MATHEMATICAL PAPERS.

London: C. J. CLAY & SONS,
CAMBRIDGE UNIVERSITY PRESS WAREHOUSE,
AVE MARIA LANE.

Cambridge: DEIGHTON, BELL AND CO.
Leipzig: F. A. BROCKHAUS.
New York: MACMILLAN AND CO.

THE COLLECTED

MATHEMATICAL PAPERS

OF

ARTHUR CAYLEY, Sc.D., F.R.S.,

SADLERIAN PROFESSOR OF PURE MATHEMATICS IN THE UNIVERSITY OF CAMBRIDGE.

VOL. IV.

CAMBRIDGE:

AT THE UNIVERSITY PRESS.

1891

CAMBRIDGE:

PRINTED BY C. J. CLAY, M.A. AND SONS,

AT THE UNIVERSITY PRESS.

ADVERTISEMENT.

THE present volume contains 77 papers numbered 223 to 299 originally published all but one of them in the years 1856 to 1862, the excepted paper, No. 265, Addition to the Memoir on an Extension of Arbogast's Method of Derivations, being now first published.

The Table for the four volumes is

Vol.	I.	Numbers	1	to	100.
„	II.	„	101	„	158.
„	III.	„	159	„	222.
„	IV.	„	223	„	299.

CONTENTS.

CLASSIFICATION.

223.

NOTE SUR UN THÉORÈME GÉNÉRAL PAR RAPPORT À L'ÉLIMINATION.

[From the *Annali di Scienze Mathematiche e Fisiche* (Tortolini), vol. VII. (1856), pp. 454—458.]

SOIENT

$$\phi = (a,\ b,\ c\ \ldots)(x,\ y)^n = 0,$$
$$\psi = (\alpha,\ \beta,\ \gamma \ldots)(x,\ y)^n = 0,$$

deux équations homogènes quelconques entre les variables x, y, et représentons par R la résultante des deux fonctions ϕ, ψ, de manière que $R=0$ sera la condition pour que les deux équations $\psi=0$, $\phi=0$ puissent avoir lieu. On sait depuis longtemps que les valeurs des variables x, y, qui satisfont à la fois aux deux équations $\psi=0$, $\phi=0$ sont données [1] par les conditions (équivalentes à une seule condition)

$$\frac{dR}{da} : \frac{dR}{db} : \frac{dR}{dc} : \text{etc.} = x^m : mx^{m-1}y : \frac{m(m-1)}{1\,.\,2}x^{m-2}y^2 : \text{etc.}$$
$$\frac{dR}{d\alpha} : \frac{dR}{d\beta} : \frac{dR}{d\gamma} : \text{etc.} = x^n : nx^{n-1}y : \frac{n(n-1)}{1\,.\,2}x^{n-2}y^2 : \text{etc.};$$

ce qui suppose cependant que les coefficients $a, b, c, \ldots\ \alpha, \beta, \gamma, \ldots$, sont des quantités absolument arbitraires: les conditions dont il s'agit peuvent aussi s'écrire sous la forme

$$\frac{dR}{da} : \frac{dR}{db} : \frac{dR}{dc} : \text{etc.} = \frac{d\phi}{da} : \frac{d\phi}{db} : \frac{d\phi}{dc} : \text{etc.}$$
$$\frac{dR}{d\alpha} : \frac{dR}{d\beta} : \frac{dR}{d\gamma} : \text{etc.} = \frac{d\psi}{d\alpha} : \frac{d\psi}{d\beta} : \frac{d\psi}{d\gamma} : \text{etc.}$$

[1] Il va sans dire que ce n'est que la valeur de $x : y$ laquelle est déterminée; mais dans la théorie des fonctions homogènes les valeurs absolues n'importent rien, et on peut dire que les valeurs x, y sont déterminées, quand $x : y$ est déterminée: on évite par cette locution des longueurs très-ennuyantes.

Or M. Schläfli dans son excellent mémoire "Ueber die Resultante eines Systemes mehrerer algebraïscher Gleichungen," *Trans. de l'Acad. de Vienne,* tom. IV. (1852), a généralisé ce théorème de la manière que voici. En considérant un nombre quelconque d'équations $\psi=0$, $\phi=0$, $\chi=0, \ldots$ entre le même nombre de variables $x=0$, $y=0$, $z=0, \ldots$ et en supposant que a, b, c, etc. soient des quantités qui entrent d'une manière quelconque dans la fonction ϕ, sans entrer dans les autres fonctions ψ, χ, etc. (il n'est nullement nécessaire que le nombre des quantités a, b, $c\ldots$ soit tel que la fonction ϕ reste absolument arbitraire, le nombre des quantités a, b, c, etc. peut même se réduire à 2) M. Schläfli fait voir que l'on a dans ce cas

$$\frac{dR}{da} : \frac{dR}{db} : \frac{dR}{dc} : \text{etc.} = \frac{d\phi}{da} : \frac{d\phi}{db} : \frac{d\phi}{dc} : \text{etc.}$$

Voici en effet le raisonnement fort simple dont se sert M. Schläfli pour établir la proposition dont il s'agit. Les équations $\psi=0$, $\phi=0$, $\chi=0$, etc. seront satisfaites par de certaines valeurs de $x:y:z$, etc. en supposant seulement que les quantités, a, b, c, etc. satisfont à la condition $R=0$. Donc les équations $\psi=0$ $\phi=0$, $\chi=0$, etc. seront encore satisfaites en donnant des variations infiniment petites quelconques δa, δb, δc, etc. aux quantités a, b, c, etc. en supposant seulement que ces variations soient telles que l'on ait

$$\delta R = \frac{dR}{da}\delta a + \frac{dR}{db}\delta b + \frac{dR}{dc}\delta c + \text{etc.} = 0.$$

Or les équations $\psi=0$, $\chi=0$, etc. qui ne contiennent pas les quantités a, b, c, etc. suffisent seules (c'est-à-dire sans l'aide de l'équation $\phi=0$) à déterminer les valeurs de $x:y:z$, etc. qui satisfont au système $\phi=0$, $\psi=0$, $\chi=0$, (on suppose toujours l'équation $R=0$): donc les nouvelles valeurs de $x:y:z$, etc. seront les mêmes qu'auparavant, et l'on doit avoir

$$\delta\phi = \frac{d\phi}{da}\delta a + \frac{d\phi}{db}\delta b + \frac{d\phi}{dc}\delta c + \text{etc.} = 0;$$

savoir cette équation aura lieu en vertu de l'équation $\delta R=0$ qui est la seule condition à laquelle on a assujetti les variations δa, δb, δc, etc., ce qui donne évidemment les conditions

$$\frac{dR}{da} : \frac{dR}{db} : \frac{dR}{dc} : \text{etc.} = \frac{d\phi}{da} : \frac{d\phi}{db} : \frac{d\phi}{dc} : \text{etc.}$$

Cela étant, on peut encore généraliser le théorème de M. Schläfli: pour cela je suppose que les quantités a, b, $c, \ldots$ entrent d'une manière quelconque dans les fonctions ϕ, ψ, χ, etc. Les équations $\phi=0$, $\psi=0$, $\chi=0$, etc. impliquent l'équation $R=0$, et en donnant aux quantités a, b, c, etc. des variations infiniment petites quelconques δa, δb, δc, etc. qui satisfont à la condition $\delta R=0$, les équations $\phi=0$, $\psi=0$, $\chi=0$, etc. seront satisfaites à la fois, cependant par des nouvelles valeurs des variables;

on peut représenter par δx, δy, δz, etc. les variations qu'il faut attribuer aux variables x, y, z, etc. Les équations $\phi=0$, $\psi=0$, $\chi=0$, etc. seront satisfaites en y variant à la fois les valeurs des variables x, y, z, etc. et des quantités a, b, c, etc.; les variations de ϕ, ψ, χ, etc. doivent donc s'évanouir: je représente de la manière que voici les conditions ainsi obtenues, savoir

$$\delta\phi + \frac{d\phi}{dx}\,\delta x + \frac{d\phi}{dy}\,\delta y + \frac{d\phi}{dz}\,\delta z + \text{etc.} = 0,$$

$$\delta\psi + \frac{d\psi}{dx}\,\delta x + \frac{d\psi}{dy}\,\delta y + \frac{d\psi}{dz}\,\delta z + \text{etc.} = 0,$$

$$\delta\chi + \frac{d\chi}{dx}\,\delta x + \frac{d\chi}{dy}\,\delta y + \frac{d\chi}{dz}\,\delta z + \text{etc.} = 0.$$

etc.

En prenant L, M, N, etc. des fonctions absolument arbitraires, et en prenant aussi

$$\delta u = - L\delta x - M\delta y - N\delta z - \text{etc.}$$

on aura l'équation identique

$$\delta u + L\delta x + M\delta y + N\delta z + \text{etc.} = 0,$$

et en éliminant les variations δx, δy, δz, etc. on obtient une équation $\square=0$; la partie de $\square$ qui contient le terme δu sera évidemment

$$\delta u \left| \begin{array}{llll} \dfrac{d\phi}{dx}, & \dfrac{d\phi}{dy}, & \dfrac{d\phi}{dz}, & \ldots \\ \dfrac{d\psi}{dx}, & \dfrac{d\psi}{dy}, & \dfrac{d\psi}{dz}, & \\ \dfrac{d\chi}{dx}, & \dfrac{d\chi}{dy}, & \dfrac{d\chi}{dz}, & \end{array} \right|$$

et le déterminant, facteur de cette expression, s'évanouit en vertu des équations $\phi=0$, $\psi=0$, $\chi=0$, etc. Cela est en effet un théorème de M. Hesse, lequel se démontre tout de suite en remarquant que l'on a

$$m\phi = x\frac{d\phi}{dx} + y\frac{d\phi}{dy} + z\frac{d\phi}{dz} + \text{etc.} = 0,$$

etc.

L'expression $\square$ ne contient donc pas de terme avec δu, et l'équation $\square=0$, peut s'écrire comme suit:

$$\left| \begin{array}{lllll} & L\,, & M\,, & N\,, & \ldots \\ \delta\phi, & \dfrac{d\phi}{dx}, & \dfrac{d\phi}{dy}, & \dfrac{d\phi}{dz}, & \\ \delta\psi, & \dfrac{d\psi}{dx}, & \dfrac{d\psi}{dy}, & \dfrac{d\psi}{dz}, & \\ \delta\chi, & \dfrac{d\chi}{dx}, & \dfrac{d\chi}{dy}, & \dfrac{d\chi}{dz}, & \end{array} \right| = 0$$

équation de la forme

$$X\delta\phi + Y\delta\psi + Z\delta\chi + \text{etc.} = 0$$

c'est-à-dire une équation entre les seules variations δa, δb, δc, etc. Or il ne peut pas y avoir entre ces variations d'autre équation que $\delta R = 0$, on doit donc avoir identiquement

$$X\delta\phi + Y\delta\psi + Z\delta\chi + \text{etc.} = k\delta R,$$

savoir cette équation sera satisfaite par les valeurs de $x : y : z$, etc. qui satisfont à $\phi = 0$, $\psi = 0$, $\chi = 0$, etc. C'est là le théorème qu'il s'agissait de démontrer; en supposant que les quantités a, b, c, etc. n'entrent que dans la fonction ϕ, on a $\delta\psi = 0$, $\delta\chi = 0$, etc., c'est-à-dire $X\delta\phi = k\delta R$, lequel est le théorème de M. Schläfli.

Je remarque que M. Schläfli a donné aussi un théorème par rapport au discriminant d'une fonction quelconque ϕ; savoir, en représentant par ∇ ce discriminant, et en supposant que les quantités a, b, c, etc. entrent d'une manière quelconque dans la fonction ϕ, les valeurs de x, y... qui satisfont aux équations

$$\frac{d\phi}{dx} = 0, \quad \frac{d\phi}{dy} = 0, \quad \frac{d\phi}{dz} = 0, \text{ etc.}$$

(lesquelles impliquent l'équation $\nabla = 0$) sont données par

$$\frac{d\nabla}{da} : \frac{d\nabla}{db} : \frac{d\nabla}{dc} : \text{etc.} = \frac{d\phi}{da} : \frac{d\phi}{db} : \frac{d\phi}{dc} : \text{etc.};$$

cela est déjà la forme la plus générale du théorème.

London, 2, *Stone Buildings*, 12 *Dec.* 1856.

224.

SUR UN THÉORÈME D'ABEL. NOTE.

[From the *Annali di Scienze Mathematiche e Fisiche* (Tortolini), vol. VIII. (1857), pp. 201—203.]

Il y a un petit mémoire d'Abel qui porte le titre "Ueber die Functionen welche der Gleichung $\phi(x)+\phi(y)=\psi\big(xf(y)+yf(x)\big)$ genugthun" (*Crelle*, tom. II. (1827), pp. 386—394). La solution du problème est contenue dans les équations que voici, savoir $f(x)$ est une fonction définie par l'équation

$$\alpha^{2n}=\big(f(x)-nx\big)^{n+\alpha'}\big(f(x)+nx\big)^{n-\alpha'},$$

et on a alors

$$\phi(x)=\frac{1}{n+\alpha'}\log C\big(f(x)+nx\big),$$

et (en réduisant un peu l'expression donné dans le mémoire)

$$\psi(x)=\frac{1}{n+\alpha'}\log C^2\alpha\left(f\left(\frac{x}{\alpha}\right)+\frac{nx}{\alpha}\right).$$

On a aussi pour $\phi(x)$ cette autre expression en forme d'intégrale indéfinie,

$$\phi(x)=\int\frac{dx}{f(x)+\alpha' x};$$

car le facteur $a\alpha$ par lequel dans le mémoire l'expression à côté droit est multiplié se réduit (comme on voit sans peine) à l'unité. En comparant les deux expressions de $\phi(x)$, on voit qu'il est permis de prendre l'intégrale depuis $x=0$, pourvu qu'on écrive $c=1$; cela donne

$$\int_0\frac{dx}{f(x)+\alpha' x}=\frac{1}{n+\alpha'}\log\big(f(x)+nx\big),$$

formule très simple pour l'intégration d'une expression algébrique laquelle ne peut pas s'exprimer à moyen de radicales.

On obtient une autre propriété de cette fonction $f(x)$ en substituant les valeurs des fonctions ϕ et ψ dans l'équation originale

$$\phi(x)+\phi(y)=\psi\left(xf(y)+yf(x)\right);$$

cela donne d'abord

$$\frac{1}{n+\alpha'}\log C^2\left(f(x)+nx\right)\left(f(x)-nx\right)=\frac{1}{n+\alpha'}\log C^2\alpha\left[f\left(\frac{xf(y)+yf(x)}{\alpha}\right)+\frac{n\,(xf(y)+yf(x))}{\alpha}\right]$$

et de là en réduisant on obtient l'équation fonctionnelle très simple

$$f(x)f(y)+n^2xy=\alpha f\left(\frac{xf(y)-yf(x)}{\alpha}\right).$$

Je remarque que l'on peut sans perte de généralité écrire $\alpha=1$, et $n=1$: je mets β au lieu de α', et j'écris aussi pour plus de simplicité $f(x)=X$, $f(y)=Y$. On a alors pour déterminer la fonction $X\left(=f(x)\right)$, l'équation

$$(X-x)^{1+\beta}(X+x)^{1-\beta}=1$$

équation dans laquelle on pourrait remplacer les exposants $1+\beta$, $1-\beta$ par deux quantités quelconques.

La formule d'intégration devient

$$\int_0\frac{dx}{X+\beta x}=\frac{1}{1+\beta}\log(X+x)=-\frac{1}{1-\beta}\log(X-x)$$

formule que l'on peut vérifier sans peine à moyen de celle-ci,

$$(X+\beta x)X'=x+\beta X,$$

que l'on obtient en différentiant l'équation pour X. L'équation fonctionnelle sera

$$XY+xy=f(xY+yX),$$

c'est-à-dire en écrivant $xY+yX=z$, $XY+xy=Z$ on doit avoir

$$(Z-z)^{1+\beta}(Z+z)^{1-\beta}=1$$

et cela se vérifie tout de suite à moyen des équations

$$Z-z=(X-x)(Y-y),\quad Z+z=(X+x)(Y+y).$$

Je remarque aussi qu'en prenant le quotient des dérivées de cette équation par rapport à x et y on obtient

$$\frac{X'-Y'}{X'Y'-1}=\left(\frac{X}{x}-\frac{Y}{y}\right)\div\left(\frac{X}{x}\cdot\frac{Y}{y}-1\right),$$

laquelle est une propriété de la fonction X et sa dérivée X'.

Londres, 17 *Juillet*, 1857.

225.

ON A CLASS OF DYNAMICAL PROBLEMS.

[From the *Proceedings of the Royal Society of London*, vol. VIII. (1857), pp. 506—511.]

THERE are a class of dynamical problems which, so far as I am aware, have not been considered in a general manner. The problems referred to (which might be designated as continuous-impact problems) are those in which the system is continually taking into connexion with itself particles of infinitesimal mass (i.e. of a mass containing the increment of time dt as a factor), so as not itself to undergo any abrupt change of velocity, but to subject to abrupt changes of velocity the particles so taken into connexion. For instance, a problem of the sort arises when a portion of a heavy chain hangs over the edge of a table, the remainder of the chain being coiled or heaped up close to the edge of the table; the part hanging over constitutes the moving system, and in each element of time dt, the system takes into connexion with itself, and sets in motion with a finite velocity, an infinitesimal length ds of the chain; in fact, if v be the velocity of the part which hangs over, then the length vdt is set in motion with the finite velocity v. The general equation of dynamics applied to the case in hand will be

$$\Sigma \left\{\left(\frac{d^2x}{dt^2} - X\right) \delta x + \left(\frac{d^2y}{dt^2} - Y\right) \delta y + \left(\frac{d^2z}{dt^2} - Z\right) \delta z\right\} dm + \Sigma \left(\Delta u \delta\xi + \Delta v \delta\eta + \Delta w \delta\zeta\right) \frac{1}{dt} d\mu = 0,$$

where the first term requires no explanation: in the second term ξ, η, ζ denote the coordinates at the time t of the particle $d\mu$ which then comes into connexion with the system; Δu, Δv, Δw are the finite increments of velocity (or, if the particle is originally at rest, then the finite velocities) of the particle $d\mu$ the instant that it has come into connexion with the system; $\delta\xi$, $\delta\eta$, $\delta\zeta$ are the virtual velocities of the same particle $d\mu$ considered as having come into connexion with and forming part of the system. The summation extends to the several particles or to the system of particles $d\mu$ which come into connexion with the system at the time t; of course, if there is only a single particle $d\mu$, the summatory sign Σ is to be omitted. The values of

Δu, Δv, Δw are $\frac{d\xi}{dt} - u$, $\frac{d\eta}{dt} - v$, $\frac{d\zeta}{dt} - w$, if by $\frac{d\xi}{dt}$, $\frac{d\eta}{dt}$, $\frac{d\zeta}{dt}$ we understand the velocities of $d\mu$ parallel to the axes, after it has come into connexion with the system; but it is to be observed, that considering ξ, η, ζ as the coordinates of the particle $d\mu$ which is continually coming into connexion with the system, then if the problem were solved and ξ, η, ζ given as functions of t (and, when there is more than one particle $d\mu$, of the constant parameters which determine the particular particle), $\frac{d\xi}{dt}$, &c., in the sense just explained, cannot be obtained by simple differentiation from such values of ξ, &c.: in fact, ξ, η, ζ so given as functions of t, belong at the time t to one particle, and at the time $t + dt$ to the next particle, but what is wanted is the increment in the interval dt of the coordinates ξ, η, ζ of one and the same particle.

Suppose as usual that x, y, z, and in like manner that ξ, η, ζ are functions of a certain number of independent variables θ, ϕ, &c., and of the constant parameters which determine the particular particle dm or $d\mu$, of which x, y, z, or ξ, η, ζ are the coordinates; parameters, that is, which vary from one particle to another, but which are constant during the motion for one and the same particle. The summations are in fact of the nature of definite integrations in regard to these constant parameters, which therefore disappear altogether from the final results. The first term,

$$\Sigma \left\{ \left(\frac{d^2x}{dt^2} - X\right) \delta x + \left(\frac{d^2y}{dt^2} - Y\right) \delta y + \left(\frac{d^2z}{dt^2} - Z\right) \delta z \right\} dm,$$

may be reduced in the usual manner to the form

$$\Theta\delta\theta + \Phi\delta\phi + \ldots$$

where, writing as usual θ', ϕ', &c. for $\frac{d\theta}{dt}$, $\frac{d\phi}{dt}$, &c., we have

$$\Theta = \frac{d}{dt}\frac{dT}{d\theta'} - \frac{dT}{d\theta} + \frac{dV}{d\theta},$$

$$\Phi = \frac{d}{dt}\frac{dT}{d\phi'} - \frac{dT}{d\phi} + \frac{dV}{d\phi}, \quad \text{\&c.,}$$

(this supposes that $Xdx + Ydy + Zdz$ is an exact differential); only it is to be observed that in the problems in hand, the mass of the system is variable, or what is the same thing, the variables θ, ϕ, &c., are introduced into T and V through the limiting conditions of the summation or definite integration, besides entering directly into T and V in the ordinary manner. And in forming the differential coefficients $\frac{d}{dt}\frac{dT}{d\theta'}$, $\frac{dT}{d\theta}$, $\frac{dV}{d\theta}$, &c., it is necessary to consider the variables θ, ϕ, &c., in so far as they enter through the limiting conditions *as exempt from differentiation*, so that the expressions just given for Θ, Φ, &c., are, in the case in hand, rather conventional representations than actual analytical values; this will be made clearer in the sequel by the consideration of the before-mentioned particular problem.

which may be written in the form

$$s\frac{ds}{dt}\,d\left(s\frac{ds}{dt}\right)=gs^2ds,$$

and the first integral is therefore

$$\frac{sds}{\sqrt{s^3-a^3}}=\sqrt{\frac{2g}{3}}\,dt,$$

where a is the length hanging over at the commencement of the motion. If $a=0$, then the equation is

$$\frac{ds}{\sqrt{s}}=\sqrt{\frac{2g}{3}}\,dt,$$

and integrating from $t=0$, $2\sqrt{s}=\sqrt{\frac{2g}{3}}\,t$, or finally $s=\frac{1}{6}gt^2$, so that the motion is the same as that of a body falling under the influence of a constant force $\frac{1}{3}g$. It is perhaps worth noticing that the differential equation may be obtained as follows:— We have, in the first instance, a mass s moving with a velocity s', and after the particle $ds\,(=s'dt)$ has been set in motion, a mass $s+s'dt$ moving say with a velocity $s'+\delta s'$, whence neglecting for the moment the effect of gravity on the mass s, the momentum of the mass in motion will be constant, or we shall have

$$ss'=(s+s'dt)(s'+\delta s')=ss'+s'^2dt+s\delta s',$$

and therefore $s\delta s'=-s'^2dt$. Hence, adding on the right-hand side the term $gsdt$ arising from gravity, and substituting $\frac{d^2s}{dt^2}\,dt$ for $\delta s'$, we have the equation, $s\frac{d^2s}{dt^2}=gs-\left(\frac{ds}{dt}\right)^2$, as before.

226.

ON PROFESSOR MAC CULLAGH'S THEOREM OF THE POLAR PLANE.

[From the *Proceedings of the Royal Irish Academy*, vol. VI. (1858), pp. 481—491.]

A RAY of polarized light, incident on the surface of an extraordinary medium, may give rise to a reflected ray and a single refracted ray; but this will be the case only for a particular position, or positions, of the plane of polarization of the incident ray. According to Professor Mac Cullagh's theory, the plane of polarization, and the relative vibrations of the three rays, are deduced from two assumed principles, which may be referred to as

1°. The principle of equivalent vibrations.

2°. The principle of equivalent moments.

And from these principles are deduced

3°. The principle of vis viva.

4°. The theorem of the polar plane.

The directions of the vibrations are completely determined by means of 4°, the theorem of the polar plane; and the relative magnitudes are then given by 1°, the principle of equivalent vibrations. The other principles, viz.:—2°, the principle of equivalent moments, and 3°, the principle of vis viva, must therefore follow as mere geometrical consequences from the first-mentioned two principles, or theorems; and I have found that the deduction depends immediately upon the following two theorems in spherical trigonometry.

Suppose (Fig. 1) that $RR'R''$ is a spherical triangle, and let W be any point in the base RR'', and N be the central point of the base; then joining WR' and

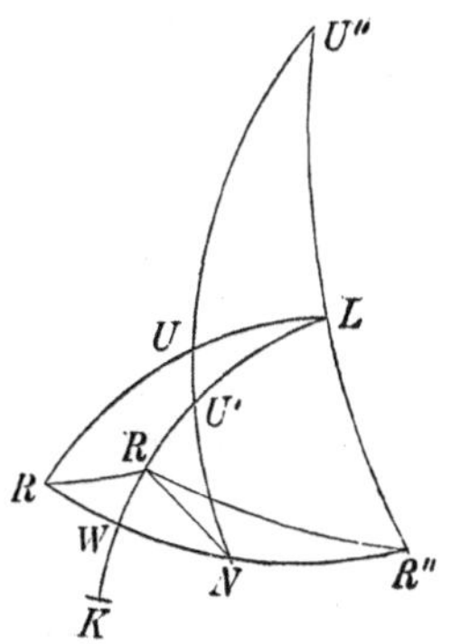

Fig. 1.

producing this arc (in the direction from W to R') to a point L, such that

$$\cot WL = \frac{\sin^2 NW}{\sin WR \sin WR'} \tan WR',$$

and joining NR', then

THEOREM I.
$$\frac{\sin^2 R''LR' - \sin^2 RLR'}{\sin^2 R''LR} = \frac{\sin NW \cos NR'}{\cos WR' \sin NR \cos NR};$$

and if we suppose also, that an arc through N, perpendicular to the base RR'', cuts LR, LR', and LR'' produced in the points U, U', U'', then

THEOREM II.

$$\sin R''LR' \cos RU \sin NUR + \sin RLR' \cos R''U'' \sin NU''R''$$
$$= \sin R''LR \frac{\cos NW}{\cos WR' \sin NR} \cos R'U' \sin NU'R'.$$

The present memoir contains the proof of the two theorems, and the application of them to the optical theory.

To prove the first theorem, I write for shortness R, R'', W to denote the angles LRR'', $LR''R$, NWR', respectively; we have then,

$$\frac{\sin^2 R''LR' - \sin^2 RLR'}{\sin^2 R''LR} = \frac{\sin (R''LR' - RLR')}{\sin R''LR}$$
$$= \frac{1}{\sin R''LR} \{\sin R''LR' \cos RLR' - \sin RLR' \cos R''LR'\}$$
$$= \frac{1}{\sin R''LR} \left\{ \frac{\sin R''W \sin R''}{\sin LW} \cdot \frac{\cos RW - \cos LR \cos LW}{\sin LR \sin LW} \right.$$
$$\left. - \frac{\sin RW \sin R}{\sin LW} \cdot \frac{\cos R''W - \cos LR'' \cos LW}{\sin LR'' \sin LW} \right\}$$

$$=\frac{1}{\sin R''LR\sin^2 LW}\left\{\frac{\sin R''}{\sin LR}\sin R''W(\cos RW-\cos LR\cos LW)\right.$$

$$\left.-\frac{\sin R}{\sin LR''}\sin RW(\cos R''W-\cos LR''\cos LW)\right\}.$$

Observing that $\frac{\sin R''}{\sin LR}$, $\frac{\sin R}{\sin LR''}$, are each equal to $\frac{\sin R''LR}{\sin R''R}$, this becomes

$$=\frac{1}{\sin R''R\sin^2 LW}\{\sin R''W(\cos RW-\cos LR\cos LW)$$

$$-\sin RW(\cos R''W-\cos LR''\cos LW)\};$$

and, substituting for $\cos LR$, $\cos LR''$, the values

$$\cos RW\cos LW-\sin RW\sin LW\cos W,$$

$$\cos R''W\cos LW+\sin R''W\sin LW\cos W,$$

the foregoing expression becomes

$$=\frac{1}{\sin R''R\sin^2 LW}\times\{\sin R''W(\cos RW\sin^2 LW+\sin RW\sin LW\cos LW\cos W)$$

$$-\sin RW(\cos R''W\sin^2 LW-\sin R''W\sin LW\cos LW\cos W)\},$$

$$=\frac{1}{\sin R''R}\{\sin R''W\cos RW-\sin RW\cos R''W+2\cot LW\sin RW\sin R''W\cos W\}$$

$$=\frac{1}{\sin R''R}\{\sin(R''W-RW)+2\cot LW\sin RW\sin R''W\cos W)\};$$

and, putting $R''W-RW=2NW$, and substituting also for $\cot WL$ its value, which gives $\cot LW\sin RW\sin R''W=\sin^2 NW\tan WR'$, the expression becomes

$$=\frac{1}{\sin R''R}\{\sin 2NW+2\sin^2 NW\tan WR'\cos W\};$$

but we have

$$\cos W=\frac{\cos NR'-\cos NW\cos WR'}{\sin NW\sin WR};$$

and therefore

$$2\sin^2 NW\tan WR'\cos W=2\frac{\sin NW}{\cos WR'}\cos NR'-\sin 2NW;$$

thus the expression becomes

$$=\frac{1}{\sin R''R}\,2\,\frac{\sin NW}{\cos WR'}\cos NR';$$

and $\sin R''R=\sin 2NR=2\sin NR\cos NR$, so that finally the expression becomes

$$=\frac{\sin NW\cos NR'}{\cos WR'\sin NR\cos NR},$$

which proves the theorem.

To prove the second theorem, take as before R, R'', W, to denote the angles LRR'', $LR''R$, NWR', respectively; and moreover, U, U', U'' to denote the angles NUR, $NU'R'$, $NU''R''$, respectively; then considering, first, the function on the left-hand side, viz.:

$$\sin R''LR' \cos RU \sin U + \sin RLR' \cos R''U'' \sin U'',$$

we have

$$\sin U = \frac{\sin NR}{\sin RU},$$

$$\cos RU \sin U = \sin NR \cot RU$$

$$= \sin NR \cos R \cot NR = \cos R \cos NR,$$

and, in like manner,

$$\sin U'' = \frac{\sin NR''}{\sin R''U''},$$

$$\cos R''U'' \sin U'' = \sin NR'' \cot R''U'' = \sin NR'' \cos R'' \cot NR''$$

$$= \cos R'' \cos NR'' = \cos R'' \cos NR;$$

the expression thus becomes

$$= \cos NR \{\sin R''LR' \cos R + \sin RLR' \cos R''\},$$

which is

$$= \cos NR \left\{\frac{\sin R''W \sin W}{\sin R''L} \cdot \frac{\cos WL - \cos RW \cos RL}{\sin RW \sin RL} - \frac{\sin RW \sin W}{\sin RL} \cdot \frac{\cos WL - \cos R''W \cos R''L}{\sin R''W \sin R''L}\right\},$$

or, substituting for $\cos RL$, $\cos R''L$ the values

$$\cos RW \cos WL - \sin RW \sin WL \cos W,$$

$$\cos R''W \cos WL + \sin R''W \sin WL \cos W,$$

the expression becomes

$$\frac{\cos NR \sin W}{\sin RL \sin R''L} \left\{ \frac{\sin R''W}{\sin RW} (\cos WL \sin^2 RW + \sin WL \sin RW \cos RW \cos W) + \frac{\sin RW}{\sin R''W} (\cos WL \sin^2 R''W - \sin WL \sin R''W \cos R''W \cos W)\right\}$$

$$= \frac{\cos NR \sin W}{\sin RL \sin R''L} \{2 \cos WL \sin RW \sin R''W + \sin WL \sin (R''W - RW) \cos W\}$$

$$= \frac{\cos NR \sin W \sin WL}{\sin RL \sin R''L} \{2 \cot WL \sin RW \sin R''W + \sin (R''W - RW) \cos W\}.$$

Hence, putting for cot WL its value, which gives

$$\cot WL \sin RW \sin R''W = \sin^2 NW \tan WR',$$

and putting also

$$\sin (R''W - RW) = \sin 2NW = 2 \sin NW \cos NW,$$

the expression becomes

$$= \frac{2 \cos NR \sin W \sin WL \sin^2 NW}{\sin RL \sin R''L} (\tan WR' + \cot NW \cos W).$$

The right-hand side of the equation to be proved is

$$\sin R''LR \frac{\sin NW}{\cos WR' \sin NR} \cos R'U' \sin U',$$

and we have

$$\sin R''LR = \frac{\sin RR'' \sin R}{\sin R''L}, \qquad \sin R = \frac{\sin WL \sin W}{\sin RL},$$

and consequently

$$\sin R''LR = \frac{\sin RR'' \sin WL \sin W}{\sin RL \sin R''L} = \frac{2 \sin NR \cos NR \sin NL \sin W}{\sin RL \sin R''L},$$

or the expression is

$$= \frac{2 \cos NR \sin WL \sin W}{\sin RL \sin R''L} \cdot \frac{\sin NW}{\cos WR'} \cos R'U' \sin U'.$$

But we have

$$\sin U' = \frac{\sin NW}{\sin W'U'};$$

and therefore

$$\begin{aligned} \frac{\sin NW}{\cos WR'} \cos R'U' \sin U' &= \sin^2 NW \frac{\cos R'U'}{\cos WR' \sin WU'} \\ &= \sin^2 NW \frac{\cos (WU' - WR')}{\cos WR' \sin WU'} \\ &= \sin^2 NW (\tan WR' + \cot WU'). \end{aligned}$$

Moreover we have $\cot WU' = \cot NW \cos W$, and the expression thus becomes

$$= \frac{2 \cos NR \sin W \sin WL \sin^2 NW}{\sin RL \sin R''L} (\tan WR' + \cot NW \cos W);$$

which is the expression previously found as the value of the left-hand side of the equation, and the theorem is therefore proved.

It is obvious that the point L might have been constructed by taking on $R'W$, produced in the direction from R' to W, a point K such that

$$\tan KW = \frac{\sin^2 NW}{\sin RW \sin R''W} \tan WR',$$

and then taking the arc KL in the reverse direction equal to 90°.

Passing now to the optical problem, it will be recollected that in Mac Cullagh's theory the direction of vibration in an extraordinary medium is perpendicular to the plane of the ray and wave normal, and that the polar plane of a refracted ray is by definition a plane through the point of incidence parallel to the direction of vibration, and also parallel to a line joining the extremity of the ray with the corresponding point on the *Index surface*,—the last-mentioned surface being the polar reciprocal of the refracted wave-surface, taken with respect to the reflected wave-surface, or wave-sphere, contemporaneously generated. We have to consider a ray of polarized light incident on the surface of an extraordinary medium, and giving rise to a reflected ray and a single refracted ray. Let the incident ray and the reflected ray be respectively produced within the medium, and let the three rays, viz., the incident ray produced, the refracted ray, and the reflected ray produced, be represented in direction (see Fig. 2) by AR, AR' and AR''; and take $AR = AR'' = 1$ as the radius of the wave-sphere and AR' as the radius of the wave-surface, corresponding at a given instant

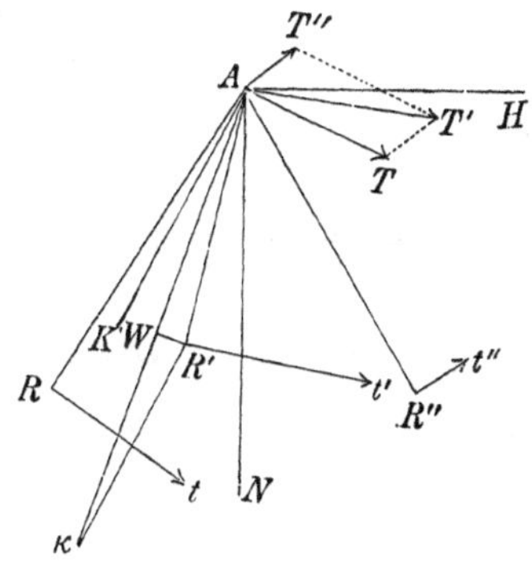

Fig. 2.

of time to the first or ordinary medium and the extraordinary medium respectively. Take also AW as the perpendicular on the tangent plane of the wave-surface at R', or 'wave-normal,' corresponding to the refracted ray AR'; and let AN represent the normal to the plane of separation of the two media, and AH the intersection of the last-mentioned plane with the plane of incidence. The lines AR, AR'', AW, AN, AH are of course all of them in the plane of incidence, the line AN bisects the angle made by the lines AR, AR'', and the lines AN, AH are at right angles to each other. The length of the wave-normal AW is given by the equation $AR \sin NAW = AW \sin NAR$, or putting, as above, $AR = 1$, and representing the two angles at A by NW, NR respectively, then, if p denote the length of the wave-normal, we have $\sin NW = p \sin NR$.

Take κ the pole of the tangent plane of the wave-surface at R' (or, what is the same thing, the image of the point W), in respect of the sphere radius AR, then κ will be the point on the index-surface corresponding to the point R' of the wave-surface; and let AK be drawn through the point A parallel to $R'\kappa$. Take AT' perpendicular to the plane WAR' (or, what is the same thing, the plane KAR') as the direction of the refracted vibration, the plane KAT' will be the polar plane; and by 4°, the theorem of the polar plane, the directions of the incident and reflected vibrations are given as the intersections of the polar plane with the wave-fronts or planes through A normal to the directions of the incident and reflected rays respectively; these intersections are represented in the figure by AT and AT''. The relative magnitudes of the vibrations are then determined by 2°, the principle of equivalent vibrations, viz., considering these vibrations as forces acting in the given directions AT', AT, AT'' respectively, the refracted vibration will be the resultant of the incident and reflected vibrations: the terminated lines AT', AT, AT'' in the figure are taken to represent in direction and magnitude the vibrations corresponding to the refracted ray and to the incident and reflected rays respectively, and the lines $R't'$, Rt, $R''t''$ are drawn through the extremities R', R, R'' of the three rays equal and parallel to AT', AT, and AT'' respectively. Let m', m, m'' denote the masses of ether set in motion by the three rays respectively, then, according to Mac Cullagh's hypothesis of equal densities, we have

$$m = m'' : m' :: AR \cos RN : \frac{AW \cos R'N}{\cos WR'},$$

(where RN, &c., denote the angles RAN, &c.); or writing as before, $AR=1$, $AW=p$, where $\sin NW = p \sin RN$, we have

$$m = m'' : m' :: \quad \cos RN : \frac{p \cos R'N}{\cos WR'} \left(= \frac{\sin NW \cos R'N}{\cos WR' \sin RN} \right).$$

This being premised, then, 3°, the principle of vis viva is that

$$m\,(Rt)^2 = m'\,(R't')^2 + m''\,(R''t'')^2;$$

or, what is the same thing,

$$\frac{Rt^2 - R''t''^2}{R't'^2} = \frac{m'}{m} = \frac{\sin NW \cos R'N}{\cos WR' \sin RN \cos RN};$$

and 2°, the principle of equivalent moments, is that the moment of $R't'$ round the axis AH, is equal to the sum of the moments of Rt and $R''t''$ round the same axis. It only remains to show that these two properties are in fact contained in the Theorems I. and II.

The point κ is the image of W in a sphere, radius unity. Hence, $A\kappa = \frac{1}{p}\,\kappa W = \frac{1}{p} - p$, and therefore

$$\tan W\kappa R' = \frac{p^2 \tan WR'}{1-p^2} = \tan KW,$$

but we have, as before, $\sin NW = p \sin RN$, and consequently,

$$\tan KW = \frac{\sin^2 NW \tan WR'}{\sin^2 RN - \sin^2 RW}$$

$$= \frac{\sin^2 NW}{\sin RW \sin R''W} \tan WR'.$$

Suppose now that the points R, R', R'', W, N, H, K, of Fig. 2, are all of them projected by radii through the centre A upon a sphere, radius unity (see Fig. 3, where the several points are represented by the same letters as in Fig. 2); and complete Fig. 3 by connecting the different points in question by arcs of great circles, and by producing KW (in the direction from K to W) to a point L, such that $KL = 90°$, and by joining LR, LR'', and drawing the arc $NU'UU''$ at right angles to $R''R$ (or, what is the same thing, with the pole H) meeting LR', LR, and LR'' produced, in the points U', U, U'' respectively. By what has preceded, the points K, L of Fig. 3

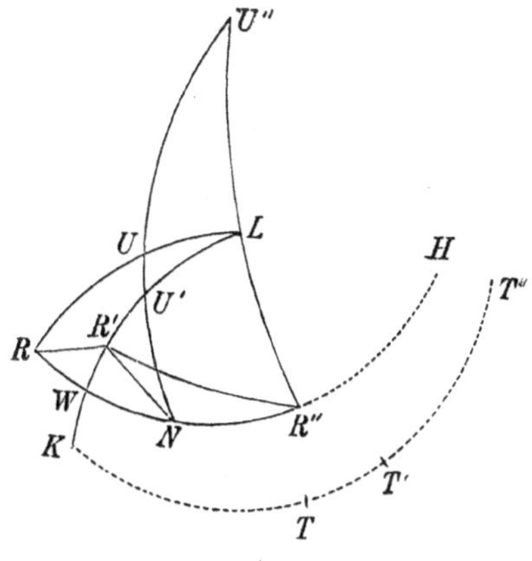

Fig. 3.

are constructed precisely in the same manner as the same points in Fig, 1, and in fact Fig. 3 is nothing else than Fig. 1 with some additional lines and points. The condition employed to determine the magnitude of the vibrations Rt, $R't'$, $R''t''$, gives that these vibrations are as

$$\sin T'T'' : \sin TT'' : \sin TT',$$

or, observing that LR, LR', LR'' are the great circles whose poles are T, T', T'' respectively, these vibrations are as

$$\sin R''LR' : \sin RLR'' : \sin RLR';$$

and, substituting these values, the equation given by the principle of vis viva becomes identical with that of Theorem I.

Proceeding to the condition given by the principle of equivalent moments, we have

$$\text{moment of } Rt \text{ round } AH$$
$$= Rt \times AR \times \cos [AR, \perp \text{dist.} (Rt, AH)] \times \sin (Rt, AH);$$

and in Fig. 3, observing that the radius through U is parallel to the perpendicular distance of (Rt, AH) (for LR has the pole T, and NU the pole H) then

$$\cos [AR, \perp \text{dist.} (Rt, AH)] = \cos RU,$$
$$\sin (Rt, AH) = \sin TH,$$

or, since T and H are the poles of LR and NW respectively, $TH = \angle NUR$, and, putting $AR = 1$, the moment is

$$= Rt \cos RU \sin NUR.$$

Similarly,

$$\text{moment of } R''t'' \text{ round } AH$$
$$= R''t'' \cos R''U'' \sin NU''R'';$$

and for the refracted ray,

$$\text{moment of } R't' \text{ round } AH$$
$$= AR' \times R't' \cos R'U' \sin NU'R'.$$

But we have

$$AR' = \frac{AW}{\cos WR'} = \frac{\sin NW}{\cos WR' \sin NR},$$

and therefore the moment is

$$= R't' \frac{\sin NW}{\cos WR' \sin NR} \cos R'U' \sin NU'R'.$$

Hence the vibrations Rt, $R''t''$, $R't'$, as before, are as

$$\sin R''LR' : \sin RLR' : \sin RLR'',$$

and thus the equation given by the principle of equivalent moments is precisely that of Theorem II.

227.

ON THE THEORY OF RECIPROCAL SURFACES.

[From the *Proceedings of the Royal Irish Academy*, vol. VII. (1862), pp. 20—28.]

THE present note is intended to be supplementary to Mr Salmon's memoir "On the Degree of a Surface reciprocal to a given one" (*Trans. R. Irish Acad.*, vol. XXI. pp. 461—488, 1857). I find that Mr Salmon's equations admit of a transformation which appears important in reference to the geometrical theory, and the object of the note is to present the system of equations under the new form.

Mr Salmon writes:

n, the order of the surface.

a, the order of the tangent cone drawn from any point to the surface.

δ, the number of the double edges of the cone.

κ, the number of its cuspidal edges.

b, the order of any double curve upon the surface.

k, the number of apparent double points of the double curve.

t, the number of triple points on the double curve.

c, the order of any cuspidal curve on the surface.

h, the number of apparent double points of the cuspidal curve.

β, the number of intersections of the double and cuspidal curves which are stationary points on the cuspidal curve.

γ, the number of intersections which are stationary points on the double curve.

i, the number of intersections which are not stationary points upon either curve.

ρ, the number of the points where the double curve is met by the curve of contact of the tangent cone.

σ, the number of the points where the cuspidal curve is met by the curve of contact.

And the accented letters denote the corresponding singularities of the reciprocal surface, or, if we choose that they should refer to the given surface, and its tangential or *class* singularities, then we have:

n', the class of the surface.

a', the class of the curve of intersection by any plane.

δ', the number of double tangents of the curve.

κ', the number of its cusps.

b', the class of the node-couple develope.

k', the number of apparent double planes of the node-couple develope.

t', the number of triple planes of the node-couple develope.

c', the class of the spinode develope.

h', the number of the apparent double planes of the spinode develope

β', the number of common tangent planes of the node-couple and spinode developes, stationary planes of the spinode develope.

γ', the number of common tangent planes, stationary planes of the node-couple develope.

i', the number of the common tangent planes which are not stationary planes of either develope.

ρ', the number of the common tangent planes of the node-couple develope, and the tangent cone.

σ', the number of the common tangent planes of the spinode develope, and the tangent cone.

The terminology made use of is that of my paper "On the Singularities of Surfaces" (*Cambridge and Dublin Mathematical Journal*, vol. VII., 1852), [106]. To explain it, I need only remark that the term node is used as synonymous with double point, and the term spinode as synonymous with cusp; a spinode plane is a tangent plane meeting the surface in a curve, having a spinode at the point of contact; and a node-couple plane is a double tangent plane, or plane meeting the surface in a curve having two nodes; the term develope is used instead of developable surface.

To collect all the formulæ, it is proper to write also:

r, the class of the cuspidal curve.

q, the class of the double curve.

r', the order of the spinode develope.

q', the order of the node-couple develope,

where q' is what Mr Salmon, who only uses it incidentally in referring to a result of Professor Schläfli's, calls (after him) A.

Mr Salmon obtains, between the twenty-eight quantities,

$$n,\ a,\ \delta,\ \kappa,\ b,\ k,\ t,\ c,\ h,\ \beta,\ \gamma,\ i,\ \rho,\ \sigma,$$

$$n',\ a',\ \delta',\ \kappa',\ b',\ k',\ t',\ c',\ h',\ \beta',\ \gamma',\ i',\ \rho',\ \sigma',$$

the twenty-one equations,

$$a = a',$$
$$a' = n(n-1) - 2b - 3c,$$
$$\kappa' = 3n(n-2) - 6b - 8c,$$
$$\delta' = \tfrac{1}{2}n(n-2)(n^2-9) - (n^2-n-6)(2b+3c) + 2b(b-1) + 6bc + \tfrac{9}{2}c(c-1),$$

$$a(n-2) = \kappa + \rho + 2\sigma,$$
$$b(n-2) = \rho + 2\beta + 3\gamma + 3t,$$
$$c(n-2) = 2\sigma + 4\beta + \gamma,$$

$$a(n-2)(n-3) = 2\delta + 2ab + 3ac - 4\rho - 9\sigma,$$
$$b(n-2)(n-3) = 4k + ab + 3bc - 9\beta - 6\gamma - 3i - 2\rho,$$
$$c(n-2)(n-3) = 6h + ac + 2bc - 6\beta - 4\gamma - 2i - 3\sigma,$$

$$n' = n(n-1)^2 - n(7b+12c) + 4b^2 + 9c^2 + 8b + 15c - 8k - 18h + 18\beta + 12\gamma + 12i - 9t,$$

$$a = n'(n'-1) - 2b' - 3c',$$
$$\kappa = 3n'(n'-2) - 6b' - 8c',$$
$${}^*\delta = \tfrac{1}{2}n'(n'-2)(n'^2-9) - (n'^2-n'-6)(2b'+3c') + 2b'(b'-1) + 6b'c' + \tfrac{9}{2}c'(c'-1),$$

$$a'(n'-2) = \kappa' + \rho' + 2\sigma',$$
$$b'(n'-2) = \rho' + 2\beta' + 3\gamma' + 3t',$$
$$c'(n'-2) = 2\sigma' + 4\beta' + \gamma',$$

$$a'(n'-2)(n'-3) = 2\delta' + 2a'b' + 3a'c' - 4\rho' - 9\sigma',$$
$$b'(n'-2)(n'-3) = 4k' + a'b' + 3b'c' - 9\beta' - 6\gamma' - 3i' - 2\rho',$$
$$c'(n'-2)(n'-3) = 6h' + a'c' + 2b'c' - 6\beta' - 4\gamma' - 2i' - 3\sigma',$$

$${}^*n = n'(n'-1)^2 - n'(7b'+12c') + 4b'^2 + 9c'^2 + 8b' + 15c' - 8k' - 18h' + 18\beta' + 12\gamma' + 12i' - 9t';$$

to which may be joined

$$q = b^2 - b - 2k - 3\gamma - 6t,$$
$$r = c^2 - c - 2h - 3\beta,$$
$$q' = b'^2 - b' - 2k' - 3\gamma' - 6t',$$
$$r' = c'^2 - c' - 2h' - 3\beta',$$

Considering the twenty-one equations, and taking as data n, b, c, β, γ, h, k, then, by means of the several equations, other than the two equations marked (*), we may express in terms of the above data a, δ, κ, t, i, ρ, σ, n', a', δ', κ', b', c', ρ', σ', $2\beta'+3\gamma'+3t'$, $4\beta'+\gamma'$, $4k'-3i'$, $6h'-2i'$; the quantities which enter into the first of the marked equations are then all given in terms of the above data, and it is clear that the equation must be satisfied identically: the quantities which enter into the second of the marked equations are given in terms of the data and of t', i', and it is not clear, *a priori*, but that the equation might lead to a relation between the data and t', i'; it will, however, appear in the sequel that the equation must be satisfied identically, independently of any particular values of t', i'. Thus, Mr Salmon's theory does not determine the values of these two quantities, nor, consequently, the values of β', γ', h', k'; it does, however, determine the values of the combinations $4\beta'+\gamma'$, $8k'-18h'$. But the twenty-one equations between the twenty-eight quantities may be replaced by seventeen equations between the twenty quantities

$$\begin{array}{l} n,\ a,\ \delta,\ \kappa,\ b,\ c,\ \rho,\ \sigma,\ 4\beta+\gamma,\ 8k-18h, \\ n',\ a',\ \delta',\ \kappa',\ b',\ c',\ \rho',\ \sigma',\ 4\beta'+\gamma',\ 8k'-18h'; \end{array}$$

this will clearly be the case if it is only shown that the equation which gives n' can by the other equations be transformed into one of the form in question; for a similar transformation will, of course, apply to the equation for n, and then we have only to reject the equation containing t, and to replace the two equations which contain i, by the equation given by the elimination of this quantity, and in like manner to reject the equation containing t', and to replace the two equations containing i', by the equation given by the elimination of this quantity, and the system will be reduced to the required form.

The reduction of the equation which gives n' is effected as follows; we have

$$\begin{aligned} (2b+3c)(n-2)(n-3) &= 8k+18h+a(2b+3c)+12bc-36\beta-24\gamma-12i-4\rho-9\sigma, \\ 3b(n-2) &= 6\beta+9\gamma+9t+3\rho; \end{aligned}$$

and thence

$$\begin{aligned} &(2b+3c)(n-2)(n-3)+3b(n-2) \\ &\quad = a(2b+3c)+12bc+8k+18h-12i+9t-\rho-9\sigma-30\beta-15\gamma \\ &\quad = a(2b+3c)+12bc+8k+18h-12i+9t-18\beta-12\gamma-\rho-9\sigma-3(4\beta+\gamma); \end{aligned}$$

and consequently

$$\begin{aligned} &-8k-18h+18\beta+12\gamma+12i-9t \\ &\quad = \{a-(n-2)(n-3)\}(2b+3c)-3b(n-2)+12bc-\rho-9\sigma-3(4\beta+\gamma), \end{aligned}$$

which (observing that the left-hand side is precisely the combination of terms which enters into the equation for n') shows that the reduction is possible; to complete it, putting for a its value $n(n-1)-2b-3c$, we have

$$\begin{aligned} &-8k-18h+18\beta+12\gamma+12i-9t \\ &\quad = b(5n-6)+c(12n-18)-4b^2-9c^2-\rho-9\sigma-3(4\beta+\gamma); \end{aligned}$$

and substituting this value in the equation for n', we obtain

$$n' = n(n-1)^2 - b(2n-2) - 3c - \rho - 9\sigma - 3(4\beta + \gamma).$$

Some of the other equations admit of simplification: the equation

$$a(n-2)(n-3) = 2\delta + a(2b + 3c) - 4\rho - 9\sigma,$$

if we put for a its value $n(n-1) - 2b - 3c$, becomes

$$(4n - 6 - 2b - 3c)(n-2)(n-3) = 2\delta - 4\rho - 9\sigma,$$

and the prescribed combination

$$(2b - 3c)(n-2)(n-3) = 8k - 18h + a(2b - 3c) - 4\rho + 9\sigma,$$

gives in like manner, putting for a its value,

$$(-n^2 + n + 4b)(n-2)(n-3) = (8k - 18h) - 4\rho + 9\sigma.$$

The system of seventeen equations then is:

$$a = a',$$
$$a' = n(n-1) - 2b - 3c,$$
$$\kappa' = 3n(n-2) - 6b - 8c,$$
$$\delta' = \tfrac{1}{2}n(n-2)(n^2-9) - (n^2 - n - 6)(2b + 3c) + 2b(b-1) + 6bc + \tfrac{9}{2}c(c-1),$$

$$a(n-2) = \kappa + \rho + 2\sigma,$$
$$c(n-2) = 2\sigma + (4\beta + \gamma),$$

$$(4n - 6 - 2b - 3c)(n-2)(n-3) = 2\delta - 4\rho - 9\sigma,$$
$$(-n^2 + n + 4b)(n-2)(n-3) = (8k - 18h) - 4\rho + 9\sigma,$$

$$n' = n(n-1)^2 - b(2n-2) - 3c - \rho - 9\sigma - 3(4\beta + \gamma),$$
$$a = n'(n'-1) - 2b' - 3c',$$
$$\kappa = 3n'(n'-2) - 6b' - 8c',$$
$$*\delta = \tfrac{1}{2}n'(n'-2)(n'^2-9) - (n'^2 - n' - 6)(2b' + 3c') + 2b'(b'-1) + 6b'c' + \tfrac{9}{2}c'(c'-1),$$

$$a'(n'-2) = \kappa' + \rho' + 2\sigma',$$
$$c'(n'-2) = 2\sigma' + (4\beta' + \gamma'),$$

$$(4n' - 6 - 2b' - 3c')(n'-2)(n'-3) = 2\delta' - 4\rho' - 9\sigma',$$
$$(-n'^2 + n' + 4b')(n'-2)(n'-3) = (8k' - 18h') - 4\rho' + 9\sigma',$$

$$*n = n'(n'-1)^2 - b'(2n'-2) - 3c' - \rho' - 9\sigma' - 3(4\beta' + \gamma').$$

We may here take as data n, b, c, $4\beta + \gamma$, $8k - 18h$, the equations exclusively of the two marked (*), then give a, δ, κ, ρ, σ, n', a', δ', κ', b', c', ρ', σ', $4\beta' + \gamma'$, $8k' - 18h'$

and then, since all the quantities entering into the two excepted equations are expressed in terms of the data, these equations are satisfied identically, and it is easy to see that this proves what was before assumed, viz., that in the system of twenty-one equations, the second of the equations marked (*) is satisfied identically.

Several of the other quantities may be expressed without difficulty in terms of the data n, b, c, $4\beta+\gamma$, $8k-18h$: we in fact have (besides a, a', κ', δ', which are originally so expressed):

$$2\sigma = (n-2)\,c - (4\beta+\gamma),$$

$$8\rho = (16n-24)\,b - (15n-18)\,c - 2\,(4b^2-9c^2) + 2\,(8k-18h) - 9\,(4\beta+\gamma),$$

$$8\kappa = 8n\,(n-1)(n-2) - (32n-56)\,b - (17n-46)\,c + 2\,(4b^2-9c^2) - 2\,(8k-18h) + 17\,(4\beta+\gamma),$$

$$2\delta = n\,(n-1)(n-2)(n-3) - (4n^2-20n+24)\,b - (6n^2-15n+18)\,c + 12bc + 18c^2 + (8k-18h) - 9\,(4\beta+\gamma),$$

$$8n' = 8n\,(n-1)^2 - (32n-40)\,b - (21n-30)\,c + 2\,(4b^2-9c^2) - 2\,(8k-18h) + 21\,(4\beta+\gamma),$$

$$c' = 4n\,(n-1)(n-2) - (16n-28)\,b - (10n-26)\,c + (4b^2-9c^2) - (8k-18h) + 10\,(4\beta+\gamma),$$

but the expressions for the remaining quantities, viz., b', ρ', σ', $4\beta'+\gamma'$, $8k'-18h'$ would be very complicated. If we suppose that b, c, $4\beta+\gamma$, $8k-18h$, vanish, or, what is the same thing, attend only to the terms which contain n alone, we have:

$$2b' = n\,(n-1)(n-2)(n^3-n^2+n-12),$$

$$\rho' = n\,(n-2)(n^3-n^2+n-12),$$

$$\sigma' = 4n\,(n-2),$$

$$4\beta'+\gamma' = 4n^2\,(n-2)(n^3-3n^2+3n-3),$$

$$8k'-18h' = n\,(n-2)(n^{10}-6n^9+16n^8-54n^7+164n^6-288n^5+403n^4-482n^3+348n^2-242n+60),$$

which agree with the values which Mr Salmon has obtained for β', γ', h', k' by means of the twenty-one equations, and the additional equations (peculiar to the case in question, of a surface of the degree n without singularities, and which are obtained by him from independent considerations), $i'=0$, and $\beta'=2n\,(n-2)(11n-24)$.

The system of seventeen equations completely accounts for the reduction of the order of the given surface considered as the reciprocal of the reciprocal surface, but the omitted equations are important for other purposes. We may by means of them express i, t in terms of the data for the system of twenty-one equations, viz., n, b, c, β, γ, h, k; and, effecting this, and annexing the corresponding values of i', t', we have the supplementary system:

$$4i = (5n-6)\,c - 6c^2 + 12h - 5\gamma,$$

$$24t = -(8n-8)\,b + (15n-18)\,c + 2\,(4b^2-9c^2) - 16k + 36h + 20\beta - 15\gamma,$$

$$4i' = (5n'-6)\,c' - 6c'^2 + 12h' - 5\gamma',$$

$$24t' = -(8n'-8)\,b' + (15n'-18c') + 2\,(4b'^2-9c'^2) - 16k' + 36h' + 20\beta' - 15\gamma,$$

to which I annex also, without transformation, the four equations for q, r, q', r', viz.:

$$q = b^2 - b - 2k - 3\gamma - 6t,$$
$$r = c^2 - c - 2h - 3\beta,$$
$$q' = b'^2 - b' - 2k' - 3\gamma' - 6t',$$
$$r' = b'^2 - c' - 2h' - 3\beta,$$

the last two of which, neglecting singularities, give

$$q' = 4n(n-2)(n-3)(n^2+2n-4),$$
$$r' = 2n(n-2)(3n-4),$$

which are the values given by Mr Salmon. I remark, in conclusion, that there is considerable difficulty in the geometrical conception of the points i and the planes i', and the subject appears to require further examination. In the case of a surface of the order n without multiple lines, we have not only $i=0$ (which is a matter of course), but also $i'=0$. In my paper before referred to, I showed, or attempted to show, by geometrical reasoning, that the common tangent planes of the spinode develope and the node-couple develope are stationary planes of the one or the other of the two developes, that is, $i'=0$, and the reasoning seems correct as far as it goes, but it was not shown how the demonstration would (as it ought to do) fail in the case of a surface having a double or cuspidal curve. I showed also that in the case where the common tangent plane is a stationary plane of the spinode develope (that is for the planes β'), the spinode curve and the node-couple curve touch instead of simply intersecting; it would seem that the tangent plane at such point is to be counted once, and not twice, in reckoning the number β' of such tangent planes: the like remark applies, of course, also to the points of intersection β of the double and cuspidal curves.

228.

SUR L'INTÉGRALE $\int_0^1 \frac{t^{\mu+\frac{1}{2}}(1-t)^{\mu-\frac{1}{2}}\,dt}{(a+bt-ct^2)^{\mu+1}}$.

Extrait d'une Lettre adressée à M. Liouville.

[From the *Journal de Mathématiques pures et appliquées* (Liouville), vol. II. (1857), pp. 49—51.]

JE suppose que le lecteur ait sous les yeux la Note de M. Liouville (tome I. [1856] page 421). Cela étant, en rétablissant les valeurs de g, h et en écrivant $i-1$ au lieu de μ, la formule de M. Liouville devient

$$\int_0^1 \frac{t^{i-\frac{1}{2}}(1-t)^{i-\frac{3}{2}}\,dt}{(a+bt-ct^2)^i} = \frac{\Gamma(\frac{1}{2})\,\Gamma(i-\frac{1}{2})}{\sqrt{a+b-c}\,\Gamma(i)\,[2a+b+2\sqrt{a(a+b-c)}]^{i-\frac{1}{2}}}, \tag{1}$$

laquelle, en y posant

$$t = \frac{x}{1+x},$$

se transforme en celle-ci:

$$\int_0^\infty \frac{x^{i-1}\,dx}{[(a+b-c)\,x^2+(2a+b)\,x+a]^i} = \frac{\Gamma(\frac{1}{2})\,\Gamma(i-\frac{1}{2})}{\sqrt{a+b-c}\;\Gamma(i)\,[2a+b+2\sqrt{a}\,(a+b-c)]^{i-\frac{1}{2}}}; \tag{2}$$

et en mettant le dénominateur de l'intégrale sous la forme

$$[(a+b-c)\,(x+\lambda)\,(x+\mu)]^i,$$

la formule devient

$$\int_0^\infty \frac{x^{i-\frac{1}{2}}\,dx}{[(x+\lambda)\,(x+\mu)]^i} = \frac{\Gamma(\frac{1}{2})\,\Gamma(i-\frac{1}{2})}{\Gamma(i)}\cdot\frac{1}{(\sqrt{\lambda}+\sqrt{\mu})^{2i-1}}: \tag{3}$$

formule qui se trouve dans mon Mémoire: "On certain formulæ for differentiation with applications to the evaluation of definite integrals," *Camb. and Dubl. Math. Journal,* tom. II., page 122, (1847), [41]. La démonstration que j'y ai donnée n'est cependant rigoureuse (à moins que l'on n'admette la théorie de la différentiation à indices quelconques) que pour le cas de $i+\frac{1}{2}$ égal à un entier positif; en effet, en écrivant

$$U_{k,i}=[(x+\lambda)(x+\mu)]^{\frac{1}{2}k}(\sqrt{x+\lambda}-\sqrt{x+\mu})^{2i},$$

je suis parti de la formule

$$\frac{d}{dx}U_{k,i}=\tfrac{1}{2}k(\lambda-\mu)^2\,U_{k-2,i-1}-(k+i)\,U_{k-1,i},$$

pour en déduire l'expression générale de $\left(\frac{d}{dx}\right)^s U_{0,i}$. Cette expression devient très-simple pour le cas $s=i+1$; on a alors

$$\frac{(-)^{i+1}}{i}\left(\frac{d}{dx}\right)^{i+1}(\sqrt{x+\lambda}-\sqrt{x+\mu})^{2i}=\frac{\Gamma(i+\frac{1}{2})}{\Gamma(\frac{1}{2})}(\lambda-\mu)^{2i}\frac{1}{[(x+\lambda)(x+\mu)]^{i+\frac{1}{2}}}, \qquad (4)$$

formule de différentiation assez singulière. En y écrivant $i-\frac{1}{2}$, au lieu de i, et en intégrant $i+\frac{1}{2}$ fois par la formule

$$\int_0^\infty x^{i-\frac{1}{2}}fx\,dx=\frac{\Gamma(i+\frac{1}{2})}{(-)^{i+\frac{1}{2}}}\left(\int_\infty d\alpha\right)^{i+\frac{1}{2}}f\alpha, \qquad \alpha=0,$$

on obtient la formule intégrale ci-devant mentionnée. Il convient d'ajouter que cette formule, représentée sous la forme

$$\int_0^\infty\frac{x^{i-\frac{1}{2}}\,dx}{(ax^2+bx+c)^i}=\frac{\Gamma(\frac{1}{2})\,\Gamma(i-\frac{1}{2})}{\Gamma(i)}\,\frac{1}{(b+2\sqrt{ac})^{i-\frac{1}{2}}}, \qquad (5)$$

est comprise dans une formule générale donnée par M. Schlömilch: "Note sur la variation des constantes arbitraires d'une intégrale définie (*Crelle*, tom. XXXIII., page 268; 1847).

Je remarque aussi que M. Donkin, en comparant ses résultats concernant les fonctions de Laplace avec ceux de M. Boole sur le même sujet, a trouvé une identité, laquelle m'a conduit à cette autre formule de différentiation

$$\left\{\frac{[(x+\lambda)(x+\mu)]^2}{2x+\lambda+\mu}\frac{d}{dx}\right\}^n\frac{(2x+\lambda+\mu)^{n+s-1}}{[(x+\lambda)(x+\mu)]^{n-\frac{1}{2}}(\sqrt{x+\lambda}-\sqrt{x+\mu})^{2n-2s}}$$
$$=\frac{[(x+\lambda)(x+\mu)]^{n+1}}{(2x+\lambda+\mu)^{n-s+1}}\left(\frac{d}{dx}\right)^n\frac{(\sqrt{x+\lambda}-\sqrt{x+\mu})^{2s}}{[(x+\lambda)(x+\mu)]^{\frac{1}{2}}}, \qquad (6)$$

formule que j'ai démontrée au moyen d'une analyse assez compliquée.

229.

NOTE SUR UNE FORMULE POUR LA RÉVERSION DES SÉRIES.

[From the *Journal für die reine und angewandte Mathematik* (Crelle), tom. LII. (1856), pp. 276—284.]

Je me propose de développer dans cette note, pour le cas de trois variables (ce qui suffit pour faire voir la loi dans le cas d'un nombre quelconque de variables) une formule qui se trouve dans le mémoire remarquable de Jacobi, "de resolutione aequationum per series infinitas," *Crelle*, t. VI. [1830] p. 257. Voici la formule dont il s'agit. Soit $f(x, y, z)$ une fonction rationnelle et entière des variables x, y, z, et mettons $u = X$, $v = Y$, $w = Z$, où X, Y, Z sont des fonctions rationnelles et entières des variables x, y, z, telles que $X - x$, $Y - y$, $Z - z$ ne contiennent que les puissances et les produits du deuxième ordre et des ordres supérieurs des variables. Cela étant, on aura

$$f(x, y, z) = \left[f(x, y, z) \frac{\partial(X, Y, Z)}{\partial(x, y, z)} \cdot \frac{1}{(X-u)(Y-v)(Z-w)} \right]_{x^{-1}y^{-1}z^{-1}},$$

où la notation $\dfrac{\partial(X, Y, Z)}{\partial(x, y, z)}$ dénote le déterminant fonctionnel ou "Jacobian" de X, Y, Z par rapport à x, y, z et la notation $[\ ]_{x^{-1}y^{-1}z^{-1}}$ signifie le coefficient de $x^{-1}y^{-1}z^{-1}$ dans le développement de la fonction en dedans des [ ]; ce développement doit s'effectuer d'une manière déterminée, savoir il faut d'abord développer les facteurs $\dfrac{1}{X-u}$ selon les puissances descendantes de x, c'est-à-dire dans la forme $\dfrac{1}{X} + \dfrac{u}{X^2} + \dfrac{u^2}{X^3} + \text{etc.}$, et puis écrivant $X = x + P$ (où P est fonction des trois variables) il faut développer les puissances de X selon les puissances descendantes du monôme x, c'est-à-dire dans la forme $X^{-m} = x^{-m} - mx^{-m-1}P + \tfrac{1}{2}m(m+1)x^{-m-2}P^2 - \text{etc.}$, ce qui est en effet un développement selon les puissances *ascendantes* des variables.

La formule donne

$$\text{Coeff. } u^{\mathrm{a}}v^{\mathrm{b}}w^{\mathrm{c}} \text{ dans } f(x, y, z) = \left[f(x, y, z)\frac{\partial(X, Y, Z)}{\partial(x, y, z)} \cdot \frac{1}{X^{\mathrm{a}+1}Y^{\mathrm{b}+1}Z^{\mathrm{c}+1}}\right]_{x^{-1}y^{-1}z^{-1}},$$

ou ce qui est la même chose

$$\text{Coeff. } u^{\mathrm{a}}v^{\mathrm{b}}w^{\mathrm{c}} \text{ dans } f(x, y, z) = \left[f(x, y, z)\frac{\partial\left(-\frac{1}{\mathrm{a}}X^{-\mathrm{a}}, -\frac{1}{\mathrm{b}}Y^{-\mathrm{b}}, -\frac{1}{\mathrm{c}}Z^{-\mathrm{c}}\right)}{\partial(x, y, z)}\right]_{x^{-1}y^{-1}z^{-1}}.$$

Soit à présent

$$\begin{aligned} X &= x\,..+A_{f,g,h}\,x^fy^gz^h + \text{etc.},\\ Y &= y\,..+B_{i,j,k}\,x^iy^jz^k + \text{etc.},\\ Z &= z\,..+C_{l,m,n}\,x^ly^mz^n + \text{etc.},\\ f(x, y, z) &= \ldots\ \Theta_{P,Q,R}\,x^Py^Qz^R + \text{etc.}; \end{aligned}$$

dans ces expressions et partout dans ce qui suit, les etc.'s se rapportent à des termes que l'on obtient en affixant des accents en nombre quelconque aux symboles indéterminés.

En employant la notation de Gauss $\Pi\alpha = 1.2.3\ldots\alpha$, on obtient pour la terme général de $-\frac{1}{\mathrm{a}}X^{-\mathrm{a}}$,

$$(-)^{r-1}\frac{\Pi(\mathrm{a}+r-1)}{\Pi\mathrm{a}\,\Pi\alpha\text{ etc.}}A^{\alpha}{}_{f,g,h}\text{ etc. } x^{-\mathrm{a}-r+F}y^Gz^H,$$

où

$$\alpha + \text{etc.} = r,\quad f\alpha + \text{etc.} = F,\quad g\alpha + \text{etc.} = G,\quad h\alpha + \text{etc.} = H.$$

De même le terme général de $-\frac{1}{\mathrm{b}}Y^{-\mathrm{b}}$ est

$$(-)^{s-1}\frac{\Pi(\mathrm{b}+s-1)}{\Pi\mathrm{b}\,\Pi\beta\text{ etc.}}B^{\beta}{}_{i,j,k}\text{ etc. } x^Iy^{-\mathrm{b}-s+J}z^K,$$

où

$$\beta + \text{etc.} = s,\quad i\beta + \text{etc.} = I,\quad j\beta + \text{etc.} = J,\quad k\beta + \text{etc.} = K,$$

et le terme général de $-\frac{1}{\mathrm{c}}Z^{-\mathrm{c}}$ est

$$(-)^{t-1}\frac{\Pi(\mathrm{c}+t-1)}{\Pi\mathrm{c}\,\Pi\gamma\text{ etc.}}C^{\gamma}{}_{l,m,n,}\text{ etc. } x^Ly^Mz^{-\mathrm{c}-t+N},$$

où

$$\gamma + \text{etc.} = t,\quad l\gamma + \text{etc.} = L,\quad m\gamma + \text{etc.} = M,\quad n\gamma + \text{etc.} = N.$$

En formant de là le terme général du Jacobian et en multipliant par le terme général de $f(x, y, z)$ on obtient pour le terme général de l'expression en dedans des [ ], la valeur que voici,

$$(-)^{r+s+t}\frac{\Pi(\mathrm{a}+r-1)\,\Pi(\mathrm{b}+s-1)\,\Pi(\mathrm{c}+t-1)}{\Pi\mathrm{a}\,\Pi\mathrm{b}\,\Pi\mathrm{c}\,\Pi\alpha\text{ etc. }\Pi\beta\text{ etc. }\Pi\gamma\text{ etc.}}\times$$

$$A^{\alpha}{}_{f,g,h}\text{ etc. }B^{\beta}{}_{i,j,k}\text{ etc. }C^{\gamma}{}_{l,m,n}\text{ etc. }\Theta_{P,Q,R}\begin{vmatrix}\mathrm{a}+r-F, & -G, & -H\\ -I, & \mathrm{b}+s-J, & -K\\ -L, & -M, & \mathrm{c}+t-N\end{vmatrix}\times$$

$$x^{-\mathrm{a}-r+F+I+L+P-1}\,.\,y^{-\mathrm{b}-s+G+J+M+Q-1}\,.\,z^{-\mathrm{c}-t+H+K+N+R-1},$$

et pour trouver le terme qui contient $x^{-1}y^{-1}z^{-1}$ on n'a qu'à écrire dans cette expression

$$F+I+L+P=\mathrm{a}+r, \quad G+J+M+Q=\mathrm{b}+s, \quad H+K+N+R=\mathrm{c}+t,$$

le coefficient de $x^{-1}y^{-1}z^{-1}$ sera alors la valeur de l'expression $[\ ]_{x^{-1}y^{-1}z^{-1}}$ qu'il s'agissait de trouver. En effectuant cela et en recapitulant les formules on obtient le théorème suivant: en posant

$$\begin{aligned} X &= x \ldots + A_{f,g,h}\, x^f y^g z^h + \text{etc.} = u, \\ Y &= y \ldots + B_{i,j,k}\, x^i y^j z^k + \text{etc.} = v, \\ Z &= z \ldots + C_{l,m,n}\, x^l y^m z^n + \text{etc.} = w, \\ f(x, y, z) &= \quad \ldots + \Theta_{P,Q,R}\, x^P y^Q z^R + \text{etc.} \end{aligned}$$

on aura pour le terme général du coeff. $u^{\mathrm{a}}v^{\mathrm{b}}w^{\mathrm{c}}$ dans $f(x, y, z)$ la valeur que voici,

$$(-)^{r+s+t}\frac{\Pi(\mathrm{a}+r-1)\,\Pi(\mathrm{b}+s-1)\,\Pi(\mathrm{c}+t-1)}{\Pi\mathrm{a}\,\Pi\mathrm{b}\,\Pi\mathrm{c}\,\Pi\alpha \text{ etc. } \Pi\beta \text{ etc. } \Pi\gamma \text{ etc.}}\times$$

$$A^{\alpha}{}_{f,g,h} \text{ etc. } B^{\beta}{}_{i,j,k} \text{ etc. } C^{\gamma}{}_{l,m,n} \text{ etc. } \Theta_{P,Q,R} \begin{vmatrix} P+I+L, & -G, & -H \\ -I, & Q+G+M, & -K \\ -L, & -M, & R+H+K \end{vmatrix}$$

dans laquelle

$$\begin{array}{lll} \alpha + \text{etc.} = r, & \beta + \text{etc.} = s, & \gamma + \text{etc.} = t, \\ f\alpha + \text{etc.} = F, & g\alpha + \text{etc.} = G, & h\alpha + \text{etc.} = H, \\ i\beta + \text{etc.} = I, & j\beta + \text{etc.} = J, & k\beta + \text{etc.} = K, \\ l\gamma + \text{etc.} = L, & m\gamma + \text{etc.} = M, & n\gamma + \text{etc.} = N, \\ P+F+I+L=\mathrm{a}+r, & Q+G+J+M=\mathrm{b}+s, & R+H+K+N=\mathrm{c}+t. \end{array}$$

Les dernières équations peuvent s'écrire sous la forme

$$\begin{array}{llll} P+(f-1)\,\alpha+\text{etc.}+ & i\beta+\text{etc.}+ & l\gamma+\text{etc.} & =\mathrm{a}, \\ Q+ \quad g\alpha+\text{etc.}+ & (j-1)\,\beta+\text{etc.}+ & m\gamma+\text{etc.} & =\mathrm{b}, \\ R+ \quad h\alpha+\text{etc.}+ & k\beta+\text{etc.}+ & (n-1)\,\gamma+\text{etc.} & =\mathrm{c}, \end{array}$$

qui sont les conditions auxquelles doivent satisfaire les valeurs de P, Q, R, α, etc., β, etc., γ, etc., f, g, h, etc., i, j, k, etc., l, m, n, etc.; en ajoutant ces équations on obtient

$$P+Q+R+(f+g+h-1)\,\alpha+\text{etc.}+(i+j+k-1)\,\beta+\text{etc.}+(l+m+n-1)\,\gamma+\text{etc.} \\ =\mathrm{a}+\mathrm{b}+\mathrm{c},$$

où les nombres $f+g+h-1$, etc. $i+j+k-1$, etc. $l+m+n-1$, etc. sont positifs, il n'y a donc qu'un nombre fini (comme cela doit être) de solutions des équations indéterminées.

Il y a une manière assez simple pour calculer le déterminant qui entre dans la formule, pour cela je représente les termes P, I, etc. par les notations symboliques $X\vartheta$, Xy, etc. de manière que le déterminant devient

$$\begin{vmatrix} X\vartheta + Xy + Xz, & -Yx, & -Zx \\ -Xy, & Y\vartheta + Yx + Yz, & -Zy \\ -Xz, & -Yz, & Z\vartheta + Zx + Zy \end{vmatrix},$$

or ce déterminant est ce que devient le produit

$$(X\vartheta + Xy + Xz)(Y\vartheta + Yz + Yx)(Z\vartheta + Zx + Zy),$$

en omettant du développement tous les termes qui contiennent un cycle tel que $Xy.Yx$, ou $Xy.Yz.Zx$. Cela donne pour le déterminant la somme des seize termes

$$\begin{aligned} X\vartheta\,.\,Y\vartheta\,.\,Z\vartheta + Y\vartheta\,.\,Z\vartheta\,(Xy + Xz) + Z\vartheta\,.\,X\vartheta\,(Yz + Yx) + X\vartheta\,.\,Y\vartheta\,(Zx + Zy) \\ + X\vartheta\,(Yx\,.\,Zx + Yx\,.\,Zy + Yz\,.\,Zx) + Y\vartheta\,(Zy\,.\,Xy + Zy\,.\,Xz + Zx\,.\,Xy) \\ + Z\vartheta\,(Xz\,.\,Yz + Xz\,.\,Yx + Xy\,.\,Yz); \end{aligned}$$

et la même chose est vraie quel que soit l'ordre du déterminant. C'est M. Sylvester qui m'a fait cette remarque.

Les formules de Jacobi s'appliquent aussi au cas où u, v, w, etc. sont données en termes de x, y, z, etc. au moyen d'équations et non pas explicitement comme auparavant; mais je ne chercherai pas à présent ce que deviennent les formules pour ce cas plus général.

On peut appliquer la formule au problème de la transformation des variables indépendantes dans le calcul différentiel. En effet, soient u, v, w des fonctions quelconques de x, y, z, et prenons ξ, η, ζ pour les incréments de x, y, z respectivement et υ, ν, ω pour les incréments de u, v, w respectivement; on aura

$$\xi = \left\{\left(\upsilon\frac{d}{du} + \nu\frac{d}{dv} + \omega\frac{d}{dw}\right) + \tfrac{1}{2}\left(\upsilon\frac{d}{du} + \nu\frac{d}{dv} + \omega\frac{d}{dw}\right)^2 + \text{etc.}\right\} x,$$

$$\eta = \left\{\left(\upsilon\frac{d}{du} + \nu\frac{d}{dv} + \omega\frac{d}{dw}\right) + \tfrac{1}{2}\left(\upsilon\frac{d}{du} + \nu\frac{d}{dv} + \omega\frac{d}{dw}\right)^2 + \text{etc.}\right\} y,$$

$$\zeta = \left\{\left(\upsilon\frac{d}{du} + \nu\frac{d}{dv} + \omega\frac{d}{dw}\right) + \tfrac{1}{2}\left(\upsilon\frac{d}{du} + \nu\frac{d}{dv} + \omega\frac{d}{dw}\right)^2 + \text{etc.}\right\} z.$$

Soit

$$\nabla = \frac{\partial(x,\ y,\ z)}{\partial(u,\ v,\ w)}$$

le Jacobian de x, y, z par rapport à u, v, w; et mettons

$$\left(\upsilon \frac{d}{du} + \nu \frac{d}{dv} + \omega \frac{d}{dw}\right) x = p,$$

$$\left(\upsilon \frac{d}{du} + \nu \frac{d}{dv} + \omega \frac{d}{dw}\right) y = q,$$

$$\left(\upsilon \frac{d}{du} + \nu \frac{d}{dv} + \omega \frac{d}{dw}\right) z = r,$$

υ, ν, ω seront des fonctions linéaires de p, q, r, et, en formant avec ces valeurs l'expression de $\upsilon \frac{d}{du} + \nu \frac{d}{dv} + \omega \frac{d}{dw}$, on trouve

$$\upsilon \frac{d}{du} + \nu \frac{d}{dv} + \omega \frac{d}{dw} = p\bar{x} + q\bar{y} + r\bar{z},$$

où $\bar{x}$, $\bar{y}$, $\bar{z}$ sont des opérations de la forme $L\frac{d}{du} + M\frac{d}{dv} + N\frac{d}{dw}$, qui peuvent être représentées symboliquement de cette manière,

$$\bar{x} = \frac{1}{\nabla} \begin{vmatrix} \frac{d}{du}, & \frac{d}{dv}, & \frac{d}{dw} \\ \frac{dy}{du}, & \frac{dy}{dv}, & \frac{dy}{dw} \\ \frac{dz}{du}, & \frac{dz}{dv}, & \frac{dz}{dw} \end{vmatrix}$$

et de même pour $\bar{y}$ et $\bar{z}$; il faut faire attention, qu'en opérant avec ces symboles, il faut traiter comme des constantes les fonctions de u, v, w qui entrent dans ces mêmes symboles.

Nous avons évidemment $\bar{x}x = 1$, $\bar{y}x = 0$, $\bar{z}x = 0$, et de même $\bar{x}y = 0$, etc.; on obtient de là

$$\xi = p + \tfrac{1}{2}(p\bar{x} + q\bar{y} + r\bar{z})^2 x + \text{etc.},$$
$$\eta = q + \tfrac{1}{2}(p\bar{x} + q\bar{y} + r\bar{z})^2 y + \text{etc.},$$
$$\zeta = r + \tfrac{1}{2}(p\bar{x} + q\bar{y} + r\bar{z})^2 z + \text{etc.}$$

Soit à présent ϑ une fonction quelconque de x, y, z, ou de u, v, w; en envisageant ϑ comme fonction de x, y, z, on trouve l'incrément de cette fonction en opérant sur ϑ avec le symbole

$$\left(\xi \frac{d}{dx} + \eta \frac{d}{dy} + \zeta \frac{d}{dz}\right) + \tfrac{1}{2}\left(\xi \frac{d}{dx} + \eta \frac{d}{dy} + \zeta \frac{d}{dz}\right)^2 + \text{etc.},$$

mais en envisageant ϑ comme fonction de u, v, w et en faisant attention à l'équation $\upsilon \frac{d}{du} + \nu \frac{d}{dv} + \omega \frac{d}{dw} = p\bar{x} + q\bar{y} + r\bar{z}$, on trouve ce même incrément en opérant sur ϑ avec le symbole

$$(p\bar{x} + q\bar{y} + r\bar{z}) + \tfrac{1}{2}(p\bar{x} + q\bar{y} + r\bar{z})^2 + \text{etc.},$$

et les deux résultats deviendront identiques en substituant pour p, q, r les valeurs de ces quantités en termes de ξ, η, ζ, valeurs qui se trouvent par la réversion des séries qui donnent ξ, η, ζ en termes de p, q, r. C'est-à-dire nous aurons

$$\left(\frac{d}{dx}\right)^{\mathrm{a}}\left(\frac{d}{dy}\right)^{\mathrm{b}}\left(\frac{d}{dz}\right)^{\mathrm{c}} = \Pi\mathrm{a}\,\Pi\mathrm{b}\,\Pi\mathrm{c}.\ \text{coeff.}\ \xi^{\mathrm{a}}\eta^{\mathrm{b}}\zeta^{\mathrm{c}}\ \text{dans}$$

$$\{(p\bar{x}+q\bar{y}+r\bar{z})+\tfrac{1}{2}(p\bar{x}+q\bar{y}+r\bar{z})^2+\text{etc.}\},$$

où

$$\xi = p+\tfrac{1}{2}(p\bar{x}+q\bar{y}+r\bar{z})^2 x+\text{etc.},$$
$$\eta = q+\tfrac{1}{2}(p\bar{x}+q\bar{y}+r\bar{z})^2 y+\text{etc.},$$
$$\zeta = r+\tfrac{1}{2}(p\bar{x}+q\bar{y}+r\bar{z})^2 z+\text{etc.}$$

C'est là le problème de la réversion des séries qui vient d'être traité; et en substituant

$$\frac{1}{\Pi f\,\Pi g\,\Pi h}x^{-f}y^{-g}z^{-h}x,\quad \frac{1}{\Pi i\,\Pi j\,\Pi k}x^{-i}y^{-j}z^{-k}y,\quad \frac{1}{\Pi l\,\Pi m\,\Pi n}x^{-l}y^{-m}z^{-n}z,\quad \frac{1}{\Pi P\,\Pi Q\,\Pi R}x^{-P}y^{-Q}z^{-R}$$

au lieu de

$$A_{f,\,g,\,h},\qquad B_{i,\,j,\,k},\qquad C_{l,\,m,\,n},\qquad \Theta_{P,\,Q,\,R},$$

on trouve le théorème suivant:

THÉORÈME. Le terme général de $\left(\frac{d}{dx}\right)^{\mathrm{a}}\left(\frac{d}{dy}\right)^{\mathrm{b}}\left(\frac{d}{dz}\right)^{\mathrm{c}}$ est

$$K\Omega\,(x^{-f}y^{-g}z^{-h}x)^{\alpha}\ \text{etc.}\ (x^{-i}y^{-j}z^{-k}y)^{\beta}\ \text{etc.}\ (x^{-l}y^{-m}z^{-n}z)^{\gamma}\ \text{etc.}\ x^{-P}y^{-Q}z^{-R},$$

expression dans laquelle

$$\begin{array}{lll} \alpha+\text{etc.}=r, & \beta+\text{etc.}=s, & \gamma+\text{etc.}=t, \\ f\alpha+\text{etc.}=F, & g\alpha+\text{etc.}=G, & h\alpha+\text{etc.}=H, \\ i\beta+\text{etc.}=I, & j\beta+\text{etc.}=J, & k\beta+\text{etc.}=K, \\ l\gamma+\text{etc.}=L, & m\gamma+\text{etc.}=M, & n\gamma+\text{etc.}=N, \end{array}$$

$$F+I+L+P=\mathrm{a}+r,\quad G+J+M+Q=\mathrm{b}+s,\quad H+K+N+R=\mathrm{c}+t;$$

$$K=(-)^{r+s+t}\times$$
$$\frac{\Pi(\mathrm{a}+r-1)\,\Pi(\mathrm{b}+s-1)\,\Pi(\mathrm{c}+t-1)}{\Pi\alpha\ \text{etc.}\ \Pi\beta\ \text{etc.}\ \Pi\gamma\ \text{etc.}\ (\Pi f\,\Pi g\,\Pi h)^{\alpha}\ \text{etc.}\ (\Pi i\,\Pi j\,\Pi k)^{\beta}\ \text{etc.}\ (\Pi l\,\Pi m\,\Pi n)^{\gamma}\ \text{etc.}\ \Pi P\,\Pi Q\,\Pi R};$$

$$\Omega = \begin{vmatrix} P+I+L, & -G, & -H \\ -I, & Q+G+M, & -K \\ -L, & -M, & R+H+K \end{vmatrix},$$

et où les nombres $f+g+h-1$, etc., $i+j+k-1$, etc., $l+m+n-1$, etc. sont tous positifs comme auparavant.

La formule contient les symboles $\bar{x}$, $\bar{y}$, $\bar{z}$ qui sont chacun une fonction linéaire de $\frac{d}{du}$, $\frac{d}{dv}$, $\frac{d}{dw}$, on pourrait donc se proposer la question de trouver le terme général en développant la formule de manière à ne contenir que des puissances et des produits de ces symboles $\frac{d}{du}$, $\frac{d}{dv}$, $\frac{d}{dw}$; c'est à quoi se rapportent les recherches très étendues que vient de faire M. Sylvester sur ce sujet, et qui embrassent aussi bien le cas où les nouvelles variables sont données explicitement que celui où les deux systèmes de variables sont liés par des équations données.

Il y a une autre manière de traiter cette question de la transformation des variables indépendantes, savoir en écrivant

$$R = \xi \frac{d}{dx} + \eta \frac{d}{dy} + \zeta \frac{d}{dz},$$

$$\rho = \xi \;\bar{x} + \eta \;\bar{y} + \zeta \;\bar{z},$$

on peut exprimer les puissances de R au moyen de ρ et de cette autre quantite symbolique

$$\Lambda = x \;\bar{x} + y \;\bar{y} + z \;\bar{z}.$$

En effet, en mettant $\chi = p\bar{x} + q\bar{y} + r\bar{z}$, on a

$$R + \tfrac{1}{2} R^2 + \tfrac{1}{6} R^3 + \text{etc.} = \chi + \tfrac{1}{2} \chi^2 + \tfrac{1}{6} \chi^3 + \text{etc.},$$

où

$$\xi = p + \tfrac{1}{2} \chi^2 x + \tfrac{1}{6} \chi^3 x + \text{etc.},$$
$$\eta = q + \tfrac{1}{2} \chi^2 y + \tfrac{1}{6} \chi^3 y + \text{etc.},$$
$$\zeta = r + \tfrac{1}{2} \chi^2 z + \tfrac{1}{6} \chi^3 z + \text{etc.}$$

En se servant de la méthode des approximations successives, on trouve comme première approximation

$$p = \xi, \quad q = \eta, \quad r = \zeta,$$

et de là

$$\chi = \rho;$$

comme seconde approximation

$$p = \xi - \tfrac{1}{2}\rho^2 x, \quad q = \eta - \tfrac{1}{2}\rho^2 y, \quad r = \zeta - \tfrac{1}{2}\rho^2 z,$$

et de là

$$\chi = \rho - \tfrac{1}{2}\rho^2 \Lambda;$$

comme troisième approximation

$$p = \xi - \tfrac{1}{2}(\rho^2 - \rho^2 \Lambda\rho)\, x - \tfrac{1}{6}\rho^3 x,$$
$$q = \eta - \tfrac{1}{2}(\rho^2 - \rho^2 \Lambda\rho)\, y - \tfrac{1}{6}\rho^3 y,$$
$$r = \zeta - \tfrac{1}{2}(\rho^2 - \rho^2 \Lambda\rho)\, z - \tfrac{1}{6}\rho^3 z,$$

et de là

$$\chi = \rho - \tfrac{1}{2}\rho^2\Lambda + \tfrac{1}{2}\rho^2\Lambda\rho\Lambda - \tfrac{1}{6}\rho^3\Lambda, \quad \chi^2 = \rho^2 - \rho^2\Lambda\rho, \quad \chi^3 = \rho^3$$

et ainsi de suite; donc, en substituant,

$$\begin{aligned} R + \tfrac{1}{2}R^2 + \tfrac{1}{6}R^3 + \text{etc.} = \rho &- \tfrac{1}{2}\rho^2\Lambda + \tfrac{1}{2}\rho^2\Lambda\rho\Lambda - \tfrac{1}{6}\rho^3\Lambda \\ &+ \tfrac{1}{2}\rho^2 \quad - \tfrac{1}{2}\rho^2\Lambda\rho \\ &\qquad\qquad + \tfrac{1}{6}\rho^3 \quad + \text{etc. etc.} \end{aligned}$$

c'est-à-dire

$$\begin{aligned} R &= \rho, \\ R^2 &= \rho^2 - \rho^2\Lambda, \\ R^3 &= \rho^3 - 3\rho^2\Lambda\rho - (\rho^3 - 3\rho^2\Lambda\rho)\,\Lambda, \\ &\text{etc.} \end{aligned}$$

formules dans lesquelles un terme tel que $\rho^2\Lambda$ signifie $\rho^2 x\,.\,\bar{x} + \rho^2 y\,.\,\bar{y} + \rho^2 z\,.\,z$, c'est-à-dire l'opération du symbole ρ^2 s'arrête aux quantités x, y, z faisant partie du symbole Λ qui vient immédiatement après le ρ^2; la même chose a lieu dans tous les cas semblables. L'équation générale est

$$R^n = \Pi - \Pi\Lambda,$$

mais pour expliquer la forme de la fonction Π il faudrait faire des développements assez longs, dans lesquels je n'entrerai pas à cette occasion; la découverte de cette équation générale est due à M. Sylvester, qui l'a établie d'une autre manière.

Londres, 2, Stone Buildings, 16 *Avril,* 1855.

230.

NOTE SUR LA MÉTHODE D'ÉLIMINATION DE BEZOUT.

[From the *Journal für die reine und angewandte Mathematik*, (Crelle), tom. LIII. (1857), pp. 366—367.]

VOICI la forme la plus simple sous laquelle on peut présenter cette méthode. Pour éliminer les variables x, y entre deux équations du $n^{\text{ième}}$ degré

$$(a, \ldots \between x, y)^n = 0,$$

$$(a', \ldots \between x, y)^n = 0,$$

on n'a qu'à former l'équation identique

$$\frac{(a, \ldots \between x, y)^n (a', \ldots \between \lambda, \mu)^n - (a', \ldots \between x, y)^n (a, \ldots \between \lambda, \mu)^n}{\mu x - \lambda y} =$$

$$\left(\begin{array}{llll} a_{0,0}, & a_{1,0}, & \ldots & a_{n-1,0} \\ a_{0,1}, & a_{1,1}, & \ldots & a_{n-1,1} \\ \vdots & & & \\ a_{0,n-1}, & a_{1,n-1}, & \ldots & a_{n-1,n-1} \end{array}\between x, y)^{n-1} (\lambda, \mu)^{n-1}$$

où l'expression qui forme le second membre représente la fonction suivante,

$$\begin{aligned} &(a_{0,0} \; x^{n-1} + a_{1,0} \; x^{n-2} y \ldots + a_{n-1,0} \; y^{n-1}) \lambda^{n-1} \\ + &(a_{0,1} \; x^{n-1} + a_{1,1} \; x^{n-2} y \ldots + a_{n-1,1} \; y^{n-1}) \lambda^{n-2} \mu \\ &\vdots \\ + &(a_{0,n-1} x^{n-1} + a_{1,n-1} x^{n-2} y \ldots + a_{n-1,n-1} y^{n-1}) \mu^{n-1}; \end{aligned}$$

le résultat de l'élimination sera

$$\begin{vmatrix} a_{0,0} & , a_{1,0} & , \dots a_{n-1,0} \\ a_{0,1} & , a_{1,1} & , \dots a_{n-1,1} \\ \vdots & & \\ a_{0,n-1}, & a_{1,n-1}, & \dots a_{n-1,n-1} \end{vmatrix} = 0.$$

Par exemple on trouve

$$\frac{(a,\ b,\ c\between x,\ y)^2(a',\ b',\ c'\between\lambda,\ \mu)^2-(a',\ b',\ c'\between x,\ y)^2(a,\ b,\ c\between\lambda,\ \mu)^2}{\mu x-\lambda y}=$$

$$\left(\begin{array}{cc} 2(ab'-a'b), & ac'-a'c \\ ac'-a'c\ , & 2(bc'-b'c) \end{array}\between x,\ y\between\lambda,\ \mu\right);$$

$$\frac{(a,\ b,\ c,\ d\between x,\ y)^3(a',\ b',\ c',\ d'\between\lambda,\ \mu)^3-(a',\ b',\ c',\ d'\between x,\ y)^3(a,\ b,\ c,\ d\between\lambda,\ \mu)^3}{\mu x-\lambda y}=$$

$$\left(\begin{array}{ccc} 3(ab'-a'b), & 3(ac'-a'c), & (ad'-a'd) \\ 3(ac'-a'c), & (ad'-a'd)+9(bc'-b'c), & 3(bd'-b'd) \\ (ad'-a'd), & 3(bd'-b'd), & 3(cd'-c'd) \end{array}\between x^2,\ xy,\ y^2\between\lambda^2,\ \lambda\mu,\ \mu^2\right);$$

d'où l'on tire immédiatement les résultats de l'élimination entre deux équations quadratiques ou cubiques.

Londres, 2 Stone Buildings, Avril, 1855.

231.

NOTE SUR L'ÉQUATION $x^2 - Dy^2 = \pm 4$, $D \equiv 5$ (mod. 8).

[From the *Journal für die reine und angewandte Mathematik*, (Crelle), tom. LIII. (1857), pp. 369—371.]

ON sait que pour un déterminant positif $D \equiv 5$ (mod. 8) le nombre des formes quadratiques proprement primitives est égal au nombre des formes improprement primitives, quand il existe une solution en nombres impairs de l'équation $x^2 - Dy^2 = 4$; mais que le nombre des formes proprement primitives est égal à trois fois le nombre des formes improprement primitives, quand il n'existe pas de solution en nombres impairs de l'équation $x^2 - Dy^2 = 4$.

Les tables de Degen font voir immédiatement, pour les nombres $D \equiv 5$ (mod. 8) qui ne sont pas plus grands que 997, s'il existe ou non une solution en nombres impairs de l'équation $x^2 - Dy^2 = 4$. À savoir si le nombre 4 se trouve dans la série des dénominateurs des quotients complets (la seconde ligne dans les tables), il existe une solution de l'équation; si non, il n'en existe pas. De plus, si la place où se trouve le nombre 4 est d'ordre pair il existe une solution tant de l'équation $x^2 - Dy^2 = +4$ que de l'équation $x^2 - Dy^2 = -4$, mais si cette place est d'ordre impair il existe seulement une solution de l'équation $x^2 - Dy^2 = 4$. On trouve la plus petite solution de ces équations au moyen de la série des quotients (la première ligne des tables), en s'arrêtant au nombre qui précède le nombre placé au-dessus du nombre 4, et en calculant la fraction continue déterminée par cette suite; par exemple, pour le nombre 61 on trouve dans les tables

61	7, 1, 4, 3, 1 (2, 2)
	1, 12, 3, 4, 9 (5, 5)
	226153980
	1766319049

Cela fait voir qu'il existe une solution de l'équation $x^2 - Dy^2 = -4$, solution que l'on obtient au moyen de la fraction continue $7 + \frac{1}{1+} \frac{1}{4}, = \frac{39}{5}$ en faisant $x = 39$, $y = 5$. La plus petite solution de l'équation $x^2 - Dy^2 = 4$ se déduit très facilement de la plus petite solution de l'équation $\tau^2 - Dv^2 = -4$ au moyen de la formule $x + y\sqrt{D} = \frac{1}{2}(\tau + v\sqrt{D})^2$, ce qui donne $x = \tau^2 + 2$, $y = \tau v$.

On obtient pareillement la plus petite solution de l'équation $x^2 - Dy^2 = -1$ au moyen de la formule $(x + y\sqrt{D}) = \frac{1}{8}(\tau + v\sqrt{D})^3$, ce qui donne

$$x = \tfrac{1}{2}(\tau^3 + 3\tau), \quad y = \tfrac{1}{2}(\tau^2 + 1)\, v.$$

De même la plus petite solution de l'équation $x^2 - Dy^2 = 1$ se déduit de la plus petite solution de l'équation $T^2 - DU^2 = 4$ au moyen de la formule $x + y\sqrt{D} = \frac{1}{8}(T + U\sqrt{D})^3$, ce qui donne $x = \frac{1}{2}(T^3 - 3T)$, $y = \frac{1}{2}(T^2 - 1)\, U$.

Je fais observer à cette occasion que suivant une remarque de Göpel ("Notiz über A. Göpel," t. XXXV. [1847] p. 315 de ce Journal) si dans l'équation $\left(\frac{x + \sqrt{y}}{p}\right)^n = P + \sqrt{Q}$, où x, y, p, n, P, Q sont des entiers, le dénominateur p est plus grand que l'unité et x, y, p n'ont pas de dénominateur commun, on aura nécessairement $p = 2$, $n = 3$ ou égal à un multiple de 3, x impair, et y de la forme $8n + 5$.

J'ai calculé au moyen des tables de Degen la table suivante des plus petites solutions en nombres impairs de l'équation $x^2 - Dy^2 = -4$ ou, si cette équation n'en admet pas, de l'équation $x^2 - Dy^2 = 4$; en d'autres termes, une table des plus petites solutions impaires de l'équation $x^2 - Dy^2 = \pm 4$, $D \equiv 5 \pmod{8}$.

Londres, 2 Stone Buildings, 10 Mars, 1855.

TABLE DES PLUS PETITES SOLUTIONS IMPAIRES DE L'ÉQUATION $x^2-Dy^2=\pm 4$, $D\equiv 5$ (MOD. 8).

D	±	x	y	D	±	x	y	D	±	x	y
5	−	1	1	341	+	277	15	677		imposs.	
13	−	3	1	349		imposs.		685	−	759	29
21	+	5	1	357	+	19	1	693	+	79	3
29	−	5	1	365	−	19	1	701		imposs.	
37		imposs.		373		imposs.		709		imposs.	
45	+	7	1	381		imposs.		717	+	241	9
53	−	7	1	389		imposs.		725	+	27	1
61	−	39	5	397	−	3447	173	733	−	27	1
69	+	25	3	405		imposs.		741	+	245	9
77	+	9	1	413	+	61	3	749	+	12945	473
85	−	9	1	421	−	444939	21685	757		imposs.	
93	+	29	3	429	+	145	7	765	+	83	3
101		imposs.		437	+	21	1	773	−	139	5
109	−	261	25	445	−	21	1	781		imposs.	
117	+	11	1	453	+	149	7	789	+	31825	1133
125	−	11	1	461	−	365	17	797	−	367	13
133	+	173	15	469	+	65	3	805	+	1447	51
141		imposs.		477	+	2599	119	813		imposs.	
149	−	61	5	485		imposs.		821	−	16189	565
157	−	213	17	493	−	111	5	829		imposs.	
165	+	13	1	501	+	28225	1261	837	+	29	1
173	−	13	1	509	−	925	41	845	−	29	1
181	−	1305	97	517	+	10573	465	853	−	27483	941
189		imposs.		525	+	23	1	861	+	1027	35
197		imposs.		533	−	23	1	869	+	49377	1675
205	+	43	3	541	−	1396425	60037	877		imposs.	
213	+	73	5	549	+	1523	65	885		imposs.	
221	+	15	1	557		imposs.		893	+	2301	77
229	−	15	1	565	−	309	13	901		imposs.	
237	+	77	5	573		imposs.		909		imposs.	
245	+	47	3	581	+	6725	279	917	+	1181	31
253	+	1177	74	589	+	4359377	179625	925		imposs.	
261	+	727	45	597	+	7949	399	933		imposs.	
269		imposs.		605	+	123	5	941	−	1135	37
277	−	2613	157	613	−	98763	3989	949	−	32685	1061
285	+	17	1	621	+	25	1	957	+	31	1
293	−	17	1	629	−	25	1	965	−	31	1
301	+	22745	1311	637	+	14159	561	973		imposs.	
309	+	5045	287	645	+	203	8	981	+	68123	2175
317	−	89	5	653	−	1661	65	989	+	103245	3283
325		imposs.		661	−	1789539	69605	997		imposs.	
333		imposs.		669	+	305285	11803				

232.

MÉMOIRE SUR LA FORME CANONIQUE DES FONCTIONS BINAIRES.

[From the *Journal für die reine und angewandte Mathematik*, (Crelle), tom. LIV. (1857), pp. 48—58.]

L'IDÉE de la forme canonique d'une fonction binaire de degré impair et la théorie de la réduction d'une pareille fonction à la forme canonique sont dues à M. Sylvester([1]), qui a, en outre, étendu sa théorie aux fonctions binaires des degrés pairs 4 et 6. Mais il n'a établi nulle part l'idée générale de la forme canonique d'une fonction binaire d'un degré pair quelconque. Je me propose de reprendre cette théorie en considérant d'abord les fonctions binaires de degré impair et ensuite les fonctions binaires de degré pair.

Soit donc proposée la fonction de degré impair

$$(a,\ b, \dots\ a' \between x,\ y)^{2m-1};$$

en désignant par le symbole Σ une somme de m termes, on voit qu'on pourra écrire

$$(a,\ b, \dots\ a' \between x,\ y)^{2m-1} = \Sigma A\,(x + \alpha y)^{2m-1},$$

l'expression qui forme le second membre de cette équation est appelée la forme canonique de la fonction donnée, et il s'agit de trouver les valeurs des quantités A, α.

En supposant que le produit des facteurs linéaires $x + \alpha y$ soit égal, à un facteur constant près, à

$$(\mathrm{a},\ \mathrm{b}, \dots \mathrm{a}' \between x,\ y)^m$$

[1] Les recherches de M. Sylvester sont contenues dans le mémoire intitulé "Sketch of a Memoir on Elimination, Transformation, and Canonical Forms," *Camb. and Dublin Mathematical Journal*, t. VI. [1851] p. 186, dans un supplément publié conjointement avec ce mémoire et intitulé "Essay on Canonical Forms by J. J. Sylvester," London, Bell, 1851, et dans le mémoire intitulé, "On a remarkable discovery in the Theory of Canonical Forms and of Hyperdeterminants," *Phil. Mag.* Nov. 1851.

la question peut s'énoncer sous la forme suivante: trouver une fonction de degré m, telle qu'en représentant $x+\alpha y$ un facteur linéaire quelconque de cette fonction, la fonction donnée de degré $2m-1$ se réduise à la forme $\Sigma A\,(x+\alpha y)^{2m-1}$. La fonction de degré m contiendra un facteur constant dont la valeur peut être prise à volonté. Mais à ce facteur près la fonction de degré m sera complètement déterminée.

Pour fixer les idées je suppose que la fonction donnée soit la fonction cubique

$$(a,\ b,\ c,\ d \between x,\ y)^3,$$

dans ce cas il s'agira de trouver une fonction quadratique

$$(\mathrm{a},\ \mathrm{b},\ \mathrm{c} \between x,\ y)^2$$

telle qu'en posant $(\mathrm{a},\ \mathrm{b},\ \mathrm{c} \between x,\ y)^2 = \mathrm{a}\,(x+\alpha y)(x+\beta y)$ on ait identiquement

$$(a,\ b,\ c,\ d \between x,\ y)^3 = A\,(x+\alpha y)^3 + B\,(x+\beta y)^3.$$

Cela se fait très-facilement par les méthodes ordinaires. En effet, on a

$$\begin{aligned} A \ + B \ &= a, \\ A\alpha + B\beta &= b, \\ A\alpha^2 + B\beta^2 &= c, \\ A\alpha^3 + B\beta^3 &= d, \end{aligned}$$

d'où il suit

$$\begin{aligned} a\xi\eta - b\,(\xi+\eta) + c &= A\ \ (\xi-\alpha)(\eta-\alpha) + B\ \ (\xi-\beta)(\eta-\beta), \\ b\,\xi\eta - c\,(\xi+\eta) + d &= A\alpha\,(\xi-\alpha)(\eta-\alpha) + B\beta\,(\xi-\beta)(\eta-\beta), \end{aligned}$$

ξ et η étant des quantités quelconques; donc, en prenant $\xi=\alpha$, $\eta=\beta$, on obtient

$$\begin{aligned} a\alpha\beta - b\,(\alpha+\beta) + c &= 0, \\ b\alpha\beta - c\,(\alpha+\beta) + d &= 0. \end{aligned}$$

Mais l'équation $(\mathrm{a},\ \mathrm{b},\ \mathrm{c} \between x,\ y)^2 = \mathrm{a}\,(x+\alpha y)(x+\beta y)$ donne

$$\alpha\beta : \alpha+\beta : 1 = \mathrm{c} : 2\mathrm{b} : \mathrm{a},$$

on a donc

$$\begin{aligned} a\mathrm{c} - 2b\mathrm{b} + c\mathrm{a} &= 0, \\ b\mathrm{c} - 2c\mathrm{b} + d\mathrm{a} &= 0, \end{aligned}$$

système qui donne les rapports a : b : c. Mais pour compléter la solution de la manière la plus élégante, il faut ajouter à ces équations, l'équation $(\mathrm{a},\ \mathrm{b},\ \mathrm{c} \between x,\ y)^2 = 0$, mise sous la forme $(\mathrm{c},\ \mathrm{b},\ \mathrm{a} \between y,\ x)^2 = 0$, ou ce qui est la même chose

$$y^2\mathrm{c} + 2yx\mathrm{b} + x^2\mathrm{a} = 0\,;$$

en éliminant a, b, c on obtient une équation de laquelle on déduit

$$(\mathrm{a},\ \mathrm{b},\ \mathrm{c}\between x,\ y)^2 = \begin{vmatrix} y^2, & -yx, & x^2 \\ a\,, & b\,, & c \\ b\,, & c\,, & d \end{vmatrix}.$$

Mais on peut trouver ce résultat d'une manière encore plus facile. Considérant l'équation

$$(a,\ b,\ c,\ d\between x,\ y)^3 = A\,(x+\alpha y)^3 + B\,(x+\beta y)^3,$$

on n'a qu'à opérer sur cette équation en se servant du symbole

$$(\mathrm{a},\ \mathrm{b},\ \mathrm{c}\between \partial_y,\ -\partial_x)^2;$$

le second membre se réduit à zéro à cause des équations $(\mathrm{a}, \mathrm{b}, \mathrm{c}\between \alpha, -1)^2 = 0$, $(\mathrm{a}, \mathrm{b}, \mathrm{c}\between \beta, -1)^2 = 0$ que donne l'équation $(\mathrm{a},\ \mathrm{b},\ \mathrm{c}\between x,\ y)^2 = \mathrm{a}\,(x+\alpha y\between x+\beta y)$. Par conséquent on a identiquement

$$(\mathrm{a},\ \mathrm{b},\ \mathrm{c}\between \partial_y,\ -\partial_x)^2\,(a,\ b,\ c,\ d\between x,\ y)^3 = 0,$$

c'est-à-dire

$$x\,(a\mathrm{c} - 2b\mathrm{b} + c\mathrm{a}) + y\,(b\mathrm{c} - 2c\mathrm{b} + d\mathrm{a}) = 0,$$

ou enfin

$$a\mathrm{c} - 2b\mathrm{b} + c\mathrm{a} = 0,$$

$$b\mathrm{c} - 2c\mathrm{b} + d\mathrm{a} = 0,$$

et de là on tire comme auparavant la valeur de la fonction $(\mathrm{a},\ \mathrm{b},\ \mathrm{c}\between x,\ y)^2$.

Les valeurs des coefficients A, B s'expriment alors en fonctions linéaires des coefficients de la fonction cubique, et des quantités α, β qui sont pour ainsi dire les racines de la fonction quadratique $(\mathrm{a},\ \mathrm{b},\ \mathrm{c}\between x,\ y)^2$.

Il est évident que le procédé suivi dans ce qui précède est tout à fait général, et que l'on obtient toujours explicitement la fonction du degré m: par exemple pour la fonction du cinquième degré

$$(a,\ b,\ c,\ d,\ e,\ f\between x,\ y)^5$$

la fonction dont il s'agit sera

$$\begin{vmatrix} y^3, & -y^2x, & yx^2, & -x^3 \\ a\,, & b\,, & c\,, & d \\ b\,, & c\,, & d\,, & e \\ c\,, & d\,, & e\,, & f \end{vmatrix}$$

et de même dans tous les autres cas. Cette fonction du degré m (laquelle est un covariant de la fonction donnée) peut être appelée le *canonisant*. Pour la fonction du cinquième degré le canonisant peut être mis sous cette autre forme

$$\begin{vmatrix} ax+by, & bx+cy, & cx+dy \\ bx+cy, & cx+dy, & dx+ey \\ cx+dy, & dx+ey, & ex+fy \end{vmatrix}$$

et on démontre facilement qu'il existe toujours une transformation semblable.

Considérons maintenant une fonction de degré pair

$$(a,\ b, \ldots a' \between x,\ y)^{2m},$$

et posons

$$(a,\ b, \ldots a' \between x,\ y)^{2m} = \Sigma A\ (x + \alpha y)^{2m} + \Lambda (\mathrm{a},\ \mathrm{b}, \ldots \mathrm{a}' \between x,\ y)^m\ (\mathrm{a}_1,\ \mathrm{b}_1, \ldots \mathrm{a}_1' \between x,\ y)^m$$

où l'on suppose $(\mathrm{a},\ \mathrm{b}, \ldots \mathrm{a}' \between x,\ y)^m = \mathrm{a}\,(x + \alpha y) \ldots$ et où $(\mathrm{a}_1,\ b_1, \ldots \mathrm{a}_1' \between x,\ y)^m$ représente un certain covariant parfaitement déterminé de $(\mathrm{a},\ \mathrm{b}, \ldots \mathrm{a}' \between x,\ y)^m$. L'expression qui forme le second membre de l'équation proposée s'appellera la forme canonique de la fonction de degré pair.

Pour fixer les idées je suppose que la fonction donnée soit la fonction du quatrième degré

$$(a,\ b,\ c,\ d,\ e \between x,\ y)^4;$$

dans ce cas, en prenant la fonction quadratique $(\mathrm{a},\ \mathrm{b},\ \mathrm{c} \between x,\ y)^2 = \mathrm{a}\,(x + \alpha y)\,(x + \beta y)$ et en représentant par $(\mathrm{a}_1,\ \mathrm{b}_1,\ \mathrm{c}_1 \between x,\ y)^2$ un covariant déterminé de $(\mathrm{a},\ \mathrm{b},\ \mathrm{c} \between x,\ y)^2$ on a

$$(a,\ b,\ c,\ d,\ e \between x,\ y)^4 = A\ (x + \alpha y)^4 + B\ (x + \beta y)^4 + \Lambda\ (\mathrm{a},\ \mathrm{b},\ \mathrm{c} \between x,\ y)^2\ (\mathrm{a}_1,\ \mathrm{b}_1,\ \mathrm{c}_1)\ (x,\ y)^2.$$

Cela posé, la condition qui sert à déterminer le covariant $(\mathrm{a}_1,\ \mathrm{b}_1,\ \mathrm{c}_1)\ (x,\ y)^2$ est la suivante

$$\tfrac{1}{2}\,(\mathrm{a},\ \mathrm{b},\ \mathrm{c} \between \partial_y,\ -\partial_x)^2 . (\mathrm{a},\ \mathrm{b},\ \mathrm{c} \between x,\ y)^2\ (\mathrm{a}_1,\ \mathrm{b}_1,\ \mathrm{c}_1 \between x,\ y)^2 = K\ (\mathrm{a},\ \mathrm{b},\ \mathrm{c} \between x,\ y)^2,$$

où K est un facteur constant dont la valeur est arbitraire.

Le premier membre sera

$$\left(\left(\begin{array}{ccc} 7\mathrm{ac} - 6\mathrm{b}^2, & -2\mathrm{ab}\ , & \mathrm{a}^2\ , \\ 2\mathrm{bc}\ , & 12\mathrm{ac} - 16\mathrm{b}^2, & 2\mathrm{ab}\ , \\ \mathrm{c}^2\ , & -2\mathrm{bc}\ , & 7\mathrm{ac} - 6\mathrm{b}^2 \end{array} \between \mathrm{a}_1,\ \mathrm{b}_1,\ \mathrm{c}_1\right)\right) (x,\ y)^2$$

en représentant par cette notation l'expression

$$\begin{array}{llll} [(7\mathrm{ac} - 6\mathrm{b}^2)\,\mathrm{a}_1 & -2\mathrm{abb}_1 & + \mathrm{a}^2\mathrm{c}_1]\ x^2 \\ +\ [\quad 2\mathrm{bca}_1 & + (12\mathrm{ac} - 16\mathrm{b}^2)\,\mathrm{b}_1 & + 2\mathrm{abc}_1]\ xy \\ +\ [\quad \mathrm{c}^2\mathrm{a}_1 & -2\mathrm{bcb}_1 & + (7\mathrm{ac} - 6\mathrm{b}^2)\,\mathrm{c}_1]\ y^2. \end{array}$$

On a donc

$$\begin{array}{llll} (7\mathrm{ac} - 6\mathrm{b}^2)\,\mathrm{a}_1 & -2\mathrm{abb}_1 & + \mathrm{a}^2\mathrm{c}_1 & = K\ \mathrm{a}, \\ 2\mathrm{bca}_1 & + (12\mathrm{ac} - 16\mathrm{b}^2)\,\mathrm{b}_1 & + 2\mathrm{abc}_1 & = K 2\mathrm{b}, \\ \mathrm{c}^2\mathrm{a}_1 & -2\mathrm{bcb}_1 & + (7\mathrm{ac} - 6\mathrm{b}^2)\,\mathrm{c}_1 & = K\ \mathrm{c}, \end{array}$$

et en ajoutant à ces équations la suivante

$$x^2\mathrm{a}_1 \qquad + 2xy\mathrm{b}_1 + y^2 \qquad \mathrm{c}_1 = \quad u,$$

on obtient, en éliminant a_1, b_1, c_1, l'équation

$$\begin{vmatrix} u\ , & x^2\ , & 2xy\ , & y^2 \\ Ka, & 7ac-6b^2, & -2ab\ , & a^2 \\ 2Kb, & 2bc\ , & 12ac-16b^2, & 2ab \\ Kc, & c^2\ , & -2bc\ , & 7ac-6b^2 \end{vmatrix} = 0,$$

laquelle peut s'écrire aussi comme il suit

$$u\begin{vmatrix} 7ac-6b^2, & -2ab\ , & a^2 \\ 2bc\ , & 12ac-16b^2, & 2ab \\ c^2\ , & -2bc\ , & 7ac-6b^2 \end{vmatrix} + K\begin{vmatrix} & x^2\ , & 2xy\ , & y^2 \\ a, & 7ac-6b^2, & -2ab\ , & a^2 \\ 2b, & 2bc\ , & 12ac-16b^2, & 2ab \\ c, & c^2\ , & -2bc\ , & 7ac-6b^2 \end{vmatrix} = 0;$$

donc, en rétablissant la valeur de u, savoir $u=(a_1, b_1, c_1)(x, y)^2$, et en prenant

$$K=\begin{vmatrix} 7ac-6b^2, & -2ab\ , & a^2 \\ 2bc\ , & 12ac-16b^2, & 2ab \\ c^2\ , & -2bc\ , & 7ac-6b^2 \end{vmatrix},$$

on obtient

$$(a_1, b_1, c_1)(x, y)^2 = -\begin{vmatrix} & x^2\ , & 2xy\ , & y^2 \\ a, & 7ac-6b^2, & -2ab\ , & a^2 \\ 2b, & 2bc\ , & 12ac-16b^2, & 2ab \\ c, & c^2\ , & -2bc\ , & 7ac-6b^2 \end{vmatrix}.$$

Ces valeurs se réduisent à des expressions très simples. En effet, on a $K=576(ac-b^2)^3$, $(a_1, b_1, c_1)(x, y)^2=-72(ac-b^2)^2(a, b, c)(x, y)^2$, de sorte qu'en supprimant le facteur constant $-72(ac-b^2)^2$ on peut prendre

$$(a_1, b_1, c_1)(x, y)^2=(a, b, c)(x, y)^2.$$

C'est ce que l'on aurait pu prévoir dès le commencement; car le procédé par lequel on obtient $(a_1, b_1, c_1)(x, y)^2$ fait voir que cette fonction est un covariant de $(a, b, c)(x, y)^2$, du second degré par rapport aux variables et du cinquième degré par rapport aux coefficients; donc cette fonction sera, à un facteur numérique près, identique avec $(ac-b^2)^2(a, b, c)(x, y)^2$. Une circonstance pareille se présente dans la réduction d'une fonction du sixième degré à la forme canonique. En effet le covariant $(a_1, b_1, c_1, d_1)(x, y)^3$ sera une fonction du troisième degré par rapport aux variables et du septième degré par rapport aux coefficients, donc à un facteur numérique près, cette fonction sera identique avec le discriminant (a^2d^2+ etc.) multiplié par le cubicovariant $(a^2d-3abc+2b^3)x^3+$ etc., ou, en supprimant le facteur constant, le covariant cherché sera tout simplement le cubicovariant de $(a, b, c, d)(x, y)^3$. De même pour la fonction du huitième degré on trouve que le covariant $(\mathrm{a}_1, \mathrm{b}_1, \mathrm{c}_1, \mathrm{d}_1, \mathrm{e}_1 \between x, y)^4$ sera tout simplement $(\mathrm{a}, \mathrm{b}, \mathrm{c}, \mathrm{d}, \mathrm{e} \between x, y)^4$. Mais rien ne nous autorise à supposer qu'une réduction de cette espèce ait lieu dans le cas général, et il paraît qu'en général le covariant $(\mathrm{a}_1, \mathrm{b}_1, \ldots \mathrm{a}_1' \between x, y)^m$ sera une fonction indécomposable du degré $2m+1$ par rapport aux coefficients; à savoir, en posant

$$(\mathrm{a}, \mathrm{b}, \ldots \mathrm{a}' \between \partial_y, -\partial_x)^m (\mathrm{a}, \mathrm{b}, \ldots \mathrm{a}' \between x, y)^m (\mathrm{a}_1, \mathrm{b}_1 \ldots \mathrm{a}_1' \between x, y)^m = K (\mathrm{a}, \mathrm{b}, \ldots \mathrm{a}' \between x, y)^m$$

le premier membre sera une fonction de la forme

$$\left(\begin{vmatrix} \mathfrak{A}, & \mathfrak{B} \ldots \\ \mathfrak{A}', & \mathfrak{B}' \ldots \\ \vdots & \end{vmatrix} \between \mathrm{a}_1, \mathrm{b}_1, \ldots \mathrm{a}_1'\right) (x, y)^m$$

où $\mathfrak{A}, \mathfrak{B}, \ldots \mathfrak{A}', \mathfrak{B}', \ldots$ etc. sont des fonctions données du second degré par rapport aux coefficients $(\mathrm{a}, \mathrm{b}, \ldots \mathrm{a}')$; cela étant, on aura

$$K = \begin{vmatrix} \mathfrak{A}, & \mathfrak{B}, \ldots \\ \mathfrak{A}', & \mathfrak{B}', \ldots \\ \vdots & \end{vmatrix},$$

et

$$(\mathrm{a}_1, \mathrm{b}_1, \ldots \mathrm{a}_1' \between x, y)^m = - \begin{vmatrix} & x^m; & m x^{m-1} y & \\ \mathrm{a}, & \mathfrak{A}, & \mathfrak{B} & \ldots \\ m\mathrm{b}, & \mathfrak{A}', & \mathfrak{B}' & \\ \vdots & & & \end{vmatrix},$$

ce qui fait voir que le covariant dont il s'agit est une fonction du degré $2m+1$ par rapport aux coefficients.

Je reviens au cas particulier, le cas général pouvant être traité précisément de la même manière. On a

$$(a, b, c, d, e \between x, y)^4 = A (x + \alpha y)^4 + B (x + \beta y)^4 + \Lambda (\mathrm{a}, \mathrm{b}, \mathrm{c} \between x, y)^2 (\mathrm{a}_1, \mathrm{b}_1, \mathrm{c}_1 \between x, y)^2.$$

J'opère sur cette équation en me servant du symbole $(\mathrm{a}, \mathrm{b}, \mathrm{c} \between \partial_y, -\partial_x)^2$ et j'obtiens

$$(\mathrm{a}, \mathrm{b}, \mathrm{c} \between \partial_y, -\partial_x)^2 (a, b, c, d, e \between x, y)^4 = 2\Lambda K (\mathrm{a}, \mathrm{b}, \mathrm{c} \between x, y)^2,$$

c'est-à-dire

$$\left.\begin{aligned} &\mathrm{a}\,(c x^2 + 2dxy + e y^2) \\ -\,&2\mathrm{b}\,(bx^2 + 2c xy + dy^2) \\ +\,&\mathrm{c}\,(ax^2 + 2bxy + c y^2) \end{aligned}\right\} = \tfrac{1}{6} \Lambda K (\mathrm{a}x^2 + 2\mathrm{b}xy + \mathrm{c}y^2);$$

en posant, pour abréger, $\frac{1}{6} \Lambda K = \lambda$, on aura

$$\left.\begin{aligned} &x^2 \,[c\mathrm{a} - 2\mathrm{b}b + \mathrm{a}(\mathrm{c} - \lambda)] \\ +\,&2xy\,[c b - 2\mathrm{b}(\mathrm{c} + \tfrac{1}{2}\lambda) + \mathrm{a}d] \\ +\,&y^2 [\mathrm{c}(\mathrm{c} - \lambda) - 2\mathrm{b}d + \mathrm{a}e] \end{aligned}\right\} = 0;$$

ce qui donne

$$\begin{aligned} &\mathrm{c}\,a \quad - 2\mathrm{b}\,b \quad + \mathrm{a}\,(c-\lambda) = 0,\\ &\mathrm{c}\,b \quad - 2\mathrm{b}\,(c+\tfrac{1}{2}\lambda) + \mathrm{a}\,d \quad = 0,\\ &\mathrm{c}\,(c-\lambda) - 2\mathrm{b}\,d \quad + \mathrm{a}\,e \quad = 0, \end{aligned}$$

et de là on tire l'équation

$$\begin{vmatrix} a\,, & b\,, & c-\lambda \\ b\,, & c+\tfrac{1}{2}\lambda, & d \\ c-\lambda, & d\,, & e \end{vmatrix} = 0$$

qui servira à la détermination de la quantité λ. Les coefficients des différentes puissances de λ seront des invariants; en effet, si on développe le déterminant, l'équation devient $2\,(ace + 2bcd - ad^2 - b^2e + c^3) + \lambda\,(ae - 4bd + 3c^2) - \lambda^3 = 0$. La quantité λ étant connue, on peut au moyen de deux quelconques des trois équations trouver les rapports a : b : c, ou ce qui revient à la même chose, on peut se servir des trois équations en introduisant les quantités arbitraires L, M, N, et l'on trouve

$$(\mathrm{a},\ \mathrm{b},\ \mathrm{c}\between x,\ y)^2 = \begin{vmatrix} & y^2\,, & -yx\,, & x^2 \\ L\,, & a\,, & b\,, & c-\lambda \\ M, & b\,, & c+\tfrac{1}{2}\lambda, & d \\ N, & c-\lambda, & d\,, & e \end{vmatrix}$$

expression qui, en réalité, ne contient rien d'indéterminé. La quantité K est une fonction connue de (a, b, c) donc $\Lambda = \dfrac{6\lambda}{K}$ sera pareillement connu, et les coefficients A, B s'expriment en fonctions linéaires des coefficients de la fonction du quatrième degré et des quantités α, β qui sont pour ainsi dire les racines de la fonction quadratique $(\mathrm{a},\ \mathrm{b},\ \mathrm{c}\between x,\ y)^2$. Comme je l'ai déjà fait observer, tout ce qui précède s'applique également à une fonction de degré pair quelconque. Le déterminant qui contient λ peut être appelé le Lambdaïque. On voit aisément quelle est sa forme générale; par exemple pour la fonction $(a,\ b,\ c,\ d,\ e,\ f,\ g\between x,\ y)^6$ ce déterminant est

$$\begin{vmatrix} a\,, & b\,, & c\,, & d-\lambda \\ b\,, & c\,, & d+\tfrac{1}{3}\lambda, & e \\ c\,, & d-\tfrac{1}{3}\lambda, & e\,, & f \\ d+\lambda, & e\,, & f\,, & g \end{vmatrix}$$

expression dont le développement ne contient que les puissances paires de λ; cela arrive, comme M. Sylvester l'a remarqué depuis longtemps, toutes les fois que m est impair, c'est-à-dire toutes les fois que le degré de la fonction dont il s'agit est un nombre de la forme $4p + 2$.

Revenons à la fonction du quatrième degré $(a,\ b,\ c,\ d,\ e \between x,\ y)^4$; en écrivant $(\mathrm{a}_1,\ \mathrm{b}_1,\ \mathrm{c}_1 \between x,\ y)^2 = (\mathrm{a},\ \mathrm{b},\ \mathrm{c} \between x,\ y)^2 = \mathrm{a}\,(x+\alpha y)\,(x+\beta y)$ on obtient pour la forme canonique

$$(a,\ b,\ c,\ d,\ e \between x,\ y)^4 = A\,(x+\alpha y)^4 + B\,(x+\beta y)^4 + \Lambda\,(x+\alpha y)^2\,(x+\beta y)^2.$$

Relativement à la fonction du sixième degré $(a,\ b,\ c,\ d,\ e,\ f,\ g \between x,\ y)^6$, je fais observer que le cubicovariant de $(x+\alpha y)\,(x+\beta y)\,(x+\gamma y)$ est

$$\{(2\alpha-\beta-\gamma)\,x-(2\beta\gamma-\gamma\alpha-\alpha\beta)\,y\}\,\{(2\beta-\gamma-\alpha)\,x-(2\gamma\alpha-\alpha\beta-\beta\gamma)\,y\}$$
$$\{(2\gamma-\alpha-\beta)\,x-(2\alpha\beta-\beta\gamma-\gamma\alpha)\,y\};$$

donc en représentant cette fonction par Φ, on a pour la forme canonique

$$(a,\ b,\ c,\ d,\ e,\ f,\ g \between x,\ y)^6$$
$$= A\,(x+\alpha y)^6 + B\,(x+\beta y)^6 + C\,(x+\gamma y)^6 + \Lambda\,(x+\alpha y)\,(x+\beta y)\,(x+\gamma y)\,\Phi.$$

Pour la fonction du huitième degré le covariant $(\mathrm{a}_1,\ \mathrm{b}_1,\ \mathrm{c}_1,\ \mathrm{d}_1,\ \mathrm{e}_1 \between x,\ y)^4$ est tout simplement égal à $(\mathrm{a},\ \mathrm{b},\ \mathrm{c},\ \mathrm{d},\ \mathrm{e} \between x,\ y)^4$ ainsi que je l'ai déjà fait observer, et l'on a pour la forme canonique

$$(a,\ b,\ c,\ d,\ e,\ f,\ g,\ h,\ i \between x,\ y)^8$$
$$= A\,(x+\alpha y)^8 + B\,(x+\beta y)^8 + C\,(x+\gamma y)^8 + D\,(x+\delta y)^8 + \Lambda\,(x+\alpha y)^2\,(x+\beta y)^2\,(x+\gamma y)^2\,(x+\delta y)^2.$$

Je n'ai pas cherché à réduire, ou à présenter sous la forme d'un déterminant, le covariant $(\mathrm{a}_1,\ \mathrm{b}_1,\ \mathrm{c}_1,\ \mathrm{d}_1,\ \mathrm{e}_1,\ \mathrm{f}_1 \between x,\ y)^5$ qui entre dans la forme canonique de la fonction du dixième degré.

On a vu que le covariant $(\mathrm{a}_1,\ \mathrm{b}_1,\ldots\ \mathrm{a}_1' \between x,\ y)^m$ s'obtient par le développement de la fonction

$$(\mathrm{a},\ \mathrm{b},\ldots\ \mathrm{a}' \between \partial_y,\ -\partial_x)^m \,.\, (\mathrm{a},\ \mathrm{b},\ldots\ \mathrm{a}_1' \between x,\ y)^m\,(\mathrm{a}_1,\ \mathrm{b}_1,\ldots\ \mathrm{a}_1' \between x,\ y)^m$$

et j'ai donné ci-dessus un exemple de ce développement; en employant pour plus de commodité les lettres italiques au lieu des lettres romaines, on peut écrire la formule dont il s'agit de la manière suivante

$$(a,\ b,\ c \between \partial_y,\ -\partial_x)^2\,(a,\ b,\ c \between x,\ y)^2\,(a_1,\ b_1,\ c_1 \between x,\ y)^2 =$$
$$2\left(\left(\begin{array}{c|c|c|} 7\,ac-6\,b^2 & -2\,ab & a^2 \\ 2\,bc & 12\,ac-16\,b^2 & 2\,ab \\ c^2 & -2\,bc & 7\,ac-6\,b^2 \end{array} \between a_1,\ b_1,\ c_1\right)\right)(x,\ y)^2.$$

Les équations analogues pour la fonction cubique etc. sont

$$(a,\ b,\ c,\ d \between \partial_y,\ -\partial_x)^3\,(a,\ b,\ c,\ d \between x,\ y)^3\,(a_1,\ b_1,\ c_1,\ d_1 \between x,\ y)^3 =$$
$$6\times\left(\left(\begin{array}{c|c|c|c|} -19\,ad+18\,bc & +39\,ac-36\,b^2 & -\ \ 3\,ab & +\ \ a^2 \\ -39\,bd+36\,c^2 & -18\,ad+27\,bc & +72\,ac-81\,b^2 & +\ \ 3\,ab \\ -\ \ 3\,cd & -72\,bd+81\,c^2 & +18\,ad-27\,bc & +39\,ac-36\,b^2 \\ -\ \ d^2 & +\ \ 3\,cd & -39\,bd+36\,c^2 & +19\,ad-18\,bc \end{array} \between a_1,\ b_1,\ c_1,\ d_1\right)\right)(x,\ y)^3$$

$$(a,\ b,\ c,\ d,\ e \between \partial_y,\ -\partial_x)^4 (a,\ b,\ c,\ d,\ e \between x,\ y)^4\ (a_1,\ b_1,\ c_1,\ d_1,\ e_1 \between x,\ y)^4 =$$

$$24 \times \left(\left(\begin{array}{l|l|l|}
+\ 71\,ae - 160\,bd + 90\,c^2 & -124\,ad + 120\,bc & +126\,ac - 120\,b^2 \\
+124\,be - 120\,cd & +160\,ae - 896\,bd + 720\,c^2 & -120\,ad + 144\,bc \\
+126\,ce - 120\,d^2 & +120\,be - 144\,cd & +180\,ae - 1440\,bd + 1296\,c^2 \\
+\ 4\,de & +240\,ce - 256\,d^2 & -120\,be + 144\,cd \\
+\ e^2 & -\ 4\,de & +126\,ce - 120\,d^2
\end{array}\right.\right.$$

$$\left.\left. \begin{array}{l|l|}
-\ 4\,ab & +\ a^2 \\
+240\,ac - 256\,b^2 & +4\,ab \\
+120\,ad - 144\,bc & +126\,ac - 120\,b^2 \\
+160\,ae - 896\,bd + 720\,c & +124\,ad - 120\,bc \\
-124\,be + 120\,cd & +\ 71\,ae - 160\,bd + 90\,c^2
\end{array} \between a_1,\ b_1,\ c_1,\ d_1,\ e_1\right)\right)(x,\ y)^4.$$

$$(a,\ b,\ c,\ d,\ e,\ f \between \partial_y,\ -\partial_x)^5 (a,\ b,\ c,\ d,\ e,\ f \between x,\ y)^5 (a_1,\ b_1,\ c_1,\ d_1,\ e_1,\ f_1 \between x,\ y)^5 =$$

$$120 \times \left(\left(\begin{array}{l|l|}
-251\,af + 600\,be - 350\,cd & +\ 655\,ae - 1700\,bd + 1050\,c^2 \\
-655\,bf + 1700\,ce - 1050\,d^2 & -\ 600\,af + 2875\,be - 2250\,cd \\
-460\,cf + 450\,de & -1700\,bf + 7750\,ce - 6000\,d^2 \\
-310\,df + 300\,e^2 & -\ 450\,cf + 500\,de \\
-\ 5\,ef & -\ 600\,df + 625\,e^2 \\
-\ e^2 & +\ 5\,ef
\end{array}\right.\right.$$

$$\begin{array}{l|l|}
-\ 460\,ad + 450\,bc & +\ 310\,ac - 300\,b^2 \\
+1700\,ae - 7750\,cd + 6000\,c^2 & -\ 450\,ad + 500\,bc \\
-\ 350\,af + 2250\,be - 2000\,cd & +2100\,ae - 12000\,bd + 10000\,c^2 \\
-2100\,bf + 12000\,ce - 10000\,d^2 & +\ 350\,af - 2250\,be + 2000\,cd \\
+\ 450\,cf - 500\,de & -1700\,bf + 7750\,ce - 6000\,d^2 \\
-\ 310\,df + 300\,e^2 & +\ 460\,cf - 450\,de
\end{array}$$

$$\left.\left. \begin{array}{l|l|}
-\ 5\,ab & +\ a^2 \\
+\ 600\,ac - 625\,b^2 & +\ 5\,ab \\
+\ 450\,ad - 500\,bc & +310\,ac - 300\,b^2 \\
+1700\,ae - 7750\,bd + 6000\,c^2 & +460\,ad - 450\,bc \\
+\ 600\,af - 2875\,be + 2250\,cd & +655\,ae - 1700\,bd + 1050\,c^2 \\
-\ 655\,bf + 1700\,ce - 1050\,d^2 & +251\,af - 600\,be + 350\,cd
\end{array} \between a_1,\ b_1,\ c_1,\ d_1,\ e_1,\ f_1\right)\right)(x,\ y)^5.$$

Les résultats qu'on vient d'obtenir pour les cas particuliers les plus simples font voir que la théorie générale doit être susceptible de développements ultérieurs. Ainsi la forme canonique de la fonction cubique $(a, b, c, d \between x, y)^3$ sera $u^3 + v^3$ si l'on pose u, v au lieu de $\sqrt[3]{A}\,(x + \alpha y)$, $\sqrt[3]{B}\,(x + \beta y)$. La méthode générale ne donne que la valeur du produit u, v à un facteur constant près; mais on peut trouver les valeurs de u et v séparément. En effet en se rappelant les résultats obtenus dans la "Note sur les covariants d'une fonction quadratique, cubique, ou biquadratique à deux indéterminées" (t. L. p. 285 de ce Journal, [135]), on reconnaît que les deux fonctions $\Phi + U\sqrt{-\square}$, $\Phi - U\sqrt{-\square}$ sont l'une et l'autre des cubes parfaits et l'on a

$$2U\sqrt{-\square} = (\Phi + U\sqrt{-\square}) - (\Phi - U\sqrt{-\square});$$

on a donc

$$u = \left\{\frac{\Phi + U\sqrt{-\square}}{2\sqrt{-\square}}\right\}^{\frac{1}{3}}, \quad v = \left\{\frac{\Phi - U\sqrt{-\square}}{2\sqrt{-\square}}\right\}^{\frac{1}{3}}.$$

De même la forme canonique de la fonction du quatrième degré $(a, b, c, d, e \between x, y)^4$ sera $u^4 + v^4 + 6\theta u^2v^2, = (u^2 - v^2)^2 + 2(1 + 3\theta)\,u^2v^2$, si l'on pose u, v au lieu de $\sqrt[4]{A}\,(x + \alpha y)$, $\sqrt[4]{B}\,(x + \beta y)$; or $IH - \varpi_1 JU$, $IH - \varpi_2 JU$, $IH - \varpi_3 JU$ étant tous les trois des carrés de fonctions quadratiques, la fonction U peut s'exprimer à moyen de deux quelconques de ces trois quantités comme une somme de deux carrés et on déduit de là les valeurs de u et de v; on trouvera le développement de cette solution dans un mémoire "Sur quelques formules pour la transformation des intégrales elliptiques", [235], qui paraîtra prochainement dans ce Journal. Quant à la fonction du cinquième degré, on pourrait écrire u, v, w au lieu de $\sqrt[5]{A}\,(x + \alpha y)$, $\sqrt[5]{B}\,(v + \beta y)$, $\sqrt[5]{C}\,(x + \gamma y)$, ce qui donnerait pour forme canonique $u^5 + v^5 + w^5$, mais il vaut mieux, en multipliant ces valeurs par des facteurs convenables, écrire $u + v + w = 0$ et prendre pour forme canonique $Au^5 + Bv^5 + Cw^5 = 0$, c'est ce qu'a fait M. Sylvester dans son mémoire ci-dessus cité. On trouve aussi des développements sur ce sujet dans le mémoire de M. Faà de Bruno, "Nota sulla teorica degli invarianti", *Tortolini, Annali di Scienze etc.* 1855.

Pour la fonction du sixième degré on peut supposer que la fonction cubique $(\mathrm{a}, \mathrm{b}, \mathrm{c}, \mathrm{d} \between x, y)^3$ soit réduite à la forme canonique $u^3 + v^3$; cela étant, le cubicovariant sera $u^3 - v^3$, et en représentant par ρ l'une des racines cubiques imaginaires de l'unité la forme canonique sera

$$A\,(u + v)^6 + B\,(u + \rho v)^6 + C\,(u + \rho^2 v)^6 + \Lambda\,(u^6 - v^6);$$

c'est encore ce qu'a fait voir M. Sylvester dans son mémoire.

Londres, 9 *Avril,* 1856.

233.

ADDITION AU MÉMOIRE SUR LA FORME CANONIQUE DES FONCTIONS BINAIRES.

[From the *Journal für die reine und angewandte Mathematik*, (Crelle), tom. LIV. (1857), p. 292.]

DANS le mémoire "Sur la forme canonique des fonctions binaires" (p. 48 de ce volume [232]) j'ai dit que M. Sylvester avait en outre étendu sa théorie aux fonctions binaires des degrés pairs 4 et 6; j'aurais dû dire, des degrés pairs 4, 6 et 8. J'ai aussi omis de faire observer que les termes *canonisant* et *lambdaïque* appartenaient à M. Sylvester. Enfin en citant dans la note les mémoires de M. Sylvester qui ont rapport à cette théorie j'ai omis de citer le mémoire "On the Calculus of Forms otherwise the Theory of Invariants, § VIII. (*Camb. and Dublin Math. Journal*, t. IX. [1854], p. 93) section qui porte le titre "On the Reduction of a Sextic Function of Two Variables to the Canonical Form."

Londres, 16 *Juillet*, 1857.

234.

DEUXIÈME NOTE SUR UNE FORMULE POUR LA RÉVERSION DES SÉRIES.

[From the *Journal für die reine und angewandte Mathematik,* (Crelle), tom. LIV. (1857), pp. 156—161: Sequel to Note t. LII. (1856), **229**.]

JE me propose de montrer, dans cette deuxième note, de quelle manière le théorème de Jacobi conduit à une formule donnée sans démonstration par M. Sylvester dans son mémoire intitulé "On the Change of Systems of Independent Variables," *Quarterly Math. Journal,* tom. I. [1857], p. 42 à 56 et 126 à 134. Pour fixer les idées je prends le cas de trois variables, et je suppose que $f(x, y, z)$ soit une fonction rationnelle et entière de x, y, z, et que ces variables soient données en fonction de u, v, w au moyen des équations $u = X$, $v = Y$, $w = Z$, où X, Y, Z sont des fonctions rationnelles et entières de x, y, z. Mais ces fonctions ne sont plus assujetties à la condition (admise dans ma première note) d'être telles que $X - x$, $Y - y$, $Z - z$ ne contiennent que les puissances et les produits du deuxième ordre et des ordres supérieurs des variables, et il s'agit de déterminer dans le cas général le développement de $f(x, y, z)$ en termes de u, v, w.

Pour résoudre ce problème j'écris

$$\begin{aligned}
X &= A_{100}\, x + A_{010}\, y + A_{001}\, z + \dots + A_{f,g,h}\, x^f y^g z^h + \text{etc.},\\
Y &= B_{100}\, x + B_{010}\, y + B_{001}\, z + \dots + B_{i,j,k}\, x^i y^j z^k + \text{etc.},\\
Z &= C_{100}\, x + C_{010}\, y + C_{001}\, z + \dots + C_{l,m,n}\, x^l y^m z^n + \text{etc.},\\
f(x, y, z) &= \qquad\qquad\qquad\qquad \dots + \Theta_{P,Q,R}\, x^P y^Q z^R + \text{etc.};
\end{aligned}$$

dans ces expressions et partout dans la suite les etc. représentent des termes qu'on obtient en donnant des accents en nombre quelconque aux symboles indéterminés. Je dois faire observer relativement au coefficient $A_{f,g,h}$ et aux coefficients semblables, que les termes qui correspondent à $f + g + h = 1$ sont écrits à part; on doit donc prendre pour les nombres f, g, h seulement les valeurs qui rendent $f + g + h > 1$.

Je pose

$$x' = A_{100}x + A_{010}y + A_{001}z,$$
$$y' = B_{100}x + B_{010}y + B_{001}z,$$
$$z' = C_{100}x + C_{010}y + C_{001}z,$$

et en représentant par

$$\begin{vmatrix} a_{100}, & b_{100}, & c_{100} \\ a_{010}, & b_{010}, & c_{010} \\ a_{001}, & b_{001}, & c_{001} \end{vmatrix}$$

la *matrice inverse* de

$$\begin{vmatrix} A_{100}, & A_{010}, & A_{001} \\ B_{100}, & B_{010}, & B_{001} \\ C_{100}, & C_{010}, & C_{001} \end{vmatrix},$$

on obtient les équations

$$x = a_{100}x' + b_{100}y' + c_{100}z',$$
$$y = a_{010}x' + b_{010}y' + c_{010}z',$$
$$z = a_{001}x' + b_{001}y' + c_{001}z'.$$

Il est presque superflu de faire observer qu'en représentant par ∇ le déterminant

$$\begin{vmatrix} A_{100}, & A_{010}, & A_{001} \\ B_{100}, & B_{010}, & B_{001} \\ C_{100}, & C_{010}, & C_{001} \end{vmatrix}$$

on a

$$a_{100} = \frac{1}{\nabla}\frac{d\nabla}{dA_{100}}, \qquad a_{010} = \frac{1}{\nabla}\frac{d\nabla}{dA_{010}}, \text{ etc.}$$

A présent, en supposant chacune des quantités u', v', w' égale à zéro, on a le système d'équations

$$u - X = 0, \qquad v - Y = 0, \qquad w - Z = 0$$
$$u' - X' = 0, \qquad v' - Y' = 0, \qquad w' - Z' = 0$$

où

$$X = x' \qquad \ldots + A_{f,g,h}\, x^f y^g z^h + \text{etc.},$$
$$Y = y' \qquad \ldots + B_{i,j,k}\, x^i y^j z^k + \text{etc.},$$
$$Z = z' \qquad \ldots + C_{l,m,n}\, x^l y^m z^n + \text{etc.},$$

$$X' = x - a_{100}x' - b_{100}y' - c_{100}z',$$
$$Y' = y - a_{010}x' - b_{010}y' - c_{010}z',$$
$$Z' = z - a_{001}x' - b_{001}y' - c_{001}z',$$

et comme auparavant

$$f(x,\ y,\ z) = \ldots + \Theta_{P,\,Q,\,R}\, x^P y^Q z^R + \text{etc.}$$

Or, en appliquant à ce nouveau système la formule de Jacobi, et en remarquant qu'il est permis de poser tout de suite $u'=0$, $v'=0$, $w'=0$, cette formule donne

$$f(x,\ y,\ z) =$$

$$\left[f(x,\ y,\ z)\,\frac{\partial(X,\ Y,\ Z,\ X',\ Y',\ Z')}{\partial(x,\ y,\ z,\ x',\ y',\ z')}\,\frac{1}{(X-u)(Y-v)(Z-w)\,X'Y'Z'}\right]_{x^{-1}y^{-1}z^{-1}x'^{-1}y'^{-1}z'^{-1}}$$

équation dans laquelle on doit d'abord développer le dernier facteur de l'expression renfermée entre crochets suivant les puissances ascendantes de u, v, w, et ensuite développer les puissances de X, Y, Z, X', Y', Z' suivant les puissances descendantes de x', y', z', x, y, z respectivement. On obtient ainsi

coeff. de $u^{\mathrm{a}}v^{\mathrm{b}}w^{\mathrm{c}}$ dans le développement de $f(x,\ y,\ z) =$

$$\left[f(x,\ y,\ z)\,\frac{\partial(X,\ Y,\ Z,\ X',\ Y',\ Z')}{\partial(x,\ y,\ z,\ x',\ y',\ z')}\,\frac{1}{X^{\mathrm{a}+1}Y^{\mathrm{b}+1}Z^{\mathrm{c}+1}X'Y'Z'}\right]_{x^{-1}y^{-1}z^{-1}x'^{-1}y'^{-1}z'^{-1}}$$

ou ce qui est la même chose

coeff. de $u^{\mathrm{a}}v^{\mathrm{b}}w^{\mathrm{c}}$ dans le développement de $f(x,\ y,\ z) =$

$$\left[f(x,\ y,\ z)\,\frac{\partial\left(-\frac{1}{\mathrm{a}}X^{-\mathrm{a}},\ -\frac{1}{\mathrm{b}}Y^{-\mathrm{b}},\ -\frac{1}{\mathrm{c}}Z^{-\mathrm{c}},\ \log X',\ \log Y',\ \log Z'\right)}{\partial(x,\ y,\ z,\ x',\ y',\ z')}\right]_{x^{-1}y^{-1}z^{-1}x'^{-1}y'^{-1}z'^{-1}}$$

Or, en posant $\Pi\mathrm{a} = 1.2.3\ldots\mathrm{a}$, etc., le terme général de $-\frac{1}{\mathrm{a}}X^{-\mathrm{a}}$ est

$$(-1)^{r-1}\,\frac{\Pi(\mathrm{a}+r-1)}{\Pi\mathrm{a}\,\Pi\alpha\ \text{etc.}}\,A^{\alpha}{}_{f,g,h}\ \text{etc.}\ x'^{-\mathrm{a}-r}\,x^F y^G z^H,$$

où

$$\alpha + \text{etc.} = r, \quad f\alpha + \text{etc.} = F, \quad g\alpha + \text{etc.} = G, \quad h\alpha + \text{etc.} = H,$$

de même le terme général de $-\frac{1}{\mathrm{b}}Y^{-\mathrm{b}}$ est

$$(-1)^{s-1}\,\frac{\Pi(\mathrm{b}+s-1)}{\Pi\mathrm{b}\,\Pi\beta\ \text{etc.}}\,B^{\beta}{}_{i,j,k}\ \text{etc.}\ y'^{-\mathrm{b}-s}x^I y^J z^K,$$

où

$$\beta + \text{etc.} = s, \quad i\beta + \text{etc.} = I, \quad j\beta + \text{etc.} = J, \quad k\beta + \text{etc.} = K$$

et le terme général de $-\frac{1}{\mathrm{c}}Z^{-\mathrm{c}}$ est

$$(-1)^{t-1}\,\frac{\Pi(\mathrm{c}+t-1)}{\Pi\mathrm{c}\,\Pi\gamma\ \text{etc.}}\,C^{\gamma}{}_{l,m,n}\ \text{etc.}\ z'^{-\mathrm{c}-t}\,x^L y^M z^N,$$

où

$$\gamma + \text{etc.} = t, \quad l\gamma + \text{etc.} = L, \quad m\gamma + \text{etc.} = M, \quad n\gamma + \text{etc.} = N.$$

Le terme général du développement de log X' est

$$-\frac{\Pi(r'-1)}{\Pi\alpha'\,\Pi\delta'\,\Pi\iota'}\,a_{100}^{\alpha'}\,b_{100}^{\delta'}\,c_{100}^{\iota'}\,x^{-r'}\,x'^{\alpha'}\,y'^{\delta'}\,z'^{\iota'}$$

où

$$r' = \alpha' + \delta' + \iota',$$

et je fais observer que pour tenir compte du terme $\log x$ que contient $\log X'$, il suffit d'attribuer à α', δ', ι', r' les valeurs $\alpha' = \delta' = \iota' = r' = 0$. En effet on n'a besoin que des coefficients différentiels de $\log X'$, et, en gardant pour le moment r' au lieu de zéro, le terme dont il s'agit sera $-\Pi(r'-1)\,x^{-r'}$, ce qui différentié par rapport à x donne $\Pi r'.x^{-r'-1}$. En faisant $r'=0$ cela devient $\frac{1}{x}$, ce qui est en effet la valeur du coefficient différentiel de $\log x$ par rapport à x.

De même le terme général de log Y' est

$$-\frac{\Pi(s'-1)}{\Pi\beta'\,\Pi\epsilon'\,\Pi\kappa'}\,a_{010}^{\beta'}\,b_{010}^{\epsilon'}\,c_{010}^{\kappa'}\,y^{-s'}\,x'^{\beta'}\,y'^{\epsilon'}\,z'^{\kappa'}$$

où

$$s' = \beta' + \epsilon' + \kappa',$$

et le terme général de $\log Z'$ est

$$-\frac{\Pi(t'-1)}{\Pi\gamma'\,\Pi\zeta'\,\Pi\lambda'}\,a_{001}^{\gamma'}\,b_{001}^{\zeta'}\,c_{001}^{\lambda'}\,z^{-t'}\,x'^{\gamma'}\,y'^{\zeta'}\,z'^{\lambda'}$$

où

$$t' = \gamma' + \zeta' + \lambda'.$$

Donc, en formant le terme général du Jacobien et en multipliant par le terme général de $f(x, y, z)$, on obtient pour le terme général de l'expression placée entre crochets la valeur suivante

$$\begin{aligned}&(-1)^{r+s+t}\frac{\Pi(\mathrm{a}+r-1)\,\Pi(\mathrm{b}+s-1)\,\Pi(\mathrm{c}+t-1)}{\Pi\mathrm{a}\,\Pi\mathrm{b}\,\Pi\mathrm{c}\,\Pi\alpha\text{ etc. }\Pi\beta\text{ etc. }\Pi\gamma\text{ etc.}}\,\frac{\Pi(r'-1)\,\Pi(s'-1)\,\Pi(t'-1)}{\Pi\alpha'\,\Pi\delta'\,\Pi\iota'\,\Pi\beta'\,\Pi\epsilon'\,\Pi\kappa'\,\Pi\gamma'\,\Pi\zeta'\,\Pi\lambda'}\\
&\times A^{\alpha}{}_{f,g,h}\text{ etc. }B^{\beta}{}_{i,j,k}\text{ etc. }C^{\gamma}{}_{l,m,n}\text{ etc. }a_{100}^{\alpha'}\,b_{100}^{\delta'}\,c_{100}^{\iota'}\,a_{010}^{\beta'}\,b_{010}^{\epsilon'}\,c_{010}^{\kappa'}\,a_{001}^{\gamma'}\,b_{001}^{\zeta'}\,c_{001}^{\lambda'}\\
&\times\Omega\\
&\times x^{F+I+L+P-r'-1}\,y^{G+J+M+Q-s'-1}\,z^{H+K+N+R-t'-1}\,x'^{-\mathrm{a}-r+\alpha'+\beta'+\gamma'-1}\,y'^{-\mathrm{b}-s+\delta'+\epsilon+\zeta-1}\,z'^{-\mathrm{c}-t+\iota'+\kappa+\lambda'-1}\end{aligned}$$

où le facteur Ω représente le déterminant numérique suivant

$$\Omega = \begin{vmatrix} F, & G, & H, & -\mathrm{a}-r, & 0, & 0 \\ I, & J, & K, & 0, & -\mathrm{b}-s, & 0 \\ L, & M, & N, & 0, & 0, & -\mathrm{c}-t \\ -r', & 0, & 0, & \alpha', & \delta', & \iota' \\ 0, & -s', & 0, & \beta', & \epsilon', & \kappa' \\ 0, & 0, & -t', & \gamma', & \zeta', & \lambda' \end{vmatrix}.$$

Pour trouver le terme qui contient $x^{-1}y^{-1}z^{-1}x'^{-1}y'^{-1}z'^{-1}$ on n'a qu'à poser

$$F+I+L+P-r'=0, \qquad G+J+M+Q-s'=0, \qquad H+K+N+R-t'=0,$$
$$-\mathrm{a}-r+\alpha'+\beta'+\gamma'=0, \qquad -\mathrm{b}-s+\delta'+\epsilon'+\zeta'=0, \qquad -\mathrm{c}-t+\iota'+\kappa'+\lambda'=0.$$

En faisant cela, et en tenant compte des formules précédentes on obtient le théorème suivant: savoir en posant

$$\begin{aligned} X &= A_{100}\,x + A_{010}\,y + A_{001}\,z \ldots + A_{f,g,h}\ x^f\,y^g\,z^h + \text{etc.} = u, \\ Y &= B_{100}\,x + B_{010}\,y + B_{001}\,z \ldots + B_{i,j,k}\ x^i\,y^j\,z^k + \text{etc.} = v, \\ Z &= C_{100}\,x + C_{010}\,y + C_{001}\,z \ldots + C_{l,m,n}\ x^l\,y^m\,z^n + \text{etc.} = w, \\ f(x,\,y,\,z) &= \qquad\qquad\qquad \ldots + \Theta^{P,\,Q,\,R}\ x^P\,y^Q\,z^R + \text{etc.} \end{aligned}$$

on trouve pour le terme général du coeff. de $u^{\mathrm{a}}v^{\mathrm{b}}w^{\mathrm{c}}$ dans le développement de $f(x, y, z)$ l'expression

$$(-1)^{r+s+t}\frac{\Pi(\mathrm{a}+r-1)\,\Pi(\mathrm{b}+s-1)\,\Pi(\mathrm{c}+t-1)}{\Pi\mathrm{a}\,\Pi\mathrm{b}\,\Pi\mathrm{c}\ \ \Pi\alpha\,\text{etc.}\ \ \Pi\beta\,\text{etc.}\ \ \Pi\gamma\,\text{etc.}}\ \frac{\Pi(r'-1)}{\Pi\alpha'\,\Pi\delta'\,\Pi\iota'}\ \frac{\Pi(s'-1)}{\Pi\beta'\,\Pi\epsilon'\,\Pi\kappa'}\ \frac{\Pi(t'-1)}{\Pi\gamma'\,\Pi\zeta'\,\Pi\lambda'}$$

$$\times\, A^{\alpha}{}_{f,g,h}\ \text{etc.}\ \ B^{\beta}{}_{i,j,k}\ \text{etc.}\ \ C^{\gamma}{}_{l,m,n}\ \text{etc.}\ \ a_{100}^{\alpha'}\,b_{100}^{\delta'}\,c_{100}^{\iota'}\,a_{010}^{\beta'}\,b_{010}^{\epsilon'}\,c_{010}^{\kappa'}\,a_{001}^{\gamma'}\,b_{001}^{\zeta'}\,c_{001}^{\lambda'}$$

$$\times\,\Omega,$$

formule dans laquelle les a_{100} etc. sont les coefficients inverses des A_{100} etc. c'est-à-dire

$$\begin{vmatrix} a_{100}, & b_{100}, & c_{100} \\ a_{010}, & b_{010}, & c_{010} \\ a_{001}, & b_{001}, & c_{001} \end{vmatrix} = \begin{vmatrix} A_{100}, & A_{010}, & A_{001} \\ B_{100}, & B_{010}, & B_{001} \\ C_{100}, & C_{010}, & C_{001} \end{vmatrix}^{-1};$$

le déterminant numérique Ω a la valeur qui vient d'être donnée, en supposant seulement que les nombres α, β, etc. qui y entrent, satisfassent aux conditions suivantes

$$\begin{array}{ccc} \alpha+\text{etc.}=r, & \beta+\text{etc.}=s, & \gamma+\text{etc.}=t, \\ f\alpha+\text{etc.}=F, & g\alpha+\text{etc.}=G, & h\alpha+\text{etc.}=H, \\ i\beta+\text{etc.}=I, & j\beta+\text{etc.}=J, & k\beta+\text{etc.}=K, \\ l\gamma+\text{etc.}=L, & m\gamma+\text{etc.}=M, & n\gamma+\text{etc.}=N, \\ \alpha'+\delta'+\iota'=r', & \beta'+\epsilon'+\kappa'=s', & \gamma'+\zeta'+\lambda'=t', \\ P+F+I+L=r', & Q+G+J+M=s', & R+H+K+N=t', \\ \alpha'+\beta'+\gamma'=\mathrm{a}+r, & \delta'+\epsilon'+\zeta'=\mathrm{b}+s, & \iota'+\kappa'+\lambda'=\mathrm{c}+t. \end{array}$$

De ces équations on tire

$$P+Q+R+F+G+H+I+J+K+L+M+N=\mathrm{a}+\mathrm{b}+\mathrm{c}+r+s+t,$$

ou, en substituant pour F, etc. leurs valeurs,

$$(f+g+h-1)\,\alpha+\text{etc.}+(i+j+k-1)\,\beta+\text{etc.}+(l+m+n-1)\,\gamma+\text{etc.}$$
$$=\mathrm{a}+\mathrm{b}+\mathrm{c}-P-Q-R,$$

les nombres $f+g+h-1$, $i+j+k-1$, $l+m+n-1$, etc. étant positifs. Il n'y a donc qu'un nombre fini de solutions des équations indéterminées, comme cela doit être.

On peut encore modifier un peu la forme du déterminant Ω; on voit d'abord que l'on peut changer les signes des quantités $-\mathrm{a}-r$, $-\mathrm{b}-s$, $-\mathrm{c}-t$, $-r'$, $-s'$, $-t'$; en faisant cela et en substituant pour ces quantités et aussi pour F, G, etc. les valeurs ci-dessus données, on obtient

$$\Omega=\left|\begin{array}{cccccc} f\alpha+\text{etc.}\;, & g\alpha+\text{etc.}\,, & h\alpha+\text{etc.}, & \alpha'+\beta'+\gamma', & 0\;, & 0 \\ i\beta+\text{etc.}\;, & j\beta+\text{etc.}\,, & k\beta+\text{etc.}, & 0\;, & \delta'+\epsilon'+\zeta', & 0 \\ l\gamma+\text{etc.}\;, & m\gamma+\text{etc.}\,, & n\gamma+\text{etc.}, & 0\;, & 0\;, & \iota'+\kappa'+\lambda' \\ \alpha'+\delta'+\iota', & 0\;, & 0\;, & \alpha'\;, & \delta'\;, & \iota' \\ 0\;, & \beta'+\epsilon'+\kappa', & 0\;, & \beta'\;, & \epsilon'\;, & \kappa' \\ 0\;, & 0\;, & \gamma'+\zeta'+\lambda', & \gamma'\;, & \zeta'\;, & \lambda' \end{array}\right|.$$

Ainsi se trouve démontré le théorème général de M. Sylvester relatif à la réversion des séries, car il est évident que les formules s'appliquent sans difficulté à un nombre quelconque de variables.

Londres, le 30 *Septembre*, 1856.

235.

SUR QUELQUES FORMULES POUR LA TRANSFORMATION DES INTÉGRALES ELLIPTIQUES.

[From the *Journal für die reine und angewandte Mathematik*, (Crelle), tom. LV. (1858), pp. 15—24.]

I.

En posant dans les formules "Fund. nova p. 15, Tab. III.," $\cos\phi = x$, on obtient

$$\frac{dy}{\sqrt{(y-\alpha)(y-\beta)(y-\gamma)(y-\delta)}} = \frac{2}{\sqrt{(\alpha-\gamma)(\beta-\delta)}+\sqrt{(\alpha-\beta)(\gamma-\delta)}}\;\frac{dx}{\sqrt{(1-x^2)(1-k^2x^2)}}$$

où

$$k^2 = \left(\frac{\sqrt{(\alpha-\gamma)(\beta-\delta)}-\sqrt{(\alpha-\beta)(\gamma-\delta)}}{\sqrt{(\alpha-\gamma)(\beta-\delta)}+\sqrt{(\alpha-\beta)(\gamma-\delta)}}\right)^2$$

$$\frac{1-x}{1+x} = \frac{(\alpha-\beta)(\beta-\delta)}{(\alpha-\gamma)(\gamma-\delta)}\;\frac{(y-\gamma)}{(\beta-y)}.$$

Maintenant soit

$$(y-\alpha)(y-\beta)(y-\gamma)(y-\delta) = (a,\ b,\ c,\ d,\ e \between y,\ 1)^4$$

et proposons-nous d'introduire dans les formules les coefficients (a, b, c, d, e) au lieu des racines α, β, γ, δ. En renvoyant d'ailleurs à la Note sur les covariants d'une fonction quadratique, cubique ou biquadratique insérée dans ce Journal, t. L. (1855), p. 285, [135], je pose pour abréger

$$(\alpha-\beta)(\gamma-\delta) = B,$$
$$(\alpha-\gamma)(\delta-\beta) = C,$$
$$(\alpha-\delta)(\beta-\gamma) = D,$$

de sorte que l'on a identiquement

$$B+C+D=0.$$

Les invariants I, J sont des fonctions rationnelles des quantités B, C, D; en effet on a

$$B^2+C^2+D^2 = 24I,$$

$$(B-C)(C-D)(D-B)=432J,$$

$$B^2C^2D^2=256(I^3-27J^2).$$

Cela étant, nous avons

$$k^2=\left(\frac{i\sqrt{C}-\sqrt{B}}{i\sqrt{C}+\sqrt{B}}\right)^2,$$

ce qui donne d'abord

$$k^4+14k^2+1=16\frac{B^2+BC+C^2}{(i\sqrt{C}+\sqrt{B})^4},$$

ou, à cause de $B+C+D=0$, d'où $B^2+C^2+D^2=2(B^2+BC+C^2)$,

$$k^4+14k^2+1=\frac{8(B^2+C^2+D^2)}{(i\sqrt{C}+\sqrt{B})^4}.$$

On trouve ensuite

$$1-k^2=\frac{4i\sqrt{BC}}{(i\sqrt{C}+\sqrt{B})^4},$$

et de là

$$k^2(1-k^2)^4=\frac{(i\sqrt{C}-\sqrt{B})^2}{(i\sqrt{C}+\sqrt{B})^2}\cdot\frac{256B^2C^2}{(i\sqrt{C}+\sqrt{B})^8},$$

ou en multipliant le numérateur et le dénominateur par $(i\sqrt{C}+\sqrt{B})^2$, et posant D^2 au lieu de $(B+C)^2$, on obtient

$$k^2(1-k^2)^4=\frac{256B^2C^2D^2}{(i\sqrt{C}+\sqrt{B})^{12}}.$$

Ces équations donnent

$$\frac{k^2(1-k^2)^4}{(k^4+14k^2+1)^3}=\frac{B^2C^2D^2}{2(B^2+C^2+D^2)^3}$$

ou, en posant

$$N=\tfrac{27}{4}\frac{1}{1-\dfrac{27J^2}{I^3}},$$

$$\frac{k^2(1-k^2)^4}{(k^4+14k^2+1)^3}=\frac{1}{16N}.$$

On trouve aussi

$$k^4+14k^2+1=\frac{192I}{(i\sqrt{C}+\sqrt{B})^4},$$

ce qui donne

$$\frac{2}{i\sqrt{C}+\sqrt{B}}=\sqrt[4]{\frac{k^4+14k^2+1}{12I}},$$

et la formule de transformation devient ainsi

$$\frac{dy}{\sqrt{(a,\ b,\ c,\ d,\ e\between y,\ 1)^4}}=\sqrt[4]{\frac{k^4+14k^2+1}{12I}}\ \frac{dx}{\sqrt{(1-x^2)(1-k^2x^2)}},$$

le module k étant déterminé par l'équation

$$\frac{k^2(1-k^2)^4}{(k^4+14k^2+1)^3}=\frac{1}{16N},$$

où

$$N=\tfrac{27}{4}\ \frac{1}{1-\dfrac{27J^2}{I^3}}.$$

Cela revient à une formule que j'ai donnée dans le mémoire intitulé, "On the Reduction of $du\div\sqrt{U}$ when U is a Function of the Fourth Order," *Camb. and Dublin Math. Journal,* t. I. (1846), p. 70, [33], et de laquelle j'ai déduit des conséquences que je vais reproduire ici. En effet l'équation en k peut s'écrire sous la forme

$$(k^4+14k^2+1)^3-16Nk^2(k^2-1)^4=0,$$

c'est-à-dire

$$\left(k^2+\frac{1}{k^2}+14\right)^3-16N\left(k-\frac{1}{k}\right)^4=0\ ;$$

en écrivant

$$k-\frac{1}{k}=\frac{4}{\sqrt{\vartheta-1}},$$

on obtient pour ϑ l'équation très-simple

$$\vartheta^3-N(\vartheta-1)=0,$$

et on a ensuite

$$k^2=\frac{7+\vartheta+4\sqrt{3+\vartheta}}{\vartheta-1},$$

ce qui donne aussi

$$k=\frac{2+\sqrt{3+\vartheta}}{\sqrt{\vartheta-1}}.$$

Soit $k=\beta^2$ l'une des valeurs de k, l'équation en k devient

$$\frac{(k^4+14k^2+1)^3}{k^2(k^2-1)^4}=\frac{(\beta^8+14\beta^4+1)^3}{\beta^4(\beta^4-1)^4},$$

équation à laquelle on satisfait, en outre, par la valeur $k=\left(\frac{1-\beta}{1+\beta}\right)^2$. En effet cette valeur donne

$$k^4+14k^2+1=\frac{16\,(\beta^8+14\beta^4+1)^3}{(1+\beta)^8},\qquad k^2-1=-\frac{8\beta\,(1+\beta^2)}{(1+\beta)^4},$$

expressions qui rendent l'équation identique. Cela fait voir que l'équation peut s'écrire sous la forme

$$(k^4+14k^2+1)^4-k^2\,(k^2-1)^4\,\frac{(\beta^8+14\beta^4+1)^3}{\beta^4\,(\beta^4-1)^4}$$

$$=(k^2-\beta^4)\left(k^2-\frac{1}{\beta^4}\right)\left(k^2-\left(\frac{1-\beta}{1+\beta}\right)^4\right)\left(k^2-\left(\frac{1+\beta}{1-\beta}\right)^4\right)\left(k^2-\left(\frac{1-\beta i}{1+\beta i}\right)^4\right)\left(k^2-\left(\frac{1+\beta i}{1-\beta i}\right)^4\right)$$

et les racines de l'équation en k^2 sont

$$\beta^4,\quad \frac{1}{\beta^4},\quad \left(\frac{1-\beta}{1+\beta}\right)^4,\quad \left(\frac{1+\beta}{1-\beta}\right)^4,\quad \left(\frac{1-\beta i}{1+\beta i}\right)^4,\quad \left(\frac{1+\beta i}{1-\beta i}\right)^4,$$

ce qui s'accorde avec un résultat obtenu par Abel (voir les œuvres d'Abel t. I. p. 310 [Ed. 2, p. 459]).

Je fais observer à présent qu'en écrivant

$$\vartheta=\frac{3\varpi}{2\varpi-3},$$

on obtient pour ϖ l'équation $(27-4N)\,\varpi^3+27N\,(\varpi-1)=0$, qui, en posant $M=\frac{-27N}{27-4N}$, ou ce qui est la même chose

$$M=\frac{I^3}{4J^2},$$

devient

$$\varpi^3-M\,(\varpi-1)=0.$$

Cette équation est précisément la même que celle à laquelle je suis parvenu dans la Note sur les covariants etc. ci-dessus citée. La quantité ϖ est liée au module k par la relation

$$k=\frac{3\sqrt{\varpi-1}+2\sqrt{2\varpi-3}}{\sqrt{\varpi+3}}.$$

On peut introduire M au lieu de N dans l'équation en k, et en combinant les formules précédemment obtenues, on trouve

$$\frac{dy}{\sqrt{(a,\,b,\,c,\,d,\,e\between y,\,1)^4}}=\sqrt[4]{\frac{k^4+14k^2+1}{12I}}\;\frac{dx}{\sqrt{(1-x^2)(1-k^2x^2)}}$$

où

$$\frac{27\,(1+14k^2+k^4)^3}{(1+k^2)^2\,(1-34k^2+k^4)^4}=4M,\quad M=\frac{I^3}{4J^2},$$

ou ce qui revient à la même chose,

$$k=\frac{3\sqrt{\varpi-1}+2\sqrt{2\varpi-3}}{\sqrt{\varpi+3}},\qquad \varpi^3-M(\varpi-1)=0.$$

De l'équation entre k et ϑ on déduit facilement la relation

$$\frac{k^4+14k^2+1}{k^4-34k^2+1}=\frac{-\vartheta}{2\vartheta-3},$$

et en substituant dans cette équation pour ϑ sa valeur on obtient

$$\frac{1-34k^2+k^4}{-3(1+14k^2+k^4)}=\frac{1}{\varpi},$$

ce qui est encore une forme de la relation entre k et ϖ.

II

On peut obtenir les résultats qui viennent d'être déduits de la formule de Jacobi, en prenant pour point de départ la transformation d'une fonction du quatrième ordre dans sa forme canonique. Je suppose d'abord que l'on a identiquement

$$\begin{aligned}(a,\ b,\ c,\ d,\ e\between x,\ y)^4&=(\lambda x+\mu y)^4+(\lambda' x+\mu' y)^4+6\theta(\lambda x+\mu y)^2(\lambda' x+\mu' y)^2,\\ &=x_1^4+y_1^4+6\theta x_1^2y_1^2.\end{aligned}$$

Cela étant, je pose $\lambda\mu'-\lambda'\mu=\Lambda$, et je forme les covariants des deux expressions; on obtient par la propriété fondamentale de ces fonctions

$$\begin{aligned}I&=\Lambda^4(1+3\theta^2),\\ J&=\Lambda^6(\theta-\theta^3),\\ U&=x_1^4+y_1^4+6\theta\, x_1^2y_1^2,\\ H&=\Lambda^2\{\theta\, x_1^4+\theta\, y_1^4+(1-3\theta^2)\, x_1^2y_1^2\},\\ \Phi&=\Lambda^3(9\theta^2-1)\, x_1\, y_1\,(x_1^4-y_1^4).\end{aligned}$$

De ces relations on tire

$$\frac{J}{I}=\Lambda^2\frac{\theta-\theta^3}{1+3\theta^2},$$

$$\frac{J^2}{I^3}=\frac{(\theta-\theta^3)^2}{(1+3\theta^2)^3},$$

de sorte qu'en posant

$$M=\frac{I^3}{4J^2},$$

on aura pour déterminer θ, l'équation

$$\frac{(1+3\theta^2)^3}{(\theta-\theta^3)^2}=4M,$$

et pour déterminer x_1, y_1 les équations

$$U=x_1^4+y_1^4+\qquad 6\theta x_1^2y_1^2,$$

$$\frac{I}{J}\,\frac{1-\theta^2}{1+3\theta^2}H=x_1^4+y_1^4+\frac{1-3\theta^2}{\theta}\,x_1^2y_1^2.$$

Je fais observer qu'en désignant par λ un coefficient tel que $U+6\lambda H$ soit un carré parfait, on obtient pour λ l'équation

$$1-9\lambda^2I-54\lambda^3J=0.$$

En effet le cubicovariant Φ d'une fonction qui est un carré parfait est identiquement égal à zéro, et le cubicovariant de la fonction $U+6\lambda H$ est $(1-9\lambda^2I-54\lambda^3J)\,\Phi$, on a donc pour λ l'équation qui vient d'être proposée; cela posé, en observant que

$$U-\frac{I}{J}\,\frac{1-\theta^2}{1+3\theta^2}\,H=\frac{9\theta^2-1}{\theta}\,x_1^2y_1^2,$$

on aura pour une des valeurs de λ,

$$\lambda=-\frac{I}{6J}\,\frac{1-\theta^2}{1+3\theta^2},$$

et en effet cette valeur donne

$$1-\frac{I^3}{J^2}\,\frac{(\theta-\theta^3)^2}{(1+3\theta^2)^3}=0,$$

ce qui est l'équation en θ. Cela étant, je pose

$$\lambda=-\frac{I}{6J}\,\frac{1}{\varpi_3};$$

alors ϖ_3 sera une racine de l'équation

$$\varpi_3-M(\varpi-1)=0,$$

on obtient

$$\varpi_3=\frac{1+3\theta^2}{1-\theta^2},$$

et de là

$$\theta^2=\frac{\varpi_3-1}{\varpi_3+3}.$$

Soient ϖ_1, ϖ_2 les deux autres racines de l'équation en ϖ, on aura

$$\varpi_1+\varpi_2=-\varpi_3,$$

$$\varpi\,\varpi_2=-\frac{\varpi_3^2}{\varpi_3-1},$$

ce qui donne

$$(\varpi_1-\varpi_2)^2=\frac{\varpi_3^{\,2}(\varpi_3+3)}{\varpi_3-1}=\frac{\varpi_3^{\,2}}{\theta^2}.$$

On pourra donc écrire

$$\theta=\frac{-\varpi_3}{\varpi_2-\varpi_1}$$

et au moyen des valeurs de U, H on obtient, par une très-simple réduction, les trois équations suivantes

$$IH-\varpi_1 JU=\quad(\varpi_3-\varpi_1)\,J\,(x_1^{\,2}+y_1^{\,2})^2,$$

$$IH-\varpi_2 JU=-\quad(\varpi_2-\varpi_3)\,J\,(x_1^{\,2}-y_1^{\,2})^2,$$

$$IH-\varpi_3 JU=-\frac{(\varpi_2-\varpi_3)(\varpi_3-\varpi_1)}{\varpi_1-\varpi_2}\,J\,.\,4x_1^{\,2}y_1^{\,2},$$

dont deux quelconques donnent les valeurs de x_1, y_1; ainsi on a obtenu la solution complète du problème de la réduction de la fonction $(a, b, c, d, e \between x, y)^4$ à la forme canonique.

Je fais observer que ces équations montrent *a posteriori* que les expressions $IH-\varpi_1 JU$, $IH-\varpi_2 JU$, $IH-\varpi_3 JU$ sont toutes les trois des carrés de fonctions quadratiques. Je fais observer, en outre, que si l'on forme la valeur de l'expression

$$(\varpi_2-\varpi_3)\sqrt{IH-\varpi_1 JU}+(\varpi_3-\varpi_1)\sqrt{IH-\varpi_2 JU}+(\varpi_1-\varpi_2)\sqrt{IH-\varpi_3 JU},$$

en posant pour un moment $\varpi_2-\varpi_3=\alpha$, $\varpi_3-\varpi_1=\beta$, $\varpi_1-\varpi_2=\gamma$, l'expression dont il s'agit sera égale, à un facteur constant près, à

$$\sqrt{\alpha}\,(x_1^{\,2}+y_1^{\,2})+i\sqrt{\beta}\,(x_1^{\,2}-y_1^{\,2})+i\sqrt{\gamma}\,.\,2x_1y_1$$

ou, ce qui est la même chose, à

$$(\sqrt{\alpha}+i\sqrt{\beta},\ i\sqrt{\gamma},\ \sqrt{\alpha}-i\sqrt{\beta}\between x_1,\ y_1)^2.$$

Or, cette fonction doit être un carré parfait, ce qu'on reconnaît en effet, en se rappelant que

$$(\sqrt{\alpha}+i\sqrt{\beta})(\sqrt{\alpha}-i\sqrt{\beta})-(i\sqrt{\gamma})^2=\alpha+\beta+\gamma=0.$$

Sa racine carrée sera, à un facteur constant près, $(\sqrt{\alpha}+i\sqrt{\beta})\,x_1+i\sqrt{\gamma}y_1$, et cette dernière fonction sera un des facteurs linéaires de $x_1^{\,4}+y_1^{\,4}+6\theta x_1^{\,2}y_1^{\,2}$. Pour vérifier cela je pose $x_1^{\,4}+y_1^{\,4}+6\theta x_1^{\,2}y_1^{\,2}=0$, ce qui donne

$$x_1^{\,2}+(3\theta+\sqrt{9\theta^2-1})\,y_1^{\,2}=0,$$

et $\quad x_1-\left(\sqrt{\dfrac{3\theta+1}{2}}+\sqrt{\dfrac{3\theta-1}{2}}\right)y_1=0.\quad$ Or $3\theta+1=\dfrac{-3\varpi_3}{\varpi_2-\varpi_1}+1=\dfrac{-3\varpi_3+\varpi_2-\varpi_1}{\varpi_2-\varpi_1}$

$=\dfrac{-2\varpi_3+2\varpi_2-(\varpi_1+\varpi_2+\varpi_3)}{\varpi_2-\varpi_1}=\dfrac{2\varpi_2-2\varpi_3}{\varpi_2-\varpi_3}=\dfrac{2\alpha}{-\gamma}$, c'est-à-dire $\sqrt{\dfrac{3\theta+1}{2}}=\dfrac{\sqrt{\alpha}}{i\sqrt{\gamma}}$, et de même $\sqrt{\dfrac{3\theta-1}{2}}=\dfrac{i\sqrt{\beta}}{i\sqrt{\gamma}}$, le facteur linéaire sera donc $i\sqrt{\gamma}\,.\,x_1-(\sqrt{\alpha}+i\sqrt{\beta})\,y_1$ ou, ce qui revient à la même chose, $(\sqrt{\alpha}+i\sqrt{\beta})\,x_1+i\sqrt{\gamma}y_1$, ce qu'il s'agissait de démontrer.

III.

Pour obtenir la formule qui sert à la transformation d'une intégrale elliptique, il faut présenter les résultats précédemment obtenus sous une forme un peu différente. J'écris

$$(a,\ b,\ c,\ d,\ e\between y,\ 1)^4 = (\lambda+\mu y)^4 - (1+k^2)(\lambda+\mu y)^2(\lambda'+\mu' y)^2 + k^2(\lambda'+\mu' y)^4$$
$$= (\lambda+\mu y)^4 . (1-x^2)(1-k^2x^2),$$

où

$$x = \frac{\lambda'+\mu' y}{\lambda+\mu y}.$$

Cela posé, en remplaçant comme auparavant $\lambda\mu'-\lambda'\mu$ par Λ, et en formant les covariants, on obtient

$$\begin{aligned}
I &= \tfrac{1}{12}\Lambda^4(1+14k^2+k^4),\\
J &= \tfrac{1}{216}\Lambda^6(1+k^2)(1-34k^2+k^4),\\
U &= (\lambda+\mu y)^4\{1-(1+k^2)x^2+k^2x^4\},\\
H &= -\tfrac{1}{12}\Lambda^2(\lambda+\mu y)^4\{2(1+k^2)-(1-10k^2+k^4)x^2+2k^2(1+k^2)x^4\},\\
\Phi &= \tfrac{1}{4}\Lambda^3(\lambda+\mu y)^6(1-k^2)^3x(1-k^2x^4).
\end{aligned}$$

On a d'abord

$$\frac{27\ (1+14k^2+k^4)^3}{(1+k^2)^2(1-34k^2+k^4)^2} = 4M,$$

en posant comme auparavant $M = \dfrac{I^3}{4J^2}$; on obtient ensuite

$$\Lambda^4 = \frac{12I}{1+14k^2+k^4},$$

et en remarquant que l'équation entre x et y donne

$$dx = \frac{(\lambda\mu'-\lambda'\mu)\,dy}{(\lambda+\mu y)^2} = \frac{\Lambda dy}{(\lambda+\mu y)^2},$$

on trouve

$$\frac{dy}{\sqrt{(a,\ b,\ c,\ d,\ e\between y,\ 1)^4}} = \frac{1}{\Lambda}\,\frac{dx}{\sqrt{(1-x^2)(1-k^2x^2)}},$$

c'est-à-dire

$$\frac{dy}{\sqrt{(a,\ b,\ c,\ d,\ e\between y,\ 1)^4}} = \sqrt[4]{\frac{k^4+14k^2+1}{12I}}\;\frac{dx}{\sqrt{(1-x^2)(1-k^2x^2)}},$$

ce qui s'accorde avec la formule ci-dessus trouvée; pour compléter la solution, je pose

$$\frac{1-34k^2+k^4}{-3(1+14k^2+k^4)} = \frac{1}{\varpi_3},$$

ϖ_3 sera une des racines de l'équation

$$\varpi_3 - M(\varpi - 1) = 0,$$

en représentant par ϖ_1, ϖ_2 les deux autres racines, on obtient

$$2\varpi_1 = -\varpi_3 \frac{1 + 6k + k^2}{1 + k^2},$$

$$2\varpi_2 = -\varpi_3 \frac{1 - 6k + k^2}{1 + k^2},$$

et ensuite

$$IH - \varpi_1 JU = J(\lambda + \mu y)^4 (\varpi_3 - \varpi_1)(1 + kx^2)^2,$$

$$IH - \varpi_2 JU = J(\lambda + \mu y)^4 (\varpi_3 - \varpi_2)(1 - kx^2)^2,$$

$$IH - \varpi_3 JU = J\frac{(\varpi_3 - \varpi_1)(\varpi_3 - \varpi_2)}{\varpi_1 - \varpi_2}(\lambda + \mu y)^4 . 4kx^2.$$

Donc en posant $\varpi_2 - \varpi_3 = \alpha$, $\varpi_3 - \varpi_1 = \beta$, $\varpi_1 - \varpi_2 = \gamma$, on a

$$(\varpi_2 - \varpi_3)\sqrt{IH - \varpi_1 JU} + (\varpi_3 - \varpi_1)\sqrt{IH - \varpi_2 JU} + (\varpi_1 - \varpi_2)\sqrt{IH - \varpi_3 JU})$$
$$= J(\lambda + \mu y)^4 \sqrt{\alpha\beta}(\sqrt{\alpha} + i\sqrt{\beta} + i\sqrt{\gamma}x\sqrt{k})^2,$$

de même

$$(\varpi_2 - \varpi_3)\sqrt{IH - \varpi_1 JU} + (\varpi_3 - \varpi_1)\sqrt{IH - \varpi_2 JU} - (\varpi_1 - \varpi_2)\sqrt{IH - \varpi_3 JU}$$
$$= J(\lambda + \mu y)^4 \sqrt{\alpha\beta}(\sqrt{\alpha} + i\sqrt{\beta} - i\sqrt{\gamma}x\sqrt{k})^2,$$

et de là enfin, en remplaçant α, β, γ par leurs valeurs,

$$\frac{(\varpi_2 - \varpi_3)\sqrt{IH - \varpi_1 JU} + (\varpi_3 - \varpi_1)\sqrt{IH - \varpi_2 JU} + (\varpi_1 - \varpi_2)\sqrt{IH - \varpi_3 JU}}{(\varpi_2 - \varpi_3)\sqrt{IH - \varpi_1 JU} + (\varpi_3 - \varpi_1)\sqrt{IH - \varpi_2 JU} - (\varpi_1 - \varpi_2)\sqrt{IH - \varpi_3 JU}}$$
$$= \left(\frac{\sqrt{\varpi_2 - \varpi_3} + i\sqrt{\varpi_3 - \varpi_1} + i\sqrt{\varpi_1 - \varpi_2}\,x\sqrt{k}}{\sqrt{\varpi_2 - \varpi_3} + i\sqrt{\varpi_3 - \varpi_1} - i\sqrt{\varpi_1 - \varpi_2}\,x\sqrt{k}}\right)^2,$$

équation dont le premier membre est le carré d'une fraction rationnelle de la forme

$$\frac{A + Bx}{C + Dx}.$$

IV.

Je terminerai ces recherches en démontrant le théorème de M. Hermite dont j'ai parlé dans la Note sur les covariants etc. ci-dessus citée. L'identité

$$JU^3 - IU^2H + 4H^3 = -\Phi^2$$

peut être mise sous la forme

$$-J+I.\frac{H}{U}-4\frac{H^3}{U^3}=\left(\frac{\Phi}{U^2}\right)^2.U.$$

Posons maintenant

$$z=\frac{H}{U},$$

formule dans laquelle je suppose qu'on ait fait $y=1$, de manière que U, H soient des fonctions rationnelles et entières de la seule variable x, savoir

$$U=(a,\ b,\ c,\ d,\ e\between x,\ 1)^4,$$

$$H=(ac-b^2,\ \tfrac{1}{2}(ad-bc),\ \tfrac{1}{6}(ae+2bd-3c^2),\ \tfrac{1}{2}(be-cd),\ ce-d^2\between x,\ 1)^4,$$

alors on aura

$$\sqrt{-J+zI-4z^3}=\frac{\Phi}{U^2}\sqrt{(a,\ b,\ c,\ d,\ e\between x,\ 1)^4},$$

et

$$dz=\frac{UdH-HdU}{U^2}.$$

En vertu de la théorie de Jacobi, $UdH-HdU$ est de la forme $M\Phi\,dx$, où M est un facteur constant; on trouve en effet très-facilement $UdH-HdU=2\Phi\,dx$, d'où l'on déduit la formule

$$\frac{dz}{\sqrt{-J+zI-4z^3}}=\frac{2dx}{\sqrt{(a,\ b,\ c,\ d,\ e\between x,\ 1)^4}},$$

et je fais observer que l'intégrale du premier membre se ramène immédiatement à une forme qui ne contient que la seule constante $M, =\frac{I^3}{4J^2}$.

Londres, 9 Avril, 1856.

236.

NOTE SUR LA COMPOSITION DU NOMBRE 47 PAR RAPPORT AUX VINGT-TROISIÈMES RACINES DE L'UNITÉ.

[From the *Journal für die reine und angewandte Mathematik*, (Crelle), tom. LV. (1858), p. 192.]

M. KUMMER a trouvé (*Journal de Liouville*, t. XII. [1847] p. 208) que le nombre 47 peut être décomposé en onze facteurs qui se déduisent du suivant $\alpha^{10}+\alpha^{13}+\alpha^{8}+\alpha^{15}+\alpha^{7}+\alpha^{16}$, α désignant une racine 23ème de l'unité, et on sait par la théorie générale qu'il doit y avoir une puissance 47^{3f} qui se décompose en vingt-deux facteurs. Le nombre 47^3 peut se décomposer en deux facteurs formés avec les demi-périodes des racines ; il était donc naturel d'essayer si le facteur $(\alpha^{10}+\alpha^{13}+\alpha^{8}+\alpha^{15}+\alpha^{7}+\alpha^{16})^3$ pourrait se décomposer de même en deux facteurs, ce qui donnerait la décomposition de 47^3 en vingt-deux facteurs. Mais on démontre très-facilement que cette décomposition n'est pas possible. En effet en posant $\alpha^\lambda=1$ (λ étant un nombre premier) et en faisant

$$A+B\alpha+\ldots K\alpha^{\lambda-1}=(a+b\alpha+\ldots k\alpha^{\lambda-1})(a+b\alpha^{\lambda-1}+\ldots k\alpha),$$

on aura $A=a^2+b^2+\ldots k^2$. Le nombre qui forme le premier membre peut se réduire au moyen de l'équation $1+\alpha+\ldots \alpha^{\lambda-1}=0$ à la forme $B'\alpha+C'\alpha^2\ldots+K'\alpha^{\lambda-1}$ et l'on aura

$$B'\alpha+C'\alpha^2\ldots+K'\alpha^{\lambda-1}=(a+b\alpha+\ldots k\alpha^{\lambda-1})(a+b\alpha^{\lambda-1}+\ldots k\alpha)-(a^2+b^2+\ldots k^2)(1+\alpha+\ldots \alpha^{\lambda-1}),$$

équation qui subsiste lorsqu'on y fait $\alpha=1$, ce qui donne

$$B'+C'+\ldots K'=(a+b+\ldots k)^2-\lambda(a^2+b^2+\ldots k^2);$$

or, la fonction qui forme le second membre, prise avec le signe négatif peut se mettre sous la forme $(a-b)^2+(a-c)^2+(b-c^2)+$ etc.; donc la décomposition n'existe pas à moins que $B' + C' + \ldots K'$ ne soit négatif. Mais, en réduisant seulement au moyen de l'équation $\alpha^{23} - 1 = 0$, on trouve la suivante

$$(\alpha^{10} + \alpha^{13} + \alpha^8 + \alpha^{15} + \alpha^7 + \alpha^{16})^3 =$$

$$\left.\begin{array}{l} 6 + 7\alpha \ + 7\alpha^2 \ + 3\alpha^3 + 6\alpha^4 + 9\alpha^5 \ + 6\alpha^6 \ + 16\alpha^7 \ + 15\alpha^8 \ + 9\alpha^9 \ + 18\alpha^{10} + 9\alpha^{11} \\ \quad + 7\alpha^{22} + 7\alpha^{21} + 3\alpha^{20} + 6\alpha^{19} + 9\alpha^{18} + 6\alpha^{17} + 16\alpha^{16} + 15\alpha^{15} + 9\alpha^{14} + 18\alpha^{13} + 9\alpha^{12} \end{array}\right\},$$

laquelle, en vertu de $1 + \alpha + \ldots \alpha^{22} = 0$, se réduit à

$$\left.\begin{array}{l} \alpha + \alpha^2 \ - 3\alpha^3 + 3\alpha^5 \ + 10\alpha^7 \ + 9\alpha^8 \ + 3\alpha^9 \ + 12\alpha^{10} + 3\alpha^{11} \\ + \alpha^{22} + \alpha^{21} - 3\alpha^{20} + 3\alpha^{18} + 10\alpha^{16} + 9\alpha^{15} + 3\alpha^{14} + 12\alpha^{13} + 3\alpha^{12} \end{array}\right\},$$

où la somme des coefficients est positive; donc la décomposition ne peut pas s'effectuer. On pourrait sans beaucoup de peine essayer de la même manière les nombres $f = 2$ ou $f = 3$, mais je ne sais pas si l'on a une idée quelconque de la grandeur du nombre f.

Londres, le 10 *Mai,* 1857.

237.

THÉORÈME SUR LES DÉTERMINANTS GAUCHES.

[From the *Journal für die reine und angewandte Mathematik*, (Crelle), tom. LV. (1858), pp. 277—278.]

DANS le mémoire intitulé "Recherches ultérieures sur les déterminants gauches," ce journal t. L. pp. 299—313 (1855), [137], j'ai donné une formule pour le développement d'un déterminant gauche bordé; mais j'ai omis de remarquer un cas particulier assez important. La formule générale se rapporte à un déterminant tel que:

$$\overline{\alpha 123 \mid \beta 123} = \begin{vmatrix} \alpha\beta, & \alpha 1, & \alpha 2, & \alpha 3 \\ 1\beta, & 11, & 12, & 13 \\ 2\beta, & 21, & 22, & 23 \\ 3\beta, & 31, & 32, & 33 \end{vmatrix},$$

que l'on obtient en bordant d'une manière quelconque la matrice gauche

$$\left(\begin{array}{|ccc|} 11, & 12, & 13 \\ 21, & 22, & 23 \\ 31, & 32, & 33 \end{array}\right).$$

On a par exemple:

$$\begin{aligned} \overline{\alpha 123 \mid \beta 123} = \; & \alpha\beta \,.\, 11 \,.\, 22 \,.\, 33 \\ & + \alpha\beta \,.\, 12 \,.\, 12 \,.\, 33 \\ & + \alpha\beta \,.\, 13 \,.\, 13 \,.\, 22 \\ & + \alpha\beta \,.\, 23 \,.\, 23 \,.\, 11 \\ & + \alpha 1 \,.\, \beta 1 \,.\, 22 \,.\, 33 \\ & + \alpha 2 \,.\, \beta 2 \,.\, 11 \,.\, 33 \\ & + \alpha 3 \,.\, \beta 3 \,.\, 11 \,.\, 22 \\ & + \alpha 123 \,.\, \beta 123, \end{aligned}$$

$$\begin{aligned}\overline{\alpha 1234 \mid \beta 1234} = \ & \alpha\beta \,.\, 11 \,.\, 22 \,.\, 33 \,.\, 44 \\ & + \alpha\beta \,.\, 12 \,.\, 12 \,.\, 33 \,.\, 44 \\ & + \ldots\ldots\ldots\ldots\ldots \\ & + \alpha\beta 1234 \,.\, 1234 \\ & + \alpha 1 \,.\, \beta 1 \,.\, 22 \,.\, 33 \,.\, 44 \\ & + \ldots\ldots\ldots\ldots\ldots \\ & + \alpha 123 \,.\, \beta 123 \,.\, 44 \\ & + \ldots\ldots\ldots\ldots\ldots \\ & + \alpha 234 \,.\, \beta 234 \,.\, 11\end{aligned}$$

où les expressions $\alpha\beta 12$, $\alpha\beta 13$ etc. sont des Pfaffiens. Cela étant, le déterminant formé en bordant d'une manière quelconque une matrice gauche et symétrique peut se nommer déterminant gauche et symétrique bordé, et la formule fait voir qu'un déterminant de cette espèce se réduit toujours au produit de deux Pfaffiens. En effet, en écrivant dans les exemples $11 = 22 = 33 = 44 = 0$, on obtient:

$$\overline{\alpha 123 \mid \beta 123} = \alpha 123 \,.\, \beta 123,$$

$$\overline{\alpha 1234 \mid \beta 1234} = \alpha\beta 1234 \,.\, 1234,$$

et de même pour un déterminant gauche et symétrique bordé quelconque, suivant que l'ordre du déterminant est pair ou impair. Je remarque à propos de cela, que dans le cas d'un déterminant d'ordre pair, le terme $\alpha\beta$ est multiplié par un mineur premier lequel (comme déterminant gauche et symétrique d'ordre impair) se réduit à zéro; le déterminant ne contient donc pas ce terme $\alpha\beta$, et sera par conséquent fonction linéo-linéaire des quantités $\alpha 1$, $\alpha 2$, etc. et 1β, 2β, etc.; de manière qu'on ne saurait être surpris de voir ce déterminant se présenter sous la forme d'un produit de deux facteurs dont l'un est fonction linéaire de $\alpha 1$, $\alpha 2$, etc. et l'autre fonction linéaire de 1β, 2β, etc. Mais pour un déterminant d'ordre impair, le coefficient du terme $\alpha\beta$ ne se réduit pas à zéro; en supposant donc que le déterminant puisse s'exprimer comme produit de deux facteurs, il est nécessaire que l'un de ces facteurs soit (comme le déterminant même) fonction linéaire de $\alpha\beta$ et linéo-linéaire de $\alpha 1$, $\alpha 2$, etc. et 1β, 2β, etc.; de cette manière on se rend compte de la différence de la forme des facteurs, qui a lieu dans les deux cas dont il s'agit.

En écrivant $\beta = \alpha$ (ce qui implique $\alpha\alpha = 0$, car on suppose toujours $\alpha\beta = -\beta\alpha$) le déterminant gauche et symétrique bordé se réduit à un déterminant gauche et symétrique ordinaire, de plus le Pfaffien $\alpha\beta 1234$ se réduit à zéro, et les équations deviennent:

$$\overline{\alpha 123 \mid \alpha 123} = (\alpha 123)^2,$$

$$\overline{\alpha 1234 \mid \alpha 1234} = 0\,;$$

savoir, quand l'ordre est pair, le déterminant se réduit au carré d'un Pfaffien, et quand l'ordre est impair, le déterminant s'évanouit; ce qui est en effet la propriété fondamentale des déterminants gauches et symétriques.

2 *Stone Buildings, Londres, le* 16 *Nov.* 1857.

238.

NOTE SUR LES NORMALES D'UNE CONIQUE.

[From the *Journal für die reine und angewandte Mathematik,* (Crelle), tom. LVI. (1859), pp. 182—185.]

On connaît les recherches très élégantes de M. Joachimsthal sur les normales d'une conique [voir le Mémoire "Ueber die Normalen der Ellipse und des Ellipsoids," *Crelle,* t. XXVI. (1843) pp. 172—180]; en particulier l'auteur a obtenu le théorème suivant: en supposant que 1, 2, 3, 4 soient des points d'une conique tels que les quatre normales se rencontrent dans un même point, on prend le pôle de la droite (1, 2) par rapport à la conique et on mène par ce pôle des perpendiculaires aux diamètres de la conique; celà étant, en prenant sur chaque diamètre dans le sens opposé un point dont la distance du centre est égale à la distance du pied de la perpendiculaire sur ce même diamètre, la droite menée par ces deux points passe par les deux points 3 et 4. Mais cette propriété peut s'énoncer d'une manière beaucoup plus simple; la droite dont il s'agit est la polaire (ou autrement dit l'harmonicale)—par rapport au triangle formé par les deux diamètres et la droite située à l'infini—du pôle de la droite (1, 2) par rapport à la conique. Or on sait que l'idée de la perpendiculaire peut être généralisée. Savoir en prenant une conique quelconque que nous appelons *la conique absolue,* deux droites harmoniques par rapport à cette conique peuvent être appelées *perpendiculaires* (et de même deux points harmoniques par rapport à la conique absolue peuvent être appelés *perpendiculaires*). Cela posé, on peut parler dans un sens plus général des normales, etc. d'une courbe quelconque. En effet, que l'on s'imagine comme auparavant (outre la conique absolue) une conique donnée quelconque et quatre points 1, 2, 3 et 4 de cette conique tels que les normales se rencontrent dans un même point. Au lieu du triangle ci-dessus mentionné on a le triangle formé par les trois axes harmoniques (ou autrement dit, *conjugués*) communs aux deux coniques, et le théorème peut s'énoncer comme suit: En prenant le pôle de la droite (1, 2) par rapport à la conique donnée et puis la polaire (l'harmonicale) de ce pôle par rapport aux axes conjugués de la conique donnée et de la conique absolue, cette polaire passe par les deux points

3 et 4. Ou (ce qui revient à la même chose) on peut considérer les points 1, 2, 3, 4 comme les angles d'un quadrilatère inscrit dans la conique donnée, les quatre tangentes à cette conique aux points dont il s'agit seront les côtés d'un quadrilatère circonscrit à la conique donnée; cela étant, les six côtés du quadrilatère inscrit seront les polaires (les harmonicales)—par rapport aux trois axes conjugués de la conique donnée et de la conique absolue—des six sommets du quadrilatère circonscrit.

Pour démontrer cela, je fais observer qu'il est permis de rapporter la conique absolue et la conique donnée aux trois axes conjugués communs, c'est-à-dire de prendre

$$x^2 + y^2 + z^2 = 0,$$

pour équation de la conique absolue, et

$$ax^2 + by^2 + cz^2 = 0,$$

pour équation de la conique donnée: cela posé (et en observant que, d'après la définition, deux droites $Ax + By + Cz = 0$, $A'x + B'y + C'z = 0$ seront perpendiculaires si $AA' + BB' + CC' = 0$) on obtient sans peine

$$\frac{x(b-c)}{x_1} + \frac{y(c-a)}{y_1} + \frac{z(a-b)}{z_1} = 0,$$

pour équation de la normale au point 1, en désignant par (x_1, y_1, z_1) les coordonnées de ce point. On a de même, en désignant par (x_2, y_2, z_2) et (x_3, y_3, z_3) les coordonnées des points 2 et 3,

$$\frac{x(b-c)}{x_2} + \frac{y(c-a)}{y_2} + \frac{z(a-b)}{z_2} = 0,$$

$$\frac{x(b-c)}{x_3} + \frac{y(c-a)}{y_3} + \frac{z(a-b)}{z_3} = 0,$$

pour les équations des normales aux points 2 et 3; et la condition qui exprime que ces trois normales se rencontrent dans un même point sera évidemment

$$\begin{vmatrix} \frac{1}{x_1}, & \frac{1}{y_1}, & \frac{1}{z_1} \\ \frac{1}{x_2}, & \frac{1}{y_2}, & \frac{1}{z_2} \\ \frac{1}{x_3}, & \frac{1}{y_3}, & \frac{1}{z_3} \end{vmatrix} = 0.$$

Mais les coordonnées (x_1, y_1, z_1) etc. satisfont à l'équation $ax^2 + by^2 + cz^2 = 0$, on a donc aussi

$$\begin{vmatrix} x_1^2, & y_1^2, & z_1^2 \\ x_2^2, & y_2^2, & z_2^2 \\ x_3^2, & y_3^2, & z_3^2 \end{vmatrix} = 0$$

et de ces deux équations on déduit la suivante

$$\left\{\begin{matrix} x_1, & y_1, & z_1 \\ x_2, & y_2, & z_2 \\ x_3, & y_3, & z_3 \end{matrix}\right\} = 0$$

en désignant par le symbole qui forme le premier membre la fonction

$$x_1y_2z_3 + x_1y_3z_2 + x_2y_3z_1 + x_2y_1z_3 + x_3y_1z_2 + x_3y_2z_1.$$

C'est ce qui résulte de l'identité

$$\left\{\begin{matrix} x_1, & y_1, & z_1 \\ x_2, & y_2, & z_2 \\ x_3, & y_3, & z_3 \end{matrix}\right\} \times \begin{vmatrix} x_1, & y_1, & z_1 \\ x_2, & y_2, & z_2 \\ x_3, & y_3, & z_3 \end{vmatrix} = \begin{vmatrix} x_1^2, & y_1^2, & z_1^2 \\ x_2^2, & y_2^2, & z_2^2 \\ x_3^2, & y_3^2, & z_3^2 \end{vmatrix} + 2x_1y_1z_1x_2y_2z_2x_3y_3z_3 \begin{vmatrix} \frac{1}{x_1}, & \frac{1}{y_1}, & \frac{1}{z_1} \\ \frac{1}{x_2}, & \frac{1}{y_2}, & \frac{1}{z_2} \\ \frac{1}{x_3}, & \frac{1}{y_3}, & \frac{1}{z_3} \end{vmatrix};$$

car le déterminant qui forme le second facteur du premier membre de l'équation ne s'évanouissant pas, c'est l'autre facteur qui devra s'évanouir en vertu des deux relations données.

L'équation de la droite (1, 2) sera

$$\begin{vmatrix} x, & y, & z \\ x_1, & y_1, & z_1 \\ x_2, & y_2, & z_2 \end{vmatrix} = 0:$$

les coordonnées du pôle de cette droite par rapport à la conique donnée $ax^2 + by^2 + cz^2 = 0$, seront

$$\frac{1}{a}(y_1z_2 - y_2z_1) : \frac{1}{b}(z_1x_2 - z_2x_1) : \frac{1}{c}(x_1y_2 - x_2y_1);$$

mais les deux équations $ax_1^2 + by_1^2 + cz_1^2 = 0$, $ax_2^2 + by_2^2 + cz_2^2 = 0$ donnent $a : b : c = y_1^2z_2^2 - y_2^2z_1^2 : z_1^2x_2^2 - z_2^2x_1^2 : x_1^2y_2^2 - x_2^2y_1^2$; par suite de cela les coordonneés du pôle deviennent

$$\frac{1}{y_1z_2 + y_2z_1} : \frac{1}{z_1x_2 + z_2x_1} : \frac{1}{x_1y_2 + x_2y_1};$$

donc l'équation de la polaire (l'harmonicale) de ce pôle par rapport aux trois droites $(x = 0,\ y = 0,\ z = 0)$ sera

$$x(y_1z_2 + y_2z_1) + y(z_1x_2 + z_2x_1) + z(x_1y_2 + x_2y_1) = 0,$$

laquelle peut être représentée comme suit

$$\left\{\begin{matrix} x, & y, & z \\ x_1, & y_1, & z_1 \\ x_2, & y_2, & z_2 \end{matrix}\right\} = 0,$$

et, en substituant (x_3, y_3, z_3) au lieu de (x, y, z), on voit que la droite dont il s'agit contient le point 3. De même cette droite contient le point 4, de sorte que le théorème se trouve démontré.

Observons encore que dans la géométrie de la sphère on peut prendre pour la conique absolue la conique imaginaire qui est l'intersection de la surface sphérique avec la surface conique imaginaire $x^2 + y^2 + z^2 = 0$. Le mot perpendiculaire aura alors la signification ordinaire, et on aura pour les coniques sphériques ce théorème très simple, savoir que les six côtés du quadrilatère inscrit seront les polaires (les harmonicales)—par rapport aux trois axes de la conique sphérique donnée—des six sommets du quadrilatère circonscrit. C'est ce que l'on reconnaît aussi par l'analyse que je viens de donner, laquelle en effet est précisément celle dont on se servirait naturellement pour les coniques sphériques.

Londres, le 10 *Mars*, 1857.

239.

ADDITION À LA NOTE SUR LA COMPOSITION DU NOMBRE 47 PAR RAPPORT AUX VINGT-TROISIÈMES RACINES DE L'UNITÉ.

[From the *Journal für die reine und angewandte Mathematik,* (Crelle), tom. LVI. (1859), pp. 186—187: Sequel to **236.**]

M. KUMMER a bien voulu m'écrire une lettre où il remarque que le cube du facteur complexe du nombre 47, *multiplié par l'unité complexe convenable,* peut effectivement se décomposer en deux facteurs réciproques; c'est-à-dire qu'il existe une unité complexe $E(\alpha)$ telle que

$$E(\alpha)(\alpha^{10}+\alpha^{13}+\alpha^{8}+\alpha^{15}+\alpha^{7}+\alpha^{16})^{3}=F(\alpha)\,F(\alpha^{-1}).$$

Pour $F(\alpha)$ M. Kummer a trouvé la valeur

$$F(\alpha)=\alpha^{4}+\alpha^{5}+\alpha^{9}+\alpha^{10}+\alpha^{16}-\alpha^{20}+\alpha^{22},$$

fonction qui par conséquent est l'un des 22 facteurs complexes de 47^3. Dans un postscriptum M. Kronecker m'a communiqué l'expression suivante de cette unité complexe

$$E(\alpha)=\frac{\alpha^{2}+\alpha^{21}}{(\alpha^{8}+\alpha^{15})(\alpha^{9}+\alpha^{14})(\alpha^{10}+\alpha^{13})^{2}(\alpha^{11}+\alpha^{12})}$$

équivalente à l'expression en fonction entière:

$$E(\alpha)=\begin{cases}-23\alpha+2\alpha^{2}-20\alpha^{3}-21\alpha^{5}-3\alpha^{6}-17\alpha^{7}-4\alpha^{8}-14\alpha^{9}-8\alpha^{10}-12\alpha^{11}\\-23\alpha^{22}+2\alpha^{21}-20\alpha^{20}-21\alpha^{18}-3\alpha^{17}-17\alpha^{16}-4\alpha^{15}-14\alpha^{14}-8\alpha^{13}-12\alpha^{12}.\end{cases}$$

En supposant que cette valeur de $E(\alpha)$ soit connue, on trouve sans peine une condition à laquelle $F(\alpha)$ doit satisfaire. La valeur que j'ai donnée pour $(\alpha^{10}+\alpha^{13}+\alpha^{8}+\alpha^{15}+\alpha^{7}+\alpha^{16})^{3}$

contient le terme constant $+6$; en y ajoutant la quantité évanouissante $-6(1+\alpha+\ldots+\alpha^{22})$ elle se réduit à

$$\begin{cases} \alpha \;+\alpha^2 -3\alpha^3 +3\alpha^5 +10\alpha^7 +9\alpha^8 +3\alpha^9 +12\alpha^{10} + 3\alpha^{11} \\ +\alpha^{22}+\alpha^{21}-3\alpha^{20}+3\alpha^{18}+10\alpha^{16}+9\alpha^{15}+3\alpha^{14}+12\alpha^{13}+3\alpha^{12} \end{cases}$$

et en multipliant cette valeur par $E(\alpha)$, le terme constant sera $+808$; donc en y ajoutant la quantité évanouissante $-808(1+\alpha+\ldots+\alpha^{22})$ on obtient

$$E(\alpha)(\alpha^{10}+\alpha^{13}+\alpha^8+\alpha^{15}+\alpha^7+\alpha^{16})^3$$

$$=\begin{cases} -5\alpha \;-8\alpha^2 -7\alpha^3-7\alpha^4 -4\alpha^5 -3\alpha^6 -7\alpha^7 -8\alpha^8 -7\alpha^9 -7\alpha^{10}-5\alpha^{11} \\ -5\alpha^{22}-8\alpha^{21}-7\alpha^{20}-7\alpha^{19}-4\alpha^{18}-3\alpha^{17}-7\alpha^{16}-8\alpha^{15}-7\alpha^{14}-7\alpha^{13}-5\alpha^{12}. \end{cases}$$

En représentant cette expression par $B'\alpha+C'\alpha^2+\ldots+K'\alpha^{22}$, j'écris

$$B'\alpha+C'\alpha^2+\ldots+K'\alpha^{22}$$
$$=(a+b\alpha+\ldots+k\alpha^{22})(a+b\alpha^{22}+\ldots+k\alpha)-(a^2+b^2+\ldots+k^2)(1+\alpha+\ldots+\alpha^{22}),$$

équation qui subsiste pour $\alpha=1$. On a donc

$$B'+C'+\ldots+K'=(a+b+\ldots+k)^2-23(a^2+b^2+\ldots+k^2),$$

ou, d'après les valeurs de B', C', ... K',

$$-136=(a+b+\ldots+k)^2-23(a^2+b^2+\ldots+k^2),$$

ce qui donne

$$(a+b+\ldots+k)^2\equiv-136 \pmod{23}.$$

On peut ajouter à $(a+b\alpha+\ldots+k\alpha^{22})$ un multiple quelconque de $(1+\alpha+\ldots+\alpha^{22})$ et changer le signe; il est donc permis de prendre $(a+b+\ldots+k)$ positif et plus petit que $\frac{23}{2}$. Cela étant la congruence donne

$$a\;+b\;+\ldots+k\;=5$$

et on obtient alors

$$a^2+b^2+\ldots+k^2=7.$$

Au moyen de cette valeur l'équation à laquelle il faut satisfaire devient

$$7+2\alpha \;-\alpha^2 \;+3\alpha^5 +4\alpha^6 -\alpha^8 +2\alpha^{11}$$
$$+2\alpha^{22}-\alpha^{19}+3\alpha^{18}+4\alpha^{17}-\alpha^{15}+2\alpha^{12}$$
$$=(a+b\alpha\ldots+k\alpha^{22})(a+b\alpha^{22}\ldots+k\alpha).$$

À cause des coefficients numériques négatifs, les coefficients cherchés $a, b, \ldots k$ ne peuvent pas être tous à la fois positifs; et cela étant il n'y a qu'une seule manière de satisfaire aux deux conditions ci-dessus écrites, savoir il faut donner à sept des coefficients $a, b, \ldots k$ les valeurs 1, 1, 1, 1, 1, 1, -1, et aux autres coefficients la valeur zéro; l'expression

$$F(\alpha)=\alpha^4+\alpha^5+\alpha^9+\alpha^{10}+\alpha^{16}-\alpha^{20}+\alpha^{22}$$

s'accorde en effet avec cette conclusion.

Londres, le 6 *Octobre,* 1858.

240.

NOTE ON A THEOREM IN SPHERICAL TRIGONOMETRY.

[From the *Philosophical Magazine*, vol. XVII. (1859), p. 151.]

I AM not aware that the following theorem has been noticed: viz., in any spherical triangle, if as usual a, b, c are the sides, and A, B, C the opposite angles, then

$$\sin b \sin c + \cos b \cos c \cos A = \sin B \sin C - \cos B \cos C \cos a,$$
$$\sin c \sin a + \cos c \cos a \cos B = \sin C \sin A - \cos C \cos A \cos b,$$
$$\sin a \sin b + \cos a \cos b \cos C = \sin A \sin B - \cos A \cos B \cos c.$$

The demonstration is very simple; in fact we have

$$\begin{aligned}
&\sin b \sin c + \cos b \cos c \cos A \\
&\quad = \sin b \sin c (\sin^2 A + \cos^2 A) + \cos b \cos c \cos A \\
&\quad = \sin b \sin c \sin^2 A + \cos A (\cos b \cos c + \sin b \sin c \cos A) \\
&\quad = \sin B \sin C \sin^2 a + \cos A \cos a \\
&\quad = \sin B \sin C (1 - \cos^2 a) + \cos A \cos a \\
&\quad = \sin B \sin C + \cos a (\cos A - \sin B \sin C \cos a) \\
&\quad = \sin B \sin C - \cos B \cos C \cos a,
\end{aligned}$$

which proves the theorem.

2, *Stone Buildings, W.C., January* 5, 1859.

A geometrical proof and interpretation are given, G. B. Airy, "Remarks on Mr Cayley's Trigonometrical Theorem, etc." *Phil. Mag.* same volume, p. 176. I transfer to this place the concluding sentence of the subsequent paper 243. "I take the opportunity of noticing that the theorem in spherical trigonometry, which I gave in the February Number, is not new, but, as pointed out by Prof. Chauvenet in the Mathematical Monthly (Cambridge, U.S.), is to be found in Cagnoli's 'Trigonometry' (1808)."

241.

ON POINSOT'S FOUR NEW REGULAR SOLIDS.

[From the *Philosophical Magazine*, vol. XVII. (1859), pp. 123—128.]

It is shown by Poinsot, in the "Mémoire sur les Polygones et les Polyèdres," *Jour. École Polyt.* vol. IV. pp. 16 to 48 (1810), that, besides the regular polyhedrons of ordinary geometry, there are (of course in an extended signification of the term) four new regular polyhedrons, viz. an icosahedron, which I will call the great icosahedron (No. 33 of the Memoir), and three dodecahedrons, which I will call the great dodecahedron (No. 37), the great stellated dodecahedron (No. 38), and the small stellated dodecahedron (No. 39). The nature of Poinsot's generalization will be best understood by conceiving, as he does, that the polyhedron is projected on a concentric sphere, so that the faces become spherical polygons. Then for the ordinary polyhedrons of geometry, the sum of the angles at a vertex $=4$ right angles; but it may, according to the more general notion, be $=e$ times 4 right angles. In like manner for the ordinary polyhedrons, the sides of a face subtend at the centre angles the sum of which is $=4$ right angles; but according to the more general notion, this sum may be (viz. if the polygons are stellated) $=e'$ times four right angles. And finally, the sum of the spherical polygons is ordinarily equal to the entire spherical surface; but according to the more general notion, it may be $=D$ times the spherical surface. (e is Poinsot's e; e' does not occur in Poinsot; and, for a reason which will appear, I have written D for Poinsot's E.)

The new polyhedra are constructed as follows:

1. *The great Icosahedron.*—Each face is made up of seven faces, or rather four faces and six half faces of the ordinary icosahedron, in the manner shown by fig. 1. There are, as in the ordinary icosahedron, five angles at each vertex; but these make up together, not four, but eight right angles, or $e=2$; but, as in the ordinary poly-

hedra, $e' = 1$; and the sum of all the faces is obviously seven times the spherical surface, or $D = 7$. (Also $E = 7$.)

Fig. 1.

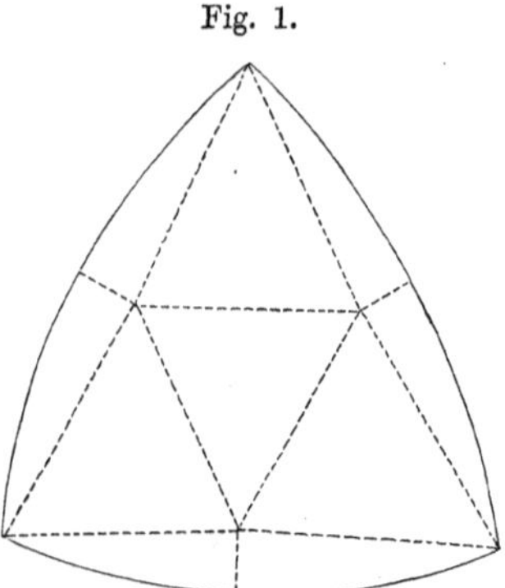

2. *The great Dodecahedron.*—Each face is made up of five faces of the ordinary icosahedron in the manner shown by the figure 2. There are five angles at each vertex, and these make up together eight right angles, or $e = 2$; but, as in ordinary polyhedra, $e' = 1$; and the sum of all the faces is obviously $12 \times \frac{5}{20}$, that is three times the spherical surface, or $D = 3$. (Also $E = 3$.)

Fig. 2.

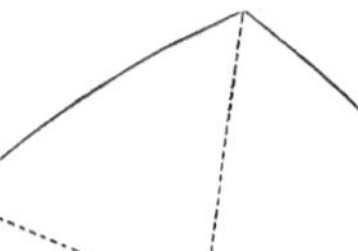

3. *The great stellated Dodecahedron.*—Each face is formed by stellating a face of the great dodecahedron in the manner shown by fig. 3. There are, as in the ordinary

Fig. 3.

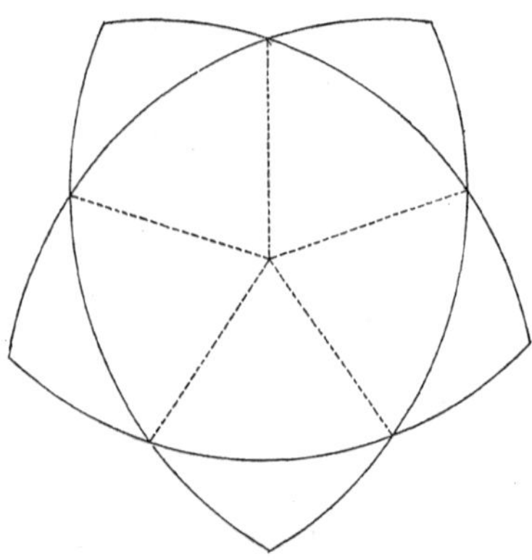

dodecahedron, three angles at each vertex, and the sum of these is simply four right angles, or $e=1$. On account of the stellation, $e'=2$. Each of the projecting parts of the face is equal $\frac{1}{3}$ of the face of the ordinary icosahedron; and if we reckon the area of the stellated pentagon to be that of the interior pentagon *plus* the projecting parts, the area of the face will be $5+\frac{5}{3}$, or $\frac{20}{3}$ of the face of the ordinary icosahedron; and the sum of the faces will be four times the spherical surface, and accordingly Poinsot writes $E=4$. If, however, what seems preferable, we reckon the area of the stellated pentagon as five times the triangle having for its vertex the centre of the face and standing upon a side (or what is the same thing, reckon the stellated pentagon as *twice* the interior pentagon *plus* the projecting parts), then the area of the face will be $10+\frac{5}{3}$ or $\frac{35}{3}$ of the face of the ordinary icosahedron, and the sum of the faces will be seven times the spherical surface, or $D=7$.

4. *The small stellated Dodecahedron.*—Each face is formed by stellating a face of the ordinary dodecahedron, as shown by fig. 4. There are five angles at each vertex; and the sum of these is four right angles, or $e=1$. On account of the stellation, $e'=2$. The area of each of the projecting parts is $\frac{1}{5}$ of the interior pentagon or face of the ordinary dodecahedron; and, according to the first mode of measurement, the

Fig. 4.

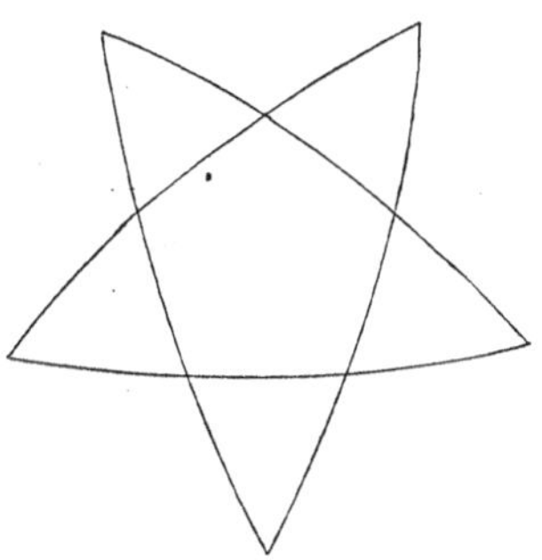

area of the stellated face is twice that of the face of the ordinary dodecahedron, and the sum of the faces is twice the spherical surface, and accordingly Poinsot writes $E=2$. But according to the second mode of measurement, the area of the stellated pentagon is three times that of the face of the ordinary dodecahedron, and the sum of the faces is three times the spherical surface, or we have $D=3$.

I form now the following Table, comprising as well the ordinary five figures as the new ones of Poinsot, and where we have

H, the number of faces.

S, the number of vertices.

A, the number of edges.

n, the number of sides to a face.

n', the number of sides (angles) at a vertex.

e, viz. the angles at a vertex make together e times four right angles.

e', viz. the angles which the sides of a face subtend at the centre of the face make together e' times four right angles.

E, viz. the faces make together E times the spherical surface, the area of a stellated face being reckoned (as by Poinsot), each portion being taken once only.

D, viz. the faces make together D times the spherical surface, the area of a stellated face being reckoned as the sum of the triangles having their vertices at the centre of the face and standing on the sides.

The Table is

Designation.	H.	S.	A.	n.	n'.	e.	e'.	D.		E.
Tetrahedron	4	4	6	3	3	1	1	1		1
Hexahedron	6	8	12	4	3	1	1	1		1
Octahedron	8	6	12	3	4	1	1	1		1
Dodecahedron	12	20	30	5	3	1	1	1		1
Icosahedron	20	12	30	3	5	1	1	1		1
Great stellated dodecahedron	12	20	30	5	3	1	2	7		4
Great icosahedron	20	12	30	3	5	2	1	7		7
Small stellated dodecahedron	12	12	30	5	5	1	2	3		2
Great dodecahedron	12	12	30	5	5	2	1	3		3

where the figures which are polar reciprocals of each other are written in pairs: viz. as is well known, the tetrahedron is its own reciprocal, the hexahedron and octahedron are reciprocals, and the dodecahedron and icosahedron are reciprocals; moreover the great stellated dodecahedron and the great icosahedron are reciprocals, and the small stellated dodecahedron and the great dodecahedron are reciprocals. The number which I have called D is reciprocal to itself; this is not the case for Poinsot's E; and I have not been able to define E in such a manner as to enable me to form the definition of a reciprocal number E': this may be possible, but in the mean time it seems better to discard E altogether, and use instead of it the number D.

Euler's well-known relation applying to ordinary polyhedra is

$$S + H = A + 2.$$

Poinsot in his memoir has (by an extension of Legendre's demonstration of Euler's theorem) obtained the more general relation,

$$eS + H = A + 2E,$$

which, however, does not apply to the two stellated figures where e' is different from unity; the general form is

$$eS + e'H = A + 2D,$$

which applies to all the nine figures. This applies to all polyhedra, regular or not, which are such that e has the same value for each vertex, and e' the same value for each face. To prove it, we have only to further extend Legendre's demonstration. If for any face, stellated or not, the sum of the angles is s, and the number of sides n, then, according to the foregoing mode of reckoning, the area of the face (measured in right angles) is

$$s + 4e' - 2n.$$

Now the sum of all the faces is D times the spherical surface, $= 8D$. But the sum of the term s is equal to the sum of the angles about each vertex, $= 4eS$; the sum of the term $4e'$ is $= 4e'H$, the sum of the term $2n$ is four times the number of edges, $= 4A$. Hence $4eS + 4e'H - 4A = 8D$, or $eS + e'H = 2D$.

I remark that the small stellated dodecahedron and the great dodecahedron are descriptively the same figures, and that, if we represent the vertices by a, b, c, d, e, f, g, h, i, j, p, q, and the faces by A, B, C, D, E, F, G, H, I, J, P, Q, then the relations of the vertices and faces is shown by either of the following Tables:

$a\ b\ c\ d\ e = P,$	$A\ C\ E\ B\ D = p,$
$p\ b\ i\ h\ e = A,$	$P\ I\ E\ B\ H = a,$
$p\ e\ j\ i\ a = B,$	$P\ J\ A\ C\ I = b,$
$p\ d\ f\ j\ b = C,$	$P\ F\ B\ D\ J = c,$
$p\ e\ g\ f\ c = D,$	$P\ G\ C\ E\ F = d,$
$p\ a\ h\ g\ d = E,$	$P\ H\ D\ A\ G = e,$
$j\ c\ d\ g\ q = F,$	$J\ D\ Q\ C\ G = f,$
$f\ d\ e\ h\ q = G,$	$F\ E\ Q\ D\ H = g,$
$g\ e\ a\ i\ q = H,$	$G\ A\ Q\ E\ I = h,$
$h\ a\ b\ j\ q = I,$	$H\ B\ Q\ A\ J = i,$
$i\ b\ c\ f\ q = J,$	$I\ C\ Q\ B\ F = j,$
$f\ g\ h\ i\ j = Q,$	$F\ H\ J\ G\ I = q,$

where it is to be noticed that in either Table each non-consecutive duad of any pentad occurs once, and only once, as a non-consecutive duad of another pentad. The restriction that a non-consecutive duad of any multiplet is *not* to occur as a duad, consecutive or non-consecutive, of any other multiplet (see my note appended to Mr Kirkman's paper "On Autopolar Polyhedra," *Phil. Trans.* 1857, p. 183 [259]), applies only to ordinary polyhedra, and not to the class here considered.

2, *Stone Buildings, W.C., January* 13, 1859.

242.

SECOND NOTE ON POINSOT'S FOUR NEW REGULAR POLYHEDRA.

[From the *Philosophical Magazine*, vol. XVII. (1859), pp. 209—210.]

THE Note on Poinsot's four new regular Polyhedra (February Number, p. 123), [241], was written without my being acquainted with Cauchy's first memoir, "Recherches sur les Polyèdres" (*Jour. Polyt.* vol. IX. pp. 68—86, 1813), the former part of which (pp. 68—76) relates to Poinsot's polyhedra. Cauchy considers the polyhedra, not as projected on the sphere, but *in solido;* and he shows, very elegantly, that all such polyhedra *must* be derived from the ordinary regular polyhedra by producing their sides or faces. The reciprocal method would be to produce the sides or join the vertices; and, adopting this reciprocal method, and projecting the figure on the sphere, we have the method employed by Poinsot, and explained and developed in my former Note. Cauchy does not at all consider Poinsot's generalized equation, $eS+H=A+2E$, nor of course my further generalization, $eS+e'H=A+2D$; but the latter part of the memoir relates to a generalization, in a different direction, of Euler's original formula, $S+H=A+2$: viz. Cauchy's theorem is—"If a polyhedron is partitioned into any number of polyhedra by taking at pleasure, in the interior of it, any number of new vertices, and if P be the total number of polyhedra thus formed, S the total number of vertices (including those of the original polyhedron), and A the total number of edges, then $S+H=A+P+1$; that is, the sum of the number of vertices and the number of faces exceeds by unity the sum of the number of edges and of the number of polyhedra."

For $P=1$, we have Euler's equation $S+H=A+2$; and for $P=0$, we have a theorem relating to the partition of a polygon; viz. if the polygon is divided into H polygons, and if S be the number of vertices, and A the number of sides, then $S+H=A+1$; from which it is easy to pass to Euler's equation, $S+H=A+2$, for polyhedra. I remark that, in the equation $S+H=A+1$, H should, in analogy with Cauchy's notation for polyhedra, be replaced by P; so that we have for a single polygon,

$$A=S;$$

and for the partitions of a polygon,

$$A=S+P-1:$$

corresponding respectively to Euler's theorem for a single polyhedron, viz.

$$S+H=A+2;$$

and to Cauchy's theorem for the partitions of a polyhedron, viz.

$$S+H=A+2+(P-1).$$

Cauchy's second memoir (pp. 87—98) contains a very beautiful demonstration of the theorem implied in the ninth definition of the eleventh book of Euclid, viz. that two convex polyhedra are equal when they are bounded by the same number of faces equal each to each.

2, *Stone Buildings, W.C., February* 1, 1859.

243.

ON THE THEORY OF GROUPS AS DEPENDING ON THE SYMBOLIC EQUATION $\theta^n = 1$. THIRD PART.

[From the *Philosophical Magazine*, vol. XVIII. (1859), pp. 34—37: Sequel to **125** and **126.**]

THE following is, I believe, a complete enumeration of the groups of 8:

I. $1, \alpha, \alpha^2, \alpha^3, \alpha^4, \alpha^5, \alpha^6, \alpha^7$ $(\alpha^8 = 1)$.

II. $1, \alpha, \alpha^2, \alpha^3, \beta, \beta\alpha, \beta\alpha^2, \beta\alpha^3$ $(\alpha^4 = 1, \beta^2 = 1, \alpha\beta = \beta\alpha)$.

III. $1, \alpha, \alpha^2, \alpha^3, \beta, \beta\alpha, \beta\alpha^2, \beta\alpha^3$ $(\alpha^4 = 1, \beta^2 = 1, \alpha\beta = \beta\alpha^3)$.

IV. $1, \alpha, \alpha^2, \alpha^3, \beta, \beta\alpha, \beta\alpha^2, \beta\alpha^3$ $(\alpha^4 = 1, \beta^2 = \alpha^2, \alpha\beta = \beta\alpha^3)$.

V. $1, \alpha, \beta, \beta\alpha, \gamma, \gamma\alpha, \gamma\beta, \gamma\beta\alpha$ $(\alpha^2 = 1, \beta^2 = 1, \gamma^2 = 1, \alpha\beta = \beta\alpha, \alpha\gamma = \gamma\alpha, \beta\gamma = \gamma\beta)$.

That the groups are really distinct is perhaps most readily seen by writing down the indices of the different terms of each group; these are

I. 1, 8, 4, 8, 2, 8, 4, 8.

II. 1, 4, 2, 4, 2, 4, 2, 4.

III. 1, 4, 2, 4, 2, 2, 2, 2.

IV. 1, 4, 2, 4, 4, 4, 4, 4.

V. 1, 2, 2, 2, 2, 2, 2, 2.

It will be presently seen why there is no group where the symbols α, β are such that $\alpha^4 = 1$, $\beta^2 = 1$, $\alpha\beta = \beta\alpha^2$. A group which presents itself for consideration is

$$1, \alpha, \alpha^2, \alpha^4, \beta, \beta\alpha, \beta\alpha^2, \beta\alpha^3 (\alpha^4 = 1, \beta^2 = \alpha^2, \alpha\beta = \beta\alpha);$$

but the indices of the different terms of this group are

$$1,\ 4,\ 2,\ 4,\ 2,\ 4,\ 2,\ 4,$$

and if we write $\beta\alpha = \gamma$, then we find $\gamma^2 = \beta\alpha\beta\alpha = \beta\beta\alpha\alpha = \alpha^4 = 1$, $\alpha\gamma = \alpha\beta\alpha = \beta\alpha\alpha = \gamma\alpha$; and the group is

$$1,\ \alpha,\ \alpha^2,\ \alpha^3,\ \gamma,\ \gamma\alpha,\ \gamma\alpha^2,\ \gamma\alpha^3\ (\alpha^4 = 1,\ \gamma^2 = 1,\ \alpha\gamma = \gamma\alpha),$$

which is the group II.

The group IV is a remarkable one; it appears to arise from the circumstance that the factors 2 and 4 of the number 8 are not prime to each other; this can only happen when the number which denotes the order of the group contains a square factor. But the nature of the group in question will be better understood by presenting it under a different form. In fact, if we write $\beta\alpha^3 = \gamma$, $\alpha^2 = \beta^2 = \vartheta$, then we find $\alpha^3 = \vartheta\alpha$, $\beta\alpha^2 = \vartheta\beta$, $\beta\alpha = \vartheta\gamma$, and the group will be

$$1,\ \alpha,\ \beta,\ \gamma,\ \vartheta,\ \vartheta\alpha,\ \vartheta\beta,\ \vartheta\gamma,$$

where the laws of combination are

$$\vartheta^2 = 1,\quad \alpha^2 = \beta^2 = \gamma^2 = \vartheta,$$

$$\beta\gamma = \alpha,\qquad \gamma\alpha = \beta,\qquad \alpha\beta = \gamma,$$

$$\gamma\beta = \alpha\vartheta = \vartheta\alpha,\quad \alpha\gamma = \beta\vartheta = \vartheta\beta,\quad \beta\alpha = \gamma\vartheta = \vartheta\gamma.$$

Observe that ϑ is a symbol of operation such that $\vartheta^2 = 1$, and that ϑ is convertible with each of the other symbols α, β, γ. It will be not so much a restrictive assumption in regard to ϑ, as a definition of -1 considered as a symbol of operation if we write $\vartheta = -1$; the group thus becomes

$$1,\ \alpha,\ \beta,\ \gamma,\ -1,\ -\alpha,\ -\beta,\ -\gamma,$$

where

$$\alpha^2 = \beta^2 = \gamma^2 = -1,$$

$$\alpha = \beta\gamma = -\gamma\beta,\quad \beta = \gamma\alpha = \alpha\gamma,\quad \gamma = \alpha\beta = \beta\alpha.$$

Hence α, β, γ combine according to the laws of the quaternion symbols i, j, k; and it is only the point of view from which the question is here considered which obliges us to consider the symbols as belonging to a group of 8, instead of (as in the theory of quaternions) a group of 4.

Suppose in general that the symbols α, β are such that

$$\alpha^m = 1,\quad \beta^n = 1,\quad \alpha\beta = \beta\alpha^s,$$

then we find

$$\alpha^u\beta^v = \beta^v\alpha^{us^v};$$

and therefore if $v=n$, $\alpha^u=\alpha^{us^n}$ or $\alpha^{u(s^n-1)}=1$, whence $u(s^n-1)\equiv 0 \,(\text{mod. } m)$; or since u is arbitrary, $s^n-1\equiv 0\,(\text{mod. } m)$, an equation which, if m, n are given, determines the admissible values of s; thus, for example, if $n=2$, and m is a prime number, then $s=1$ or $s=m-1$. The equation $\alpha^u\beta^v=\beta^v\alpha^{us^v}$ shows that any combination whatever of the symbols α, β can be expressed in the form $\beta^q\alpha^p$ (or, if we please, in the form $\alpha^p\beta^q$). It is proper to show that the assumed law is consistent with the associative law, viz. that the expression

$$\beta^b\alpha^a\,.\,\beta^d\alpha^c\,.\,\beta^f\alpha^e$$

can be transformed in one way only into the form $\beta^q\alpha^p$. We in fact have

$$\beta^b\alpha^a\,.\,\beta^d\alpha^c=\beta^b\,.\,\alpha^a\beta^d\,.\,\alpha^c=\beta^b\,.\,\beta^d\alpha^{as^d}.\,\alpha^c=\beta^{b+d}\alpha^{as^d+c};$$

and multiplying this by the remaining factor $\beta^f\alpha^e$, we have

$$\beta^{b+d}\,.\,\alpha^{as^d+c}\beta^f\,.\,\alpha^e,$$

which is equal to

$$\beta^{b+d}\,.\,\beta^f\alpha^{as^{d+f}+cs^f}.\,\alpha^e,$$

or finally to

$$\beta^{b+d+f}\alpha^{as^{d+f}+cs^f+e};$$

and the result would have been precisely the same if, instead of thus combining together the first and second factors and the product with the third factor, we had combined the first factor with the product of the second and third factors, so that the associative law is satisfied.

It is now easy to see that if, as before,

$$\alpha^m=1,\quad \beta^n=1,\quad \alpha\beta=\beta\alpha^s,$$

conditions which it has been shown imply $s^n\equiv 1\,(\text{mod. } m)$, then the symbols $\beta^q\alpha^p$ (or, if we please, $\alpha^p\beta^q$), where p has the values $0, 1, 2, \ldots, m-1$, and q the values $0, 1, 2, \ldots, n-1$, form a group of mn terms. In particular, as already noticed, if $n=2$ and m is prime, then $s=1$ or $s=m-1$; the two groups so obtained are essentially distinct from each other. If $n=2$, but m is not prime, then s has in general more than two values: thus for $m=12$, $s^2\equiv 1$ (mod. 12), which is satisfied by $s=1$, 5, 7 and 11; the group corresponding to $s=1$ is distinct from that for any other value of s, but I have not ascertained whether the values other than unity do, or do not, give groups distinct from each other.

For the sake of an observation to which it gives rise, I write down an example of a group corresponding to $n=2$, $s=m-1$, say $m=5$, and therefore $s=4$, so that we have

$$\alpha^5=1,\quad \beta^2=1,\quad \alpha\beta=\beta\alpha^4,$$

and the group is

$$1,\ \alpha,\ \alpha^2,\ \alpha^3,\ \alpha^4,\ \beta,\ \beta\alpha,\ \beta\alpha^2,\ \beta\alpha^3,\ \beta\alpha^4,$$

the indices of the several terms being

$$1,\ 5,\ 5,\ 5,\ 5,\ 2,\ 2,\ 2,\ 2,\ 2.$$

The group is here expressed by means of the symbols α, β, having the indices 5 and 2 respectively, but it may be expressed by means of two symbols having each of them the index 2. Thus putting $\beta\alpha = \gamma$, we find $\beta^2 = 1$, $\gamma^2 = 1$, $(\beta\gamma)^5 = 1$, which is equivalent to $(\gamma\beta)^5 = 1$, and the group may be represented in the form

$$1,\ \beta,\ \gamma,\ \beta\gamma,\ \gamma\beta,\ \beta\gamma\beta,\ \gamma\beta\gamma,\ \beta\gamma\beta\gamma,\ \gamma\beta\gamma\beta,\ \beta\gamma\beta\gamma\beta = \gamma\beta\gamma\beta\gamma,$$

the equality of the last two symbols being an obvious consequence of the equation $(\beta\gamma)^5 = 1$. It is clear that for any even number $2p$ whatever, there is always a group which can be expressed in this form.

2, *Stone Buildings, W.C., June* 9, 1859.

244.

ON AN ANALYTICAL THEOREM RELATING TO THE DISTRIBUTION OF ELECTRICITY UPON SPHERICAL SURFACES.

[From the *Philosophical Magazine*, vol. XVIII. (1859), pp. 119—127.]

THERE is contained in Plana's "Mémoire sur la distribution de l'électricité à la surface de deux sphères conductrices complètement isolées" (*Mém. de Turin*, vol. VII. 1845), an identical relation which is remarkable, as well in itself as because by means of it the author corrects an error into which Poisson had fallen in his researches on the same subject. The development of a certain definite integral is obtained in the form (equation 165)

$$y = -2\,.\,3\frac{M_1 h}{b^2}\sin^2\tfrac{1}{2}\theta + \frac{3\,.\,4\,.\,5}{1\,.\,2}\,\frac{M_2 h}{b^2}\sin^4\tfrac{1}{2}\theta + \&\text{c}.$$

Poisson had in effect shown that $M_1 = 0$; and he thence inferred that, θ being small, the function in question $\propto \sin^4\frac{1}{2}\theta$, or what is the same thing, $\propto (1-\cos\theta)^2$. In the former part of the memoir, Plana shows that this is not the true form of the development; the foregoing development must therefore be illusory; and Plana in fact shows, by a laborious induction carried as far as M_7, that all the coefficients M vanish identically. The identical equation $M_i = 0$, where i is any positive integer whatever, constitutes the analytical theorem above referred to. Plana's expression for the function M_i is as follows:

B_1, B_3, B_5, &c. denote Bernoulli's numbers as given by the equation

$$\frac{t}{e^t-1} = 1 - \tfrac{1}{2}t + B_1\frac{t^2}{1\,.\,2} - B_3\frac{t^4}{1\,.\,2\,.\,3\,.\,4} + \&\text{c}.$$

($B_1 = \frac{1}{6}$, $B_3 = \frac{1}{30}$, $B_5 = \frac{1}{42}$, $B_7 = \frac{1}{30}$, &c. I have, in conformity with the usual practice written the equations so as to make these numbers all positive; with Plana they are alternately positive and negative). And in the equation 162, writing for k its value $\frac{1}{1+b}$, we have, λ being any positive integer,

$$(1+b)^\lambda G_\lambda = \frac{1}{\lambda+1} - \tfrac{1}{2}(1+b) + \frac{\lambda.\lambda-1}{1.2} B_1 \frac{(1+b)^2}{\lambda-1}$$

$$- \frac{\lambda.\lambda-1.\lambda-2.\lambda-3}{1.2.3.4} B_3 \frac{(1+b)^4}{\lambda-3}$$

$$+ \frac{\lambda.\lambda-1.\lambda-2.\lambda-3.\lambda-4.\lambda-5}{1.2.3.4.5.6} B_5 \frac{(1+b)^6}{\lambda-5},$$

$$\pm \text{ \&c.},$$

where the series is continued for so long as the factor in the denominator is positive. It should be observed that this factor really divides out, and that the rule just mentioned amounts to this, viz. that when λ is odd, the finite series on the right-hand side is to be continued to its last term; but when λ is even, the series is to be continued only to the last term but one. And G_λ being thus defined, the expression for M_i (see equation 164, in which I have written for k its value $\frac{1}{1+b}$) is

$$M_i = (1+b)^{i-1} \left\{ G_i + (i+2)\frac{1}{1}\frac{1+b}{b} G_{i+1} + (i+3)\frac{i}{1.2}\left(\frac{1+b}{b}\right)^2 G_{i+2} \right.$$

$$\left. + (i+4)\frac{i.i-1}{1.2.3}\left(\frac{1+b}{b}\right)^3 G_{i+3} + \text{\&c.} \right\},$$

where on the right-hand side the finite series is continued up to its last term, the value of which is obviously

$$(1+2i+1)\frac{1}{i+1}\left(\frac{1+b}{b}\right)^{i+1} G_{2i+1}, \text{ that is, } 2\left(\frac{1+b}{b}\right)^{i+1} G_{2i+1}.$$

But the form of this equation may be somewhat simplified. We in fact have

$$(1+b)M_i = 1\left\{\frac{1}{i+1} - \tfrac{1}{2}(1+b) + B_1\frac{i}{1.2}(1+b)^2 - B_3\frac{i.i-1.i-2}{1.2.3.4}(1+b)^4 + \text{\&c.}\right\}$$

$$+ \frac{i+2}{1}\frac{1}{b}\left\{\frac{1}{i+2} - \tfrac{1}{2}(1+b) + B_1\frac{i+1}{1.2}(1+b)^2 - B_3\frac{i+1.i.i-1}{1.2.3.4}(1+b)^4 + \text{\&c.}\right\}$$

$$+ \frac{i+3.i}{1.2}\frac{1}{b^2}\left\{\frac{1}{i+3} - \tfrac{1}{2}(1+b) + B_1\frac{i+2}{1.2}(1+b)^2 - B_3\frac{i+2.i+1.i}{1.2.3.4}(1+b)^4 + \text{\&c.}\right\},$$

which is at once changed into

$$(1+b)M_i =$$

$$\frac{1}{i+1}\left\{1 - \tfrac{1}{2}\frac{i+1}{1}(1+b) + B_1\frac{i+1.i}{1.2}(1+b)^2 - B_3\frac{i+1.i.i-1.i-2}{1.2.3.4}(1+b)^4 + \text{\&c.}\right\}$$

$$+\tfrac{1}{1}\frac{1}{b}\left\{1-\tfrac{1}{2}\frac{i+2}{1}(1+b)+B_1\frac{i+2\,.\,i+1}{1\,.\,2}(1+b)^2-B_3\frac{i+2\,.\,i+1\,.\,i\,.\,i-1}{1\,.\,2\,.\,3\,.\,4}(1+b)^4+\&c.\right\}$$

$$+\frac{i}{1\,.\,2}\frac{1}{b^2}\left\{1-\tfrac{1}{2}\frac{i+3}{1}(1+b)+B_1\frac{i+3\,.\,i+2}{1\,.\,2}(1+b)^2-B_3\frac{i+3\,.\,i+2\,.\,i+1\,.\,i}{1\,.\,2\,.\,3\,.\,4}(1+b)^4+\&c.\right\}$$

$+\ \&c.$;

or multiplying by $i+1$, and then putting $i-1$ in the place of i, we have

$$i(1+b)\,M_{i-1}=\Theta_i+\frac{i}{1}\,\frac{1}{b}\,\Theta_{i+1}+\frac{i\,.\,i-1}{1\,.\,2}\,\frac{1}{b^2}\,\Phi_{i+2}+\&c.=0,$$

where

$$\Theta_i=1-\tfrac{1}{2}\,\frac{i}{1}(1+b)+B_1\frac{i\,.\,i-1}{1\,.\,2}(1+b)^2-B_3\frac{i\,.\,i-1\,.\,i-2\,.\,i-3}{1\,.\,2\,.\,3\,.\,4}(1+b)^4\pm\&c.$$

In this last equation the finite series on the right-hand side is, when i is *even*, to be continued up to its last term, but when i is odd, then only up to the last term but one. The equation to be proved is

$$0=\Theta_i+\frac{i}{1}\,\frac{1}{b}\,\Theta_{i+1}+\frac{i\,.\,i-1}{1\,.\,2}\,\frac{1}{b^2}\,\Theta_{i+2}+\&c.,$$

where on the right-hand side the finite series is to be continued up to its last term: and the equation holds for any integer value of i which is > 2. This is the simplest form of Plana's theorem.

We have

$$\Theta_i=i\,(1+b)^{i-1}\,G_{i-1};$$

or writing this equation under the form $\Theta_i=i\,(1+b)\dfrac{G_{i-1}}{k^{i-2}}$, and comparing with Plana's developed expressions for $\dfrac{G_{i-1}}{k^{i-2}}$ (which are continued by him as far as G_{17}), we find

$$\begin{aligned}
\Theta_2 &= -\ b,\\
\Theta_3 &= -\tfrac{1}{2}\,b+\tfrac{1}{2}\,b^2,\\
\Theta_4 &= b^2,\\
\Theta_5 &= \tfrac{1}{6}\,b+\tfrac{2}{3}\,b^2-\tfrac{3}{2}\,b^3-\tfrac{1}{6}\,b^4,\\
\Theta_6 &= -\tfrac{1}{2}\,b^2-2\,b^3-\tfrac{1}{2}\,b^4,\\
\Theta_7 &= -\tfrac{1}{6}\,b-b^2-\tfrac{4}{3}\,b^3+\tfrac{4}{3}\,b^4+b^5+\tfrac{1}{6}\,b^6,\\
\Theta_8 &= \tfrac{2}{3}\,b^2+4\,b^3+\tfrac{23}{3}\,b^4+4\,b^5+\tfrac{2}{3}\,b^6,\\
\Theta_9 &= \tfrac{3}{10}\,b+\tfrac{12}{5}\,b^2+\tfrac{32}{5}\,b^3+\tfrac{24}{5}\,b^4-\tfrac{24}{5}\,b^5-\tfrac{32}{5}\,b^6-\tfrac{12}{5}\,b^7-\tfrac{3}{10}\,b^8,\\
\Theta_{10} &= -\tfrac{3}{2}\,b^2-12b^3-37b^4-54b^5-37\,b^6-12\,b^7-\tfrac{3}{2}\,b^8,\\
&\&c.,
\end{aligned}$$

which are of course the results obtained by developing the foregoing expression for Θ_i, in powers of b, and collecting the terms. The formulæ put in evidence a remarkable symmetry which does not exist in the original expression in powers of $1+b$.

It would be now easy to verify, for moderately small values of the suffix, the equations

$$\Theta_2 + \frac{2}{b}\Theta_3 + \frac{1}{b^2}\Theta_4 = 0,$$

$$\Theta_3 + \frac{3}{b}\Theta_4 + \frac{3}{b^2}\Theta_5 + \frac{1}{b^3}\Theta_6 = 0$$

&c.

This is, in fact, Plana's process, which, however, as the suffixes increase, becomes a very laborious one, and the law of the terms which destroy each other is not in anywise exhibited thereby.

I have succeeded in obtaining a complete demonstration, founded on Herschel's theorem for the development of a function of e^t, and the expression thereby given for Bernoulli's numbers. The theorem in question [See Herschel's *Collection of Examples in the Calculus of Finite Differences*, Cambridge, 1820, p. 70, where the theorem is given in the form $f\{(1+\Delta)^n\}\, 0^x = n^x . f(1+\Delta)\, 0^x$] is, that for any function of e^t which admits of development in positive integer powers of t,

$$f(e^t) = f(1+\Delta) e^{t\cdot 0},$$

where the right-hand side denotes the series the general term whereof is

$$\frac{t^n}{1.2.3\ldots n} f(1+\Delta)\, 0^n;$$

and $f(1+\Delta)$ is of course to be developed in powers of Δ, and the different terms Δ, Δ^2, Δ^3 &c., applied to the symbol 0^n (viz. $\Delta 0^n = 1^n - 0^n$, $\Delta^2 0^n = 2^n - 2.1^n + 0^n$, &c.). This gives

$$\frac{t}{e^t - 1} = \frac{\log(1+\Delta)}{\Delta} e^{t\cdot 0},$$

and comparing the development of the right-hand side with the development

$$1 - \tfrac{1}{2}t + B_1 \frac{t^2}{1.2} - B_3 \frac{t^4}{1.2.3.4} + \&c.$$

we find

$$\frac{\log(1+\Delta)}{\Delta} 0^0 = 1,$$

$$\frac{\log(1+\Delta)}{\Delta} 0^1 = -\tfrac{1}{2},$$

$$\frac{\log(1+\Delta)}{\Delta} 0^{2x} = (-)^{x+1} B_{2x-1},$$

$$\frac{\log(1+\Delta)}{\Delta} 0^{2x-1} = 0, \ (x > 1).$$

It is now easy to obtain the equation

$$\Theta_i = \frac{\log(1+\Delta)}{\Delta}\{(1+0(1+b))^i - (-0(1+b))^i\}.$$

In fact, the first two terms of the development of the expression on the right-hand side agree with those of the foregoing expression for Θ_i. For any even power $2x$ (except, when i is even, the power $2x = i$) the term is

$$\frac{[i]^{2x}}{[2x]^{2x}}(1+b)^{2x}\frac{\log(1+\Delta)}{\Delta}0^{2x},$$

which agrees; and when i is even, then for the power $2x = i$ there are two equal and opposite terms which destroy each other, and the whole term in Θ_i is, as it ought to be, zero. For any odd power $2x-1$, $(x>1)$, (including, when i is odd, the power $2x-1=i$), the term vanishes as containing an evanescent factor. The expression for Θ_i is thus shown to be true.

I write for shortness,

$$\Theta_i = \frac{\log(1+\Delta)}{\Delta}\{X^i - Y^i\},$$

where

$$X = 1 + 0(1+b),$$

$$Y = \quad -0(1+b).$$

Forming the expression for $i(1+b)M_{i-1}$,

$$= \Theta_i + \frac{i}{1}\frac{1}{b}\Theta_{i+1} + \frac{i.i-1}{1.2}\frac{1}{b^2}\Theta_{i+2} + \&c.,$$

this is

$$i(1+b)M_{i-1} = \frac{\log(1+\Delta)}{\Delta}\left\{\left(X\left(1+\frac{X}{b}\right)\right)^i - \left(Y\left(1+\frac{Y}{b}\right)\right)^i\right\},$$

and we have

$$X \quad = (1+0)(1+b) - b,$$

and therefore

$$1+\frac{X}{b} \quad = (1+0)\frac{1+b}{b},$$

$$\left(X\left(1+\frac{X}{b}\right)\right)^i = \left(\frac{1+b}{b}\right)^i ((1+0)\{(1+0)(1+b)-b\})^i,$$

$$Y \quad = -0(1+b),$$

$$1+\frac{Y}{b} = 1 - 0\frac{1+b}{b} = -\frac{1}{b}\{0(1+b)-b\},$$

$$\left(Y\left(1+\frac{Y}{b}\right)\right)^i = \left(\frac{1+b}{b}\right)^i (0\{0(1+b)-b\})^i.$$

We see that the expression for $\left(X\left(1+\frac{X}{b}\right)\right)^i$ is deduced from that of $\left(Y\left(1+\frac{Y}{b}\right)\right)^i$ by writing therein $1+0$ in the place of 0; we have therefore

$$\left(X\left(1+\frac{X}{b}\right)\right)^i=(1+\Delta)\left(Y\left(1+\frac{Y}{b}\right)\right)^i,$$

and consequently

$$\left(X\left(1+\frac{X}{b}\right)\right)^i-\left(Y\left(1+\frac{Y}{b}\right)\right)^i=\Delta\left(Y\left(1+\frac{Y}{b}\right)\right)=\left(\frac{1+b}{b}\right)\Delta\left(0\left\{0\left(1+b\right)-b\right\}\right)^i;$$

whence also

$$i(1+b)M_{i-1}=\left(\frac{1+b}{b}\right)^i\log(1+\Delta)\left(0\left\{0\left(1+b\right)-b\right\}\right)^i.$$

We have by the general theorem,

$$t=\log e^t=\log(1+\Delta)\,e^{t\cdot 0};$$

and consequently whenever $n \nless 2$,

$$\log(1+\Delta)\,0^n=0.$$

But $i \nless 2$, and the function $\left(0\left\{0\left(1+b\right)-b\right\}\right)^i$ contains only 0^i and the superior powers; it is therefore reduced to zero by the operation $\log(1+\Delta)$, and we have

$$i(1+b)M_{i-1}=\Theta_i+\frac{i}{1}\frac{1}{b}\Theta_{i+1}+\frac{i\,.\,i-1}{1\,.\,2}\frac{1}{b^2}\Theta_{i+2}+\&\text{c.}=0;$$

and the theorem in question is thus proved. The foregoing expressions for Θ_2, Θ_3, &c. show that these functions all divide by b, and moreover that when i is even and greater than 2, then that Θ_i divides by b^2. The equation

$$\Theta_i=\frac{\log(1+\Delta)}{\Delta}\left\{(1+0\,(1+b))^i-(-0\,(1+b))^i\right\}$$

gives generally for the term in Θ_i involving b^α, the expression

$$\frac{[i]^\alpha}{[\alpha]^\alpha}\,b^\alpha\,\frac{\log(1+\Delta)}{\Delta}\left\{(1+0)^{i-\alpha}\,0^\alpha-(-0)^i\right\};$$

and it is to be shown, first, that the coefficient vanishes for $\alpha=0$; and next, that when i is even and >2, the coefficient also vanishes for $\alpha=1$. Putting $\alpha=0$, the coefficient is

$$\frac{\log(1+\Delta)}{\Delta}\left\{(1+0)^i-(-0)^i\right\},$$

which is equal to

$$\frac{\log(1+\Delta)}{\Delta}\left\{1+\Delta-(-)^i\,1\right\}0^i,$$

or to

$$\log(1+\Delta)\,0^i+\left\{1-(-)^i\,1\right\}\frac{\log(1+\Delta)}{\Delta}\,0^i,$$

where, since $i \nless 2$, the former term vanishes, as above remarked; and the latter term, when i is even, vanishes on account of the factor $1-(-)^i 1$; and when i is odd, on account of the other factor. Hence the coefficient vanishes for $\alpha = 0$.

Next, if i is even, and $\alpha = 1$, the coefficient becomes

$$\frac{\log(1+\Delta)}{\Delta}\{(1+0)^{i-1}0 - 0^i\},$$

which, writing $(1+0)-1$ for 0, becomes

$$\frac{\log(1+\Delta)}{\Delta}\{(1+0)^i - (1+0)^{i-1} - 0^i\},$$

which, since $(1+0)^i - 0^i = \Delta 0^i$, $(1+0)^{i-1} = (1+\Delta)\,0^{i-1}$, is equal to

$$\log(1+\Delta)\,0^i - \frac{(1+\Delta)\log(1+\Delta)}{\Delta}\,0^{i-1},$$

or since the first term vanishes, to

$$-\frac{(1+\Delta)\log(1+\Delta)}{\Delta}\,0^{i-1}.$$

But this function is to a numerical factor *près* the coefficient of t^{i-1} in $\dfrac{e^t \log e^t}{e^t - 1}$, or (what is the same thing) in $\dfrac{-t}{1-e^{-t}}$; and if in the expression for $\dfrac{t}{e^t-1}$ we write $-t$ in the place of t, we find

$$-\frac{t}{1-e^{-t}} = 1 + \tfrac{1}{2}t + B_1\frac{t^2}{1.2} - B_3\frac{t^4}{1.2.3.4} + \&c.,$$

so that, i being even and greater than 2, the function in question vanishes. Hence in the case under consideration the coefficient vanishes for $\alpha = 1$.

Writing β for $i-\alpha$, or assuming $\alpha+\beta = i$, the symmetry of the foregoing expressions for Θ_2, Θ_3, &c. shows that we ought to have

$$\frac{\log(1+\Delta)}{\Delta}\{(1+0)^\alpha 0^\beta - (-0)^{\alpha+\beta}\} = \pm\frac{\log(1+\Delta)}{\Delta}\{(1+0)^\beta 0^\alpha - (-0)^{\alpha+\beta}\},$$

where the upper or under sign is to be taken according as $\alpha+\beta$ is even or odd. Or separating the two cases, we find

$$\frac{\log(1+\Delta)}{\Delta}\{(1+0)^\alpha 0^\beta - (1+0)^\beta 0^\alpha\} = 0, \quad \alpha+\beta \text{ even},$$

and

$$\frac{\log(1+\Delta)}{\Delta}\{(1+0)^\alpha 0^\beta + (1+0)^\beta 0^\alpha + 2.0^{\alpha+\beta}\} = 0, \quad \alpha+\beta \text{ odd}.$$

I have not attempted to verify *à posteriori* these elegant formulæ.

2, *Stone Buildings, W.C., June* 18, 1859.

245.

ON AN ANALYTICAL THEOREM CONNECTED WITH THE DISTRIBUTION OF ELECTRICITY ON SPHERICAL SURFACES. SECOND PART.

[From the *Philosophical Magazine*, vol. XVIII. (1859), pp. 193—202: continuation of **244**.]

THE theorem is certainly true; but its existence gives rise to a difficulty to which I shall advert in the sequel. I propose, in the first instance, to give a demonstration which starts from the expression for fx given by Plana's equation (115), instead of the deduced equation which was the basis of my former proof. It will be proper to explain the origin and meaning of the formulæ. We have two conducting spherical surfaces, radii 1 and b, in contact with each other (so that the distance between the centres is $1+b$): and then, if x is the distance from the centre of the sphere, radius 1, of an exterior point, and $\mu\,(=\cos\theta)$ the cosine of the inclination of this distance to the line from the centre to the centre of the other sphere, the potential $\phi(\mu, x)$ of the sphere, radius 1, at the point whose coordinates are (x, μ) is deduced from the potential fx of a point in the axis; that is, if

$$fx = A_0 \quad + A_1 x \quad + A_2 x^2 \quad + \&\text{c.},$$

then

$$\phi(\mu, x) = A_0 P_0 + A_1 P_1 x + A_2 P_2 x^2 + \&\text{c.},$$

where P_0, P_1, P_2, &c. are Legendre's functions, viz. the functions of μ which are the coefficients of the successive powers of x in the development of $(1-2\mu x+x^2)^{-\frac{1}{2}}$ in ascending powers of x. And the electrical thickness y at any point of the surface of the sphere, radius 1, is given by the formula

$$y = x\,\frac{d\phi(\mu, x)}{dx} + 2\phi(\mu, x)$$

where, after the differentiation, $x=1$.

The problem consequently depends on the determination of the potential fx for a point on the axis; and this is determined by the functional equation

$$fx - \frac{b}{1+2b-(1+b)x} f\left(\frac{1+b-x}{1+2b-(1+b)x}\right) = h - \frac{bh}{1+b-x}$$

(Plana's equation (G), in which I have written for β, γ, H their values, and substituted also for g its value $=h$). The solution of this equation is (equation (H), writing therein $g=h$)

$$fx = \frac{P}{1-x} + bh\Sigma_{n_0}^{\infty} \frac{1}{b+n(1+b)-n(1+b)x} - bh\Sigma_{n_0}^{\infty} \frac{1}{(n+1)(1+b)-\left(1+n(1+b)\right)x},$$

where P is an arbitrary constant *quoad* the functional equation, viz. it is any function whatever which has the property of remaining unaltered when x is changed into $\frac{1+b-x}{1+2b-(1+b)x}$. Poisson, and Plana after him, arrive at the conclusion that in the physical problem $P=0$. It appears to me that there is ground for holding that this is only true *sub modo*, and that $\frac{P}{(1-x)^2}$ for $x=1$ (which, if P were retained, would be a term occurring in the expression for the thickness at the point of contact) is not of necessity zero. But the term, if it exists, can be replaced at the conclusion; and I write therefore

$$fx = bh\Sigma_{n_0}^{\infty} \frac{1}{b+n(1+b)-n(1+b)x} - bh\Sigma_{n_0}^{\infty} \frac{1}{(n+1)(1+b)-\left(1+n(1+b)\right)x}.$$

According to the process by which the solution of the functional equation was obtained, this is the true form of the solution; for although the series are non-convergent, and the two sums are in fact each of them infinite, there is nothing to show a relation between the number of terms which must be taken in each series. However, nothing immediately turns upon this, as the expression is only used for obtaining an expression for fx in the form of a definite integral, viz., equation (36),

$$fx = bh \int_0^1 \frac{dt\, t^{b-1}(1-t^{1-x})}{1-t^{(1+b)(1-x)}};$$

or, equation (39),

$$fx = \frac{bh}{(1+b)(1-x)} \int_0^1 \frac{dt\,(t^{-\frac{1}{1+b}}-1)\, t^{\frac{bx}{(1+b)(1-x)}}}{1-t};$$

the latter of which gives (equation (115), in which I have written for a its value $\frac{b}{1+b}$)

$$fx = \frac{hb}{(1+b)(1-x)} \left\{ Z'\left(\frac{1+b-x}{(1+b)(1-x)}\right) - Z'\left(\frac{b}{1+b-x}\right)\right\},$$

where $Z'(p)$ is Legendre's function $\frac{d}{dp}\log \Gamma p$, which is developable in the form

$$Z'p = \log p - \frac{1}{2p} - \frac{B_1}{2p^2} + \frac{B_3}{4p^4} - \frac{B_5}{6p^6} + \&\text{c.},$$

where B_1, B_3, &c. are Bernoulli's numbers.

This is the starting-point of the present investigation; and attending to the equations

$$\frac{\log(1+\Delta)}{\Delta}\,0^1 \quad = -\tfrac{1}{2},$$

$$\frac{\log(1+\Delta)}{\Delta}\,0^{2x-1} = 0, \ \ (x>1),$$

$$\frac{\log(1+\Delta)}{\Delta}\,0^{2x} \quad = B_{2x-1},$$

we see that the development of $Z'p$ becomes

$$Z'p = \log p + \frac{\log(1+\Delta)}{\Delta}\left(\frac{0}{p} - \frac{0^2}{2p^2} + \frac{0^3}{2p^3} - \&c.\right) = \log p + \frac{\log(1+\Delta)}{\Delta}\log\left(1+\frac{0}{p}\right),$$

which, observing that

$$\left(\frac{\log(1+\Delta)}{\Delta} - 1\right)\log p = 0,$$

can be expressed under the more simple form

$$Z'p = \frac{\log(1+\Delta)}{\Delta}\log(p+0).$$

We deduce hence

$$fx = \frac{hb}{(1-x)(1+b)}\,\frac{\log(1+\Delta)}{\Delta}\left\{\log\left(\frac{1+b-x}{(1-x)(1+b)}+0\right) - \log\left(\frac{b}{(1-x)(1+b)}+0\right)\right\};$$

or what is the same thing,

$$fx = \frac{hb}{(1-x)(1+b)}\,\frac{\log(1+\Delta)}{\Delta}\left\{\log\big(1+b-x+(1-x)(1+b)\,0\big) - \log\big(b+(1-x)(1+b)\,0\big)\right\},$$

which may be converted into

$$fx = \frac{hb}{1+b}\,\frac{\log(1+\Delta)}{\Delta}\int_0^1 \frac{dt}{b+t(1-x)}$$
$$+\frac{hb\log(1+\Delta)}{\Delta}\,0\left\{\int_0^1 \frac{dt}{1-x+b+(1-x)(1+b)\,t0} - \int_0^1 \frac{dt}{b+(1-x)(1+b)\,t0}\right\};$$

or what is the same thing,

$$fx = \frac{hb}{1+b}\,\frac{\log(1+\Delta)}{\Delta}\int_0^1 \frac{dt}{b+t-tx}$$
$$+\frac{hb\log(1+\Delta)}{\Delta}\,0\left\{\int_0^1 \frac{dt}{(1+b)(1+t0) - x\big(1+(1+b)\,t0\big)} - \int_0^1 \frac{dt}{b+(1+b)\,t0 - x(1+b)\,t0}\right\},$$

the object of the transformation being to express fx so that x may only enter under the form $\frac{1}{A-Bx}$. The factor $\frac{\log(1+\Delta)}{\Delta}$ which multiplies the first of the three definite integrals, might be reduced to unity, but it is more convenient not to make this change.

Now if a fraction $\frac{1}{A-Bx}$ be operated upon by expanding in ascending powers of x, and multiplying the successive terms of the development by P_0, P_1, P_2, &c., it is converted into

$$\frac{1}{(A^2-2AB\mu x+B^2x^2)^{\frac{1}{2}}}.$$

Hence from the foregoing expression for fx we pass at once to the expression for $\phi(\mu, x)$; that is, we have

$$\phi(\mu, x) = \frac{hb}{1+b}\frac{\log(1+\Delta)}{\Delta}0\int_0^1 \frac{dt}{(A^2-2AB\mu x+B^2x^2)^{\frac{1}{2}}}$$

$$+\frac{hb\log(1+\Delta)}{\Delta}0\left\{\int_0^1 \frac{dt}{(A'^2-2A'B'\mu x+B'^2x^2)^{\frac{1}{2}}}-\int_0^1 \frac{dt}{(A''^2-2A''B''\mu x+B''^2x^2)^{\frac{1}{2}}}\right\};$$

where for shortness,

$$A=b+t,\quad A'=(1+b)(1+t0),\quad A''=b+(1+b)t0,$$
$$B=\quad t,\quad B'=1+(1+b)t0,\quad B''=\quad (1+b)t0;$$

and it may be remarked that

$$A'=1+b+B'',\quad B'=1+B'',\quad A''=b+B''.$$

We thence obtain

$$x\frac{d\phi(\mu, x)}{dx}+2\phi(\mu, x)=\frac{hb}{1+b}\frac{\log(1+\Delta)}{\Delta}\int_0^1 \frac{(A^2-B^2x^2)\,dt}{(A^2-2AB\mu x+B^2x^2)^{\frac{3}{2}}}$$

$$+\frac{hb\log(1+\Delta)}{\Delta}0\left\{\int_0^1 \frac{(A'^2-B'^2x^2)\,dt}{(A'^2-2A'B'\mu x+B'^2x^2)^{\frac{3}{2}}}-\int_0^1 \frac{(A''^2-B''^2x^2)\,dt}{(A''^2-2A''B''\mu x+B''^2x^2)^{\frac{3}{2}}}\right\},$$

and writing $x=1$,

$$y=\frac{hb}{1+b}\frac{\log(1+\Delta)}{\Delta}\int_0^1 \frac{(A^2-B^2)\,dt}{(A^2-2AB\mu+B^2)^{\frac{3}{2}}}$$

$$+\frac{hb\log(1+\Delta)}{\Delta}0\left\{\int_0^1 \frac{(A'^2-B'^2)\,dt}{(A'^2-2A'B'\mu+B'^2)^{\frac{3}{2}}}-\int_0^1 \frac{(A''^2-B'')\,dt}{(A''^2-2A''B''\mu+B''^2)^{\frac{3}{2}}}\right\},$$

the integrals in the foregoing expression are of the form

$$\int_0^1 \frac{(G+Ht)\,dt}{(L+2Mt+Nt^2)^{\frac{3}{2}}};$$

the value of the indefinite integral is

$$\frac{1}{LN-M^2}\,\frac{(NG-MH)\,t+MG-LH}{(L+2Nt+Mt^2)^{\frac{1}{2}}},$$

from which the value of the definite integral can be at once found. It is easy, by means of the values to be presently given, to verify that, in each of the three definite integrals, $NG-MH=0$; and the expression for the definite integral is therefore

$$\frac{MG-LH}{LN-M^2}\left\{\frac{1}{(L+2M+N)^{\frac{1}{2}}}-\frac{1}{L^{\frac{1}{2}}}\right\}.$$

In the first integral we have

$$G=b^2,\qquad L=b^2,$$
$$H=2b,\qquad M=b\,(1-\mu),$$
$$N=2\,(1-\mu),$$

whence

$$LN-M^2=b^2\,(1-\mu)\,(1+\mu),\qquad MG-LH=-b^3\,(1+\mu),$$
$$L+2M+N=b^2+2\,(1-\mu)\,(1+b),\qquad L=b^2;$$

and the integral is

$$\frac{-b}{1-\mu}\left\{\frac{1}{\sqrt{b^2+2\,(1-\mu)\,(1+b)}}-\frac{1}{b}\right\}.$$

For the second integral we have

$$G=b^2+2b,\qquad L=b^2+2\,(1-\mu)\,(1+b),$$
$$H=2b\,(1+b)\,0,\qquad M=(1-\mu)\,(2+b)\,(1+b)\,0,$$
$$N=2\,(1-\mu)\,(1+b)^2\,0^2;$$

and thence

$$LN-M^2=(1-\mu)\,(1+\mu)\,b^2\,(1+b)^2\,0^2,\qquad MG-LH=-(1+\mu)\,b^3\,(1+b)\,0,$$
$$L+2M+N=b^2+2\,(1-\mu)\,(1+b)\,\{(1+0)^2+b\,(0+0^2)\};\qquad L=b^2+2\,(1-\mu)\,(1+b);$$

and the value of the integral is

$$-\frac{b}{(1+b)\,(1-\mu)\,0}\left\{\frac{1}{\sqrt{b^2+2\,(1-\mu)\,(1+b)\,\big((1+0)^2+b\,(0+0^2)\big)}}-\frac{1}{\sqrt{b^2+2\,(1-\mu)\,(1+b)}}\right\}.$$

For the third integral,

$$G=b^2,\qquad L=b^2,$$
$$H=2b\,(1+b)\,0,\qquad M=(1-\mu)\,b\,(1+b)\,0,$$
$$N=2\,(1-\mu)\,(1+b)^2\,0^2,$$

and thence

$$LN-M^2=(1-\mu)\,(1+\mu)\,b^2\,(1+b)^2\,0^2,\qquad MG-LH=-(1+\mu)\,b^3\,(1+b)\,0,$$
$$L+2M+N=b^2+2\,(1-\mu)\,(1+b)\,\big(0^2+b\,(0+0^2)\big),\qquad L=b^2;$$

and the value of the integral is

$$\frac{-b}{(1+b)(1-\mu)0}\left\{\frac{1}{\sqrt{b^2+2(1-\mu)(1+b)\left(0^2+b(0+0^2)\right)}}-\frac{1}{b}\right\}.$$

Hence the expression for y is

$$y=\frac{-hb^2}{(1-\mu)(1+b)}\frac{\log(1+\Delta)}{\Delta}\left\{\frac{1}{\sqrt{b^2+2(1+\mu)(1+b)}}-\frac{1}{b}\right\},$$

$$-\frac{hb^2}{(1-\mu)(1+b)}\frac{\log(1+\Delta)}{\Delta}\times$$

$$\left\{\frac{1}{\sqrt{b^2+2(1-\mu)(1+b)\left((1+0)^2+b(0+0^2)\right)}}-\frac{1}{\sqrt{b^2+2(1-\mu)(1+b)}}\right\},$$

$$+\frac{hb^2}{(1-\mu)(1+b)}\frac{\log(1+\Delta)}{\Delta}\times$$

$$\frac{1}{\sqrt{b^2+2(1-\mu)(1+b)\left(0^2+b(0+0^2)\right)}}\quad-\quad\frac{1}{b}\quad\Bigg\};$$

the top line is destroyed by the second terms of the other two lines, and we have

$$y=\frac{-hb^2}{(1-\mu)(1+b)}\frac{\log(1+\Delta)}{\Delta}\times$$

$$\left\{\frac{1}{\sqrt{b^2+2(1-\mu)(1+b)\left((1+0)^2+b(0+0^2)\right)}}-\frac{1}{\sqrt{b^2+2(1-\mu)(1+b)\left(0^2+b(0+0^2)\right)}}\right\}.$$

This expression admits of expansion in positive integer powers of $1-\mu$; and when so expanded the result ought, according to Plana's theorem, to be identically equal to zero. And I proceed to show that this is in fact the case. The coefficient of $(1-\mu)^{m-1}$ is to a factor *près* of the form

$$\frac{\log(1+\Delta)}{\Delta}\{((1+0)^2-b(0+0^2))^m-\left(0^2+b(0+0^2)\right)^m\},$$

which is the sum of a series of terms each of the form

$$\frac{\log(1+\Delta)}{\Delta}\{(1+0)^{2m-2n}-0^{2m-2n}\}(0+0^2)^n;$$

this is equal to

$$\frac{\log(1+\Delta)}{\Delta}\{(1+0)^{2m-n}0^n-0^{2m-n}(1+0)^n\},$$

which is of the form

$$\frac{\log(1+\Delta)}{\Delta}\{(1+0)^\alpha 0^\beta-(1+0)^\beta 0^\alpha\},$$

where $\alpha+\beta=2m$ is even, or what is the same thing, $\alpha-\beta$ is even; and, as remarked in the first part of the present paper, such expression is in fact equal to zero. The demonstration, which is very simple, will be given in a note; but assuming for the moment the truth of the proposition, the coefficient of $(1-\mu)^{m-1}$ is the sum of a finite number of evanescent terms, and it is therefore identically equal to zero.

I consider this demonstration as identical *in principle* with that given by Plana; the same function is, by two processes, different indeed from each other, but which cannot but lead to the same result, developed in an infinite series of positive integer powers of $1-\mu$; and it is shown that the coefficient of each power of $1-\mu$ is equal to zero. But the difficulty I find is that the investigation *proves too much*, viz. it appears to prove that y is actually equal to zero. There are undoubtedly functions such as the function $e^{-\frac{1}{x^2}}$ (noticed by Cauchy and Sir W. R. Hamilton), which *in a sense* have the property in question, viz. that if we attempt to develope them in positive integer powers of x, the coefficients are found to be all of them zero; and it would appear that y is, in regard to $1-\mu$, a function of this nature. But it cannot be asserted *simpliciter* that $e^{-\frac{1}{x^2}}$ and its differential coefficients do in fact vanish for $x=0$; they only vanish for $x=0$ considered as the limit of an indefinitely small *real* positive or negative quantity. (This is quite consistent with a remarkable theorem of Cauchy's, by which it appears *à priori* that $e^{-\frac{1}{x^2}}$ cannot be expanded in positive integer powers of x, because it is discontinuous for the modulus zero.) And if, instead of a direct application of Maclaurin's theorem, we first expand $e^{-\frac{1}{x^2}}$, say in positive powers of $1-x$, and then develope the several terms in powers of x, we obtain for the coefficient of x^0, or any other power of x, an infinite series, which I apprehend is not convergent, and which can only be equal to zero in the same conventional sense in which $e^{-\frac{1}{x^2}}$ is equal to zero for $x=0$. This appears to be something very different from finding for the coefficient of x^0, or of any other power of x, an expression composed of a finite number of finite terms the sum whereof is identically equal to zero.

Plana has given for the calculation of y when μ is nearly equal to 1, an expression (equation (127)) which is deduced from the same development of $Z'p$ which is here made use of; but it appears to me that this expression is, for the following reason, open to objection. The expression referred to contains explicitly positive and integer powers of μ, and also powers of the radical $\sqrt{b^2+2(1-\mu)(1+b)}$: it would be, for anything that appears to the contrary, allowable to develope as well the positive and integer powers of μ as also the powers of the radical in question, in a series of positive and integer powers of $1-\mu$; but if this were done, we should obtain as a mere transformation of Plana's expression (127), an expression for y developed in a series of positive integer powers of $1-\mu$; and for consistency with the before-mentioned result, the coefficients of the different powers of $1-\mu$ must be each equal to zero. But if this be so, it does not appear how the original expression (127) can be anything else than zero. The difficulty is, I think, a real one; and I do not see how it is to be got over: it seems to render necessary a more careful study of the effect of the multiplication of the successive terms of the development of a function fx by Legendre's functions P_0, P_1, P_2, &c., so as to pass from fx to the function of two variables $\phi(\mu, x)$, as well generally as when this transformation is performed upon the as yet imperfectly studied transcendental function Z'.

I remark that the original expression for fx is of the form

$$fx = hb\Sigma_{n_0}^{\infty}\frac{1}{p-qx} - hb\Sigma_{n_0}^{\infty}\frac{1}{p'-q'x};$$

and this gives (Plana's equation (131))

$$y = hb\Sigma_{n_0}^{\infty}\frac{p^2-q^2}{(p^2-2pq\mu+q^2)^{\frac{3}{2}}} - hb\Sigma_{n_0}^{\infty}\frac{p'^2-q'^2}{(p'^2-2p'q'\mu+q'^2)^{\frac{3}{2}}},$$

the values of p, q, p', q' being

$$p = b+n(1+b),\quad p'=(n+1)(1+b),$$
$$q = \quad n(1+b),\quad q' = 1+n\,(1+b);$$

so that

$$p-q=b=p'-q', \text{ and } p'+q'=2+b+2n(1+b)=p+q+2.$$

Hence, putting $\mu=1$, we find

$$y = hb\Sigma_{n_0}^{\infty}\left(\frac{p+q}{b^2}-\frac{p'+q'}{b^2}\right) = -\frac{2h}{b^2}\Sigma_{n_0}^{\infty}1 = -\infty,$$

which is inconsistent with the expression $y=0$, deduced from the definite integral. If, however, it is assumed that fx contains the term $\frac{P}{1-x}$, then the corresponding term of y will be

$$\frac{P(1-x^2)}{(1-2\mu x+x^2)^{\frac{3}{2}}},$$

which, when $\mu=1$, becomes $\frac{P(1+x)}{(1-x)^2}$; and if P be put equal to zero, then it is conceivable that, for $x=1$, $\frac{P}{1-x}$ may be equal to zero, but $\frac{P(1+x)}{(1-x)^2}$, or what will be the same thing, $\frac{2P}{(1-x)^2}$ may be finite or even infinite. This is perhaps the explanation of the apparent contradiction.

Note on the demonstration of the Theorem

$$\frac{\log(1+\Delta)}{\Delta}\{0^\alpha(1+0)^\beta - 0^\beta(1+0)^\alpha\} = 0, \quad \alpha-\beta \text{ even.}$$

Consider the function

$$\frac{e^t(t+z)}{e^{t+z}-1} = \phi(t, z),$$

which, it is clear, admits of expansion in positive integer powers of t and z. Changing the signs of t, z, we have

$$\frac{e^{-t}(-t-z)}{e^{-t-z}-1} = \phi(-t, -z),$$

or, what is the same thing,

$$\frac{e^z(t+z)}{e^{t+z}-1} = \phi(-t, -z),$$

and thence

$$\frac{(e^t-e^z)(t+z)}{e^{t+z}-1} = \phi(t, z) - \phi(-t, -z);$$

so that the development in positive integer powers of t, z, of the function on the right-hand side does not contain any term $t^\alpha z^\beta$ for which $\alpha-\beta$ is even. Writing the function under the form

$$\frac{e^t(t+z)}{e^{t+z}-1} - \frac{e^z(t+z)}{e^{t+z}-1},$$

and considering the two parts separately, then by Herschel's theorem extended to two variables, the coefficient of $t^\alpha z^\beta$ in the first term is

$$\frac{(1+\Delta_1)\log\{(1+\Delta_1)(1+\Delta_2)\}}{(1+\Delta_1)(1+\Delta_2)-1}\, 0_1{}^\alpha\, 0_2{}^\beta,$$

which is equal to

$$\frac{\log\{(1+\Delta_1)(1+\Delta_2)\}}{(1+\Delta_1)(1+\Delta_2)-1}\,(1+0_1)^\alpha\, 0_2{}^\beta,$$

or, what is the same thing,

$$\frac{\log(1+\Delta)}{\Delta}(1+0)^\alpha\, 0^\beta;$$

and forming in like manner the expression for the coefficient of $t^\alpha z^\beta$ in the second term, this is

$$\frac{\log(1+\Delta)}{\Delta}\, 0^\alpha(1+0)^\beta;$$

the difference of the two expressions therefore vanishes when $\alpha-\beta$ is even, which is the above-mentioned theorem. It would be easy to obtain a variety of similar theorems.

2, *Stone Buildings, W.C., June* 29, 1859.

246.

ON CONTOUR AND SLOPE LINES.

[From the *Philosophical Magazine*, vol. XVIII. (1859), pp. 264—268.]

It is, I think, interesting as a question of topography, to consider the general configuration of a system of contour lines and steepest or slope lines (*lignes de niveau* and *lignes de la plus grande pente*). Imagine, to fix the ideas, a mountainous island, the exterior or sea-level contour line being consequently a closed curve; the case where any contour line is a curve cutting itself is an important one, which will be considered; but disregarding it for the moment, and excluding (as I do throughout) a curve which cuts itself from the notion of a closed curve, the entire contour line corresponding to a given elevation will be either a single closed curve, or it will consist of two or more separate closed curves, in the latter case each of these may be considered as being by itself a contour line, and we may therefore say that the contour line is in general a closed curve. It may happen that the elevation of a given contour line is a maximum or minimum; in other words, that the consecutive curve without the given contour line and that within it are each of them higher or each of them lower than the given contour line; but this is a speciality which need not be particularly attended to; in general the consecutive curve without the given contour line will be lower, and that within it higher than the given contour line, in which case the tract bounded by the contour line is an elevation (hill, table-land, or mountain, as the case may be); or else the consecutive curve without the given contour line is higher, and that within it lower than the given contour line; in which case the tract bounded by the contour line is a depression. But there may be within the contour line bounding an elevation, spaces lower than the bounding line, and within the contour line bounding a depression, spaces higher than the bounding line. A depression usually contains water, and indeed is filled so as to overflow, in which case there is a lake with an outlet; if the depression is not filled to overflowing, the lake will have no outlet. The contour line bounding an elevation may become indefinitely small and ultimately reduce itself to a point, which is a *summit;* the contour line bounding a depression may in like manner become indefinitely small, and ultimately reduce itself to a point, which is what I call an *immit.* A summit is a point of maximum elevation (though of course there may be summits, or even immits, which are higher); an immit is a point of minimum elevation. But there are besides, as at the heads of passes, points where the surface is horizontal, but where the elevation is neither a maximum nor a minimum; you descend backwards and forwards, but ascend right and left: I will for the present purpose call this kind of point a *knot.* And this leads to the consideration of a contour line which cuts itself: the point where this happens is in fact a knot, or geometrically the knot is a node or double point on the contour line. It may be assumed that the contour line through a knot does not pass through any other knot; for although there may be neighbouring passes of precisely the same elevation, yet the general configuration of the country will not be altered by giving a slight difference of elevation to such passes: the effect of this alteration is to distribute among contour lines of slightly different elevations (one to each line) the different knots which would otherwise occur upon one and the same contour line. The contour line through a knot cuts itself therefore at this point only: such contour line is either a figure of eight, or as I will term it, an *outloop* curve; or else it is the figure formed by the union (so as to give rise to a node or double point) of two closed curves, one of which lies within the other of them; this I call an *inloop* curve. An outloop curve consists of two loops; the spaces within these may also be spoken of as the loops. An inloop curve consists of an outer and an inner loop; the space within the inner loop may be spoken of as the inner loop, that between the two loops as the *lune.* It usually happens, and to fix the ideas I will assume, that for an outloop curve each of the loops is an elevation: this is the case of two mountain summits connected by a ridge or col, the lowest point whereof, or head of the pass, is the knot on the outloop contour line through this point. And in like manner, that for an inloop curve the lune is an elevation, the inner loop a depression; and that the outer loop, considered as a portion of the contour line, is higher than the consecutive exterior contour line. This is the case of a lake having an outlet; if the lake were dry, the passage up stream into the bed of it would be over a ridge, col, or barrier, the lowest point whereof, or point of outlet for the water of the lake, is the knot on the inloop contour line passing through this point, the shore of the lake being of course the inner loop of this contour line, and the waters being retained by means of the raised ground within the lune between the two loops of the contour line.

The slope lines cut at right angles the contour lines; and this property applies also to the projections of the two systems of lines; so that the two systems of lines delineated *in plano* intersect at right angles. Consider the contour lines which are closed curves surrounding a given summit or immit; the exterior contour line is intersected at each of its points by a slope line; and all these slope lines must, it is clear, intersect all the interior contour lines, and ultimately unite at the interior summit or immit. In order to see more distinctly the form of the system of slope lines, it is to be noticed that, if (as is in general the case) the indicatrix at the summit or immit be an ellipse, the contour lines in the immediate neighbourhood thereof will be a system of similar and similarly situated concentric ellipses, the major and minor axes whereof correspond respectively with the directions of least and greatest curvature; the equation of any orthogonal trajectory of the ellipses, if a, b are the semi-axes, major and minor, of any one of them, is $y^{b^2} = Cx^{a^2}$; and unless $C = \infty$, the curve represented by this equation touches the axis of x, which is the direction of least curvature; if however $C = \infty$, then the equation becomes $x = 0$, and the curve touches the axis of y, which is the direction of greatest curvature. Hence in general at a summit or immit the slope curves all, except one (which is a limiting case) touch the line which is the direction of least curvature. The only exception is when the summit or immit is an umbilicus—the indicatrix is then a circle; the contour lines in the immediate neighbourhood of this point are concentric circles, and the slope lines pass in all directions through the summit or immit.

The indicatrix at a knot is in general a hyperbola, and consequently the contour lines in the neighbourhood of a knot are similar and similarly situated concentric hyperbolas; and if a, b are the semi-axes of one of these hyperbolas, the equation of an orthogonal trajectory is $x^{a^2}y^{b^2} = C$: and when this passes through the knot, $C = 0$; and therefore either $x = 0$ or else $y = 0$; there are consequently through the knot only two slope lines, which bisect the angles made by the two branches of the contour line and intersect each other at right angles. The slope lines through a knot may be termed ridge and course lines: and for one of these—the ridge line—the knot is a point of minimum elevation; for the other of them—the course line—the knot is a point of maximum elevation. But this requires some further development. To fix the ideas, consider the case where the contour line is an outloop curve, the loops being each of them elevations. The slope line through the knot, and which lies within the two loops, would be, according to the definition, a ridge line. Suppose that the contour lines within one of the loops are closed curves surrounding a summit, the ridge line will, it is clear, cut all these curves and ultimately arrive at the summit. But if the contour lines within the loop are not all of them closed curves; if, for instance, they are first closed curves, then an outloop curve, and within each of the loops of this, closed curves surrounding a summit, then it may happen that the above-mentioned ridge line will pass through the knot of the inner outloop curve: and with respect to this knot, it will be, not a ridge line, but a course line; so that the slope line in question cannot be spoken of *simpliciter* either as a ridge line or as a course line, but it is the one or the other *quoad* the knot in reference to which it is considered; and, considered by itself, it can only be spoken of as a ridge-or-course line. The case just referred to is, however, an exceptional one; in general the slope line in question would not pass through the knot of the inner outloop curve, but would cut one of the loops of this curve, and then cutting all the contour lines within such loop, arrive at last at the summit within such loop. And when the ridge line has once arrived at a summit, there is little meaning in continuing it further, and it may be considered as ending there; in fact there are through the summit an infinity of slope lines, all of them (except in the case where the summit is an umbilicus) coincident in direction with the ridge line, and consequently the ridge line may, without graphical discontinuity, be considered as proceeding along any one of these lines indifferently;

and although, when the surface is a geometrical one capable of being represented by an equation, there would be geometrically one of these slope lines which could be identified as the continuation of the ridge line, there would be no advantage in making this identification. Hence it may be considered that in general a ridge line passes from summit to summit, through a single intervening knot which is a point of minimum elevation on the ridge line; and in like manner, that in general a course line passes from immit to immit through a single intervening knot which is a point of maximum elevation on the course line; it need not be considered as an exception when, as is frequently the case, the course line arrives at the sea-level contour line without previously reaching an immit. It is to be noticed that a ridge line or a course line may commence and terminate at one and the same summit or immit, and thus form a closed curve.

The ridge lines, as above defined, determine the watershed. In the case of an isolated conical or dome-shaped mountain, and in general when the contour lines are all of them closed curves, there is no definable watershed; but in the case of a chain of mountain summits, the watershed runs from summit to summit through the heads of the passes over the connecting cols, i.e. it is made up of a series of ridge lines each extending from a summit to a summit through an intervening knot. And the course lines are, as nearly as may be, the beds of the streams which flow from the heads of the passes down the lateral valleys. The ridge line and the course line respectively are, I believe, the so-called *ligne de faîte* and *ligne de thalweg.*

2, *Stone Buildings, W.C., July* 20, 1859.

247.

ON THE ANALYTICAL FORMS CALLED TREES. SECOND PART.

[From the *Philosophical Magazine*, vol. XVIII. (1859), pp. 374—378. Continuation of **203.**]

THE following class of "trees" presented itself to me in some researches relating to functional symbols; viz., attending only to the terminal knots, the trees with one knot, two knots, three knots, and four knots respectively are shown in the figures 1, 2, 3 and 4:

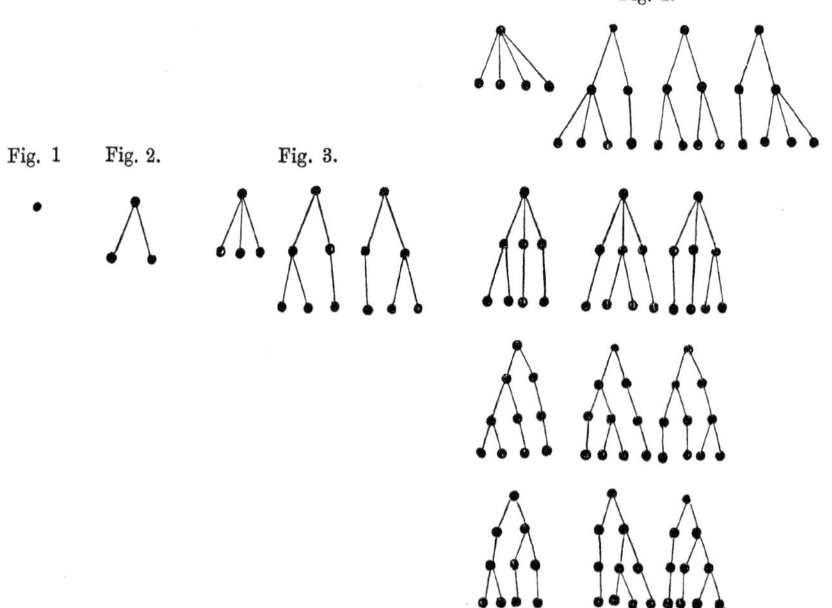

and similarly for any number of knots. The trees with four knots are formed first from those of one knot by attaching thereto in every possible way (one way only) four knotted branches; secondly, from those with two knots by attaching thereto in every possible way (three different ways) four knotted branches; and thirdly, from those with three knots by attaching thereto in every possible way (three different ways) four knotted branches,—the original knots of the trees of one knot and two and three knots, being no longer terminal knots, are disregarded. The total numbers of trees with one knot and with two and three knots being respectively 1, 1, 3; the total number of trees with four knots is $1.1+3.1+3.3=13$. And in general, if the number of trees with m knots is ϕm, then it is easy to see that we have

$$\phi m = \phi 1 + \frac{m-1}{1}\phi 2 + \frac{m-1\,.\,m-2}{1\,.\,2}\phi 3 \ldots + \frac{m-1}{1}\phi(m-1);$$

or what is the same thing,

$$2\phi m = \phi 1 + \frac{m-1}{1}\phi 2 + \frac{m-1\,.\,m-2}{1\,.\,2}\phi 3 \ldots + \frac{m-1}{1}\phi(m-1) + \phi m.$$

Hence if

$$u = \phi 1 + \frac{x}{1}\phi 2 + \frac{x^2}{1\,.\,2}\phi 2 + \ldots,$$

we obtain

$$\begin{aligned} e^x\,.\,u = & \quad \phi 1 \\ & + x\ (\phi 1 + \phi 2) \\ & + \frac{x^2}{1\,.\,2}(\phi 1 + 2\phi 2 + \phi 3) \\ & + \quad \&c. \\ = & \quad 2\phi 1 - 1 \\ & + x^2\,.\,2\phi 2 \\ & + \frac{x^2}{1\,.\,2}\,.\,2\phi 3 \\ & + \quad \&c.; \end{aligned}$$

that is,

$$e^x u = 2u - 1,$$

and thence

$$u = \frac{1}{2-e^x},$$

which gives for ϕm the expression

$$\phi m = 1\,.\,2\,.\,3 \ldots (m-1) \text{ coeff. } x^{m-1} \text{ in } \frac{1}{2-e^x};$$

and the value of ϕm might easily be obtained in an explicit form in terms of the differences of the powers of zero. The values of ϕm are, for

$$\begin{array}{lllllllll} m = 1, & 2, & 3, & 4, & 5, & 6, & 7, & 8, & \&c. \\ \phi m = 1, & 1, & 3, & 13, & 75, & 541, & 4683, & 47293. \end{array}$$

In the foregoing problem, the number of branches descending from a non-terminal knot is one, two, or more. But assume that the number of branches descending from a non-terminal knot is always two; so that attending, as before, only to the terminal knots, the trees with two knots, three knots, four knots respectively are shown in the figures, 5, 6, and 7.

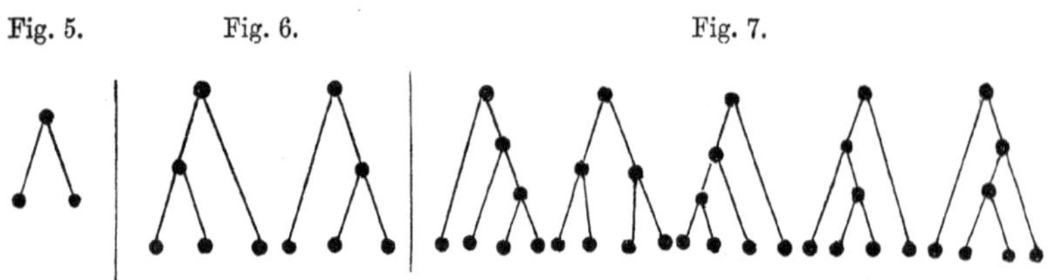
Fig. 5. Fig. 6. Fig. 7.

This corresponds to the following problem in the theory of symbols; viz. if A, B, C, D, &c. are symbols capable of successive binary combinations, but do not satisfy the associative law, what is the number of the different significations of the ambiguous expressions ABC, $ABCD$, $ABCDE$, &c. respectively? For instance, AB has only one meaning; ABC may mean either $A.BC$ or $AB.C$. In like manner $ABCD$ may mean $A(B.CD)$, or $AB.CD$, or $(AB.C)D$, or $(A.BC)D$, or $A(BC.D)$; the numbers, 1, 2, 5 being those of the trees in the last three figures respectively; and similarly for any greater number of symbols.

Let ϕm be the required value corresponding to the number m; then we may in any manner whatever separate the number m into two parts m', m'', and then combining *inter se* the m' knots (or symbols) and the m'' knots (or symbols) respectively, ultimately combine the two combinations; hence a part of ϕm is $\phi m' . \phi m''$. The assumed definition of ϕm does not apply to the case $m=1$; but if we write $\phi 1 = 1$, then the foregoing consideration shows that we have

$$\begin{aligned} \phi m = \quad & \phi 1 \phi (m-1) \\ & + \phi 2 \phi (m-2) \\ & \quad \vdots \\ & + \phi (m-1) \phi 1 ; \end{aligned}$$

from which it is easy to calculate

$$\phi 1 = 1,\ \phi 2 = 1,\ \phi 3 = 2,\ \phi 4 = 5,\ \phi 5 = 14,\ \phi 6 = 42,\ \phi 7 = 132,\ \&c.$$

But to obtain the law, consider the generating function

$$u = \phi 1 + \quad x\phi 2 + \quad x^2 \phi 3 + \&c.;$$

we have

$$u^2 = \phi 1\ \phi 1 + x(\phi 1\ \phi 2 + \phi 2\ \phi 1) + x^2(\phi 1\ \phi 3 + \phi 2\ \phi 2 + \phi 3\ \phi 1) + \&c.,$$

which is

$$= \phi 2 + \quad x\phi 3 + \quad x^2 \phi 4 + \&c.;$$

and we have therefore

$$xu^2 = u - 1,$$

and consequently

$$u = \frac{1 - \sqrt{1 - 4x}}{2x}.$$

But

$$\sqrt{1 - 4x} = 1 - \tfrac{1}{2}\, 4x + \frac{\tfrac{1}{2} \cdot -\tfrac{1}{2}}{1 \cdot 2} (4x)^2 - \frac{\tfrac{1}{2} \cdot -\tfrac{1}{2} \cdot -\tfrac{3}{2}}{1 \cdot 2 \cdot 3} (4x)^3 + \&\text{c.}$$

$$= 1 - 2x - 2x^2 - 4x^3 - 10x^4 + \&\text{c.},$$

and therefore

$$u = \frac{1 - \sqrt{1 - 4x}}{2x} = 1 + 1x + 2x^2 + 5x^3 + \&\text{c.},$$

the series of coefficients 1, 1, 2, 5, &c. agreeing with the values already found. The expression for the general term is at once seen to be

$$\phi m = \frac{1 \cdot 3 \cdot 5 \ldots 2m - 3}{1 \cdot 2 \cdot 3 \ldots m}\, 2^{m-1},$$

which is a remarkably simple form.

2, *Stone Buildings, W.C., June* 9, 1859.

248.

SKETCH OF A PROOF OF THE THEOREM THAT EVERY ALGEBRAIC EQUATION HAS A ROOT.

[From the *Philosophical Magazine*, vol. XVIII. (1859), pp. 436—439.]

I HAVE referred to the theorem as usually stated; for it is an easy consequence of the existence of a single root of an equation of any order, that for an equation of the nth order there are n roots: the proof here proposed goes, however, to show directly the existence of the n roots: it is in form a geometrical one, and was suggested to me some months ago by a letter from Prof. De Morgan, containing the remark made in his memoir, "A Proof of the Existence of a Root in every Algebraic Equation," &c. (*Camb. Phil. Trans.* vol. X. 1858), viz. "that the curves $P=0$, $Q=0$, the intersections whereof determine the root-points, are such that two branches, one of each curve, cannot enclose a space." The proof which occurred to me was in character somewhat similar to that given by the Astronomer Royal in the paper, "Suggestion of a Proof of the Theorem that every Algebraic Equation has a Root" (*Camb. Phil. Trans.* vol. X. 1858), and which was suggested to him by Prof. De Morgan's memoir. I have since varied my proof by considering therein cones in the place of plane curves. It will be obvious, upon reading it, that the proof is closely connected with Cauchy's well-known theorem for the number of roots within a given circuit; the circuit being in this case infinity, and the number of roots included within it consequently equal to the order of the equation.

The curve represented by an equation of the nth degree between the coordinates (x, y) is by definition a curve of the nth order; and a cone standing on any such curve (taking the vertex for origin) is represented by a homogeneous equation of the nth degree between the coordinates (x, y, z), and is by definition a cone of the nth order. It is very easy to show that an equation of the nth degree cannot have more than n roots; and we have thence the geometrical theorems, that a curve of the nth order is not intersected by a line in more than n points, and that a cone of the nth order is not intersected by a plane (I speak throughout of planes through the vertex) in more than n lines. I assume that an algebraic curve is always a continuous curve, viz. that it consists of a branch or branches, no one of which is a *courbe pointillée*, or a branch terminating abruptly in a point; an algebraic cone will be in the like sense a continuous surface. An algebraic curve cannot be an indefinite spiral, for in that case there would be lines meeting it in an infinity of points; and in like manner an algebraic cone cannot be an indefinite spiral surface: an algebraic cone consists, therefore, of a closed sheet or sheets. An algebraic curve may indeed have conjugate or isolated points, and an algebraic cone have conjugate or isolated lines: this is a circumstance which will be adverted to in the sequel. It will fix the ideas as to the general form of an algebraic cone, to remark that it may comprise twin-pair sheets, such as the sheet of a cone of the second order (this is properly spoken of as a twin-pair sheet, each of the two opposite portions of it being called, for distinction, a twin-sheet); and of single sheets, such as one at least of the sheets of a cone of the third or any other odd order (see the annexed "Note upon Cones of the Third Order," [249]). The advantage of the consideration of cones instead of plane curves, is that we have only closed sheets, and thus get rid of the distinction which exists for plane curves between infinite branches and the branches which are closed curves.

My proof depends on the following lemma, viz. "Consider two algebraic cones with the same vertex, each of them of the order n; then *if* there be some one plane meeting the first cone in n lines, and the second cone in n lines, such that the lines of each set occur alternately, the two cones intersect in at least n lines."

The truth of this lemma is, I conceive, a matter of intuition, depending only on the notion of the continuity of the sheets of the surface. Thus, if we have *in plano*, through a point O, the lines A, A' and B, B', such that, A, α being opposite points on the same line, and so for the other lines, the order round O is A, B, A', B', α, β, α', β', it is obvious that we cannot through the lines A, A' draw a cone, and through the lines B, B' draw a cone, without making these cones intersect in at least two lines: and in like manner for two sets, each of n lines. I have, in the enunciation of the lemma, said that the cones are each of them of the order n; this was necessary in order to exclude a case which might otherwise have happened, viz. a line of intersection of the plane with either of the cones might have been a conjugate or isolated line without any sheet through it; and if this were so, we could not infer the existence of the n lines of intersection of the two cones. But if a plane meet an algebraic cone of the nth order in n lines, no one of these can be a conjugate or isolated line; for such line is to be considered as two or more coincident lines, and there would be in all more than n lines of intersection of the plane and cone.

Consider now the equation

$$\phi u = 0,$$

where ϕu is a rational and integral function of u with (in general) imaginary coefficients, and write

$$\phi(x+y\sqrt{-1})=P+Q\sqrt{-1},$$

P, Q being real functions of (x, y), each of them of the degree n; if (x, y) are rectangular coordinates, then $P=0$, $Q=0$ are real curves each of the order n. And to each point of intersection of the two curves there corresponds a root of the equation. The two curves do not intersect in more than n points (for if they did, the equation $\phi u=0$ would have more than n roots); hence if it be shown that the two curves intersect in at least n points, they will intersect in precisely n points, and the equation will have n roots. Take any point as the common vertex of two cones standing upon the curves $P=0$, $Q=0$ respectively; each point of intersection of the two curves corresponds to a line of intersection of the two cones, and it is only necessary to show that the two cones intersect in at least n lines Take for the vertex a point in the perpendicular at the origin of (x, y) to the plane of the two curves, and at a distance unity from such origin, viz. a point such that, treating it as the origin of the coordinates (x, y, z), the coordinates in respect thereto of the origin (x, y) are $x=0$, $y=0$, $z=1$. The equations of the cones are at once deduced from those of the curves by writing therein $\left(\frac{x}{z}, \frac{y}{z}\right)$ in the place of (x, y) and, to render the equation integral, multiplying by z^n; or if $P'=0$, $Q'=0$ are the equations of the cones, we have

$$z^n\phi\left(\frac{x+y\sqrt{-1}}{z}\right)=P'+Q'\sqrt{-1}.$$

Consider the section by the plane through the vertex parallel to the plane of the two curves: the equation of this plane is $z=0$; and it is clear that, to obtain the intersections of this cone with the plane in question, we have only in $\phi\left(\frac{x+y\sqrt{-1}}{z}\right)$ to disregard all the terms after the first. Suppose that

$$\phi u=(a+b\sqrt{-1})\,u^n+\&c.;$$

then putting

$$(a+b\sqrt{-1})(x+y\sqrt{-1})^n=P_0'+Q_0'\sqrt{-1},$$

the equations $P_0'=0$, $Q_0'=0$ determine the intersections of the plane $z=0$ with the cones $P=0$, $Q=0$ respectively. But writing

$$a+b\sqrt{-1}=A(\cos\alpha+\sqrt{-1}\sin\alpha),$$

$$x+y\sqrt{-1}=r(\cos\theta+\sqrt{-1}\sin\theta),$$

we have

$$Ar^n\{\cos(n\theta+\alpha)+\sqrt{-1}\sin(n\theta+\alpha)\}=P_0'+Q_0'\sqrt{-1},$$

so that

$$P_0' = Ar^n \cos(n\theta + \alpha), \quad Q_0' = Ar^n \sin(n\theta + \alpha),$$

or the intersections with the cone $P = 0$ are the n lines given in direction by the equation

$$n\theta + \alpha = (m + \tfrac{1}{2})\,\pi,$$

and the intersections with the cone $Q = 0$ are the n lines given in direction by the equation

$$n\theta + \alpha = m\pi\,;$$

in each of which equations m is any integer number from 0 to $n - 1$. Hence the plane $z = 0$ meets the cones in two sets of lines succeeding each other alternately, as required by the lemma, and the two cones intersect in at least n lines. And it is thus shown that the given equation of the nth degree has n roots.

2, *Stone Buildings, W.C., September* 26, 1859.

249.

NOTE ON CONES OF THE THIRD ORDER.

[From the *Philosophical Magazine*, vol. XVIII. (1859), pp. 439—442.]

THE distinction adverted to in the preceding paper between the twin-pair sheets and single sheets of an algebraic cone is made (with respect to spherical curves, which is the same thing) by Möbius, in the interesting Memoir "Ueber die Grundformen der Linien der dritten Ordnung," *Abh. der K. Sächs. Ges. zu Leipzig*, vol. I. (1849) [and *Werke*, vol. II. pp. 86—176]. Consider the generating line POP' of a cone, vertex O, and let pOp' be any position of this line, the points P, P', and in like manner the points p, p', being on opposite sides of the vertex; then if OP originally coincides with Op (and therefore OP' with Op'), and if, in the course of the generation of the surface, OP (without having first come to coincide with Op') comes to coincide with Op, at the same time OP' (without having first come to coincide with Op) will come to coincide with Op', and we have a *twin-pair sheet*, viz. one twin-sheet generated by OP, and the other twin-sheet generated by OP'. This is the ordinary case of a cone of the second order, and requires no further explanation. It is proper to remark that, for cones of superior orders, the conical angle of each twin-sheet is not (as for a cone of the second order) necessarily less than 360°. But suppose that OP, starting from the position Op, and before it again comes to coincide therewith, comes to coincide with Op', then at the same time OP' (without having first come again to coincide with Op') will come to coincide with Op; the generation is here complete, and we have a *single sheet*, which, if the motion were continued until OP came to coincide with Op, would only be generated over again. The conical angle of a single sheet is necessarily greater than 360°; for OP in coming to coincide with Op' must describe an angle greater than 180°, and OP' describing an equal angle, the entire angle is therefore greater than 360°; in the limiting case, where the entire angle is precisely 360°, the conical surface is a plane. It is easy to cut out in paper and join together two sectors of a circle so as to form therewith a sector the angle whereof exceeds 360°; such a sector can then, by joining together the two radial edges, be converted into a cone of a single sheet; the generating lines being all finite lines equal in length, the curve formed by the circular edge is, it is clear, the spherical curve which is the intersection of the cone by a concentric sphere. It is shown by Möbius (stating his result with respect to cones instead of spherical curves) that a cone of an odd order must have at least one single sheet; a cone of the third order consists (1) of a single sheet, or else (2) of a single sheet and a twin-pair sheet. These are the two general forms of cones of the third order. But there are two special forms and one subspecial form, making in all five forms: viz., the two special forms are, (3) the cone has a nodal line; (4) the cone has an isolated line; and the subspecial form is, (5) the cone has a cuspidal line. The relation of the different forms may be explained as follows.

Starting from the form (1), as the constants of the equation change, the cone gathers itself up together so as to have a nodal line; this is the form (3). The loops of this form then detach themselves so as to form a twin-pair sheet, the remaining part of the surface reverting to a form similar to that of (1); we have thus a single sheet and twin-pair sheet, which is the form (2). The twin-pair sheet then dwindles away into an isolated line, giving the form (4); and lastly, the isolated line disappears and the cone resumes the form (1): these four forms constitute, there-

fore, a complete cycle. The constants may be such that the loop of the form (3) is evanescent, or, what is the same thing, that the forms (3) and (4) arise simultaneously; there is in this case a cuspidal line, or we have the form (5). It may be added that for the general forms (1) and (2) there are always three lines of inflexion. This is also the case with the form (4), where there is an isolated line; but in the form (3), where there is a nodal line, there is but one line of inflexion; and in the form (5), where there is a cuspidal line, there is not any line of inflexion: the equivalent theorem for the spherical curves is given by Möbius [1]. It was remarked long ago by Sir I. Newton, that all curves of the third order could be generated as the shadows of the five cubical parabolas; these are, in fact, sections in a particular manner of the above-mentioned five forms of cones of the third order: the existence of five essentially distinct forms of cones of the third order is noticed by M. Chasles in the *Aperçu Historique*, 1837. The analytical distinction between the forms (1) and (2) depends on the sign of the function $1-\frac{64S^3}{T^2}$, where S, T are the quartinvariant and sextinvariant of the cubic form. I annex stereoscopic representations of the cones of the third order of the general form (1), and of the form with a nodal line (3). The generating lines are finite lines of equal length, and the curved contours shown in the figures are consequently the spherical curves which are the intersections of the cones by concentric spheres. The figures are intended to be looked at with the glasses of a Reeves's book stereoscope.

[1] It is hardly necessary to mention that, according to the general theory of cones of the third order, there are always nine lines of inflexion,—three real and six imaginary. Six of the lines of inflexion disappear when there is a double line, viz., in the case of a nodal line, two real and four imaginary lines of inflexion; but in the case of an isolated line, the six imaginary lines of inflexion. When there is a cuspidal line, eight lines of inflexion, viz. two real lines and the six imaginary lines, disappear.

2, *Stone Buildings, W.C., September* 26, 1859.

250.

SUR LA SURFACE QUI EST L'ENVELOPPE DES PLANS CONDUITS PAR LES POINTS D'UN ELLIPSOIDE PERPENDICULAIREMENT AUX RAYONS MENÉS PAR LE CENTRE.

[From the *Annali di Matematica pura ed applicata* (Tortolini), tom. II. (1859), pp. 3—14.]

La considération de la surface dont il s'agit me fut suggérée, il y a quelques années par le Prof. Stokes, mais il convient de remarquer que la courbe enveloppe des droites menées par les points d'une ellipse perpendiculaires aux rayons conduits par le centre, est mentionnée en passant dans le mémoire de M. Tortolini "Sulle relazioni ec." *Annali*, tom. VI. pp. 433—446, 1855 (voir p. 461), où il trouve que l'équation de la courbe est

$$[4(a^4+b^4-a^2b^2)-3(a^2x^2+b^2y^2)]^3$$
$$=[9a^2(2b^2-a^2)x^2+9b^2(2a^2-b^2)y^2-4(a^2+b^2)(2a^2-b^2)(2b^2-a^2)]^2,$$

équation qui était trouvée en égalant à zéro le discriminant d'une fonction quadratique. L'auteur remarque que cette équation fut trouvée par lui en 1846 dans la *Raccolta Scientifica di Roma*, et il remarque que la courbe est connue sous le nom de la courbe de Talbot.

Selon ma méthode l'équation de la courbe est trouvée en égalant à zéro le discriminant d'une fonction cubique, et l'équation de la surface est trouvée en égalant à zéro le discriminant d'une fonction biquadratique. Comme on a besoin de la courbe pour la discussion de la surface, je commence par la considération de la courbe.

On voit sans peine qu'en s'imaginant un cercle passant par le centre de l'ellipse et touchant l'ellipse à l'un quelconque de ses points, le point correspondant de la courbe est situé à l'extrémité du diamètre du cercle, lequel diamètre passe par le centre de l'ellipse, ou ce qui est la même chose (prenant pour origine le centre de l'ellipse) les coordonnées du point de la courbe sont respectivement les doubles des coordonnées du centre du cercle. Prenez X, Y pour les coordonnées du point de l'ellipse, et x, y pour celles du point de la courbe, l'équation de l'ellipse sera

$$\frac{X^2}{a^2}+\frac{Y^2}{b^2}=1$$

et, de la construction géométrique dont je viens de parler, on obtient sans peine

$$x=X\left(2-\frac{1}{a^2}(X^2+Y^2)\right), \qquad y=Y\left(2-\frac{1}{b^2}(X^2+Y^2)\right)$$

lesquelles sont les expressions de x, y en X, Y.

Ces expressions font voir que les points selon lesquels l'ellipse est coupée par le cercle $X^2+Y^2=2a^2$ (c'est-à-dire les points

$$(b^2-a^2)\,X^2=a^2(b^2-2a^2), \qquad (b^2-a^2)\,Y^2=a^2b^2,$$

qui sont des points imaginaires de l'ellipse) donnent pour la courbe des points doubles sur l'axe des Y (ces points sont toujours imaginaires) et de même les points selon lesquels l'ellipse est coupée par le cercle $X^2+Y^2=2b^2$ (c'est-à-dire les points

$$(a^2-b^2)\,X^2=a^2b^2, \qquad (a^2-b^2)\,Y^2=b^2(a^2-2b^2),$$

qui sont des points réels de l'ellipse en supposant $a^2>2b^2$) donnent pour la courbe des points doubles sur l'axe des X. Les coordonnées de ces points doubles sont données par $a^2x^2=4b^2(a^2-b^2)$, $y^2=0$, et elles sont ainsi réelles, soit pour $a^2<2b^2$, soit pour $a^2>2b^2$, mais dans le premier cas les points doubles sont des points conjugués.

En formant les équations

$$dx=\left(2-\frac{1}{a^2}(3X^2+Y^2)\right)dX-\frac{2XY}{a^2}dY=0,$$

$$dy=-\frac{2XY}{b^2}dX+\left(2-\frac{1}{b^2}(X^2+3Y^2)\right)dY=0,$$

et en éliminant dX, dY, on obtient l'équation

$$\left(2-\frac{1}{a^2}(3X^2+Y^2)\right)\left(2-\frac{1}{b^2}(X^2+3Y^2)\right)-\frac{4X^2Y^2}{a^2b^2}=0,$$

laquelle au moyen de l'équation de l'ellipse se réduit à

$$-2(X^2+Y^2)\left(\frac{1}{a^2}+\frac{1}{b^2}\right)+3(X^2+Y^2)^2\frac{1}{a^2b^2}=0.$$

On voit que les points de l'ellipse lesquels donnent lieu à des points de rebroussement de la courbe doivent être situés sur la courbe donnée par l'équation dernièrement mentionnée. Cette équation se divise en deux facteurs: l'équation $X^2+Y^2=0$ combinée avec l'équation de l'ellipse donne les points $X^2=-\frac{a^2b^2}{a^2-b^2}$, $Y^2=\frac{a^2b^2}{a^2-b^2}$, lesquels sont des points imaginaires de l'ellipse; mais on peut se convaincre sans peine que ces points *ne donnent pas lieu* à des points de rebroussement de la courbe. L'autre facteur, savoir $X^2=Y^2=\frac{2}{3}(a^2+b^2)$, combiné avec l'équation de l'ellipse donne les points

$$(a^2-b^2)\,X^2=\tfrac{1}{3}\,a^2\,(2a^2-b^2),\qquad (a^2-b^2)\,Y^2=\tfrac{1}{3}\,b^2\,(a^2-2b^2),$$

lesquels pour $a^2>2b^2$ sont des points réels de l'ellipse, et qui donnent lieu à des points de rebroussement de la courbe. On trouve alors pour les coordonnées des points de rebroussement

$$(a^2-b^2)\,a^2x^2=\tfrac{4}{27}\,(2a^2-b^2)^3,\qquad (b^2-a^2)\,b^2y^2=\tfrac{4}{27}\,(2b^2-a^2)^3$$

lesquels sont aussi des points réels pour $a^2>2b^2$.

On trouve sans peine les points dans lesquels la courbe coupe l'ellipse; en effet ces points se dérivent des points où l'ellipse est coupée par le cercle

$$X^2+Y^2=\frac{3a^2b^2}{a^2+b^2},$$

lesquels sont des points réels de l'ellipse si $a^2>2b^2$, et les coordonnées sont données par

$$(a^4-b^4)\,X^2=a^2b^2\,(2a^2-b^2),\qquad (b^4-a^4)\,Y^2=a^2b^2\,(2b^2-a^2),$$

de là on obtient, pour les coordonnées des points où l'ellipse est coupée par la courbe,

$$(a^2+b^2)^3\,(a^2-b^2)\,x^2=a^2b^2\,(2a^2-b^2)^3,\qquad (a^2+b^2)^3\,(b^2-a^2)\,y^2=a^2b^2\,(2b^2-a^2)^3,$$

lesquels sont de même des points réels si $a^2>2b^2$. A présent il est facile d'apercevoir la forme de la courbe; en effet on voit d'abord que la courbe est symétrique par rapport à chacune des deux axes et qu'elle touche l'ellipse aux extrémités des deux axes: pendant que $a^2<2b^2$ (c'est-à-dire pendant que l'ellipse n'est pas trop excentrique) la courbe est une ovale située entièrement en dedans de l'ellipse. Seulement il y a deux points conjugués sur l'axe de x. En supposant $a^2=2b^2$, on n'a plus les deux points conjugués, mais l'apparence de la courbe n'est pas autrement changée: cependant les points aux extrémités de l'axe majeur de l'ellipse sont ici des points singuliers d'une espèce particulière; car chacun de ces points est formé par l'union et l'amalgamation de deux points de rebroussement et d'un point double. Mais, pour $a^2>2b^2$, la forme de la courbe est entièrement changée: aux extrémités de l'axe majeur, la courbe est située en dehors de l'ellipse, avec sa convexité dans le sens opposé à celle de l'ellipse; la courbe va de chaque côté jusqu'à un point de rebroussement où elle change de direction, et puis elle coupe l'ellipse, coupe aussi l'axe majeur dans un point en dedans de l'ellipse (ce point est l'intersection de deux branches de la courbe, et ainsi un point double), et enfin la courbe arrive à l'extrémité de l'axe mineur où elle se réunit avec la branche située de l'autre côté de l'axe mineur; la construction géométrique par points donne le moyen de tracer la courbe avec exactitude; et dans le cas $a^2 > 2b^2$ on pourrait aussi par les valeurs ci-devant données pour les coordonnées construire les points doubles, et les points de rebroussement. Pour trouver l'équation de la courbe il faut éliminer X, Y entre l'équation de l'ellipse, et les équations trouvées pour x, y; pour faire cela j'écris $X^2 + Y^2 = \theta$, on a

$$x = X\left(2 - \frac{\theta}{a^2}\right), \qquad y = Y\left(2 - \frac{\theta}{b^2}\right),$$

et l'équation $X^2 + Y^2 = \theta$ et l'équation de l'ellipse donnent alors

$$\theta = \frac{x^2}{\left(2 - \frac{\theta}{a^2}\right)^2} + \frac{y^2}{\left(2 - \frac{\theta}{b^2}\right)^2}, \qquad 1 = \frac{x^2}{a^2\left(2 - \frac{\theta}{a^2}\right)^2} + \frac{y^2}{b^2\left(2 - \frac{\theta}{b^2}\right)^2},$$

entre lesquelles il faut éliminer θ; en multipliant la première équation par 2, et la seconde équation par $-\theta$ et en ajoutant, on trouve

$$\theta = \frac{x^2}{2 - \frac{\theta}{a^2}} + \frac{y^2}{2 - \frac{\theta}{b^2}},$$

et la seconde équation est l'équation dérivée de celle-ci par rapport à θ; c'est-à-dire l'équation de la courbe sera trouvée en éliminant θ entre cette équation et sa dérivée par rapport à θ: en écartant les dénominateurs, l'équation devient

$$\theta^3 - \theta^2(2a^2 + 2b^2) + \theta(a^2x^2 + b^2y^2 + 4a^2b^2) - 2a^2b^2(x^2 + y^2) = 0,$$

et le discriminant de cette fonction cubique de θ, égalé à zéro donne l'équation de la courbe.

Je représente la fonction cubique par

$$(A, B, C, D \between \theta, 1)^3,$$

de manière que

$$\begin{aligned} A &= 1, \\ B &= -\tfrac{2}{3}(a^2 + b^2), \\ C &= \tfrac{1}{3}(a^2x^2 + b^2y^2 + 4a^2b^2), \\ D &= -2a^2b^2(x^2 + y^2). \end{aligned}$$

En représentant le discriminant par $\square$, dans le cas actuel où $A = 1$, il sera convenable de se servir de la forme

$$A^2\square = 4(AC - B^2)^3 - (3ABC - A^2D - 2B^3)^2 = 0;$$

l'on a

$$AC - B^2 = \tfrac{1}{9}(3a^2x^2 + 3b^2y^2 - 4a^4 + 4a^2b^2 - 4b^4),$$

$$AD - BC = \tfrac{2}{9}[(a^2 - 8b^2)\,a^2x^2 + (b^2 - 8a^2)\,b^2y^2 + 4\,(a^2 + b^2)\,a^2b^2],$$

$$BD - C^2 = -\tfrac{1}{9}[a^4x^4 + 2a^2b^2x^2y^2 + b^2y^4 - 4\,(a^2 + 3b^2)\,a^2b^2x^2 - 4\,(b^2 + 3a^2)\,a^2b^2y^2 + 16a^4b^4],$$

et de là

$$\begin{aligned} 3ABC - A^2D - 2B^3 &= 2B\,(AC - B^2) - A\,(AD - BC) \\ &= -\tfrac{2}{27}[9\,(a^2 - 2b^2)\,a^2x^2 + 9\,(b^2 - 2a^2)\,b^2y^2 - 8a^6 + 12a^4b^2 + 12a^2b^4 - 8b^6], \end{aligned}$$

et l'équation de la courbe est ainsi trouvée sous la forme

$$\begin{aligned} &(3a^2x^2 + 3b^2y^2 - 4a^4 + 4a^2b^2 - 4b^4)^3 \\ &\qquad + [9\,(a^2 - 2b^2)\,a^2x^2 + 9\,(b^2 - 2a^2)\,b^2y^2 - 8a^6 + 12a^4b^2 + 12a^2b^4 - 8b^6]^2 = 0 \end{aligned}$$

laquelle est en effet la forme ci-devant mentionnée; j'ajoute que l'on trouverait les expressions déjà données pour les coordonnées des points de rebroussement en égalant à zéro les deux fonctions quadratiques qui entrent dans l'équation.

Je considère à présent la surface qui est l'enveloppe des plans menés par les points d'un ellipsoïde perpendiculairement aux rayons conduits par le centre. En s'imaginant une sphère qui passe par le centre de l'ellipsoïde et touche l'ellipsoïde dans l'un quelconque de ses points, le point correspondant de la surface sera à l'extrémité d'un diamètre de la sphère, lequel diamètre passe par le centre de l'ellipsoïde: ou ce qui est la même chose (prenant le centre de l'ellipsoïde pour origine) les coordonnées du point de la surface sont respectivement les doubles des coordonnées du centre de la sphère. Prenez X, Y, Z pour les coordonnées du point de l'ellipsoïde, et x, y, z pour les coordonnées du point de la surface, l'équation de l'ellipsoïde sera

$$\frac{X^2}{a^2} + \frac{Y^2}{b^2} + \frac{Z^2}{c^2} = 1,$$

et la construction qui vient d'être mentionnée donne

$$x = X\left[2 - \frac{1}{a^2}(X^2 + Y^2 + Z^2)\right],\quad y = Y\left[2 - \frac{1}{b^2}(X^2 + Y^2 + Z^2)\right],\quad z = Z\left[2 - \frac{1}{c^2}(X^2 + Y^2 + Z^2)\right],$$

pour les expressions de x, y, z en fonction de X, Y, Z.

La discussion de la surface peut être effectuée au moyen de ces expressions. Il y a un assez grand nombre de différentes formes de la surface; mais le seul cas que je vais considérer (lequel apparemment est celui qui présente le plus grand nombre des singularités réelles) est le cas où $a^2 > 2b^2$, $b^2 > 2c^2$; en effet dans ce cas, les courbes du sixième ordre, ou sextiques, correspondantes aux sections principales de l'ellipsoïde ont chacune des points doubles et des points de rebroussement.

Les courbes selon lesquelles l'ellipsoïde est coupé par les sphères concentriques

$$X^2+Y^2+Z^2=2a^2, \qquad X^2+Y^2+Z^2=2b^2, \qquad X^2+Y^2+Z^2=c^2,$$

donnent lieu dans la surface à des courbes *nodales* (courbes doubles) dans les plans yz, zx, xy respectivement. La première de ces courbes d'intersection, et la courbe nodale dans le plan de zy qui y correspond, sont l'une et l'autre imaginaires: les deux autres courbes d'intersection, et les courbes nodales dans les plans de zx et xy qui y correspondent, sont réelles; la courbe selon laquelle l'ellipsoïde est coupé par la sphère

$$X^2+Y^2+Z^2=2b^2,$$

laquelle courbe est une espèce d'ovale autour de l'extrémité de l'axe le plus grand de l'ellipsoïde (il va sans dire que les ovales dont je parle sont des courbes à double courbure), cette courbe d'intersection, je dis, donne lieu à une courbe nodale hyperbolique sur le plan de ZX. Pour en trouver l'équation on a

$$\frac{X^2}{a^2}+\frac{Y^2}{b^2}+\frac{Z^2}{c^2}=1, \qquad X^2+Y^2+Z^2=2b^2,$$

$$x=2X\left(1-\frac{b^2}{a^2}\right), \qquad y=0, \qquad z=2Z\left(1-\frac{b^2}{c^2}\right),$$

et de là, en éliminant X, Y, Z, on trouve

$$\frac{a^2x^2}{4b^2(a^2-b^2)}-\frac{c^2z^2}{4b^2(b^2-c^2)}=1$$

pour l'équation de la courbe nodale hyperbolique dans le plan de zx; on s'assure sans peine que cette courbe passe par les points doubles de la courbe sextique dans le plan de xy.

De même la courbe selon laquelle l'ellipsoïde est coupé par le sphère

$$X^2+Y^2+Z^2=2c^2,$$

laquelle courbe est une espèce d'ovale autour de l'extrémité de l'axe le plus petit, donne lieu à une courbe nodale elliptique dans le plan de xy, et l'on a d'une manière semblable

$$\frac{a^2x^2}{4c^2(a^2-c^2)}+\frac{b^2y^2}{4c^2(b^2-c^2)}=1$$

pour l'équation de cette courbe nodale elliptique dans le plan de xy: cette courbe passe par les points doubles des courbes sextiques dans les plans de yz et zy respectivement. La section de la surface par un plan principal de l'ellipsoïde est composée de la courbe sextique donnée par la section principale de l'ellipsoïde et de la courbe nodale (ellipse hyperbole, ou conique imaginaire) considérée comme deux coniques coïncidentes; les sections principales de la surface sont ainsi des courbes de l'ordre 10, et (ce qui sera montré plus complètement dans la suite) la surface elle-même est de l'ordre 10.

A présent je vais chercher la courbe sur l'ellipsoïde laquelle donne lieu à une courbe *cuspidale* (courbe de rebroussement) sur la surface. Pour cela je forme les équations:

$$dx = \left\{2 - \frac{1}{a^2}(3X^2 + Y^2 + Z^2)\right\} dX \qquad - \frac{2XY}{a^2} dY \qquad - \frac{2XZ}{a^2} dZ = 0,$$

$$dy = \qquad - \frac{2YX}{b^2} dX + \left\{2 - \frac{1}{b^2}(X^2 + 3Y^2 + Z^2)\right\} dY \qquad - \frac{2YZ}{b^2} dZ = 0,$$

$$dz = \qquad - \frac{2ZX}{c^2} dX \qquad - \frac{2ZY}{c^2} dY + \left\{2 - \frac{1}{c^2}(X^2 + Y^2 + Z^2)\right\} dZ = 0,$$

et en éliminant dX, dY, dZ et réduisant au moyen de l'équation de l'ellipsoïde on trouve

$$4\left(\frac{X^2}{a^4} + \frac{Y^2}{b^4} + \frac{Z^2}{c^4}\right) - 2(X^2 + Y^2 + Z^2)\left(\frac{1}{b^2c^2} + \frac{1}{c^2a^2} + \frac{1}{a^2b^2}\right) + 3(X^2 + Y^2 + Z^2)^2 \frac{1}{a^2b^2c^2} = 0,$$

laquelle est l'équation d'une surface du quatrième ordre qui coupe l'ellipsoïde suivant une courbe qui donne lieu à une courbe cuspidale sur la surface. La courbe sur l'ellipsoïde rencontre chacune des sections principales dans les points qui donnent lieu à la courbe sextique qui correspond à la section principale, et dans les points selon lesquels la section principale est coupée par la courbe qui donne lieu à une courbe nodale: ainsi les points d'intersection de la surface du quatrième ordre avec la section principale

$$\frac{X^2}{a^2} + \frac{Z^2}{c^2} = 1,\ Y = 0,$$

sont les points d'intersection de cette ellipse avec les courbes

$$X^2 + Z^2 = \tfrac{2}{3}(a^2 + c^2) \text{ et } X^2 + Z^2 = 2b^2,$$

lesquels points sont réels; les points d'intersection avec la section principale

$$\frac{Y^2}{b^2} + \frac{Z^2}{c^2} = 1,\ X = 0,$$

sont les points d'intersection de cette ellipse avec le cercle

$$Y^2 + Z^2 = \tfrac{2}{3}(b^2 + c^2),$$

lesquels points sont réels, et avec le cercle

$$Y^2 + Z^2 = 2a^2,$$

lesquels points sont imaginaires; les points d'intersection avec la section principale

$$\frac{X^2}{a^2} + \frac{Y^2}{b^2} = 1,\ Z = 0,$$

sont les points d'intersection de cette ellipse avec le cercle

$$X^2 + Y^2 = \tfrac{2}{3}(a^2 + b^2),$$

lesquels points sont réels, et avec le cercle

$$X^2 + Y^2 = 2c^2,$$

lesquels points sont imaginaires. La courbe sur l'ellipsoïde qui donne lieu à la courbe cuspidale rencontre les courbes sur l'ellipsoïde qui donnent lieu aux courbes nodales, seulement dans les points où ces dernières courbes sont rencontrées par les sections principales respectivement; et les points où cela arrive sont des points de contact des courbes dont il s'agit; les seuls points réels de contact sont les points où la section principale

$$\frac{X^2}{a^2} + \frac{Z^2}{c^2} = 1, \quad Y = 0,$$

est coupée par le cercle $X^2 + Z^2 = 2b^2$. Pour fixer les idées je suppose $a^2 + b^2 > 3b^2$, on a alors $2b^2 < \frac{2}{3}(a^2 + b^2)$ et les points dont je viens de parler sont situés plus près des extrémités de l'axe le plus petit que ne sont les points où cette même ellipse est coupée par le cercle

$$X^2 + Z^2 = \tfrac{2}{3}(a^2 + c^2).$$

La courbe sur l'ellipsoïde que donne lieu à la courbe cuspidale est composée de deux paires d'ovales: les ovales de l'une de ces paires sont situés autour des extrémités de l'axe le plus petit, et les ovales de l'autre paire autour des extrémités de l'axe le plus grand; et par la remarque déjà faite, les ovales de la première paire touchent la courbe sur l'ellipsoïde qui donnent lieu à la courbe hyperbolique nodale, les coordonnées des points de contact étant données par

$$\frac{X^2}{a^2} + \frac{Z^2}{c^2} = 1, \quad X^2 + Y^2 = 2b^2, \quad Y = 0,$$

les coordonnées des points correspondants sur la surface sont données par

$$x^2 = \frac{4(a^2 - b^2)(2b^2 - c^2)}{a^2(a^2 - c^2)}, \qquad y^2 = \frac{4(b^2 - c^2)(a^2 - 2b^2)}{c^2(a^2 - c^2)}.$$

Nous avons jusqu'ici parlé de la courbe nodale hyperbolique comme étant une hyperbole complète, mais il convient de remarquer que la courbe réelle entière sur l'ellipsoïde ne donne lieu qu'à des parties finies de cette hyperbole, savoir les parties finies comprises entre les points dont je viens de donner les coordonnées, et que pour ces parties de l'hyperbole la courbe nodale est une courbe nodale proprement dite, savoir il y a pour ces parties deux nappes réelles, mais pour les autres parties de l'hyperbole il n'y a aucune nappe réelle de la surface, ou autrement dit, la courbe nodale est une courbe conjuguée: les points dont il s'agit (lesquels pour plus de commodité je vais appeler les *points critiques*[1]) seront évidemment des points cuspidaux sur la courbe nodale (c'est-à-dire des points où les deux plans tangents deviennent identiques). Mais ces points sont ici des points d'une singularité plus élevée, savoir ils sont des

[1] Dans le mémoire anglais j'ai dit "points of cesser" ce qui était un terme plus expressif. A. C.

points de rebroussement sur la courbe cuspidale, et de plus la courbe nodale hyperbolique, la courbe cuspidale, et la courbe sextique dans le plan zx ont à chacun de ces points une tangente commune. Car, comme je l'ai auparavant mentionné, les courbes sur l'ellipsoïde qui donnent lieu à la courbe nodale hyperbolique et la courbe cuspidale respectivement, se touchent l'une l'autre aux points qui donnent lieu aux points critiques; de là il suit que la courbe nodale hyperbolique et la courbe cuspidale se touchent l'une l'autre aux points critiques, et par cela seulement que la tangente de la courbe cuspidale est, à l'un quelconque des points critiques, dans le plan de zx, (puisque la courbe cuspidale est symétrique par rapport à ce plan) on voit que le point critique sera un point de rebroussement sur la courbe cuspidale. Il ne reste qu'à montrer que la courbe nodale hyperbolique et la courbe sextique ont une tangente commune au point critique. La tangente de la courbe nodale hyperbolique est donnée par l'équation

$$\frac{a^2x\,dx}{a^2-b^2}-\frac{c^2z\,dz}{b^2-c^2}=0,$$

où

$$x=X\left[2-\frac{1}{a^2}(X^2+Y^2+Z^2)\right],\quad z=Z\left[2-\frac{1}{c^2}(X^2+Y^2+Z^2)\right],$$

et

$$\frac{X^2}{a^2}+\frac{Z^2}{c^2}=1,\quad X^2+Z^2=2b^2,\quad Y=0,$$

c'est-à-dire que l'on a

$$x=\frac{2X}{a^2}(a^2-b^2),\qquad z=-\frac{2Z}{c^2}(b^2-c^2)$$

et l'équation devient ainsi $Xdx+Zdz=0$, ce qui fait voir que la tangente est perpendiculaire au rayon mené par le centre de l'ellipsoïde au point qui donne lieu au point critique, et ainsi la tangente de la courbe nodale hyperbolique coïncide avec celle de la courbe sextique dans le plan de zx. Et l'on peut à présent expliquer plus particulièrement la forme de la courbe cuspidale (le cas dont il s'agit est celui où l'on a non seulement $a^2>2b^2$, $b^2>2c^2$ mais aussi $a^2+c^2>3b^2$) en effet la courbe cuspidale est composée en premier lieu de deux ovales en forme d'œil situés symétriquement de chaque côté du plan de xy, et qui passent par les points critiques, et les points de rebroussement de la courbe sextique dans le plan de yz, et en second lieu de deux ovales ordinaires situés symétriquement de chaque côté du plan de yz, et qui passent par les points de rebroussement des courbes sextiques dans les plans de xy et zx respectivement: tous les ovales dont il s'agit étant, il va sans dire, des courbes à double courbure. Les relations des différentes courbes sur l'ellipsoïde seront mieux comprises au moyen de la figure 1[a], où les sections principales de l'ellipsoïde sont tracées, non pas en perspective, mais développées sur le plan de la figure; et les courbes qui donnent lieu aux courbes nodales et cuspidales respectivement sont seulement marquées par des courbes menées par les points dans les sections principales par lesquels points passent respectivement les courbes dont il s'agit. On aura une idée de la forme de la surface au moyen de la figure 2[a] (laquelle est dessinée en perspective orthogonale au moyen de la première figure) et qui fait voir la forme des sections principales de la surface (y compris les courbes nodales) et la forme de la courbe cuspidale. Les numéros et lettres dans les deux figures font voir quels points des différentes courbes sur l'ellipsoïde donnent lieu respectivement aux points des différentes courbes sur la surface. En particulier il convient de remarquer que dans la figure 1ª le point 4 qui donne lieu à un point critique est situé entre les points 3 et 5 qui donnent lieu respectivement au point double et au point de rebroussement de la courbe sextique dans le plan de zx, et cela fait voir très évidemment dans quelle partie de la courbe sextique est située le point critique, où cette courbe est touchée par la courbe nodale hyperbolique. Je n'ai pas cherché à dessiner d'autres sections de la surface, cela donnerait trop de complication à la figure, et il serait encore plus difficile d'apercevoir la forme de la surface, laquelle peut à peine être représentée sinon par un modèle ou au moins une stéréographie. On peut trouver l'équation de la surface de la même manière que celle de la courbe sextique, à savoir il faut éliminer X, Y, Z entre l'équation

$$\frac{X^2}{a^2}+\frac{Y^2}{b^2}+\frac{Z^2}{c^2}=1,$$

et les équations

$$x=X\left[2-\frac{1}{a^2}(X^2+Y^2+Z^2)\right],\ y=Y\left[2-\frac{1}{b^2}(X^2+Y^2+Z^2)\right],\ z=Z\left[2-\frac{1}{c^2}(X^2+Y^2+Z^2)\right],$$

qui donnent les points de la surface.

En écrivant $\theta=X^2+Y^2+Z^2$ on trouve

$$\theta=\frac{x^2}{\left(2-\frac{\theta}{a^2}\right)^2}+\frac{y^2}{\left(2-\frac{\theta}{b^2}\right)^2}+\frac{z^2}{\left(2-\frac{\theta}{c^2}\right)^2},$$

$$1=\frac{x^2}{a^2\left(2-\frac{\theta}{a^2}\right)^2}+\frac{y^2}{b^2\left(2-\frac{\theta}{b^2}\right)^2}+\frac{z^2}{c^2\left(2-\frac{\theta}{c^2}\right)^2},$$

et en multipliant la seconde équation par 2, et la première équation par $-\theta$ et en ajoutant on trouve

$$\theta=\frac{x^2}{2-\frac{\theta}{a^2}}+\frac{y^2}{2-\frac{\theta}{b^2}}+\frac{z^2}{2-\frac{\theta}{c^2}}$$

laquelle a pour dérivée par rapport à θ la seconde équation. Cette équation peut s'écrire

$$\begin{aligned}&\theta^4\\&+\theta^3[-2(a^2+b^2+c^2)]\\&+\theta^2[4(b^2c^2+c^2a^2+a^2b^2)+a^2x^2+b^2y^2+c^2z^2]\\&+\theta\ [-8a^2b^2c^2-2(b^2+c^2)a^2x^2-2(c^2+a^2)b^2y^2-2(a^2+b^2)c^2z^2]\\&+\quad 4a^2b^2c^2(x^2+y^2+z^2)=0,\end{aligned}$$

et l'équation de la surface sera trouvée en égalant à zéro le discriminant de cette fonction de θ. En représentant l'équation par

$$\tfrac{1}{6}(A,\ B,\ C,\ D,\ E)\ (\theta,\ 1)^4 = 0,$$

on aura

$$\begin{aligned}
A &= \quad 6,\\
B &= -\ 3\,(a^2+b^2+c^2),\\
C &= \quad 4\,(b^2c^2+c^2a^2+a^2b^2)+a^2x^2+b^2y^2+c^2z^2,\\
D &= -\ 12a^2b^2c^2-3\,(b^2+c^2)\,a^2x^2-3\,(c^2+a^2)\,b^2y^2-3\,(a^2+b^2)\,c^2z^2,\\
E &= \quad 24a^2b^2c^2\,(x^2+y^2+z^2),
\end{aligned}$$

et l'équation de la surface sera

$$(AE-4BD+3C^2)^3-27\,(ACE+2BCD-AD^2-B^2E-C^3)^2=0.$$

Il convient de remarquer que C^6, qui serait du douzième ordre en x, y, z n'entre pas dans l'équation, et que tous les autres termes sont au plus de l'ordre 10, la surface est donc (comme j'ai déjà remarqué) de l'ordre 10. Il ne m'a pas paru nécessaire de développer plus complètement l'équation[1].

[1] Mi sia lecito di richiamare qui per un' istante in seguito alla bella ed elegante Memoria dell' illustre Geometra, come più volte questa superficie di decimo ordine derivata dall' ellissoide sia stato fin da molti anni per me il soggetto di ricerche relativo al calcolo integrale per ciò che riguarda specialmente la sua quadratura dipendente da trascendenti ellittici di prima e seconda specie; così in un breve articolo inserito nella *Raccolta Scientifica* di Roma Anno 2° No. 9 in data del 2 Aprile 1846 ritrovai le coordinate della nuova superficie in funzione delle coordinate corrispondenti dell' ellissoide, e determinai ancora l' integrale definito duplicato di forma razionale per la quadratura della stessà superficie: più estesamente mi occupai dello stesso argomento in una Memoria inserita nel tom. 34° del Sig. *Crelle* in data 15 Aprile 1846: più completamente poi risolvei la stessa questione in altro articolo pubblicato nella citata *Raccolta Scientifica* dello stesso Anno 2° No. 21 con la data del 23 Ottobre 1846. Infine ripresi a risolvere lo stesso problema con altri metodi in una Memoria *Sulla riduzione di alcuni integrali definiti ai trascendenti ellittici* inserita pel mese di Agosto e Settembre 1848 nel giornale arcadico di Roma, e precisamente nel No. 18 e seguenti. Sovente ancora nei privati miei studii tentai di giungere all' equazione algebrica della superficie, il che dipendeva da un' eliminazione, ma proponendomi la questione sotto un' aspetto complicato, mi trovava arrestato dalla lunghezza dei calcoli. Il maraviglioso ripiego analitico ritrovato dal Sig. Cayley riduce quest' eliminazione ad una simplicità inaspettata, e rende inutile qualunque altra ricerca, che si volesse intraprendere sullo stesso soggetto. Termino questa breve nota col fare i miei distinti ringraziamenti al chiarissimo Autore, che si è compiaciuto dietro un mio suggerimento comunicatogli dal Sig. D. *Hirst* fare la traduzione della Memoria per essere inserita negli *Annali*.—*B. T.*

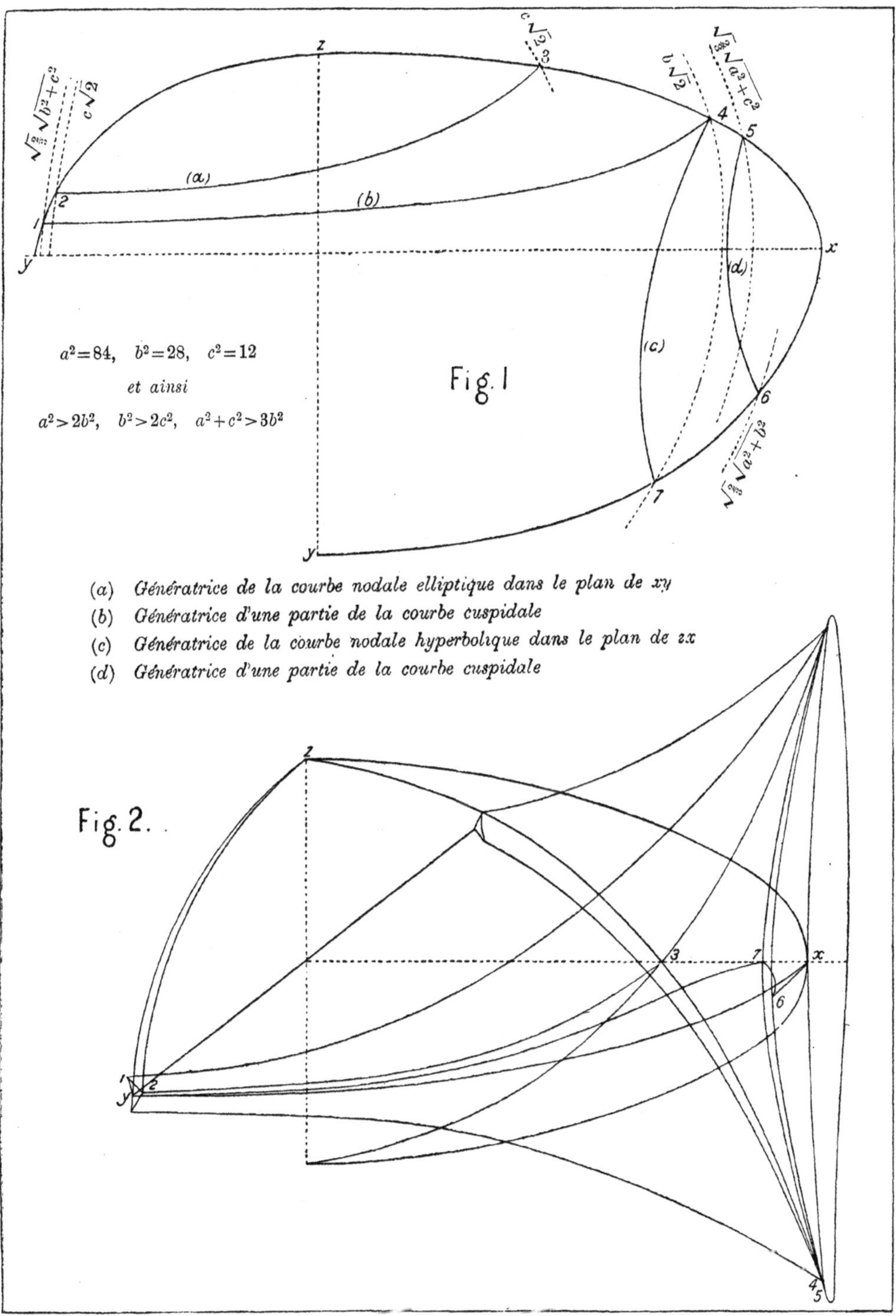

$a^2=84, \quad b^2=28, \quad c^2=12$
et ainsi
$a^2>2b^2, \quad b^2>2c^2, \quad a^2+c^2>3b^2$
Fig. 1
(a) Génératrice de la courbe nodale elliptique dans le plan de xy
(b) Génératrice d'une partie de la courbe cuspidale
(c) Génératrice de la courbe nodale hyperbolique dans le plan de zx
(d) Génératrice d'une partie de la courbe cuspidale
Fig. 2.

251.

SUR QUELQUES FORMULES POUR LA DIFFÉRENTIATION.

[From the *Annali di Matematica pura ed applicata* (Tortolini) tom. II. (1859), pp. 214—230. Traduction par l'auteur d'un mémoire présenté à la Société Royale de Londres le 26 Novembre, 1857.]

En cherchant une formule dans la théorie des intégrales définies multiples je fus conduit il y a plusieurs années à chercher les coefficients différentiels successifs de l'expression $(\sqrt{x+\lambda}-\sqrt{x+\mu})^{2i}$, et les résultats que j'ai trouvés sont donnés dans le Mémoire "On certain formulæ for differentiation with applications to the evaluation of definite integrals," *Camb. and Dublin Mathematical Journal*, tom. II. pp. 122—128 (1847), [41]. J'ai depuis cherché les coefficients différentiels successifs de l'expression plus générale $[(x+\lambda)(x+\mu)]^{\frac{1}{2}k}(\sqrt{x+\lambda}-\sqrt{x+\mu})^{2i}$, mais l'investigation n'était pas achevée. Mon attention fut rappelée à ce sujet par deux identités remarquables trouvées par le Prof. Donkin dans son Mémoire "On the equation of Laplace's functions etc." *Phil. Trans.* tom. CXLVII. (1857), pp. 43—57, au moyen de la comparaison de ses résultats avec ceux du Prof. Boole; identités qui appartenaient, je l'aperçus, à la classe des formules ci-dessus mentionnées: la première de ces deux identités se déduit en effet assez facilement d'une formule exposée dans mon mémoire; la démonstration de la seconde identité est beaucoup plus difficile, et je n'ai réussi à l'établir qu'en la faisant dépendre de l'établissement de l'égalité des coefficients numériques de deux expressions de la même forme. Je suis depuis revenu aux investigations incomplètes dont j'ai parlé ci-dessus et les résultats que j'ai trouvés sont donnés dans le présent mémoire. Je remarque qu'en écrivant pour abréger

$$P=2x+\lambda+\mu,\quad Q=\sqrt{(x+\lambda)(x+\mu)},\quad R=(\sqrt{x+\lambda}-\sqrt{x+\mu})^2,$$

le sujet auquel appartiennent tous les résultats est la différentiation de l'expression $P^\alpha Q^\beta R^\gamma$: l'expression ci-dessus mentionnée $[(x+\lambda)(x+\mu)]^{\frac{1}{2}k}(\sqrt{x+\lambda}-\sqrt{x+\mu})^{2i}$ est de la forme dont il s'agit, et la question qui s'y rapporte est celle d'obtenir le développement de $D_x^\mu P^\alpha Q^\beta R^\gamma$, où $\alpha=0$.

La question suggérée par la seconde identité du Prof. Donkin est celle d'obtenir le développement de $(P^{-1}Q^4D_x)^\mu P^\alpha Q^\beta R^\gamma$, où $\alpha=\gamma-\beta$. Comme la démonstration de ces identités est l'un des objets de ce Mémoire, j'ai donné dans la première section la réduction des identités à la forme sous laquelle je les ai depuis considérées. La seconde section se rapporte au développement de l'expression $D_x^\mu P^\alpha Q^\beta R^\gamma$, où $\alpha=0$; la troisième section à celui de l'expression $(P^{-1}Q^4D_x)^\mu P^\alpha Q^\beta R^\gamma$, où $\alpha=\gamma-\beta$: enfin la quatrième section contient l'application des deux identités, et quelques autres applications des formules.

§ I.

1. La première des deux identités du Prof. Donkin est

$$(\sin\theta D_\theta \sin\theta)^n (\text{tang}\, \tfrac{1}{2}\theta)^n = 1\,.\,3\,.\,5\,\ldots(2n-1)(\sin\theta)^{2n},$$

laquelle est l'équation (27) article No. 14, de son Mémoire[1].

En écrivant $\cot\theta=t$, l'équation devient

$$(-)^n D_t^n \frac{(\sqrt{1+t^2}-t)^n}{\sqrt{1+t^2}} = \frac{1\,.\,3\,.\,5\,\ldots(2n-1)}{(1+t^2)^{n+\frac{1}{2}}},$$

et en posant comme à l'ordinaire $i=\sqrt{-1}$ on obtient

$$\sqrt{1+t^2}-t=-\tfrac{1}{2}(\sqrt{t+i}-\sqrt{t-i})^2,$$

et en écrivant aussi

$$1\,.\,3\,.\,5\ldots(2n-1)=2^n\,.\,\tfrac{1}{2}\,.\,\tfrac{3}{2}\,.\,\tfrac{5}{2}\ldots(n-\tfrac{1}{2})=2^n\,[n-\tfrac{1}{2}]^n,$$

la formule devient

$$D_t^n \frac{(\sqrt{t+i}-\sqrt{t-i})^{2n}}{\sqrt{1+t^2}} = \frac{2^n\,[n-\frac{1}{2}]^n}{(1+t^2)^{n+\frac{1}{2}}}.$$

Ceci est un cas particulier de

$$D_x^n \frac{(\sqrt{x+\lambda}-\sqrt{x+\mu})^{2n}}{\sqrt{(x+\lambda)(x+\mu)}} = \frac{(-)^n(R-\mu)^{2n}\,[n-\frac{1}{2}]^n}{[(x+\lambda)(x+\mu)]^{n+\frac{1}{2}}},$$

ou, en écrivant comme auparavant

$$P=2x+\lambda+\mu,\quad Q=\sqrt{(x+\lambda)(x+\mu)},\quad R=[\sqrt{x+\lambda}-\sqrt{x+\mu}]^2,$$

la formule est

$$D_x^n Q^{-1}R^n=(-)^n(\lambda-\mu)^{2n}\,[n-\tfrac{1}{2}]\,Q^{-2n-1}.$$

[1] Vedi la nota in fine. [This Note by the Editor, Professor Tortolini, relating merely to the transformation $(\sin\theta D_\theta \sin\theta)^n=(\sin\theta)^{-1}(\sin^2\theta D_\theta)^n \sin\theta$, is not here reproduced.]

2. (1) La comparaison mentionnée article No. 5 du Mémoire du Prof. Donkin donne

$$(\sin\theta)^{-n}(\sin\theta\, D_\theta \sin\theta)^n \left\{ f\left[e^{\phi\sqrt{-1}} \operatorname{tang}\tfrac{1}{2}\theta + F\left(\frac{e^{\phi\sqrt{-1}}}{\operatorname{tang}\frac{1}{2}\theta}\right)\right]\right\}$$

$$= \mu^n \left(D_\mu \frac{1}{\mu}\right)^n \left\{ (\mu+\mu^2)^n f\left(\frac{\mu}{1+\mu}\,\frac{\sqrt{1-\mu'^2}}{\mu'}\, e^{\phi\sqrt{-1}}\right)\right.$$

$$\left. + (-)^n (\mu-\mu^2)^n F\left(\frac{\mu}{1-\mu}\,\frac{\sqrt{1-\mu'^2}}{\mu'}\, e^{\phi\sqrt{-1}}\right)\right\},$$

où $\mu = \cos\theta$, et après les différentiations $\mu' = \cos\theta$. Les parties qui contiennent les fonctions indéterminées f et F doivent être égales, chacune prise à part de l'autre. Les parties qui contiennent f seront égales, si cette égalité subsiste pour $fx = x^s$, où s est un indice quelconque; c'est-à-dire l'égalité subsistera si

$$(\sin\theta)^{-n}(\sin\theta\, D_\theta \sin\theta)^n (\operatorname{tang}\tfrac{1}{2}\theta)^s$$

$$= \mu^n \left(D_\mu \frac{1}{\mu}\right)^n (\mu+\mu^2)^n \frac{\mu^s}{(1+\mu)^s}\left(\frac{\sqrt{1-\mu'^2}}{\mu'}\right)^s,$$

$$= \mu^n \left(\frac{\sqrt{1-\mu'^2}}{\mu'}\right)^s \left(D_\mu \frac{1}{\mu}\right)^n (\mu+\mu^2)^n \frac{\mu^s}{(1+\mu)^s},$$

ou enfin en écrivant μ, au lieu de μ', si

$$(\sin\theta)^{-n}(\sin\theta\, D_\theta \sin\theta)^n (\operatorname{tang}\tfrac{1}{2}\theta)^s = \mu^{n-s}(1-\mu^2)^{\frac{1}{2}s}\left(D_\mu \frac{1}{\mu}\right)^n \mu^{n+s}(1+\mu)^{n-s},$$

où $\mu = \cos\theta$; c'est en effet la seconde des deux identités du Prof. Donkin. L'égalité des parties qui contiennent F dépend de la même manière de l'équation

$$(\sin\theta)^{-n}(\sin\theta\, D_\theta \sin\theta)^n (\operatorname{tang}\tfrac{1}{2}\theta)^s = (-)^n (1-\mu^2)^{\frac{1}{2}s}\left(D_\mu \frac{1}{\mu}\right)^2 \mu^{n+s}(1-\mu)^{n-s},$$

laquelle se déduit de l'autre équation en y écrivant $180° - \theta$ au lieu de θ.

3. La seconde identité (voir article No. 16) est

$$(\sin\theta)^{-n}(\sin\theta\, D_\theta \sin\theta)^n (\operatorname{tang}\tfrac{1}{2}\theta)^s = \mu^{n-s}(1-\mu^2)^{\frac{1}{2}s}\left(D_\mu \frac{1}{\mu}\right)^n \mu^{n+s}(1+\mu)^{n-s},$$

où comme auparavant $\mu = \cos\theta$. En écrivant $\cot\theta = t$, et en faisant attention que le côté gauche de l'équation peut s'écrire sous la forme

$$(\sin\theta)^{-n-1}(\sin^2\theta\, D_\theta)^n \sin\theta\, (\operatorname{tang}\tfrac{1}{2}\theta)^n,$$

et que l'on a

$$\sin\theta = \frac{1}{\sqrt{1+t^2}},\quad \operatorname{tang}\tfrac{1}{2}\theta = \sqrt{1+t^2} - t,\quad \sin^2\theta\, D_\theta = -D_t,$$

[1] Le lecteur pourrait omettre cet article, qui ne fait que montrer qu'une certaine formule du Professeur Donkin se réduit à son identité seconde.

le côté gauche devient

$$(-)^n (1+t^2)^{\frac{1}{2}(n+1)} D_t^n \frac{(\sqrt{1+t^2}-t)^s}{\sqrt{1+t^2}}.$$

Le côté droit peut s'écrire sous la forme

$$\mu^{n-s+1}(1-\mu^2)^{\frac{1}{2}s}\left(\frac{1}{\mu}D_\mu\right)^n \mu^{n+s-1}(1+\mu)^{n+s},$$

et en observant que

$$\mu = \frac{t}{\sqrt{1+t^2}}, \quad \frac{1}{\mu}D_\mu = \frac{(1+t^2)^2}{t}D_t,$$

le côté droit devient

$$\frac{t^{n-s+1}}{(1+t^2)^{\frac{1}{2}(n+1)}}\left(\frac{(1+t^2)^2}{t}D_t\right)^2 \frac{t^{n+s-1}}{(1+t^2)^{n-\frac{1}{2}}(\sqrt{1+t^2}-t)^{n-s}};$$

en comparant les deux expressions on obtient

$$(-)^n \frac{(1+t^2)^{n+1}}{t^{n-s+1}} D_t^n \frac{(\sqrt{1+t^2}-t)^s}{\sqrt{1+t^2}} = \left(\frac{(1+t^2)^2}{t}D_t\right)^n \frac{t^{n+s-1}}{(1+t^2)^{n-\frac{1}{2}}(\sqrt{1+t^2}-t)^{n-s}},$$

et de là, en écrivant comme auparavant $i=\sqrt{-1}$, et

$$\sqrt{1+t^2}-t = -\tfrac{1}{2}(\sqrt{t+i}-\sqrt{t-i})^2$$

la formule devient

$$\left(\frac{(1+t^2)^2}{2t}D_t\right)^n \frac{(2t)^{n+s-1}}{(1+t^2)^{n-\frac{1}{2}}(\sqrt{1+i}-\sqrt{t-i})^{2n-2s}} = \frac{(1+t^2)^{n+1}}{(2t)^{n-s+1}} D_t^n \frac{(\sqrt{t+i}-\sqrt{t-i})^{2s}}{\sqrt{1+t^2}},$$

laquelle est un cas particulier de

$$\left[\frac{[(x+\lambda)(x+\mu)]^2}{2x+\lambda+\mu}D_x\right]^n \frac{(2x+\lambda+\mu)^{n+s-1}}{[(x+\lambda)(x+\mu)]^{n-\frac{1}{2}}(\sqrt{x+\lambda}-\sqrt{x+\mu})^{2n-2s}}$$

$$= \frac{[(x+\lambda)(x+\mu)]^{n+1}}{(2x+\lambda+\mu)^{n-s+1}} D_x^n \frac{(\sqrt{x+\lambda}-\sqrt{x+\mu})^{2s}}{\sqrt{(x+\lambda)(x+\mu)}}$$

c'est-à-dire de

$$(P^{-1}Q^4 D_x)^n (P^{n+s-1}Q^{-2n+1}R^{-n+s}) = P^{-n+s-1}Q^{2n+2}D_x^n(Q^{-1}R^s).$$

§ II.

4. En écrivant comme auparavant

$$P = 2x+\lambda+\mu, \quad Q=\sqrt{(x+\lambda)(x+\mu)}, \quad R=(\sqrt{x+\lambda})-\sqrt{x+\mu})^2,$$

on a $R=P-2Q$, et en écrivant pour abréger $(\lambda-\mu)^2=\Lambda$, on obtient aussi

$$\Lambda = P^2 - 4Q^2, \quad \frac{\Lambda}{R} = P + 2Q;$$

on trouve de plus

$$D_x P = 2, \quad D_x Q = \tfrac{1}{2}\frac{P}{Q}, \quad D_x R = -\frac{R}{Q}.$$

5. Les dernières formules donnent

$$D_x P^\alpha Q^\beta R^\gamma = \tfrac{1}{2}\beta P^{\alpha+1} Q^{\beta-2} R^\gamma - \gamma P^\alpha Q^{\beta-1} R^\gamma + 2\alpha P^{\alpha-1} Q^\beta R^\gamma,$$

formule dont on pourrait se servir pour chercher $D_x{}^r P^\alpha Q^\beta R^\gamma$; on a par exemple

$$\begin{aligned}D_x{}^2 P^\alpha Q^\beta R^\gamma &= \tfrac{1}{4}\beta(\beta-2) P^{\alpha+2} Q^{\beta-4} R^\gamma - \tfrac{1}{2}\gamma(2\beta-1) P^{\alpha+1} Q^{\beta-3} R^\gamma \\ &+ [\beta(2\alpha+1)+\gamma^2] P^\alpha Q^{\beta-2} R^\gamma - 4\alpha\gamma P^{\alpha-1} Q^{\beta-1} R^\gamma + 4(\alpha-1)\gamma P^{\alpha-2} Q^\beta R^\gamma,\end{aligned}$$

et ainsi de suite. Et de même

$$P^{-1}Q^4 D_x P^\alpha Q^\beta R^\gamma = \tfrac{1}{2}\beta P^\alpha Q^{\beta+2} R^\gamma - \gamma P^{\alpha-1} Q^{\beta+3} R^\gamma + 2\alpha P^{\alpha-2} Q^{\beta+4} R^\gamma,$$

et de là

$$\begin{aligned}(P^{-1}Q^4 D_x)^2 P^\alpha Q^\beta R^\gamma &= \tfrac{1}{4}\beta(\beta+2) P^\alpha Q^{\beta+4} R^\gamma - \tfrac{1}{2}(2\beta+3)\gamma P^{\alpha-1} Q^{\beta+5} R^\gamma \\ &+ [2\alpha(\beta+2)+\gamma^2] P^{\alpha-2} Q^{\beta+6} R^\gamma - 2(2\alpha-1)\gamma P^{\alpha-3} Q^{\beta+7} R^\gamma \\ &+ 4\alpha(\alpha-2) P^{\alpha-4} Q^{\beta+8} R^\gamma,\end{aligned}$$

et ainsi de suite. Mais il serait difficile d'obtenir de cette manière l'expression de la r-ième répétition de l'opération D_x, ou $P^{-1}Q^4 D_x$, sur $P^\alpha Q^\beta R^\gamma$ et je ne poursuis pas la question.

6. Je vais à présent chercher le développement de $D_x{}^r P^\alpha Q^\beta R^\gamma$ $(\alpha = 0)$ ou ce qui est la même chose $D_x{}^r Q^\beta R^\gamma$. On a

$$D_x Q^\beta R^\gamma = \tfrac{1}{2}\beta P Q^{\beta-2} R^\gamma - \gamma Q^{\beta-1} R^\gamma,$$

ou en substituant pour P la valeur

$$P = \frac{\Lambda}{R} - 2Q,$$

on a

$$D_x Q^\beta R^\gamma = -(\beta+\gamma) Q^{\beta-1} R^\gamma - \tfrac{1}{2}\beta \Lambda Q^{\beta-2} R^{\gamma-1}.$$

La répétition de l'opération D_x donne évidemment une expression de la forme

$$\begin{aligned}D_x{}^r Q^\beta R^\gamma = \quad & (-)^r \quad L_r \quad \Lambda^0 Q^{\beta-r} \quad R^\gamma \\ & \vdots \\ + & (-)^{r-\theta} \frac{1}{2^\theta} L_{r,\theta} \quad \Lambda^\theta Q^{\beta-r-\theta} R^{\gamma-\theta} \\ & \vdots \\ + & \quad \frac{1}{2^r} L_{r,\mu r} . \Lambda^r Q^{\beta-2r} \quad R^{\gamma-r},\end{aligned}$$

et on obtient pour $L_{r,\theta}$ l'équation aux différences

$$L_{r+1,\,\theta+1}-(\beta+\gamma-r-2\theta-2)\,L_{r,\,\theta+1}-(\beta-r-\theta)\,L_{r,\,\theta}=0,$$

laquelle, avec les conditions particulières

$$L_{r,\,-1}=0,\quad L_{r,\,r+1}=0,\quad L_{0,\,0}=1,$$

suffit pour déterminer les coefficients $L_{r,\theta}$ de la formule.

7. Avant d'aller plus loin, je vais considérer le cas particulier $\beta=0$. L'équation aux différences est ici

$$L_{r+1,\,\theta+1}-(\gamma-r-2\theta-2)\,L_{r,\,\theta+1}+(r+\theta)\,L_{r,\,\theta}=0,$$

et l'on satisfait à cette équation en posant

$$L_{r,\,\theta}=\frac{(-)^{\theta}\,[r+\theta-1]^{2\theta}\,[\gamma-\theta-1]^{r-\theta-1}}{2^{\theta}\,[\theta]^{\theta}},$$

ou, ce qui est la même chose,

$$L_{r,\,\theta}=\frac{(-)^{\theta}\,2^{\theta}\gamma\,[\theta-\tfrac{1}{2}]^{\theta}\,[r+\theta-1]^{r-\theta-1}\,[\gamma-\theta-1]^{r-\theta-1}}{[r-\theta-1]^{r-\theta-1}}.$$

En effet on déduit de la première expression

$$\begin{aligned}L_{r+1,\,\theta+1}&-(\gamma-r-2\theta-2)\,L_{r,\,\theta+1}\\&=\frac{(-)^{\theta+1}\gamma}{2^{\theta+1}\,[\theta+1]^{\theta+1}}\,(r+\theta)\,[r+\theta-1]^{2\theta}\,[\gamma-\theta-2]^{r-\theta-2}\\&\qquad\times[(r+\theta+1)(\gamma-r)-(\gamma-r-2\theta-2)(r-\theta-1)],\end{aligned}$$

et le facteur en [ ] est $2(\theta+1)(\gamma-\theta-1)$, de manière que l'expression devient

$$=\frac{-(-)^{\theta}\gamma}{2^{\theta}\,[\theta]^{\theta}}\,(r+\theta)\,[r+\theta-1]^{2\theta}\,[\gamma-\theta-1]^{r-\theta-1},$$

ce qui est

$$=-(r+\theta)\,L_{r,\,\theta}.$$

L'équation aux différences est donc satisfaite ; les conditions pour les limites sont aussi satisfaites, et la valeur qui vient d'être donnée est donc celle de $L_{r,\theta}$ dans le cas particulier dont il s'agit où $\beta=0$.

8. Il sera convenable d'écrire

$$\frac{(-)^{\theta}}{2^{\theta}\gamma}\,L_{r,\,\theta}=\frac{[r+\theta-1]^{2\theta}\,[\gamma-\theta-1]^{r-\theta-1}}{2^{2\theta}\,[\theta]^{\theta}}=\frac{[\theta-\tfrac{1}{2}]^{\theta}\,[r+\theta-1]^{r-\theta-1}\,[\gamma-\theta-1]^{r-\theta-1}}{[r-\theta-1]^{r-\theta-1}}=L'_{r,\theta},$$

et alors en observant que $L_{r,\theta}$, ou $L'_{r,\theta}$ se réduit à zéro pour $\theta=r$, on obtient

$$\frac{(-)^{r}}{\gamma}\,D_x^{\,r}R^{\gamma}=L'_{r,\,0}\,Q^{-r}\,R^{\gamma}\ldots+L'_{r,\,\theta}\,\Lambda^{\theta}\,Q^{-r-\theta}\,R^{\gamma-\theta}\ldots+L'_{r,\,r-1}\,\Lambda^{r-1}\,Q^{-2r+1}\,R^{\gamma-r-1}.$$

9. Cela est en effet sous une forme un peu modifiée la formule fondamentale de mon mémoire dans le *Camb. and Dub. Math. Journal;* la valeur qu'on y trouve du coefficient $K_{r,\theta}$ (en mettant γ au lieu de i est)

$$K_{r,\theta}=\frac{\Gamma(r-\frac{1}{2}-\theta)\,\Gamma(2r-1-\theta)\,\Gamma(\gamma-r-\theta-1)}{\Gamma(\frac{1}{2})\,\Gamma(\theta+1)\,\Gamma(2r-1-2\theta)\,\Gamma(\gamma-r+1)},$$

et le coefficient $L_{r,\theta}\,(\beta=0)$ du présent mémoire est lié avec $K_{r,\theta}$ par l'équation

$$\frac{(-)^\theta}{2^\theta\gamma}L_{r,\theta}=L'_{r,\theta}=K_{r,\,r-1-\theta}.$$

La valeur de $L'_{r,\,r-1}$ est $[r-\frac{3}{2}]^{r-1}$, et pour $r=\gamma+1$ le coefficient devient $[\gamma-\frac{1}{2}]^\gamma$, et tous les autres coefficients se réduisent à zéro: la formule devient donc tout simplement

$$\frac{(-)^{\gamma+1}}{\gamma}D_x{}^{\gamma+1}R^\gamma=[\gamma-\tfrac{1}{2}]^\gamma\,\Lambda^\gamma\,Q^{-2\gamma-1}.$$

10. Dans le cas $r>\gamma+1$ il convient de modifier la forme de l'équation. J'écris $r=\gamma+1+s$, puis en faisant attention que les coefficients $L'_{\gamma+1+s,\,\theta}$ se réduisent à zéro pour $\theta<\gamma$, on peut écrire $\gamma+\theta$ au lieu de θ, et dans la formule nouvelle étendre θ depuis $\theta=0$ jusqu'à $\theta=s$. On a ainsi

$$\begin{aligned}\frac{(-)^{\gamma+1+s}}{\gamma}D_x{}^{\gamma+1+s}R^\gamma = &\ L'_{\gamma+1+s,\,\gamma}\ \Lambda^\gamma\ Q^{-2\gamma-1-s}\ R^0\\ &\ \vdots\\ &+L'_{\gamma+1+s,\,\gamma+\theta}\,\Lambda^{\gamma+\theta}\,Q^{-2\gamma-1-s-\theta}\,R^{-\theta}\\ &\ \vdots\\ &+L'_{\gamma+1+s,\,\gamma+s}\,\Lambda^{\gamma+s}\,Q^{-2\gamma-1-2s}\ R^{-s},\end{aligned}$$

et dans cette équation le côté gauche peut s'exprimer sous la forme

$$L'_{\gamma+1,\,\gamma}\,\Lambda^\gamma\,(-)^s\,D_x{}^s\,Q^{-2\gamma-1}.$$

On obtient ainsi

$$\begin{aligned}D_x{}^s\,Q^{-2\gamma-1} = &\ L''_{s,\,0}\qquad Q^{-2\gamma-1-s}\\ &\ \vdots\\ &+L''_{s,\,\theta}\,\Lambda^\theta\,Q^{-2\gamma-1-s-\theta}\,R^{-\theta}\\ &\ \vdots\\ &+L''_{s,\,s}\,\Lambda^s\,Q^{-2\gamma-1-2s}\ R^{-s},\end{aligned}$$

où $L''_{s,\,\theta}=(-)^s\,L'_{\gamma+1+s,\,\gamma+\theta}\div L'_{\gamma+1,\,\gamma}$, et en substituant pour les coefficients L' leurs valeurs, et en faisant attention que $[-\theta-1]^{s-\theta}=(-)^{s-\theta}\,[s]^{s-\theta}$,

on trouve

$$L''_{s,\,\theta}=\frac{(-)^\theta\,[2\gamma+s+\theta]^{2\gamma+2\theta}\,[s]^{s-\theta}}{2^{2\theta}\,[2\gamma]^{2\theta}\,[\gamma+\theta]^\theta}=\frac{(-)^\theta\,[\gamma+\theta-\frac{1}{2}]^\theta\,[2\gamma+s+\theta]^{s-\theta}\,[s]^{s-\theta}}{[s-\theta]^{s-\theta}};$$

la formule qui vient d'être trouvée est donnée sous une forme différente dans mon mémoire déjà cité.

11. Je résume l'équation générale

$$L_{r+1,\,\theta+1}-(\beta+\gamma-r-2\theta-2)\,L_{r,\,\theta+1}-(\beta-r-\theta)\,L_{r,\,\theta}=0,$$

les conditions aux limites étant comme auparavant

$$L_{r,\,-1}=0,\quad L_{r,\,r+1}=0,\quad L_{0,\,0}=1.$$

On a en particulier

$$\begin{aligned}
&L_{r+1,\,0}-(\beta+\gamma-r\quad)\,L_{r,\,0} &&=0,\\
&L_{r+1,\,1}-(\beta+\gamma-r-2)\,L_{r,\,1}-(\beta-\ r\quad)\,L_{r,\,0}&&=0,\\
&L_{r+1,\,2}-(\beta+\gamma-r-4)\,L_{r,\,2}-(\beta-\ r-1)\,L_{r,\,1}&&=0,\\
&\vdots\\
&L_{r+1,\,r+1}\,\cdots\cdots\qquad\qquad -(\beta-2r\quad)\,L_{r,\,r}&&=0.
\end{aligned}$$

De là on voit tout de suite que les valeurs de $L_{r,\,0}$, $L_{r,\,r}$ sont

$$L_{r,\,0}=[\beta+\gamma]^r,\quad L_{r,\,r}=[\beta,\ -2]^r,$$

où dans la dernière équation la notation au côté droit dénote une factorielle à différence -2.

12. Les autres coefficients $L_{r,\,1}$, $L_{r,\,2}$, etc. peuvent s'obtenir successivement par une intégration directe des équations; et quoiqu'on ne les obtienne pas de cette manière sous la forme la plus commode, cependant il convient de donner l'investigation. Pour $L_{r,\,1}$ on peut écrire

$$L_{r,\,1}=[\beta+\gamma-2]^r\,M_{r,\,1},$$

l'équation pour $M_{r,\,1}$ sera

$$M_{r+1,\,1}-M_{r,\,1}=\frac{(\beta-r)\,[\beta+\gamma]^r}{[\beta+\gamma-2]^{r+1}},\ =\frac{(\beta-r)\,(\beta+\gamma)\,(\beta+\gamma-1)}{(\beta+\gamma-r)\,(\beta+\gamma-r-1)\,(\beta+\gamma-r-2)},$$

laquelle peut aussi s'écrire

$$M_{r+1,\,1}-M_{r,\,1}=1+\frac{-\frac{1}{2}(r+2)\,(r+1)\,(\gamma-2)}{\beta+\gamma-r-2}+\frac{(r+1)\,r\,(\gamma-1)}{\beta+\gamma-r-1}+\frac{-\frac{1}{2}r\,(r-1)\,\gamma}{\beta+\gamma-r}.$$

Cette équation a une solution

$$M_{r,\,1}=r+\frac{A_r}{\beta+\gamma-r-1}+\frac{B_r}{\beta+\gamma-r},$$

cela donne en effet

$$M_{r+1,\,1}-M_{r,1}=1+\frac{A_{r+1}}{\beta+\gamma-r-2}+\frac{B_{r+1}-A_r}{\beta+\gamma-r-1}+\frac{-B_r}{\beta+\gamma-r},$$

et l'on doit ainsi avoir

$$A_{r+1}=-\tfrac{1}{2}(r+2)\,(r+1)\,(\gamma-2),\quad B_{r+1}-A_r=(r+1)\,r\,(\gamma-1),\quad -B_r=-\tfrac{1}{2}r\,(r-1)\,\gamma,$$

équations qui sont satisfaites par

$$A_r=-\tfrac{1}{2}(r+1)\,r\,(\gamma-2),\quad B_r=\tfrac{1}{2}r\,(r-1)\,\gamma.$$

Les conditions aux limites sont aussi satisfaites, et en formant l'expression de L_{r+1} on obtient

$$L_{r,\,1}=[\beta+\gamma-2]^r\left(r+\frac{-\frac{1}{2}(r+1)\,r\,(\gamma-2)}{\beta+\gamma-r-1}+\frac{\frac{1}{2}r\,(r-1)\,\gamma}{\beta+\gamma-r}\right).$$

13. La valeur de $L_{r,2}$ peut s'obtenir par un procédé semblable, et l'on a

$$L_{r,2}=[\beta+\gamma-4]^r\left(\tfrac{1}{2}r(r-1)+\frac{A_r}{\beta+\gamma-r-3}+\frac{B_r}{\beta+\gamma-r-2}+\frac{C_r}{\beta+\gamma-r-1}+\frac{D_r}{\beta+\gamma-r}\right),$$

où

$$\begin{aligned}
A_r &= \tfrac{1}{48}(r+1)\,r\left((r^2+5r-6)\gamma^2+(-11r^2-31r+42)\gamma+(24r^2+48r-72)\right),\\
B_r &= -\tfrac{3}{48}\,r(r-1)\left((r^2+3r-2)\gamma^2+(-7r^2-13r+6)\gamma+8r^2+16r\right),\\
C_r &= \tfrac{3}{48}(r-1)(r-2)\left((r^2+r)\gamma^2+(-3r^2-3r)\gamma\right),\\
D_r &= -\tfrac{1}{48}(r-2)(r-3)\left((r^2-r)\gamma^2+(r^2-r)\gamma\right),
\end{aligned}$$

et l'on peut remarquer que les suites des coefficients numériques (1, 1, 1, 1), (5, 3, 1, −1), (−6, −2, 0, 0) ont respectivement les premières, secondes, et troisièmes différences égales à zéro, les suites (−11, −7, −3, 1), (−31, −13, −3, −1) ont respectivement les secondes, et troisièmes différences égales à zéro, et la suite (24, 8, 0, 0) a les troisièmes différences égales à zéro. Je remarque que pour $r=1$, $r=2$ les expressions de $L_{r,1}$, $L_{r,2}$ donnent

$$L_{1,1}=(\beta+\gamma-2)\left(1-\frac{(\gamma-2)}{\beta+\gamma-2}\right),\ =\beta,$$

$$L_{2,2}=(\beta+\gamma-4)(\beta+\gamma-5)\left(1+\frac{\gamma^2-8\gamma+15}{\beta+\gamma-5}+\frac{-\gamma^2+6\gamma-8}{\beta+\gamma-4}\right),\ =\beta(\beta-2),$$

lesquelles s'accordent avec les résultats donnés par l'expression générale de $L_{r,r}$.

14. L'expression de $L_{r,1}$ peut se transformer en

$$L_{r,1}=r\beta\,[\beta+\gamma-2]^{r-1}-\tfrac{1}{2}r(r-1)[\beta+\gamma-2]^{r-2},$$

et l'expression de $L_{r,2}$ peut de même, avec beaucoup plus de peine, se transformer en

$$L_{r,2}=\tfrac{1}{2}r(r-1)\left\{\begin{array}{l}\beta(\beta-2)[\beta+\gamma-4]^{r-2}\\ -\ (r-2)\beta(\beta-1)[\beta+\gamma-4]^{r-3}\\ +\tfrac{1}{2}(r-2)(r-3)\tfrac{1}{2}(\beta+1)\beta[\beta+\gamma-4]^{r-4},\end{array}\right.$$

et la forme de ces équations et d'autres considérations m'ont conduit à assumer en général

$$\begin{array}{rll}
L_{r,\theta}= & \dfrac{1}{[\theta]^\theta} & A_{\theta,0}[r]^\theta\quad[\beta+\gamma-2\theta]^{r-\theta}\\
 & \vdots & \\
 & -\dfrac{1}{[\theta-1]^{\theta-1}[1]^1} & \tfrac{1}{2}A_{\theta,1}[r]^{\theta+1}\,[\beta+\gamma-2\theta]^{r-\theta-1}\\
 & \vdots & \\
 & +\dfrac{(-)^\sigma}{[\theta-\sigma]^{\theta-\sigma}[\sigma]^\sigma}\cdot & \dfrac{1}{2^\sigma}A_{\theta,\sigma}[r]^{\theta+\sigma}[\beta+\gamma-2\theta]^{r-\theta-\sigma}\\
 & \vdots & \\
 & \dfrac{(-)^\theta}{[\theta]^\theta} & A_{\theta,\theta}[r]^{2\theta}\quad[\beta+\gamma-2\theta]^{r-2\theta}.
\end{array}$$

15. En posant pour plus de commodité $\theta-1$ au lieu de θ dans l'équation aux différences, cette équation devient

$$L_{r+1,\,\theta}-(\beta+\gamma-r-2\theta)\,L_{r,\,\theta}=(\beta-r-\theta+1)\,L_{r,\,\theta-1},$$

et puis substituant dans cette équation la valeur assumée de $L_{r,\,\theta}$, et les valeurs correspondantes de $L_{r+1,\,\theta}$ et $L_{r,\,\theta-1}$, le terme général au côté gauche sera

$$\frac{(-)^\sigma}{[\sigma]^\sigma\,[\theta-\sigma]^{\theta-\sigma}}\,\frac{1}{2^\sigma}\,[r]^{\theta+\sigma-1}\,[\beta+\gamma-2\theta]^{r-\theta-\sigma}\,A_{\theta,\,\sigma}$$
$$\times\,[(r+1)(\beta+\gamma-r-\theta-\sigma)-(\beta+\gamma-r-2\theta)(r-\theta-\sigma+1)]\,;$$

l'expression en [ ] est $(\theta+\sigma)(\beta+\gamma-2\theta+1)$, et en substituant cette valeur le terme général au côté gauche devient

$$\frac{(-)^\sigma}{[\sigma]^\sigma\,[\theta-\sigma]^{\theta-\sigma}}\cdot\frac{1}{2^\sigma}\,(\theta+\sigma)\,[r]^{\theta+\sigma-1}\,[\beta+\gamma-2\theta+1]^{r+1-\theta-\sigma}\,A_{\theta,\,\sigma}.$$

Le terme général au côté droit est

$$\frac{(-)^\sigma}{[\sigma]^\sigma\,[\theta-1-\sigma]^{\theta-1-\sigma}}\,\frac{1}{2^\sigma}\,[r]^{\theta-1-\sigma}\,[\beta+\gamma-2\theta+1]^{r-\theta-\sigma}\,A_{\theta-1,\,\sigma}$$
$$\times\,(\beta-r-\theta+1)\,(\beta+\gamma-2\theta+2),$$

dont le dernier facteur est égal à

$$(\beta+\gamma-r-\theta+1+\sigma)\,(\beta-2\theta+2-\sigma)-(r-\theta+1-\sigma)\,(\gamma+\sigma);$$

et en substituant cette valeur le terme devient

$$\frac{(-)^\sigma}{[\sigma]^\sigma\,[\theta-1-\sigma]^{\theta-1-\sigma}}\,\frac{1}{2^\sigma}\,[r]^{\theta-1+\sigma}\,[\beta+\gamma-2\theta-1]^{r-\theta-\sigma+1}\,(\beta-2\theta+2-\sigma)\,A_{\theta-1,\,\sigma}$$
$$-\,\frac{(-)^\sigma}{[\sigma]^\sigma\,[\theta-1-\sigma]^{\theta-1-\sigma}}\cdot\frac{1}{2^\sigma}\,[r]^{\theta+\sigma}\,[\beta+\gamma-2\theta+1]^{r-\theta-\sigma}\,(\gamma+\sigma)A_{\theta-1,\,\sigma}.$$

En écrivant dans la seconde ligne (ce qui est permis) $\sigma-1$ au lieu de σ, cette ligne devient

$$-\,\frac{(-)^{\sigma-1}}{[\sigma-1]^{\sigma-1}\,[\theta-\sigma]^{\theta-\sigma}}\cdot\frac{1}{2^{\sigma-1}}\,[r]^{\theta+\sigma-1}\,[\beta+\gamma-2\theta+1]^{r-\theta-\sigma+1}\,(\gamma+\sigma-1)\,A_{\theta-1,\,\sigma-1},$$

et les deux lignes ensemble seront

$$\frac{(-)^\sigma}{[\sigma]^\sigma\,[\theta-\sigma]^{\theta-\sigma}}\,\frac{1}{2^\sigma}\,[r]^{\theta+\sigma-1}\,[\beta+\gamma-2\theta+1]^{r-\theta-\sigma+1}$$
$$\times\,[(\theta-\sigma)\,(\beta-2\theta+2-\sigma)\,A_{\theta-1,\,\sigma}+2\sigma\,(\gamma+\sigma-1)\,A_{\theta-1,\,\sigma-1}],$$

ce qui est le terme général au côté droit. En comparant cela avec l'expression ci-dessus trouvée pour le terme général au côté gauche on obtient

$$(\theta+\sigma)\,A_{\theta,\,\sigma}=(\theta-\sigma)(\beta-2\theta+2-\sigma)\,A_{\theta-1,\,\sigma}+2\sigma\,(\gamma+\sigma-1)\,A_{\theta-1,\,\sigma-1},$$

pour l'équation aux différences à laquelle doit satisfaire le coefficient $A_{\theta,\,\sigma}$; en y écrivant

$$A_{\theta,\,\sigma}=[\gamma+\sigma-1]^{\sigma}\,B_{\theta,\,\sigma},$$

on aura pour $B_{\theta,\,\sigma}$ l'équation

$$(\theta+\sigma)\,B_{\theta,\,\sigma}=(\theta-\sigma)(\beta-2\theta+2-\sigma)\,B_{\theta-1,\,\sigma}+2\sigma\,B_{\theta-1,\,\sigma-1},$$

et on voit sans peine que les conditions aux limites sont

$$B_{\theta,\,-1}=0,\quad B_{\theta,\,\theta+1}=0,\quad B_{0,\,0}=1.$$

16. On a en particulier

$$B_{\theta,\,0}=(\beta-2\theta+2)\,B_{\theta-1,\,0},\quad B_{\theta,\,\theta}=B_{\theta-1,\,\theta-1},$$

et de là

$$B_{\theta,\,0}=[\beta,\,-2]^{\theta},\quad B_{\theta,\,\theta}=1.$$

Les autres fonctions $B_{\theta,\,\sigma}$ sont des fonctions rationnelles et entières de β, du degré $\theta-\sigma$, données par l'équation aux différences; mais les coefficients numériques des différentes puissances de β n'admettent, à ce qu'il paraît, aucune expression simple; les fonctions $B_{\theta,\,\sigma}$ sont, pour ainsi dire, une espèce de transcendantes rationnelles propres pour exprimer la valeur de $L_{r,\,\theta}$, et il sera suffisant de tabuler les valeurs de $B_{\theta,\,\sigma}$ sans pousser plus loin la recherche de la loi de ces valeurs. On a en effet pour $B_{0,\,0}$, $B_{1,\,0}$, etc.... $B_{1,\,1}$ les valeurs

$$\begin{array}{cccccc}
1, & \beta, & \beta(\beta-2), & \beta(\beta-2)(\beta-4), & \beta(\beta-2)(\beta-4)(\beta-6), & \text{etc.}\ldots\\
 & 1, & \beta-1, & \beta^2-4\beta+\tfrac{5}{2}, & \beta^3-9\beta^2+\tfrac{43}{2}\beta-\tfrac{21}{2}, & \\
 & & 1\;, & \beta-2\;, & \beta^2-6\beta+7, & \\
 & & & 1\;, & \beta-3, & \\
 & & & & 1. &
\end{array}$$

17. Puis en substituant pour $A_{\theta,\,\sigma}$ la valeur $[\gamma+\sigma-1]^{\sigma}\,B_{\theta,\,\sigma}$ on trouve

$$\begin{aligned}
L_{r,\theta}= &\ \frac{1}{[\theta]^{\theta}}\qquad B_{\theta,\,0}\qquad [r]^{\theta}\qquad [\beta+\gamma-2\theta]^{r-\theta}\\
&+\frac{(-)}{[\theta-1]^{\theta-1}\,[1]^{1}}\ \tfrac{1}{2}\,B_{\theta,\,1}\ [\gamma]^{1}\,[r]^{\theta+1}\qquad [\beta+\gamma-2\theta]^{r-\theta-1}\\
&\quad\vdots\\
&+\frac{(-)^{\sigma}}{[\theta-\sigma]^{\theta-\sigma}\,[\sigma]^{\sigma}}\ \frac{1}{2^{\sigma}}\,B_{\theta,\,\sigma}\,[\gamma+\sigma-1]^{\sigma}\,[r]^{\theta+\sigma}\,[\beta+\gamma-2\theta]^{r-\theta-\sigma}\\
&\quad\vdots\\
&+\quad\frac{(-)^{\theta}}{[\theta]^{\theta}}\qquad \frac{1}{2^{\theta}}\,B_{\theta,\,\theta}\,[\gamma+\theta-1]^{\theta}\quad [r]^{2\theta}\quad [\beta+\gamma-2\theta]^{r-2\theta},
\end{aligned}$$

ce qui est l'expression de $L_{r,\theta}$ dans la formule

$$\begin{aligned} D_x{}^r Q^\beta B^\gamma = (-)^{-r} \quad & L_{r,0} \quad Q^{\beta-\gamma} \quad R^\gamma \\ & \vdots \\ + (-)^{r-\theta} & \frac{1}{2^\theta} L_{r,\theta} \Lambda^\theta Q^{\beta-r-\theta} R^{\gamma-\theta} \\ & \vdots \\ + \quad & \frac{1}{2^r} L_{r,r} \Lambda^r Q^{\beta-2r} R^{\gamma-r}. \end{aligned}$$

§ III.

18. Je passe à présent au développement de

$$(P^{-1} Q^4 D_x)^r P^\alpha \quad Q^\beta R^\gamma, \quad (\alpha = \gamma - \beta),$$

ou, ce qui est la même chose,

$$(P^{-1} Q^4 D_x)^r P^{\gamma-\beta} Q^\beta R^\gamma.$$

On a comme auparavant

$$P^{-1} Q^4 D_x P^\alpha Q^\beta R^\gamma = \tfrac{1}{2}\beta P^\alpha Q^{\beta+2} R^\gamma - \gamma P^{\alpha-1} Q^{\beta+3} R^\gamma + 2\alpha P^{\alpha-2} Q^{\beta+4} R^\gamma,$$

et de là

$$\begin{aligned} P^{-1} Q^4 D_x P^{\gamma-\beta} Q^\beta R^\gamma &= \tfrac{1}{2}\beta P^{\gamma-\beta} Q^{\beta+2} R^\gamma - \gamma P^{\gamma-\beta-1} Q^{\beta+3} R^\gamma + 2(\gamma-\beta) P^{\gamma-\beta-2} Q^{\beta+4} R^\gamma, \\ &= \tfrac{1}{2}\beta (P-2Q) P^{\gamma-\beta-1} Q^{\beta+2} R^\gamma - (\gamma-\beta)(P-2Q) P^{\gamma-\beta-2} Q^{\beta+3} R^\gamma, \end{aligned}$$

c'est-à-dire

$$P^{-1} Q^4 D_x P^{\gamma-\beta} Q^\beta R^\gamma = \tfrac{1}{2}\beta P^{\gamma-\beta-1} Q^{\beta+2} R^{\gamma+1} - (\gamma-\beta) P^{\gamma-\beta-2} Q^{\beta+3} R^{\gamma+1}.$$

On peut donc écrire

$$\begin{aligned} (P^{-1} Q^4 D_x)^r P^{\gamma-\beta} Q^\beta R^\gamma = \quad & \frac{1}{2^r} N_{r,0} P^{\gamma-\beta-r} Q^{\beta+2r} R^{\gamma+r} \\ & \vdots \\ + & \frac{(-)^\theta}{2^{r-\theta}} N_{r,\theta} P^{\gamma-\beta-r-\theta} Q^{\beta+2r+\theta} R^{\beta+r} \\ & \vdots \\ + & (-)^r N_{r,r} P^{\gamma-\beta-2r} Q^{\beta+2r} R^{\gamma+r}, \end{aligned}$$

et on trouve pour $N_{r,\theta}$ l'équation aux différences

$$N_{r+1,\theta+1} - (\beta + 2r + \theta + 1) N_{r,\theta+1} - (\gamma - \beta - r - \theta) N_{r,\theta} = 0,$$

laquelle avec les conditions aux limites

$$N_{r,-1} = 0, \quad N_{r,r+1} = 0, \quad N_{0,0} = 1,$$

détermine le coefficient $N_{r,\theta}$.

On a, en particulier

$$N_{r+1,\,0} = (\beta + 2r) \qquad N_{r,\,0},$$
$$N_{r+1,\,1} = (\beta + 2r + 1) \quad N_{r,\,1} + (\gamma - \beta - r) \qquad N_{r,\,0},$$
$$N_{r+1,\,2} = (\beta + 2r + 2) \quad N_{r,\,2} + (\gamma - \beta - r - 1)\, N_{r,\,1},$$
$$\vdots$$
$$N_{r+1,\,r+1} = (\gamma - \beta - 2r)\, N_{r,\,r},$$

lesquelles donnent tout de suite

$$N_{r,\,0} = [\beta + 2r - 2,\ -2]^r, \quad N_{r,\,r} = [\gamma - \beta,\ -2]^r.$$

19. L'équation ressemble à celle pour $L_{r,\theta}$, et l'on pourrait croire que l'intégration s'accomplirait d'une manière semblable, mais cela n'est pas ainsi; car en considérant la seconde équation de la suite, c'est-à-dire

$$N_{r+1,\,1} - (\beta + 2r + 1)\, N_{r,\,1} = (\gamma - \beta - r)\, [\beta + 2r - 2,\ -2]^r,$$

en y mettant

$$N_{r,\,1} = [\beta + 2r - 1,\ -2]^r\, M_{r,\,1},$$

on obtient

$$[\beta + 2r + 1,\ -2]^{r+1}\, (M_{r+1,\,1} - M_{r,\,1}) = (\gamma - \beta - r)\, [\beta + 2r - 2,\ -2]^r,$$

et de là

$$M_{r+1,\,1} - M_{r,\,1} = \frac{(\gamma - \beta - r)\, [\beta + 2r - 2,\ -2]^r}{[\beta + 2r + 1,\ -2]^{r+1}};$$

mais les facteurs du numérateur, et du dénominateur ne sont pas ici (ce qui arrivait dans l'équation pour la quantité dénotée auparavant par le même symbole $M_{r,\,1}$) de la même forme, et il n'y a pas de simplification dans la forme de la fraction. En écrivant successivement $r-1$, $r-2, \ldots 2, 1, 0$ pour r on trouve

$$M_{r,\,1} = \gamma - \beta + \frac{(\gamma - \beta - 1)\,\beta}{\beta + 3} + \frac{(\gamma - \beta - 2)(\beta + 2)\,\beta}{(\beta + 5)(\beta + 3)}$$
$$\ldots + \frac{(\gamma - \beta - r + 1)(\beta + 2r - 4)\ldots(\beta + 2)\,\beta}{(\beta + 2r - 1)\ldots(\beta + 5)(\beta + 3)},$$

et de là

$$N_{r,\,1} = (\beta + 2r - 1)\ldots(\beta + 5)(\beta + 3)\left(\gamma - \beta + \frac{(\gamma - \beta - 1)\,\beta}{\beta + 3} + \frac{(\gamma - \beta - 2)(\beta + 2)\,\beta}{(\beta + 5)(\beta + 3)}\right.$$
$$\left.\ldots + \frac{(\gamma - \beta - r + 1)(\beta + 2r - 4)\ldots(\beta + 2)\,\beta}{(\beta + 2r - 1)\ldots(\beta + 5)(\beta + 3)}\right).$$

Il ne paraît pas que la suite en [ ] puisse être additionnée, et cela m'empêche de pousser plus loin la recherche de la forme des coefficients $N_{r,\,\theta}$; la solution, telle que je l'ai trouvée, est donc donnée par l'expression de $(P^{-1} Q^4 D_x)^r\, P^{\gamma - \beta} Q^\beta R^\gamma$ en termes des coefficients numériques $N_{r,\,\theta}$, et par l'équation aux différences et conditions aux limites qui déterminent les coefficients $N_{r,\,\theta}$.

§ IV.

20. La première des identités du Prof. Donkin est

$$D_x^n Q^{-1} R^n = (-)^n \Lambda^n [n - \tfrac{1}{2}]^n Q^{-2n-1};$$

mais par une forme ci-dessus trouvée No. 9 en y écrivant n au lieu de γ on a

$$\frac{(-)^{n+1}}{n} D_x^{n+1} \quad R^n = \Lambda^n [n - \tfrac{1}{2}]^n Q^{-2n-1},$$

et, en remarquant que $D_x R^n = -nQ^{-1} R^n$, on voit que les deux formules sont identiques, et la vérité du théorème est ainsi démontrée.

21. La seconde des deux identités est

$$P^{-n+s-1} Q^{2n+2} D_x^n Q^{-1} R^s = (P^{-1} Q^4 D_x)^n P^{n+s-1} Q^{-2n+1} R^{-n+s}.$$

On a par la formule (No. 10) en y mettant s au lieu de γ et $n+1$ au lieu de r,

$$\begin{aligned}\frac{(-)^{n+1}}{s} D_x^{n+1} R^s = &\quad L'_{n+1,0} \quad Q^{-n-1} \quad R^s \\ &\quad \vdots \\ &+ L'_{n+1,\theta} \Lambda^\theta \, Q^{-n-1-\theta} R^{s-\theta} \\ &\quad \vdots \\ &+ L'_{n+1,n} \Lambda^n Q^{-2n-1} \quad R^{s-n},\end{aligned}$$

où

$$L'_{n+1,\theta} = \frac{[\theta - \tfrac{1}{2}]^\theta [n+\theta]^{n-\theta} [s - \theta - 1]^{n-\theta}}{[n-\theta]^{n-\theta}}.$$

22. En renversant l'ordre des termes et mettant aussi $L'_{n+1,\theta} = V_{n-\theta}$, de manière que

$$V_\theta = \frac{[n - \tfrac{1}{2} - \theta]^{n-\theta} [2n - \theta]^\theta [s - n + \theta - 1]^\theta}{\lfloor \theta \rfloor^\theta},$$

on obtient

$$\begin{aligned}\frac{(-)^{n+1}}{s} D_x^{n+1} R^s = &\quad V_0 \Lambda^n \quad Q^{-2n-1} \quad R^{s-n} \\ &\quad \vdots \\ &+ V^\theta \Lambda^{n-\theta} Q^{-2n-1+\theta} R^{s-n+\theta} \\ &\quad \vdots \\ &+ V_n \qquad Q^{-n-1} \quad R^s,\end{aligned}$$

et en remarquant que $D_x R^s = -sQ^{-1}R^s$, et que l'on a aussi $\Lambda = P^2 - 4Q^2$ et $R = P - 2Q$, de manière que $\Lambda R^{-1} = (P+2Q)$, l'équation peut s'écrire sous la forme

$$\begin{aligned}Q^{2n+2} P^{-n+s-1} D_x^n Q^{-1} R = (-)^n \times &\; V_0 (P+2Q)^n \quad P^{-n+s-1} Q R^s \\ &\quad \vdots \\ &+ V_\theta (P+2Q)^{n-\theta} P^{-n+s-1} Q^{1+\theta} R^s \\ &\quad \vdots \\ &+ V_n \qquad\qquad P^{-n+s-1} Q^{1+n} R^s,\end{aligned}$$

et en développant les binomiels on obtient pour $Q^{2n+2} P^{-n+s-1} D_x{}^n Q^{-1} R^s$ la valeur $(-)^n \times$

$$(V_0) P^{s-1} QR^s + \left(\frac{[n]^1}{[1]^1} 2^1 V_0 + V_1\right) P^{s-2} Q^2 R^s + \left(\frac{[n]^2}{[2]^2} 2^2 V_0 + \frac{[n-1]^1}{[1]^1} 2^1 V_1 + V_2\right) P^{s-3} Q^3 R^s$$

$$\ldots + (2^n V_0 + 2^{n-1} V_1 \ldots + 2^1 V_{n-1} + V_n) P^{s-n-1} Q^{n+1} R^s.$$

Cela devrait être égal à

$$(P^{-1} Q^4 D_x)^n P^{n+s-1} Q^{-2n+1} R^{-n+s}$$

et le développement de cette dernière expression se déduit de celui (No. 18) de

$$(P^{-1} Q^4 D_x)^r P^{\gamma-\beta} Q^\gamma R^\beta$$

en y écrivant n, $-2n+1$, $-n+s$ au lieu de r, β, γ; la valeur sera ainsi

$$= \frac{1}{2^n} N_{n,0} P^{s-1} QR^s - \frac{1}{2^{n-1}} N_{n,1} P^{s-2} Q^2 R^s + \frac{1}{2^{n-2}} N_{n,2} P^{s-3} Q^3 R^s \ldots + (-)^n N_{n,n} P^{s-n-1} Q^{n+1} R^s,$$

laquelle a la même forme que le développement de la fonction au côté gauche de l'équation, et l'identité des deux expressions dépend des équations

$$\frac{1}{2^n} N_{n,0} = (-)^n V_0,$$

$$\frac{1}{2^{n-1}} N_{n,1} = (-)^{n-1} \left(\frac{[n]^1}{[1]^1} 2^1 V_0 + V_1\right),$$

$$\frac{1}{2^{n-2}} N_{n,2} = (-)^{n-2} \left(\frac{[n]^2}{[2]^2} 2^2 V_0 + \frac{[n-1]^1}{[1]^1} 2^1 V_1 + V_2\right),$$

$$\vdots$$

$$N_{n,n} = (2^n V_0 + 2^{n-1} V_1 \ldots + 2 V_{n-1} + V_n);$$

mais comme on ne sait pas la forme générale des coefficients $N_{n,\theta}$ je n'ai pas pu vérifier complètement ces équations. On a cependant en mettant n, $-2n+1$, $-n+s$ pour r, β, γ

$$N_{n,0} = [-1, +2]^n, = (-)^n [2n-1, -2]^n, = (-1)^n 2^n [n-\tfrac{1}{2}]^n,$$
$$N_{n,1} = s[-3, +2]^{n-1}, = (-)^{n-1} s [2n-1, -2]^{n-1}, = (-)^{n-1} 2^{n-1} s [n-\tfrac{1}{2}]^{n-1},$$

et les deux premières équations seront

$$[n-\tfrac{1}{2}]^n = [n-\tfrac{1}{2}]^n; \quad s[n-\tfrac{1}{2}]^{n-1} = \frac{n}{1} 2 [n-\tfrac{1}{2}]^n + [n-\tfrac{3}{2}]^{n-1} (2n-1)(s-n),$$

où dans la seconde équation l'expression au côté droit est

$$[2n + 2(s-n)] [n-\tfrac{1}{2}]^n = 2s [n-\tfrac{1}{2}]^n = s [n-\tfrac{1}{2}]^{n-1},$$

comme cela devrait être. La dernière équation de la suite est

$$N_{n,n} = 2^n V_0 + 2^{n-1} V_1 \ldots + 2 V_{n-1} + V_n$$

et comme on a $N_{n,n} = [n+s-1, -2]^n$, cette dernière équation peut aussi se vérifier.

252.

NOTE SUR L'ÉQUATION DES DIFFÉRENCES POUR UNE ÉQUATION DONNÉE DE DEGRÉ QUELCONQUE.

[From the *Annali di Matematica pura ed applicata* (Tortolini), tom. II. (1859), pp. 365, 366.]

Il s'agit de trouver l'équation qui a pour racines les carrés des différences des racines d'une équation donnée

$$(*)\,(v,\ 1)^n = 0.$$

En représentant cette équation par $\phi v = 0$, soient x, y deux racines différentes quelconques, on a non seulement $\phi x = 0$, $\phi y = 0$; mais aussi

$$\phi x + \phi y = 0,\ \frac{\phi x - \phi y}{x - y} = 0,$$

et en écrivant dans ces équations

$$x + y = 2s,\quad (x - y)^2 = 4\theta,$$

(ou ce qui est la même chose $x = s + \sqrt{\theta}$, $y = s - \sqrt{\theta}$) on obtient deux équations rationnelles en s, et θ, et en éliminant s on obtient l'équation qui donne

$$\theta = \tfrac{1}{4}(x - y)^2.$$

Il convient de changer un peu la forme des équations; en effet la première équation est du degré n, la seconde du degré $n - 1$ pour rapport à s, mais en écrivant les deux équations sous la forme

$$n(\phi x + \phi y) - (x + y)\frac{\phi x - \phi y}{x - y} = 0,\qquad \frac{\phi x - \phi y}{x - y} = 0,$$

l'une et l'autre équation sera du degré $n-1$ par rapport à s. La forme sous laquelle j'ai présenté la méthode de Bezout s'applique au problème; en représentant les deux équations par $Fs=0$, $Gs=0$ j'écris pour le moment

$$\phi(s+\sqrt{\theta})=A, \quad \phi(s-\sqrt{\theta})=B,$$

on a alors

$$Fs=n(A+B)-s\frac{A-B}{\sqrt{\theta}}, \quad Gs=\frac{A-B}{\sqrt{\theta}},$$

et en écrivant

$$\phi(s'+\sqrt{\theta})=A', \quad \phi(s'-\sqrt{\theta})=B',$$

on a de même

$$Fs'=n(A'+B')-\frac{s'(A'-B')}{\sqrt{\theta}}, \quad Gs'=\frac{A'-B'}{\sqrt{\theta}}.$$

On obtient de là

$$\Phi(s, s')=\frac{Fs\,Gs'-Fs'Gs}{s-s'},$$

$$=\left[\left(n(A+B)-\frac{s(A-B)}{\sqrt{\theta}}\right)\frac{A'-B'}{\sqrt{\theta}}-\left(n(A'+B')-\frac{s'(A'-B')}{\sqrt{\theta}}\right)\frac{(A-B)}{\sqrt{\theta}}\right]\frac{1}{s-s'},$$

ou, en réduisant,

$$-\Phi(s, s')=\frac{2n(AB'-A'B)}{(s-s')\sqrt{\theta}}+\frac{(A-B)(A'-B')}{\theta}.$$

Donc, en rétablissant les valeurs de A, B, A', B', on a la fonction

$$\frac{2n[\phi(s+\sqrt{\theta})\,\phi(s'-\sqrt{\theta})-\phi(s'+\sqrt{\theta})\,\phi(s-\sqrt{\theta})]}{(s-s')\sqrt{\theta}}$$
$$+\frac{[\phi(s+\sqrt{\theta})-\phi(s-\sqrt{\theta})]\,[\phi(s'+\sqrt{\theta})-\phi(s'-\sqrt{\theta})]}{\theta}$$

laquelle sera de la forme

$$\left(\begin{array}{l} a_{0,0},\ a_{0,1},\ \ldots\ a_{0,n-2} \\ a_{1,0},\ a_{1,1},\ \ldots \\ \vdots \\ a_{n-2,0},\ \ldots \end{array}\right)(s,\ 1)^{n-2}(s',\ 1)^{n-2},$$

où les coefficients a sont des fonctions rationnelles de θ, et en égalant à zero le déterminant formé avec ces coefficients on a l'équation qu'il s'agissait de trouver. Quoique cette solution soit analytiquement la plus simple, j'ai une autre méthode nouvelle plus adaptée au calcul, laquelle j'ai appliquée à trouver l'équation des différences pour l'équation quintique

$$(a,\ b,\ c,\ d,\ e,\ f\between v,\ 1)^5=0.$$

Londres, 4 *Fév.* 1860.

253.

SUR LA COURBE PARALLÈLE À L'ELLIPSE.

[From the *Annali di Matematica pura ed applicata* (Tortolini), tom. III. (1860), pp. 311—316.]

IL fut remarqué par Cauchy (*Comptes Rendus*, tom. XIII. [1841], p. 1063) que l'équation de cette courbe pourrait se trouver en éliminant θ entre les équations

$$\frac{a^2x^2}{(\theta+a^2)^2}+\frac{b^2y^2}{(\theta+b^2)^2}=1, \qquad \frac{\theta^2x^2}{(\theta+a^2)^2}+\frac{\theta^2y^2}{(\theta+b^2)^2}=k^2,$$

et cette élimination fut effectuée, et le résultat trouvé sous la forme la plus simple par M. Catalan (*Terquem*, tom. III. 1844, p. 553); mais pour faire l'élimination de la manière la plus facile je remarque que ces deux équations donnent

$$\frac{x^2}{\theta+a^2}+\frac{y^2}{\theta+b^2}=1+\frac{k^2}{\theta}, \qquad \frac{x^2}{(\theta+a^2)^2}+\frac{y^2}{(\theta+b^2)^2}=\frac{k^2}{\theta^2},$$

dont la seconde est la dérivée, par rapport à θ, de la première; or cette première équation est

$$(\theta+k^2)(\theta+a^2)(\theta+b^2)-x^2\,\theta\,(\theta+b^2)-y^2\,\theta\,(\theta+a^2)=0,$$

et, en égalant à zéro le discriminant de la fonction cubique, on aura l'équation de la courbe; en posant

$$A=x^2+y^2-k^2-a^2-b^2, \quad B=b^2x^2+a^2y^2-a^2k^2-b^2k^2-a^2b^2, \quad C=a^2b^2k^2,$$

la fonction cubique, multipliée par 3, sera

$$(3,\,-A,\,-B,\,3C\between\theta,\,1)^3;$$

on aura donc

$$4\,(A^2+3B)\,(B^2+3AC)-(AB-9C)^2=0,$$

ou ce qui est la même chose

$$A^2B^2+4B^3+4A^3C+18ABC-27C^2=0,$$

c'est-à-dire en substituant les valeurs de A, B, C, l'équation de la courbe sera

$$\begin{aligned}&(x^2+y^2-k^2-a^2-b^2)^2(b^2x^2+a^2y^2-a^2k^2-b^2k^2-a^2b^2)^2\\&+4(b^2x^2+a^2y^2-a^2k^2-b^2k^2-a^2b^2)^3\\&+4a^2b^2k^2(x^2+y^2-k^2-a^2-b^2)^3\\&+18a^2b^2k^2(x^2+y^2-k^2-a^2-b^2)(b^2x^2+a^2y^2-a^2k^2-b^2k^2-a^2b^2)\\&-27a^4b^4k^4=0,\end{aligned}$$

laquelle est en effet l'équation trouvée par M. Catalan.

En écrivant $k=0$ l'équation devient

$$0=(b^2x^2+a^2y^2-a^2b^2)\left[(x^2+y^2)^2-2(a^2-b^2)(x^2+y^2)+(a^2-b^2)^2\right],$$

ce qui équivaut à l'équation de l'ellipse

$$b^2x^2+a^2y^2-a^2b^2=0$$

deux fois répétée, et aux équations

$$(x\pm ae)^2+y^2\ \ =0,$$

(où comme à l'ordinaire $a^2e^2=a^2-b^2$) lesquelles appartiennent aux droites menées chacune par un foyer réel et un foyer imaginaire de l'ellipse; ces droites sont aussi les tangentes menées à l'ellipse par les quatre foyers. En effet en prenant sur l'ellipse le point imaginaire dont les coordonnées sont

$$x=\frac{a}{e},\quad y=ia\left(\frac{1}{e}-e\right),\ (i=\sqrt{-1})$$

l'équation du cercle, rayon 0, ayant pour centre le point dont il s'agit, sera

$$\left(x-\frac{a}{e}\right)^2+\left[y-ia\left(\frac{1}{e}-e\right)\right]^2=0\,;$$

ce cercle se réduit donc à deux droites, savoir la droite

$$x-\frac{a}{e}+i\left[y-ia\left(\frac{1}{e}-e\right)\right]=0,$$

ou, ce qui est la même chose,

$$x-ae+iy=0,$$

laquelle est tangente à l'ellipse, et la droite

$$\left(x-\frac{a}{e}\right)-i\left[y-ia\left(\frac{1}{e}-e\right)\right]=0,$$

ou, ce qui est la même chose,

$$x - a\left(\frac{2}{e} - e\right) - iy = 0,$$

laquelle est la droite menée par le point de contact à l'autre point circulaire à l'infini (celui qui n'est pas situé sur la tangente $x - ae + iy = 0$). Le cercle, comme courbe composée d'une droite tangente à l'ellipse et d'une autre droite menée par le point de contact, a même avec l'ellipse une intersection à trois points réunis. Ces considérations font voir pourquoi au cas $k = 0$ les quatre droites $(x \pm ae)^2 + y^2 = 0$ font partie de la courbe.

En supposant que k est quelconque, mais que l'on a $a = b$, l'équation devient

$$a^4 (x^2 + y^2)^2 [(x^2 + y^2 - a^2)^2 - 2k^2 (x^2 + y^2 + a^2) + k^4] = 0,$$

où, en écartant le facteur $(x^2 + y^2)^2$ et le facteur constant a^4, l'équation se réduit à

$$(x^2 + y^2 - a^2)^2 - 2k^2 (x^2 + y^2 + a^2) + k^4 = 0,$$

c'est-à-dire

$$[x^2 + y^2 - (a + k)^2] [x^2 + y^2 - (a - k)^2] = 0;$$

ainsi la courbe se réduit à l'ensemble des deux droites $(x^2 + y^2) = 0$ (chacune deux fois répétée, et des deux cercles

$$x^2 + y^2 - (a + k)^2 = 0, \quad x^2 + y^2 - (a - k)^2 = 0$$

comme évidemment cela doit être.

Par rapport aux singularités de la courbe, la forme de l'équation montre à premier coup d'œil qu'il y a quatre points doubles à l'infini; savoir les deux points circulaires à l'infini (points où l'infini est rencontré par les droites $x^2 + y^2 = 0$), et les deux points où l'infini est rencontré par les droites $b^2x^2 + a^2y^2 = 0$. Il y a de plus deux points doubles sur chacun des axes: en effet en écrivant dans l'équation de la courbe $y = 0$ l'équation résultante, toute réduction faite, devient

$$[(x - a)^2 - k^2] [(x + a)^2 + k^2] [b^2x^2 - (a^2 - b^2) (b^2 - k^2)]^2 = 0,$$

ce qui fait voir qu'il y a sur l'axe de x les deux points doubles dont les coordonnées sont données par l'équation

$$b^2x^2 - (a^2 - b^2) (b^2 - k^2) = 0.$$

Mais pour parvenir à cette conclusion il convient de considérer que l'axe de x rencontre la courbe dans des points tels que pour chacun la distance normale à l'ellipse est égale à k: il y a quatre distances normales k, mesurées le long de l'axe; et quatre distances normales k mesurées à des points situés symétriquement par rapport à l'axe, lesquels se rencontrent deux à deux sur l'axe aux points dont les coordonnées sont $b^2x^2 = (a^2 - b^2) (b^2 - k^2)$; c'est là l'origine des deux points doubles sur l'axe de x. On doit aussi remarquer que les coordonnées des points sur l'ellipse qui correspondent à ces points doubles sont données par les équations

$$x^2 = \frac{a^4}{b^2}\frac{b^2-k^2}{a^2-b^2}, \quad y^2 = \frac{a^2k^2-b^4}{a^2-b^2};$$

on voit de là que les points sur l'ellipse qui donnent lieu aux deux points doubles sur l'axe de x, ne seront réels à moins que $k > \frac{b^2}{a^3}$, $< b$; les deux points doubles seront cependant réels pour toute valeur $k > 0 < b$. J'ajoute que pour les valeurs $k \overline{>} 0$, $< \frac{b^2}{a}$, les deux points doubles seront des points conjugués ou isolés; pour $k = \frac{b^2}{a}$, chacun de ces points se réunit à la courbe, et à deux points de rebroussement, mais il n'y a pas de singularité visible dans la forme de la courbe; pour $k > \frac{b^2}{a} < b$ il y a deux points doubles avec des branches réelles; et pour $k = b$, ces deux points viennent se réunir au centre de l'ellipse et il y a deux branches ayant pour tangente commune à ce point l'axe de x; enfin pour $k > b$, les deux points doubles sur l'axe de x deviennent imaginaires.

Des considérations pareilles s'appliquent aux points doubles sur l'axe de y; pour $x = 0$ l'équation de la courbe se transforme en

$$[(y-b)^2 - k^2]\,[(y+b)^2 + k^2]\,[a^2y^2 - (a^2-b^2)(k^2-a^2)]^2 = 0;$$

il y a donc sur l'axe de y deux points doubles dont les coordonnées sont données par

$$a^2y^2 - (a^2-b^2)(k^2-a^2) = 0,$$

ces points correspondent à des points sur l'ellipse dont les coordonnées sont données par

$$x^2 = \frac{a^4-b^2k^2}{a^2-b^2}, \quad y^2 = \frac{b^4}{a^2}\frac{k^2-a^2}{a^2-b^2};$$

les points sur l'ellipse ne seront donc réels à moins que $k > \frac{a^2}{b}$; les deux points doubles seront cependant réels pour toute valeur $k > a$. Pour $k < a$ les deux points doubles seront imaginaires. Pour $k = a$ le centre est point de réunion de deux points doubles et il y a deux branches qui ont à ce point pour tangente commune l'axe de y: pour $k = \frac{a^2}{b}$, chacun des deux points doubles se réunit à deux points de rebroussement; mais il n'y a pas de singularité visible dans la forme de la courbe; enfin pour $k > \frac{a^2}{b}$ les deux points doubles se détachent de la courbe et deviennent des points conjugués ou isolés. En résumé, il y a huit points doubles, savoir quatre points doubles à l'infini, et quatre points doubles sur les deux axes.

Il y a de plus 12 points de rebroussement; ces points sont en effet les centres de courbure aux points de l'ellipse pour lesquels le rayon de courbure est égal à k;

les douze points seront tous imaginaires à moins que k n'ait une valeur entre les limites $\frac{b^2}{a}$, $\frac{a^2}{b}$: pour une telle valeur de k les douze points seront 4 réels, et 8 imaginaires. Pour $k=\frac{b^2}{a}$ les quatre points réels se réunissent deux à deux aux deux points doubles sur l'axe de x; pour $k=\frac{a^2}{b}$, ils se réunissent deux à deux aux deux points doubles sur l'axe de y; cela s'accorde avec ce qui est dit ci-dessus par rapport aux points doubles. On peut encore trouver que le nombre des points de rebroussement est 12 à moyen des équations

$$A^2+3B=0,\quad B^2+3AC=0,\quad AB-9C=0,$$

lesquelles appartiennent chacune à une courbe du quatrième ordre qui passe par les points de rebroussement; la forme des équations fait voir que ces courbes ont en commun 12 points, et seulement 12 points, d'intersection. C'est là en effet la manière la plus simple de trouver les coordonnées des points de rebroussement; car ces équations donnent tout de suite

$$A=-3C^{\frac{1}{3}},\qquad B=-3C^{\frac{2}{3}},$$

c'est-à-dire on a les équations

$$\begin{aligned} x^2+\ \ y^2 &= a^2\ \ +\ b^2\ +\ \ \ k^2-3\,(abk)^{\frac{2}{3}},\\ b^2x^2+a^2y^2 &= a^2b^2+(a^2+b^2)\,k^2-3\,(abk)^{\frac{4}{3}} \end{aligned}$$

qui donnent les coordonnées des douze points de rebroussement.

Pour vérifier que ces points correspondent aux points de l'ellipse pour lesquels le rayon de courbure est égal à k; je prends $a\cos\theta$, $b\sin\theta$ pour les coordonnées d'un point sur l'ellipse, les coordonnées du centre de courbure seront données par

$$ax=(a^2-b^2)\cos^3\theta,\qquad by=-(a^2-b^2)\sin^3\theta,$$

et en supposant que le rayon de courbure soit égal à k, on a

$$k=\frac{1}{ab}\,(a^2\sin^2\theta+b^2\cos^2\theta)^{\frac{3}{2}},$$

et de là

$$a^2\sin^2\theta+b^2\cos^2\theta=(abk)^{\frac{2}{3}}.$$

Cela donne

$$(a^2-b^2)\cos^2\theta=a^2-(abk)^{\frac{2}{3}},\qquad -(a^2-b^2)\sin^2\theta=b^2-(abk)^{\frac{2}{3}},$$

et de là on déduit

$$x^2+y^2=\frac{1}{a^2-b^2}\left[\frac{1}{a^2}\left(a^2-(abk)^{\frac{2}{3}}\right)^3-\frac{1}{b^2}\left(b^2-(abk)^{\frac{2}{3}}\right)^3\right]$$

$$b^2x^2+a^2y^2=\frac{1}{a^2-b^2}\left[\frac{b^2}{a^2}\left(a^2-(abk)^{\frac{2}{3}}\right)^3-\frac{a^2}{b}\left(b^2-(abk)^{\frac{2}{3}}\right)^3\right]$$

valeurs qui s'accordent en effet avec les valeurs ci-dessus trouvées.

La courbe parallèle à l'ellipse est de l'ordre 8 et il y a 8 points doubles, et 12 points de rebroussement; elle sera donc de la classe $56-16-36=4$. En effet pour mener d'un point donné une tangente à la courbe parallèle on peut décrire avec ce point comme centre un cercle rayon k, pour mener une tangente commune au cercle et à l'ellipse; la droite, menée par le point donné, parallèle à la tangente commune sera la tangente cherchée: on peut donc par un point donné mener 4 tangentes à la courbe, ou la courbe parallèle est de la classe 4. Au reste on peut trouver très facilement l'équation de la courbe parallèle en coordonnées tangentielles. Car en représentant par θ l'inclination d'une tangente quelconque de l'ellipse à l'axe de x, l'équation de cette tangente est

$$x\cos\theta+y\sin\theta \qquad -\sqrt{a^2\cos^2\theta+b^2\sin^2\theta}=0,$$

et l'équation de la tangente parallèle de la courbe sera ainsi

$$x\cos\theta+y\sin\theta-k-\sqrt{a^2\cos^2\theta+b^2\sin^2\theta}=0;$$

donc en représentant cette équation par $\xi x+\eta y+\zeta=0$ on aura

$$\xi:\eta:\zeta=\cos\theta:\sin\theta:k+\sqrt{a^2\cos^2\theta+b^2\sin^2\theta},$$

et de là

$$\zeta+k\sqrt{\xi^2+\eta^2}+\sqrt{a^2\xi^2+b^2\eta^2}=0,$$

laquelle est l'équation dont il s'agit; la forme rationnelle est

$$(a^2-k^2)^2\xi^4+(b^2-k^2)^2\eta^4+2(a^2-k^2)(b^2-k^2)\xi^2\eta^2-2(a^2+k^2)\zeta^2\xi^2-2(b^2+k^2)\zeta^2\eta^2+\zeta^4=0,$$

équation du quatrième ordre comme cela devait être.

J'ai cru qu'il n'était pas nécessaire de faire voir comment on pourrait à moyen de l'équation tracer la courbe; en effet on trouve les différentes formes assez facilement par des considérations géométriques, en considérant la courbe comme la développée de l'évolute de l'ellipse: on se rend compte à ce moyen de ce qui est déjà dit par rapport aux points doubles et aux points de rebroussement.

Je remarque enfin que l'équation d'une normale quelconque de l'ellipse est

$$ax\sin\theta-by\cos\theta=(a^2-b^2)\sin\theta\cos\theta$$

(où θ est un paramètre arbitraire): donc en considérant la trajectoire orthogonale de ce système de droites, on obtient

$$(xdx+ydy)\sqrt{b^2dx^2+a^2dy^2}+(a^2-b^2)\,dx\,dy=0$$

comme équation différentielle de la courbe parallèle à l'ellipse. L'intégrale de cette équation est donc l'équation du huitième ordre (contenant k comme constante arbitraire) qu'on a ci-dessus trouvée.

Londres, 2 *Stone Buildings*, 22 Oct. 1860.

254.

SUR LA SURFACE PARALLÈLE À L'ELLIPSOIDE.

[From the *Annali di Matematica pura ed applicata*, (Tortolini), tom. III. (1860), pp. 345—352.]

Il y a pour la surface parallèle à l'ellipsoïde une solution tout à fait semblable à celle pour la courbe parallèle à l'ellipse. En effet soit

$$\frac{X^2}{a^2}+\frac{Y^2}{b^2}+\frac{Z^2}{c^2}=1$$

l'équation de l'ellipsoïde; k la distance de la surface parallèle, en prenant cette distance sur la normale extérieure au point (X, Y, Z) et en écrivant pour abréger

$$\sqrt{\frac{X^2}{a^4}+\frac{Y^2}{b^4}+\frac{Z^2}{c^4}}=\frac{k}{\lambda},$$

on trouve pour les coordonnées (x, y, z) de l'extrémité de la normale

$$x=X\left(1+\frac{\lambda}{a^2}\right),\quad y=Y\left(1+\frac{\lambda}{b^2}\right),\quad z=Z\left(1+\frac{\lambda}{c^2}\right),$$

et, en substituant pour X, Y, Z les valeurs données par ces équations, on obtient

$$\frac{a^2x^2}{(a^2+\lambda)^2}+\frac{b^2y^2}{(b^2+\lambda)^2}+\frac{c^2z^2}{(c^2+\lambda)^2}=1,$$

$$\frac{\lambda^2x^2}{(a^2+\lambda)^2}+\frac{\lambda^2y^2}{(b^2+\lambda)^2}+\frac{\lambda^2z^2}{(c^2+\lambda)^2}=k^2.$$

Or ces équations donnent

$$\frac{x^2}{a^2+\lambda}+\frac{y^2}{b^2+\lambda}+\frac{z^2}{c^2+\lambda}=1+\frac{k^2}{\lambda},$$

$$\frac{x^2}{(a^2+\lambda)^2}+\frac{y^2}{(b^2+\lambda)^2}+\frac{z^2}{(c^2+\lambda)^2}=\frac{k^2}{\lambda^2},$$

et la seconde équation est la dérivée par rapport à λ de la première. C'est-à-dire en obtient l'équation de la surface parallèle en égalant à zéro le discriminant par rapport à λ de l'équation

$$\frac{x^2}{a^2+\lambda}+\frac{y^2}{b^2+\lambda}+\frac{z^2}{c^2+\lambda}=1+\frac{k^2}{\lambda},$$

ou autrement dit, la surface parallèle est l'enveloppe par rapport à λ de cette équation. En multipliant par $\lambda(a^2+\lambda)(b^2+\lambda)(c^2+\lambda)$, on aura une équation du quatrième ordre, dont le discriminant égalé à zéro donne l'équation de la surface: on voit sans peine que cette équation sera de l'ordre 10.

J'avais à peine trouvé cette méthode, quand j'ai appris que le problème était déjà résolu par MM. Salmon et W. Roberts. M. Salmon considère la surface comme lieu des centres des sphères, rayon k, qui touchent à l'ellipsoïde $\frac{X^2}{a^2}+\frac{Y^2}{b^2}+\frac{Z^2}{c^2}=1$, ce qui lui donne tout de suite le théorème que voici; savoir on obtient l'équation de la surface parallèle, en égalant à zéro le discriminant par rapport à λ du discriminant par rapport à (X, Y, Z) de l'équation

$$\lambda\left(\frac{X^2}{a^2}+\frac{Y^2}{b^2}+\frac{Z^2}{c^2}-1\right)+(X-x)^2+(Y-y)^2+(Z-z)^2=0;$$

cette solution est donnée en passant dans une Note publiée dans le No. de Septembre 1860 des Nouvelles Annales (Terquem et Gerono). En formant le discriminant par rapport à X, Y, Z on trouve l'équation en λ ci-dessus donnée.

La solution de M. W. Roberts n'avait pas été publiée: il a bien voulu me permettre d'en faire part aux géomètres. En considérant une section circulaire de l'ellipsoïde, l'enveloppe des sphères, rayon k, ayant leurs centres sur cette section, sera une surface annulaire, et l'enveloppe des surfaces annulaires qui correspondent aux différentes

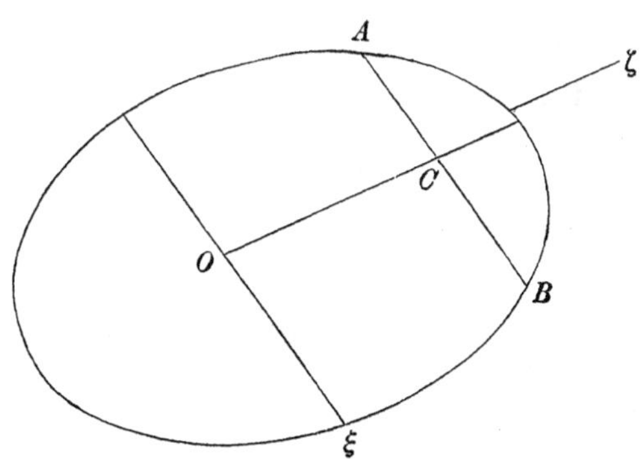

sections circulaires de l'ellipsoïde sera la surface parallèle. Dans la figure (qui est censée située dans le plan des axes le plus grand et le plus petit) soit $O\xi$ la trace de la section circulaire centrale, $O\zeta$ le diamètre perpendiculaire, AB la trace d'une section quelconque parallèle à la section circulaire centrale: soit de plus $AB=2\delta$, et (α, γ)

les coordonnées du point C qui est le centre de la section AB. En prenant pour un moment ce point C pour origine, et l'axe de η parallèle à l'axe moyen, la surface annulaire sera l'enveloppe de la sphère

$$(\xi-\delta\cos\phi)^2+(\eta-\delta\sin\phi)^2+\zeta^2=k^2,$$

(où ϕ est un paramètre arbitraire); donc l'équation de la surface annulaire sera

$$4\delta^2(\xi^2+\eta^2)=(\xi^2+\eta^2+\zeta^2+\delta^2-k^2)^2,$$

ou, en prenant le centre de l'ellipsoïde pour origine, cette équation sera

$$4\delta^2(\xi^2+\eta^2-2\alpha\xi+\alpha^2)=(\xi^2+\eta^2+\zeta^2+\delta^2-k^2-2\alpha\xi-2\gamma\zeta-\alpha^2-\gamma^2)^2.$$

Mais par les propriétés de l'ellipse on a $\gamma=l\alpha$ et $\delta^2=m+n\alpha^2$, où l, m, n sont des quantités constantes, fonctions des axes a, b, c de l'ellipsoïde. L'équation de la surface annulaire devient ainsi

$$4(m+n\alpha^2)(\xi^2+\eta^2-2\alpha\xi+\alpha^2)=[\xi^2+\eta^2+\zeta^2+m-k^2-2\alpha(\xi+l\zeta)+\alpha^2(n+1+l^2)]^2,$$

équation qui contient le paramètre variable α au quatrième degré, et en égalant à zéro le discriminant par rapport à α de cette équation, on obtiendrait l'équation de la surface parallèle. À propos de cette solution je remarque qu'en considérant la surface annulaire, enveloppe des sphères, rayon k, ayant leurs centres sur l'ellipse

$$\frac{X^2}{a^2}+\frac{Y^2}{b^2}-1=0,$$

on obtient sans peine le système d'équations

$$\frac{a^2x^2}{(a^2+\lambda)^2}+\frac{b^2y^2}{(b^2+k)^2}=1,\qquad \frac{\lambda^2x^2}{(a^2+\lambda)^2}+\frac{\lambda^2y^2}{(c^2+\lambda)^2}=k^2-z^2,$$

et de là l'équation de la surface annulaire s'obtient en égalant à zéro le discriminant par rapport à λ de l'équation

$$\frac{x^2}{a^2+\lambda}+\frac{y^2}{b^2+\lambda}=1+\frac{k^2-z^2}{\lambda},$$

c'est-à-dire en égalant à zéro le discriminant d'une fonction *cubique* de λ. Mais en supposant que l'ellipse devient un cercle, l'on n'aura qu'une fonction quadratique de λ; c'est là pourquoi la surface annulaire considérée par M. Roberts s'exprime sous une forme assez simple pour qu'on puisse s'en servir pour trouver l'équation de la surface parallèle. Soit à présent l'ellipse, section de l'ellipsoïde

$$\frac{X^2}{a^2}+\frac{Y^2}{b^2}+\frac{Z^2}{c^2}=1$$

par un plan quelconque

$$lX+mY+nZ=p;$$

il est clair que l'équation de la surface annulaire doit se trouver de même à moyen d'une fonction cubique de λ. Pour vérifier cela, je remarque que l'on a à trouver l'enveloppe de la sphère $(X-x)^2+(Y-y)^2+(Z-z)^2=k^2$, où les paramètres X, Y, Z sont liés par les deux équations qui viennent d'être mentionnées: cela donne tout de suite pour la surface annulaire le système des équations

$$x - X - \lambda \frac{X}{a^2} - \mu l \;= 0,$$

$$y - Y - \lambda \frac{Y}{b^2} - \mu m = 0,$$

$$z - Z - \lambda \frac{Z}{c^2} - \mu n \;= 0,$$

où λ, μ sont des paramètres arbitraires: ces équations donnent

$$X = \frac{a^2(x-\mu l)}{a^2+\lambda}, \quad Y = \frac{b^2(y-\mu m)}{b^2+\lambda}, \quad Z = \frac{c^2(z-\mu n)}{c^2+\lambda},$$

et de là, en mettant pour abréger

$$\frac{la^2x}{a^2+\lambda} + \frac{mb^2y}{b^2+\lambda} + \frac{nc^2z}{c^2+\lambda} - p = P,$$

$$\frac{l^2a^2}{a^2+\lambda} + \frac{m^2b^2}{b^2+\lambda} + \frac{n^2c^2}{c^2+\lambda} \qquad = Q,$$

on trouve à moyen de l'équation linéaire

$$Q\mu - P = 0, \text{ où } \mu = \frac{P}{Q},$$

et les deux autres équations donnent alors

$$\frac{a^2(x-\mu l)^2}{(a^2+\lambda)^2} + \frac{b^2(y-\mu m)^2}{(b^2+\lambda)^2} + \frac{c^2(z-\mu n)^2}{(c^2+\lambda)^2} = 1,$$

$$\frac{(\lambda x + a^2\mu l)^2}{(a^2+\lambda)^2} + \frac{(\lambda y + b^2\mu m)^2}{(b^2+\lambda)^2} + \frac{(\lambda z + c^2\mu n)^2}{(c^2+\lambda)^2} = k^2,$$

où μ est censé dénoter la valeur $\frac{P}{Q}$. Je déduis de là les équations

$$\frac{x^2}{a^2+\lambda} + \frac{y^2}{b^2+\lambda} + \frac{z^2}{c^2+\lambda} + \frac{\mu^2}{\lambda} Q = 1 + \frac{k^2}{\lambda},$$

$$\begin{aligned} &\frac{x^2}{(a^2+\lambda)^2} + \frac{y^2}{(b^2+\lambda)^2} + \frac{z^2}{(c^2+\lambda)^2} \\ &+ 2\left(\frac{a^2lx}{(a^2+\lambda)^2} + \frac{b^2my}{(b^2+\lambda)^2} + \frac{c^2nz}{(c^2+\lambda)^2}\right)\frac{\mu}{\lambda} \\ &+ \left(\frac{a^4l^2}{(a^2+\lambda)^2} + \frac{b^4m^2}{(b^2+\lambda)^2} + \frac{c^4n^2}{(c^2+\lambda)^2}\right) \qquad = \frac{k^2}{\lambda^2}. \end{aligned}$$

Soient P', Q' les dérivées de P, Q par rapport à λ, on a

$$-P' = \frac{la^2x}{(a^2+\lambda)^2} + \frac{mb^2y}{(b^2+\lambda)^2} + \frac{nc^2z}{(a^2+\lambda)^2},$$

$$-Q' = \frac{l^2a^2}{(a^2+\lambda)^2} + \frac{m^2b^2}{(b^2+\lambda)^2} + \frac{n^2c^2}{(c^2+\lambda)^2},$$

et de là aussi

$$Q+\lambda Q' = \frac{a^4l^2}{(a^2+\lambda)^2} + \frac{b^4m^2}{(b^2+\lambda)^2} + \frac{c^4n^2}{(c^2+\lambda)^2}.$$

A moyen de ces équations, et en substituant pour μ la valeur $\frac{Q}{P}$, les deux équations ci-dessus données deviennent

$$\left(\frac{x^2}{a^2+\lambda} + \frac{y^2}{b^2+\lambda} + \frac{z^2}{c^2+\lambda} - 1 - \frac{k^2}{\lambda}\right) Q + \frac{P^2}{\lambda} = 0,$$

$$\left(\frac{x^2}{(a^2+\lambda)^2} + \frac{y^2}{(b^2+\lambda)^2} + \frac{z^2}{(c^2+\lambda)^2} - \frac{k^2}{\lambda^2}\right) Q - 2\frac{PP'}{\lambda} + (Q+\lambda Q')\frac{P^2}{Q\lambda^2} = 0;$$

en différentiant la première équation par rapport à λ on obtient

$$-\left(\frac{x^2}{(a^2+\lambda)^2} + \frac{y^2}{(b^2+\lambda)^2} + \frac{z^2}{(c^2+\lambda)^2} - \frac{k^2}{\lambda^2}\right) Q$$
$$+\left(\frac{x^2}{a^2+\lambda} + \frac{y^2}{b^2+\lambda} + \frac{z^2}{c^2+\lambda} - 1 - \frac{k^2}{\lambda}\right) Q' + \frac{2PP'}{\lambda} - \frac{P^2}{\lambda^2} = 0,$$

et cette équation se réduit à une identité à moyen des deux équations: donc l'équation de la surface annulaire s'obtient en égalant à zéro le discriminant par rapport à λ de l'équation

$$\left(\frac{x^2}{a^2+\lambda} + \frac{y^2}{b^2+\lambda} + \frac{z^2}{c^2+\lambda} - 1 - \frac{k^2}{\lambda}\right) Q + \frac{P^2}{\lambda} = 0,$$

c'est-à-dire, en substituant pour P, Q leurs valeurs, l'équation sera

$$\left(\frac{x^2}{a^2+\lambda} + \frac{y^2}{b^2+\lambda} + \frac{z^2}{c^2+\lambda} - 1 - \frac{k^2}{\lambda}\right)\left(\frac{a^2l^2}{a^2+\lambda} + \frac{b^2m^2}{b^2+\lambda} + \frac{c^2n^2}{c^2+\lambda}\right)$$
$$+\frac{1}{\lambda}\left(\frac{a^2lx}{a^2+\lambda} + \frac{b^2my}{b^2+\lambda} + \frac{c^2nz}{c^2+\lambda} - p\right)^2 = 0.$$

Dans cette équation, en réunissant les termes

$$\frac{x^2}{a^2+\lambda}\,\frac{a^2l^2}{a^2+\lambda} + \frac{1}{\lambda}\,\frac{a^4l^2x^2}{(a^2+\lambda)^2},$$

on voit que ces termes se réduisent à

$$\frac{(a^2+\lambda)\,\frac{a^4l^2x^2}{\lambda}}{(a^2+\lambda)}, = \frac{a^4l^2x^2}{\lambda\,(a^2+\lambda)},$$

et de même pour les termes semblables en y, z. Donc en multipliant par

$$\lambda(a^2+\lambda)(b^2+\lambda)(c^2+\lambda)$$

les dénominateurs disparaissent, et l'équation est réellement de l'ordre 3 par rapport à λ.

La section sera circulaire en supposant

$$l=\sqrt{\frac{1}{b^2}-\frac{1}{a^2}},\quad m=0,\quad n=\sqrt{\frac{1}{c^2}-\frac{1}{b^2}};$$

cela donne

$$\frac{a^2l^2}{a^2+\lambda}+\frac{b^2m^2}{b^2+\lambda}+\frac{c^2n^2}{c^2+\lambda}=\frac{\frac{a^2}{b^2}-1}{a^2+\lambda}+\frac{1-\frac{c^2}{b^2}}{c^2+\lambda}=\frac{(a^2-c^2)(b^2+\lambda)}{b^2(a^2+\lambda)(c^2+\lambda)},$$

l'équation en λ devient ainsi

$$\left(\frac{x^2}{a^2+\lambda}+\frac{y^2}{b^2+\lambda}+\frac{z^2}{c^2+\lambda}-1-\frac{k^2}{\lambda}\right)(a^2-c^2)(b^2+\lambda)$$

$$+\frac{b^2(a^2+\lambda)(c^2+\lambda)}{\lambda}\left(\frac{a^2x\sqrt{\frac{1}{b^2}-\frac{1}{a^2}}}{a^2+\lambda}+\frac{c^2z\sqrt{\frac{1}{c^2}-\frac{1}{b^2}}}{c^2+\lambda}-p\right)^2=0,$$

laquelle se réduit à

$$\begin{aligned}x^2[\lambda(a^2-c^2)+c^2(a^2-b^2)]+y^2\lambda(a^2-c^2)+z^2[\lambda(a^2-c^2)+a^2(b^2-c^2)]\\+2acxz\sqrt{a^2-b^2}\,.\sqrt{b^2-c^2}\\-2bp[ax(c^2+\lambda)\sqrt{a^2-b^2}+by(a^2+\lambda)\sqrt{b^2-c^2}]\\-2(a^2-c^2)(b^2+\lambda)(c^2+\lambda)+b^2p^2(a^2+\lambda)(c^2+\lambda)=0,\end{aligned}$$

équation de l'ordre 2 par rapport à (x, y, z), par rapport à λ, et par rapport à p. Donc en égalant à zéro le discriminant par rapport à λ, on obtiendrait une équation du quatrième ordre par rapport aux coordonnées (x, y, z), et par rapport à p; on aurait ainsi l'équation de la surface annulaire de M. Roberts, rapportée aux axes de l'ellipsoïde, et en termes des quantités a, b, c, k et du paramètre p qui donne la position du plan

$$X\sqrt{\frac{1}{b^2}-\frac{1}{a^2}}+Y\sqrt{\frac{1}{c^2}-\frac{1}{b^2}}-p=0$$

de la section circulaire.

M. Roberts a donné (*Comptes Rendus* Nov. 14, 1859) le théorème que voici qui se rapporte à une surface primitive quelconque:—"Soit D la surface semblable à la primitive en multipliant par 2 ses rayons vecteurs, et soit P la surface parallèle ou équidistante de D par la longueur constante k, l'équation qui résulte de la substitution de $\sqrt{x^2+y^2+z^2}$ au lieu de k dans l'équation de P, coïncide avec l'équation de la première dérivée négative de la surface primitive"—ou je rappelle que dans la théorie des surfaces dérivées de M. Hirst la *première dérivée négative* est la surface enveloppe des plans conduits par les points de la surface primitive perpendiculairement aux rayons menés par un point quelconque. Pour démontrer le théorème, prenons $f(X, Y, Z)=0$ pour l'équation de la surface primitive: en supposant que (X, Y, Z) soient les coordonnées d'un point de cette surface, $(2X, 2Y, 2Z)$ seront les coordonnées d'un point de la surface D, et la surface parallèle à D sera l'enveloppe des sphères

$$(x-2X)^2+(y-2Y)^2+(z-2Z)^2=k^2,$$

avec la relation $f(X, Y, Z)=0$ entre les paramètres. Or on peut *avant d'effectuer l'élimination* écrire $\sqrt{x^2+y^2+z^2}$ au lieu de k, l'équation devient ainsi

$$X(x-X)+Y(y-Y)+Z(z-Z)=0,$$

avec cette même relation $f(X, Y, Z)=0$ entre les paramètres; or cette équation est celle d'un plan conduit par le point (X, Y, Z) perpendiculairement au rayon mené par l'origine des coordonnées: et ce dernier système donne ainsi l'équation de la surface dérivée.

L'équation de la surface parallèle de l'ellipsoïde est trouvée en égalant à zéro le discriminant par rapport à λ de l'équation

$$\frac{x^2}{a^2+\lambda}+\frac{y^2}{b^2+\lambda}+\frac{z^2}{c^2+\lambda}=1+\frac{k^2}{\lambda};$$

donc, en écrivant dans cette équation $x^2+y^2+z^2$ au lieu de k, et $4a^2$, $4b^2$, $4c^2$ au lieu de a^2, b^2, c^2, on doit obtenir l'équation de la dérivée de l'ellipsoïde, les rayons étant menés par le centre. Pour me conformer à la notation de mon Mémoire "Sur la surface qui est l'enveloppe des plans conduits par les points d'un ellipsoïde perpendiculairement aux rayons menés par le centre" (*Journal*, tom. II. 1859 [250]) j'écris -2θ au lieu de λ: l'équation en λ devient ainsi

$$\frac{x^2}{4a^2-2\theta}+\frac{y^2}{4b^2-2\theta}+\frac{z^2}{4c^2-2\theta}=1-\frac{x^2+y^2+z^2}{2\theta},$$

laquelle se réduit tout de suite à

$$\frac{x^2}{2-\frac{\theta}{a^2}}+\frac{y^2}{2-\frac{\theta}{b^2}}+\frac{z^2}{2-\frac{\theta}{c^2}}=\theta,$$

et la surface dérivée est l'enveloppe de cette équation en θ, résultat trouvé dans le mémoire que je viens de mentionner.

Mais je dois à M. Roberts la remarque que réciproquement l'équation en λ pour la surface parallèle peut se déduire des formules de ce mémoire. En effet en prenant $a\cos\alpha$, $b\cos\beta$, $c\cos\gamma$ pour les coordonnées d'un point de l'ellipsoïde (on a comme à l'ordinaire $\cos^2\alpha+\cos^2\beta+\cos^2\gamma=1$) l'équation du plan perpendiculaire au rayon par le centre sera

$$ax\cos\alpha+by\cos\beta+cz\cos\gamma=a^2\cos^2\alpha+b^2\cos^2\beta+c^2\cos^2\gamma,$$

et l'enveloppe de cette équation est l'enveloppe par rapport à θ de l'équation

$$\frac{x^2}{2-\frac{\theta}{a^2}}+\frac{y^2}{2-\frac{\theta}{b^2}}+\frac{z^2}{2-\frac{\theta}{c^2}}=\theta.$$

Donc *en général* l'enveloppe de

$$l\cos\alpha+m\cos\beta+n\cos\gamma=p\cos^2\alpha+q\cos^2\beta+r\cos^2\gamma,$$

(où comme auparavant $\cos^2\alpha+\cos^2\beta+\cos^2\gamma=1$) sera l'enveloppe par rapport à θ de l'équation

$$\frac{l^2}{2p-\theta}+\frac{m^2}{2q-\theta}+\frac{n^2}{2r-\theta}=\theta.$$

Or la surface parallèle à l'ellipsoïde est l'enveloppe de

$$(X-x)^2+(Y-y)^2+(Z-z)^2=k^2,$$

ou, en écrivant $a\cos\alpha$, $b\cos\beta$, $c\cos\gamma$ au lieu de X, Y, Z, cette surface est l'enveloppe de

$$l\cos\alpha+m\cos\beta+n\cos\gamma=p\cos^2\alpha+q\cos^2\beta+r\cos^2\gamma,$$

en posant $\rho=x^2+y^2+z^2$, et

$$l=2ax,\ m=2by,\ n=2cz,$$

$$p=\rho+a^2-k^2,\ q=\rho+b^2-k^2,\ r=\rho+c^2-k^2.$$

Donc cette surface sera l'enveloppe par rapport à θ de l'équation

$$\frac{4a^2x^2}{2(\rho+a^2-k^2)-\theta}+\frac{4b^2y^2}{2(\rho+b^2-k^2)-\theta}+\frac{4c^2z^2}{2(\rho+c^2-k^2)-\theta}=\theta,$$

et en écrivant $(2\rho-k^2)-\theta=2\lambda$ cette équation devient

$$\frac{a^2x^2}{a^2+\lambda}+\frac{b^2y^2}{b^2+\lambda}+\frac{c^2z^2}{c^2+\lambda}=\rho-k^2-\lambda,$$

c'est-à-dire

$$x^2+y^2+z^2-\lambda\left(\frac{x^2}{a^2+\lambda}+\frac{y^2}{b^2+\lambda}+\frac{z^2}{c^2+\lambda}\right)=\rho-k^2-\lambda,$$

ou enfin

$$\frac{x^2}{a^2+\lambda}+\frac{y^2}{b^2+\lambda}+\frac{z^2}{c^2+\lambda}=1+\frac{k^2}{\lambda},$$

ce qu'il s'agissait de faire voir. Je réserve à une autre occasion la discussion de la forme, et des singularités de la forme, et des singularités de la surface parallèle à l'ellipsoïde.

2, *Stone Buildings, W.C., Londres*, 7 *Nov.* 1860.

255.

ON A PROBLEM OF DOUBLE PARTITIONS.

[From the *Philosophical Magazine,* vol. XX. (1860), pp. 337—341.]

IF $a+b+c+\ldots=m$, $\alpha+\beta+\gamma+\ldots=\mu$ (the quantities being all positive integer numbers, not excluding zero), then $(a, \alpha)+(b, \beta)+(c, \gamma)+\ldots$ is considered as a partition of (m, μ): and the partible quantity (m, μ), and parts (a, α), (b, β), &c. being each of them composed of two elements, such partition is said to belong to the theory of *Double Partitions.* The subject (so far as I am aware) has hardly been considered except by Professor Sylvester, and it is greatly to be regretted that only an outline of his valuable researches has been published: the present paper contains the demonstration of a theorem, due to him, by which (subject to certain restrictions) the question of Double Partitions is made to depend upon the ordinary theory of Single Partitions.

Let the question be proposed, "In how many ways can (m, μ) be made up of the given parts (a, α), (b, β), (c, γ), &c." under the following conditions (which are, it will be seen, necessary in the demonstration of the theorem constituting the solution), viz.

$$\frac{a}{\alpha},\ \frac{b}{\beta},\ \frac{c}{\gamma},\ \text{\&c. are unequal fractions, each in its least terms,}$$

and

$$\alpha,\ \beta,\ \gamma,\ \text{\&c., are each less than } \mu+2.$$

The number of partitions is

$$=\text{coeff. } x^m y^\mu \text{ in } \frac{1}{(1-x^a y^\alpha)(1-x^b y^\beta)(1-x^c y^\gamma)\ldots}$$

the fraction being developed in ascending powers of x, y.

Considering the fraction as a function of y, it may be expressed as a sum of partial fractions in the form

$$\frac{A(x, y)}{1 - x^a y^\alpha} + \frac{B(x, y)}{1 - x^b y^\beta} + \frac{C(x, y)}{1 - x^c y^\gamma} + \dots,$$

where

$A(x, y)$ is rational in x, rational and integral of degree $\alpha - 1$ in y,
$B(x, y)$ " " " $\beta - 1$ "
$C(x, y)$ " " " $\gamma - 1$ "
&c.

To find $A(x, y)$ we have, when $y = x^{-\frac{a}{\alpha}}$,

$$A(x, y) = \frac{1}{(1 - x^b y^\beta)(1 - x^c y^\gamma)\dots};$$

or what is the same thing, we have

$$A(x, x^{-\frac{a}{\alpha}}) = \frac{1}{(1 - x^{b - \frac{a\beta}{\alpha}})(1 - x^{c - \frac{a\gamma}{\alpha}})\dots}.$$

This in fact determines $A(x, y)$; for the right-hand side of the equation may be reduced to the form

$$\frac{A_0 + A_1 x^{-\frac{a}{\alpha}} \dots + A_{\alpha-1} x^{-\frac{(\alpha-1)a}{\alpha}}}{(1 - x^{\alpha b - a\beta})(1 - x^{\alpha c - a\gamma})\dots},$$

where $A_0, A_1 \dots A_{\alpha-1}$ are rational functions of x: to do this, it is only necessary (taking ω an imaginary α-th root of unity) to multiply the numerator and denominator by

$$\Pi(1 - \omega x^{b - \frac{a\beta}{\alpha}})\,\Pi(1 - \omega x^{c - \frac{a\gamma}{\alpha}})\dots,$$

where Π denotes the product of the factors corresponding to the $\alpha - 1$ values of ω; the denominator is thus converted into

$$(1 - x^{\alpha(b - \frac{a\beta}{\alpha})})(1 - x^{\alpha(c - \frac{a\gamma}{\alpha})})\dots,$$

which is of the form in question; and the numerator becomes a rational function of x and $x^{-\frac{a}{\alpha}}$, integral as regards $x^{-\frac{a}{\alpha}}$, and therefore at once expressible in the form in question. And the equation, viz.

$$A(x, x^{-\frac{a}{\alpha}}) = \frac{A_0 + A_1 x^{-\frac{a}{\alpha}} \dots + A_{\alpha-1} x^{-\frac{(\alpha-1)a}{\alpha}}}{(1 - x^{\alpha b - a\beta})(1 - x^{\alpha c - a\gamma})\dots},$$

remains true if instead of $x^{-\frac{a}{\alpha}}$ we write $\omega x^{-\frac{a}{\alpha}}$; in fact, instead of writing in the first instance $y = x^{-\frac{a}{\alpha}}$, it would have been allowable to write $y = \omega x^{-\frac{a}{\alpha}}$, ω being any α-th root, real or imaginary, of unity. Hence recollecting that $A(x, y)$ is a rational and integral function of the degree $\alpha - 1$ in y, the equation

$$A(x, y) = \frac{A_0 + A_1 y \ldots + A_{\alpha-1} y^{\alpha-1}}{(1 - x^{ab-a\beta})(1 - x^{ac-a\gamma})\ldots},$$

which is true for the α values $\omega x^{-\frac{a}{\alpha}}$ of y, must be true identically; or this equation gives the value of $A(x, y)$. And the values of $B(x, y)$, $C(x, y)$, &c. are of course of the like form.

Now consider the term

$$\frac{A(x, y)}{1 - x^a y^\alpha},$$

where $A(x, y)$ is a rational and integral function of the degree $\alpha - 1$ in y, and $\frac{a}{\alpha}$ is by hypothesis a fraction in its least terms. The coefficient therein of $x^m y^\mu$ (the fraction being developed in ascending powers of x, y) is

$$= \text{coeff. } x^m y^\mu \text{ in } \frac{A(x, y)}{1 - x^{\frac{a}{\alpha}} y}$$

(the fraction being developed in ascending powers of x, y). In fact the two fractions only differ by a wholly *irrational* function of x, as is at once obvious by developing $\frac{1}{1 - x^{\frac{a}{\alpha}} y}$ in ascending powers of y. We have, separating the integral part,

$$\frac{A(x, y)}{1 - x^{\frac{a}{\alpha}} y} = U + \frac{A(x, x^{-\frac{a}{\alpha}})}{1 - x^{\frac{a}{\alpha}} y},$$

where U is a rational and integral function of the degree $\alpha - 2$ in y. But α being by hypothesis $< \mu + 2$, or what is the same thing, $\alpha - 2 < \mu$, U does not contain any term of the form $x^m y^\mu$, and therefore

$$\text{coeff. } x^m y^\mu \text{ in } \frac{A(x, y)}{1 - x^{\frac{a}{\alpha}} y}$$

$$= \quad \text{do.} \quad \text{in } \frac{A(x, x^{-\frac{a}{\alpha}})}{1 - x^{\frac{a}{\alpha}} y};$$

and this last is

$$= \text{coeff. } x^m \text{ in } x^{\frac{\mu a}{\alpha}} A(x, x^{-\frac{a}{\alpha}}),$$

$$= \text{coeff. } x^{m - \frac{\mu a}{\alpha}} \text{ in } A(x, x^{-\frac{a}{\alpha}}),$$

$$= \text{coeff. } x^{\alpha m - a\mu} \text{ in } A(x^\alpha, x^{-a});$$

and from the foregoing equation

$$A(x, x^{-\frac{a}{\alpha}}) = \frac{1}{(1-x^{b-\frac{a\beta}{\alpha}})(1-x^{c-\frac{a\gamma}{\alpha}})\ldots},$$

this is

$$= \text{coeff. } x^{\alpha m - a\mu} \text{ in } \frac{1}{(1-x^{\alpha b - a\beta})(1-x^{\alpha c - a\gamma})\ldots}.$$

The last-mentioned expression is thus the value of

$$\text{coeff. } x^m y^\mu \text{ in } \frac{A(x, y)}{1 - x^a y^\alpha};$$

and hence, *Theorem*,

$$\text{coeff. } x^m y^\mu \text{ in } \frac{1}{(1-x^a y^\alpha)(1-x^b y^\beta)(1-x^c y^\gamma)\ldots}$$

$$= \text{coeff. } x^{\alpha m - a\mu} \text{ in } \frac{1}{(1-x^{\alpha b - a\beta})(1-x^{\alpha c - a\gamma})\ldots}$$

$$+ \text{coeff. } x^{\beta m - b\mu} \text{ in } \frac{1}{(1-x^{\beta a - b\alpha})(1-x^{\beta c - b\gamma})\ldots}$$

$$+ \text{coeff. } x^{\gamma m - c\mu} \text{ in } \frac{1}{(1-x^{\gamma a - c\alpha})(1-x^{\gamma b - c\beta})\ldots}$$

$$+ \&c.,$$

the fraction on the left-hand side being expanded in ascending powers of x, y, and those on the right-hand side being expanded in ascending powers of x, and the data satisfying the above-mentioned conditions. The number of partitions of (m, μ) is thus found to be equal to the expression on the right-hand side. It is to be noticed that on the right-hand side, when any of the indices $\alpha m - a\mu$, $\beta m - b\mu$, ... is negative, the corresponding coefficient vanishes; and that when the index of the power of x in any factor of a denominator is negative, e.g. if $\alpha b - a\beta = -p$, then (in order to develope in ascending powers of x) we must in the place of $\frac{1}{1-x^{\alpha b - a\beta}} = \frac{1}{1-x^{-p}}$ write $\frac{x^p}{x^p - 1}$, $= -\frac{x^p}{1-x^p}$, and develope in the form $-(x^p + x^{2p} + x^{3p} + \ldots)$. The right-hand side is thus seen to be the sum of a series of positive or negative numbers, each of which *taken positively* denotes the number of the single partitions of a given partible number into given parts.

If, using a term of Professor Sylvester's, we say that

$$\text{coeff. } x^m \text{ in } \frac{1}{(1-x^a)(1-x^b)\ldots}$$

(where m, a, b, c ... are positive or negative integers, and the fraction is developed in ascending powers of x) is

$$= \text{Denumerant of } m \text{ in respect to the elements } (a, b, c, \ldots), \text{ say}$$

$$= \text{Denumerant } (m;\ a, b, c\ldots),$$

then when m, a, b, c... are positive, but not otherwise, Denumerant $(m;\ a, b, c\ldots)$ denotes the number of ways in which m can be made up of the parts a, b, c...; and the foregoing result shows that the number of ways in which (m, μ) can be made up of the parts (a, α), (b, β), (c, γ), &c. is equal to the sum

$$\begin{aligned}
&\text{Denumerant } (\alpha m - a\mu;\ \alpha b - a\beta,\ \alpha c - a\gamma, \ldots) \\
&+ \text{Denumerant } (\beta m - b\mu;\ \beta a - b\alpha,\ \beta c - b\gamma, \ldots) \\
&+ \text{Denumerant } (\gamma m - c\mu;\ \gamma a - c\alpha,\ \gamma b - c\beta, \ldots) \\
&+ \&\text{c.}
\end{aligned}$$

But, as appears from what precedes, a denumerant may be equal to zero, or may denote a number of partitions taken negatively; and it is not allowable, in the place, e.g., of the first denumerant, to write *simpliciter*, number of partitions of $\alpha m - a\mu$ in respect of $\alpha b - a\beta$, $\alpha c - a\gamma$, &c. The notion of a Denumerant is, in fact, an important generalization of the notion of a number of partitions.

2, *Stone Buildings, W.C., October* 4, 1860.

256.

ON A SYSTEM OF ALGEBRAIC EQUATIONS.

[From the *Philosophical Magazine*, vol. XX. (1860), pp. 341, 342.]

THE determination of a, b, c from the system of equations

$$a^2 + bc = \lambda,$$
$$b^2 + ca = \mu,$$
$$c^2 + ab = \nu,$$

in the case where λ, μ, ν have the values 16, 17, and 18 respectively, is the problem known as Colonel Titus's Arithmetical Problem. See Masères' *Tracts on the Resolution of affected Algebraic Equations*, Lond. 1800. If for shortness we put

$$\sigma = \nu - c^2,$$

then the third equation gives $b = \frac{\sigma}{a}$; and substituting this value of b in the two other equations, we have

$$a^2 + \frac{\sigma c}{a} = \lambda,$$

$$\frac{\sigma^2}{a^2} + ca = \mu;$$

or what is the same thing,

$$a^3 - \lambda a + \sigma c = 0,$$
$$ca^3 - \mu a^2 + \sigma^2 = 0;$$

and from these equations, eliminating a, we have

$$\sigma^4 - 3c^2\sigma^3 + (3c^4 - 2\lambda\mu)\,\sigma^2 - c^2(c^4 - \lambda\mu)\,\sigma + c^4\lambda\mu - c^2(\lambda^3 + \mu^3) + \lambda^2\mu^2 = 0,$$

where $\sigma = \nu - c^2$. The equation in c^2 is thus of the fourth order; and in like manner, if instead of c^2 we take σ as the unknown quantity, and substitute therefore for c^2 its value $\nu - \sigma$, the equation in σ will be also of the fourth order: and effecting the reduction, this equation is

$$8\sigma^4 - 12\nu\sigma^3 + (6\nu^2 - 2\lambda\mu)\sigma^2 + (\lambda^3 + \mu^3 - \nu^3 - \lambda\mu\nu)\sigma + (\nu\lambda - \mu^2)(\nu\mu - \lambda^2) = 0.$$

It may be remarked that if $\sigma = 0$, then a or b vanishes; and therefore, from the original equations, $\nu\lambda - \mu^2 = 0$, or $\nu\mu - \lambda^2 = 0$, which agrees with the result afforded by the foregoing equation in σ. Again, if $\sigma = \nu$, then $c = 0$; and therefore, from the original equations, $\nu^2 - \lambda\mu = 0$. The left-hand side of the equation in σ, writing therein $\sigma = \nu$, should therefore contain the factor $\nu^2 - \lambda\mu$; its value in fact is $\nu^4 - 2\lambda\mu\nu^2 + \lambda^2\mu^2$, or $(\nu^2 - \lambda\mu)^2$.

If in the original equations we write $a = \frac{x}{z}$, $b = \frac{y}{z}$, the equations become

$$x^2 + cyz - \lambda z^2 = 0,$$

$$y^2 + czx - \mu z^2 = 0,$$

$$(c^2 - \nu)z^2 + xy = 0,$$

which are three homogeneous equations of the second order; from which, if the variables x, y, z are eliminated, we have the required equation in c. And it would not, I think, be difficult, from the known formula for the general case, to deduce the foregoing result corresponding to the very particular case which is here in question.

2, *Stone Buildings, W.C., September* 25, 1860.

257.

ON THE CUBIC CENTRES OF A LINE WITH RESPECT TO THREE LINES AND A LINE.

[From the *Philosophical Magazine*, vol. XX. (1860), pp. 418—423.]

CONSIDER a line L in relation to the three lines X, Y, Z and the line I: through the point of intersection of the lines X, L, draw any line meeting the lines I, Y, Z, and let the harmonic of the intersection with I, in relation to the intersections with Y, Z, be ξ; then the locus of the point ξ is a conic passing through the points YI, ZI, YZ.

If, in like manner, through the point of intersection of the lines Y, L, there is drawn any line meeting the lines I, Z, X, and the harmonic of the intersection with I, in relation to the intersections with Z, X, is called η, the locus of the point η is a conic passing through the points ZI, XI, ZX.

And so, if through the point of intersection of the lines Z, L there is drawn any line meeting the lines I, X, Y, and the harmonic of the intersection with I, in relation to the intersections with X, Y, is called ζ, then the locus of ζ is a conic passing through the points XI, YI, XY.

The pairs of conics, viz. the second and third, third and first, first and second conics, have obviously in common the points XI, YI, ZI respectively. They besides intersect all three of them in three points, which may be termed the *cubic centres* of the line L in relation to the lines X, Y, Z and the line I.

The line L may be such that two of the three cubic centres coincide; the locus of the coincident centres is in this case a conic which touches the lines X, Y, Z harmonically in regard to the line I; that is, it touches each of the three lines in the point which is the harmonic of its intersection with I in relation to its intersections with the other two lines.

Except that the line I is there taken to be infinity, the foregoing theorems occur in Plücker's *System der analytischen Geometrie* (Berlin, 1835), p. 177 *et seq.*; and they play an important part in his classification of curves of the third order (see p. 220 *et seq.*). It is, I think, an omission that he has not sought for the curve which is the envelope of the line L in the above-mentioned case of the two coincident centres: I find that the envelope is a curve of the fourth order, having four-pointic contact with the lines X, Y, Z harmonically in regard to the line I; viz., if the equations of the lines X, Y, Z are $x=0$, $y=0$, $z=0$ respectively, and the equation of the line I is $x+y+z=0$, then the equation of the envelope in question is

$$\sqrt[4]{x}+\sqrt[4]{y}+\sqrt[4]{z}=0,$$

a result which is also interesting as exhibiting a geometrical construction of the curve represented by this equation.

The investigation of the series of theorems is as follows; taking

$$\begin{array}{rcr} x=0 & \text{for the equation of} & X, \\ y=0 & ,, & Y, \\ z=0 & ,, & Z, \\ x+\ y+\ z=0 & ,, & I, \\ \lambda x+\mu y+\nu z=0 & ,, & L, \end{array}$$

then, first, in order to find the curve which is the locus of ξ, the coordinates of the point XL are given by $x:y:z=0:\nu:-\mu$; or if, as it is convenient to do, we take X, Y, Z (instead of x, y, z) for current coordinates, by $X:Y:Z=0:\nu:-\mu$. Hence taking x, y, z as the coordinates of ξ, the equation of the line through XL, ξ is

$$\left|\begin{array}{ccc} X, & Y, & Z \\ x, & y, & z \\ 0, & \nu, & -\mu \end{array}\right|=0,$$

viz.

$$X(\mu y+\nu z)-x(\mu Y+\nu Z)=0;$$

and at the point where this line meets the line I, the equation whereof is

$$X+Y+Z=0,$$

we have

$$(Y+Z)(\mu y+\nu z)+x(\mu Y+\nu Z)=0;$$

that is

$$Y(\mu x+\mu y+\nu z)+Z(\nu x+\mu y+\nu z)=0.$$

Hence this line, and the line

$$Yz-Zy=0,$$

with the lines

$$Y=0,\ Z=0,$$

are the lines which pass through the point YZ and the four harmonic points, and they form therefore a harmonic pencil; or we have

$$y(\mu x+\mu y+\nu z)-z(\nu x+\mu y+\nu z)=0,$$

or, what is the same thing,

$$(\mu y-\nu z)(x+y+z)+2yz(\nu-\mu)=0,$$

as the locus of the point ξ: the locus is therefore a conic passing through the points YI, ZI, YZ.

The equations of the conics which are the loci of X, Y, Z respectively, are therefore

$$\begin{aligned} U&=(\mu y-\nu z)(x+y+z)+2yz(\nu-\mu)=0,\\ V&=(\nu z-\lambda x)(x+y+z)+2zx(\lambda-\nu)=0,\\ W&=(\lambda x-\mu y)(x+y+z)+2xy(\mu-\lambda)=0;\end{aligned}$$

and the identical equation,

$$U\lambda x+V\mu y+W\nu z=0,$$

shows that these conics have three points of intersection in common. The three equations, and a fourth one to which they give rise, may be written

$$\frac{\mu}{z}-\frac{\nu}{y}+\frac{2(\nu-\mu)}{x+y+z}=0,$$

$$\frac{\nu}{x}-\frac{\lambda}{z}+\frac{2(\lambda-\nu)}{x+y+z}=0,$$

$$\frac{\lambda}{y}-\frac{\mu}{x}+\frac{2(\mu-\lambda)}{x+y+z}=0,$$

$$\frac{\nu-\mu}{x}+\frac{\lambda-\nu}{y}+\frac{\mu-\lambda}{z}=0;$$

and each of these is the equation of a conic passing through the three cubic centres.

If two of the three centres coincide, then the conics all touch at the coincident centres. Consider the first and second conics: these intersect at the point $z=0$ $x+y+z=0$; and the line $x+y+z=2kz$, if k be properly determined, or what is the same thing, the line $x+y+z=\dfrac{2(\theta+\nu)}{\theta}z$, if θ is properly determined, will be a line passing through the last-mentioned point and one of the other points of intersection: k or θ will of course be determined by a cubic equation; and if this has a pair of equal roots, the conics will touch. But the equation of the line, combined with those of the two conics, gives

$$x : y : z = \frac{1}{\theta+\lambda} : \frac{1}{\theta+\mu} : \frac{1}{\theta+\nu};$$

and substituting these values in the equation of the line, we have

$$\frac{1}{\theta+\lambda} + \frac{1}{\theta+\mu} + \frac{1}{\theta+\nu} - \frac{2}{\theta} = 0,$$

which is (as it should be) a cubic equation in θ.

If the equation in θ has equal roots, then

$$\frac{1}{(\theta+\lambda)^2} + \frac{1}{(\theta+\mu)^2} + \frac{1}{(\theta+\nu)^2} - \frac{2}{\theta^2} = 0;$$

and putting in these two equations,

$$x = \frac{m}{\theta+\lambda}, \quad y = \frac{m}{\theta+\mu}, \quad z = \frac{m}{\theta+\nu},$$

we have

$$x + y + z - \frac{2m}{\theta} = 0,$$

$$x^2 + y^2 + z^2 - \frac{2m^2}{\theta^2} = 0;$$

whence eliminating m,

$$(x+y+z)^2 = 2(x^2+y^2+z^2);$$

that is

$$x^2 + y^2 + z^2 - 2yz - 2zx - 2xy = 0;$$

or, what is the same thing,

$$\sqrt{x} + \sqrt{y} + \sqrt{z} = 0,$$

for the equation of the locus of the coincident centres: such locus is therefore a conic touching the lines $x=0$, $y=0$, $z=0$, in the points of intersection with the lines $y-z=0$, $z-x=0$, $x-y=0$ respectively; it is a conic touching the lines X, Y, Z harmonically in regard to the line I.

To find the envelope of the line L, the most convenient course is to take the equation in θ in the reduced form

$$\theta^3 - \theta(\mu\nu + \nu\lambda + \lambda\mu) - 2\lambda\mu\nu = 0;$$

this will have a pair of equal roots if

$$(\mu\nu + \nu\lambda + \lambda\mu)^3 - 27\lambda^2\mu^2\nu^2 = 0;$$

that is, if

$$\mu\nu + \nu\lambda + \lambda\mu - 3(\lambda\mu\nu)^{\frac{2}{3}} = 0;$$

or if

$$\frac{1}{\lambda} + \frac{1}{\mu} + \frac{1}{\nu} - 3\frac{1}{(\lambda\mu\nu)^{\frac{1}{3}}} = 0;$$

or finally if

$$\lambda^{-\frac{1}{3}} + \mu^{-\frac{1}{3}} + \nu^{-\frac{1}{3}} = 0,$$

which is the relation between λ, μ, ν in order that the line

$$\lambda x + \mu y + \nu z = 0$$

may have two coincident centres; this gives at once for the equation of the envelope

$$\sqrt[4]{x} + \sqrt[4]{y} + \sqrt[4]{z} = 0,$$

which is the equation of a curve of the fourth order having four-pointic contact with the lines $x=0$, $y=0$, $z=0$, at the points of intersection with the lines $y-z=0$, $z-x=0$, $x-y=0$ respectively, i.e. it has four-pointic contact with the lines X, Y, Z harmonically in regard to the line I.

It may be noticed that the rationalized form of the equation $\sqrt[4]{x} + \sqrt[4]{y} + \sqrt[4]{z} = 0$ is

$$x^4 + y^4 + z^4 - 4(yz^3 + y^3z + zx^3 + z^3x + xy^3 + x^3y)$$
$$+ 6(y^2z^2 + z^2x^2 + x^2y^2) - 124(x^2yz + y^2zx + z^2xy) = 0.$$

If, to fix the ideas, the signs of the coordinates x, y, z are so determined that a point *within* the triangle $x=0$, $y=0$, $z=0$ has its coordinates positive (in which case the line $x+y+z=0$ will cut the three sides *produced*), the curve $\sqrt[4]{x} + \sqrt[4]{y} + \sqrt[4]{z} = 0$ will lie wholly within the triangle, and will be of the form shown by the annexed

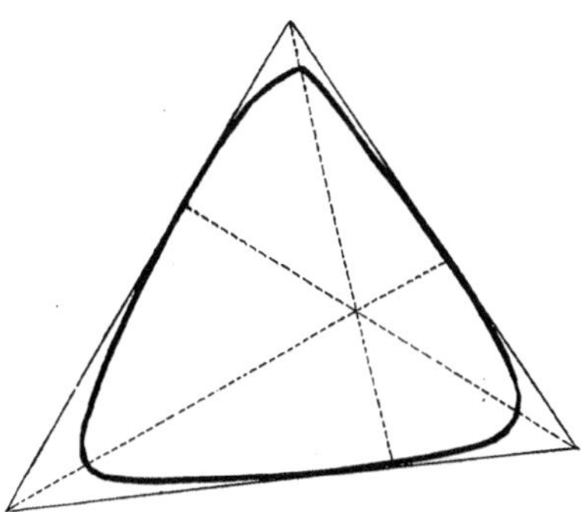

figure. This is, in fact, the form of the curve in the case considered by Plücker, where the line I is at infinity, the points of contact being the middle points of the sides. And his five groups of curves, α, β, γ, δ, ϵ, and two subdivisions of the group

β (see pp. 221—224), correspond to the following positions of the line in regard to the triangle and curve, viz.

α. The line cuts the three sides produced.

β. It passes through an angle, (a) cutting, or (b) not cutting the curve.

γ. It cuts two sides and a side produced, but does not cut or touch the curve.

δ. It cuts two sides and a side produced, and touches the curve.

ϵ. It cuts two sides and a side produced, and cuts the curve.

It is hardly necessary to remark that, in the general case, the tangential equation of the curve is

$$\xi^{-\frac{1}{3}} + \eta^{-\frac{1}{3}} + \zeta^{-\frac{1}{3}} = 0\,;$$

or what is the same thing,

$$(\eta\zeta + \zeta\xi + \xi\eta)^3 - 27\xi^2\eta^2\zeta^2 = 0\,;$$

and that the curve is therefore of the sixth class.

2, *Stone Buildings, W.C., October* 16, 1860.

258.

ON A RELATION BETWEEN TWO TERNARY CUBIC FORMS.

[From the *Philosophical Magazine*, vol. xx. 1860, pp. 512—514.]

THE cubic form

$$x^3 + y^3 + z^3 + 6lxyz$$

is in general linearly transformable into the form

$$(X + Y + Z)^3 + 27kXYZ;$$

in fact, writing

$$X = 2lx - y - z,$$
$$Y = 2ly - z - x,$$
$$Z = 2lz - x - y,$$

we have identically

$$(1 - 2l + 4l^2)(X + Y + Z)^3 + 24(l - 1)^3 XYZ = 8(2l + 1)^2 (l - 1)^3 (x^3 + y^3 + z^3 + 6lxyz);$$

and the value of k consequently is

$$k = -\frac{8(l-1)^3}{9(1 - 2l + 4l^2)}.$$

If, however, $l = 1$ or $l = -\frac{1}{2}$, the transformation fails. In the former case, viz. for $l = 1$, the equations for the linear transformation become

$$X = 2x - y - z,$$
$$Y = 2y - z - x,$$
$$Z = 2z - x - y,$$

which give $X+Y+Z=0$, so that X, Y, Z are no longer independent; and the formula of transformation becomes

$$(X+Y+Z)^3=0.$$

It may be noticed that the invariant S of the form

$$x^3+y^3+z^3+6lxyz$$

is $S=-l+l^4$, so that $l=1$ is one of the values which make S vanish. And the above transformation is not applicable to the cubic form $x^3+y^3+z^3+6xyz$, which is a form for which S vanishes. The transformation, however, holds good for $l=0$, which is another value which makes S vanish; or it does apply to the form $x^3+y^3+z^3$, for which S vanishes. The transformation, in fact, is

$$(X+Y+Z)^3+24XYZ=-8(x^3+y^3+z^3),$$

with the linear equations

$$X=-y-z,$$
$$Y=-z-x,$$
$$Z=-x-y.$$

The above two forms for which S vanishes, viz.

$$x^3+y^3+z^3+6xyz,$$
$$x^3+y^3+z^3,$$

are, notwithstanding, equivalent to each other, as appears by the identical equation

$$(x+y+z)^3+(x+\omega y+\omega^2 z)^3+(x+\omega^2 y+\omega z)^3=3(x^3+y^3+z^3+6xyz),$$

where ω is an imaginary cube root of unity. In the latter of the two cases of failure, viz. for $l=-\frac{1}{2}$, the equations for the linear transformations are

$$X=Y=Z=-x-y-z;$$

so that X, Y, Z are not only not independent, but they are connected by *two* linear relations. And the formula of transformation becomes

$$(X+Y+Z)^3-27XYZ=0,$$

which is, in fact, true in virtue of the equations $X=Y=Z$.

The two forms of equation,

$$x^3+y^3+z^3+6lxyz=0,$$
$$(x+y+z)^3+27kxyz=0,$$

represent each of them equally well a curve of the third order without a double point. In the first form the three real points of inflexion are given by

$$(x=0,\ y+z=0),\ (y=0,\ z+x=0),\ (z=0,\ x+y=0);$$

or what is the same thing, the points in question are the intersections of the lines $x=0$, $y=0$, $z=0$ with the line $x+y+z=0$; or we have $x+y+z=0$ for the equation of the line through the three points of inflexion; and the equations of the tangents at the points of inflexion are

$$2lx - y - z = 0, \quad 2ly - z - x = 0, \quad 2lz - x - y = 0.$$

For the second form it is obvious that the points of inflexion are the intersections of the lines $x=0$, $y=0$, $z=0$ with the line $x+y+z=0$; and, moreover, that the lines $x=0$, $y=0$, $z=0$ are the tangents at the point of inflexion.

The first of the above-mentioned forms, however, cannot represent a curve with a double point. In fact the condition for its doing so would be $1+8l^3=0$; but when this condition is satisfied, the left-hand side breaks up into linear factors, and the equation represents, not a proper curve of the third order, but a system of three lines. The second form *can* represent a curve having a double point; viz. if $k=-1$, the curve will have a conjugate or isolated point at the point $x=y=z$. It is clear *à priori* that ($x=0$, $y=0$, $z=0$ being real lines) neither of the forms can represent a curve of the third order having a double point with two real branches through it, since in this case the curve has only one real point of inflexion.

I have elsewhere used the word "node" to denote a double point, and I take the opportunity of suggesting the employment of the words "crunode" (*crus*) and "acnode" (*acus*) to denote respectively a double point with two real branches through it, and a conjugate or isolated point.

2, *Stone Buildings*, *W.C.*, *October* 19, 1860.

259.

THE PROBLEM OF POLYHEDRA.

[From the *Philosophical Transactions of the Royal Society of London*, vol. CXLVII. (for the year 1857), pp. 183—185: printed as a Note to Mr Kirkman's Memoir "On Autopolar Polyhedra," pp. 183—215.]

LET a, b, c, d, e, f, g, h, &c. represent the vertices of a polyhedron, then a face will be represented, e.g. by *abcde*, where the contiguous duads, viz. *ab, bc, cd, de, ea* are the edges of the face; and calling the face K, we may write

$$K = abcde. \tag{1}$$

It is to be noticed that the letters of a face-symbol may be taken forwards or backwards from any letter without altering the meaning of the symbol. Thus, *abcde*, *bcdea*, &c., *edcba*, &c. might any of them be taken to denote the face K. The diagonal of a face cannot be either an edge or a diagonal of any other face, i.e. a non-contiguous duad such as *ac* in a face-symbol K cannot be a duad, contiguous or non-contiguous, of any other face-symbol. But each edge of a face must be an edge of one and only one other face, i.e. each contiguous duad such as *ab* in the face-symbol K must be a contiguous duad of one and only one other face-symbol L. And moreover two faces cannot have more than a single edge in common, i.e. two face-symbols cannot contain more than a single contiguous duad, the same in each symbol.

The face K contains the edges *ab, ac*, i.e. the edge *ab* is contained in the face K; it will also be contained in one and only one other face, suppose L; this face will contain another edge through the vertex a, suppose the edge *af*, and so on, until we arrive at a face containing the edge *ae*; we have, for example,

$$
\begin{aligned}
K &= eabcd,\\
L &= baf\ldots,\\
M &= fag\ldots,\\
N &= gah\ldots,\\
P &= hai\ldots,\\
Q &= iae\ldots;
\end{aligned}
$$

and we thence derive the vertex-symbol

$$a = KLMNPQ, \qquad (2)$$

where the contiguous duads KL, LM, MN, NP, PQ, QK represent in order the edges through the vertex a. The remarks before made with respect to the face-symbols apply to the vertex-symbols. A non-contiguous duad such as KN of the vertex-symbol a cannot be a duad, contiguous or non-contiguous, of any other vertex-symbol; but each contiguous duad such as KL of the vertex-symbol a must be a contiguous duad of one and only one other vertex-symbol b. And the symbols of two vertices cannot contain more than one contiguous duad, the same in each symbol.

Any edge of the polyhedron admits of a double representation; it is the junction of two vertices, or the intersection of two faces. Thus ab and KL will represent the same edge, or we may write

$$ab = KL. \qquad (3)$$

It is to be remarked that in this system, to each equation $ab = KL$ there corresponds one and only one equation of the form $ae = KQ$, i.e. to an edge considered as drawn from a given vertex in a given face there corresponds one and only one other edge from the same vertex in the same face.

It has been shown how the system of face-symbols (1) leads to the system of vertex-symbols (2), and the system of edge symbols (3); and generally, any one of the three systems leads to the other two; and the three systems conjointly, or each system by itself, is a complete representation of the polyhedron. As an example, take the hexaedron: the three systems are:

$K = abcd,$ (1)	$a = LPK,$ (2)
$L = abfe,$	$b = LMK,$
$M = bfge,$	$c = MNK,$
$N = gcdh,$	$d = NPK,$
$P = dhea,$	$e = LPQ,$
$Q = hefg,$	$f = LMQ,$
	$g = MNQ,$
	$h = NPQ,$

$$ab = KL, \quad ae = PL, \quad ef = LQ, \quad (3)$$
$$bc = KM, \quad bf = LM, \quad fg = MQ,$$
$$cd = KN, \quad cg = MN, \quad gh = NQ,$$
$$de = KP, \quad dh = NP, \quad he = PQ.$$

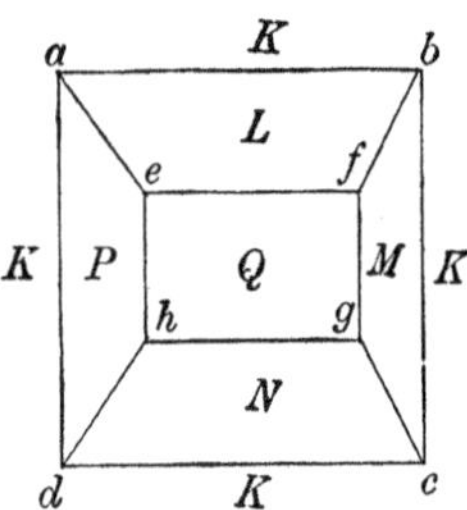

Consider, now, two polyhedra having the same number of vertices and also the same number of faces. And let the vertices and faces of the first polyhedron taken in any order be represented by

$$abcde \ldots KLM \ldots,$$

and the vertices and faces of the second polyhedron taken in a certain order be represented by

$$a'b'c'd'e' \ldots K'L'M' \ldots;$$

then, forming the substitution symbol

$$a'b'c'd'e' \ldots K'L'M' \ldots abcde \ldots KLM \ldots,$$

which denotes that a' is to be written for a, b' for $b \ldots K'$ for K, &c., if operating with this upon the symbol system of the first polyhedron, we obtain the symbol system of the second polyhedron, the second polyhedron will be syntypic with the first. It should be noticed, that there may be several modes of arrangement of the vertices and faces of the second polyhedron, which will render it syntypic according to the foregoing definition with the first polyhedron, i.e. the second polyhedron may be syntypic in several different ways with the first polyhedron. This is, in fact, the same as saying that a polyhedron may be syntypic with itself in several different ways. Suppose, next, that the number of vertices of the second polyhedron is equal to the number of faces of the first polyhedron, and the number of faces of the second polyhedron is equal to the number of vertices of the first polyhedron; and let the vertices and faces of the first polyhedron in any order be represented by

$$abcde \ldots KLM \ldots,$$

and the faces and vertices of the second polyhedron in a certain order be represented by

$$A'B'C'D'E' \ldots k'l'm' \ldots.$$

Then, forming the substitution symbol

$$A'B'C'D'E' \ldots k'l'm' \ldots abcde \ldots KLM \ldots,$$

if, operating with this upon the symbol system of the first polyhedron, we obtain the symbol system of the second polyhedron, the second polyhedron is said to be polar-syntypic with the first; and, as in the case of syntypicism, this may happen in several different ways.

Lastly, if there be a polyhedron having the same number of vertices and faces, and if the vertices and faces in any order be represented by

$$abcd \ldots KLMN \ldots,$$

and the faces and vertices in a certain order be represented by

$$ABCD \ldots klmn \ldots;$$

then, forming the substitution symbol

$$ABCD \ldots klmn \ldots abcd \ldots KLMN \ldots,$$

if, operating with this upon the symbol system of the polyhedron, we reproduce such symbol system, i.e. in fact, if the polyhedron be polar-syntypic with itself, the polyhedron is said to be *autopolar;* and in accordance with a preceding remark, this may happen in several different ways. It is clear that the substitution symbol, operating on the symbol system of the vertices, must give the symbol system of the faces, and conversely; but operating on the symbol system of the edges, it must reproduce such symbol system of the edges: and this last condition will by itself suffice to make the polyhedron autopolar, i.e. the polyhedron will be autopolar if the substitution symbol, operating on the symbol system of the edges, reproduces such symbol system.

260.

ON THE DOUBLE TANGENTS OF A PLANE CURVE.

[From the *Philosophical Transactions of the Royal Society of London*, vol. CXLIX. *for the year* 1859, pp. 193—212. Received March 17,—Read April 14, 1859.]

IT was first shown by Plücker on geometrical principles, that the number of the double tangents of a plane curve of the order m was $\frac{1}{2}m(m-2)(m^2-9)$: see the note, "Solution d'une question fondamentale concernant la théorie générale des Courbes," *Crelle*, t. XII. pp. 105—108 (1834), and the "Theorie der algebraischen Curven" (1839). The memoir by Hesse, "Ueber die Wendepuncte der Curven dritter Ordnung," *Crelle*, t. XXVIII. pp. 97—107 (1844), contains the analytical solution of the allied easier problem of the determination of the points of inflexion of a plane curve. In the memoir, "Recherches sur l'élimination et sur la théorie des Courbes," *Crelle*, t. XXXIV. (1847), pp. 30—45, [53], I showed how the problem of double tangents admitted of an analytical solution, viz. if $U=0$ is the equation of the curve, L, M, N the first derived functions of U, and

$$D=\alpha(M\partial_x-N\partial_y)+\beta(N\partial_y-L\partial_z)+\gamma(L\partial_z-M\partial_x)$$

(where α, β, γ are arbitrary), then the points of contact of the double tangents are given as the intersections of the curve $U=0$, with a curve the equation whereof is in the first instance obtained under the form $[Y]=0$; $[Y]$ being a given function of

$$D^2U,\ D^3U, \ldots D^mU,$$

of the degree m^2-m-6 in respect of (α, β, γ), the degree $m^3-2m^2-10m+12$ in respect of (x, y, z), and the degree m^2+m-12 in respect of the coefficients of U. It was necessary, in order that the points of intersection should be independent of the arbitrary quantities (α, β, γ), that we should have identically

$$[Y]=\Lambda\,.\,U+N\,.\,\Pi U,$$

N being of the degree m^2-m-6 in (α, β, γ), and consequently ΠU a function of (x, y, z) without (α, β, γ). Guided by Hesse's investigation for the points of inflexion, I asserted that it was probable that N was of the form $(\alpha x+\beta y+\gamma z)^{m^2-m-6}$; which being so, ΠU would be of the degree $(m-2)(m^2-9)$ in respect of (x, y, z), and the degree m^2+m-12 in respect of the coefficients, and I was thus led to the theorem, "On trouve les points de contact des tangentes doubles en combinant avec l'équation de la courbe une équation $\Pi U=0$, de l'ordre $(m-2)(m^2-9)$ par rapport aux variables et de l'ordre m^2+m-12 par rapport aux coefficients—c'est-à-dire, puisqu'il correspond deux points de contact à une tangente double, le nombre de ces tangentes est égal à $\frac{1}{2}m(m-2)(m^2-9)$: théorème démontré indirectement par M. Plücker."

Hesse, in the memoir "Ueber Curven dritter Ordnung u.s.w.," *Crelle*, t. XXXVI. pp. 143—176 (1848), showed how the components $D^2U, D^3U, \ldots D^mU$ of $[Y]$ could each of them be expressed in a simplified form, and he thus effected the actual reduction of $[Y]$ to the form $\Lambda . U+(\alpha x+\beta y+\gamma z)^{4(m-3)}R$, where R still contained the arbitrary quantities (α, β, γ) in the degree $(m-2)(m-3)$. In particular for a quartic curve, the equation $R=0$ was shown to be

$$3Q_2Q_4-Q_3^2=0,$$

where the left-hand side is of the degree 2 in (α, β, γ) and the degree 16 in (x, y, z); and which should therefore by means of the equation $U=0$ be reducible so as to contain the factor $(\alpha x+\beta y+\gamma z)^2$.

Jacobi's paper, "Beweis des Satzes, dass eine Curve n-ten Grades im allgemeinen $\frac{1}{2}n(n-2)(n^2-9)$ Doppeltangenten hat," *Crelle*, t. XL. pp. 237—260 (1850), did not, I think, materially advance the solution of the question. In a letter to Jacobi, dated the 30th December, 1849, published at the conclusion of the last-mentioned paper, Hesse gave the equation of the curve of the 14th order for the points of contact of the double tangents of a quartic, viz. in my notation,

$$(\mathfrak{A}, \mathfrak{B}, \mathfrak{C}, \mathfrak{F}, \mathfrak{G}, \mathfrak{H}\between\partial_xH, \partial_yH, \partial_zH)^2-H(\mathfrak{A}, \mathfrak{B}, \mathfrak{C}, \mathfrak{F}, \mathfrak{G}, \mathfrak{H}\between\partial_x, \partial_y, \partial_z)^2H=0,$$

and the demonstration is given in Hesse's paper, "Ueber die ganzen homogenen Functionen von der dritten und vierten Ordnung zwischen drei Variabeln," *Crelle*, t. XLI. pp. 285—292 (1851), and is reproduced in Mr Salmon's Treatise on the Higher Plane Curves (1852). Two very interesting memoirs by Hesse and Steiner, *Crelle*, t. XLIX. (1855), relate to the *geometrical* theory of the double tangents of a quartic, and it is not necessary to refer to them more particularly. It is to be observed that the curve which determines the points of contact of the double tangents is not absolutely determinate; for we may, it is clear, in the place of $\Pi U=0$, write $\Pi U+M.U=0$, where M is an arbitrary function of the proper degree: a very elegant transformation in the case of the quartic is given in Hesse's paper, "Transformation der Gleichung der Curven 14ten Grades, welche eine gegebene Curve 4ten Grades in den Berührungspuncten ihrer Doppeltangenten schneiden," *Crelle*, t. LII. pp. 97—103 (1856).

Mr Salmon's work above referred to, contains the fundamental theorem of the *tangential* of a cubic, viz. a tangent to a cubic meets the cubic in a third point which lies on the second or line polar of the point of contact with respect to the Hessian. In my "Memoir on Curves of the Third Order," *Phil. Trans.* vol. CXLVII. (1857), pp. 415—446, art. No. 37, [146], I gave an identical equation relating to the tangential of a cubic, but which is not there exhibited in its proper form; this was afterwards effected by Mr Salmon, in the paper "On Curves of the Third Order," *Phil. Trans.* vol. CXLVIII. (1858), pp. 535—541. The equation, as given by Mr Salmon, is in the notation of the present memoir,

$$-\mathfrak{H} \,.\, U + \tfrac{1}{3}\mathfrak{D}\mathfrak{H} \,.\, DU - \tfrac{1}{3}DH \,.\, \mathfrak{D}\Upsilon + H \,.\, \Upsilon = 0,$$

an equation which in fact puts in evidence the last-mentioned theorem for the tangential of a cubic.

The idea occurred to me of considering, in the case of the higher plane curves, the *tangentials* of a given point of the curve, viz. the points in which the tangent again meets the curve; for by expressing that two of these tangentials were coincident, we should have the condition that the given point is the point of contact of a double tangent. But I was not able to complete the solution.

Finally, Mr Salmon discovered the equation of a curve of the order $m-2$, which by its intersections with the tangent at the given point determines the tangentials, and by expressing that the curve in question is touched by the tangent, he was led to a complete solution of the Double-tangent problem. Mr Salmon's result is given in the note, "On the Double Tangents to Plane Curves," in the *Philosophical Magazine* for October 1858. The discovery just referred to led me to the investigations of the present memoir, in which it will be seen that I obtain, for a curve of any order whatever, the identical equation corresponding to the before-mentioned equation obtained by Mr Salmon in the case of a cubic; which identical equation puts in evidence the theorem as to the tangentials of the curve, and may thus be considered as containing in itself the solution of the Double-tangent problem: the identical equation is besides interesting for its own sake, as a part of the theory of ternary quantics.

1. Mr Salmon's solution of the problem of double tangents is based upon the following analytical determination of the tangentials of any point of the curve.

Let

$$\Upsilon = (*\between X,\ Y,\ Z)^n = 0$$

be the equation of the given curve, $(X,\ Y,\ Z)$ being current coordinates; and let $(x,\ y,\ z)$ be the coordinates of a point on the curve, so that we have

$$U = (*\between x,\ y,\ z)^n = 0,$$

a condition satisfied by the coordinates of the point in question.

Then the tangent

$$V = (X\partial_x + Y\partial_y + Z\partial_z)\, U = 0$$

at the point $(x,\ y,\ z)$, meets the curve besides in $(n-2)$ points, which are the tangentials of the given point $(x,\ y,\ z)$, and which are determined as the intersections of the tangent $V=0$ with a certain curve,

$$\Omega = (\dagger\between x,\ y,\ z)^{n-2} = 0.$$

2. To express the equation of this curve, let U_1, U_2, ... be the successive emanants of U, taken with the facients of emanation $(x_{,}, y_{,}, z_{,})$, viz.

$$U_1 = \frac{1}{n} \left[\text{should be } \frac{1}{n-2}\right] (x_{,}\partial_x + y_{,}\partial_y + z_{,}\partial_z)\, U,$$

$$U_2 = \frac{1}{n(n-1)} \left[\text{should be } \frac{1}{(n-2)(n-3)}\right] (x_{,}\partial_x + y_{,}\partial_y + z_{,}\partial_z)^2 U,$$

$$\vdots$$

where it should be noticed that the numerical determination is such, that putting (x, y, z) for $(x_{,}, y_{,}, z_{,})$, then U_1, U_2, ... become respectively equal to U. [The numerical determination should have been and in the latter part of the memoir is assumed to be such as to render H_1, H_2, &c. equal to H, on making the substitution in question: the correction was made in a later memoir "On the double tangents of a curve of the fourth order."] Suppose also that H, H_1, H_2, ... are the Hessians of U, U_1, U_2, ..., viz. H is the determinant formed with the second derived functions of U with respect to (x, y, z), H_1 the like determinant with the second derived functions of U_1 with respect to the same quantities (x, y, z); and so on. Moreover let $D^{n-2}H, = (X\partial_x + Y\partial_y + Z\partial_z)^{n-2}H$, denote the $(n-2)$thic emanant of H with respect to the current coordinates (X, Y, Z) as facients of emanation; and similarly let $D^{n-2}H_1$, $D^{n-2}H_2$, ... denote the $(n-2)$thic emanants of H_1, H_2, ... in respect to the same facient of emanation—it being understood that in all these functions, $(x_{,}, y_{,}, z_{,})$ are after the differentiations to be replaced by (x, y, z). It is to be observed that U_r is of the degree $(n-r)$ in (x, y, z), and consequently H_r of the degree $3(n-2-r)$; hence $D^{n-2}H_r$ is of the degree $3(n-2-r)-(n-2)$, $= 2(n-2)-3r$, which implies that $r \not> \tfrac{2}{3}(n-2)$, for otherwise $D^{n-2}H_r$ would be identically equal to zero. Upon replacing $(x_{,}, y_{,}, z_{,})$ by (x, y, z), $D^{n-2}H_r$ (r satisfying the above condition) becomes of the degree $2(n-2)$ in (x, y, z), and it is obviously of the degree 3 in the coefficients of U, and of the degree $(n-2)$ in the current coordinates (X, Y, Z).

3. This being premised, we have

$$\Omega = (\dagger \between X, Y, Z)^{n-2},$$

$$= D^{n-2}H - \frac{n-1}{1} D^{n-2}H_1 + \&c. = 0,$$

for the equation of the curve of the order $(n-2)$, which by its intersection with the tangent gives the tangentials of the given point; the numerical coefficients are the binominal coefficients of the order $(n-1)$ taken with the signs + and − alternately, and the series is continued as long as the terms do not vanish, that is, if as before r denote the suffix of H, for so long as $r \not> \tfrac{2}{3}(n-2)$; but of course the value will not be altered by continuing the series to $r = n-1$. In particular, for the quartic we have

$$\Omega = D^2H - 3D^2H_1,$$

for the quintic

$$\Omega = D^3H - 4D^3H_1 + 6D^3H_2,$$

and so on. The function Ω, like the several component terms, is of course of the degree 3 in the coefficients of U, and of the degree $2(n-2)$ in (x, y, z).

4. It is to be remarked that the formula applies to a cubic; we have here simply $\Omega = DH$, which agrees with a result already mentioned. It may be noticed also that in the general case the formula gives at once the condition for the points of inflexion; in fact, if the point (x, y, z) be a point of inflexion, then one of the tangentials must coincide with this point, or the equation $\Omega = 0$ will be satisfied by writing therein (x, y, z) for (X, Y, Z); but when this is done $D^{n-2}H$, $D^{n-2}H_1$ &c. reduce themselves (to numerical factors *près*) to H, and the equation becomes simply $H = 0$, which is the well-known condition for the points of inflexion.

5. If two of the tangentials coincide, or what is the same thing, if the tangent $V = 0$ touches the curve $\Omega = 0$, then the point (x, y, z) will be the point of contact of a double tangent. The equation which expresses the condition in question, treating therein (x, y, z) as current coordinates, is consequently that of a curve, intersecting the given curve (now represented by $U = 0$) in the points of contact of the double tangents. The process leads to a determinate form $\Pi U = 0$, of the curve in question, but of course any curve whatever, $\Pi U + M . U = 0$, will intersect the curve $U = 0$ in the points of contact of the double tangents.

6. I write for the moment

$$\begin{aligned} \Omega &= (A, \ldots \between X, Y, Z)^{n-2} = 0, \\ V &= \xi X + \eta Y + \zeta Z \qquad = 0, \end{aligned}$$

for the two equations; the coefficients $(A, \ldots)$, as already mentioned, are of the degree $2(n-2)$ in (x, y, z) and of the degree 3 in the coefficients of U; or as we may express it,

$$A, \ldots = (a, \ldots)^3 (x, y, z)^{2(n-2)}.$$

In like manner ξ, η, ζ are of the degree $(n-1)$ in (x, y, z), and the degree 1 in the coefficients of U, or we may write

$$\xi, \eta, \zeta = (a, \ldots)^1 (x, y, z)^{n-1}.$$

7. The equation which expresses that the line $V = 0$ touches the curve $\Omega = 0$, is $\mathsf{F}\Omega = 0$, where the facients of the Reciprocant $\mathsf{F}\Omega$ are the coefficients (ξ, η, ζ) of the linear function. This equation is of the form

$$(A, \ldots)^{2(n-3)} (\xi, \eta, \zeta)^{(n-2)(n-3)} = 0;$$

or attending to the forms of $(A, \ldots)$ and (ξ, η, ζ), it is of the form

$$(a, \ldots)^{6(n-3) + (n-2)(n-3)} (x, y, z)^{4(n-2)(n-3) + (n-1)(n-2)(n-3)} = 0,$$

or what is the same thing, the form

$$(a, \ldots)^{(n+4)(n-3)} \qquad (x, y, z)^{(n-2)(n^2-9)} \qquad = 0,$$

viz. the curve through the points of contact of the double tangents is a curve of the order $(n-2)(n^2-9)$, and its equation contains the coefficients of the equation $U=0$ of the given curve in the degree $(n+4)(n-3)$. And since each double tangent corresponds to two points of contact, the number of double tangents is $\frac{1}{2}n(n-2)(n^2-9)$. This agrees with the before-mentioned results.

8. The whole problem is thus reduced to the demonstration of Mr Salmon's expression for the curve $\Omega=0$. To fix the ideas, consider the case of a quartic curve $\Upsilon=(*\between X, Y, Z)^4=0$, and let the function $U=(*\between x, y, z)^4$ (or as for shortness we may write it, $U=(x, y, z)^4$) and certain of its emanants be represented as follows, viz.—

$$\begin{aligned}
a &= (x, y, z)^4 \qquad\qquad\qquad\qquad ,\\
b &= (x, y, z)^3(X, Y, Z),\\
c &= (x, y, z)^2(X, Y, Z)^2,\\
d &= (x, y, z)\,(X, Y, Z)^3,\\
e &= \ldots\ldots\ldots\,(X, Y, Z)^4,\\[1ex]
a' &= (x, y, z)^3 \ldots\ldots\ldots\,(X', Y', Z'),\\
b' &= (x, y, z)^2(X, Y, Z)\,(X', Y', Z'),\\
c' &= (x, y, z)\,(X, Y, Z)^2(X', Y', Z'),\\
d' &= \ldots\ldots\,(X, Y, Z)^3(X', Y', Z'),\\
a'' &= (x, y, z)^2 \ldots\ldots\ldots\,(X', Y', Z')^2,\\
b'' &= (x, y, z)\,(X, Y, Z)\,(X', Y', Z')^2,\\
c'' &= \ldots\ldots\ldots\,(X, Y, Z)^2(X', Y', Z')^2,
\end{aligned}$$

where (X', Y', Z') are new arbitrary facients; but, as before, (X, Y, Z) are taken to be current coordinates, and (x, y, z) the coordinates of the given point on the curve:

$e=0$ is the equation of the curve;

$d=0$, the equation of the first or cubic polar of the point (x, y, z);

$b=0$, the equation of the last or line polar of the point (x, y, z), or what is the same thing (the point being on the curve), the tangent of the curve at this point;

$a=0$, the condition which expresses that the point is on the curve.

9. Imagine now an identical equation,

$$a\text{I}+b\text{II}+d\text{III}+e\text{IV}=0;$$

then, since $a=0$, we have

$$b\text{II}+d\text{III}+e\text{IV}=0;$$

and if in this equation we write $b=0$, $e=0$, it becomes $\text{III}d=0$, that is, the points of intersection of the curve $e=0$ and the tangent $b=0$ lie on one or other of the curves $d=0$, $\text{III}=0$. But the points in question do not lie on the curve $d=0$, consequently they lie on the curve $\text{III}=0$.

10. To explain the law of formation of the multipliers I, II, III, IV, I form the matrix

$$\left(\begin{array}{cccc|ccc} a, & b, & c, & d; & a', & b', & c' \\ b, & c, & d, & e; & b', & c', & d' \\ a', & b', & c', & d'; & a'', & b'', & c'' \end{array}\right),$$

and then we have

$$\text{I}=-\begin{vmatrix} d, & c, & b' \\ e, & d, & c' \\ d', & c', & b'' \end{vmatrix} + \begin{vmatrix} d, & b, & c' \\ e, & c, & d' \\ d', & b', & d'' \end{vmatrix},$$

$$\text{II}=-\begin{vmatrix} d, & c, & a' \\ e, & d, & b' \\ d', & c', & a'' \end{vmatrix} - \begin{vmatrix} d, & b, & b' \\ e, & c, & c' \\ d', & b', & b'' \end{vmatrix} - \begin{vmatrix} d, & a, & c' \\ e, & b, & d' \\ d', & a', & c'' \end{vmatrix},$$

$$\text{III}=-\begin{vmatrix} a, & b, & c' \\ b, & c, & d' \\ a', & b', & c'' \end{vmatrix} - \begin{vmatrix} a, & c, & b' \\ b, & d, & c' \\ a', & c', & b'' \end{vmatrix} - \begin{vmatrix} a, & d, & a' \\ b, & e, & b' \\ a', & d', & a'' \end{vmatrix},$$

$$\text{IV}=\begin{vmatrix} a, & b, & b' \\ b, & c, & c' \\ a', & b', & b'' \end{vmatrix} + \begin{vmatrix} a, & c, & a' \\ b, & d, & b' \\ a', & c', & a'' \end{vmatrix},$$

values which, as I proceed to show, satisfy the identical equation

$$a\text{I}+b\text{II}+d\text{III}+e\text{IV}=0.$$

11. We have in fact

$$\begin{aligned} \text{I}= &\ \ d\,(db''-c'^2+cc''-b'd') \\ &+e\,(b'c'-b''c+b'c'-bc'') \\ &+d'(cc'-db'+bd'-cc'), \end{aligned}$$

where the last line is $=bd'^2-db'd'$;

$$\begin{aligned} \text{II}= &\ \ d\,(b'c'-a''d+b'c'-b''c+a'd'-bc'') \\ &+e\,(ca''-a'c'+bb''-b'^2+ac''-a'c') \\ &+d'(a'd-b'c+b'c-bc'+bc'-ad'), \end{aligned}$$

where the last line is $da'd'-ad'^2$;

$$\begin{aligned} \text{III} = \; & a\,(-b'd' + cc'' + c'^2 - b''d + b'd' - ea'') \\ & + b\,(\;bc'' - b'c' + b''c - b'c' + a''d - a'd') \\ & + a'(\;cc' - bd' + b'd - cc' + a'e - b'd\;), \end{aligned}$$

where the last line is $-ba'd' + ea'^2$; and

$$\begin{aligned} \text{IV} = \; & a\,(b''c - b'c' + da'' - b'c') \\ & + b\,(b'^2 - bb'' + a'c' - a''c) \\ & + a'(bc' - b'c + cb' - a'd), \end{aligned}$$

where the last line is $ba'c' - a'^2d$.

These values may be expressed as follows:

$$\begin{array}{rll} \text{I} = & a(\;0\;) & \\ & +b(\;d'^2\;) & (12) \\ & +d(\;b''d + c''c - b'd' - c'c' - d'b'\;) & (13) \\ & +e(-b''c - c''b + b'c' + c'b'\;), & (14) \end{array}$$

$$\begin{array}{rll} \text{II} = & a(-\;d'^2\;) & (21) \\ & +b(\;0\;) & \ldots \\ & +d(-a''d - b''c - c''b + a'd' + b'c' + c'b' + d'a') & (23) \\ & +e(\;a''c + b''b + c''a - a'c' - b'b' - c'a'\;), & (24) \end{array}$$

$$\begin{array}{rll} \text{III} = & a(-b''d - c''d + b'd' + c'c' + d'b'\;) & (31) \\ & +b(\;a''d + b''c + c''b - a'd' - b'c' - c'b' - d'a') & (32) \\ & +d(\;0\;) & \ldots \\ & +e(-a''a + a'a'\;), & (34) \end{array}$$

$$\begin{array}{rll} \text{IV} = & a(\;b''c + c''b - b'c' - c'b'\;) & (41) \\ & +b(-a''c - b''b - c''a + a'c' + b'b' + c'a'\;) & (42) \\ & +d(\;a''a - a'a'\;) & (43) \\ & +e(\;0\;), & \ldots \end{array}$$

which are of the form

$$\begin{aligned}
\text{I} &= a\ \ 0\ \ + b\,(12) + d\,(13) + e\,(14),\\
\text{II} &= a\,(21) + b\ \ 0\ \ + d\,(23) + e\,(24),\\
\text{III} &= a\,(31) + b\,(32) + d\ \ 0\ \ + e\,(34),\\
\text{IV} &= a\,(41) + b\,(42) + d\,(43) + e\ \ 0\ \ ,
\end{aligned}$$

where $(12) = -(21)$, &c., and which therefore satisfy the equation

$$a\text{I} \quad + b\text{II} \quad + d\text{III} \quad + e\text{IV} = 0.$$

12. The equation of the curve which by its intersection with the tangent gives the tangentials, is

$$\text{III} = -\begin{vmatrix} a, & b, & c' \\ b, & c, & d' \\ a', & b', & c'' \end{vmatrix} - \begin{vmatrix} a, & c, & b' \\ b, & d, & c' \\ a', & c', & b'' \end{vmatrix} - \begin{vmatrix} a, & d, & a' \\ b, & e, & b' \\ a', & d', & a'' \end{vmatrix} = 0,$$

the degrees of which are

in the coefficients of U, 3,
in (x, y, z) 6,
in (X, Y, Z) 4,
in (X', Y', Z') 2;

and it only remains to divest this equation of a factor which it contains,

$$\begin{vmatrix} x, & y, & z \\ X, & Y, & Z \\ X', & Y', & Z' \end{vmatrix}^2,$$

which being thrown out, the equation will be independent of (X', Y', Z') and will be of the degrees

in the coefficients of U, 3,
in (x, y, z) 4,
in (X, Y, Z) 2,

and will in fact be the before-mentioned equation $\Omega = D^2H - 3D^2H_1 = 0$.

13. Write for shortness,

$$\begin{vmatrix} x, & y, & z \\ X, & Y, & Z \\ X', & Y', & Z' \end{vmatrix} = \Lambda,$$

it is to be shown that

$$\text{III} = -\tfrac{1}{2}\Lambda^2 (D^2H - 3D^2H_1).$$

14. To effect this I remark that we have identically

$$\begin{vmatrix} a, & b, & a' \\ b, & c, & b' \\ a', & b', & a'' \end{vmatrix} = \Lambda^2 H;$$

and I proceed to operate upon this equation with $D = X\partial_x + Y\partial_y + Z\partial_z$.

I notice that

$$a, \quad b, c, d, e; \qquad a', \quad b', c', d'; \qquad a'', \quad b'', c''$$

are in regard to (x, y, z) of the degrees

$$4, \quad 3, 2, 1, 0; \qquad 3, \quad 2, 1, 0; \qquad 2, \quad 1, 0;$$

or what is the same thing, since for the case in hand $n = 4$, of the degrees

$$n, n-1, \ldots; \qquad n-1, n-2, \ldots; \quad n-2, n-3, \ldots$$

and we have

$$Da = nb,\ Db = (n-1)c, \ldots \quad Da' = (n-1)b',\ Db' = (n-2)c', \ldots \quad Da'' = (n-2)b'', \ldots$$

15. In the determinant

$$\begin{vmatrix} a, & b, & a' \\ b, & c, & b' \\ a', & b', & a'' \end{vmatrix}, = \Lambda^2 H,$$

the degrees of the terms (other than each top term, the degree of which is higher by unity) in the several columns are $n-1$, $n-2$, $n-2$; if then we operate on the determinant with D, and as regards the top terms we write

$$\begin{aligned} Da &= b + (n-1)b, \\ Db &= c + (n-2)c, \\ Da' &= b' + (n-2)b', \end{aligned}$$

we have in the first place a term

$$\begin{vmatrix} b, & c, & b' \\ b, & c, & b' \\ a', & b', & a'' \end{vmatrix}$$

which vanishes, and next the terms

$$(n-1)\begin{vmatrix} b, & b, & a' \\ c, & c, & b' \\ b', & b', & a'' \end{vmatrix} + (n-2)\begin{vmatrix} a, & c, & a' \\ b, & d, & b' \\ a', & c', & a'' \end{vmatrix} + (n-2)\begin{vmatrix} a, & b, & b' \\ b, & c, & c' \\ a', & b', & b'' \end{vmatrix}$$

the first of which vanishes. On the right-hand side $D\Lambda = 0$ identically, and therefore $D.\Lambda^2H = \Lambda^2DH$, or we have

$$(n-2)\begin{vmatrix} a, & c, & a' \\ b, & d, & b' \\ a', & c', & a'' \end{vmatrix} + (n-2)\begin{vmatrix} a, & b, & b' \\ b, & c, & c' \\ a', & b', & b'' \end{vmatrix} = \Lambda^2DH.$$

16. I repeat the operation D: we have

$$\left.\begin{array}{l} (n-2)(n-1)\begin{vmatrix} b, & c, & a' \\ c, & d, & b' \\ b', & c', & a'' \end{vmatrix} + (n-2)(n-1)\begin{vmatrix} b, & b, & b' \\ c, & c, & c' \\ b', & b', & b'' \end{vmatrix} \\ + (n-2)(n-3)\begin{vmatrix} a, & d, & a \\ b, & c, & b \\ a', & d', & a'' \end{vmatrix} + (n-2)(n-2)\begin{vmatrix} a, & c, & b' \\ b, & d, & c' \\ a', & c', & b'' \end{vmatrix} \\ + (n-2)(n-2)\begin{vmatrix} a, & c, & b' \\ b, & d, & c' \\ a', & c', & b'' \end{vmatrix} + (n-2)(n-3)\begin{vmatrix} a, & b, & c' \\ b, & c, & d' \\ a', & b', & c'' \end{vmatrix} \end{array}\right\} = \Lambda^2D^2H;$$

or, collecting the different terms,

$$(n-2)(n-3)\begin{vmatrix} a, & b, & c' \\ b, & c, & d' \\ a', & b', & c'' \end{vmatrix} + 2(n-2)^2\begin{vmatrix} a, & c, & b' \\ b, & d, & c' \\ a', & c', & b'' \end{vmatrix} + (n-2)(n-3)\begin{vmatrix} a, & d, & a' \\ b, & e, & b' \\ a', & d', & a'' \end{vmatrix} + (n-2)(n-1)\begin{vmatrix} b, & c, & a' \\ c, & d, & b' \\ b', & c', & a'' \end{vmatrix} = \Lambda^2D^2H$$

17. A little consideration will show that in this equation we may write $n-1$ for n, and H_1 for H. In fact, putting for a moment $\delta = x_{\prime}\partial_x + y_{\prime}\partial_y + z_{\prime}\partial_z$, we have corresponding to the equation

$$\begin{vmatrix} a, & b, & a' \\ b, & c, & b' \\ a', & b', & a'' \end{vmatrix} = \Lambda^2H,$$

this other equation,

$$\begin{vmatrix} \delta a, & \delta b, & \delta a' \\ \delta b, & \delta c, & \delta b' \\ \delta a', & \delta b', & \delta a'' \end{vmatrix} = \Lambda^2H_1,$$

where ultimately $(x_{\prime}, y_{\prime}, z_{\prime})$ are to be replaced by (x, y, z). We may operate upon this equation with D, D^2, ... as before, the only difference being that in the first instance δa, δb, &c. are as regards (x, y, z) of degrees lower by unity than a, b, &c., that is $n-1$ must be substituted throughout in the place of n; and when at the end of the process $(x_{,}, y_{,}, z_{,})$ are replaced by (x, y, z), then δa, δb, &c. become equal to a, b, &c., from which the truth of the asserted proposition is manifest.

18. Hence writing $n=4$, we have

$$2\begin{vmatrix} a, & b, & c' \\ b, & c, & d' \\ a', & b', & c'' \end{vmatrix} + 8\begin{vmatrix} a, & d, & a' \\ b, & e, & b' \\ a', & d', & a'' \end{vmatrix} + 2\begin{vmatrix} a, & d, & a' \\ b, & e, & b' \\ a', & d', & a'' \end{vmatrix} + 6\begin{vmatrix} b, & c, & a' \\ c, & d, & b' \\ b', & c', & a'' \end{vmatrix} = \Lambda^2 D^2 H,$$

$$2\begin{vmatrix} a, & d, & a' \\ b, & e, & b' \\ a, & d', & a'' \end{vmatrix} + 2\begin{vmatrix} b, & c, & a' \\ c, & d, & b' \\ b', & c', & a'' \end{vmatrix} = \Lambda^2 D^2 H_1;$$

and hence

$$2\left\{\begin{vmatrix} a, & b, & c' \\ b, & c, & d' \\ a', & b', & c'' \end{vmatrix} + \begin{vmatrix} a, & c, & b' \\ b, & d, & c' \\ a', & c', & b'' \end{vmatrix} + \begin{vmatrix} a, & d, & a' \\ b, & e, & b' \\ a', & d', & a'' \end{vmatrix}\right\} = \Lambda^2(D^2H - 3D^2H_1),$$

which is the required equation

$$\text{III} = -\tfrac{1}{2}\Lambda^2(D^2H - 3D^2H_1).$$

19. It is to be added, that the equation for $\Lambda^2 DH$ gives $\text{IV} = \frac{1}{2}\Lambda^2 DH$; the values of II and I are at once obtained from those of III and IV by interchanging (x, y, z) and (X, Y, Z). Hence if we represent by $\mathfrak{H}$, $\mathfrak{D}$, &c. the values which H, D, &c. assume by this interchange, we may write

$$\begin{aligned} \text{I} &= -\tfrac{1}{2}\Lambda^2\mathfrak{D}\mathfrak{H}, \\ \text{II} &= +\tfrac{1}{2}\Lambda^2(\mathfrak{D}^2\mathfrak{H} - 3\mathfrak{D}^2\mathfrak{H}_1), \\ \text{III} &= -\tfrac{1}{2}\Lambda^2(D^2H - 3D^2H_1), \\ \text{IV} &= +\tfrac{1}{2}\Lambda^2 DH; \end{aligned}$$

and the identical equation,

$$a\text{I} + b\text{II} + d\text{III} + e\text{IV} = 0,$$

gives therefore

$$-\mathfrak{D}\mathfrak{H}\,.\,U + \tfrac{1}{4}(\mathfrak{D}^2\mathfrak{H} - 3\mathfrak{D}^2\mathfrak{H}_1)DU - \tfrac{1}{4}(D^2H - 3D^2H_1)\mathfrak{D}\Upsilon + DH\,.\,\Upsilon = 0,$$

which is of itself sufficient to put in evidence the property that the curve $D^2H - 3D^2H_1 = 0$ gives by its intersections with the tangent $DU=0$, the tangentials of the point (x, y, z). The last-mentioned equation is the equation for a quartic corresponding to Mr Salmon's equation

$$-\mathfrak{H}\,.\,U + \tfrac{1}{3}\mathfrak{D}\mathfrak{H}\,.\,DU - \tfrac{1}{3}DH\,.\,\mathfrak{D}\Upsilon + H\,.\,\Upsilon = 0$$

for the cubic $U=0$.

20. It is worth while to give the investigation of the equation for the cubic; the matrix is

$$\left(\begin{array}{ccc|cc} a, & b, & c; & a', & b' \\ b, & c, & d; & b', & c' \\ a', & b', & c'; & a'', & b'' \end{array}\right)$$

and the identical equation is

$$a\mathrm{I} + b\mathrm{II} + c\mathrm{III} + d\mathrm{IV} = 0,$$

where

$$\mathrm{I} = \begin{vmatrix} c, & b, & b' \\ d, & c, & c' \\ c', & b', & b'' \end{vmatrix},$$

$$\mathrm{II} = -\begin{vmatrix} c, & b, & a' \\ d, & c, & b' \\ c', & b', & a'' \end{vmatrix} - \begin{vmatrix} c, & a, & b' \\ d, & b, & c' \\ c', & a', & b'' \end{vmatrix},$$

$$\mathrm{III} = -\begin{vmatrix} a, & b, & b' \\ b, & c, & c' \\ a', & b', & b'' \end{vmatrix} - \begin{vmatrix} a, & c, & a' \\ b, & d, & b' \\ a', & c', & a'' \end{vmatrix},$$

$$\mathrm{IV} = \begin{vmatrix} a, & b, & a' \\ b, & c, & b' \\ a', & b', & a'' \end{vmatrix},$$

or, as we may express them,

$$\begin{array}{rll} \mathrm{I} = & a(\ 0\ ) & \\ & +b(\ c'^2\ ) & (12) \\ & +c(b''c - b'c' - c'b') & (13) \\ & +d(-b''b + b'b'\ ), & (14) \end{array}$$

$$\begin{array}{rll} \mathrm{II} = & a(\ -c'^2\ ) & (21) \\ & +b(\ 0\ ) & \ldots \\ & +c(-a''c - b''b + a'c' + b'b' + c'a') & (23) \\ & +d(\ a''b + b''a - a'b' - b'a'\ ), & (24) \end{array}$$

$$\begin{array}{rll} \mathrm{III} = & a(-b''c + b'c' + c'b'\ ) & (31) \\ & +b(\ a''c + b''b - a'c' - b'b' - c'a') & (32) \\ & +c(\ 0\ ) & \ldots \\ & +d(-aa'' \quad + a'^2\ ), & (34) \end{array}$$

$$\begin{array}{rll} \mathrm{IV} = & a(b''b \quad - b'b'\ ) & (41) \\ & +b(-a''b - b''a + a'b' + b'a'\ ) & (42) \\ & +c(\ a''a \quad - a'a'\ ) & (43) \\ & +d(\ 0\ ), & \ldots \end{array}$$

which verify the identical equation. We have $\text{III} = -\Lambda^2 DH$, $\text{IV} = \Lambda^2 H$, and thence $\text{II} = +\Lambda^2 \mathfrak{D}\mathfrak{H}$, $\text{I} = -\Lambda^2 \mathfrak{H}$; hence the equation in question

$$-\mathfrak{H}\,.\,U + \tfrac{1}{3}\mathfrak{D}\mathfrak{H}\,.\,DU - \tfrac{1}{3}DH\,.\,\mathfrak{D}\Upsilon + H\,.\,\Upsilon = 0.$$

21. One other example will be sufficient to render manifest the law of the formation of the multipliers I, II, III, IV.

In the case of a sextic curve we have the matrix

$$\begin{pmatrix} a, & b, & c, & d, & e, & f; & a', & b', & c', & d', & e' \\ b, & c, & d, & e, & f, & g; & b', & c', & d', & e', & f' \\ a', & b', & c', & d', & e', & f'; & a'', & b'', & c'', & d'', & e'' \end{pmatrix},$$

the identical equation is

$$a\text{I} + b\text{II} + f\text{III} + g\text{IV} = 0;$$

and the expressions for the multipliers I, II, III, IV are:

$$\text{I} = \begin{vmatrix} f, & e, & b' \\ g, & f, & c' \\ f', & e', & b'' \end{vmatrix} + \begin{vmatrix} f, & d, & c' \\ g, & e, & d' \\ f', & d', & c'' \end{vmatrix} + \begin{vmatrix} f, & c, & d \\ g, & d, & e \\ f', & c', & d'' \end{vmatrix} + \begin{vmatrix} f, & b, & e' \\ g, & c, & f' \\ f', & b', & e'' \end{vmatrix},$$

$$\text{II} = -\begin{vmatrix} f, & e, & a' \\ g, & f, & b' \\ f', & e', & a'' \end{vmatrix} - \begin{vmatrix} f, & d, & b' \\ g, & e, & c' \\ f', & d', & b'' \end{vmatrix} - \begin{vmatrix} f, & c, & c' \\ g, & d, & d' \\ f', & c', & d'' \end{vmatrix} - \begin{vmatrix} f, & b, & d' \\ g, & c, & e' \\ f', & b', & d'' \end{vmatrix} - \begin{vmatrix} f, & a, & e' \\ g, & b, & f' \\ f', & a', & e'' \end{vmatrix},$$

$$\text{III} = -\begin{vmatrix} a, & b, & e' \\ b, & c, & f' \\ a', & b', & e'' \end{vmatrix} - \begin{vmatrix} a, & c, & d' \\ b, & d, & e' \\ a', & c', & e'' \end{vmatrix} - \begin{vmatrix} a, & d, & c' \\ b, & e, & d' \\ a', & d', & e'' \end{vmatrix} - \begin{vmatrix} a, & e', & b' \\ b, & f', & c' \\ a', & e'', & b'' \end{vmatrix} - \begin{vmatrix} a, & f, & a' \\ b, & g, & b' \\ a', & f', & a'' \end{vmatrix},$$

$$\text{IV} = \begin{vmatrix} a, & b, & d' \\ b, & c, & e' \\ a', & b', & d'' \end{vmatrix} + \begin{vmatrix} a, & c, & c' \\ b, & d, & d' \\ a', & d', & c'' \end{vmatrix} - \begin{vmatrix} a, & d, & b' \\ b, & e, & c' \\ a', & d', & b'' \end{vmatrix} - \begin{vmatrix} a, & e, & a' \\ b, & f, & b' \\ a', & e', & a'' \end{vmatrix}.$$

22. We have in fact

$$\begin{aligned} \text{I} = \ & a(0 &&) \\ & +b(-ge'' && +ff' &&) \quad (12) \\ & +f(\ b''f + c''e + d''d + e''c && -b'f' - c'e' - d'd' - e'c' - f'b' &&) \quad (13) \\ & +g(-b''e - c''d - d''c && +b'e' + c'd' + d'c' + e'b' &&), \quad (14) \end{aligned}$$

$$\begin{aligned} \text{II} = \ & a(\ e''g && -f'f' &&) \quad (21) \\ & +b(0 && &&) \quad \ldots \\ & +f(-a''f - b''e - c''d - d''e - e''b + a'f' + b'e' + c'd' + d'c' + e'b' + f'a') && && \quad (23) \\ & +g(\ a''e + b''d + c''c + d''b && -a'e' - b'd' - c'c' - d'b' - e'a' &&), \quad (24) \end{aligned}$$

$$\begin{aligned}
\text{III} = &\ a(-b''f - c''e - d''d - e''c \quad + b'f' + c'e' + d'd' + e'c' + f'b' \quad) \quad (31)\\
&+ b(+a''f + b''e + c''d + d''c + e''b - a'f' - b'e' - c'd' - d'c' - e'b' - f'a') \quad (32)\\
&+ f\ 0 \quad \ldots\\
&+ g(-a''a \quad + a'^2 \quad), \quad (34)\\
\text{IV} = &\ a(\ b''e + c''d + d''c \quad - b'e' - c'd' - d'c - e'b' \quad) \quad (41)\\
&+ b(-a''e - b''d - c''c - d''b \quad + a'e' + b'd' + c'c' + d'b' + e'a' \quad) \quad (42)\\
&+ f(\ a''a \quad - a'^2 \quad) \quad (43)\\
&+ g(0 \quad), \quad \ldots
\end{aligned}$$

which are of the form

$$\begin{aligned}
\text{I} &= a\ \ 0 \ \ + b(12) + f(13) + g(14),\\
\text{II} &= a(21) + b\ \ 0\ \ + f(23) + g(24),\\
\text{III} &= a(31) + b(32) + f\ \ 0\ \ + g(34),\\
\text{IV} &= a(41) + b(42) + f(43) + g\ \ 0\ ,
\end{aligned}$$

where $(12) = -(21)$ &c., and the equation

$$a\text{I} \quad + b\text{II} \quad + f\text{III} \quad + g\text{IV} = 0$$

is consequently satisfied.

23. The expression

$$\text{III} = -\begin{vmatrix} a, & b, & e' \\ b, & c, & f' \\ a', & b', & e'' \end{vmatrix} - \begin{vmatrix} a, & c, & d' \\ b, & d, & e' \\ a', & c', & d'' \end{vmatrix} - \begin{vmatrix} a, & d, & c' \\ b, & e, & d' \\ a', & d', & c'' \end{vmatrix} - \begin{vmatrix} a, & f, & a' \\ b, & g, & b' \\ a', & f', & a'' \end{vmatrix}$$

leads to

$$\text{III} = -\Lambda^2 (D^4H - 5D^4H_1 + 10D^4H_2),$$

and consequently the equation of the curve which by its intersections with the tangent determines the tangentials of a point of a sextic, is

$$D^4H - 5D^4H_1 + 10D^4H_2 = 0.$$

24. In the general case of a curve of the order n the matrix is

$$\left(\begin{array}{llll} a_0, & a_1, & a_2 \ldots a_{n-1}; & a_0', \ a_1' \ldots a'_{n-2} \\ a_1, & a_2, & a_3 \ldots a_n\ ; & a_1', \ a_2' \ldots a'_{n-1} \\ a_0', & a_1', & a_2' \ldots a'_{n-1}; & a_0'', \ a_1'' \ldots a''_{n-2} \end{array}\right)$$

where, in analogy with what precedes,

$$\begin{aligned}
a_0 &= (X,\ Y,\ Z)^n \quad ,\\
a_1 &= (X,\ Y,\ Z)^{n-1}(x,\ y,\ z) \quad ,\\
&\ \vdots\\
a_{n-1} &= (X,\ Y,\ Z)\ \ (x,\ y,\ z)^{n-1},\\
a_n &= \qquad\qquad\quad (x,\ y,\ z)^n \quad ,
\end{aligned}$$

and similarly for the accented letters, so that

$a_0 \;\; = 0$ is the equation of the curve;

$a_1 \;\; = 0$ is the equation of the first or $(n-1)$thic polar;

$a_{n-1} = 0$ is the equation of the last or line-polar, or what is the same thing, since (x, y, z) is a point on the curve, the tangent at this point;

$a_n \;\; = 0$, the condition which expresses that (x, y, z) is a point of the curve;

and we have to form the identical equation

$$a_0 \mathrm{I} + a_1 \mathrm{II} + a_{n-1} \mathrm{III} + a_n \mathrm{IV} = 0.$$

25. If, for shortness, the columns of the last-mentioned matrix are represented by

$$1,\ 2,\ 3 \ldots n,\ (1),\ (2) \ldots (n-1),$$

and the determinants formed with these columns respectively by a corresponding notation $\{1, 2, (1)\}$, $\{1, 2, (2)\}$, &c., then the expressions for the multipliers I, II, III, IV are as follows, viz.

$$\begin{aligned}
\mathrm{I} &= \{n, n-1, (2)\} + \{n, n-2, (3)\} \ldots + \{n, 2, (n-1)\} \\
\mathrm{II} &= -\{n, n-1, (1)\} - \{n, n-2, (2)\} \ldots - \{n, 2, (n-2)\} - \{n, 1, (n-1)\}, \\
\mathrm{III} &= -\{1, 2, (n-1)\} - \{1, 3, (n-2)\} \ldots - \{1, n-1, (2)\} - \{1, n, (1)\}, \\
\mathrm{IV} &= \{1, 2, (n-2)\} + \{1, 3, (n-1)\} \ldots + \{1, n-1, (1)\};
\end{aligned}$$

the truth of the identical equation being shown, as in the foregoing special cases, by the transformation of the multipliers into the form

$$\begin{aligned}
\mathrm{I} &= a_0\ 0 \ + a_1(12) + a_{n-1}(13) + a_n(14), \\
\mathrm{II} &= a_0(21) + a_1\ 0 \ + a_{n-1}(23) + a_n(24), \\
\mathrm{III} &= a_0(31) + a_1(32) + a_{n-1}\ 0 \ + a_n(34), \\
\mathrm{IV} &= a_0(41) + a_1(42) + a_{n-1}(43) + a_n\ 0\ ,
\end{aligned}$$

where $(12) = -(21)$, &c.: the required expressions may be written down without difficulty.

26. Proceeding then to reduce the equation

$$\mathrm{III} = -\{1, 2, (n-1)\} - \{1, 3, (n-2)\} \ldots - \{1, n-1, (2)\} - \{1, n, (1)\},$$

we have the equation

$$\{1, 2, (1)\} = \Lambda^2 H,$$

which is to be successively operated on with D. The degrees (less unity) of the columns

$$1, \quad 2, \quad \ldots n-1, \quad n, \quad (1) \quad (2), \quad \ldots n-1,$$

are

$$n-1, \ n-2, \ldots \ 1, \quad 0, \ n-2, \ n-3, \ldots \ 0;$$

and the rule is to operate on each column of the determinant, multiplying by the degree less unity, and increasing the symbolical number by unity. Thus

$$\begin{aligned}
D\{1, 2, (1)\} &= (n-1)\{2, 2, (1)\} + (n-2)\{1, 3, (1)\} + (n-2)\{1, 2, (2)\}, \\
&= \qquad\qquad\qquad\qquad (n-2)\{1, 3, (1)\} + (n-2)\{1, 2, (2)\},
\end{aligned}$$

since $\{2, 2, (1)\}$ vanishes identically. The following Table shows the mode of effecting the operations:

H	1			= 1	$[n-2]^0$	$[n-2]^0$		12(1)
DH	1			= 1	$[n-2]^0$	$[n-2]^1$		12(2)
		1		1	$[n-2]^1$	$[n-2]^0$		13(1)
D^2H	1			= 1	$[n-2]^0$	$[n-2]^2$		12(3)
	1	1		2	$[n-2]^1$	$[n-2]^1$		13(2)
		1		1	$[n-2]^2$	$[n-2]^0$		14(1)
			1	1	$[n-2]^1$	$[n-2]^0$	$[n-1]^1$	23(1)
D^3H	1			1	$[n-2]^0$	$[n-2]^3$		12(4)
	2	1		3	$[n-2]^1$	$[n-2]^2$		13(3)
	1	2		3	$[n-2]^2$	$[n-2]^1$		14(2)
		1		1	$[n-2]^3$	$[n-2]^0$		15(1)
	1		2	3	$[n-2]^1$	$[n-2]^1$	$[n-1]^1$	23(2)
		1	1	2	$[n-2]^2$	$[n-2]^0$		24(1)
D^4H	1			= 1	$[n-2]^0$	$[n-2]^4$		12(5)
	3	1		4	$[n-2]^1$	$[n-2]^3$		13(4)
	3	3		6	$[n-2]^2$	$[n-2]^2$		14(3)
	1	3		4	$[n-2]^3$	$[n-2]^1$		15(2)
		1		1	$[n-2]^4$	$[n-2]^0$		16(1)
	3		3	6	$[n-2]^1$	$[n-2]^2$	$[n-1]^1$	23(3)
	2	3	3	8	$[n-2]^2$	$[n-2]^1$		24(2)
		2	1	3	$[n-2]^3$	$[n-2]^0$		25(1)
			2	2	$[n-2]^2$	$[n-2]^0$	$[n-1]^2$	34(1)
D^5H	1			= 1	$[n-2]^0$	$[n-2]^5$		12(6)
	4	1		5	$[n-2]^1$	$[n-2]^4$		13(5)
	6	4		10	$[n-2]^2$	$[n-2]^3$		14(4)
	4	6		10	$[n-2]^3$	$[n-2]^2$		15(3)
	1	4		5	$[n-2]^4$	$[n-2]^1$		16(2)
		1		1	$[n-2]^5$	$[n-2]^0$		17(1)
	6		4	10	$[n-2]^1$	$[n-2]^4$	$[n-1]^1$	23(4)
	8	6	6	20	$[n-2]^2$	$[n-2]^3$		24(3)
	3	8	4	15	$[n-2]^3$	$[n-2]^2$		25(2)
		3	1	4	$[n-2]^4$	$[n-2]^1$		26(1)
	2		8	10	$[n-2]^2$	$[n-2]^1$	$[n-1]^2$	34(2)
		2	3	5	$[n-2]^3$	$[n-2]^0$		35(1)
D^6H	1			= 1	$[n-2]^0$	$[n-2]^6$		12(7)
	5	1		6	$[n-2]^1$	$[n-2]^5$		13(6)
	10	5		15	$[n-2]^2$	$[n-2]^4$		14(5)
	10	10		20	$[n-2]^3$	$[n-2]^3$		15(4)
	5	10		15	$[n-2]^4$	$[n-2]^2$		16(3)
	1	5		6	$[n-2]^5$	$[n-2]^1$		17(2)
		1		1	$[n-2]^6$	$[n-2]^0$		18(1)
	10		5	15	$[n-2]^1$	$[n-2]^4$	$[n-1]^1$	23(5)
	20	10	10	40	$[n-2]^2$	$[n-2]^3$		24(4)
	15	20	10	45	$[n-2]^3$	$[n-2]^2$		25(3)
	4	15	5	24	$[n-2]^4$	$[n-2]^1$		26(2)
		4	1	5	$[n-2]^5$	$[n-2]^0$		27(1)
	10		20	30	$[n-2]^2$	$[n-2]^2$	$[n-1]^2$	34(3)
	5	10	15	30	$[n-2]^3$	$[n-2]^1$		35(2)
		5	4	9	$[n-2]^4$	$[n-2]^0$		36(1)
			5	5	$[n-2]^3$	$[n-2]^0$	$[n-1]^3$	45(1)
D^7H	&c.							

where the first three columns show the numbers which give, by the addition of the numbers in the same horizontal line, the numerical coefficients of the factorials which multiply the different terms of H, DH, &c., and where in the last column 12(1), &c. are written for shortness in the place of $\{1, 2, (1)\}$, &c.

27. It is clear that we have in general

$$\begin{aligned}
\Lambda^2 D^r H = \quad & 1\,[n-2]^0[n-2]^r\ \{1,\ 2,\ (r+1)\}\\
+\ & \frac{r}{1}\,[n-2]^1[n-2]^{r-1}\{1,\ 3,\ (r)\}\\
+\ & \frac{r\,.\,r-1}{1\,.\,2}\,[n-2]^2[n-2]^{r-2}\{1,\ 4,\ (r-1)\}\\
& \vdots\\
+\ & 1\,[n-2]^r[n-2]^0\ \{1,\ r+2,\ (1)\}\\[6pt]
+[n-1]^1 & \left\{\begin{array}{l} +R_0'\ [n-2]^1\ [n-2]^{r-2}\ \{2,\ 3,\ (r-1)\}\\ +R_1'\ [n-2]^2\ [n-2]^{r-3}\ \{2,\ 4,\ (r-2)\}\\ \vdots\\ +R'_{r-2}[n-2]^{r-1}[n-2]^0\ \{2,\ r+1,\ (1)\}\end{array}\right.\\[6pt]
+[n-1]^2 & \left\{\begin{array}{l} +R_0''\ [n-2]^2\ [n-2]^{r-4}\ \{3,\ 4,\ (r-3)\}\\ \vdots\\ +R''_{r-4}[n-2]^{r-2}[n-2]^0\ \{3,\ r,\ (1)\}\end{array}\right.\\
& \vdots\\
+[n-1]^{\frac{1}{2}r} & \left\{\begin{array}{l} +R_0^{\frac{1}{2}r}\ [n-2]^{\frac{1}{2}r}\ [n-2]^0\{\tfrac{1}{2}r+1\,,\ \tfrac{1}{2}r+2\,,\ (1)\}\\ r \text{ even; or } r \text{ odd}\end{array}\right.\\[6pt]
+[n-1]^{\frac{1}{2}(r-1)} & \left\{\begin{array}{l} +R_0^{\frac{1}{2}(r-1)}[n-2]^{\frac{1}{2}(r-1)}[n-2]^1\{\tfrac{1}{2}(r+1),\ \tfrac{1}{2}(r+3),\ (2)\}\\ +R_0^{\frac{1}{2}(r+1)}[n-2]^{\frac{1}{2}(r+1)}[n-2]^0\{\tfrac{1}{2}(r+1),\ \tfrac{1}{2}(r+5),\ (1)\};\end{array}\right.
\end{aligned}$$

and the general term is

$$[n-1]^\delta_{\cdot}\ R_s^\delta[n-2]^{\delta+s}[n-2]^{r-2\delta-s}\{\delta+1,\ \delta+2+s,\ (r-2\delta+1-s)\},$$

where s extends from $s=0$ to $s=r-2\delta$, and δ from $\delta=0$ to $\delta=\frac{1}{2}r$ or $\frac{1}{2}(r-1)$, according as r is even or odd. The expression for the coefficients R_s^0 is

$$R_s^0 = \frac{[r]^s}{[s]^s},$$

and that of the other coefficients $R_{s\delta}$ ($\delta = \text{or} < 1$) is not required for the present purpose.

28. According to a remark already made, the expressions for D^rH_1, D^rH_2, &c. are at once obtained from that for D^rH by merely writing $n-1$, $n-2$, &c. in the place of n: it is however to be noticed, that the quantity within the [ ] must not be negative, and that on its becoming so, the factorial is to be omitted.

29. I write now

$$+s=\alpha, \quad r-2\delta-s=\beta,$$

and I consider the expression

$$D^r H - \frac{n-1}{1} D^r H_1 + \&c.;$$

the general term of which is

$$R_s^\delta\{\delta+1,\ \delta+2+s,\ (r-2\delta+1-s)\}$$
$$\times \left\{ \begin{array}{ll} & [n-1]^\delta [n-2]^\alpha [n-2]^\beta \\ -\dfrac{n-1}{1} & [n-2]^\delta [n-3]^\alpha [n-3]^\beta \\ + & \&c. \end{array} \right\},$$

or, as this may be written, putting $q=n-\delta-1$,

$$R_s^\delta\{\delta+1,\ \delta+2+s,\ (r-2\delta+1-s)\}$$
$$\times [n-1]^\delta \left\{ \begin{array}{l} [n-2]^\alpha [n-2]^\beta \\ -\dfrac{q}{1} [n-3]^\alpha [n-3]^\beta \\ + \ \&c. \end{array} \right\}.$$

30. I assume $r \not> n-2$, we have then $\alpha+\beta=r-\delta \not> n-\delta-2$, and therefore $\alpha+\beta<q$. The general term of the series in $\{\}$ is

$$(-)^\vartheta \frac{[q]^\vartheta}{[\vartheta]^\vartheta} [n-2-\vartheta]^\alpha [n-2-\vartheta]^\beta,$$

where the terms for which $n-2-\vartheta$ is negative are to be excluded, or what is the same thing, the series is not to be continued beyond $\vartheta=n-2$. But observing that $[q]^\vartheta$ vanishes for $\vartheta>q$, that is, $\vartheta>n-\delta-1$, it is in fact the same thing whether the series is continued indefinitely or only to the term for which $\vartheta=n-\delta-1$, and we may consistently with the condition $\vartheta \not> n-2$, continue the series as far as $\vartheta=n-\delta-1$, except in the case $\delta=0$, when by doing so we include the term corresponding to $\vartheta=n-1$, which in virtue of the condition ought to be excluded. The expression for the term in question is $(-)^{n-1}[-1]^\alpha[-1]^\beta$; hence if the sum of the series continued to the proper point is S, the sum continued indefinitely (in the particular case $\delta=0$) is $S+(-)^{n-1}[-1]^\alpha[-1]^\beta$, but in every other case the sum continued indefinitely is simply S. And by a well-known theorem in finite differences, the sum continued indefinitely is in fact zero. That is, except in the case $\delta=0$, we have $S=0$, but in the excepted case

$$S+(-)^{n-1}[-1]^\alpha[-1]^\beta=0;$$

or observing that $\alpha+\beta(=r-\delta)$ is in this case $=r$, and transforming the factorials, we have

$$S=(-)^{n-r}[\alpha]^\alpha[\beta]^\beta,$$

or substituting for α and β their values,

$$S=(-)^{n-r}[s]^s\,[r-s]^{r-s}.$$

31. Hence the general term of

$$D^r H - \frac{n-1}{1} D^r H_1 + \&c.$$

vanishes except for $\delta = 0$, but when $\delta = 0$, its value is

$$R_s^0\{1,\ 2+s,\ (r+1-s)\} \times (-)^{n-r}[s]^s[r-s]^{r-s};$$

or observing that R_s^0 is equal to $[r]^r \div [s]^s[r-s]^{r-s}$, the value is simply

$$(-)^{n-r}[r]^r\{1,\ 2+s,\ (r+1-s)\},$$

that is, we have

$$D^r H - \frac{n-1}{1} D^r H_1 + \&c.$$

$$= (-)^{n-r}[r]^r\, S_s\{1,\ 2+s,\ (r+1-s)\},$$

the summation in respect to s extending from $s=0$ to $s=r$. In particular, giving to r the values $n-2$ and $n-1$, and attending to the expressions for III and IV, we find

$$\Lambda^2\left(D^{n-2}H - \frac{n-1}{1} D^{n-2}H_1 + \&c. \ldots\right) = -[n-2]^{n-2}\ \text{III},$$

$$\Lambda^2\left(D^{n-3}H - \frac{n-1}{1} D^{n-3}H_1 + \&c. \ldots\right) = -[n-3]^{n-3}\ \text{IV}.$$

32. The equation III $=0$ belongs to the curve which by its intersections with the tangent, gives the tangentials of a point of the curve $U=0$. Hence the equation of the curve in question is

$$D^{n-2}H - \frac{n-1}{1} D^{n-2}H_1 + \&c. = 0,$$

which is Mr Salmon's theorem, leading to the solution of the problem of double tangents.

33. The expressions for I and II are obtained from those of IV and III by interchanging $(X,\ Y,\ Z)$ and $(x,\ y,\ z)$, and reversing the sign. Hence if, as before, $\mathfrak{H}$, $\mathfrak{D}$, &c. denote the values which H, D, &c. assume by this interchange, we have

$$\Lambda^2\left(\mathfrak{D}^{n-2}\mathfrak{H} - \frac{n-1}{1} \mathfrak{D}^{n-2}H_1 + \&c. \ldots\right) = [n-2]^{n-2}\ \text{II},$$

$$\Lambda^2\left(\mathfrak{D}^{n-3}\mathfrak{H} - \frac{n-1}{1} \mathfrak{D}^{n-3}H_1 + \&c. \ldots\right) = [n-3]^{n-3}\ \text{I},$$

and the identical equation

$$a_0\text{I} + a_1\text{II} + a_{n-1}\text{III} + a_n\text{IV} = 0$$

becomes therefore

$$\left.\begin{array}{l} \left(\mathfrak{D}^{n-3}\mathfrak{H} - \dfrac{n-1}{1}\mathfrak{D}^{n-3}\mathfrak{H}_1 + \&c.\right) U \\ + \dfrac{1}{n(n-2)}\left(\mathfrak{D}^{n-2}\mathfrak{H} - \dfrac{n-1}{1}\mathfrak{D}^{n-2}\mathfrak{H}_1 + \&c.\right) DU \\ - \dfrac{1}{n(n-2)}\left(D^{n-2}H - \dfrac{n-1}{1} D^{n-2}H_1 + \&c.\right)\mathfrak{D}\Upsilon \\ \left(D^{n-3}H - \dfrac{n-1}{1} D^{n-3}H_1 + \&c.\right)\Upsilon \end{array}\right\} = 0,$$

which is the general identical equation referred to in the introduction to the present memoir.

34. It is to be noticed that for $n=3$, the equation is

$$(\mathfrak{H} - 2\mathfrak{H}_1) U + \tfrac{1}{3}\mathfrak{D}\mathfrak{H} \,.\, DU - \tfrac{1}{3}DH \,.\, \mathfrak{D}\Upsilon - (H - 2H_1)\Upsilon = 0.$$

But we have $H_1 = H$, and in like manner $\mathfrak{H}_1 = \mathfrak{H}$, and the equation thus becomes

$$-\mathfrak{H} U + \tfrac{1}{3}\mathfrak{D}\mathfrak{H} \,.\, D\Upsilon - \tfrac{1}{3}DH \,.\, \mathfrak{D}\Upsilon + H\Upsilon = 0.$$

And so also for $n=4$, the equation is

$$(\mathfrak{D}\mathfrak{H} - 3\mathfrak{D}\mathfrak{H}_1) U + \tfrac{1}{8}(\mathfrak{D}^2\mathfrak{H} - 3\mathfrak{D}^2\mathfrak{H}_1) DU - \tfrac{1}{8}(D^2H - 3D^2H_1)\mathfrak{D}\Upsilon - (DH - 3DH_1)\Upsilon = 0.$$

But we have in general $DH_1 = \dfrac{n-3}{n-2} DH$, and therefore in the present case $DH_1 = \tfrac{1}{2}DH$, and consequently $\mathfrak{D}\mathfrak{H}_1 = \tfrac{1}{2}\mathfrak{D}\mathfrak{H}$, and the equation thus becomes

$$-\mathfrak{D}\mathfrak{H} \,.\, U + \tfrac{1}{4}(\mathfrak{D}^2\mathfrak{H} - 3\mathfrak{D}^2\mathfrak{H}_1) DU - \tfrac{1}{4}(D^2H - 3D^2H_1)\mathfrak{D}\Upsilon + DH \,.\, \Upsilon = 0,$$

which agree with the results previously obtained for the two particular cases.

261.

ON THE CONIC OF FIVE-POINTIC CONTACT AT ANY POINT OF A PLANE CURVE.

[From the *Philosophical Transactions of the Royal Society of London,* vol. CXLIX. (for the year 1859), pp. 371—400. Received March 1,—Read March 24, 1859.]

THE tangent is a line passing through two consecutive points of a plane curve, and we may in like manner consider the conic which passes through five consecutive points of a plane curve; and as there are certain singular points, viz. the points of inflexion where three consecutive points of the curve lie in a line, so there are singular points where six consecutive points of the curve lie in a conic. In the particular case where the given curve is a cubic, the last-mentioned species of singular points have been considered by Plücker and Steiner, and in the same particular case, the theory of the conic of five-pointic contact has recently been established by Mr Salmon. But the general case, where the curve is of any order whatever, has not so far as I am aware been hitherto considered;—the establishment of this theory is the object of the present memoir.

I. *Investigation of the Equation of the Conic of Five-pointic Contact.*

1. I take (X, Y, Z) as current coordinates, and I represent the equation of the given curve by

$$\Upsilon = (*\between X, Y, Z)^m = 0.$$

Let (x, y, z) be the coordinates of a given point on the curve, and let $U = (*\between x, y, z)^m$ be what Υ becomes when (x, y, z) are written in the place of (X, Y, Z); we have therefore $U=0$ as a condition satisfied by the coordinates of the point in question.

2. Write for shortness

$$D\,U = (X\partial_x + Y\partial_y + Z\partial_z)\ U,$$
$$D^2 U = (X\partial_x + Y\partial_y + Z\partial_z)^2\ U,$$

and let $\Pi = aX + bY + cZ = 0$ be the equation of a line. It is easy to see that

$$D^2 U - \Pi\,.\,DU = 0$$

will be the equation of a conic having an ordinary (two-pointic) contact with the curve at the point (x, y, z). In fact the equation $DU=0$ is that of the tangent at the point in question, and the equation $D^2U=0$ is that of the penultimate polar (or polar conic) of the point, which conic is touched by the tangent; the assumed equation represents therefore a conic having an ordinary (two-pointic) contact with the polar conic, and therefore with the curve. It may be added that the two conics intersect besides in a pair of points, and that the line joining these, or common chord of the two conics, is the line represented by the equation $\Pi=0$; and this being so, the constants (a, b, c) of the line $\Pi=0$ can be so determined as to give rise to a five-pointic contact.

3. Consider the coordinates of a point of the curve as functions of a single variable parameter; then for the present purpose the coordinates of a point consecutive to (x, y, z) may be taken to be

$$\begin{aligned} &x + dx + \tfrac{1}{2}\, d^2x + \tfrac{1}{6}\, d^3x + \tfrac{1}{24}\, d^4x,\\ &y + dy + \tfrac{1}{2}\, d^2y + \tfrac{1}{6}\, d^3y + \tfrac{1}{24}\, d^4y,\\ &z + dz + \tfrac{1}{2}\, d^2z + \tfrac{1}{6}\, d^3z + \tfrac{1}{24}\, d^4z,\end{aligned}$$

values which, substituted for X, Y, Z, must satisfy the equations

$$\Upsilon = 0, \quad D^2U - \Pi\,.\,DU = 0.$$

4. I write for shortness

$$\begin{aligned} \partial_1 &= dx\,\partial_x + dy\,\partial_y + dz\,\partial_z,\\ \partial_2 &= d^2x\,\partial_x + d^2y\,\partial_y + d^2z\,\partial_z,\\ \partial_3 &= d^3x\,\partial_x + d^3y\,\partial_y + d^3z\,\partial_z,\\ \partial_4 &= d^4x\,\partial_x + d^4y\,\partial_y + d^4z\,\partial_z,\end{aligned}$$

then the consecutive value of Υ is

$$\text{exp.}\,(\partial_1 + \tfrac{1}{2}\,\partial_2 + \tfrac{1}{6}\,\partial_3 + \tfrac{1}{24}\,\partial_4)\, U$$

(Read exp. z, exponential of z, $=e^z$), which is

$$= \left.\begin{array}{l} (1 + \partial_1 + \tfrac{1}{2}\partial_1^2 + \tfrac{1}{6}\partial_1^3 + \tfrac{1}{24}\partial_1^4)\\ \times(1 \quad + \tfrac{1}{2}\partial_2 \quad + \tfrac{1}{8}\,\partial_2^2)\\ \times(1 \quad + \tfrac{1}{6}\partial_3 \quad)\\ \times(1 \quad + \tfrac{1}{24}\partial_4\,) \end{array}\right\} U$$

$$= \left.\begin{array}{l} 1 + \partial_1 + \tfrac{1}{2}\partial_1^2 + \tfrac{1}{6}\partial_1^3 + \tfrac{1}{24}\partial_1^4\\ \quad + \tfrac{1}{2}\partial_2 + \tfrac{1}{2}\partial_1\partial_2 + \tfrac{1}{4}\,\partial_1^2\partial_2\\ \quad + \tfrac{1}{8}\,\partial_2^2\\ \quad + \tfrac{1}{6}\partial_3 + \tfrac{1}{6}\,\partial_1\,\partial_3\\ \quad + \tfrac{1}{24}\partial_4 \end{array}\right\} U$$

$$\begin{aligned} =\; & U\\ &+ \partial_1 U\\ &+ \tfrac{1}{2}\,(\partial_1^2 + \partial_2)\, U\\ &+ \tfrac{1}{6}\,(\partial_1^3 + 3\partial_1\partial_2 + \partial_3)\, U\\ &+ \tfrac{1}{24}\,(\partial_1^4 + 6\partial_1^2\partial_2 + 4\partial_1\partial_3 + 3\partial_2^2 + \partial_4)\, U,\end{aligned}$$

the several terms of which must respectively vanish, and we have therefore

$$\begin{aligned} U &= 0, \\ \partial_1 U &= 0, \\ \partial_2 U &= -\ \partial_1^2 U, \\ \partial_3 U &= -(\partial_1^3 + 3\partial_1\partial_2)\, U, \\ \partial_4 U &= -(\partial_1^4 + 6\partial_1^2\partial_2 + 4\partial_1\partial_3 + 3\partial_2^2)\, U. \end{aligned}$$

5. Next, preparing to substitute in the equation

$$D^2U - \Pi\,.\,DU = 0,$$

the consecutive value of DU is

$$\begin{aligned} &(x + dx + \tfrac{1}{2}d^2x + \tfrac{1}{6}d^3x + \tfrac{1}{24}d^4x)\,\partial_x U + \&c. \\ &= (\partial_0 + \partial_1 + \tfrac{1}{2}\partial_2 + \tfrac{1}{6}\partial_3 + \tfrac{1}{24}\partial_4)\ \ U, \end{aligned}$$

where

$$\partial_0 U = (x\partial_x + y\partial_y + z\partial_z)\, U = mU.$$

Reducing by the above results, the consecutive value of DU is

$$= -\tfrac{1}{2}\partial_1^2 U - \tfrac{1}{6}(\partial_1^3 + 3\partial_1\partial_2)\, U - \tfrac{1}{24}(\partial_1^4 + 6\partial_1^2\partial_2 + 4\partial_1\partial_3 + 3\partial_2^2)\, U.$$

6. Hence also writing

$$\begin{aligned} P &= ax + by + cz\ , \\ \partial_1 P &= adx + bdy + cdz\,, \\ \partial_2 P &= ad^2x + bd^2y + cd^2z, \end{aligned}$$

the consecutive value of $-\Pi DU$ is $-(P + \partial_1 P + \tfrac{1}{2}\partial_2 P)$ multiplied into the consecutive value of DU, and the product is

$$\begin{aligned} = \ & P\,.\,\tfrac{1}{2}\,\partial_1^2 U \\ & + P\,.\,\tfrac{1}{6}\,(\partial_1^3 + 3\partial_1\partial_2)\, U \qquad + \partial_1 P\,.\,\tfrac{1}{2}\partial_1^2 U \\ & + P\,.\,\tfrac{1}{24}(\partial_1^4 + 6\partial_1^2\partial_2 + 4\partial_1\partial_3 + 3\partial_2^2)\, U + \partial_1 P\,.\,\tfrac{1}{6}(\partial_1^3 + 3\partial_1\partial_2)\, U + \tfrac{1}{2}\partial_2 P\,.\,\tfrac{1}{2}\partial_1^2 U. \end{aligned}$$

7. The consecutive value of D^2U is

$$= (x + dx + \tfrac{1}{2}d^2x + \tfrac{1}{6}d^3x + \tfrac{1}{24}d^4x)^2\,\partial_x^2 U + \&c.$$

$$= \left.\begin{array}{l} x^2 \\ +\ 2x\,dx \\ +\ x^2\,dx + (dx)^2 \\ +\ \tfrac{1}{3}\,x\,d^3x + dx\,d^2x \\ +\ \tfrac{1}{12}x\,d^4x + \tfrac{1}{3}dx\,d^3x + \tfrac{1}{4}\,(d^2x)^2 \end{array}\right\}\partial_x^2 U + \&c.,$$

which is

$$=\left.\begin{array}{l} \partial_0^2 \\ +\ 2\partial_0\partial_1 \\ +\ \partial_0\partial_2+\partial_1^2 \\ +\tfrac{1}{3}\partial_0\partial_3+\partial_1\partial_2 \\ +\tfrac{1}{12}\partial_0\partial_4+\tfrac{1}{3}\partial_1\partial_3+\tfrac{1}{4}\partial_2^2 \end{array}\right\} U;$$

and observing that

$$\begin{aligned} \partial_0^2\ U &= m(m-1)\ U, \\ \partial_0\partial_1 U &= (m-1)\,\partial_1 U, \\ \partial_0\partial_2 U &= (m-1)\,\partial_2 U, \\ \partial_0\partial_3 U &= (m-1)\,\partial_3 U, \\ \partial_0\partial_4 U &= (m-1)\,\partial_4 U, \end{aligned}$$

and reducing as before, the consecutive value of D^2U is

$$\begin{aligned} =-\ &(m-2)\,\partial_1^2 U \\ -\tfrac{1}{3}\ &[(m-1)\,\partial_1^3+3(m-2)\,\partial_1\partial_2]\,U \\ -\tfrac{1}{12}\,&[(m-1)(\partial_1^4+6\partial_1^2\partial_2)+(m-2)(4\partial_1\partial_3+3\partial_2^2)]\,U. \end{aligned}$$

8. Substituting in the equation $D^2U-\Pi\,.\,DU=0$, we obtain as the conditions of a five-pointic contact

$$-(m-2)\,\partial_1^2U+P\,.\,\tfrac{1}{2}\partial_1^2U=0,$$

$$-\tfrac{1}{3}\ [(m-1)\ \partial_1^3+3(m-2)\,\partial_1\partial_2]\ U+P\,.\,\tfrac{1}{6}(\partial_1^3+3\partial_1\partial_2)\ U+\partial_1 P\,.\,\tfrac{1}{2}\partial_1^2U=0,$$

$$\begin{aligned} &-\tfrac{1}{12}\,[(m-1)(\partial_1^4+6\partial_1^2\partial_2)+(m-2)(4\partial_1\partial_3+3\partial_2^2)]\ U \\ &+P\,.\,\tfrac{1}{24}\,[\partial_1^4+6\partial_1^2\partial_2+4\partial_1\partial_3+3\partial_2^2]\ U+\partial_1 P\,.\,\tfrac{1}{6}\,[\partial_1^3+3\partial_1\partial_2]\ U+\tfrac{1}{2}\partial_2 P\,.\,\tfrac{1}{2}\partial_1^2U=0\,; \end{aligned}$$

or reducing

$$P=2(m-2),$$

$$\partial_1 P=\tfrac{2}{3}\frac{\partial_1^3U}{\partial_1^2U},$$

$$\partial_2 P=\tfrac{1}{2}\,\frac{[\partial_1^4+6\partial_1^2\partial_2]\ U}{\partial_1^2U}-\tfrac{4}{9}\frac{\partial_1^3U}{\partial_1^2U}\,\frac{[\partial_1^3+3\partial_1\partial_2]\ U}{\partial_1^2U},$$

which are the conditions of a five-pointic contact: it is to be remarked that if only the first and second conditions are satisfied, we have a four-pointic contact, and if only the first condition is satisfied, a three-pointic contact.

9. We have to reduce the last-mentioned equations; suppose that A, B, C are the first derived functions of U, then the equation $\partial_1 U=0$ may be written

$$Adx+Bdy+Cdz=0,$$

and this will be satisfied identically if

$$dx = B\nu - C\mu,$$
$$dy = C\lambda - A\nu,$$
$$dz = A\mu - B\lambda,$$

where λ, μ, ν are arbitrary multipliers, which may be taken to be constants. We have therefore $\partial_1 = \mathbf{D}$, where

$$\mathbf{D} = (B\nu - C\mu)\,\partial_x + (C\lambda - A\nu)\,\partial_y + (A\mu - B\lambda)\,\partial_z.$$

10. The resulting expressions for $\partial_1^2 U$, $\partial_1^3 U$, $\partial_1^4 U$ may be exhibited in the reduced forms given by Hesse, viz. if $\vartheta = \lambda x + \mu y + \nu z$, we have

$$\partial_1^2 U = P_2 U - Q_2 \vartheta^2,$$
$$\partial_1^3 U = P_3 U - Q_3 \vartheta^2,$$
$$\partial_1^4 U = P_4 U - Q_4 \vartheta^2,$$

where the values of P_2, P_3, P_4; Q_2, Q_3, Q_4 are as follows, viz. if (a, b, c, f, g, h) are the second derived functions of U, and if

$$H = \begin{vmatrix} a, & h, & g \\ h, & b, & f \\ g, & f, & c \end{vmatrix}$$

be the Hessian; if, moreover,

$$\Phi = -\begin{vmatrix} & \lambda, & \mu, & \nu \\ \lambda & a, & h, & g \\ \mu & h, & b, & f \\ \nu & g, & f, & c \end{vmatrix}$$

be the bordered Hessian (we may also write $\Phi = (\mathfrak{A}, \mathfrak{B}, \mathfrak{C}, \mathfrak{F}, \mathfrak{G}, \mathfrak{H} \between \lambda, \mu, \nu)^2$, where $(\mathfrak{A}, \mathfrak{B}, \mathfrak{C}, \mathfrak{F}, \mathfrak{G}, \mathfrak{H})$ are the inverse coefficients of (a, b, c, f, g, h), viz. $\mathfrak{A} = (bc - f^2)$, &c.); and finally, if for shortness we write

$$\Pi = \partial_x\Phi \,.\, \partial_\lambda\Phi + \partial_y\Phi \,.\, \partial_\mu\Phi + \partial_z\Phi \,.\, \partial_\nu\Phi,$$
$$\square = \partial_x H \,.\, \partial_\lambda\Phi + \partial_y H \,.\, \partial_\mu\Phi + \partial_z H \,.\, \partial_\nu\Phi,$$

then we have

$$P_2 = \frac{m}{m-1}\,\Phi, \qquad Q_2 = \frac{1}{(m-1)^2}\,H,$$

$$P_3 = \frac{m}{m-1}\,\mathbf{D}\Phi, \qquad Q_3 = \frac{1}{(m-1)^2}\,\mathbf{D}H,$$

$$P_4 = \frac{m}{m-1}\,\mathbf{D}^2\Phi - \frac{m\vartheta}{(m-1)^2}\,\Pi, \qquad Q_4 = \frac{1}{(m-1)^2}\,\mathbf{D}^2 H - \frac{\vartheta}{(m-1)^3}\,\square + \frac{3(m-2)}{(m-1)^3}\,H\Phi.$$

In the present case $U=0$, and we have

$$\partial_1^2 U = -Q_2 \vartheta^2,$$
$$\partial_1^3 U = -Q_3 \vartheta^2,$$
$$\partial_1^4 U = -Q_4 \vartheta^2.$$

11. Hence, substituting for Q_2 and Q_3 their values, the first and second of the equations for the five-pointic contact give

$$P = 2(m-2),$$
$$\partial_1 P = \tfrac{2}{3}\frac{\mathbf{D}H}{H},$$

and observing that Π is a linear function of (X, Y, Z), and consequently that P, $\partial_1 P$ denote simply the values which Π assumes when (x, y, z), (dx, dy, dz) are respectively substituted for (X, Y, Z), we see at once that these two conditions will be satisfied if we put

$$\Pi = \tfrac{2}{3}\frac{1}{H} DH + \Lambda \,.\, DU,$$

where Λ is an arbitrary constant, or, what is the same thing, an arbitrary function of (x, y, z). We have thus the general equation of a conic of four-pointic contact.

12. The above value of Π gives

$$\partial_2 P = \tfrac{2}{3}\frac{1}{H}\partial_2 H + \Lambda\, \partial_2 U,$$

and the third equation of the system of conditions for a five-pointic contact is therefore

$$\tfrac{2}{3}\frac{1}{H}\partial_2 H + \Lambda\,\partial_2 U = \tfrac{1}{3}\frac{[\partial_1^4 + 6\partial_1^2\partial_2]\,U}{\partial_1^2 U} - \tfrac{4}{9}\frac{\partial_1^3 U}{\partial_1^2 U}\frac{[\partial_1^3 + 3\partial_1\partial_2]\,U}{\partial_1^2 U},$$

which leads to the value of Λ.

13. We have in general

$$\partial_1^2 U = \frac{m}{m-1}\Phi\, U - \frac{1}{(m-1)^2} H\,\vartheta^2,$$

$$\partial_1^3 U = \frac{m}{m-1}\mathbf{D}\Phi \,.\, U - \frac{1}{(m-1)^2}\mathbf{D}H \,.\, \vartheta^2,$$

$$(\partial_1^3 + 3\partial_1\,\partial_2)\, U = \frac{m}{m-1}\mathbf{D}\Phi \,.\, U - \frac{1}{(m-1)^2}\mathbf{D}H \,.\, \vartheta^2,$$

$$(\partial_1^4 + 6\partial_1^2\partial_2)\, U = \frac{m}{m-1}\left(2\partial_2\Phi + \mathbf{D}^2\Phi + \frac{\vartheta}{m-1}\Pi\right) U$$
$$- \frac{1}{(m-1)^2}\left(2\partial_2 H + \mathbf{D}^2 H + \frac{\vartheta}{m-1}\square - \frac{3(m-2)}{m-1} H\Phi\right)\vartheta^2,$$

the last two of which have not yet been demonstrated. The value of $\partial_2 H$ (which, however, is not required for the present purpose) is

$$\partial_2 H = \frac{-3(m-3)}{m-1} H\Phi + \frac{\frac{1}{2}\square}{m-1}\vartheta,$$

which also is not yet demonstrated.

14. Putting $U=0$, we have

$$\partial_1{}^2\, U = -\frac{1}{(m-1)^2} \quad H\, \vartheta^2,$$

$$\partial_1{}^3\, U = -\frac{1}{(m-1)^2} \quad \mathbf{D}H\,.\,\vartheta^2,$$

$$(\partial_1{}^3 + 3\partial_1\partial_2)\, U = -\frac{1}{(m-1)^2} \quad \mathbf{D}H\,.\,\vartheta^2,$$

$$(\partial_1{}^4 + 6\partial_1{}^2\partial_2)\, U = -\frac{1}{(m-1)^2}\left(2\partial_2 H + \mathbf{D}^2 H + \frac{\vartheta}{m-1}\square - \frac{3(m-2)}{m-1} H\Phi\right)\vartheta^2,$$

and substituting,

$$\tfrac{2}{3}\frac{1}{H}\partial_2 H + \frac{1}{(m-1)^2} H\vartheta^2\Lambda$$

$$= \tfrac{1}{3}\frac{1}{H}\left(2\partial_2 H + \mathbf{D}^2 H + \frac{\vartheta}{m-1}\square - \frac{3(m-2)}{m-1} H\Phi\right) - \tfrac{4}{9}\left(\frac{\mathbf{D}H}{H}\right)^2,$$

where the term involving $\partial_2 H$ disappears; the equation may be written

$$\frac{9H^3\vartheta^2}{(m-1)^2}\Lambda = 3H\left(\mathbf{D}^2 H + \frac{\vartheta}{m-1}\square - \frac{3(m-2)}{m-1} H\Phi\right) - 4(\mathbf{D}H)^2,$$

which I will represent by

$$\frac{9H^3\vartheta^2}{(m-1)^2}\Lambda = 3R_2R_4 - 4R_3{}^2,$$

the values of R_2, R_3, R_4 being

$$\begin{cases} R_2 = H, \\ R_3 = \mathbf{D}H, \\ R_4 = \mathbf{D}^2 H + \dfrac{\vartheta}{m-1}\square - \dfrac{3(m-2)}{m-1} H\Phi. \end{cases}$$

15. We have $R_2 = H$, and it will be shown that

$$\begin{aligned}(m-1)^2 R_3{}^2 = &- 9(m-2)^2 H^2\Phi \\ &+ 3(m-2) H\square\vartheta \\ &- \Psi\vartheta^2,\end{aligned}$$

$$\begin{aligned}(m-1)^2 R_4 = &-12(m-2)^2 H\Phi \\ &+ 4(m-2)\square\vartheta \\ &- \Omega\vartheta^2,\end{aligned}$$

where for shortness

$$\Psi = (\mathfrak{A}, \mathfrak{B}, \mathfrak{C}, \mathfrak{F}, \mathfrak{G}, \mathfrak{H} \between \partial_x H, \partial_y H, \partial_z H)^2,$$

$$\Omega = (\mathfrak{A}, \mathfrak{B}, \mathfrak{C}, \mathfrak{F}, \mathfrak{G}, \mathfrak{H} \between \partial_x\ , \partial_y\ , \partial_z\ )^2 H,$$

and hence, writing

$$9H^3 \vartheta^2 \Lambda = 3R_2 (m-1)^2 R_4 - 4(m-1)^2 R_3^2,$$

we have

$$9H^3\Lambda = -3\Omega H + 4\Psi;$$

or replacing Ω, Ψ by their values,

$$\Lambda = \frac{1}{9H^3}\left\{\begin{array}{l} -3(\mathfrak{A}, \mathfrak{B}, \mathfrak{C}, \mathfrak{F}, \mathfrak{G}, \mathfrak{H} \between \partial_x\ , \partial_y\ , \partial_z\ )^2 H \,.\, H \\ +4(\mathfrak{A}, \mathfrak{B}, \mathfrak{C}, \mathfrak{F}, \mathfrak{G}, \mathfrak{H} \between \partial_x H, \partial_y H, \partial_z H)^2 \end{array}\right\},$$

and Λ having this value, the equation of the five-pointic conic is

$$D^2U - \left(\tfrac{2}{3}\frac{1}{H}DH + \Lambda DU\right)DU = 0,$$

where it will be recollected that the current coordinates are (X, Y, Z), and that D denotes $X\partial_x + Y\partial_y + Z\partial_z$.

16. I remark, in passing, that the problem of finding the circle of curvature at a given point of a plane curve, is in fact that of determining the conic having with the curve at the given point a three-pointic contact, and besides passing through two given points. The equation of a conic having an ordinary contact, is

$$D^2U - \Pi DU = 0,$$

where

$$\Pi = aX + bY + cZ,$$

and the condition of a three-pointic contact is

$$ax + by + cz = 2(m-2).$$

Let the coordinates of the two given points be

$$(x_1, y_1, z_1), \quad (x_2, y_2, z_2),$$

and let $(D^2U)_1$ &c. be the corresponding values of D^2U, &c., then we have

$$aX + bY + cZ = \frac{D^2U}{DU},$$

$$ax_1 + by_1 + cz_1 = \left(\frac{D^2U}{DU}\right)_1,$$

$$ax_2 + by_2 + cz_2 = \left(\frac{D^2U}{DU}\right)_2;$$

and if from the four equations we eliminate a, b, c, we find

$$\begin{vmatrix} XDU, & YDU, & ZDU, & D^2U \\ x\ , & y\ , & z\ , & 2(m-2) \\ x_1\ , & y_1\ , & z_1\ , & \left(\dfrac{D^2U}{DU}\right)_1 \\ x_2\ , & y_2\ , & z_2\ , & \left(\dfrac{D^2U}{DU}\right)_2 \end{vmatrix} = 0$$

for the equation of the conic in question; x, y, z being the coordinates of the point of contact, and X, Y, Z current coordinates.

II. *Demonstration of Identities assumed in the preceding section.*

Proof of the expressions for $(\partial_1^3 + 3\partial_1\partial_2)\,U$ and $(\partial_1^4 + 6\partial_1^2\partial_2)\,U$:

17. It will be remembered that ∂_1, ∂_2 stand originally for

$$d\,x\,\partial_x + d\,y\,\partial_y + d\,z\,\partial_z,$$
$$d^2x\,\partial_x + d^2y\,\partial_y + d^2z\,\partial_z,$$

and that A, B, C being the first derived functions of U, dx, dy, dz are changed into

$$B\nu - C\mu,\quad C\lambda - A\nu,\quad A\mu - B\lambda,$$

and that the resulting value of ∂_1, viz.

$$(B\nu - C\mu)\,\partial_x + (C\lambda - A\nu)\,\partial_y + (A\mu - B\lambda)\,\partial_z,$$

is also represented by $\mathbf{D}$, so that $\partial_1 = \mathbf{D}$. The corresponding values of d^2x, d^2y, d^2z are

$$d^2x = \nu dB - \mu dC,$$
$$d^2y = \lambda dC - \nu dA,$$
$$d^2z = \mu dA - \lambda dB,$$

where we have

$$dA = a dx + h dy + g dz, \text{ \&c.},$$

in which dx, dy, dz are to be replaced by the values

$$B\nu - C\mu,\quad C\lambda - A\nu,\quad A\mu - B\lambda;$$

and ∂_2 really denotes what

$$d^2x\,\partial_x + d^2y\,\partial_y + d^2z\,\partial_z$$

becomes when the above values are substituted for d^2x, d^2y, d^2z. But in the expressions ∂_1^2U, $\partial_1\partial_2U$ &c., the symbols ∂_x, ∂_y, ∂_z contained in ∂_1 and ∂_2 operate only on U, and not on the variable quantities A, B, C, &c. contained in ∂_1 and ∂_2.

18. If now we treat ∂_1 as an operand, that is, perform the differentiations on the variable quantities A, B, C which enter into ∂_1, we obtain

$$\partial_1 . \partial_1 = \partial_2,$$

or, what is the same thing, operating on $\partial_1 U$ with ∂_1, the result is

$$\partial_1 . \partial_1 U = \partial_1^2 U + \partial_2 U = (\partial_1^2 + \partial_2)\, U,$$

and in like manner

$$\partial_1 . \partial_1^2 U = (\partial_1^3 + 2\partial_1\, \partial_2)\, U,$$

$$\partial_1 . \partial_1^3 U = (\partial_1^4 + 3\partial_1^2 \partial_2)\, U.$$

It is, in fact, upon these principles that Hesse's values of $\partial_1^2 U$, $\partial_1^3 U$, &c. were obtained, and we may by means of them obtain the other expressions assumed in the preceding section.

19. In fact, starting from Hesse's equation,

$$\partial_1^2 U = \frac{m}{m-1} \Phi U - \frac{1}{(m-1)^2} H \vartheta^2,$$

we have

$$(\partial_1^3 + 2\partial_1 \partial_2)\, U = \frac{m}{m-1} (U \mathbf{D}\Phi + \Phi \mathbf{D} U) - \frac{1}{(m-1)^2} (\mathbf{D}H . \vartheta^2 + H . 2\vartheta \mathbf{D}\vartheta).$$

But we have identically $\mathbf{D}U = 0$, $\mathbf{D}\vartheta = 0$, and this equation becomes therefore

$$(\partial_1^3 + 2\partial_1 \partial_2)\, U = \frac{m}{m-1} U \mathbf{D}\Phi - \frac{1}{(m-1)^2} \mathbf{D}H . \vartheta^2;$$

this is precisely Hesse's value of $\partial_1^3 U$, and thus we have $\partial_1 \partial_2 U = 0$, and therefore

$$(\partial_1^3 + 3\partial_1 \partial_2)\, U = \partial_1^3\, U = \frac{m}{m-1} U \mathbf{D}\Phi - \frac{1}{(m-1)^2} \mathbf{D}H . \vartheta^2.$$

20. In like manner, starting from the expression of $\partial_1^3 U$, we have

$$(\partial_1^4 + 3\partial_1^2 \partial_2)\, U = \frac{m}{m-1} (\mathbf{D}U . \mathbf{D}\Phi + U\mathbf{D} . \mathbf{D}\Phi) - \frac{1}{(m-1)^2} (\mathbf{D}H . 2\vartheta \mathbf{D}\vartheta + \vartheta^2 \mathbf{D} . \mathbf{D}H);$$

or since $\mathbf{D}U$ and $\mathbf{D}\vartheta$ vanish identically, and the values of $\mathbf{D}.\mathbf{D}\Phi$, $\mathbf{D}.\mathbf{D}H$ are $\partial_2 \Phi + \mathbf{D}^2 \Phi$, $\partial_2 H + \mathbf{D}^2 H$, we have

$$(\partial_1^4 + 3\partial_1^2 \partial_2)\, U = \frac{m}{m-1} U (\partial_2 \Phi + \mathbf{D}^2 \Phi) - \frac{1}{(m-1)^2} \vartheta^2 (\partial_2 H + \mathbf{D}^2 H);$$

and if from the double of this equation we subtract Hesse's equation,

$$\partial_1^4 U = \frac{m}{m-1} U \left(\mathbf{D}^2 \Phi - \frac{\vartheta}{m-1} \Pi \right) - \frac{1}{(m-1)^2} \vartheta^2 \left(\mathbf{D}^2 H - \frac{\vartheta \Box}{m-1} + \frac{3(m-2)}{m-1} H\Phi \right),$$

we find the required relation,

$$(\partial_1^4 + 6\partial_1^2\partial_2)\, U = \frac{m}{m-1}\left(2\partial_2\Phi + \mathbf{D}^2\Phi + \frac{\vartheta}{m-1}\,\Pi\right) U$$

$$- \frac{1}{(m-1)^2}\left(2\partial_2 H + \mathbf{D}^2 H + \frac{\vartheta\Box}{m-1} - \frac{3(m-2)}{m-1}\, H\Phi\right)\vartheta^2.$$

Proof of the expression for $\partial_2 H$:

21. We have

$$\partial_2 H = \partial_x H\,(\nu dB - \mu dC) + \partial_y H\,(\lambda dC - \nu dB) + \partial_z H\,(\mu dA - \lambda dB),$$

where

$$dA = adx + hdy + gdz,$$
$$dB = hdx + bdy + fdz,$$
$$dC = gdx + fdy + cdz,$$

in which dx, dy, dz are to be replaced by their values: we have therefore

$$\begin{aligned}\partial_2 H = \partial_x H\,\{\quad &\nu\,[(bC - fB)\lambda + (fA - hC)\mu + (hB - bA)\nu] \\ &- \mu\,[(fC - cB)\lambda + (cA - gC)\mu + (gB - fA)\nu]\}, \\ &+ \&c.,\end{aligned}$$

where the coefficient of $\partial_x H$ is

$$0\lambda^2 + (gC - cA)\mu^2 + (hB - bA)\nu^2 + (2fA - gB - hC)\mu\nu + (bC - fB)\nu\lambda + (cB - fC)\lambda\mu\,;$$

or, since we have

$$(m-1)A = ax + hy + gz,$$
$$(m-1)B = hx + by + fz,$$
$$(m-1)C = gx + fy + cz,$$

the coefficient, omitting the factor $\dfrac{1}{m-1}$, which will be afterwards restored, is

$$\begin{aligned}&0\lambda^2 \\ &+[\ g(gx + fy + cz) - c(ax + hy + gz)]\,\mu^2 \\ &+[\ h(hx + by + fz) - b(ax + hy + gz)]\,\nu^2 \\ &+[2f(ax + hy + gz) - g(hx + by + fz) - h(gx + fy + cz)]\,\mu\nu \\ &+[\ b(gx + fy + cz) - f(hx + by + fz)]\,\nu\lambda \\ &+[\ c(hx + fy + cz) - f(gx + fy + cz)]\,\lambda\mu\end{aligned}$$

$$\begin{aligned}=\ &0\lambda^2 \\ &+(-\mathfrak{B}x + \mathfrak{H}y)\,\mu^2 \\ &+(-\mathfrak{C}x + \mathfrak{G}z)\,\nu^2 \\ &+(-2\mathfrak{F}x + \mathfrak{G}y + \mathfrak{H}z)\,\mu\nu \\ &+(-\mathfrak{G}x + \mathfrak{A}z)\,\nu\lambda \\ &+(-\mathfrak{H}x + \mathfrak{A}y)\,\lambda\mu,\end{aligned}$$

which is equal to

$$-(\mathfrak{A}\lambda^2+\mathfrak{B}\mu^2+\mathfrak{C}\nu^2+2\mathfrak{F}\mu\nu+2\mathfrak{G}\nu\lambda+2\mathfrak{H}\lambda\mu)\,x$$
$$+(\mathfrak{A}\lambda\ +\mathfrak{H}\mu+\mathfrak{G}\nu)(\lambda x+\mu y+\nu z).$$

The coefficients of $\partial_y H$, ∂H_z have a similar form; and uniting the three terms, observing that $x\partial_x H+y\partial_y H+z\partial_z H$ is equal to $3(m-2)H$, and attending to the values of Φ, $\square$, ϑ, we have, restoring the omitted factor $\frac{1}{m-1}$,

$$\partial_2 H=\frac{-3(m-2)}{m-1}H\Phi+\frac{\frac{1}{2}\square}{m-1}\vartheta.$$

Proof of the expressions for $(\mathbf{D}H)^2$ and $\mathbf{D}^2H$:

22. These are obtained (for the particular case $m=4$, which makes but little difference) in Mr Salmon's *Higher Plane Curves*, pp. 88 and 89, and I merely reproduce his investigation; we have

$$(\mathbf{D}H)^2=\{\ (B\nu-C\mu)\,\partial_x H+\ (C\lambda-A\nu)\,\partial_y H+(A\mu-B\lambda)\,\partial_z H\}^2,$$

or, what is the same thing,

$$(\mathbf{D}H)^2=\{\lambda\,(C\partial_y-B\partial_z)\,H+\mu\,(A\partial_z-C\partial_x)\,H+\nu\,(B\partial_x-A\partial_z)\,H\}^2;$$

and if we consider first the term which contains λ^2, the coefficient is

$$\{(C\partial_y-B\partial_z)\,H\}^2.$$

Now making use of the equations

$$(m-1)\,A=ax+hy+gz,$$
$$(m-1)\,B=hx+by+fz,$$
$$(m-1)\,C=gx+fy+cz,$$

and

$$m\,(m-1)\,U=ax^2+by^2+cz^2+2fyz+2gzx+2hxy=0,$$

we have

$$(m-1)^2\ \ C^2=(gx+fy+cz)^2-c\,(ax^2+by^2+cz^2+2fyz+2gzx+2hxy)$$
$$=-\mathfrak{B}x^2+2\mathfrak{H}xy-\mathfrak{A}y^2,$$
$$(m-1)^2\,BC=(hx+by+fz)\,(gx+fy+cz)-f\,(ax^2+by^2+cz^2+2fyz+2gzx+2hxy)$$
$$=\mathfrak{F}x^2-\mathfrak{G}xy-\mathfrak{H}xz+\mathfrak{A}yz,$$
$$(m-1)^2\,B^2=(hx+by+fz)^2-b\,(ax^2+by^2+cz^2+2fyz+2gzx+2hxy)$$
$$=-\mathfrak{C}x^2+2\mathfrak{G}xz-\mathfrak{A}z^2;$$

and hence

$$\begin{aligned}(m-1)^2\left((C\partial_y-B\partial_z)\,H\right)^2=&\ \ (-\mathfrak{B}x^2+2\mathfrak{H}xy\qquad-\mathfrak{A}y^2)\,(\partial_y H)^2\\&+2\,(-\mathfrak{F}x^2+\mathfrak{G}xy+\mathfrak{H}xz-\mathfrak{A}yz)\,\partial_y H\,.\,\partial_z H\\&+\ (-\mathfrak{C}x^2+\qquad 2\mathfrak{G}xz-\mathfrak{A}z^2)\,(\partial_z H)^2\\=&-\ x^2\,(\mathfrak{B},\ \mathfrak{F},\ \mathfrak{C}\between\partial_y H,\ \partial_z H)^2\\&+2\,x\,(y\partial_y H+z\partial_z H)\,(\mathfrak{H}\partial_y H+\mathfrak{G}\partial_z H)\\&-\ \mathfrak{A}\,(y\partial_y H+z\partial_z H)^2,\end{aligned}$$

and

$$y\partial_y H + z\partial_z H = 3(m-2)H - z\partial_z H,$$

so that the term is

$$\begin{aligned}&- x^2 (\mathfrak{B}, \mathfrak{F}, \mathfrak{C} \between \partial_y H, \partial_x H)^2 \\ &+ 2x \left(3(m-2)H - z\partial_z H\right)(\mathfrak{H}\partial_y H + \mathfrak{G}\partial_z H) \\ &- \mathfrak{A}\left(3(m-2)H - z\partial_z H\right)^2;\end{aligned}$$

or, reducing, we have

$$\begin{aligned}(m-1)^2 \{(C\partial_y - B\partial_z) H\}^2 \\ = - \qquad & x^2 (\mathfrak{A}, \mathfrak{B}, \mathfrak{C}, \mathfrak{F}, \mathfrak{G}, \mathfrak{H} \between \partial_x H, \partial_y H, \partial_z H)^2 \\ &+ 6(m-2) Hx (\mathfrak{A}\partial_x H + \mathfrak{H}\partial_y H + \mathfrak{G}\partial_z H) \\ &- 9(m-2)^2 \mathfrak{A} H^2.\end{aligned}$$

23. The other terms may be obtained in a similar manner; and it is easy to see that, collecting all the terms, the sum will be

$$\begin{aligned}&-(\lambda x + \mu y + \nu z)^2 (\mathfrak{A}, \mathfrak{B}, \mathfrak{C}, \mathfrak{F}, \mathfrak{G}, \mathfrak{H} \between \partial_x H, \partial_y H, \partial_z H)^2 \\ &+ 6(m-2) H (\lambda x + \mu y + \nu z) \{(\mathfrak{A}\lambda + \mathfrak{H}\mu + \mathfrak{G}\nu)\partial_x H + (\mathfrak{H}\lambda + \mathfrak{B}\mu + \mathfrak{F}\nu)\partial_y H + (\mathfrak{G}\lambda + \mathfrak{F}\mu + \mathfrak{C}\nu)\partial_z H\} \\ &- 9(m-2)^2 H^2 (\mathfrak{A}, \mathfrak{B}, \mathfrak{C}, \mathfrak{F}, \mathfrak{G}, \mathfrak{H} \between \lambda, \mu, \nu)^2;\end{aligned}$$

or, attending to the signification of the symbols ϑ, Φ, $\square$, Ψ, we have

$$\begin{aligned}(m-1)^2 (\mathbf{D}H)^2 = &- 9(m-2)^2 H^2 \Phi \\ &+ 3(m-2) H \square \vartheta \\ &- \Psi \vartheta^2.\end{aligned}$$

24. Next

$$\mathbf{D}^2 H = \left((B\nu - C\mu)\partial_x + (C\lambda - A\nu)\partial_y + (A\mu - B\lambda)\partial_z\right)^2 H,$$

or, what is the same thing,

$$\mathbf{D}^2 H = \{\lambda(C\partial_y - B\partial_z) + \mu(A\partial_z - C\partial_x) + \nu(B\partial_x - A\partial_z)\}^2 H;$$

and if we attend first to the term which contains λ^2, the coefficient is

$$(C\partial_y - B\partial_z)^2 H.$$

Now substituting for C^2, CB, B^2 as before, we have

$$\begin{aligned}(m-1)^2 (C\partial_y - B\partial_z)^2 H = &\quad (-\mathfrak{B}x^2 + 2\mathfrak{H}xy - \mathfrak{A}y^2)\partial_y^2 H \\ &+ 2(-\mathfrak{F}x^2 + \mathfrak{G}xy + \mathfrak{H}xz - \mathfrak{A}yz)\partial_y\partial_z H \\ &+ (-\mathfrak{C}x^2 + 2\mathfrak{G}xz - \mathfrak{A}z^2)\partial_z^2 H \\ = &- x^2 (\mathfrak{B}, \mathfrak{F}, \mathfrak{C} \between \partial_y, \partial_z)^2 H \\ &+ 2x(y\partial_y + z\partial_z)(\mathfrak{H}\partial_y + \mathfrak{G}\partial_z) H \\ &- \mathfrak{A}(y\partial_y + z\partial_z)^2 H,\end{aligned}$$

where it is to be observed that the symbols of differentiation affect H only. We have

$$\begin{aligned}(y\partial_y + z\partial_z)^2 H = &\quad (x\partial_x + y\partial_y + z\partial_z)^2 H \\ &- 2(x\partial_x + y\partial_y + z\partial_z)\, x\partial_x H \\ &+ \ (x\partial_x)^2 H,\end{aligned}$$

where H is a homogeneous function of the degree $3m-6$; $x\partial_x H$, since the x is not affected by the differentiation, must be treated as of the degree $3m-7$, and $(x\partial_x)^2 H$, for the like reason, stands for $x^2\partial_x{}^2 H$; we have therefore

$$\begin{aligned}(y\partial_y + z\partial_z)^2 H = &\quad (3m-6)(3m-7) H \\ &- 2(3m-7)\, x\partial_x H \\ &+ x^2\partial_x{}^2 H.\end{aligned}$$

In like manner,

$$\begin{aligned}(y\partial_y + z\partial_z)(\mathfrak{H}\partial_y + \mathfrak{G}\partial_z) H = &\quad (x\partial_x + y\partial_y + z\partial_z)(\mathfrak{H}\partial_y + \mathfrak{G}\partial_z) H - x\partial_x(\mathfrak{H}\partial_y + \mathfrak{G}\partial_z) H \\ = &\ (3m-7)(\mathfrak{H}\partial_y + \mathfrak{G}\partial_z) H - x\partial_x(\mathfrak{H}\partial_y + \mathfrak{G}\partial_z) H;\end{aligned}$$

and hence

$$\begin{aligned}(m-1)^2 (C\partial_y - B\partial_z)^2 H = &- x^2 (\mathfrak{B}, \mathfrak{F}, \mathfrak{C}\between \partial_y, \partial_z)^2 H \\ &+ 2x\,[(3m-7)(\mathfrak{H}\partial_y + \mathfrak{G}\partial_z) H - x\partial_x(\mathfrak{H}\partial_y + \mathfrak{G}\partial_z) H] \\ &- \mathfrak{A}\ [(3m-6)(3m-7) H - 2(3m-7)\, x\partial_x H + x^2\partial_x{}^2 H] \\ = &- x^2 (\mathfrak{A}, \mathfrak{B}, \mathfrak{C}, \mathfrak{F}, \mathfrak{G}, \mathfrak{H}\between \partial_x, \partial_y, \partial_z)^2 H \\ &+ 2(3m-7)\, x\, (\mathfrak{A}\partial_x + \mathfrak{H}\partial_y + \mathfrak{G}\partial_z) H \\ &- (3m-6)(3m-7)\, \mathfrak{A} H.\end{aligned}$$

25. The other terms are formed in a similar manner; and collecting all the terms, we have

$$\begin{aligned}(m-1)^2 \mathbf{D}^2 H = &-(3m-6)(3m-7)(\mathfrak{A}, \mathfrak{B}, \mathfrak{C}\ \mathfrak{F}, \mathfrak{G}, \mathfrak{H}\between \lambda, \mu, \nu)^2 H \\ &+ 2(3m-7)(\lambda x + \mu y + \nu z)((\mathfrak{A}\lambda + \mathfrak{H}\mu + \mathfrak{G}\nu)\,\partial_x H + (\mathfrak{H}\lambda + \mathfrak{B}\mu + \mathfrak{F}\nu)\,\partial_y H + (\mathfrak{G}\lambda + \mathfrak{F}\mu + \mathfrak{C}\nu)\,\partial_z H) \\ &- (\lambda x + \mu y + \nu z)^2 (\mathfrak{A}, \mathfrak{B}, \mathfrak{C}, \mathfrak{F}, \mathfrak{G}, \mathfrak{H}\between \partial_x, \partial_y, \partial_z)^2 H,\end{aligned}$$

or, attending to the signification of the symbols ϑ, Φ, $\square$, Ω, this is

$$\begin{aligned}(m-1)^2 \mathbf{D}^2 H = &-(3m-6)(3m-7)\, H\Phi \\ &+ (3m-7)\, \vartheta\square \\ &- \Omega\vartheta^2.\end{aligned}$$

Proof of the expressions for $(m-1)^2R_3^2$, $(m-1)^2R_4$:

26. We have

$$\begin{aligned}(m-1)^2R_3^2 &= (m-1)^2(\mathbf{D}H)^2\\ &= -9(m-2)^2H^2\Phi\\ &\quad +3(m-2)H\square\vartheta\\ &\quad -\Psi\vartheta^2,\end{aligned}$$

$$\begin{aligned}(m-1)^2R_4 &= (m-1)^2\mathbf{D}^2H+(m-1)\vartheta\square-(3m-6)(m-1)\Phi H\\ &= -12(m-2)^2H\Phi\\ &\quad +\ 4(m-2)\square\vartheta\\ &\quad -\Omega\vartheta^2,\end{aligned}$$

which are the expressions required.

Proof of corresponding expressions for $(m-1)^6Q_3^2$, $(m-1)^4Q_4$:

27. We have

$$\begin{aligned}(m-1)^6Q_3^2 &= (m-1)^2(\mathbf{D}H)^2\\ &= -9(m-2)^2H^2\Phi\\ &\quad +3(m-2)H\square\vartheta\\ &\quad -\Psi\vartheta,\end{aligned}$$

$$\begin{aligned}(m-1)^4Q_4 &= (m-1)^2\mathbf{D}^2H-(m-1)\vartheta\square+(3m-6)(m-1)H\Phi\\ &= -6(m-2)(m-3)H\Phi\\ &\quad +2(m-3)\square\vartheta\\ &\quad -\Omega\vartheta^2;\end{aligned}$$

to which I join

$$(m-1)^2Q_2 = H.$$

28. We have consequently

$$\begin{aligned}(m-1)^6(3Q_2Q_4-Q_3^2) &= -9(m-2)(m-4)H^2\Phi\\ &\quad +3(m-4)H\square\vartheta\\ &\quad +\ (-3\Omega H+\Psi)\vartheta^2,\end{aligned}$$

and

$$(m-1)^6\left(3(m-2)Q_2Q_4-2(m-3)Q_3^2\right) = \left(-3(m-2)\Omega H+2(m-3)\Psi\right)\vartheta^2,$$

which for $m=4$ becomes

$$729(3Q_2Q_4-Q_3^2) = (-3\Omega H+\Psi)\vartheta^2.$$

29. In the case $m=4$, we have Hesse's theorem, that the equation $3Q_2Q_4-Q_3=0$ gives a curve of the 14th order, which passes through the points of contact of the double tangents, viz. substituting for Ω, Ψ their values, the equation of this curve is

$$\begin{aligned}&-3H\,(\mathfrak{A},\ \mathfrak{B},\ \mathfrak{C},\ \mathfrak{F},\ \mathfrak{G},\ \mathfrak{H}\between\partial_x\ ,\ \partial_y\ ,\ \partial_z)^2H\\ &+\quad (\mathfrak{A},\ \mathfrak{B},\ \mathfrak{C},\ \mathfrak{F},\ \mathfrak{G},\ \mathfrak{H}\between\partial_xH,\ \partial_yH,\ \partial_zH)^2=0.\end{aligned}$$

I have added these remarks for the sake of pointing out the striking resemblance of the expressions which occur in the double-tangent problem for $m=4$, and in the present theory of the five-pointic conic for any value whatever of m. It has not hitherto been shown what the expressions $3(m-2)Q_2Q_4-2(m-3)Q_3^2$ and $-3(m-2)\Omega H+2(m-3)\Psi$ respectively denote, except in the particular case $m=4$.

III. *Application of the Formulæ to the Cubic.*

30. I shall apply the formula for the five-pointic conic to a cubic; to avoid confusion as to a numerical factor, I write U', H' in the place of U, H, so that we have

$$\Lambda=\frac{1}{9H'^3}\left\{\begin{array}{l}-3(\mathfrak{A},\ \mathfrak{B},\ \mathfrak{C},\ \mathfrak{F},\ \mathfrak{G},\ \mathfrak{H})(\partial_x\ ,\ \partial_y\ ,\ \partial_z)^2H'\,.\,H'\\+4(\mathfrak{A},\ \mathfrak{B},\ \mathfrak{C},\ \mathfrak{F},\ \mathfrak{G},\ \mathfrak{H})(\partial_xH',\ \partial_yH',\ \partial_zH')^2\end{array}\right\},$$

and then the equation of the five-pointic conic is

$$D^2U'-\tfrac{2}{3}\left(\frac{1}{H'}DH'+\Lambda DU'\right)DU'=0.$$

I take as the equation of the cubic,

$$U=x^3+y^3+z^3+6lxyz=0\,;$$

the formulæ Table No. 70 of my Third Memoir on Quantics, *Phil. Trans.*, vol. CXLVI. (1856), pp. 627—647, [144], putting H for HU, give

$$H=l^2(x^3+y^3+z^3)-(1+2l^3)\,xyz.$$

Hence writing

$$U'=\tfrac{1}{6}(x^3+y^3+z^3+6lxyz),$$

the first derived functions are

$$\tfrac{1}{2}(x^2+2lyz),\quad \tfrac{1}{2}(y^2+2lzx),\quad \tfrac{1}{2}(z^2+2lxy)\,;$$

the second derived functions, or $(a,\ b,\ c,\ f,\ g,\ h)$, are $(x,\ y,\ z,\ lx,\ ly,\ lz)$, whence $H'=-H$; the inverse coefficients $(\mathfrak{A},\ \mathfrak{B},\ \mathfrak{C},\ \mathfrak{F},\ \mathfrak{G},\ \mathfrak{H})$ are

$$(yz-l^2x^2,\ zx-l^2y^2,\ xy-l^2z^2,\ l^2yz-lx^2,\ l^2zx-ly^2,\ l^2xy-lz^2)\,;$$

and putting $U'=\frac{1}{6}U$ and $H'=-H$, we have

$$\Lambda=-\frac{1}{9H^3}\left\{\begin{array}{l}-3(\mathfrak{A},\ \mathfrak{B},\ \mathfrak{C},\ \mathfrak{F},\ \mathfrak{G},\ \mathfrak{H})(\partial_x\ ,\ \partial_y\ ,\ \partial_z)^2H\,.\,H\\+4(\mathfrak{A},\ \mathfrak{B},\ \mathfrak{C},\ \mathfrak{F},\ \mathfrak{G},\ \mathfrak{H})(\partial_xH,\ \partial_yH,\ \partial_zH)^2\end{array}\right\};$$

and the equation of the five-pointic conic is

$$D^2U-\left(\tfrac{2}{3}\frac{1}{H}DH+\tfrac{1}{6}\Lambda DU\right)DU=0.$$

31. We have

$$
\begin{aligned}
&(\mathfrak{A}, \mathfrak{B}, \mathfrak{C}, \mathfrak{F}, \mathfrak{G}, \mathfrak{H} \between \partial_x, \partial_y, \partial_z)^2 H \\
= \;& \quad (yz - l^2x^2). 6l^2x \\
&+ (zx - l^2y^2). 6l^2y \\
&+ (xy - l^2z^2). 6l^2z \\
&+ 2(l^2yz - lx^2). -(1+2l^3)\,x \\
&+ 2(l^2zx - ly^2). -(1+2l^3)\,y \\
&+ 2(l^2xy - lz^2). -(1+2l^3)\,z,
\end{aligned}
$$

which is

$$
\begin{aligned}
= \;& 18l^2xyz - 6l^4(x^3+y^3+z^3) \\
& - 6l^2(1+2l^3)xyz + 2l(1+2l^3)(x^3+y^3+z^3), \\
= \;& (12l^2 - 12l^5)xyz + (2l - 2l^4)(x^3+y^3+z^3), \\
= \;& 2(l-l^4)(x^3+y^3+z^3+6lxyz),
\end{aligned}
$$

or we have

$$(\mathfrak{A}, \mathfrak{B}, \mathfrak{C}, \mathfrak{F}, \mathfrak{G}, \mathfrak{H} \between \partial_x, \partial_y, \partial_z)^2 H = -2S.U,$$

where S is the quartinvariant (see the Table No. 70). For the present purpose $U=0$, and consequently

$$(\mathfrak{A}, \mathfrak{B}, \mathfrak{C}, \mathfrak{F}, \mathfrak{G}, \mathfrak{H} \between \partial_x, \partial_y, \partial_z)^2 H = 0.$$

32. Next,

$$
\begin{aligned}
&(\mathfrak{A}, \mathfrak{B}, \mathfrak{C}, \mathfrak{F}, \mathfrak{G}, \mathfrak{H} \between \partial_x H, \partial_y H, \partial_z H) \\
= \;& \quad (yz - l^2x^2)\,[3l^2x^2 - (1+2l^3)\,yz]^2 \\
&+ (zx - l^2y^2)\,[3l^2y^2 - (1+2l^3)\,zx]^2 \\
&+ (xy - l^2z^2)\,[3l^2z^2 - (1+2l^3)\,xy]^2 \\
&+2(l^2yz - lx^2)\,[3l^2y^2 - (1+2l^3)\,zx]\,[3l^2z^2 - (1+2l^3)\,xy] \\
&+2(l^2zx - ly^2)\,[3l^2z^2 - (1+2l^3)\,xy]\,[3l^2x^2 - (1+2l^3)\,yz] \\
&+2(l^2xy - lz^2)\,[3l^2x^2 - (1+2l^3)\,yz]\,[3l^2y^2 - (1+2l^3)\,zx],
\end{aligned}
$$

the first three lines of which are

$$
\begin{aligned}
&9l^4\,xyz\,(x^3+y^3+z^3) - 18l^2\,(1+2l^3)\,x^2y^2z^2 + (1+2l^3)^2(y^3z^3+z^3x^3+x^3y^3) \\
&-9l^6\,(x^6+y^6+z^6) + 6l^4\,(1+2l^3)\,xyz\,(x^3+y^3+z^3) - 3l^2\,(1+2l^3)^2x^2y^2z^2\,;
\end{aligned}
$$

or collecting and reducing,

$$
\begin{aligned}
&(-9l^6 \qquad\qquad\qquad\qquad)\,(x^6+y^6+z^6) \\
&+(1+4l^3+4l^6 \qquad\qquad)\,(y^3z^3+z^3x^3+x^3y^3) \\
&+(\quad 15l^4+12l^7 \qquad\quad)\,(x^3+y^3+z^3)\,xyz \\
&+(-21l^2-48l^5-12l^8)\,x^2y^2z^2\,;
\end{aligned}
$$

the second three lines are

$$18l^6(y^3z^3+z^3x^3+x^3y^3)-12l^4(1+2l^3)\,xyz\,(x^3+y^3+z^3)+6l^2(1+2l^3)^2x^2y^2z^2$$
$$-54l^5x^2y^2z^2+12l^3(1+2l^3)(y^3z^3+z^3x^3+x^3y^3)-2l(1+2l^3)^2xyz\,(x^3+y^3+z^3)\,;$$

or collecting and reducing,

$$\begin{aligned}&(12l^3+42l^6)\,(y^3z^3+z^3x^3+x^3y^3)\\&+(-2l-20l^4-32l^7)\,(x^3+y^3+z^3)\,xyz\\&+(6l^2-30l^5+24l^8)\,x^2y^2z^2.\end{aligned}$$

Hence in the first part replacing the top line by

$$-9l^6(x^3+y^3+z^3)^2+18l^6(y^3z^3+z^3x^3+x^3y^3),$$

and uniting the two parts, we find

$$\begin{aligned}&(\mathfrak{A},\ \mathfrak{B},\ \mathfrak{C},\ \mathfrak{F},\ \mathfrak{G},\ \mathfrak{H}\between\partial_xH,\ \partial yH,\ \partial zH)^2\\=\;&(1+8l^3)^2\,(y^3z^3+z^3x^3+x^3y^3)\\&+(-9l^6)\,(x^3+y^3+z^3)^2\\&+(-2l-5l^4-20l^7)\,(x^3+y^3+z^3)\,xyz\\&+(-15l^2-78l^5+12l^8)\,x^2y^2z^2\,;\end{aligned}$$

and referring to the Table No. 70, and writing Θ for ΘU, we have

$$(\mathfrak{A},\ \mathfrak{B},\ \mathfrak{C},\ \mathfrak{F},\ \mathfrak{G},\ \mathfrak{H}\between\partial_xH,\ \partial_yH,\ \partial_zH)^2=\Theta,$$

where Θ is the first of the three functions which may be chosen to represent the octicovariant of the cubic.

33. We have thus

$$\Lambda=-\tfrac{4}{9}\frac{\Theta}{H^3},$$

and thence

$$D^2U-\left(\tfrac{2}{3}\frac{1}{H}DH-\tfrac{2}{27}\frac{\Theta}{H^3}DU\right)DU=0$$

as the equation of the five-pointic conic: the investigation has been conducted by means of the canonical form of the equation of the cubic, but the form of the result shows that it applies to the equation of the cubic in any form whatever.

34. If, however, we continue to represent the cubic by the canonical equation

$$x^3+y^3+z^3+6lxyz=0,$$

the result may be further reduced. We have, putting $U=0$, or writing $x^3+y^3+z^3=-6lxyz$,

$$H=-(1+8l^3)\,xyz\,;$$

moreover, putting $U=0$, the Table No. 70 gives

$$\Theta=(1+8l^3)^2(y^3z^3+z^3x^3+x^3y^3)-3l^2H^2;$$

or substituting for H the last-mentioned value, and putting for shortness

$$Q=y^3z^3+z^3x^3+x^3y^3-3l^2x^2y^2z^2,$$

we have

$$\Theta=(1+8l^3)^2\,Q;$$

and with these values of H and Q, the equation of the five-pointic conic is

$$D^2U+\left(\tfrac{2}{3}\frac{1}{(1+8l^3)\,xyz}DH-\tfrac{2}{27}\frac{Q}{(1+8l^3)\,x^3y^3z^3}DU\right)DU=0,$$

where

$$DU=3\{(x^2+2lyz)X+(y^2+2lzx)Y+(z^2+2lxy)Z\}$$

$$D^2U=6\{(X^2+2lYZ)x+(Y^2+2lZX)y+(Z^2+2lXY)z\},$$

or, as it will be convenient to write it,

$$=6(x,\ y,\ z,\ lx,\ ly,\ lz\between X,\ Y,\ Z)^2,$$

$$DH=\left(3l^2x^2-(1+2l^3)yz\right)X+\left(3l^2y^2-(1+2l^3)zx\right)Y+\left(3l^2z^2-(1+2l^3)xy\right)Z;$$

whence, finally, the equation of the five-pointic conic of the cubic

$$X^3+Y^3+Z^3+6lXYZ=0$$

at the point $(x,\ y,\ z)$ is

$$\begin{aligned}&9(1+8l^3)\,x^3y^3z^3(x,\ y,\ z,\ lx,\ ly,\ lz\between X,\ Y,\ Z)^2\\&\qquad+\{(x^2+2lyz)X+(y^2+2lzx)Y+(z^2+2lxy)Z\}\times\\&\left\{\begin{array}{l}3x^2y^2z^2\left[\left(3l^2x^2-(1+2l^3)yz\right)X+\left(3l^2y^2-(1+2l^3)zx\right)Y+\left(3l^2z^2-(1+2l^3)xy\right)Z\right]\\-Q\left[(\ x^2+\ 2l\ yz)X+(\ y^2+\ 2l\ zx)Y+(\ z^2+\ 2l\ xy)Z\right]\end{array}\right\},\end{aligned}$$

a result which I had previously obtained by a special method.

35. But the expression may be exhibited in a different form by a transformation suggested by a geometrical theorem of Mr Salmon's. In fact the tangent at the point $(x,\ y,\ z)$ meets the cubic in the tangential of this point, and the coordinates of the tangential are $x(y^3-z^3)$, $y(z^3-x^3)$, $z(x^3-y^3)$. Calling these ξ, η, ζ, the equation of the tangent to the cubic at the tangential is

$$(\xi^2+2l\eta\zeta)X+(\eta^2+2l\zeta\xi)Y+(\zeta^2+2l\xi\eta)Z=0.$$

Now we have identically,

$$\begin{aligned}&3x^2y^2z^2\left\{\left(3l^2x^2-(1+2l^3)yz\right)X+\left(3l^2y^2-(1+2l^3)yz\right)Y+\left(3l^2z^2-(1+2l^3)zx\right)Z\right\}\\&\quad-Q\{(x^2+2lyz)X+(y^2+2lzx)Y+(z^2+2lxy)Z\}\\&=\quad\{(\xi^2+2l\eta\zeta)X+(\eta^2+2l\zeta\xi)Y+(\zeta^2+2l\xi\eta)Z\}\\&\quad+U\{x^2(-y^3-z^2+2lyz)X+y^2(-z^3-x^3+2lzx)Y+z^2(-x^3-y^3+2lxy)Z\}.\end{aligned}$$

In fact this relation will be true if only

$$3x^2y^2z^2\left(3l^2x^2-(1+2l^3)\,yz\right)-Q\,(x^2+2lyz)-(\xi^2+2l\eta\zeta)=Ux^2\,(-y^3-z^3+2lyz)\,;$$

and substituting for ξ, η, ζ and Q their values, the left-hand side is

$$\begin{aligned}&3x^2y^2z^2\left(3l^2x^2-(1+2l^3)\,yz\right)\\&-(x^2+2lyz)\,(y^3z^3+z^3x^3+x^3y^3-3l^2x^2y^2z^2)\\&-\ x^2\,(y^3-z^3)^2\\&-\ 2lyz\,(z^3-x^3)\,(x^3-y^3)\,;\end{aligned}$$

and expanding and reducing, the result is

$$x^2\left(-(y^3+z^3)^2-x^3\,(y^3+z^3)\right)+2lyz\left(-2x^3\,(y^3+z^3)+x^6\right)+12l^2x^4y^2z^2\,;$$

whence, dividing by x^2, the equation becomes

$$\begin{aligned}-(y^3+z^3)\,(x^3+y^3+z^3)+2lxyz\left(x^3-2\,(y^3+z^3)\right)+12l^2x^2y^2z^2&\\=(-y^3-z^3+2lyz)\,(x^3+y^3+z^3+6lxyz),&\end{aligned}$$

which is identically true.

36. Hence in the identical equation putting $U=0$, we see that the equation of the five-pointic conic may be written

$$\begin{aligned}&9\,(1+8l^3)\,x^3y^3z^3\,(x,\ y,\ z,\ lx,\ ly,\ lz\between X,\ Y,\ Z)^2\\&+\{(x^2+2lyz)\,X+(y^2+2lzx)\,Y+(z^2+2lxy)\,Z\}\times\\&\quad\{(\xi^2+2l\eta\zeta)\,X+(\eta^2+2l\zeta\xi)\,Y+(\zeta^2+2l\xi\eta)\,Z\}\\&=0,\end{aligned}$$

where ξ, η, ζ stand for $x\,(y^3-z^3)$, $y\,(z^3-x^3)$, $z\,(x^3-y^3)$, the coordinates of the tangential of the given point, and which puts in evidence the geometrical theorem above referred to, viz.

THEOREM.—The common chord of the five-pointic conic and the polar conic is the tangent to the cubic at the tangential of the given point.

37. The five-pointic conic meets the cubic in the point of contact, considered as five coincident points, and in a remaining sixth point or point of simple intersection. The process by which I originally obtained the equation of the five-pointic conic, led also to the equation of the line joining the point of contact with the point of simple intersection: the equation of this line is

$$\begin{aligned}&9x^3y^3z^3\left\{\begin{aligned}&Xx\left((1+8l^3)\,x^4+(4l+41l^4)\,x^2yz+(-2l^2+2l^5)\,y^2z^2\right)\\&+Yy\left((1+8l^3)\,y^4+(4l+41l^4)\,y^2zx+(-2l^2+2l^5)\,z^2x^2\right)\\&+Zz\left((1+8l^3)\,z^4+(4l+41l^4)\,z^2xy+(-2l^2+2l^5)\,x^2y^2\right)\end{aligned}\right\}\\&-6Qx^2y^2z^2\left\{X\left(3l^2x^2-(1+2l^3)\,yz\right)+Y\left(3l^2y^2-(1+2l^3)\,zx\right)+Z\left(3l^2z^2-(1+2l^3)\,xy\right)\right\}\\&+\ Q^2\qquad\{X\,(x^2+2lyz)+Y\,(y^2+2lzx)+Z\,(z^2+2lxy)\}=0.\end{aligned}$$

38. If the conic meet the cubic in six coincident points, that is, if the point of contact be a singular point of the kind already spoken of, or, as we may term it, a sextactic point, then the last-mentioned line must coincide with the tangent at the point. Represent for a moment the equation of the line by

$$AX + BY + CZ = 0,$$

then this line is to coincide with the line

$$(x^2 + 2lyz)\,X + (y^2 + 2lzx)\,Y + (z^2 + 2lxy)\,Z = 0,$$

or we must have

$$\begin{aligned} B\,(z^2 + 2lxy) - C\,(y^2 + 2lzx) &= 0,\\ C\,(x^2 + 2lyz) - A\,(z^2 + 2lxy) &= 0,\\ A\,(y^2 + 2lzx) - B\,(x^2 + 2lyz) &= 0, \end{aligned}$$

which must be equivalent to a single condition. The terms A, B, C, which contain

$$x^2 + 2lyz, \quad y^2 + 2lzx, \quad z^2 + 2lxy$$

respectively, may, it is clear, be omitted, and omitting also a factor $3x^2y^2z^2$, we may write

$$\begin{aligned} A &= 3x^2yz\left((1 + 8l^3)\,x^4 + (4l + 41l^4)\,x^2yz + (-2l^2 + 2l^5)\,y^2z^2\right) - 2Q\left(3l^2x^2 - (1 + 2l^3)\,yz\right),\\ B &= 3xy^2z\left((1 + 8l^3)\,y^4 + (4l + 41l^4)\,y^2zx + (-2l^2 + 2l^5)\,z^2x^2\right) - 2Q\left(3l^2y^2 - (1 + 2l^3)\,zx\right), \end{aligned}$$

and the like value for C. The last of the three equations is

$$\begin{aligned} &3xyz\left\{\begin{array}{l} \quad(y^2 + 2lzx)\left((1 + 8l^3)\,x^5 + (4l + 41l^4)\,x^3yz + (-2l^2 + 2l^5)\,xy^2z^2\right)\\ -(x^2 + 2lyz)\left((1 + 8l^3)\,y^5 + (4l + 41l^4)\,y^3zx + (-2l^2 + 2l^5)\,yz^2x^2\right)\end{array}\right\}\\ &-2Q\{\ (y^2 + 2lzx)\left(3l^2x^2 - (1 + 2l^3)\,yz\right) - (x^2 + 2lyz)\left(3l^2y^2 - (1 + 2l^3)\,zx\right)\}\\ &= 0, \end{aligned}$$

where the function on the left hand is

$$\begin{aligned} &= 3xyz\left\{\begin{array}{l} \quad(1 + 8l^3)\left(x^5y^2 - x^2y^5 + 2l\,(x^6z - y^6z)\right)\\ +\,(4l + 41l^4)\,2l\,(x^4yz^2 - xy^4z^2)\\ +\,(-2l^2 + 2l^5)\,(xy^4z^2 - x^4yz^2)\end{array}\right\}\\ &\quad -2Q\,\{(1 + 2l^3)\,(y^3z - x^3z) + 6l^2\,(x^3z - y^3z)\}, \end{aligned}$$

or, what is the same thing, throwing out the factor z, it is

$$\begin{aligned} &= \{3\,(1 + 8l^3)\,x^3y^3 + 6l\,(1 + 8l^3)\,xyz\,(x^3 + y^3) + 30l^2\,(1 + 8l^3)\,x^2y^2z^2\}\,(x^3 - y^3)\\ &\qquad -2Q\,(1 + 8l^3)\,(x^3 - y^3)\,; \end{aligned}$$

or throwing out the factor $(1 + 8l^3)$ and substituting for Q its value, it is

$$= \{3x^3y^3 + 6lxyz\,(x^3 + y^3) + 30l^2x^2y^2z^2 - 2\,(y^3z^3 + z^3x^3 + x^3y^3 - 3l^2x^2y^2z^2)\}\,(x^3 - y^3).$$

The first factor, reducing by the equation $x^3+y^3+z^3+6lxyz=0$, is

$$\begin{aligned}
&= 3x^3y^3 - (x^3+y^3)(x^3+y^3+z^3) + (x^3+y^3+z^3)^2 - 2(y^3z^3+z^3x^3+x^3y^3),\\
&= 3x^3y^3 + z^3(x^3+y^3+z^3) - 2(y^3z^3+z^3x^3+x^3y^3),\\
&= z^6 - z^3(x^3+y^3) + x^3y^3,\\
&= (z^3-x^3)(z^3-y^3).
\end{aligned}$$

39. Hence putting for the moment

$$M = (y^3-z^3)(z^3-x^3)(x^3-y^3),$$

it appears that the last of the three equations is $Mz=0$; the first and second are of course $Mx=0$ and $My=0$, and the required condition is $M=0$, that is,

$$(y^3-z^3)(z^3-x^3)(x^3-y^3)=0,$$

the equation which, combined with the equation of the curve

$$x^3+y^3+z^3+6lxyz=0,$$

gives the sextactic points. There are consequently twenty-seven such points, and it is at once seen that these are the points of contact of the tangents to the cubic from the points of inflexion, or, what is the same thing, that the twenty-seven sextactic points form nine groups of three each, such that the three points of a group have for their common tangential one of the nine points of inflexion. In fact, let ω be a cube root (real or imaginary) of unity, the three sextactic points of one of the groups will be given by

$$\left\{\begin{aligned} x - \omega y \qquad &= 0,\\ x^3+y^3+z^3+6lxyz &= 0.\end{aligned}\right.$$

Now consider the tangential of any one of these points, its coordinates are

$$\begin{aligned}
x_1 &= x(y^3-z^3),\\
y_1 &= y(z^3-x^3),\\
z_1 &= z(x^3-y^3);
\end{aligned}$$

or, reducing by the equation $x-\omega y=0$, $x_1=\omega y(x^3-z^3)$, $y_1=-y(x^3-z^3)$, $z_1=0$, or, what is the same thing, $x_1+\omega y_1=0$, $z_1=0$; that is, the point (x_1, y_1, z_1) is one of the points of inflexion. This is the construction of the sextactic points obtained by Plücker and Steiner.

40. Reverting to the equation of the line joining the point of contact of the five-pointic conic with the point of simple intersection, this meets the cubic in a third point, and Mr Salmon has shown that this third point is in fact the second tangential (tangential of the tangential) of the point of contact, or, what is the same thing:

THEOREM. The point of simple intersection of the cubic and the five-pointic conic is the third point of intersection with the cubic, of the line joining the point of contact with the second tangential of this point.

41. I have not sought to verify this theorem by my formulæ. I remark, that combining it with the before-mentioned theorem, the five-pointic conic is completely determined as follows; viz.

THEOREM. The five-pointic conic touches the conic at the point of contact (two conditions); it passes through the two points in which the polar conic is intersected by the tangent to the cubic at the tangential of the point of contact (two conditions); and it passes through the point which is the third point of intersection with the cubic of the line joining the point of contact with its second tangential.

42. The construction for the point of simple intersection leads at once to that for the sextactic points; in fact, consider a point having for its tangential a point of inflexion: a point of inflexion is its own tangential, and the second tangential of the first-mentioned point is therefore the point of inflexion: the line joining the point with the second tangent is therefore the tangent at the point, and the point of simple intersection coincides with the point itself, that is, the point in question is a sextactic point.

43. I represent the equation of the five-pointic conic by

$$(a,\ b,\ c,\ f,\ g,\ h \between X,\ Y,\ Z)^2 = 0;$$

the value of a is

$$\begin{aligned}&= 9\,(1+8l^3)\,x^4y^3z^3\\ &\quad + 3x^2y^2z^2\left(3l^2x^2-(1+2l^3)\,yz\right)(x^2+2lyz)\\ &\quad - Q\,(x^2+2lyz)^2,\end{aligned}$$

in which equation

$$Q = y^3z^3 + z^3x^3 + x^3y^3 - 3l^2x^2y^2z^2;$$

or, reducing by the equation $x^3+y^3+z^3+6lxyz=0$,

$$Q = y^3z^3 + x^3\,(-x^3-6lxyz) - 3l^2x^2y^2z^2,$$

that is,

$$-Q = x^6 + 6lx^4yz + 3l^2x^2y^2z^2 - y^3z^3,$$

and we have

$$\begin{aligned}a &= 9\,(1+8l^3)\,x^4y^3z^3\\ &\quad + 3x^2y^2z^2\left(3l^2x^4+(-1+4l^3)\,x^2yz-2l\,(1+2l^3)\,y^2z^2\right)\\ &\quad + (x^6+6lx^4yz+3l^2x^2y^2z^2-y^3z^3)\,(x^4+4xl^2yz+4l^2y^2z^2).\end{aligned}$$

We have in like manner

$$\begin{aligned}2f &= 18\,(1+8l^3)\,lx^4y^3z^3\\ &\quad + 3x^2y^2z^2\left\{\left(3l^2y^2-(1+2l^3)\,zx\right)(z^2+2lxy)+\left(3l^2z^2-(1+2l^3)\,xy\right)(y^2+2lzx)\right\}\\ &\quad - 2Q\,(y^2+2lzx)\,(z^2+2lxy);\end{aligned}$$

the coefficient of $3x^2y^2z^2$ in the second line is

$$= 6l^2y^2z^2 + (-1+4l^3)\,x\,(y^3+z^3) - 4l\,(1+2l^3)\,x^2yz;$$

or reducing by the equation of the curve,

$$= (1-4l^3)\,x^4 + (2l-32l^4)\,x^2yz + 6l^2y^2z^2;$$

and the coefficient of $-2Q$ is

$$= y^2z^2 + 2lx(y^3+z^3) + 4l^2x^2yz;$$

or reducing by the equation of the curve,

$$= -2lx^4 - 8l^2x^2yz + y^2z^2.$$

We thus have

$$\begin{aligned} 2f = {} & 18l(1+8l^3)x^4y^3z^3 \\ & + 3x^2y^2z^2\left((1-4l^3)x^4 + (2l-32l^4)x^2yz + 6l^2y^2z^2\right) \\ & + 2(x^6 + 6lx^4yz + 3l^2x^2y^2z^2 - y^3z^3)(-2lx^4 - 8l^2x^2yz + y^2z^2). \end{aligned}$$

Reducing the expressions of a and $2f$, we find for the coefficients (a, b, c, f, g, h),

$$\left\{\begin{aligned} a = {} & x^{10} + 10lx^8yz + 40l^2x^6y^2z^2 + (5+120l^3)x^4y^3z^3 - 10lx^2y^4z^4 - 4l^2y^5z^5, \\ & \vdots \\ 2f = {} & -4lx^{10} - 40l^2x^8yz + (5-120l^3)x^6y^2z^2 + 40lx^4y^3z^3 + 8l^2x^2y^4z^4 - 2y^5z^5, \\ & \vdots \end{aligned}\right.$$

which gives the completely developed form of the equation of the five-pointic conic.

44. I investigate the coordinates of the point of simple intersection of the cubic and the five-pointic conic as follows: the equations of the two curves are

$$X^3 + Y^3 + Z^3 + 6lXYZ = 0,$$
$$(a, b, c, f, g, h \between X, Y, Z)^2 = 0;$$

or if we write

$$\begin{aligned} \alpha &= 1, & A &= c, \\ \gamma &= 6lXY, & B &= 2(gX + fY), \\ \delta &= X^3 + Y^3, & C &= aX^2 + 2hXY + bY^2, \end{aligned}$$

then the two equations are

$$\alpha Z^3 + \gamma Z + \delta = 0,$$
$$AZ^2 + BZ + C = 0,$$

and the result of the elimination of Z will be

$$(\alpha Z_1^3 + \gamma Z_1 + \delta)(\alpha Z_2^3 + \gamma Z_2 + \delta) = 0,$$

where Z_1, Z_2 are the roots of the equation $AZ^2 + BZ + C = 0$; that is, we have

$$\left.\begin{aligned} & \alpha^2\ C^3 \\ & + \alpha\gamma\ C(B^2 - 2AC) \\ & + \alpha\delta\ (-B^3 + 2ABC) \\ & + \gamma^2\ CA^2 \\ & - \gamma\delta\ BA^2 \\ & + \delta^2\ A^3 \end{aligned}\right\} = 0;$$

and substituting for A, B, C their values, but attending only to the terms which involve X^6 and Y^6, the result is

$$(a^3+c^3-8g^3+6acg)\,X^6+\ldots+(b^3+c^3-8f^3+6bcf)\,Y^6=0.$$

45. But the result of the elimination must obviously be

$$(Xy-Yx)^5\,(Xy_1-Yx_1)=0,$$

if (x_1, y_1, z_1) are the coordinates of the point of simple intersection. Comparing the two results, and forming the analogous third equation, we may write

$$\begin{aligned} x^5x_1 &= b^3+c^3-8f^3+6bcf,\\ y^5y_1 &= c^3+a^3-8g^3+6cag,\\ z^5z_1 &= a^3+b^3-8h^3+6abh, \end{aligned}$$

where the value of x^5x_1 may also be written $(b+c-2f)\,(b+c\omega-2\omega^2 f)\,(b+c\omega^2-2\omega f)$, ω being an imaginary cube root of unity, and so for the other two terms. The factors of x^5x_1 might be calculated from the identical equation

$$\begin{aligned} bY^2+2fYZ+cZ^2 = {} & 9\,(1+8l^3)\,x^3y^3z^3\,(yY^2+zZ^2+2lxYZ)\\ & +\{(y^2+2lzx)\,Y+(z^2+2lxy)\,Z\}\times\\ & \left\{\begin{array}{l} 3x^2y^2z^2\left[\left(3l^2y^2-(1+2l^3)\,zx\right)Y+\left(3l^2z^2-(1+2l^3)\,xy\right)Z\right]\\ -Q\,[(y^2+2lzx)\,Y+(z^2+2lxy)\,Z] \end{array}\right\}. \end{aligned}$$

I remark, that putting $x=0$, we have

$$bY^2+2fYZ+cZ^2=-\,y^3z^3\,(y^2Y+z^2Z)^2,$$

and hence writing 1 for Y, and -1, $-\omega$, $-\omega^2$ for Z, we have

$$b+c-2f=-\,y^3z^3\,(y^2-z^2)^2,\;\; b+c\omega-2\omega^2 f=-\,y^3z^3\,(y^2-\omega^4z^2)^2,\;\; b+c\omega^2-2\omega f=-\,y^3z^3\,(y^2-\omega^2z^2)^2;$$

and hence the product of the three factors is $-\,y^9z^9\,(y^2-z^2)^2\,(y^2-\omega^2z^2)^2\,(y^2-\omega^4z^2)^2$, which is equal to $-\,y^9z^9\,(y^3-z^3)^2\,(y^3+z^3)^2$, which vanishes in virtue of the assumed question $x=0$. This shows that the function $b^3+c^3-8f^3+6bcf$ contains the factor x. I have not verified *à posteriori*, but I assume it to be true, that it contains in fact the factor x^6, and consequently that the expressions for x_1, y_1, z_1 are rational and integral functions of (x, y, z) of the degree 25, and containing respectively the factors x, y, z.

46. In the theory of the cubic, a point which depends linearly upon a given point may be termed a derivative of such point. According to a very beautiful theorem of Professor Sylvester's, the coordinates of a derivative point are necessarily rational and integral functions *of a square degree* of the coordinates (x, y, z) of the given point; and moreover, there is but one derivative point having its coordinates of any given square degree m^2, or, as we may express it, only one derivative point of the degree m^2. The successive tangentials are derivative points of the degrees 4, 16, 64, &c.; the third point of intersection with the cubic, of the line joining two

derivative points of the degrees m^2 and n^2 respectively, is a derivative point of the degree $(m \pm n)^2$. Thus the third point of intersection with the cubic, of the line joining the given point with its second tangential, is a derivative point of the degree $(4 \pm 1)^2$, and it is easy to see that the degree is not 9; it is therefore 25. The point of simple intersection of the five-pointic conic is a derivative point of the degree 25; it is therefore, according to Professor Sylvester's general theory, identical with the point given by the former construction; this agrees with the before-mentioned theorem of Mr Salmon.

IV. *Independent investigation for the Cubic.*

47. The following is, in substance, the method by which I first obtained the equation of the five-pointic conic, for the cubic

$$X^3 + Y^3 + Z^3 + 6lXYZ = 0.$$

Write for shortness

$$\begin{aligned}
U &= x^3 + y^3 + z^3 + 6lxyz,\\
V &= (x^2 + 2lyz)\,X + (y^2 + 2lzx)\,Y + (z^2 + 2lxy)\,Z,\\
W &= (X^2 + 2lYZ)\,x + (Y^2 + 2lXZ)\,y + (Z^2 + 2lXY)\,z,\\
\Upsilon &= X^3 + Y^3 + Z^3 + 6lXYZ,\\
\mathrm{P} &= ax + by + cz,\\
\Pi &= aX + bY + cZ;
\end{aligned}$$

then, X, Y, Z being current coordinates, and x, y, z the coordinates of a point of the cubic (so that $U = 0$), the equation of the cubic will be

$$\Upsilon = 0,$$

and the equation of a conic having with it an ordinary (two-pointic) contact at the point (x, y, z), will be (1)

$$2W - \Pi V = 0.$$

48. Now imagine from the point of contact lines drawn to the other four intersections of the two curves; in the case of the five-pointic conic, three of these lines will coincide with the tangent $V = 0$, and the remaining line will be the line joining the point of contact with the point of simple intersection. The equations of the lines in question can be found by Joachimsthal's theorem, viz. if (x, y, z) be the coordinates of a given point, and (X, Y, Z) current coordinates, then if in the equations of any two curves we substitute for the coordinates, $\lambda x + \mu X$, $\lambda y + \mu Y$, $\lambda z + \mu Z$, and between the equations so obtained eliminate λ, μ, the resulting equation will be that

1 I have introduced the factor 2, to make this correspond with the form $D^2U - \Pi DU = 0$, in the case in question, $m = 3$.

of the lines drawn from the point (x, y, z) to the points of intersection of the two curves. The point (x, y, z) is any point whatever, and it may therefore be a point of intersection, or, as in the present instance, a point of contact of the two curves; the only difference is, that in either case the degree of each equation as regards (λ, μ) is reduced by unity, and the degree of the resulting equation in X, Y, Z is also reduced by unity: in the case of a point of simple intersection this is the only reduction; but in the case of a point of contact, the resulting equation contains the equation of the tangent as a factor, and rejecting this factor, the reduction in degree is two units.

49. Applying the method to the two equations, $\Upsilon=0$, $2W-\Pi V=0$, and substituting therein for the original current coordinates X, Y, Z the values $\lambda x+\mu X$, $\lambda y+\mu Y$, $\lambda z+\mu Z$, the equations become

$$\lambda^3 U+3\lambda^2\mu V+3\lambda\mu^2 W+\mu^3\Upsilon=0,$$
$$2(\lambda^2 U+2\lambda\mu V+\mu^2 W)-(\lambda P+\mu\Pi)(\lambda U+\mu V)=0;$$

or writing $U=0$, and omitting from each equation the factor μ, the equations become

$$\lambda^2 . 3V+\lambda\mu . 3W+\mu^2\Upsilon \quad =0,$$
$$\lambda(4-P)V+\mu(2W-\Pi V)=0;$$

and putting in the first equation $\lambda=2W-\Pi V$, $\mu=-(4-P)V$, the result of the elimination contains the factor V, rejecting which it becomes

$$3(2W-\Pi V)^2-3(4-P)(2W-\Pi V)W+(4-P)^2 V\Upsilon=0,$$

which is of the fourth degree in (X, Y, Z), as it should be, and represents therefore the lines drawn from the point of contact to the other four points of intersection of the conic and cubic.

50. The equation may be written

$$-3(2W-\Pi V)((2-P)W+\Pi V)+(4-P)^2 V\Upsilon=0,$$

and we obtain at once the condition that this may contain the factor V, viz. this condition is

$$P=2;$$

and if this be satisfied the conic will have a three-pointic conic, and there will be three other points of intersection. And writing $P=2$, and throwing out the factor V, we find

$$3\Pi^2 V-6\Pi W+4\Upsilon=0$$

for the equation of the lines from the point of contact to the three points of intersection. And we have now to determine Π so that the function on the left hand may divide by V^2.

51. I simplify my original method by the use of a theorem of Mr Salmon's, viz. writing

$$\begin{array}{l|l} \Upsilon = X^3 + Y^3 + Z^3 + 6lXYZ, & \mathfrak{H} = l^2(X^3+Y^3+Z^3) - (1+2l^3)XYZ, \\ W = (X^2 + 2lYZ)x + .. , & \mathfrak{H}_1 = (3l^2X^2 - (1+2l^3)YZ)x + .. , \\ V = (x^2 + 2lyz)X + .. , & H_1 = (3l^2x^2 - (1+2l^3)yz)X + .. , \\ U = x^3 + y^3 + z^3 + 6lxyz , & H = l^2(x^3+y^3+z^3) - (1+2l^3)xyz , \end{array}$$

we have identically

$$\Upsilon H - U\mathfrak{H} = WH_1 - V\mathfrak{H}_1,$$

and in the present case, since $U=0$,

$$\Upsilon H = WH_1 - V\mathfrak{H}_1.$$

Hence, multiplying by H and substituting this value of ΥH, the equation becomes

$$3H\Pi^2V - 6H\Pi W + 4WH_1 - 4V\mathfrak{H}_1 = 0,$$

or, as we may write it,

$$V(3H\Pi^2 - 4\mathfrak{H}_1) + 2W(2H_1 - 3H\Pi) = 0;$$

and we can at once make the equation divide by V, viz. by assuming

$$\Pi = \tfrac{2}{3}\frac{H_1}{H} + \tfrac{1}{2}\Lambda V,$$

where Λ is arbitrary: we have thus a four-pointic contact. And substituting for Π, and throwing out the factor V, the equation becomes

$$3H\left(\tfrac{2}{3}\frac{H_1}{H} + \tfrac{1}{2}\Lambda V\right)^2 - 4\mathfrak{H}_1 + 2W(\tfrac{3}{2}H\Lambda) = 0;$$

or reducing,

$$16(H_1^2 - 3H\mathfrak{H}_1) - 36H^2\Lambda W + 24HH_1\Lambda V + 9H^2\Lambda^2V^2 = 0,$$

which is the equation of the lines drawn from the point of contact to the remaining two points of intersection.

52. I write for greater convenience

$$\Lambda = -\tfrac{4}{9}\frac{\Theta}{H^3},$$

Θ being as yet indeterminate; the equation is thus reduced to

$$9H^3\{H(H_1^2 - 3H\mathfrak{H}_1) + \Theta W\} - 6H^2H_1\Theta V + \Theta^2V^2 = 0,$$

and we have then to determine Θ so that the left-hand side may divide by V; or, what is the same thing, we must determine Θ so that

$$H(H_1^2 - 3H\mathfrak{H}_1) + \Theta W$$

may divide by V. This implies the existence of an identical equation,

$$H(H_1^2 - 3H\mathfrak{H}_1) + \Theta W = MU + NV,$$

which for $U=0$ would give the decomposition in question; but I have not investigated the values of M and N. I assume at the outset $U=0$, and putting as before

$$Q = y^3z^3 + z^3x^3 + x^3y^3 - 3l^2x^2y^2z^2,$$

and writing also

$$\begin{aligned} S = \;& Xx\,[(1+8l^3)\,x^4 + (4l+41l^4)\,x^2yz + (-2l^2+2l^5)\,y^2z^2] \\ &+ Yy\,[(1+8l^3)\,y^4 + (4l+41l^4)\,y^2zx + (-2l^2+2l^5)\,z^2x^2] \\ &+ Zz\,[(1+8l^3)\,z^4 + (4l+41l^4)\,z^2xy + (-2l^2+2l^5)\,x^2y^2], \end{aligned}$$

I remark that for $U=0$ we have

$$-xyzH_1^2 + VS + (1+8l^3)(QW - 3x^2y^2z^2\mathfrak{H}_1) = 0,$$

an equation which, observing that

$$H = -(1+8l^3)\,xyz,$$

and assuming also

$$\Theta = \quad (1+8l^3)^2\,Q,$$

may be written

$$H(H_1^2 - 3H\mathfrak{H}_1) + \Theta W = -(1+8l^3)\,VS,$$

which gives the required decomposition, viz., Θ having this value, the conic will have a five-pointic contact. Reducing by the last equation, and throwing out the factor V, we find

$$-9H^3(1+8l^3)\,V - 6H^2\Theta H_1 + \Theta^2 V = 0$$

for the equation of the line joining the point of contact with the point of simple intersection. And if in this equation we write $H = -(1+8l^3)\,xyz$, and $\Theta = (1+8l^3)^2\,Q$, we obtain, finally, for the equation of the line in question,

$$9x^3y^3z^3S - 6x^2y^2z^2QH_1 + Q^2V = 0,$$

which is the before-mentioned result.

53. It only remains to verify the assumed equation

$$-xyzH_1^2 + VS + (1+8l^3)(QW - 3x^2y^2z^2\mathfrak{H}_1) = 0.$$

We may write

$$-Q = x^6 + 6lx^4yz + 3l^2x^2y^2z^2 - y^3z^3,$$

and then observing that

$$\begin{aligned} W &= \quad xX^2 + \ldots + \qquad 2lxYZ + \ldots, \\ \mathfrak{H}_1 &= 3l^2xX^2 + \ldots - (1+2l^3)\,xYZ - \ldots, \end{aligned}$$

we find at once

$$(1+8l^3)(QW-3x^2y^2z^2\mathfrak{H}_1)=$$

$$-(1+8l^3)\left\{\begin{array}{l}(x^7+6lx^5yz+12l^2x^3yz\;-xy^3z^3)\,X^2\\ \quad\vdots\\ +(2lx^7+12l^2x^5yz-3x^3y^2z^2-2lxy^3z^3)\,YZ\\ \quad\vdots\end{array}\right.$$

Next writing

$$H_1=\left(3l^2x^2-(1+2l^3)\,yz\right)X+\ldots,$$
$$V\;=(x^2+2lyz)\,X+\ldots,$$
$$S\;=x\left\{(1+8l^3)\,x^4+(4l+41l^4)\,x^2yz+(-2l^2+2l^5)\,y^2z^2\right\}X+\ldots,$$

and forming the expression for

$$-xyzH_1^{\,2}+VS,$$

the coefficient of X^2 is

$$\begin{array}{l}-xyz\qquad\qquad\{3l^2x^2-(1+2l^3)\,yz\}^2\\ +(x^2+2lyz)\,x\,\{(1+8l^3)\,x^4+(4l+41l^4)\,x^2yz+(-2l^2+2l^5)\,y^2z^2\},\end{array}$$

which is

$$=(1+8l^3)\,(x^7+6lx^5yz+12l^2x^3y^2z^2-xy^3z^5)\,;$$

the coefficient of YZ is

$$\begin{array}{l}-2xyz\left(3l^2y^2-(1+2l^3)\,zx\right)\left(3l^2z^2-(1+2l^3)\,xy\right)\\ +(y^2+2lzx)\,z\,\{(1+8l^3)\,z^4+(4l+41l^4)\,z^2xy+(-2l^2+2l^5)\,x^2y^2\}\\ +(z^2+2lxy)y\,\{(1+8l^3)\,y^4+(4l+41l^4)\,y^2zx+(-2l^2+2l^5)\,x^2y^2\},\end{array}$$

which is

$$=(1+8l^3)\,\{y^2z^2\,(y^3+z^3)+8lxy^3z^3+12l^2x^2yz\,(y^3+z^3)+2lx\,(y^6+z^6)-2x^3y^2z^2\}\,;$$

or substituting for y^3+z^3 and y^6+z^6 the values $-x^3-6lxyz$ and $(x^3+6lxyz)^2-2y^3z^3$ respectively, this is

$$=(1+8l^3)\,(2lx^7+12l^2x^5yz-3x^3y^2z^2-2lxy^3z^3).$$

We have thus

$$-xyzH_1^{\,2}+VS$$

$$=(1+8l^3)\left\{\begin{array}{l}(x^7+6lx^5yz+12l^2x^3yz-xy^3z^3)\,X^2\\ \quad\vdots\\ +(2lx^7+12l^2x^5yz-3x^3y^2z^2-2lxy^3z^3)\,YZ\\ \quad\vdots\end{array}\right.$$

and the equation

$$-xyzH_1^{\,2}+VS+(1+8l^3)\,(QW-3x^2y^2z^2\mathfrak{H}_1)=0$$

is thus verified.

ADDITION.—The foregoing memoir was communicated to Mr Salmon, and I am indebted to him for two notes, containing the extension to a curve of any order, of the preceding investigation for the case of a cubic; I reproduce this extension in the following section.

V. *Extension of the last preceding method to a curve of any order.*

54. Consider the curve of the m-th order $\Upsilon = 0$, and in the place of the coordinates write $\lambda x + \mu X$, $\lambda y + \mu Y$, $\lambda z + \mu Z$, where, as before, (x, y, z) are the coordinates of the point of the curve, and X, Y, Z current coordinates; the term involving λ^m vanishes, and dividing out the factor μ the equation becomes

$$\lambda^{m-1} DU + \tfrac{1}{2}\lambda^{m-2}\mu D^2U + \tfrac{1}{6}\lambda^{m-3}\mu^2 D^3U + \tfrac{1}{24}\lambda^{m-4}\mu^3 D^4U + \&c. = 0.$$

Making the like substitution in $D^2U - \Pi DU = 0$, the assumed equation of the five-pointic conic, the factor μ divides out and the equation becomes

$$2\lambda (m-1) DU + \mu D^2U - (\lambda P + \mu\Pi) D^2U = 0,$$

or, what is the same thing,

$$\lambda DU \big(2(m-1) - P\big) + \mu (D^2U - \Pi DU) = 0;$$

and if from the two equations we eliminate λ, μ, the result, throwing out the factor DU, is

$$(D^2U - \Pi DU)^{m-1} - \tfrac{1}{2}\big(2(m-1) - P\big) D^2U (D^2U - \Pi DU)^{m-2} + \&c. = 0,$$

where all the terms after the second contain the factor DU; the condition in order that the equation may divide by DU, is consequently $2(m-1) - P = 2$, or $P = 2(m-2)$, the condition of a three-pointic contact. Substituting this value, and dividing by DU, the equation becomes

$$-\Pi (D^2U - \Pi DU)^{m-2} + \tfrac{2}{3} D^3U (D^2U - \Pi DU)^{m-3} - \tfrac{1}{3} D^4U \,.\, DU (D^2U - \Pi DU)^{m-4} + \&c. = 0,$$

which will be divisible by DU if $-\Pi D^2U + \tfrac{2}{3} D^3U$ is divisible by DU, and the condition for this is found to be, as before,

$$\Pi = \tfrac{2}{3}\frac{1}{H} DH + \Lambda DU,$$

where Λ is arbitrary; we have thus the conditions of a four-pointic contact.

55. Substituting this value of Π, we see that $D^3U - \dfrac{1}{H} DH \,.\, D^2U$ divides by DU, viz. there exists an identical equation,

$$D^3U - \frac{1}{H} DH \,.\, D^2U = IU + J \,.\, DU;$$

and hence if $U = 0$,

$$\frac{1}{DU}\left(D^3U - \frac{1}{H} DH \,.\, D^2U\right) = J,$$

where J is a quadric function of (X, Y, Z). I do not know the general form of this function, but Mr Salmon has obtained a result which may be generalized as follows, viz. writing for X, Y, Z the values $B\nu - C\mu$, $C\lambda - A\nu$, $A\mu - B\lambda$ (where, as before, A, B, C are the first derived functions of U and λ, μ, ν are arbitrary), the expression for J is

$$J = \frac{1}{DU}\left(D^3U - \frac{1}{H}DH \,.\, D^2U\right) = \frac{3(m-2)}{(m-1)}\Phi - \frac{1}{(m-1)H}\square\vartheta;$$

a formula which will be presently useful.

56. The foregoing equation may be written

$$(D^2U)^{m-3}(-\Pi D^2U + \tfrac{2}{3}D^3U)$$
$$+ (D^2U)^{m-4}DU\{(m-2)\Pi^2D^2U - \tfrac{2}{3}(m-3)\Pi D^3U - \tfrac{1}{3}D^4U\} + \&c.\ (DU)^2 \ldots = 0;$$

and the term $-\Pi D^2U + \frac{2}{3}D^3U$ is equal to

$$\tfrac{2}{3}\left(D^3U - \frac{1}{H}DH \,.\, D^2U\right) - \Lambda DU \,.\, D^2U, = \tfrac{2}{3}JDU - \Lambda DU \,.\, D^2U.$$

Substituting this value the equation divides by DU, and throwing out this factor it becomes

$$(D^2U)^{m-3}(\tfrac{2}{3}J - \Lambda D^2U)$$
$$+ (D^2U)^{m-4}\{(m-2)\Pi^2D^2U - \tfrac{2}{3}(m-3)\Pi D^3U - \tfrac{1}{3}D^4U\} + \&c.\ DU = 0;$$

or observing that $\frac{2}{3}\Pi D^3U = \Pi^2D^2U +$ term containing DU, this may be written

$$(D^2U)^{m-3}(\tfrac{2}{3}J - \Lambda D^2U) + (D^2U)^{m-4}(\Pi^2D^2U - \tfrac{1}{3}D^4U) + \&c.\ DU = 0.$$

57. If the equation divides by DU we shall have a five-pointic contact; the condition for this is that

$$-\Lambda(D^2U)^2 + \tfrac{2}{3}JD^2U + \Pi^2D^2U - \tfrac{1}{3}D^4U$$

may divide by DU, or more simply that

$$-\Lambda(D^2U)^2 + \tfrac{2}{3}JD^2U + \tfrac{4}{9}\left(\frac{1}{H}DH\right)^2 D^2U - \tfrac{1}{3}D^4U$$

may divide by DU, or, what is the same thing, the function in question must vanish in virtue of the substitution of the values $B\nu - C\mu$, $C\lambda - A\nu$, $A\mu - B\lambda$ in the place of X, Y, Z. The expression for J has just been given; we have besides

$$\left(\frac{1}{H}DH\right)^2 = \frac{(m-1)^4}{H^2}Q_3^2,\quad D^4U = -Q_4\vartheta^2,$$

where the values of Q_3^2, Q_4 are given (*ante*, No. 27); we have thus

58. $$\frac{1}{DU}\left(D^3U - \frac{1}{H}DH \,.\, D^2U\right) = \frac{3(m-2)}{m-1}\Phi - \frac{1}{(m-1)H}\square\vartheta,$$

$$\tfrac{4}{9}\left(\frac{1}{H}DH\right)^2 = -\frac{4(m-2)^2}{(m-1)^2}\Phi + \tfrac{4}{3}\frac{(m-2)^2}{(m-1)^2}\frac{\square}{H}\vartheta - \tfrac{4}{9}\frac{1}{(m-1)^2}\frac{1}{H^2}\Psi\vartheta^2,$$

$$-\tfrac{1}{3}D^4U = \vartheta^2\left\{-\frac{2(m-2)(m-3)}{(m-1)^4}\Phi H + \tfrac{4}{3}\frac{(m-3)}{(m-1)^4}\square\vartheta - \tfrac{1}{3}\frac{1}{(m-1)^4}\Omega\vartheta^2\right\},$$

and

$$D^2U = -\frac{1}{(m-1)^2} H\vartheta^2,$$

whence

$$\Lambda \frac{1}{(m-1)^4} H^2\vartheta^4 =$$

$$-\tfrac{2}{3}\frac{1}{(m-1)^2} H\vartheta^2 \left\{\frac{3(m-2)}{m-1}\Phi - \frac{1}{(m-1)H}\square\vartheta\right\}$$

$$-\frac{1}{(m-1)^2} H\vartheta^2 \left\{\frac{-4(m-2)^2}{(m-1)^2}\Phi + \tfrac{4}{3}\frac{(m-2)}{(m-1)^2 H}\square\vartheta - \tfrac{4}{9}\frac{1}{(m-1)^2}\frac{1}{H^2}\Psi\vartheta^2\right\}$$

$$+\vartheta^2 \left\{-\frac{2(m-2)(m-3)}{(m-1)^4}\Phi H + \tfrac{2}{3}\frac{(m-3)}{(m-1)^4}\square\vartheta - \tfrac{1}{3}\frac{1}{(m-1)^4}\Omega\vartheta^2\right\},$$

$$= \frac{1}{9(m-1)^4 H}(4\Psi - 3\Omega H),$$

and consequently

$$\Lambda = \frac{1}{9H^3}(4\Psi - 3\Omega H),$$

which agrees with the result before obtained, and thus the present method gives the complete solution of the problem.

262.

ON THE EQUATION OF DIFFERENCES FOR AN EQUATION OF ANY ORDER, AND IN PARTICULAR FOR THE EQUATIONS OF THE ORDERS TWO, THREE, FOUR, AND FIVE.

[From the *Philosophical Transactions of the Royal Society of London,* vol. CL. (for the year 1860), pp. 93—112. Received March 2,—Read March 29, 1860.]

THE term, *equation of differences,* denotes the equation for the squared differences of the roots of a given equation; the equation of differences afforded a means of determining the number of real roots, and also limits for the real roots, of a given numerical equation, and was upon this account long ago sought for by geometers. In the *Philosophical Transactions* for 1763, Waring gives, but without demonstration or indication of the mode of obtaining it, the equation of differences for an equation of the fifth order wanting the second term: the result was probably obtained by the method of symmetric functions. This method is employed in the *Meditationes Algebraicæ* (1782), where the equation of differences is given for the equations of the third and fourth orders wanting the second terms; and in p. 85 the before-mentioned result for the equation of the fifth order wanting the second term, is reproduced. The formulæ for obtaining by this method the equation of differences, are fully developed by Lagrange in the *Traité des Equations Numériques* (1808); and he finds by means of them the equation of differences for the equations of the orders two and three, and for the equation of the fourth order wanting the second term; and in Note III. he gives, after Waring, the result for the equation of the fifth order wanting the second term. It occurred to me that the equation of differences could be most easily calculated by the following method. The coefficients of the equation of differences, *quâ* functions of the differences of the roots of the given equation, are leading coefficients of covariants, or (to use a shorter expression) they are "Seminvariants [1]," that is, each of them is a function of the coefficients which is reduced to zero by one of the two operators which reduce an invariant to zero. In virtue of this property they can be calculated, when their values are known for the particular case in which one of the coefficients of the given equation is zero. To fix the ideas, let the given equation be $(*\between v, 1)^n = 0$; then, when the last coefficient or constant term vanishes, the equation breaks up into $v=0$ and into an equation of the degree $n-1$, which I call the reduced equation; the equation of differences will break up into two equations, one of which is the equation of differences for the reduced equation, the other is the equation for the squares of the roots of the same reduced equation. This hardly requires a proof; let the roots of the given equation be α, β, γ, δ, &c., those of the equation of differences are $(\alpha-\beta)^2$, $(\alpha-\gamma)^2$, $(\alpha-\delta)^2$, &c., $(\beta-\gamma)^2$, $(\beta-\delta)^2$, $(\gamma-\delta)^2$, &c.; but in putting the constant term equal to zero, we in effect put one of the roots, say α, equal to zero; the roots of the equation of differences thus become β^2, γ^2, δ^2 &c., $(\beta-\gamma)^2$, $(\beta-\delta)^2$, $(\gamma-\delta)^2$, &c. The equation for the squares of the roots can be found without the slightest difficulty; hence if the equation of differences for the reduced equation of the order $n-1$ is known, we can, by combining it with the equation for the squares of the roots, form the equation of differences for the given equation with the constant term put equal to zero, and thence by the above-mentioned property of the Seminvariancy of the coefficients, find the equation of differences for the given equation. The present memoir shows the application of the process to equations of the orders two, three, four, and five: part of the calculation for the equation of the fifth order was kindly performed for me by the Rev. R. Harley. It is to be noticed that the best course is to apply the method in the first instance to the forms $(a, b, \ldots \between v, 1)^n = 0$, without numerical coefficients (or, as they may be termed, the *denumerate forms*), and to pass from the results so obtained to those which belong to the forms $(a, b, \ldots \between v, 1)^n = 0$, or *standard forms*. The equation of differences, for $(\alpha-\beta)^2$, &c., the coefficients of which are seminvariants, naturally leads to the consideration of a more general equation having for its roots $(\alpha-\beta)^2(x-\gamma y)^2(x-\delta y)^2 \ldots$, &c., the coefficients of which are covariants; and in fact, when, as for equations of the orders two, three, and four, all the covariants are known, such covariant equation can be at once formed from the equation of differences; for equations of the fifth order, however, where the covariants are not calculated beyond a certain degree, [they are now all calculated, see 141 and 143], only a few of the coefficients of the covariant equation can be thus at once formed. At the conclusion of the memoir, I show how the equation of differences for an equation of the order n can be obtained by the elimination of a single quantity from two equations each of the order $n-1$; and by applying to these two equations the simplification which I have made in Bezout's abridged method of elimination, I exhibit the equation of differences for the given equation of the order n, in a compendious form by means of a determinant; the first-mentioned method is, however, that which is best adapted for the actual development of the equation of differences for the equation of a given order.

[1] The term "Seminvariant" seems to me preferable to M. Brioschi's term "Peninvariant."

The equations successively considered are

$$(a,\ b,\ c \between v,\ 1)^2 = 0,$$
$$(a,\ b,\ c,\ d \between v,\ 1)^3 = 0,$$
$$(a,\ b,\ c,\ d,\ e \between v,\ 1)^4 = 0,$$
$$(a,\ b,\ c,\ d,\ e,\ f \between v,\ 1)^5 = 0.$$

The equation of differences for the quadric, and that for the squares of the roots, are considered to be known, and the other results are derived from them: it will be convenient to write down in the first instance the results for the quadric, the cubic, and the quartic equations, and then explain the process of obtaining them.

For the quadric equation,

Equation of differences is, 0 =

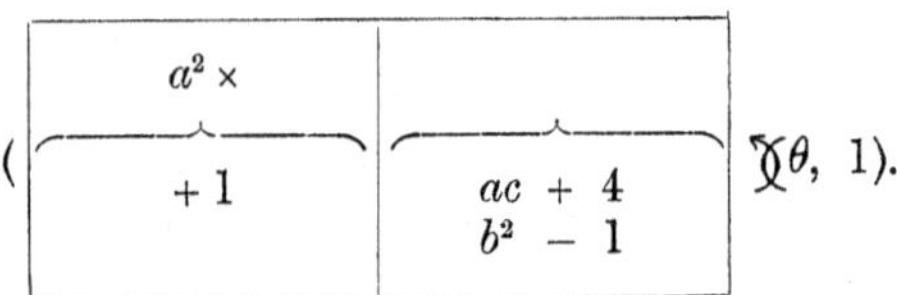

$a^2 \times$	
$+1$	$ac\ +4$ $b^2\ -1$

$(\ \ldots \between \theta,\ 1).$

Equation for the squares of the roots is, 0 =

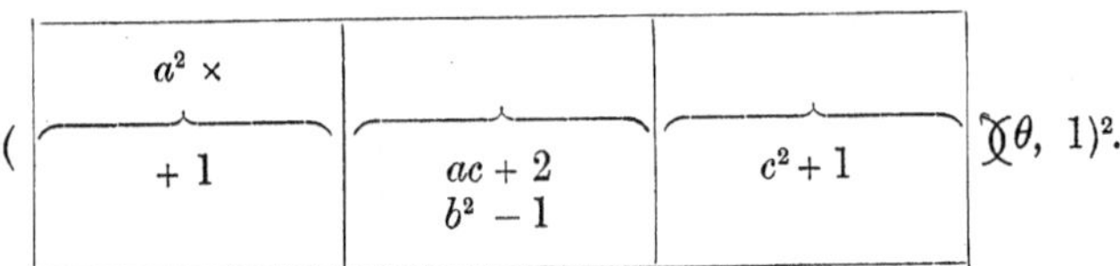

$a^2 \times$		
$+1$	$ac+2$ $b^2\ -1$	c^2+1

$(\ \ldots \between \theta,\ 1)^2.$

For the cubic equation,

Equation of differences is, 0 =

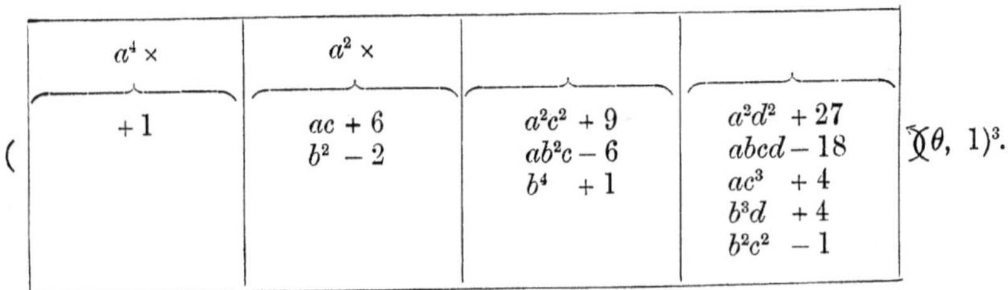

$a^4 \times$	$a^2 \times$		
$+1$	$ac\ +6$ $b^2\ -2$	$a^2c^2\ +9$ ab^2c-6 $b^4\ \ +1$	$a^2d^2\ +27$ $abcd-18$ $ac^3\ \ +4$ $b^3d\ \ +4$ $b^2c^2\ \ -1$

$(\ \ldots \between \theta,\ 1)^3.$

Equation for the squares of the roots is, 0 =

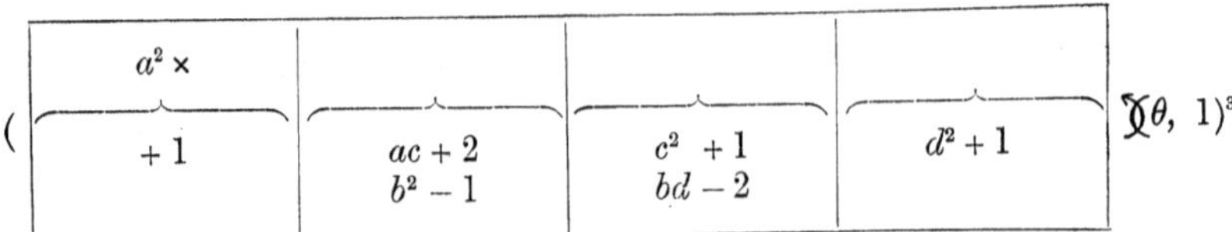

$a^2 \times$			
$+1$	$ac+2$ b^2-1	$c^2\ +1$ $bd\ -2$	d^2+1

$(\ \ldots \between \theta,\ 1)^3.$

For the quartic equation,

Equation of differences is, $0 =$

(

$a^6 \times$	$a^4 \times$	$a^2 \times$				
$+1$	$ac + 8$	$a^3e + 8$	$a^4ce + 16$	$a^4e^2 - 112$	$a^3ce^2 - 192$	$a^3e^3 + 256$
	$b^2 - 3$	$a^2bd - 2$	$a^4d^2 + 26$	$a^3bde + 56$	$a^3d^2e + 216$	$a^2bde^2 - 192$
		$a^2c^2 + 22$	$a^3b^2e - 6$	$a^3c^2e + 24$	$a^2b^2e^2 + 72$	$a^2c^2e^2 - 128$
		$ab^2c - 16$	$a^3bcd - 30$	$a^3cd^2 + 48$	$a^2bcde - 120$	$a^2cd^2e + 144$
		$b^4 + 3$	$a^3c^3 + 28$	$a^2b^2ce - 32$	$a^2bd^3 - 54$	$a^2d^4 - 27$
			$a^2b^3d + 8$	$a^2b^2d^2 - 25$	$a^2c^3e + 32$	$ab^2ce^2 + 144$
			$a^2b^2c^2 - 24$	$a^2bc^2d - 54$	$a^2c^2d^2 + 18$	$ab^2d^2e - 6$
			$ab^4c + 8$	$a^2c^4 + 17$	$ab^3de + 18$	$abc^2de - 80$
			$b^6 - 1$	$ab^4e + 6$	$ab^2c^2e - 6$	$abcd^3 + 18$
				$ab^3cd + 38$	$ab^2cd^2 + 42$	$ac^4e + 16$
				$ab^2c^3 - 12$	$abc^3d - 26$	$ac^3d^2 - 4$
				$b^5d - 6$	$ac^5 + 4$	$b^4e^2 - 27$
				$b^4c^2 + 2$	$b^4d^2 - 9$	$b^3cde + 18$
					$b^3c^2d + 6$	$b^3d^3 - 4$
					$b^2c^4 - 1$	$b^2c^3e - 4$
						$b^2c^2d^2 + 1$

$)(\theta, 1)^6$.

Equation for squares of the roots is, $0 =$

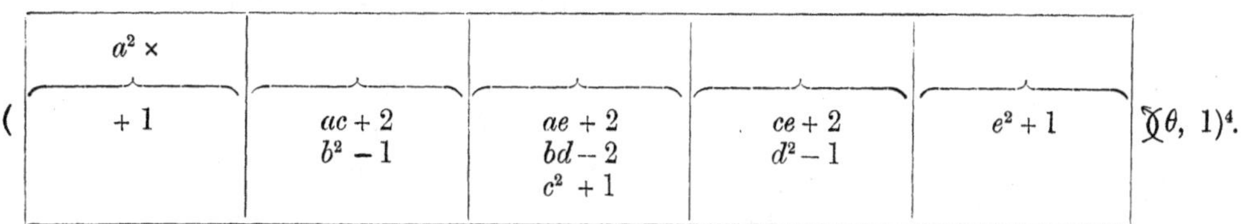

(

$a^2 \times$				
$+1$	$ac + 2$	$ae + 2$	$ce + 2$	$e^2 + 1$
	$b^2 - 1$	$bd - 2$	$d^2 - 1$	
		$c^2 + 1$		

$)(\theta, 1)^4$.

The multiplication of the equation of differences and the equation of the squares of the roots of the quadric equation, gives the equation, $0 =$

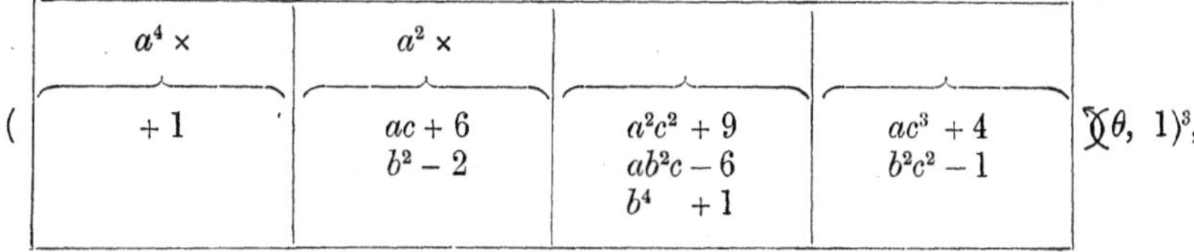

(

$a^4 \times$	$a^2 \times$		
$+1$	$ac + 6$	$a^2c^2 + 9$	$ac^3 + 4$
	$b^2 - 2$	$ab^2c - 6$	$b^2c^2 - 1$
		$b^4 + 1$	

$)(\theta, 1)^3$,

where all the coefficients except the last are reduced to zero by the operator

$$3a\partial_b + 2b\partial_c + c\partial_d,$$

and they are consequently (without any alteration) coefficients of the equation of differences of the cubic equation: the last coefficient is not reduced to zero by the operator, and requires therefore to be completed by the adjunction of the terms in d (the series, here and in every other case, is of course a finite one, the number of terms might easily be calculated *à priori*). Let the value be $L_0 + L_1 d + L_2 d^2 + \&c.$, we have $L_0 = +4ac^3 - 1b^2c^2$; and putting for shortness $\nabla' = 3a\partial_b + 2b\partial_c$, the operator which reduces this to zero is $\nabla' + c\partial_d$; we ought therefore to have

$$0 = \begin{array}{c|c|c} \nabla' L_0 + d & \nabla' L_1 + d^2 & \nabla' L_2 + \dots \\ + \;\; cL_1 & 2cL_2 & 3cL_3 \end{array}$$

and consequently

$$L_1 = -\frac{1}{c}\nabla' L_0, \;\; L_2 = -\frac{1}{2c}\nabla' L_1, \;\; L_3 = -\frac{1}{3c}\nabla' L_2, \;\; \&c.,$$

giving

$$L_1 = -\frac{1}{c}\{(-6+24=)\,18abc^2 - 4b^3c\} = -18abc + 4b^3,$$

$$L_2 = -\frac{1}{2c}\{-54a^2c + (36-36=)\,0\,ab^2\} = +27a^2,$$

$$L_3 = 0, \;\; \&c.,$$

and consequently for the last coefficient the value above written down; it will be presently seen how in more complicated cases the calculations should be arranged.

Again, multiplying together the equation of differences and the equation for the squares of the roots of the cubic equation, we obtain an equation which it is not necessary to write down, as it can be at once formed by putting $e=0$ in the equation of differences for the quartic equation. And from the equation so obtained, by the adjunction of the terms in e, we find the equation of differences for the quartic equation, viz. each coefficient is of the form $L_0 + L_1 e + L_2 e^2 + \&c.$, where L_0 is known, and such coefficient is reduced to zero by the operator

$$4a\partial_b + 3b\partial_c + 2c\partial_d + d\partial_e\,;$$

or putting for shortness $\nabla' = 4a\partial_b + 3b\partial_c + 2c\partial_d$, the operator $\nabla' + d\partial_e$. We have therefore

$$L_1 = -\frac{1}{d}\nabla' L_0, \quad L_2 = -\frac{1}{2d}\nabla' L_1, \quad L_3 = -\frac{1}{3d}\nabla' L_2, \quad \&c.$$

It is to be observed that the last coefficient of the equation of differences is the discriminant, and that the above method of calculating the coefficients of the equation of differences, as applied to the last coefficient, is nothing else than the method of calculating the discriminant given in my Fourth Memoir on Quantics, [155].

The multiplication of the equation of differences, and the equation for the squares of the roots of the quartic equation, gives, in like manner, the equation of differences for the quintic equation, except as to the terms involving f; and these are obtained as above, viz. each coefficient is of the form $L_0 + L_1 f + L_2 f^2 + \&c.$, where L_0 is known; such coefficient is reduced to zero by the operator

$$5a\partial_b + 4b\partial_c + 3c\partial_d + 2d\partial_e + e\partial_f;$$

or putting for shortness $\nabla' = 5a\partial_b + 4b\partial_c + 3c\partial_d + 2d\partial_e$, the operator $\nabla' + e\partial_f$. We have therefore

$$L_1 = -\frac{1}{e}\nabla' L_0, \quad L_2 = -\frac{1}{2e}\nabla' L_1, \quad L_3 = -\frac{1}{3e}\nabla' L_2, \ \&c.,$$

which give L_1, L_2, &c. by means of L_0. For the calculation of $\nabla' L$ (where L is any one of the coefficients L_0, L_1, &c.), it is proper to separate the terms involving the different powers of e, and write $L = M_0 + M_1 e + M_2 e^2 + \&c.$, $\nabla' = \nabla'' + 2d\partial_e$, where $\nabla'' = 5a\partial_b + 4b\partial_c + 3c\partial_d$. We have then

$$\nabla L = \begin{array}{l|l|l} \nabla'' M_0 \ + e & \nabla'' M_1 \ + e^2 & \nabla'' M_2 + \&c., \\ + 2dM_1 & + 4dM_2 & + 6dM_3, \end{array}$$

or, what is the same thing,

$$\frac{1}{e}\nabla L = \begin{array}{l|l} \nabla'' M_1 \ + e & \nabla'' M_2 + \ldots \\ + 4dM_2 & + 4dM_3; \end{array}$$

and as an equation which should be satisfied identically, and which would therefore serve as a verification,

$$\nabla'' M_0 + 2dM_1 = 0;$$

but, since a verification was obtained by other means, the equations of this kind were not used for the purpose. It may be interesting to give the actual calculation of one of the coefficients, say of coefficient θ^2 (which, with coefficient θ, was calculated by Mr Harley).

Calculation of coefficient θ^2 in equation of differences for the quintic equation

$$(a,\ b,\ c,\ d,\ e,\ f \between v,\ 1)^5 = 0.$$

Calculation of L_0.

	$2ae$	$-2bd$	$+c^2$	$2ce$	$-d^2$	e^2	
	Coeff. θ^0.			Coeff. θ^1.		Coeff. θ^2.	In eq. of diff. for quartic.
a^4e^4	+512					−112 =	+400
a^3bde^3	−384	−512				+56	−840
$a^3c^2e^3$	−256		+256	−384		+24	−360
$a^3cd^2e^2$	+288			+432	+192	+48	+960
a^3d^4e	−54				−216		−270
$a^2b^2ce^3$	+288			+144		−32	+400
$a^2b^2d^2e^2$	−12	+384			−72	−25	+275
$a^2bc^2de^2$	−160	+256	−192	−240		−54	−390
a^2bcd^3e	+36	−288		−108	+120		−240
abd^5		+54			+54		+108
$a^2c^4e^2$	+32		−128	+64		+17	−15
$a^2c^3d^2e$	−8		+144	+36	−32		+140
$a^2c^2d^4$			−27		−18		−45
ab^4e^3	−54					+6	−48
ab^3cde^2	+36	−288		+36		+38	−178
ab^3d^3e	−8	+12			−18		−14
$ab^2c^3e^2$	−8		+144	−12		−12	+112
$ab^2c^2d^2e$	+2	+160	−6	+84	+6		+246
ab^2cd^4		−36			−42		−78
abc^4de		−32	−80	−52			−164
abc^3d^3		+8	+18		+26		+52
ac^6e			+16	+8			+24
ac^5d^2			−4		−4		−8
b^5de^2		+54				−6	+48
$b^4c^2e^2$			−27			+2	−25
b^4cd^2e		−36		−18			−54
b^4d^4		+8			+9		+17
b^3c^3de		+8	+18	+12			+38
$b^3c^2d^3$		−2	−4		−6		−12
b^2c^5e			−4	−2			−6
$b^2c^4d^2$			+1		+1		+2

$L_0 = A + Be + Ce^2 + De^3 + Ee^4$, suppose, then

$-B=$		$-C=$		$-D=$		$-E=$	
a^2d^3	+270	a^3cd^2	−960	a^3bd	+840	a^4	−400
a^2bcd	+240	$a^2b^2d^2$	−275	a^3c^2	+360		
$a^2c^3d^2$	−140	a^2bc^2d	+390	a^2b^2c	−400		
ab^3d^3	+14	a^2c^4	+15	ab^4	+48		
$ab^2c^2d^2$	−246	ab^3cd	+178				
abc^4d	+164	ab^2c^3	−112				
ac^6	−24	b^5d	−48				
b^4cd^2	+54	b^4c^2	+25				
b^3c^3d	−38						
b^2c^5	+6						

Calculation of L_1.

Terms of $L_1 f$ not involving e.

$-f(\nabla'' B + 4dC)$ viz.	$5a\partial_b$	$+4b\partial_c$	$+3c\partial_d$	$4d$		
		$-B$		$-C$		
a^3cd^3f	+ 1200		+ 3240	− 3840	=	+ 600
$a^2b^2d^3f$	+ 210	+ 960		− 1100		+ 70
$a^2bc^2d^2f$	− 2460	− 1680	+ 2160	+ 1560		− 420
a^2c^4df	+ 820		− 840	+ 60		+ 40
ab^3cd^2f	+ 1080	− 1968	+ 126	+ 712		− 50
ab^2c^3df	− 570	+ 2624	− 1476	− 448		+ 130
abc^5f	+ 60	− 576	+ 492			− 24
b^5d^2f		+ 216		− 192		+ 24
b^4c^2df		− 456	+ 324	+ 100		− 32
b^3c^4f		+ 120	− 114			+ 6

Terms of $L_1 f$ involving e^1.

$-ef(\nabla'' C + 6dD)$ viz.	$5a\partial_b$	$+4b\partial_c$	$+3c\partial_d$	$6d$		
		$-C$		$-D$		
a^3bd^2ef	− 2750	− 3840		+ 5040	=	− 1550
a^3c^2def	+ 1950		− 5760	+ 2160		− 1650
a^2b^2cdef	+ 2670	+ 3120	− 1650	− 2400		+ 1740
a^2bc^3ef	− 1120	+ 240	+ 1170			+ 290
ab^4def	− 1200	+ 712		+ 288		− 200
ab^3c^2ef	+ 500	− 1344	+ 534			− 310
b^5cef		+ 200	− 144			+ 56

Terms of $L_1 f$ involving e^2.

$-e^2f(\nabla'' D + 8dE)$ viz.	$5a\partial_b$	$4b\partial_c$	$3c\partial_d$	$8d$		
		$-D$		$-E$		
a^4de^2f	+ 4200			− 3200	=	+ 1000
a^3bce^2f	− 4000	+ 2880	+ 2520			+ 1400
$a^2b^3e^2f$	+ 960	− 1600				− 640

Calculation of L_2.

$L_1 = A + Be + Ce^2$, suppose, where

$-\frac{1}{2}B =$		$-\frac{1}{2}C =$	
a^3bd^2	+ 775	a^4d	− 500
a^3c^2d	+ 825	a^3bc	− 700
a^2b^2cd	− 870	a^2b^3	+ 320
a^2bc^3	− 145		
ab^4d	+ 100		
ab^3c^2	+ 155		
b^5c	− 28		

Terms of L_2f^2 not involving e.

$-f^2(\nabla''\frac{1}{2}B+4d\frac{1}{2}C)$ viz.	$5a\partial_b$	$4b\partial_c$	$3c\partial_d$	$4d$		
	$-\frac{1}{2}B$			$-\frac{1}{2}C$		
$a^4d^2f^2$	+ 3875			− 2000	=	+ 1875
a^3bcdf^2	− 8700	+ 6600	+ 4650	− 2800		− 250
$a^3c^3f^2$	− 725		+ 2475			+ 1750
$a^2b^3df^2$	+ 2000	− 3480		+ 1280		− 200
$a^2b^2c^2f^2$	+ 2325	− 1740	− 2610			− 2025
ab^4cf^2	− 700	+ 1240	+ 300			+ 840
ab^6f^2		− 112				− 112

Terms of L_2f^2 involving e^1.

$-ef^2\Delta''\frac{1}{2}C$ viz.	$5a\partial_b$	$4b_{dc}$	$3c\partial_d$		
	$-\frac{1}{2}C$				
a^4cef^2	− 3500		− 1500	=	− 5000
$a^3b^2ef^2$	+ 4800	− 2800			+ 2000

Calculation of L_3 (gives $L_3 = 0$).

$$L_2 = A + Be,$$

$-\frac{1}{3}B =$	
a^4c a^3b^2	$+\ 1666\frac{2}{3}$ $-\ \ 666\frac{2}{3}$

Terms of L_3f^3.

$-f^3\nabla''\frac{1}{3}B$	viz. $\underbrace{5a\partial_b \quad 4b\partial_c \quad 3c\partial_d}_{-\frac{1}{3}B}$			
a^4bf^3	$-\ 6666\frac{2}{3}$	$+\ 6666\frac{2}{3}$	=	0

and the required coefficient of θ^2 is

$$L_0 + L_1f + L_2f^2.$$

All the coefficients were calculated in this manner, except the last coefficient, which was deduced from the known value of the discriminant for the standard form. And we have thus the complete expression of the equation of differences for the general quintic equation under the denumerate form $(a, b, c, d, e, f \between v, 1)^5 = 0$, viz.

Equation of differences for $(a, b, c, d, e\ f \between v, 1)^5 = 0$ is

θ^{10}	θ^{9}	θ^{8}	θ^{7}	θ^{6}	θ^{5}	θ^{4}
$a^8 \times$	$a^6 \times$	$a^4 \times$	$a^2 \times$			
$+1$	$ac + 10$	$a^3e + 10$	$a^4ce + 50$	$a^6df + 200$	$a^6f^2 + 625$	$a^5cf^2 + 1750$
	$b^2 - 4$	$a^2bd - 4$	$a^4d^2 + 25$	$a^6e^2 - 95$	$a^5bef - 250$	$a^5def - 950$
		$a^2c^2 + 39$	$a^3b^2e - 20$	$a^5bcf - 120$	$a^5cdf + 400$	$a^5e^3 + 40$
		$ab^2c - 30$	$a^3bcd - 50$	$a^5bde + 36$	$a^5ce^2 - 360$	$a^4b^2f^2 - 700$
		$b^4 + 6$	$a^3c^3 + 80$	$a^5c^2e + 124$	$a^5d^2e + 260$	$a^4bcef - 130$
			$a^2b^3d + 16$	$a^5cd^2 + 92$	$a^4b^2df - 110$	$a^4bd^2f + 380$
			$a^2b^2c^2 - 81$	$a^4b^3f + 32$	$a^4b^2e^2 + 169$	$a^4bde^2 + 142$
			$ab^4c + 30$	$a^4b^2ce - 98$	$a^4bc^2f - 240$	$a^4c^2df + 240$
			$b^6 - 4$	$a^4b^2d^2 - 44$	$a^4bcde - 104$	$a^4c^2e^2 - 522$
				$a^4bc^2d - 160$	$a^4bd^3 - 104$	$a^4cd^2e + 708$
				$a^4c^4 + 95$	$a^4c^3e + 196$	$a^4d^4 - 53$
				$a^3b^4e + 18$	$a^4c^2d^2 + 118$	$a^3b^3ef + 128$
				$a^3b^3cd + 116$	$a^3b^3cf + 150$	$a^3b^2ce^2 + 388$
				$a^3b^2c^3 - 104$	$a^3b^3de - 10$	$a^3b^2cdf - 394$
				$a^2b^5d - 20$	$a^3b^2c^2e - 180$	$a^3b^2d^2e - 378$
				$a^2b^4c^2 + 45$	$a^3b^2cd^2 + 20$	$a^3bc^3f - 144$
				$ab^6c - 10$	$a^3bc^3d - 220$	$a^3bc^2de - 480$
				$b^8 + 1$	$a^3c^5 + 66$	$a^3bcd^3 - 156$
					$a^2b^5f - 24$	$a^3c^4e + 194$
					$a^2b^4ce + 66$	$a^3c^3d^2 + 52$
					$a^2b^4d^2 - 3$	$a^2b^4df + 66$
					$a^2b^3c^2d + 192$	$a^2b^4e^2 - 84$
					$a^2b^2c^4 - 66$	$a^2b^3c^2f + 194$
					$ab^6e - 8$	$a^2b^3cde + 330$
					$ab^5cd - 66$	$a^2b^3d^2 + 92$
					$ab^4c^3 + 24$	$a^2b^2c^3e - 152$
					$b^7d + 8$	$a^2b^2c^2d^2 + 174$
					$b^6c^2 - 3$	$a^2bc^4d - 140$
						$a^2c^6 + 25$
						$ab^5cf - 70$
						$ab^5de - 42$
						$ab^4c^2e + 32$
						$ab^4cd^2 - 144$
						$ab^3c^3d + 100$
						$ab^2c^5 - 18$
						$b^7f + 8$
						$b^6ce - 2$
						$b^6d^2 + 22$
						$b^5c^2d - 16$
						$b^4c^4 + 3$

$(\mathfrak{D}\theta,\ 1)^{10}=0$, viz. the function in θ is

θ^3		θ^2		θ		θ^0	
a^5ef^2	− 3750	a^4cef^2	− 5000	a^4df^3	− 6250	a^4f^4	+ 3125
a^4bdf^2	+ 1500	$a^4d^2f^2$	+ 1875	$a^4e^2f^2$	+ 5000	a^3bef^3	− 2500
a^4be^2f	+ 1500	a^4de^2f	+ 1000	a^3bcf^3	+ 3750	a^3cdf^3	− 3750
$a^4c^2f^2$	+ 2500	a^4e^4	+ 400	a^3bdef^2	− 250	$a^3ce^2f^2$	+ 2000
a^4cdef	− 2150	$a^3b^2ef^2$	+ 2000	a^3be^3f	− 2000	$a^3d^2ef^2$	+ 2250
a^4ce^3	− 80	a^3bcdf^2	− 250	$a^3c^2ef^2$	− 3750	a^3de^3f	− 1600
a^4d^3f	+ 700	a^3bce^2f	+ 1400	$a^3cd^2f^2$	+ 3000	a^3e^5	+ 256
$a^4d^2e^2$	+ 570	a^3bd^2ef	− 1550	a^3cde^2f	+ 200	$a^2b^2df^3$	+ 2000
$a^3b^2cf^2$	− 2450	a^3bde^3	− 840	a^3ce^4	+ 320	$a^2b^2e^2f^2$	− 50
a^3b^2def	− 40	$a^3c^3f^2$	+ 1750	a^3d^3ef	− 450	$a^2bc^2f^3$	+ 2250
$a^3b^2e^3$	− 118	a^3c^2def	− 1650	$a^3d^2e^3$	− 40	a^2bcdef^2	− 2050
a^3bc^2ef	+ 290	$a^3c^2e^3$	− 360	$a^2b^3f^3$	− 1000	a^2bce^3f	+ 160
a^3bcd^2f	− 400	a^3cd^3f	+ 600	$a^2b^2cef^2$	+ 1950	$a^2bd^3f^2$	− 900
a^3bcde^2	− 158	$a^3cd^2e^2$	+ 960	$a^2b^2d^2f^2$	− 1150	$a^2bd^2e^2f$	+ 1020
a^3bd^3e	− 596	a^3d^4e	− 270	$a^2b^2de^2f$	+ 1170	a^2bde^4	− 192
a^3c^3df	+ 80	$a^2b^3df^2$	− 200	$a^2b^2e^4$	+ 72	$a^2c^3ef^2$	− 900
$a^3c^3e^2$	− 308	$a^2b^3e^2f$	− 640	$a^2bc^2df^2$	− 2100	$a^2c^2d^2f^2$	+ 825
$a^3c^2d^2e$	+ 612	$a^2b^2c^2f^2$	− 2025	$a^2bc^2e^2f$	+ 1380	$a^2c^2de^2f$	+ 560
a^3cd^4	− 102	a^2b^2cdef	+ 1740	a^2bcd^2ef	− 550	$a^2c^2e^4$	− 128
$a^2b^4f^2$	+ 490	$a^2b^2ce^3$	+ 400	a^2bcde^3	− 504	a^2cd^3ef	− 630
a^2b^3cef	+ 180	$a^2b^2d^3f$	+ 70	a^2bd^4f	+ 180	$a^2cd^2e^3$	+ 144
$a^2b^3d^2f$	+ 112	$a^2b^2d^2e^2$	+ 275	$a^2bd^3e^2$	+ 138	a^2d^5f	+ 108
$a^2b^3de^2$	+ 92	a^2bc^3ef	+ 290	$a^2c^4f^2$	+ 675	$a^2d^4e^2$	− 27
$a^2b^2c^2df$	+ 86	$a^2bc^2d^2f$	− 420	a^2c^3def	− 330	ab^3cf^3	− 1600
$a^2b^2c^2e^2$	+ 388	$a^2bc^2de^2$	− 390	$a^2c^3e^3$	− 224	ab^3def^2	+ 160
$a^2b^2cd^2e$	+ 150	a^2bcd^3e	− 240	$a^2c^2d^3f$	+ 60	ab^3e^3f	− 36
$a^2b^2d^4$	+ 160	a^2bd^5	+ 108	$a^2c^2d^2e^2$	+ 434	$ab^2c^2ef^2$	+ 1020
a^2bc^4f	− 48	a^2c^4df	+ 40	a^2cd^4e	− 198	$ab^2cd^2f^2$	+ 560
a^2bc^3de	− 504	$a^2c^4e^2$	− 15	a^2d^6	+ 27	ab^2cde^2f	− 746
a^2c^5e	+ 106	$a^2c^3d^2e$	+ 140	ab^4ef^2	− 240	ab^2ce^4	+ 144
$a^2c^4d^2$	− 7	$a^2c^2d^4$	− 45	ab^3cdf^2	+ 1320	ab^2d^3ef	+ 24
ab^5ef	− 68	ab^4cf^2	+ 840	ab^3ce^2f	− 1230	$ab^2d^2e^3$	− 6
ab^4cdf	− 86	ab^4def	− 200	ab^3d^2ef	− 12	abc^3df^2	− 630
ab^4ce^2	− 178	ab^4e^3	− 48	ab^3de^3	+ 18	abc^3e^2f	+ 24
ab^4d^2e	− 54	ab^3c^2ef	− 310	$ab^2c^3f^2$	− 450	abc^2d^2ef	+ 356
ab^3c^2de	+ 234	ab^3cd^2f	− 50	ab^2c^2def	+ 594	abc^2de^3	− 80
ab^3c^3f	+ 34	ab^3cde^2	− 178	$ab^2c^2e^3$	+ 282	$abcd^4f$	− 72
ab^3cd^3	− 148	ab^3d^3e	− 14	ab^2cd^3f	− 154	$abcd^3e^2$	+ 18
ab^2c^4e	− 56	ab^2c^3df	+ 130	$ab^2cd^2e^2$	− 114	ac^5f^2	+ 108
$ab^2c^3d^2$	+ 112	$ab^2c^3e^2$	+ 112	ab^2d^4e	+ 6	ac^4def	− 72
abc^5d	− 34	$ab^2c^2d^2e$	+ 246	abc^4ef	− 72	ac^4e^3	+ 16
ac^7	+ 4	ab^2cd^4	− 78	abc^3d^2f	+ 24	ac^3d^3f	+ 16
b^6df	+ 16	abc^5f	− 24	abc^3de^2	− 186	$ac^3d^2e^2$	− 4
b^6e^2	+ 26	abc^4de	− 164	abc^2d^3e	+ 116	b^5f^3	+ 256
b^5c^2f	− 6	abc^3d^3	+ 52	$abcd^5$	− 18	b^4cef^2	− 192
b^5cde	− 30	ac^6e	+ 24	ac^5e^2	+ 36	$b^4d^2f^2$	− 128
b^5d^3	+ 28	ac^5d^2	− 8	ac^4d^2e	− 24	b^4de^2f	+ 144
b^4c^3e	+ 8	b^6f^2	− 112	ac^3d^4	+ 4	b^4e^4	− 27
$b^4c^2d^2$	− 24	b^5cef	+ 56	b^5df^2	− 192	$b^3c^2df^2$	+ 144
b^3c^4d	+ 8	b^5d^2f	+ 24	b^5e^2f	+ 216	$b^3c^2e^2f$	− 6
b^2c^6	− 1	b^5de^2	+ 48	$b^4ce^2f^2$	+ 72	b^3cd^2ef	− 80
		b^4c^2df	− 32	b^4cdef	− 120	b^3cde^3	+ 18
		$b^4c^2e^2$	− 25	b^4ce^3	− 54	b^3d^4f	+ 16
		b^4cd^2e	− 54	b^4d^3f	+ 32	$b^3d^3e^2$	− 4
		b^4d^4	+ 17	$b^4d^2e^2$	+ 18	$b^2c^4f^2$	− 27
		b^3c^4f	+ 6	b^3c^3ef	+ 18	b^2c^3def	+ 18
		b^3c^3de	+ 38	$b^3c^2d^2f$	− 6	$b^2c^3e^3$	− 4
		$b^3c^2d^3$	− 12	$b^3c^2de^2$	+ 42	$b^2c^2d^3f$	− 4
		b^2c^5e	− 6	b^3cd^3e	− 26	$b^2c^2d^2e^2$	+ 1
		$b^2c^4d^2$	+ 2	b^3d^5	+ 4		
				$b^2c^4e^2$	− 9		
				$b^2c^3d^2e$	+ 6		
				$b^2c^2d^4$	− 1		

$(\mathfrak{D}\theta,\ 1)^{10}$

It may be remarked, that if ω is an imaginary cube root of unity, then the roots of the equation $(1, 1, 1, 1, 1, 1 \between v, 1)^5 = 0$ are $-1, \omega, \omega^2, -\omega, -\omega^2$; the differences of the roots are

$$\begin{array}{cccccccccc} -1-\omega, & -1-\omega^2, & -1+\omega, & -1+\omega^2, & \omega-\omega^2, & 2\omega, & \omega+\omega^2, & \omega^2+\omega, & 2\omega^2, & -\omega+\omega^2 \\ = \omega^2, & \omega, & -1+\omega, & -1+\omega^2, & \omega-\omega^2, & 2\omega, & -1, & -1, & 2\omega^2, & -\omega+\omega^2, \end{array}$$

and the squares of the differences are

$$\omega, \quad \omega^2, \quad -3\omega, \quad -3\omega^2, \quad -3, \quad 4\omega^2, \quad 1, \quad 1, \quad 4\omega, \quad -3,$$

from which the equation of differences is found to be

$$(\theta^2+\theta+1)(\theta^2-3\theta+9)(\theta^2+4\theta+16)(\theta^2-2\theta+1)(\theta^2+6\theta+9) = 0;$$

or multiplying out, it is

$$(1, 6, 21, 46, 108, 546, 493, -1410, -567, -540, +1296 \between \theta, 1)^{10} = 0;$$

which is what the preceding expression of the equation of differences becomes upon writing therein $a=b=c=d=e=f=1$. Moreover, upon passing (as will presently be done) to the standard form, and then writing $a=b=c=d=e=f=1$, all the coefficients (except the first coefficient, which is equal to unity) should become equal to zero; these two tests afford a complete verification of the result.

The following corrections have to be made in Waring's result, as given by himself and Lagrange (Waring, *Phil. Trans.* 1762).

Waring, *Meditationes Algebraicæ*, p. 85:

for $+\ 169\,q^3s$ *read* $+196\,q^3s$ (in coefficient w^5).

Lagrange, *Equations Numériques*, p. 108:

for $+1200\,CE$ *read* $+200\,CE$ (in d)
for $-\ 169\,B^3D$ *read* $-196\,B^3D$ (in e)
for $-\ 25\,B^6$ *read* $+\ 25\,B^6$ (in f)
for $+\ 27\,C^4D^2$ *read* $-\ 27\,C^4D^2$ (in k).

It may be noticed, that if in the coefficients of the several powers of θ (as they are written down in the columns, without regarding the power of a which multiplies the entire column), we attend only to the terms independent of a, we have the series

$$\begin{array}{llllllll} 1, & b^2-4, & b^4+6, & b^6-4, & b^8+1, & b^7d+8, & b^7f\ +\ 8, & \&c. \\ & & & & & b^6c^2-3, & b^6ce\ -\ 2 & \\ & & & & & & b^6d^2\ +22 & \\ & & & & & & b^5c^2d-16 & \\ & & & & & & b^4c^4\ +\ 3, & \end{array}$$

the law of the first terms of which, up to the term $+1b^8$, is obvious; but the term $+1b^8$, which is the last term of this initial series, is also the first term of a terminal series, the terms of which are deduced from the coefficients in the equation of differences for the quartic equation $(a, b, c, d\ e \between v, 1)^4 = 0$, viz. these coefficients are

$a^6 \times$	$a^4 \times$	$a^2 \times$	&c.
$+1$	$ac + 8$	$a^3e \ + \ 8$	
	$b^2 - 3$	$a^2bd - \ 2$	
		$a^2c^2 + 22$	
		$ab^2c - 16$	
		$b^4 \ + \ 3$;	

and by writing b, c, d, e, f in the place of a, b, c, d, e respectively, and multiplying by b^2, we have the above-mentioned series,

$$b^8 + 1, \quad b^7d + 8, \ \&c.$$
$$b^6c^2 - 3.$$

It is easy to see, *à priori*, in the case of an equation of any order, that this property holds good.

Passing now to the standard forms:

For the quadric $(a, b, c \between v, 1)^2 = 0$, the equation of differences is, $0 =$

$a^2 \times$	$4 \times$
$+1$	$ac + 1$
	$b^2 - 1$

$$(\ \ldots \ \between \theta, 1).$$

For the cubic equation $(a, b, c, d \between v, 1)^3 = 0$, the equation of differences is, $0 =$

$a^4 \times$	$18\, a^2 \times$	$81 \times$	$27 \times$
$+1$	$ac + 1$	$a^2c^2 + 1$	$a^2d^2 + 1$
	$b^2 - 1$	$ab^2c - 2$	$abcd - 6$
		$b^4 \ + 1$	$ac^3 \ + 4$
			$b^3d \ + 4$
			$b^2c^2 \ - 3$

$$(\ \ldots \ \between \theta, 1)^3.$$

For the quartic equation $(a, b, c, d, e \between v, 1)^4 = 0$, the equation of differences is, $0 =$

$a^6 \times$	$48a^4 \times$	$8a^2 \times$	$32 \times$	$16 \times$	$1152 \times$	$256 \times$
$+1$	$ac + 1$	$a^3e \ + \ 1$	$a^4ce \ + \ 3$	$a^4e^2 \ - \ 7$	$a^3ce^2 \ - \ 1$	$a^3e^3 \ + \ 1$
	$b^2 - 1$	$a^2bd - \ 4$	$a^4d^2 \ + \ 13$	$a^3bde \ + \ 56$	$a^3d^2e \ + \ 3$	$a^2bde^2 - \ 12$
		$a^2c^2 + \ 99$	$a^3b^2e \ - \ 3$	$a^3c^2e \ + \ 54$	$a^2b^2e^2 \ + \ 1$	$a^2c^2e^2 \ - \ 18$
		$ab^2c - 192$	$a^3bcd \ - \ 90$	$a^3cd^2 \ + \ 288$	$a^2bcde - 10$	$a^2cd^2e + \ 54$
		$b^4 \ + \ 96$	$a^3c^3 \ + 189$	$a^2b^2ce \ - \ 192$	$a^2bd^3 \ - 12$	$a^2d^4 \ - \ 27$
			$a^2b^3d \ + \ 64$	$a^2b^2d^2 - \ 400$	$a^2c^3e \ + \ 6$	$ab^2ce^2 + \ 54$
			$a^2b^2c^2 \ - 432$	$a^2bc^2d - 1944$	$a^2c^2d^2 \ + \ 9$	$ab^2d^2e - \ 6$
			$ab^4c \ + 384$	$a^2c^4 \ + 1377$	$ab^3de \ + \ 4$	$abc^2de - 180$
			$b^6 \ - 128$	$ab^4e \ + \ 96$	$ab^2c^2e \ - \ 3$	$abcd^3 + 108$
				$ab^3cd \ + 3648$	$ab^2cd^2 + 56$	$ac^4e \ + \ 81$
				$ab^2c^3 \ - 2592$	$abc^3d \ - 78$	$ac^3d^2 \ - \ 54$
				$b^5d \ - 1536$	$ac^5 \ + 27$	$b^4e^2 \ - \ 27$
				$b^4c^2 \ + 1152$	$b^4d^2 \ - 32$	$b^3cde + 108$
					$b^3c^2d \ + 48$	$b^3d^3 \ - \ 64$
					$b^2c^2 \ - 18$	$b^2c^3e \ - \ 54$
						$b^2c^2d^2 + \ 36$

$$(\ \ldots \ \between \theta, 1)^6.$$

For the quintic equation $(a, b, c, d, e, f \between v, 1)^5 = 0$, the equation of differences is,

θ^{10}	θ^9	θ^8	θ^7	θ^6	θ^5	θ^4
$a^8 \times$	$100\,a^6 \times$	$50\,a^4 \times$	$2500\,a^2 \times$	$125 \times$	$625 \times$	$2500 \times$
$+1$	$ac + 1$	$a^3e + 1$	$a^4ce + 1$	$a^6df + 16$	$a^6f^2 + 1$	$a^5cf^2 + 7$
	$b^2 - 1$	$a^2bd - 4$	$a^4d^2 + 1$	$a^6e^2 - 19$	$a^5bef - 10$	$a^5def - 19$
		$a^2c^2 + 78$	$a^3b^2e - 1$	$a^5bcf - 48$	$a^5cdf + 64$	$a^5e^3 + 2$
		$ab^2c - 150$	$a^3bcd - 10$	$a^5bde + 72$	$a^5ce^2 - 144$	$a^4b^2f^2 - 7$
		$b^4 + 75$	$a^3c^3 + 32$	$a^5c^2e + 496$	$a^5d^2e + 208$	$a^4bcef - 13$
			$a^2b^3d + 8$	$a^5cd^2 + 736$	$a^4b^2df - 44$	$a^4bd^2f + 76$
			$a^2b^2c^2 - 81$	$a^4b^3f + 32$	$a^4b^2e^2 + 169$	$a^4bde^2 + 71$
			$ab^4c + 75$	$a^4b^2ce - 980$	$a^4bc^2f - 192$	$a^4c^2df + 96$
			$b^6 - 25$	$a^4b^2d^2 - 880$	$a^4bcde - 416$	$a^4c^2e^2 - 522$
				$a^4bc^2d - 6400$	$a^4bd^3 - 832$	$a^4cd^2e + 1416$
				$a^4c^4 + 7600$	$a^4c^3e + 1568$	$a^4d^4 - 212$
				$a^3b^4e + 450$	$a^4c^2d^2 + 1888$	$a^3b^3ef + 32$
				$a^3b^3cd + 11600$	$a^3b^3cf + 300$	$a^3b^2ce^2 + 970$
				$a^3b^2c^3 - 20800$	$a^3b^3de - 100$	$a^3b^2cdf - 394$
				$a^2b^5d - 5000$	$a^3b^2c^2e - 3600$	$a^3b^2d^2e - 1890$
				$a^2b^4c^2 + 22500$	$a^3b^2cd^2 + 800$	$a^3bc^3f - 288$
				$ab^6c - 12500$	$a^3bc^3d - 17600$	$a^3bc^2de - 4800$
				$b^8 + 3125$	$a^3c^5 + 10560$	$a^3bcd^3 - 3120$
					$a^2b^5f - 120$	$a^3c^4e + 3880$
					$a^2b^4ce + 3300$	$a^3c^3d^2 + 2080$
					$a^2b^4d^2 - 300$	$a^2b^4df + 165$
					$a^2b^3c^2d + 38400$	$a^2b^4e^2 - 525$
					$a^2b^2c^4 - 26400$	$a^2b^3c^2f + 970$
					$ab^6e - 1000$	$a^2b^3cde + 8250$
					$ab^5cd - 33000$	$a^2b^3d^3 + 4600$
					$ab^4c^3 + 24000$	$a^2b^2c^3e - 7600$
					$b^7d + 10000$	$a^2b^2c^2d^2 + 17400$
					$b^6c^2 - 7500$	$a^2bc^4d - 28000$
						$a^2c^6 + 10000$
						$ab^5cf - 875$
						$ab^5de - 2625$
						$ab^4c^2e + 4000$
						$ab^4cd^2 - 36000$
						$ab^3c^3d + 50000$
						$ab^2c^5 - 18000$
						$b^7f + 250$
						$b^6ce - 625$
						$b^6d^2 + 13750$
						$b^5c^2d - 20000$
						$b^4c^4 + 7500$
$+1$	± 1	± 154	± 117	± 46627	± 91258	± 125515

$0 = (\text{☊}\theta,\ 1)^{10} = 0$, viz. the function in θ is

θ^3		θ^2		θ		θ^0		
6250 ×		62500 ×		62500 ×		3125 ×		
a^5ef^2	− 3	a^4cef^2	− 4	a^4df^3	− 1	a^4f^4	+ 1	
a^4bdf^2	+ 12	$a^4d^2f^2$	+ 3	$a^4e^2f^2$	+ 2	a^3bef^3	− 20	
a^4be^2f	+ 30	a^4de^2f	+ 4	a^3bcf^3	+ 3	a^3cdf^3	− 120	
$a^4c^2f^2$	+ 40	a^4e^4	+ 4	a^3bdef^2	− 1	$a^3ce^2f^2$	+ 160	
a^4cdef	− 172	$a^3b^2ef^2$	+ 4	a^3be^3f	− 20	$a^3d^2ef^2$	+ 360	
a^4ce^3	− 16	a^3bcdf^2	− 2	$a^3c^2ef^2$	− 30	a^3de^3f	− 640	
a^4d^3f	+ 112	a^3bce^2f	+ 28	$a^3cd^2f^2$	+ 48	a^3e^5	+ 256	
$a^4d^2e^2$	+ 228	a^3bd^2ef	− 62	a^3cde^2f	+ 8	$a^2b^2df^3$	+ 160	
$a^3b^2cf^2$	− 98	a^3bde^3	− 84	a^3ce^4	+ 32	$a^2b^2e^2f^2$	− 10	
a^3b^2def	− 8	$a^3c^3f^2$	+ 28	a^3d^3ef	− 36	$a^2bc^2f^3$	+ 360	
$a^3b^2e^3$	− 59	a^3c^2def	− 132	$a^3d^2e^3$	− 8	a^2bcdef^2	− 1640	
a^3bc^2ef	+ 116	$a^3c^2e^3$	− 72	$a^2d^3f^3$	− 2	a^2bce^3f	+ 320	
a^3bcd^2f	− 320	a^3cd^3f	+ 96	$a^2b^2cef^2$	+ 39	$a^2bd^3f^2$	− 1440	
a^3bcde^2	− 316	$a^3cd^2e^2$	+ 384	$a^2b^2d^2f^2$	− 46	$a^2bd^2e^2f$	+ 4080	
a^3bd^3e	− 2384	a^3d^4e	− 216	$a^2b^2de^2f$	+ 117	a^2bde^4	− 1920	
a^3c^3df	+ 128	$a^2b^3df^2$	− 4	$a^2b^2e^4$	+ 18	$a^2c^3ef^2$	− 1440	
$a^3c^3e^2$	− 1232	$a^2b^3e^2f$	− 32	$a^2bc^2df^2$	− 168	$a^2c^2d^2f^2$	+ 2640	
$a^3c^2d^2e$	+ 4896	$a^2b^2c^2f^2$	− 81	$a^2bc^2e^2f$	+ 276	$a^2c^2de^2f$	+ 4480	
a^3cd^4	− 1632	a^2b^2cdef	+ 348	a^2bcd^2ef	− 220	$a^2c^2e^4$	− 2560	
$a^2b^4f^2$	+ 49	$a^2b^2ce^3$	+ 200	a^2bcde^3	− 504	a^2cd^3ef	− 10080	
a^2b^3cef	+ 180	$a^2b^2d^3f$	+ 28	a^2bd^4f	+ 144	$a^2cd^2e^3$	+ 5760	
$a^2b^3d^2f$	+ 224	$a^2b^2d^2e^2$	+ 275	$a^2bd^3e^2$	+ 276	a^2d^5f	+ 3456	
$a^2b^3de^2$	+ 460	a^2bc^3ef	+ 116	$a^2c^4f^2$	+ 108	$a^2d^4e^2$	− 2160	
$a^2b^2c^2df$	+ 344	$a^2bc^2d^2f$	− 336	a^2c^3def	− 264	ab^3cf^3	− 640	
$a^2b^2c^2e^2$	+ 3880	$a^2bc^2de^2$	− 780	$a^2c^3e^3$	− 448	ab^3def^2	+ 320	
$a^2b^2cd^2e$	+ 3000	a^2bcd^3e	− 960	$a^2c^2d^3f$	+ 96	ab^3e^3f	− 180	
$a^2b^2d^4$	+ 6400	a^2bd^5	+ 864	$a^2c^2d^2e^2$	+ 1736	$ab^2c^2ef^2$	+ 4080	
a^2bc^4f	− 384	a^2c^4df	+ 64	a^2cd^4e	− 1584	$ab^2cd^2f^2$	+ 4480	
a^2bc^3de	− 20160	$a^2c^4e^2$	− 60	a^2d^6	+ 432	ab^2cde^2f	− 14920	
a^2c^5e	+ 8480	$a^2c^3d^2e$	+ 1120	ab^4ef^2	− 12	ab^2ce^4	+ 7200	$(\text{☊}\theta,\ 1)^{10}$
$a^2c^4d^2$	− 1120	$a^2c^2d^4$	− 720	ab^3cdf^2	+ 264	ab^2d^3ef	+ 960	
ab^5ef	− 170	ab^4cf^2	+ 84	ab^3ce^2f	− 615	$ab^2d^2e^3$	− 600	
ab^4cdf	− 860	ab^4def	− 100	ab^3d^2ef	− 12	abc^3df^2	− 10080	
ab^4ce^2	− 4450	ab^4e^3	− 60	ab^3de^3	+ 45	abc^3e^2f	+ 960	
ab^4d^2e	− 2700	ab^3c^2ef	− 310	$ab^2c^3f^2$	− 180	abc^2d^2ef	+ 28480	
ab^3c^2de	+ 23400	ab^3cd^2f	− 100	ab^2c^2def	+ 1188	abc^2de^3	− 16000	
ab^3c^3f	+ 680	ab^3cde^2	− 890	$ab^2c^2e^3$	+ 1410	$abcd^4f$	− 11520	
ab^3cd^3	− 29600	ab^3d^3e	− 140	ab^2cd^3f	− 616	$abcd^3e^2$	+ 7200	
ab^2c^4e	− 11200	ab^2c^3df	+ 520	$ab^2cd^2e^2$	− 1140	ac^5f^2	+ 3456	
$ab^2c^3d^2$	+ 44800	$ab^2c^3e^2$	+ 1120	ab^2d^4e	+ 120	ac^4def	− 11520	
abc^5d	− 27200	$ab^2c^2d^2e$	+ 4920	abc^4ef	− 288	ac^4e^3	+ 6400	
ac^7	+ 6400	ab^2cd^4	− 3120	abc^3d^2f	+ 192	ac^3d^3f	+ 5120	
b^6df	+ 400	abc^5f	− 192	abc^3de^2	− 3720	$ac^3d^2e^2$	− 3200	
b^6e^2	+ 1625	abc^4de	− 6560	abc^2d^3e	+ 4640	b^5f^3	+ 256	
b^5c^2f	− 300	abc^3d^3	+ 4160	$abcd^5$	− 1440	b^4cef^2	− 1920	
b^5cde	− 7500	ac^6e	+ 1920	ac^5e^2	+ 1440	$b^4d^2f^2$	− 2560	
b^5d^3	+ 14000	ac^5d^2	− 1280	ac^4d^2e	− 1920	b^4de^2f	+ 7200	
b^4c^3e	+ 4000	b^6f^2	− 28	ac^3d^4	+ 640	b^4e^4	− 3375	
$b^4c^2d^2$	− 24000	b^5cef	+ 140	b^5df^2	− 96	$b^3c^2df^2$	+ 5760	
b^3c^4d	+ 16000	b^5d^2f	+ 120	b^5e^2f	+ 270	$b^3c^2e^2f$	− 600	
b^2c^6	− 4000	b^5de^2	+ 600	$b^4c^2f^2$	+ 72	b^3cd^2ef	− 16000	
		b^4c^2df	− 320	b^4cdef	− 600	b^3cde^3	+ 9000	
		$b^4c^2e^2$	− 625	b^4ce^3	− 675	b^3d^4f	+ 6400	
		b^4cd^2e	− 2700	b^4d^3f	+ 320	$b^3d^3e^2$	− 4000	
		b^4d^4	+ 1700	$b^4d^2e^2$	+ 450	$b^2c^4f^2$	− 2160	
		b^3c^4f	+ 120	b^3c^3ef	+ 180	b^2c^3def	+ 7200	
		b^3c^3de	+ 3800	$b^3c^2d^2f$	− 120	$b^2c^3e^3$	− 4000	
		$b^3c^2d^3$	− 2400	$b^3c^2de^2$	+ 2100	$b^2c^2d^3f$	− 3200	
		b^2c^5e	− 1200	b^3cd^3e	− 2600	$b^2c^2d^2e^2$	+ 2000	
		$b^2c^4d^2$	+ 800	b^3d^5	+ 800			
				$b^2c^4e^2$	− 900			
				$b^2c^3d^2e$	+ 1200			
				$b^2c^2d^4$	− 400			
	± 139884		± 23570		± 18666		± 128505	

[where in each column after the first the sum of the positive coefficients is equal to that of the negative coefficients, and I have inserted at the foot this sum with the sign +. A single coefficient − 7500 in place of − 3750 has been corrected.]

The coefficients in the preceding equations of differences are functions of the seminvariants of the quantics to which they belong; for instance, in the case of the quartic, the coefficient of θ^4 is

$$8a^2\{a^2(ae-4bd+3c^2)+96(ac-b^2)^2\},$$

that of θ^3 is

$$32\{-13a^3(ace-ad^2-b^2e+2bcd-c^3)+16a^2(ac-b^2)(ae-4bd+3c^2)+128(ac-b^2)^3\},$$

and so for the other coefficients; and by replacing each seminvariant by the covariant to which it belongs, we pass from the solution of the original problem of finding the equation for $\theta=(\alpha-\beta)^2$, to that of the problem of finding the equation for

$$\theta=\frac{(\alpha-\beta)^2}{(x-\alpha y)^2(x-\beta y)^2}.$$

The results are as follows:

For the quadric $(a, b, c\between x, y)^2$, the equation in θ is, $0=$

$$(U^2,\ 4\square\between\theta,\ 1),$$

where U is the quadric, $\square$ the discriminant.

For the cubic $(a, b, c, d\between x, y)^3$, the equation in θ is, $0=$

$$(U^4,\ 18U^2H,\ 81H^2,\ 27\square\between\theta,\ 1)^3,$$

where U is the cubic, H the Hessian, $\square$ the discriminant.

For the quartic $(a, b, c, d, e\between x, y)^4$, the equation in θ is, $0=$

$$\left(\begin{array}{l} U^6, \\ 48U^4H, \\ 8U^2(U^2I+96H^2), \\ 32(-13U^3J+16U^2HI+128H^3), \\ 16(-\ 7U^2I^2-288UHJ+384H^2I), \\ 1152(-\ 3UIJ+2HI^2), \\ 256(\ I^3\ -27J^2), \end{array}\right)\between\theta,\ 1)^6,$$

where U is the quartic, H the Hessian, I and J the quadrinvariant and the cubinvariant respectively.

For the quintic $(a, b, c, d, e, f \between x, y)^5$, the equation in θ, as far as it can be expressed in terms of known covariants, is, $0 =$

$$\left(\begin{array}{l} U^8, \\ 100\ U^6\,(\text{Tab. No. }15), \\ 50\ U^4\,[U^2\,(\text{Tab. No. }14) + 75\ (\text{Tab. No. }15)^2], \\ \vdots \\ 3125\ \text{Discriminant}\ (= \text{Tab. No. }26) \end{array}\right\between \theta, 1)^{10},$$

where the Tables referred to are those in my Second Memoir on Quantics, [141; for the completed equation see *post* p. 261].

The form of the preceding results may be modified by writing $\theta = \vartheta \div U^2$; we have thus the equations for

$$\vartheta = a^2 (\alpha - \beta)^2 (x - \gamma y)^2 (x - \delta y)^2 \ldots;$$

thus for example, in the case of the cubic $(a, b, c, d \between x, y)^3$,

the equation for $\vartheta\, [= a^2 (\alpha - \beta)^2 (x - \gamma y)^2]$ is

$$0 = (1,\ 18H,\ 81H^2,\ 27\square\, U^2 \between \vartheta, 1)^3;$$

this equation may be written

$$(\vartheta + 9H)^2 \vartheta + 27 \square\, U^2 = 0;$$

or putting $v = \sqrt{\vartheta}$, we have

$$v^3 + 9Hv + U\sqrt{-27\square} = 0,$$

an equation the roots of which are

$$a(\alpha - \beta)(x - \gamma y), \qquad a(\beta - \gamma)(x - \alpha y), \qquad a(\gamma - \alpha)(x - \beta y),$$

and which leads to the formula, given in my fifth Memoir on Quantics, [156], for the solution of a cubic equation. But this decomposition of the equation in ϑ is peculiar to the cubic.

The equation of differences for an equation of any order may be found by the following entirely distinct method. Let the proposed equation $(* \between v, 1)^n = 0$, be for shortness represented by $\phi v = 0$, and let x, y be any two distinct roots; we have not only $\phi x = 0$, $\phi y = 0$, but also $\phi x + \phi y = 0$, $\dfrac{\phi x - \phi y}{x - y} = 0$. Writing $\theta = (x - y)^2$, $s = x + y$, we have

$$x = \tfrac{1}{2}(s + \sqrt{\theta}), \qquad y = \tfrac{1}{2}(s - \sqrt{\theta}),$$

values which are to be substituted for x, y in the equations

$$\phi x + \phi y = 0, \qquad \frac{\phi x - \phi y}{x - y} = 0.$$

We have thus two equations rational in s and θ, and the elimination between them of the quantity s leads to the required equation in θ. But it is proper to modify the form of the system; in fact the two equations are, as regards s, the first of them of the degree n, the second of the degree $n-1$; but if we write

$$n(\phi x+\phi y)-\frac{(x+y)(\phi x-\phi y)}{x-y}=0, \qquad \frac{\phi x-\phi y}{x-y}=0,$$

then each of the equations will be of the same degree $n-1$ in s.

For instance, let $\phi v=(a,\ b,\ c,\ d \between v,\ 1)^3$, then $x=\frac{1}{2}(s+\sqrt{\theta})$, $y=\frac{1}{2}(s+\sqrt{\theta})$; the equations $\phi x+\phi y=0$, $\dfrac{\phi x-\phi y}{x-y}=0$, are

$$\begin{aligned} s^3 . a+3s^2 . 2b+3s(4c+a\theta)+8d+6b\theta&=0,\\ 3s^2 . a+3s\ . 4b+\quad 12c+a\theta\qquad &=0; \end{aligned}$$

and multiplying the first equation by 3 and the second by $-s$, adding and dividing by 2, we have an equation

$$s^2 . 3b+s(12c+4a\theta)+12d+9b\theta=0.$$

The second equation and this equation may be written

$$\begin{aligned} (3a,\ 12b\qquad\quad 12c+a\theta \between s,\ 1)^2&=0,\\ (3b,\ 12c+4a\theta,\ \ 12d+9b\theta \between s,\ 1)^2&=0, \end{aligned}$$

and the elimination of s from these equations gives the required equation in θ. The result may be obtained under either of the two forms,

$$\{\ a^2\theta^2+(15ac-27b^2)\,\theta-36(bd-c^2)\}\,\{a^2\theta+\ \ 3(ac-b^2)\}+3\,\{2ab\theta+3(ad-bc)\}^2=0$$

and

$$\{4a^2\theta^2+(24ac-27b^2)\,\theta-36(bd-c^2)\}\,\{a^2\theta+12(ac-b^2)\}+3\,\{\ ab\theta+6(ad-bc)\}^2=0,$$

the expansions of which respectively coincide with the before-mentioned result.

In the case of the quartic equation $\phi v=(a,\ b,\ c,\ d \between v,\ 1)^4=0$, we have

$$\begin{aligned} s^4 . a+4s^3 . 2b+6s^2(\ 4c+a\theta)+4s(8d+6b\theta)+16e+24\,c\theta+a\theta^2&=0,\\ 4s^3 . a+6s^2 . 4b+4s\ (12c+a\theta)+\quad 32d+8b\theta\qquad\qquad &=0, \end{aligned}$$

from which we derive another cubic equation; the two cubic equations thus are

$$\begin{aligned} (4a,\ \ 24b\qquad\ \ ,\ 48c+\ 4a\theta,\ \ 32d+\ 8b\theta\qquad\ \between s,\ 1)^3&=0,\\ (4b,\ \ 24c+10a\theta,\ \ 48d+44b\theta,\ \ 32e+48c\theta+2a\theta^2 \between s,\ 1)^3&=0, \end{aligned}$$

from which, if s be eliminated, we have the equation in θ.

Similarly, for the quintic equation $\phi v=(a,\ b,\ c,\ d,\ e,\ f\between v,\ 1)^5=0$, we have

$$s^5.a+\ 5s^4.2b+10s^3(\ 4c+a\theta)+10s^2(\ 8d+6b\theta)+5s(16e+24c\theta+a\theta^2)+32f+80d\theta+10b\theta^2=0,$$

$$5s^4.a+10s^3.4b+10s^2(12c+a\theta)+\ 5s\ (32d+8b\theta)+\ (80e+40c\theta+a\theta^2)\qquad =0,$$

from which we derive another quartic equation; the two equations are

$$(5a,\quad 40b\quad,\quad 120c+\ 10a\theta,\quad 160d+\ 40b\theta\quad,\quad 80e+\ 40c\theta+\ a\theta^2\between s,\ 1)^4=0,$$

$$(5b,\quad 40c+20a\theta,\quad 120d+130b\theta,\quad 160e+280c\theta+12a\theta^2,\quad 80f+200d\theta+25b\theta^2\between s,\ 1)^4=0,$$

from which, if s be eliminated, we have the equation in θ.

But we may apply Bezout's method to the two equations each of the order $n-1$, which result from the equation of the nth order $\phi v=(*\between v,\ 1)^n=0$. The process is as follows: Suppose, in general, that s is to be eliminated from the two equations

$$Fs=0,\quad Gs=0,$$

each of the order $n-1$; it is only necessary to form the expression

$$\frac{FsGs'-Fs'Gs}{s-s'},$$

which will be a function of s, s' of the form

$$\left(\begin{array}{l|} a_{0,0}\ ,\quad a_{0,1},\ldots a_{0,n-2}\\ a_{1,0}\ ,\\ \cdot\\ \cdot\\ a_{n-2,0},\end{array}\between s,\ 1)^{n-2}(s',\ 1)^{n-2}\right.$$

where the coefficients are such that $a_{l,m}=a_{m,l}$; and by equating to zero the determinant formed with these coefficients, we have the result of the elimination.

In the present case, writing for a moment $\phi\frac{1}{2}(s+\sqrt{\theta})=A$, $\phi\frac{1}{2}(s-\sqrt{\theta})=B$, and in like manner $\phi\frac{1}{2}(s'+\sqrt{\theta})=A'$, $\phi\frac{1}{2}(s'-\sqrt{\theta})=B'$, we have

$$Fs=n(A+B)-\frac{s(A-B)}{\sqrt{\theta}},\qquad Gs=\frac{A-B}{\sqrt{\theta}},$$

$$Fs'=n(A'+B')-\frac{s'(A'-B')}{\sqrt{\theta}},\qquad Gs'=\frac{A'-B'}{\sqrt{\theta}},$$

and therefore

$$FsGs'-Fs'Gs=n\frac{(A+B)(A'-B')-(A-B)(A'+B')}{\sqrt{\theta}}-\frac{(s-s')(A-B)(A'-B')}{\theta};$$

or reducing and dividing by $s-s'$,

$$-\frac{FsGs'-Fs'Gs}{s-s'}=2n\frac{AB'-A'B}{\sqrt{\theta}(s-s')}+\frac{(A-B)(A'-B')}{\theta}.$$

Hence, substituting for A, B, A', B' these values, we have the expression

$$2n\,\frac{\phi\tfrac{1}{2}(s+\sqrt{\theta})\,\phi\tfrac{1}{2}(s'-\sqrt{\theta})-\phi\tfrac{1}{2}(s-\sqrt{\theta})\,\phi\tfrac{1}{2}(s'+\sqrt{\theta})}{(s-s')\sqrt{\theta}}$$

$$+\frac{\{\phi\tfrac{1}{2}(s+\sqrt{\theta})-\phi\tfrac{1}{2}(s-\sqrt{\theta})\}\,\{\phi\tfrac{1}{2}(s'+\sqrt{\theta})-\phi\tfrac{1}{2}(s'-\sqrt{\theta})\}}{\theta},$$

which is of the form

$$\left(\begin{array}{l|} a_{0,0}\;,\; a_{0,1},\ldots a_{0,\,n-2} \\ a_{1,0}\;, \\ \vdots \\ a_{n-2,\,0}, \end{array}\right.\!\!\between s,\ 1)^{n-2}(s,\ 1)^{n-2}$$

and equating to zero the determinant formed with the coefficients, we have an equation in θ which is the equation of differences of the given equation $\phi v=0$. For instance, if the given equation is $\phi v=(a,\ b,\ c,\ d\between v,\ 1)^3=0$, then we have

$$\begin{aligned}8\phi\tfrac{1}{2}(s+\sqrt{\theta})&=(a,\ 2b+a\sqrt{\theta},\ 4c+4b\sqrt{\theta}+a\theta,\ 8d+12c\sqrt{\theta}+6b\theta+a\theta\sqrt{\theta}\between s,\ 1)^3\\&=(A,\ B,\ C,\ D\between s,\ 1)^3,\end{aligned}$$

$$\begin{aligned}8\phi\tfrac{1}{2}(s-\sqrt{\theta})&=(a,\ 2b-a\sqrt{\theta},\ 4c-4b\sqrt{\theta}+a\theta,\ 8d-12c\sqrt{\theta}+6b\theta-a\theta\sqrt{\theta}\between s,\ 1)^3\\&=(A',\ B',\ C',\ D'\between s,\ 1)^3;\end{aligned}$$

and the function in s, s' is

$$\frac{6}{\sqrt{\theta}}\left\{\begin{array}{l} 3\,(AB'-A'B)\,s^2s'^2 \\ +3\,(AC'-A'C)\,ss'\,(s+s') \\ +\ \ (AD'-A'D)\,(s^2+ss'+s'^2) \\ +9\,(BC'-B'C)\,ss' \\ +3\,(BD'-B'D)\,(s+s') \\ +3\,(CD'-C'D) \end{array}\right\}$$

$$+\frac{1}{\theta}(A-A',\ B-B',\ C-C',\ D-D'\between s,\ 1)^3\,(A-A',\ B-B',\ C-C',\ D-D'\between s',\ 1)^3,$$

which is equal to

$$12\left\{\begin{array}{l} -3a^2s^2s'^2 \\ -12ab\ ss'\,(s+s') \\ +(-12ac-a^2\theta)\,(s^2+ss'+s'^2) \\ +(36ac-72b^2+9a^2\theta)\,ss' \\ +(24ad-72bc+12ab\theta)\,(s+s') \\ +96bd-144c^2+(-48ac+72b^2)\,\theta-3a^2\theta^2 \end{array}\right\}$$

$$+4\,(3as^2+12bs+12c+a\theta)\,(3as'^2+12bs'+12c+a\theta);$$

or reducing and dividing by 32, this is

$$\{9(ac-b^2)+3a^2\theta\}\,ss'+\{9(ad-bc)+6ab\theta\}(s+s')+36(bd-c^2)+(-15ac+27b^2)\theta-a^2\theta^2,$$

the terms in s^2, s'^2 disappearing, as they should do. Writing this under the form

$$\left(\begin{array}{ll} 9(ac-b^2)+3a^2\theta, & 9(ad-bc)+6ab\theta \\ 9(ad-bc)+6ab\theta, & 36(bd-c^2)+(-15ac+27b^2)\theta-a^2\theta^2 \end{array}\between s,\ 1)(s',\ 1)\right.$$

and equating the determinant to zero, we have the required equation in θ: the form is that obtained by the ordinary process of applying Bezout's method to the two equations $(3a,\ 12b,\ 12c+a\theta\between s,\ 1)^2=0$, $(3b,\ 12c+4a\theta,\ 12d+8b\theta\between s,\ 1)^2=0$, being in fact the before-mentioned equation

$$\left(a^2\theta^2+(15ac-27b^2)\theta-36(bd-c^2)\right)\left(a^2\theta+3(ac-b^2)\right)+3\left(2ab\theta+3(ad-bc)\right)^2=0.$$

But, as already remarked, this elimination process is less convenient for the complete development of the result, than the method first explained in the present memoir.

[The equation p. 257, changing the notation and inserting the omitted coefficients, becomes $0=$

θ^{10}	$\theta^9 . 100\times$	$\theta^8 . 50\times$	$\theta^7 . 2500\times$	$\theta^6 . 125\times$	$\theta^5 . 625\times$	$\theta^4 . \frac{1}{9}2500\times$
A^8+1	A^6C+1	$A^6B\ +\ 1$	$A^4BC\ +\ 1$	$A^4H\ +\ 16$	$A^4G\ +\ 1$	$A^3J\ -\ 212$
		A^4C^2+75	$A^2F^2\ +\ 1$	$A^4B^2\ -\ 19$	$A^2B^2C-\ 160$	$A^2BH-\ 1545$
			$A^2C^3\ +29$	$A^2BC^2+\ 450$	$AFI\ +\ 60$	$A^2B^3\ +\ 230$
				$CF^2\ +\ 800$	$AC^2D-1140$	$AEI\ +\ 275$
				$C^4\ +6325$	$BF^2\ +\ 280$	$BEF\ +\ 1225$
					$BC^3\ +3620$	$B^2C^2\ +11650$
						$C^2H\ +\ 1125$

$\theta^3 . \frac{1}{9}6250\times$	$\theta^2 . 62500\times$	$\theta . 62500\times$	$\theta^0 . 3125\times$
$A^2M\ -\ 336$	$ABJ\ -20$	$BM\ -16$	$Q'+1$
$A^2BG-\ 27$	$ADG\ -\ 3$	$B^2G\ +\ 2$	
$AB^2D-2400$	$B^4\ +24$	$GH\ -\ 1$	
$ACJ\ -1520$	$BCG\ +19$	$DJ\ -48$	
$B^3C\ +3200$	$B^2H\ -36$		
$BCH\ -2800$	$CM\ -28$		
$C^2G\ +1625$			

where the capital letters denote the covariants of the quantic as explained **141** and **143**.]

263.

DEMONSTRATION OF A THEOREM IN FINITE DIFFERENCES.

[From the *Philosophical Transactions of the Royal Society of London*, vol. CL. (for the year 1860), pp. 321—323: printed as a note to Sir J. W. F. Herschel's Memoir "On the Formulæ investigated by Dr BRINKLEY for the general Term in the Development of LAGRANGE'S Expression for the Summation of Series and for Successive Integrations, pp. 319—321.]

THE formula (B) [of Sir J. W. F. Herschel's Memoir], substituting therein for A_x the value $\frac{1}{[x]}\nabla^{-n}0^x$, becomes

$$\nabla^{-n}0^x = -\frac{1}{[n-1]}\left\{\frac{[x+n]}{[x-1]}\frac{1}{[n+1]}2.3..n\nabla - \frac{[x+n]}{[x-2]}\frac{1}{[n+2]}3.4\ldots\overline{n+1}\,\nabla^2 - \&c.\right\}0^x;$$

or, as this may be written,

$$\nabla^{-n}0^x = \frac{[x+n]}{[n-1]}\left\{\qquad -\frac{1}{[1][x-1](n+1)}\nabla + \frac{1}{[2][x-2](n+2)}\nabla^2 - \&c.\right\}0^x;$$

or, inserting a first term which vanishes except in the case $x=0$, and which is required in order that the formula may hold good for this particular value,

$$\nabla^{-n}0^x = \frac{[x+n]}{[n-1]}\left\{\frac{1}{[0][x]n}\nabla^0 - \frac{1}{[1][x-1](n+1)}\nabla + \frac{1}{[2][x-2](n+2)}\nabla^2 - \&c.\right\}0^x;$$

where the series on the right-hand side need only be continued up to the term containing $\nabla^x 0^x$, since the subsequent terms vanish. [In these formulæ and throughout the present paper $[x]$ is written to denote the factorial $[x]^x$ or $\Pi(x)$, and so in other cases.]

Now $\nabla^{-n}0^x$, or $\left(\frac{\Delta}{\log(1+\Delta)}\right)^{-n}0^x$, is equal to $[x]\times$ coef. t^x in $\left(\frac{e^t-1}{t}\right)^{-n}$, and so $\nabla^q 0^x$, or $\left(\frac{\Delta}{\log(1+\Delta)}\right)^q 0^x$, is equal to $[x]\times$ coef. t^x in $\left(\frac{e^t-1}{t}\right)^q$.

Hence, putting $R = \frac{e^t - 1}{t}$, the last-mentioned formula will be true if, as regards the term which contains t^x, we have

$$R^{-n} = \frac{[x+n]}{[n-1]} \left\{ \frac{1}{[0]\,[x]\,n} R^0 - \frac{1}{[1]\,[x-1]\,(n+1)} R^1 + \frac{1}{[2]\,[x-2]\,(n+2)} R^2 - \&c. \right\},$$

the series on the right-hand side being continued up to the term in R^x. This formula is, in fact, true if R, instead of being restricted to denote $\frac{e^t-1}{t}$, denotes any function whatever of the form $1 + bt + ct^2 + \&c.$, and it is true not only for the term in t^x, but for all the powers of t not higher than t^x. And, moreover, R^{-n} may denote any positive or negative integral or fractional power of R. In fact, the formula (assuming for a moment the truth of it) shows that the expansion of any power whatever of a series of the form in question, can be obtained by means of the expansions of the successive positive integer powers of the same series: the existence of such a formula (at least for negative powers) was indicated by Eisenstein, *Crelle*, t. XXXIX. p. 181 (1850), and the formula itself, in a slightly different form, was obtained in a very simple manner by Professor Sylvester in his paper, "Development of an idea of Eisenstein," *Quart. Math. Journ.* t. I. p. 199 (1855); the demonstration was in fact as follows, viz. writing

$$R^n = (1 + \overline{R-1})^n = 1 + \frac{n}{1}(R-1) + \frac{n\,.\,n-1}{1\,.\,2}(R-1)^2 + \&c.:$$

if we attend only to the terms involving powers of t not higher than t^x, the series on the right-hand side needs only to be continued up to the term involving $(R-1)^x$, and the right side being thus converted into a rational and integral function of R, it may be developed in a series of powers of R (the highest power being of course R^x), and the coefficients of the several powers are finite series which admit of summation; this gives the required formula. But there is an easier method; the process shows that the series on the right-hand side, continued as above up to the term involving t^x, is, as regards n, a rational and integral function of the degree x; and by Lagrange's interpolation formula, any rational and integral function of n of the degree x, can be at once expressed in terms of the values corresponding to $x+1$ particular values of n. The investigation will be as follows:—Let R denote a series of the form $1 + bt + ct^2 + \&c.$, and let R^n denote the development of the nth power of R, continued up to the term containing t^x, the terms involving higher powers of t being rejected: R^0, R^1, R^2, &c., and generally R^s, will in like manner denote the developments of these powers up to the term involving t^x, or what is the same thing, they will be the values of R^n, corresponding to $n = 0, 1, 2, \ldots s$. By what precedes R^n is a rational and integral function of n of the degree x, and it can therefore be expressed in terms of the values $R^0, R^1, R^2, \ldots R^x$, which correspond to $n = 0, 1, 2, \ldots x$. Let s have any one of the last-mentioned values, then the expression

$$\frac{n\,.\,n-1\,.\,n-2\ldots n-x}{n-s}\;\frac{1}{s\,.\,s-1\ldots 2\,.\,1\,.\,-1\,.\,-2\ldots -(x-s)},$$

which, as regards n, is a rational and integral function of the degree x (the factor

$n-s$, which occurs in the numerator and in the denominator being of course omitted), vanishes for each of the values $n=0, 1, 2, \ldots x$, except only for the value $n=s$, in which case it becomes equal to unity. The required formula is thus seen to be

$$R^n = \Sigma \left\{ \frac{n \,.\, n-1 \,.\, n-2 \ldots n-x}{n-s} \; \frac{1}{s \,.\, s-1 \ldots 2 \,.\, 1 \,.\, -2 \ldots -(x-s)} R^s \right\},$$

where the summation extends to the several values $s = 0, 1, 2, \ldots x$; or, what is the same thing, it is

$$R^n = \Sigma \left\{ \frac{n \,.\, n-1 \,.\, n-2 \,.. \, n-x}{n-s} \; \frac{(-)^{x-s}}{1 \,.\, 2 \ldots s \,.\, 1 \,.\, 2 \ldots (x-s)} R^s \right\};$$

or changing the sign of n, it is

$$R^{-n} = \Sigma \left\{ \frac{n \,.\, n+1 \,.\, n+2 \,..\, n+x}{n+s} \; \frac{(-)^s}{1 \,.\, 2 \ldots s \,.\, 1 \,.\, 2 \ldots x-s} R^s \right\},$$

or, as this may be written,

$$R^{-n} = \frac{[x+n]}{[n-1]} \Sigma \left\{ \frac{(-)^s}{[s]\,[x-s]\,(n+s)} R^s \right\}$$

or substituting for s the values $0, 1, 2, \ldots x$, the formula is

$$R^{-n} = \frac{[x+n]}{[n-1]} \left\{ \frac{1}{[0]\,[x]\,n} R^0 - \frac{1}{[1]\,[x-1]\,(n+1)} R^1 + \frac{1}{[2]\,[x-2]\,(n+2)} R^2 - \ldots \right\}$$

continued up to the term involving R^x, which is the theorem in question.

264.

ON AN EXTENSION OF ARBOGAST'S METHOD OF DERIVATIONS.

[From the *Philosophical Transactions of the Royal Society of London,* vol. CLI. (for the year 1861, pp. 37—43, Received October 18,—Read December 13, 1860.]

ARBOGAST'S Method of Derivations was devised by him with a view to the development of a function $\phi(a+bx+cx^2+\dots)$, but it is at least as useful for the formation of only the literal parts of the coefficients, or, what is the same thing, the combinations of a given degree and weight in the letters $(a, b, c, d, \dots)$, the weights of the successive letters being 0, 1, 2, 3, &c. Thus instead of applying the method to finding the coefficients

$$a^4,\ 4a^3b,\ 4a^3c+6a^2b^2,\ \&c.,$$

we may apply it merely to finding the sets of terms

$$\begin{array}{llll} a^4, & a^3b, & a^3c, & \&c. \\ & & a^2b^2. & \end{array}$$

To derive any column from the one which immediately precedes it, we operate on a letter by changing it into its immediate successor in the alphabet, and we must in each term operate on the last letter, and also, when the last but one letter in the term is the immediate antecessor in the alphabet of the last letter (but in this case only), operate on the last but one letter. Thus a^3c gives a^3d, but a^2b^2 gives a^2bc and ab^3, and the next succeeding column is therefore

$$\begin{array}{l} a^3d \\ a^2bc \\ ab^3. \end{array}$$

If the series of letters is finite, and the last letter of the term is also the last letter of the series, then it is impossible to operate on the last letter of the term, but the last but one letter (when the foregoing rule applies to it) is still to be operated on; and if the rule does not apply, then the term does not give rise to a term in the succeeding column; the operations will at length terminate, and a complete series of columns be obtained. Thus, if the letters are (a, b, c, d), and the operations are (as before) performed upon a^4, the entire series of columns is

a^4	a^3b	a^3c	a^3d	a^2bd	a^2cd	a^2d^2	abd^2	acd^2	ad^3	bd^3	cd^3	c^4
		a^2b^2	a^2bc	a^2c^2	ab^2d	$abcd$	ac^2d	b^2d^2	bcd^2	c^2d^2		
			ab^3	ab^2c	abc^2	ac^3	b^2cd	bc^2d	c^3d			
				b^4	b^3c	b^3d	bc^3	c^4				
						b^2c^2						

Nothing can be more convenient than the process when the entire series of columns is required; but it is very desirable to have a process for the formation of any column apart from the others; and the object of the present memoir is to investigate a rule for the purpose. But as Arbogast's rule, applied as above, depends upon very similar principles, I will commence by showing how this rule is to be demonstrated. If we take any combination b^3d and operate backwards on the last letter (viz. by changing it into its immediate antecessor in the alphabet), we obtain b^3c, which is a term in the next preceding column; b^3d is therefore obtainable from a term in the next preceding column, viz. b^3c, and the process is to operate on the last letter. If, instead of b^3d, the term is abc^2 (the last letter here entering as a power), the operation backwards on the last letter gives ab^2c, which is also a term of the preceding column; and it is to be noticed that the last but one letter b is here the immediate antecessor of the last letter c (and would have been so even if b had not entered into the given term abc^2, thus ac^2 operated on backwards would have given abc). Hence abc^2 is obtained by operating on a term in the next preceding column, viz. the term ab^2c, but in this case the operation is performed on the last but one letter. *Every* term is thus obtained from the next preceding column, viz. the terms are obtained by operating on the last letter, and (when the last but one letter is the immediate antecessor in the alphabet of the last letter) then also on the last but one letter, of each term of the next preceding column, and the correctness of the rule is thus demonstrated. It is to be observed that the terms are operated upon in order, the operation on the last but one letter (when it is operated on) being made immediately after that upon the last letter of the same term, and that the terms of a column are thus obtained in the proper alphabetical order.

I pass now to the above mentioned question of the formation of a single column by itself; it will be convenient, by way of illustration, to write down the columns

a^2e^3	,	ade^3
$abde^2$		bce^3
ac^2e^2		bd^2e^2
acd^2e		c^2de^2
ad^4		cd^3e
b^2ce^2		d^5
b^2d^2e		
bc^2de		
bcd^3		
c^4e		
c^3d^2		

which belong to the set (a, b, c, d, e), and are of the degree 5 and the weights 12 and 15 respectively.

Some definitions and explanations are required. I speak of the first and last letters *of the set*, simply as the first and the last letter: there is frequent occasion to speak of the last letter; and to avoid confusion, the last letter *of a term* will be spoken of as the ultimate letter; it is necessary also to consider the penultimate, antepenultimate, and pro-antepenultimate letters of the term. It will be convenient to distinguish between the ultimate letter and the ultimate, which may be either the ultimate letter or a power of such letter; and similarly for the penultimate, &c. Thus in the term bcd^3, the ultimate letter is d and the ultimate is d^3, the penultimate and the penultimate letter are each of them c; of course the ultimate, penultimate, &c. letters are always distinct from each other. We have also to consider the pairs of letters contained in a term; cd^2e contains the pairs (c, d), (c, e), (d, d), (d, e), and so in other cases; the letters of a pair are taken in the natural order. A pair not containing either the first letter or the last letter is *expansible;* thus the set being as before (a, b, c, d, e), the pairs (b, c), (d, d) are expansible: they are *expanded* by retreating the prior and advancing the posterior letter each one step; thus the just mentioned pairs (b, c), (d, d), are expanded into (a, d) and (c, e) respectively.

A pair composed of two distinct non-contiguous letters is *contractible;* it is *contracted* by advancing the prior and retreating the posterior letter each one step: thus (a, d), (c, e) are contractible pairs, and they give by contraction the pairs (b, c), (d, d) respectively: the processes of expansion and contraction are obviously converse to each other.

The expression the *last expansible pair* of a term hardly requires explanation; (d, d) is the last expansible pair of the term bcd^3, or of the term $b^2d^2e^2$, (c, d) the last expansible pair of the term c^2de (the set being always (a, b, c, d, e)), and so in all other cases. The expression the *last expansion*, in regard to any term, means the expansion of the last expansible pair of such term.

The expansion or contraction of any pair of a term leaves unaltered as well the weight as the degree, the resulting term belongs therefore to the same column as the original one. But the effect of an expansion is to diminish, and that of a contraction to increase the alphabetical rank, or rank in the column, the ranks being reckoned as the first or lowest, second, third, and so on, up to the last or highest rank. In particular, by performing upon any term the last expansion, we diminish the rank in the column, and by a succession of last expansions we bring the term up to the head of the column. Such term is not susceptible of any further expansion; it may therefore contain the first and last letters or either of them, and it may also contain a single intermediate letter in the first power only; thus the first and last letters being a, e, the head term of a column is of the form a^me^q or a^mce^q, c being some intermediate letter, and the powers a^m, e^q being both or either of them omitable. In like manner, by a succession of contractions of any term we obtain the term at the foot of the column; such term is not susceptible of any further contraction, and it must therefore be composed either of a single letter or of two contiguous letters, that is it must be of the form c^m, or of the form $c^m d^q$, where c, d are any two contiguous letters, not excluding the case where the single letter or each or either of the contiguous letters is a first or a last letter.

It is to be observed that the passage up from any term to the term at the head of the column (or top term) by means of a succession of last expansions, is a perfectly unique one; but as no selection has been made of a like unique process of contraction, this is not the case with respect to the passage down from any term to the term at the foot of the column (or bottom term) by a succession of contractions.

Every term gives by the last expansion a term above it; it can therefore be obtained from such term above it by means of a contraction. But the contraction of the upper term is by hypothesis such that, performing upon the contracted term the last expansion, we obtain the upper term; a contraction, which, performed on any term, gives a lower term which by performing upon it the last expansion reproduces the first mentioned or upper term, may be called *a reversible contraction.* And it is clear that if we perform on the top term all the reversible contractions, and on each of the resulting terms all the reversible contractions, and so on as long as the process is possible, we obtain without repetition all the terms of the column. The column is in fact similar to a genealogical tree in the male line, each lower term issuing from a single upper term, while each upper term generates a lower term or terms, or does not generate any such term, and the top term being the common origin of the entire series. It may be added that when the order of the reversible contractions of the same term is duly fixed, the alphabetical arrangement of the terms in the column agrees with the order as of primogeniture (an elder son and his issue male preceding all the younger sons and their respective issue male) in the genealogical tree.

It only remains then to inquire under what conditions a contraction is reversible. Now as regards any term, in order that a contraction performed on it may be reversible, it is necessary and sufficient that the pair produced by the contraction should be the last expansible pair of the contracted term. There are several cases to be considered. First, if the contraction affects the ultimate and penultimate letters of the term: this implies that the ultimate and penultimate letters are not contiguous. Let the term terminate in $e^m h$, the contracted term will terminate in $e^{m-1}fg$, and (f, g), the pair produced by the contraction, is the last expansible pair of the contracted term; the contraction is in this case reversible. If, however, the term terminate in $e^m h^p$ $(p>1)$, the contracted term will terminate in $e^{m-1}fgh^{p-1}$, and the last expansible pair is not as before (f, g), but it is (according as $p=2$ or $p>2$) (g, h) or (h, h): unless indeed h is the last letter, in which case (f, g) remains the last expansible pair of the contracted term. In the example e^m has been written, but the case $m=1$ is not excluded; moreover, the penultimate letter e is removed three steps from the ultimate letter h, but the result would have been the same if instead of e we had had any preceding letter, or had had the letter f; by hypothesis it is not g, the letter contiguous to the ultimate letter h. The conclusion is that a contraction on the ultimate and penultimate letters (these being non-contiguous) is reversible if the ultimate is a simple letter, or if, being a power, it is a power of the last letter.

Next let the contraction affect the ultimate and antepenultimate letters. The two letters are here separated by the penultimate letter, and are therefore not contiguous. Suppose that the termination is $f^l g^m k$ (f and g contiguous), the contracted term terminates in $f^{l-1} g g^m j$, and (g, j) the pair arising from the contraction is the last expansible pair of the contracted term; the contraction is therefore reversible. In the example f^l has been written, but the case $l=1$ is not excluded; moreover the ultimate letter k is taken non-contiguous to the penultimate letter g; but if the ultimate letter had been the contiguous letter h, the only difference is that the pair would be (g, g), and the conclusion is not altered. But suppose the termination of the term is $e^l g^m k$ (e, g, non-contiguous), then the contracted term terminates in $e^{l-1} f g^m j$, where the pair arising from the contraction is (f, j), but the last expansible pair is (g, j); the contraction therefore is not reversible. The case $l=1$ is not excluded; nor is it necessary that the ultimate letter should be non-contiguous to the penultimate; if the ultimate letter had been h, the pair arising from the contraction would have been (f, g), but the last expansible pair (g, g), and the contraction is still not reversible. In each of the cases considered the ultimate has been a simple letter; if in the first case the ultimate had been k^p $(p > 1)$, then the contracted pair would terminate in jk^{p-1}, and (according as $p=2$ or $p<2$) the last expansible pair would be (j, k) or (k, k), which is not the pair (g, j) produced by the contraction; the contraction is therefore not reversible, unless indeed k is the last letter, in which case it continues reversible. In the second case the contraction, which is not reversible when the ultimate is the simple letter k, remains not reversible when the ultimate is a power of such letter. The conclusion is that a contraction on the ultimate and antepenultimate letters is reversible, if the penultimate and antepenultimate letters are contiguous, and the ultimate is a simple letter, or if, being a power, it is a power of the last letter.

A contraction on the ultimate and pro-antepenultimate letters, or on the ultimate letter and any letter preceding the pro-antepenultimate letter, is never reversible. To show this, it will be sufficient to consider the case where the term terminates in $efgh$, the contracted term terminates in $ffgg$, the pair arising from the contraction being (f, g), and the last expansible pair being (g, g); and *à fortiori*, if the letters or any of them occur as powers, or are non-contiguous.

Consider, next, a contraction on the penultimate and antepenultimate letters; this assumes that these letters are non-contiguous. Such a contraction may be reversible if only the ultimate is the last letter or a power thereof; and the condition is then similar to that in the case of the ultimate and penultimate letters; only as the penultimate cannot be a power of the last letter, it must be a simple letter. And the conditions in order that the contraction may be reversible then are that the ultimate is the last letter or a power thereof, and the penultimate a simple letter.

The next case is that of a contraction on the penultimate and pro-antepenultimate letters. Such contraction may be reversible if the ultimate is the last letter or a power thereof; and the condition is then similar to that for the case of the ultimate and antepenultimate letters; only as the penultimate cannot be a power of the last letter, it must be a simple letter. The conditions, in order that the contraction may be reversible, are that the ultimate may be the last letter or a power thereof, the penultimate a simple letter, and the antepenultimate and pro-antepenultimate letters contiguous.

A contraction on the penultimate letter and on any letter preceding the pro-antepenultimate letter, or upon any two letters preceding the penultimate letter, is never reversible. If the ultimate, penultimate, antepenultimate and pro-antepenultimate letters are denoted by U, P, A, P' respectively, then by what precedes, the following contractions, viz. UP, UA, PA, PP', may be reversible, and they will be so under the conditions shown in the annexed Table. It is to be noticed that the conditions for UA, PA are mutually exclusive, and consequently that the number of reversible contractions to be performed upon any term is at most 3. The Table is

	P	A	P'
U	U, P, non-contiguous letters. The ultimate a simple letter, or a power of the last letter.	P, A, contiguous letters. The ultimate a simple letter, or a power of the last letter.	
P		P, A, non-contiguous letters. Penultimate a simple letter. Ultimate the last letter, or a power thereof.	A, P', contiguous letters. Penultimate a simple letter. Ultimate the last letter, or a power thereof.

The contractions are to be applied in the order UP, UA, PA, PP', but all the contracted terms originating in a prior contraction of a given term are to be exhausted before proceeding to a posterior contraction of the same term. As an example of the process, the set being (a, b, c, d, e), I will take the column

The top term abe^3 is given. The contractions applicable to it are UP, UA. And UP gives $acde^2$. The only contraction applicable to this is UA, giving ad^3e. And

there is not any contraction applicable to ad^3e. We revert therefore to UA on abe^3, this gives b^2de^2. The only applicable contraction is PA, giving bc^2e^2. The contractions applicable to this are UP, UA. And UP gives bcd^2e. The only contraction applicable to this is UA, giving bd^4, and there is not any contraction applicable to bd^4. We revert therefore to UA on bc^2e^2, this gives c^3de, and the only contraction applicable to this is UA, giving c^2d^3. There is not any contraction applicable to this, and the process is therefore complete. It may be remarked that the example presents no instance of the contraction PP'; in fact the only terms having a pro-antepenultimate are $acde^2$, in which A, P' are not contiguous letters, and bcd^2e, in which the penultimate is not a simple letter, so that the contraction PP' is in each case excluded. It must be confessed that a considerable amount of practice would be required before the process could be readily made use of.

265.

ADDITION TO THE MEMOIR ON AN EXTENSION OF ARBOGAST'S METHOD OF DERIVATIONS.

[Now first published, 1891.]

THE process explained in the foregoing Memoir is not easily workable, and I have in fact never used it. I have devised a new process theoretically less complete, and of a somewhat mixed character consisting, as it does, in the analytical reduction of the problem to the same problem for a smaller number of letters. I have however found it very convenient for obtaining the literal terms in the theory of the sextic, viz. where we have the seven letters (a, b, c, d, e, f, g); and I propose to explain the process by applying it to the determination of the terms of the discriminant, degree $= 6$, weight $= 18$.

Here writing a, b, c, d, e, f, g to denote the indices of these letters respectively in a term such as $a^\alpha b^\beta c^\gamma\ldots$ (that is, writing for convenience $a, b, c, \ldots$ instead of $\alpha, \beta, \gamma, \ldots$) we have

$$a + b + c + d + e + f + g = 6,$$

$$b + 2c + 3d + 4e + 5f + 6g = 18,$$

and thence

$$6a + 5b + 4c + 3d + 2e + f = 18.$$

I separate off from the others the first three letters a, b, c. The equation gives $a = 3$ at most. And then

if $a = 3$, then $5b + 4c + 3d + 2e + f = 18$, $b = 0$, $c = 0$;

if $a = 2$, $5b + 4c + 3d + 2e + f = 6$, $b = 1$ at most;

if $b = 1$, $4c + 3d + 2e + f = 1$, $c = 0$,

„ $b = 0$, $4c + 3d + 2e + f = 6$, $c = 1$ at most,

if	$a = 1$,	$5b + 4c + 3d + 2e + f = 12$,	$b = 2$ at most;
	if $b = 2$,	$4c + 3d + 2e + f = 2$,	$c = 0$,
	„ $b = 1$,	$4c + 3d + 2e + f = 7$,	$c = 1$ at most,
	„ $b = 0$,	$4c + 3d + 2e + f = 12$,	$c = 3$ „ „
if	$a = 0$,	$5b + 4c + 3d + 2e + f = 18$,	$b = 3$ at most;
	if $b = 3$,	$4c + 3d + 2e + f = 3$,	$c = 0$,
	„ $b = 2$,	$4c + 3d + 2e + f = 8$,	$c = 2$ at most,
	„ $b = 1$,	$4c + 3d + 2e + f = 13$,	$c = 3$ „ „
	„ $b = 0$,	$4c + 3d + 2e + f = 18$,	$c = 4$ „ „

Hence considering the several terms as arranged in alphabetical order and writing down only the factors in a, b, c, we obtain col. 1 of the following diagram;

Col. 1.	Col. 2.	Col. 3.	
a^3	$(d^3)^9$	1	1
a^2b	$(d^3)^8$	1	1
a^2b^0c	$(d^3)^7$	2	
c^0	$(d^4)^6$	5	7
$a\,b^2c^0$	$(d^3)^7$	2	
$b\,c^1$	$(d^3)^6$	3	
c^0	$(d^4)^5$	4	9
$a\,b^0c^3$	$(d^2)^6$	1	
c^2	$(d^3)^5$	3	
c^1	$(d^4)^4$	4	
c^0	$(d^5)^3$	3	11
$a^0b^3c^0$	$(d^3)^6$	3	3
b^2c^2	$(d^2)^6$	1	
c^1	$(d^3)^5$	3	
c^0	$(d^4)^4$	4	8
b^1c^3	$(d^2)^5$	1	
c^2	$(d^3)^4$	3	
c^1	$(d^4)^3$	3	
c^0	$(d^5)^2$	2	9
b^0c^4	$(d^2)^4$	2	
c^3	$(d^3)^3$	3	
c^2	$(d^4)^2$	2	
c^1	$(d^5)^1$	1	
c^0	$(d^6)^0$	1	9
			58

We then form col. 2 by annexing to each term a term $(d^\theta)^\phi$ which denotes the derivative ϕ of d^θ; the value of θ being such that the whole term may be of the proper degree 6, and the value of ϕ being such that the whole term may be of the proper weight 18. Thus

$$a^3 (d^3)^9, \text{ degree is } 3 + 3, = 6; \text{ weight is } 0 + 9 + 9, = 18,$$
$$ab^2c^0 (d^3)^7, \quad „ \quad 1 + 2 + 3, = 6; \quad „ \quad „ \quad 0 + 2 + 9 + 7, = 18,$$

viz. d^3 being of weight 9, then $(d^3)^9$ is of weight $9+9$, and $(d^3)^7$ of weight $9+7$.

The derivatives of any power of d are given by those of the indefinite power d^m, which, omitting therein the several powers of d, may be tabulated thus. I remark that the table is carried up to the column 21 in order that it may be applicable to the calculation of the literal terms of the degree 15 and weight 45 which belong to the highest invariant of the sextic.

SUBSIDIARY TABLE OF THE DERIVATIVES OF d^m.

	1	1	2	3	4	5	7	8	10	12	14	16	19	21	24	27	30	33	37	40	44	48
	0	1	2	3	4	5	6	7	8	9	10	11	12	13	14	15	16	17	18	19	20	21
1	1	e	f	g	eg	fg	g^2	eg^2	fg^2	g^3	eg^3	fg^3	g^4	eg^4	fg^4	g^5	eg^5	fg^5	g^6	eg^6	fg^6	g^7
2			e^2	ef	f^2	e^2g	efg	f^2g	e^2g^2	efg^2	f^2g^2	e^2g^3	efg^3	f^2g^3	e^2g^4	efg^4	f^2g^4	e^2g^5	efg^5	f^2g^5	e^2g^6	efg^6
3				e^3	e^2f	ef^2	f^3	e^2fg	ef^2g	f^3g	e^2fg^2	ef^2g^2	f^3g^2	e^2fg^3	ef^2g^3	f^3g^3	e^2fg^4	ef^2g^4	f^3g^4	e^2fg^5	ef^2g^5	f^3g^5
4					e^4	e^3f	e^3g	ef^3	f^4	e^3g^2	ef^3g	f^4g	e^3g^3	ef^3g^2	f^4g^2	e^3g^4	ef^3g^3	f^4g^3	e^3g^5	ef^3g^4	f^4g^4	e^3g^6
5						e^5	e^2f^2	e^4g	e^3fg	e^2f^2g	f^5	e^3fg^2	$e^2f^2g^2$	f^5g	e^3fg^3	$e^2f^2g^3$	f^5g^2	e^3fg^4	$e^2f^2g^4$	f^5g^3	e^3fg^5	$e^2f^2g^5$
6							e^4f	e^3f^2	e^2f^3	ef^4	e^4g^2	e^2f^3g	ef^4g	e^4g^3	$e^2f^3g^2$	ef^4g^2	e^4g^4	$e^2f^3g^3$	ef^4g^3	e^4g^5	$e^2f^3g^4$	ef^4g^4
7							e^6	e^5f	e^5g	e^4fg	e^3f^2g	ef^5	f^6	$e^3f^2g^2$	ef^5g	f^6g	$e^3f^2g^3$	ef^5g^2	f^6g^2	$e^3f^2g^4$	ef^5g^3	f^6g^3
8								e^7	e^4f^2	e^3f^3	e^2f^4	e^5g^2	e^4fg^2	e^2f^4g	f^7	e^4fg^3	$e^2f^4g^2$	f^7g	e^4fg^4	$e^2f^4g^3$	f^7g^2	e^4fg^5
9									e^6f	e^6g	e^5fg	e^4f^2g	e^3f^3g	ef^6	e^5g^3	$e^3f^3g^2$	ef^6g	e^5g^4	$e^3f^3g^3$	ef^6g^2	e^5g^5	$e^3f^3g^4$
10									e^8	e^5f^2	e^4f^3	e^3f^4	e^2f^5	e^5fg^2	$e^4f^2g^2$	e^2f^5g	f^8	$e^4f^2g^3$	$e^2f^5g^2$	f^8g	$e^4f^2g^4$	$e^2f^5g^3$
11										e^7f	e^7g	e^6fg	e^6g^2	e^4f^3g	e^3f^4g	ef^7	e^5fg^3	$e^3f^4g^2$	ef^7g	e^5fg^4	$e^3f^4g^3$	ef^7g^2
12										e^9	e^6f^2	e^5f^3	e^5f^2g	e^3f^5	e^2f^6	e^6g^3	$e^4f^3g^2$	e^2f^6g	f^9	$e^4f^3g^3$	$e^2f^6g^2$	f^9g
13											e^8f	e^8g	e^4f^4	e^7g^2	e^6fg^2	$e^5f^2g^2$	e^3f^5g	ef^8	e^6g^4	$e^3f^5g^2$	ef^8g	e^6g^5
14											e^{10}	e^7f^2	e^7fg	e^6f^2g	e^5f^3g	e^4f^4g	e^2f^7	e^6fg^3	$e^5f^2g^3$	e^2f^7g	f^{10}	$e^5f^2g^4$
15												e^9f	e^6f^3	e^5f^4	e^4f^5	e^3f^6	e^7g^3	$e^5f^3g^2$	$e^4f^4g^2$	ef^9	e^6fg^4	$e^4f^4g^3$
16												e^{11}	e^9g	e^8fg	e^8g^2	e^7fg^2	$e^6f^2g^2$	e^4f^5g	e^3f^6g	e^7g^4	$e^5f^3g^3$	$e^3f^6g^2$
17													e^8f^2	e^7f^3	e^7f^2g	e^6f^3g	e^5f^4g	e^3f^7	e^2f^8	$e^6f^2g^3$	$e^4f^5g^2$	e^2f^8g
18													$e^{10}f$	$e^{10}g$	e^6f^4	e^5f^5	e^4f^6	e^8g^3	e^7fg^3	$e^5f^4g^2$	e^3f^7g	ef^{10}
19													e^{12}	e^9f^2	e^9fg	e^9g^2	e^8fg^2	$e^7f^2g^2$	$e^6f^3g^2$	e^4f^6g	e^2f^9	e^7fg^4
20														$e^{11}f$	e^8f^3	e^8f^2g	e^7f^3g	e^6f^4g	e^5f^5g	e^3f^8	e^8g^4	$e^6f^3g^3$
21														e^{13}	$e^{11}g$	e^7f^4	e^6f^5	e^5f^6	e^4f^7	e^8fg^3	$e^7f^2g^3$	$e^5f^5g^2$
22															$e^{10}f^2$	$e^{10}fg$	$e^{10}g^2$	e^9fg^2	e^9g^3	$e^7f^3g^2$	$e^6f^4g^2$	e^4f^7g
23															$e^{12}f$	e^9f^3	e^9f^2g	e^8f^3g	$e^8f^2g^2$	e^6f^5g	e^5f^6g	e^3f^9
24															e^{14}	$e^{12}g$	e^8f^4	e^7f^5	e^7f^4g	e^5f^7	e^4f^8	e^9g^4
25																$e^{11}f^2$	$e^{11}fg$	$e^{11}g^2$	e^6f^6	$e^{10}g^3$	e^9fg^3	$e^8f^2g^3$
26																$e^{13}f$	$e^{10}f^3$	$e^{10}f^2g$	$e^{10}fg^2$	$e^9f^2g^2$	$e^8f^3g^2$	$e^7f^4g^2$
27																e^{15}	$e^{13}g$	e^9f^4	e^9f^3g	e^8f^4g	e^7f^5g	e^6f^6g
28																	$e^{12}f^2$	$e^{12}fg$	e^8f^5	e^7f^6	e^6f^7	e^5f^8
29																	$e^{14}f$	$e^{11}f^3$	$e^{12}g^2$	$e^{11}fg^2$	$e^{11}g^3$	$e^{10}fg^3$
30																	e^{16}	$e^{14}g$	$e^{11}f^2g$	$e^{10}f^3g$	$e^{10}f^2g^2$	$e^9f^3g^2$
31																		$e^{13}f^2$	$e^{10}f^4$	e^9f^5	e^9f^4g	e^8f^5g
32																		$e^{15}f$	$e^{13}fg$	$e^{13}g^2$	e^8f^6	e^7f^7
33																		e^{17}	$e^{12}f^3$	$e^{12}f^2g$	$e^{12}fg^2$	$e^{12}g^3$
34																			$e^{15}g$	$e^{11}f^4$	$e^{11}f^3g$	$e^{11}f^2g^2$
35																			$e^{14}f^2$	$e^{14}fg$	$e^{10}f^5$	$e^{10}f^4g$
36																			$e^{16}f$	$e^{13}f^3$	$e^{14}g^2$	e^9f^6
37																			e^{18}	$e^{16}g$	$e^{13}f^2g$	$e^{13}fg^2$
38																				$e^{15}f^2$	$e^{12}f^4$	$e^{12}f^3g$
39																				$e^{17}f$	$e^{15}fg$	$e^{11}f^5$
40																				e^{19}	$e^{14}f^3$	$e^{15}g^2$
41																					$e^{17}g$	$e^{14}f^2g$
42																					$e^{16}f^2$	$e^{13}f^4$
43																					$e^{18}f$	$e^{16}fg$
44																					e^{20}	$e^{15}f^3$
45																						$e^{18}g$
46																						$e^{17}f^2$
																						$e^{19}f$
																						e^{21}

This means for instance that the derivatives of d^2 are

0	1	2	3	4	5	6
d^2	de	df e^2	dg ef	eg f^2	fg	g^2

those of d^3 are

0	1	2	3	4	5	6	7	8	9
d^3	d^2e	d^2f de^2	d^2g def e^3	deg df^2 e^2f	dfg e^2g ef^2	dg^2 efg f^3	eg^2 f^2g	fg^2	g^3

and so on; viz. to form any column we affix to the terms of the corresponding column of the subsidiary table the proper power of d, so that the degree in all the letters may be 2, 3, &c. as the case may be, using from each column of the subsidiary table only the terms which are not of too high a degree; for instance for col. 4 of the derivatives of d^3, only the terms eg, f^2, e^2f of the complete column eg, f^2, e^2f, e^3.

It is to be observed that these tables of the derivatives of d^2, d^3, &c. do not need to be actually formed; any column which is wanted can be written down at once, *currente calamo,* from the column of the subsidiary table. And if we require only the number of terms, then these numbers are at once taken out from the Subsidiary Table. Thus for the numerical column of the diagram, the Subsidiary Table, col. 9, contains but 1 term g^3 of degree 3, and thus the number set opposite to $(d^3)^9$ is 1; so col. 8 contains but 1 term fg^2 of the degree 3, and so the number opposite to $(d^3)^8$ is 1; col. 7 contains 2 terms eg^2, f^2g of degree 3, and so the number opposite to $(d^3)^7$ is 2; and so on: we have in this way the numbers 1, 1, 2, 5, 2, 3, 4, &c. the partial sums of which are 1, 1, 7, 9, 11, 3, 8, 9, 9 giving the total sum 58: viz. this is the number of the terms degree 6, weight 18, which can be formed with the seven letters (a, b, c, d, e, f, g).

When the literal terms are required, it is proper in the first instance as a safeguard against accidental omissions in copying, to take the numbers in this manner; and we can then take out the terms themselves, viz. these are

$$\begin{array}{ll} a^3 & g^3 \\ a^2b & fg^2 \\ a^2c & eg^2 \\ & f^2g \\ a^2 & d^2g^2 \\ & defg \\ & df^3 \\ & e^3g \\ & e^2f^2, \end{array}$$

and so on, the whole number being $=58$ as just mentioned.

266.

ON THE EQUATION FOR THE PRODUCT OF THE DIFFERENCES OF ALL BUT ONE OF THE ROOTS OF A GIVEN EQUATION.

[From the *Philosophical Transactions of the Royal Society of London*, vol. CLI. for the year 1861, pp. 45—59. Received November 30, 1860,—Read January 10, 1861.]

It is easy to see that for an equation of the order n, the product of the differences of all but one of the roots will be determined by an equation of the order n, the coefficients of which are alternately rational functions of the coefficients of the original equation, and rational functions multiplied by the square root of the discriminant. In fact, if the equation be $\phi v = (a, \ldots \between v, 1)^n = a(v-\alpha)(v-\beta)\ldots$, then putting for the moment $a=1$, and disregarding numerical factors, $\sqrt{\Box}$, the square root of the discriminant, is equal to the product of the differences of the roots, and $\phi'\alpha$ is equal to $(\alpha-\beta)(\alpha-\gamma)\ldots$, consequently the product of the differences of the roots, all but α, is equal to $\sqrt{\Box} \div \phi'\alpha$, and the expression $\dfrac{1}{\phi'\alpha}$ is the root of an equation of the order n, the coefficients of which are rational functions of the coefficients of the original equation. I propose in the present memoir to determine the equation in question for equations of the orders three, four, and five: the process employed is similar to that in my memoir "On the Equation of Differences for an Equation of any Order, and in particular for Equations of the Orders Two, Three, Four, and Five," *Phil. Trans.*, vol. CL. (1860), [262], viz. the last coefficient of the given equation is put equal to zero, so that the given equation breaks up into $v=0$ and into an equation of the order $n-1$ called the reduced equation; and this being so, the required equation breaks up into an equation of the order $n-1$ (which however is not, as for the equation of differences, that which corresponds to the reduced equation) and into a linear equation; the equation of the order $n-1$ is calculated by the method of symmetric functions; and combining it with the linear equation, which is known, we have the required equation, except as regards the terms involving the last coefficient, which terms are found by the consideration that the coefficients of the required equation are seminvariants. The solution leads immediately to that of a more general question; for if the product of the differences of all the roots except α, of the given equation

$$\phi v = (* \between v,\ 1)^n = a(v-\alpha)(v-\beta)\ldots = 0$$

(which product is a function of the degree $n-2$ in regard to each of the roots $\beta, \gamma, \delta\,..$), is multiplied by $(x-\alpha y)^{n-2}$, the function so obtained will be the root of an equation of the order n, the coefficients of which are covariants of the quantic $(* \between x,\ y)^n$, and these coefficients can be at once obtained by writing, in the place of the seminvariants of the former result, the covariants to which they respectively belong. In the case of the quintic equation, one of these covariants is, in regard to the coefficients, of the degree 6, which exceeds the limit of the tabulated covariants, [the covariants are all tabulated, 141 and 143], the covariant in question has therefore to be now first calculated. The covariant equations for the cubic and the quartic might be deduced from the formulæ Nos. 119 and 142 of my Fifth memoir on Quantics *Phil. Trans.*, vol. CXLVIII. (1858), [156]; they are in fact the bases of the methods which are there given for the solution of the cubic and the quartic equations respectively; and it was in this way that I was led to consider the problem which is here treated of.

1. The notation $\zeta(\alpha, \beta, \gamma, ..)$ is used (after Professor Sylvester) to denote the product of the squared differences of $(\alpha, \beta, \gamma, ..)$, and the notation $\zeta^{\frac{1}{2}}(\alpha, \beta, \gamma, ..)$ to denote the product of the differences taken in a determinate order, viz.

$$\begin{aligned}\zeta^{\frac{1}{2}}(\alpha, \beta, \gamma, \delta, ..) = (\alpha-\beta)(\alpha-\gamma)(\alpha-\delta)&\,..\\ (\beta-\gamma)(\beta-\delta)&\,..\\ (\gamma-\delta)&\,..\\ &\,..\end{aligned}$$

2. The product of the differences of the roots of an equation depends, as already noticed, on the square root of the discriminant; and in order to fix the numerical factors and signs, it will be convenient, in regard to the equations

$$(a,\ b,\ c \between v,\ 1)^2 = 0,$$

$$(a,\ b,\ c,\ d \between v,\ 1)^3 = 0,$$

$$(a,\ b,\ c,\ d,\ e \between v,\ 1)^4 = 0,$$

$$(a,\ b,\ c,\ d,\ e,\ f \between v,\ 1)^5 = 0,$$

to write as follows:

$$\zeta^{\frac{1}{2}}(\alpha,\ \beta) = \frac{1}{a}\sqrt{-(4ac-b^2)} \qquad = \frac{1}{a}\sqrt{-\square},$$

$$\zeta^{\frac{1}{2}}(\alpha,\ \beta,\ \gamma) = \frac{1}{a^2}\sqrt{-(27a^2d^2+4ac^3+\ldots)} = \frac{1}{a^2}\sqrt{-\square},$$

$$\zeta^{\frac{1}{2}}(\alpha,\ \beta,\ \gamma,\ \delta) = -\frac{1}{a^3}\sqrt{256a^3e^3-27a^2d^4+\ldots} \quad = -\frac{1}{a^3}\sqrt{\square},$$

$$\zeta^{\frac{1}{2}}(\alpha,\ \beta,\ \gamma,\ \delta,\ \epsilon) = -\frac{1}{a^4}\sqrt{3125a^4f^4+256a^3e^5+\ldots} = -\frac{1}{a^4}\sqrt{\square},$$

where it is to be observed, for example, that writing in the last equation $\epsilon=0$, and therefore $f=0$, we have $\zeta^{\frac{1}{2}}(\alpha,\ \beta,\ \gamma,\ \delta,\ 0) = -\frac{e}{a^4}\sqrt{256a^3e^3+\ldots}$, which agrees with the equation $\zeta^{\frac{1}{2}}(\alpha,\ \beta,\ \gamma,\ \delta,\ 0) = \alpha\beta\gamma\delta\ \zeta^{\frac{1}{2}}(\alpha,\ \beta,\ \gamma,\ \delta) = \frac{e}{a}\zeta^{\frac{1}{2}}(\alpha,\ \beta,\ \gamma,\ \delta)$, if for $\zeta^{\frac{1}{2}}(\alpha,\ \beta,\ \gamma,\ \delta)$ we substitute the value given by the last equation but one.

For the cubic equation $(a,\ b,\ c,\ d\between v,\ 1)^3 = 0$;

3. We have to find the equation for $\theta = \zeta^{\frac{1}{2}}(\alpha,\ \beta) = \alpha-\beta$; the roots are

$$\theta_1 = \beta-\gamma,\quad \theta_2 = \gamma-\alpha,\quad \theta_3 = \alpha-\beta.$$

To apply the method above explained, write $\gamma=0$, and therefore also $d=0$; the roots thus become

$$\theta_1 = \beta,\quad \theta_2 = -\alpha,\quad \theta_3 = \alpha-\beta,$$

and we have the quadric and linear equations

$$(\theta+\alpha)(\theta-\beta)=0,\quad \theta-(\alpha-\beta)=0,$$

where $(\alpha,\ \beta)$ are the roots of the equation

$$(a,\ b,\ c\between v,\ 1)^2 = 0.$$

Hence, writing

$$Z = 4ac-b^2,$$

we have

$$\alpha-\beta = \frac{1}{a}\sqrt{-Z},$$

and the two equations become

$$\theta^2a+\theta\sqrt{-Z}-c=0,\quad \theta a-\sqrt{-Z}=0;$$

or multiplying the two equations together,

$$\theta^3a^2+\theta^2 0+\theta(3ac-b^2)+c\sqrt{-Z}=0,$$

which is what the required equation becomes, on putting therein $d=0$; the coefficients of the complete equation are seminvariants, and the terms in d are to be inserted by means of this property. The coefficient $3ac-b^2$ is reduced to zero by the operator

$$3a\partial_b + 2b\partial_c + c\partial_d,$$

it is therefore a seminvariant, and remains unaltered. The coefficient $c\sqrt{-Z}$ is what $\sqrt{-\square}$ becomes ($\square$ being the discriminant of the cubic equation) on putting therein $d=0$, it is therefore to be changed into $\sqrt{-\square}$. Hence

4. For the cubic equation $(a, b, c, d \between v, 1)^3$ the equation for $\theta\left(=\zeta^{\frac{1}{2}}(\alpha, \beta)\right)$ is $0=$

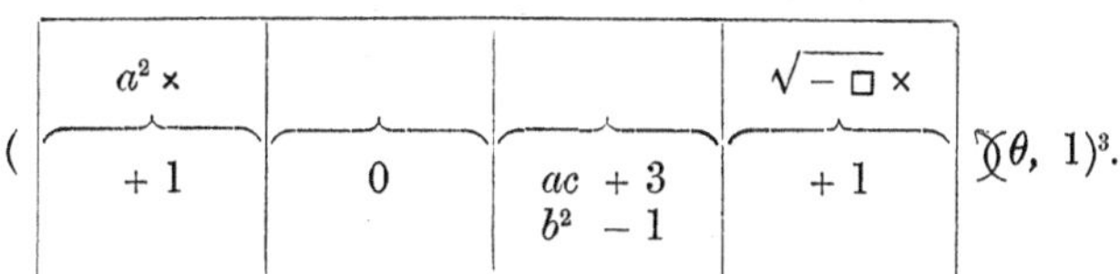

$a^2 \times$			$\sqrt{-\square}\times$
$+1$	0	ac $+3$ b^2 -1	$+1$

$\between \theta, 1)^3.$

5. For the quartic equation $(a, b, c, d, e \between v, 1)^4=0$;

$$\theta=-\zeta^{\frac{1}{2}}(\alpha, \beta, \gamma)=-(\alpha-\beta)(\alpha-\gamma)(\beta-\gamma),$$

the roots are

$$\begin{aligned}
\theta_1 &= \ \ \zeta^{\frac{1}{2}}(\beta, \gamma, \delta),\\
\theta_2 &= -\zeta^{\frac{1}{2}}(\gamma, \delta, \alpha),\\
\theta_3 &= \ \ \zeta^{\frac{1}{2}}(\delta, \alpha, \beta),\\
\theta_4 &= -\zeta^{\frac{1}{2}}(\alpha, \beta, \gamma),
\end{aligned}$$

the signs being in this case (and indeed for an equation of any even order) alternately positive and negative; in fact, if the equation is represented by $\phi v=0$, then the roots divided by $\zeta^{\frac{1}{2}}(\alpha, \beta, \gamma, \delta)$ should respectively be $\phi'\alpha$, $\phi'\beta$, $\phi'\gamma$, $\phi'\delta$, and this will be the case if the signs are taken as above.

6. Putting now $\delta=0$ (and therefore $e=0$) the roots become

$$\begin{aligned}
\theta_1 &= \beta\gamma(\beta-\gamma),\\
\theta_2 &= \gamma\alpha(\gamma-\alpha),\\
\theta_3 &= \alpha\beta(\alpha-\beta),\\
\theta_4 &= -\zeta^{\frac{1}{2}}(\alpha, \beta, \gamma),
\end{aligned}$$

where (α, β, γ) are the roots of $(a, b, c, d \between v, 1)^3=0$. Let Z denote the discriminant of the cubic function, then $\zeta^{\frac{1}{2}}(\alpha, \beta, \gamma)=\frac{1}{a^2}\sqrt{-Z}$, and we have thus the linear equation; the cubic equation is

$$\Pi_3\{\theta-\beta\gamma(\beta-\gamma)\}=0,$$

the coefficients of which can be calculated by the method of symmetric functions (see Annex No. 1).

7. The cubic equation being thus obtained, we have the two equations

$$\begin{array}{l|l} \theta^3 \,.\, a^4 & \theta \,.\, a^2 \\ +\,\theta^2 \,.\, -a^2\sqrt{-Z} & +\sqrt{-Z} \\ +\,\theta\ \,.\, -9a^2d^2 + 4abcd - b^3d & \\ +\qquad d^2\sqrt{-Z} & \\ =0 & =0\,; \end{array}$$

and multiplying these together, the resultant equation is

$$\begin{array}{l} \theta^4 \,.\, a^6 \\ +\,\theta^3 \,.\, 0 \\ +\,\theta^2 \,.\, a^2\,(-9a^2d^2 + 4abcd - b^3d + Z) \\ +\,\theta\ \,.\quad (-8a^2d + 4abc - b^3)\, d\sqrt{-Z} \\ \qquad\quad -d^2Z = 0, \end{array}$$

where the coefficients have to be completed by adding the terms which contain e. We have $\sqrt{\square}$ in the place of $d\sqrt{-Z}$, and $\square$ in the place of $-d^2Z$. The coefficient $-8a^2d + 4abc - b^3$ is a seminvariant, and requires no alteration. The coefficient

$$-\ 9a^2d^2 + \ 4abcd - \ b^3d + Z$$

is

$$\begin{array}{l} -\ \ 9a^2d^2 + \ \ 4abcd \qquad\qquad\qquad - b^3d \\ +\,27a^2d^2 - 18abcd + 4ac^3 + 4b^3d - b^2c^2\,; \end{array}$$

that is,

$$\begin{array}{ll} a^2d^2 & +18 \\ abcd & -14 \\ ac^3 & +\ 4 \\ b^3d & +\ 3 \\ b^2c^2, & -\ 1 \end{array}$$

and the terms in e to be added to this, in order to make it a seminvariant, are easily found to be

$$\begin{array}{ll} a^2ce & -16 \\ ab^2e & +\ 6 \end{array}$$

8. Hence, for the quartic equation $(a, b, c, d \between v, 1)^4$, the equation for $\theta\,(=\zeta^{\frac{1}{4}}(\alpha, \beta, \gamma))$ is $0 =$

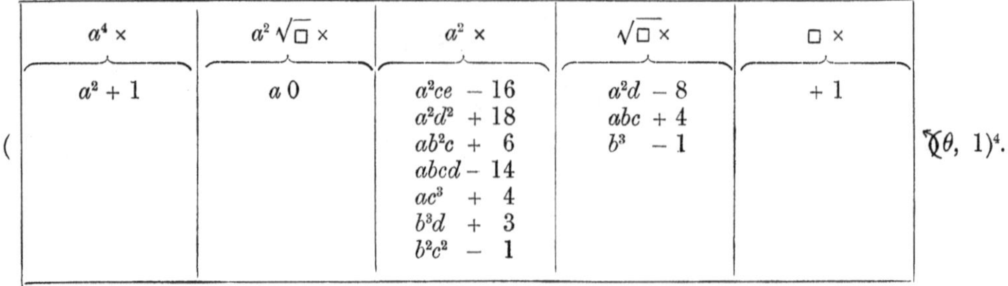

$a^4 \times$	$a^2\sqrt{\square}\times$	$a^2\times$	$\sqrt{\square}\times$	$\square\times$
$a^2 + 1$	$a\ 0$	$a^2ce\ -16$ $a^2d^2\ +18$ $ab^2c\ +\ 6$ $abcd-\ 14$ $ac^3\ +\ 4$ $b^3d\ +\ 3$ $b^2c^2\ -\ 1$	$a^2d\ -8$ $abc\ +4$ $b^3\ -1$	$+1$

$\between \theta, 1)^4$.

For the quintic equation $(a, b, c, d, e, f \between v, 1)^5 = 0$;

9. We have $\theta = \zeta^{\frac{1}{2}}(\alpha, \beta, \gamma, \delta)$, the roots being

$$\begin{array}{lll}
\theta_1 = \zeta^{\frac{1}{2}}(\beta, \gamma, \delta, \epsilon), & \text{which for } \epsilon = 0 \text{ becomes} & \beta\gamma\delta\zeta^{\frac{1}{2}}(\beta, \gamma, \delta), \\
\theta_2 = \zeta^{\frac{1}{2}}(\gamma, \delta, \epsilon, \alpha), & \quad ,, \qquad\quad ,, & -\gamma\delta\alpha\zeta^{\frac{1}{2}}(\gamma, \delta, \alpha), \\
\theta_3 = \zeta^{\frac{1}{2}}(\delta, \epsilon, \alpha, \beta), & \quad ,, \qquad\quad ,, & \delta\alpha\beta\zeta^{\frac{1}{2}}(\delta, \alpha, \beta), \\
\theta_4 = \zeta^{\frac{1}{2}}(\epsilon, \alpha, \beta, \gamma), & \quad ,, \qquad\quad ,, & -\alpha\beta\gamma\zeta^{\frac{1}{2}}(\alpha, \beta, \gamma), \\
\theta_5 = \zeta^{\frac{1}{2}}(\alpha, \beta, \gamma, \delta), & \quad ,, \qquad\quad ,, & \zeta^{\frac{1}{2}}(\alpha, \beta, \gamma, \delta).
\end{array}$$

10. The linear equation is $\theta a^3 + \sqrt{Z} = 0$; the quartic equation may be written $\Pi_4(\theta - \theta_1) = 0$, for the determination of which see Annex No. 2. The two equations are

$$\begin{array}{l|l}
\theta^4 . \; a^9 & \theta . a^3 \\
+\theta^3 - a^6\sqrt{Z} & +\sqrt{Z} \\
+\theta^2 . \; a^3 Me & \\
+\theta . \; Ne^2\sqrt{Z} & \\
+ \quad e^3 Z & \\
= 0 & = 0;
\end{array}$$

and multiplying these together, the resulting equation is

$$\begin{array}{l}
\theta^5 . a^{12} \\
+\theta^4 . 0 \\
+\theta^3 . a^6 (Me - Z) \\
+\theta^2 . a^3 (Ne + M) e\sqrt{Z} \\
+\theta . \quad (N + a^3 e) e^2 Z \\
+ \qquad e^3 Z\sqrt{Z} = 0,
\end{array}$$

where the coefficients have to be completed by the addition of the terms in f. We have $\sqrt{\Box}$ in the place of $e\sqrt{Z}$, and therefore $\Box$ in the place of e^2Z.

11. The value of $Me - Z$ is

a^3e^3	+ 96	− 256 =	− 160
a^2bde^2	− 60	+ 192 =	+ 132
$a^2c^2e^2$	− 40	+ 128 =	+ 88
a^2cd^2e	+ 27	− 144 =	− 177
a^2d^4		+ 27 =	+ 27
ab^2ce^2	+ 47	− 144 =	− 97
ab^2d^2e		+ 6 =	+ 6
abc^2de	− 18	+ 80 =	+ 62
$abcd^3$		− 18 =	− 18
ac^4e	+ 4	− 16 =	− 12
ac^3d^3		+ 4 =	+ 4
b^4e^2	− 9	+ 27 =	+ 18
b^3cde	+ 4	− 18 =	− 14
b^3d^3		+ 4 =	+ 4
b^2c^3e	− 1	+ 4 =	+ 3
$b^2c^2d^2$		− 1 =	− 1

and the terms in f are found to be

a^3def	$+300$
a^2bcef	-130
a^2bd^2f	-120
a^2c^2df	$+40$
ab^3ef	$+28$
ab^2cdf	$+66$
abc^3f	-24
b^4df	-16
b^3c^2f	$+6$
a^3cf^2	-125
$a^2b^2f^2$	$+50$

12. The value of $Ne+M$ is

a^3e^2	-16	$+96=$	$+80$
a^2bde	$+6$	$-60=$	-54
a^2c^2e	$+4$	$-40=$	-36
a^2cd^2		$+27=$	$+27$
ab^2ce	-5	$+47=$	$+42$
abc^2d		$-18=$	-18
ac^4		$+4=$	$+4$
b^4e	$+1$	$-9=$	-8
b^3cd		$+4=$	$+4$
b^2c^3		$-1=$	-1

and the terms in f to be added thereto are found to be

a^3df	-50
a^2bcf	$+30$
ab^3f	-8

13. The value of $N+a^3e$ is

a^3e	-15
a^2bd	$+6$
a^2c^2	$+4$
ab^2c	-5
b^4	$+1$

which is a seminvariant, and requires no addition.

14. Hence, for the quintic equation

$$(a,\ b,\ c,\ d,\ e,\ f \between v,\ 1)^5 = 0,$$

the equation for $\theta \left(= \zeta^{\frac{1}{2}} (\alpha,\ \beta,\ \gamma,\ \delta)\right)$ is $0 =$

(

$a^{12} \times$		$a^6 \times$	$a^3 \sqrt{\square} \times$	$\square \times$	$\square \sqrt{\square} \times$
$+1$	0	$a^3cf^2 \quad -125$	$a^3df \quad -50$	$a^3e \quad -15$	$+ \quad 1$
		$a^3def \quad +300$	$a^3e^2 \quad +80$	$a^2bd \quad +6$	
		$a^3e^3 \quad -160$	$a^2bcf \quad +30$	$a^2c^2 \quad +4$	
		$a^2b^2f^2 \quad +50$	$a^2bde \quad -54$	$ab^2c \quad -5$	
		$a^2bde^2 \quad +132$	$a^2c^2e \quad -36$	$b^4 \quad +1$	
		$a^2bcef \quad -130$	$a^2cd^2 \quad +27$		
		$a^2bd^2f \quad -120$	$ab^3f \quad -8$		
		$a^2c^2df \quad +40$	$ab^2ce \quad +42$		
		$a^2c^2e^2 \quad +88$	$abc^2d \quad -18$		
		$a^2cd^2e \quad -117$	$ac^4 \quad +4$		
		$a^2d^4 \quad +27$	$b^4e \quad -8$		
		$ab^3ef \quad +28$	$b^3cd \quad +4$		
		$ab^2ce^2 \quad -97$	$b^2c^3 \quad -1$		
		$ab^2d^2e \quad +6$			
		$abc^2de \quad +62$			
		$ab^2cdf \quad +66$			
		$abcd^3 \quad -18$			
		$abc^3f \quad -24$			
		$ac^4e \quad -12$			
		$ac^3d^2 \quad +4$			
		$b^4df \quad -16$			
		$b^4e^2 \quad +18$			
		$b^3cde \quad -14$			
		$b^3d^3 \quad +4$			
		$b^3c^2f \quad +6$			
		$b^2c^3e \quad +3$			
		$b^2c^2d^2 \quad -1$			

$\between \theta,\ 1)^5.$

15. As a verification of this result, I remark that, taking for the quintic equation $v^5 + v^4 + v^3 + v^2 + v + 1 = 0$, the roots of this equation are $-1,\ \omega,\ \omega^2,\ -\omega,\ -\omega^2$, where ω is an imaginary cube root of unity $(\omega^2 + \omega + 1 = 0)$. We ought to have $\zeta^{\frac{1}{2}} (\alpha,\ \beta,\ \gamma,\ \delta,\ \epsilon)$ $= -\sqrt{\square} = -36$; and this will be the case if, for instance, $\alpha,\ \beta,\ \gamma,\ \delta,\ \epsilon$ are respectively $-1,\ \omega,\ \omega^2,\ -\omega^2,\ -\omega$. We have then

$$\zeta^{\frac{1}{2}} (\alpha,\ \beta,\ \gamma,\ \delta) = -1-\omega \,.\, -1-\omega^2 \,.\, -1+\omega^2 \,.\, \omega-\omega^2 \,.\, \omega+\omega^2 \,.\, 2\omega^2 \quad = \quad 6,$$

$$\zeta^{\frac{1}{2}} (\beta,\ \gamma,\ \delta,\ \epsilon) = \quad \omega-\omega^2 \,.\, \omega+\omega^2 \,.\, 2\omega \,.\, 2\omega^2 \,.\, \omega^2+\omega \,.\, -\omega^2+\omega \quad = -12,$$

$$\zeta^{\frac{1}{2}} (\gamma,\ \delta,\ \epsilon,\ \alpha) = \quad 2\omega^2 \,.\, \omega^2+\omega \,.\, \omega^2+1 \,.\, -\omega^2+\omega \,.\, -\omega^2+1 \,.\, -\omega+1 \quad = +6\,(\omega-\omega^2),$$

$$\zeta^{\frac{1}{2}} (\delta,\ \epsilon,\ \alpha,\ \beta) = -\omega^2+\omega \,.\, -\omega^2+1 \,.\, -\omega^2-\omega \,.\, -\omega+1 \,.\, -2\omega \,.\, -1-\omega = -6\,(\omega-\omega^2),$$

$$\zeta^{\frac{1}{2}} (\epsilon,\ \alpha,\ \beta,\ \gamma) = -\omega+1 \,.\, -2\omega \,.\, -\omega-\omega^2 \,.\, -1-\omega \,.\, -1-\omega^2 \,.\, \omega-\omega^2 \quad = \quad 6.$$

The equation in θ is thus $(\theta-6)^2(\theta+12)(\theta^2+108)=0$, or multiplying out it is

$$(1,\ 0,\ 0,\ +432,\ -11664,\ +46656 \between \theta,\ 1)^4=0,$$

which in fact (observing that $\sqrt{\square}=36$) is what the preceding formula becomes for the equation $(1,\ 1,\ 1,\ 1,\ 1,\ 1 \between v,\ 1)^5=0$.

The analogous verifications for the cubic and the quartic equations are as follows:

16. For the cubic, if the assumed equation is $v^3+v^2+v+1=0$, the roots whereof are $-1,\ i,\ -i\ (i^2=-1)$, then we should have $\zeta^{\frac{1}{2}}(\alpha,\ \beta,\ \gamma)=\sqrt{-\square}=4i$, which will be the case if $\alpha,\ \beta,\ \gamma=-1,\ i,\ -i$, respectively, and the roots $\beta-\gamma,\ \gamma-\alpha,\ \alpha-\beta$ of the equation in θ then are $2i,\ -i+1,\ -i-1$, so that the equation in θ is $(\theta^2+2i\theta-2)(\theta-2i)=0$, or

$$(1,\ 0,\ 2,\ 4i \between \theta,\ 1)^3=0,$$

which (observing that $\sqrt{\square}=4i$) is what the formula for the equation in θ becomes for the equation $(1,\ 1,\ 1,\ 1 \between v,\ 1)^3=0$.

17. For the quartic equation, taking this to be $v^4+v^3+v^2+v+1=0$, the roots are $\omega,\ \omega^2,\ \omega^3,\ \omega^4$, where ω is an imaginary fifth root of unity $(\omega^4+\omega^3+\omega^2+\omega+1=0)$, and putting $\alpha,\ \beta,\ \gamma,\ \delta$ equal to $\omega,\ \omega^2,\ \omega^3,\ \omega^4$ respectively, we have

$$-\sqrt{\square}=\zeta^{\frac{1}{2}}(\alpha,\ \beta,\ \gamma,\ \delta)=-5(\omega+\omega^4-\omega^2-\omega^3),$$

giving, as it should do, $\square=125$. The equation in θ is therefore by the formula

$$(1,\ 0,\ 0,\ -25(\omega+\omega^4-\omega^2-\omega^3),\ 125 \between \theta,\ 1)^4=0.$$

But the roots are

$$\begin{aligned}
\theta_1 &= \zeta^{\frac{1}{2}}(\beta,\ \gamma,\ \delta)=-.\,\omega^2-\omega^3.\,\omega^2-\omega^4.\,\omega^3-\omega^4= 2-\omega+\omega^2-2\omega^3= 2-X,\\
\theta_2 &=-\zeta^{\frac{1}{2}}(\gamma,\ \delta,\ \alpha)=-.\,\omega^3-\omega^4.\,\omega^3-\omega\,.\,\omega^4-\omega =-1+3\omega+2\omega^2+\omega^3=-1+Y,\\
\theta_3 &= \zeta^{\frac{1}{2}}(\delta,\ \alpha,\ \beta)= .\,\omega^4-\omega\,.\,\omega^4-\omega^2.\,\omega-\omega^2=-4-3\omega-2\omega^2-\omega^3=-4-Y,\\
\theta_4 &=-\zeta^{\frac{1}{2}}(\alpha,\ \beta,\ \gamma)=-.\,\omega-\omega^2.\,\omega-\omega^3.\,\omega^2-\omega^3= 3+\omega-\omega^2+2\omega^3= 3+X,
\end{aligned}$$

if, for shortness,

$$X=\omega-\omega^2+2\omega^3,\quad Y=3\omega+2\omega^2+\omega^3.$$

The equation in θ is therefore

$$(\theta-2+X)(\theta+1-Y)(\theta+4+Y)(\theta-3-X)=0,$$

where the left-hand side is the product of the factors

$$(\theta-2+X)(\theta-3-X)=\theta^2-5\theta+6-X-X^2=\theta^2-5\theta+10-5(\omega+\omega^4)$$

and

$$(\theta+1-Y)(\theta+4+Y)=\theta^2+5\theta+4-3Y-Y^2=\theta^2+5\theta+10-5(\omega^2+\omega^3)$$

and the equation in θ is, therefore, as it should be,

$$(1,\ 0,\ 0,\ -25(\omega+\omega^4-\omega^2-\omega^3),\ 125 \between \theta,\ 1)^4=0.$$

Passing from the denumerate to the standard forms:

18. For the cubic equation $(a, b, c, d \between v, 1)^3 = 0$, the equation for $\theta \left(= \zeta^{\frac{1}{2}} (\alpha, \beta)\right)$ is $0 =$

$a^2 \times$		$9 \times$	$\sqrt{-27\square} \times$	
$+1$	0	$ac + 1$ $b^2 - 1$	$+1$	$\between \theta, 1)^3$.

19. For the quartic equation $(a, b, c, d, e \between v, 1)^4 = 0$, the equation for $\theta \left(= \zeta^{\frac{1}{2}} (\alpha, \beta, \gamma)\right)$ is $0 =$

$a^6 \times$		$96a^2 \times$	$512\sqrt{\square} \times$	$256\square \times$	
$+1$	0	$a^2ce \;\; - 1$ $a^2d^2 \;\; + 3$ $ab^2e \;\; + 1$ $abcd \;\; - 14$ $ac^3 \;\; + 9$ $b^3d \;\; + 8$ $b^2c^2 \;\; - 6$	$a^2d \;\; - 1$ $abc \;\; + 3$ $b^3 \;\; - 2$	$+1$	$\between \theta, 1)^4$.

20. For the quintic equation $(a, b, c, d, e, f \between v, 1)^5 = 0$, the equation for $\theta \left(= \zeta^{\frac{1}{2}} (\alpha, \beta, \gamma, \delta)\right)$ is $0 =$

$a^{12} \times$		$625\, a^6 \times$	$12500\sqrt{\square}a^3 \times$	$15625\square \times$	$76125\square\sqrt{\square}$	
$+1$	0	$a^3cf^2 \;\; - 2$ $a^3def \;\; + 24$ $a^3e^3 \;\; - 32$ $a^2b^2f^2 \;\; + 2$ $a^2bde^2 \;\; + 264$ $a^2bcef \;\; - 52$ $a^2bd^2f \;\; - 96$ $a^2c^2df \;\; + 64$ $a^2c^2e^2 \;\; + 352$ $a^2cd^2e \;\; - 936$ $a^2d^4 \;\; + 432$ $ab^3ef \;\; + 28$ $ab^2ce^2 \;\; - 970$ $ab^2d^2e \;\; + 120$ $ab^2cdf \;\; + 264$ $abc^2de \;\; + 2480$ $abcd^3 \;\; - 1440$ $abc^3f \;\; - 192$ $ac^4e \;\; - 960$ $ac^3d^2 \;\; + 640$ $b^4df \;\; - 160$ $b^4e^2 \;\; + 450$ $b^3cde \;\; 1400$ $b^0d^0 \;\; + 800$ $b^3c^2f \;\; + 120$ $b^2c^3e \;\; + 600$ $b^2c^2d^2 \;\; - 400$	$a^3df \;\; - 1$ $a^3e^2 \;\; + 4$ $a^2bcf \;\; + 3$ $a^2bde \;\; - 27$ $a^2c^2e \;\; - 36$ $a^2cd^2 \;\; + 54$ $ab^3f \;\; - 2$ $ab^2ce \;\; + 105$ $abc^2d \;\; - 180$ $ac^4 \;\; + 80$ $b^4e \;\; - 50$ $b^3cd \;\; + 100$ $b^2c^3 \;\; - 50$	$a^3e \;\; - 3$ $a^2bd \;\; + 12$ $a^2c^2 \;\; + 16$ $ab^2c \;\; - 50$ $b^4 \;\; + 25$	$+1$	$\between \theta, 1)^5$.

21. I remark, with respect to the equation in θ, for the cubic, that it leads at once to the equation of differences. In fact we have

$$a^2\theta^3 + 9(ac - b^2)\theta + \sqrt{-27\square} = \Pi_3\{\theta - (\alpha - \beta)\},$$

whence changing the sign of θ,

$$a^2\theta^3 + 9(ac - b^2)\theta - \sqrt{-27\square} = \Pi_3\{\theta + (\alpha - \beta)\};$$

or multiplying the two equations and putting u for θ^2,

$$u\{a^2u + 9(ac - b^2)\}^2 + 27\square = \Pi_3\{u - (\alpha - \beta)^2\},$$

that is, the equation of differences is

$$a^4u^3 + 18(ac - b^2)a^2u^2 + 81(ac - b^2)^2u + 27\square = 0;$$

but this mode of composition is peculiar to the case of the cubic.

If in the several equations in θ we substitute for the seminvariants the covariants to which they respectively belong, we obtain as follows:

22. For the cubic equation $(a, b, c, d \between v, 1)^3 = 0$, the equation for $(\vartheta = (\beta - \gamma)(x - \alpha y))$ is

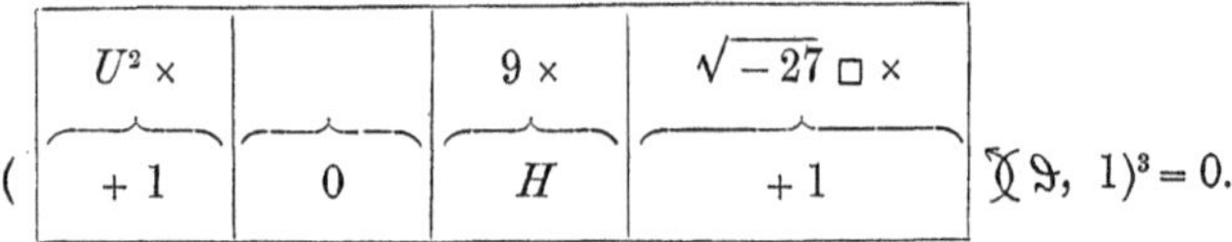

23. For the quartic equation $(a, b, c, d, e \between v, 1)^4 = 0$, the equation for

$\vartheta\,(= (\beta - \gamma)(\gamma - \delta)(\delta - \beta)(x - \alpha y)^2)$ is

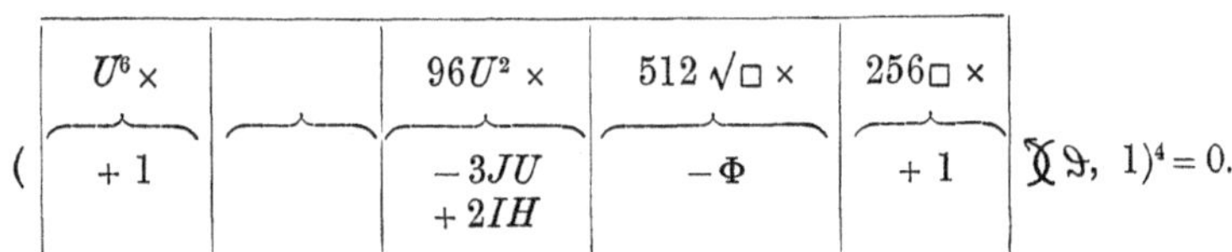

24. And for the quintic equation $(a, b, c, d, e, f \between v, 1)^5 = 0$, [denoting the covariants of the quintic as in 141, A the quintic itself, &c.; and completing the expression for the coefficient of ϑ^3] the equation for

$\vartheta\,(= (\beta - \gamma)(\beta - \delta)(\beta - \epsilon)(\gamma - \delta)(\gamma - \epsilon)(\delta - \epsilon)(x - \alpha y)^3)$ is

$$\left(\begin{array}{l} A^{12}, \\ 0\ , \\ 625A^6(48AJ - 80B^3 + 120BH - 50CG, \\ 12500\sqrt{\square}\ A^3\{4AB^2 - AH - 50CD\}, \\ 15625\ \ \square\ \ \{3A^2B + 25C^2\}, \\ 76125\ \ \square\sqrt{\square} \end{array}\right\between \vartheta, 1)^5 = 0,$$

where the covariant which enters into the coefficient of ϑ^3 being of the sixth degree in the coefficients, is not given in the Tables.

Its value (completed for me, from the first term, by Mr Davis) is

$a^3cf^2 - 2$	$a^3df^2 + 18$	$a^3ef^2 - 30$	$a^3f^3 - 10$	$a^2bf^3 - 30$	$a^2cf^3 + 18$	$a^2df^3 - 2$
$a^3def + 24$	$a^3e^2f - 48$	$a^2bdf^2 + 210$	$a^2bef^2 + 30$	$a^2cef^2 + 210$	$a^2def^2 - 66$	$a^2e^2f^2 + 2$
$a^3e^3 - 32$	$a^2bcf^2 - 66$	$a^2be^2f + 60$	$a^2cdf^2 + 680$	$a^2d^2f^2 + 180$	$a^2e^3f + 48$	$abcf^3 + 24$
$a^2b^2f^2 + 2$	$a^2bdef + 348$	$a^2c^2f^2 + 180$	$a^2ce^2f - 320$	$a^2de^2f - 840$	$ab^2f^3 - 48$	$abdef^2 - 52$
$a^2bde^2 + 264$	$a^2be^3 + 48$	$a^2cdef + 120$	$a^2d^2ef - 1320$	$a^2e^4 + 480$	$abcef^2 + 348$	$abe^3f + 28$
$a^2bcef - 52$	$a^2c^2ef + 624$	$a^2ce^3 - 480$	$a^2de^3 + 960$	$ab^2ef^2 + 60$	$abd^2f^2 + 624$	$ac^2ef^2 - 96$
$a^2bd^2f - 96$	$a^2cd^2f - 936$	$a^2d^3f - 1080$	$ab^2df^2 - 320$	$abcdf^2 + 120$	$abe^4 + 720$	$acd^2f^2 + 64$
$a^2c^2df + 64$	$a^2cde^2 - 576$	$a^2d^2e^2 + 1080$	$ab^2e^2f + 290$	$abce^2f - 540$	$abde^2f - 1596$	$acde^2f + 264$
$a^2c^2e^2 + 352$	$a^2d^3e + 648$	$ab^2cf^2 - 840$	$abc^2f^2 - 1320$	$abd^2ef + 960$	$ac^2df^2 - 936$	$ace^4 - 160$
$a^2cd^2e - 936$	$ab^3f^2 + 48$	$ab^2def - 540$	$abcdef + 4160$	$abde^3 - 600$	$ac^2e^2f - 48$	$ad^3ef - 192$
$a^2d^4 + 432$	$ab^2cef - 1596$	$ab^2e^3 + 450$	$abce^3 - 3200$	$ac^3f^2 - 1080$	$acd^2ef + 3504$	$ad^2e^3 + 120$
$ab^3ef + 28$	$ab^2de^2 + 210$	$abc^2ef + 960$	$abd^3f + 960$	$ac^2def + 4560$	$ad^4f - 1728$	$b^3f^3 - 32$
$ab^2ce^2 - 970$	$ab^2d^2f - 48$	$abcd^2f + 4560$	$abd^2e^2 - 600$	$ac^2e^3 - 2400$	$acde^3 - 1920$	$b^2cef^2 + 264$
$ab^2d^2e + 120$	$abc^2df + 3504$	$abcde^2 - 4200$	$ac^3ef + 960$	$acd^3f - 2880$	$ad^3e^2 + 1080$	$b^2d^2f^2 + 352$
$ab^2cdf + 264$	$abc^2e^2 + 720$	$ac^3df - 2880$	$ac^2d^2f - 2560$	$acd^2e^2 + 1800$	$b^3ef^2 + 48$	$b^2de^2f - 970$
$abc^2de + 2480$	$abcd^2e - 2160$	$ac^3e^2 + 2400$	$ac^2de^2 + 1600$	$b^3df^2 - 480$	$b^2cdf^2 - 576$	$b^2e^4 + 450$
$abc^3f - 192$	$ac^4f - 1728$	$b^4f^2 + 480$	$bc^3f^2 + 960$	$b^3e^2f + 450$	$b^2ce^2f + 210$	$bc^2df^2 - 936$
$abcd^3 - 1440$	$ac^3de + 960$	$b^3cef - 600$	$b^3def - 3200$	$b^2c^2f^2 + 1080$	$b^2d^2ef + 720$	$bcd^2ef + 2480$
$ac^4e - 960$	$b^4ef + 720$	$b^3d^2f - 2400$	$b^3e^3 + 2250$	$b^2cdef - 4200$	$b^2de^3 - 450$	$bc^2e^2f + 120$
$ac^3d^2 + 640$	$b^3cdf - 1920$	$b^3de^2 + 2250$	$b^2c^2ef - 600$	$b^2ce^3 + 2250$	$bc^3f^2 + 648$	$bcde^3 - 1400$
$b^4df - 160$	$b^3ce^2 - 450$	$b^2c^2df + 1800$	$b^2cd^2f + 1600$	$b^2d^3f + 2400$	$bc^2def - 2160$	$bd^4f - 960$
$b^4e^2 + 450$	$b^3d^2e + 1200$	$b^2c^2e^2 - 1500$	$b^2cde^2 - 1000$	$b^2d^2e^2 - 1500$	$bc^2e^3 + 1200$	$bd^3e^2 + 600$
$b^3cde - 1400$	$b^2c^3f + 1080$				$bcd^3f + 960$	$c^4f^2 + 432$
$b^3c^2f + 120$	$b^2c^2de - 600$				$bcd^2e^2 - 600$	$c^3def - 1440$
$b^3d^3 + 800$						$c^3e^3 + 800$
$b^2c^3e + 600$						$c^2d^3f + 640$
$b^2c^2d^2 - 400$						$c^2d^2e^2 - 400$

$\big)(x, y)^6.$

[viz. this is $= 48AJ - 80B^3 + 120BH - 50CG$ as above].

In the following two Annexes, the notation of the symmetric functions is the same as in my "Memoir on the Symmetric Functions of the Roots of an Equation," *Phil. Trans.* vol. CXLVII. (1857), [147] and the values of the symmetric functions are taken from that memoir, the powers of a being restored by the principle of homogeneity. The suffixes of the Σ indicate the number of terms in the sum; thus in the first Annex

$$\Sigma_3 \gamma (\beta - \gamma)(\gamma - \alpha) = \Sigma_3 (\beta\gamma^2 - \alpha\beta\gamma - \gamma^3 + \alpha\gamma^2);$$

the terms $\Sigma_3 (\beta\gamma^2 + \alpha\gamma^2)$ are equal to $\Sigma_6 \alpha^2\beta$, the complete symmetric function; the correct result will be obtained (though of course neither of these equations is true) by writing $\Sigma_3 \beta\gamma^2 = \frac{1}{2}\Sigma_6 \alpha^2\beta$, $\Sigma_3 \alpha\gamma^2 = \frac{1}{2}\Sigma_6 \alpha^2\beta$, and so in similar cases; the insertion of the suffix to the Σ very much facilitates the calculation, and is a check on its accuracy.

Annex No. 1, containing the calculation of the equation $\Pi_3 (\theta - \theta_1) = 0$, where

$$\theta_1 = \beta\gamma (\beta - \gamma), \ \theta_2 = \gamma\alpha (\gamma - \alpha), \ \theta_3 = \alpha\beta (\alpha - \beta),$$

α, β, γ being the roots of the cubic equation $(a, b, c, d \between v, 1)^3 = 0$.

We have

$$\Sigma_3 \theta_1 = \Sigma \beta\gamma (\beta - \gamma) = -(\alpha - \beta)(\beta - \gamma)(\gamma - \alpha) = \zeta^{\frac{1}{2}} (\alpha, \beta, \gamma) = a^{-2}\sqrt{-Z},$$

where $Z = 27a^2d^2 + \&c.$ is the discriminant of the cubic.

$$\Sigma_3\theta_1\theta_2 = \Sigma_3\beta\gamma(\beta-\gamma)\gamma\alpha(\gamma-\alpha) = \alpha\beta\gamma\Sigma_3\gamma(\beta-\gamma)(\gamma-\alpha),$$

where $\alpha\beta\gamma = -a^{-1}d$ and

$$\Sigma_3\gamma(\beta-\gamma)(\gamma-\alpha) = \Sigma_3(\beta\gamma^2 - \alpha\beta\gamma - \gamma^3 + \alpha\gamma^3)$$

$$\begin{array}{lll|l} = -\Sigma_3\alpha^3 & = - & (3) = a^{-2} & 3a^2d - 3abc + 1b^3 \\ + \Sigma_6\alpha^2\beta & + & (21) & +3a^2d - 1abc \\ -3\alpha\beta\gamma & & -3(13) & +3a^2d \end{array}$$

$$= a^{-2}(9a^2d - 4abc + 1b^3),$$

and therefore

$$\Sigma_3\theta_1\theta_2 = -a^{-3}(9a^2d^2 - 4abcd + 1b^3d).$$

And lastly,

$$\Sigma_1\theta_1\theta_2\theta_3, \text{ or } \theta_1\theta_2\theta_3 = \alpha^2\beta^2\gamma^2(\alpha-\beta)(\beta-\gamma)(\gamma-\alpha)$$

$$= -\alpha^2\beta^2\gamma^2\zeta^{\frac{1}{2}}(\alpha,\ \beta,\ \gamma)$$

$$= -a^{-4}d^2\sqrt{-Z}$$

so that the equation is the one given above, No. 7.

Annex No. 2, containing the calculation of the equation $\Pi_4(\theta-\theta_1)=0$, where

$\theta_1 = \beta\gamma\delta\zeta^{\frac{1}{2}}(\beta,\ \gamma,\ \delta)$, $\theta_2 = -\gamma\delta\alpha\zeta^{\frac{1}{2}}(\gamma,\ \delta,\ \alpha)$, $\theta_3 = \delta\alpha\beta\zeta^{\frac{1}{2}}(\delta,\ \alpha,\ \beta)$, and $\theta_4 = -\alpha\beta\gamma\zeta^{\frac{1}{2}}(\alpha,\ \beta,\ \gamma)$,

α, β, γ, δ being the roots of the quartic equation $(a,\ b,\ c,\ d,\ e \between v,\ 1)^4 = 0$.

$$\Sigma_4\theta_1 = (^1)\,\Sigma_4\beta\gamma\delta\zeta^{\frac{1}{2}}(\beta,\ \gamma,\ \delta) = -(\alpha-\beta)(\alpha-\gamma)(\alpha-\delta)(\beta-\gamma)(\beta-\delta)(\gamma-\delta)$$

$$= -\zeta^{\frac{1}{2}}(\alpha,\ \beta,\ \gamma,\ \delta)$$

$$= -a^{-3}$$

where $Z = 256a^3e^3 + \&c.$ is the discriminant of the quartic.

$$\Sigma_6\theta_1\theta_2 = \Sigma_6\theta_3\theta_4 = \Sigma_6\delta\alpha\beta\,\zeta^{\frac{1}{2}}(\delta,\ \alpha,\ \beta) \times -\alpha\beta\gamma\,\zeta^{\frac{1}{2}}(\alpha,\ \beta,\ \gamma)$$

$$= \Sigma_6\delta\alpha\beta(\delta-\alpha)(\delta-\beta)(\alpha-\beta) \times -\alpha\beta\gamma(\alpha-\beta)(\alpha-\gamma)(\beta-\gamma)$$

$$= -\alpha\beta\gamma\delta\,\Sigma_6\alpha\beta(\alpha-\beta)^2(\alpha-\gamma)(\alpha-\delta)(\beta-\gamma)(\beta-\delta),$$

where $\alpha\beta\gamma\delta = -a^{-1}e$, and

[1] The signs of θ_1, θ_2, θ_3, θ_4 are taken account of implicitly.

$$\Sigma_6\alpha\beta(\alpha-\beta)^2(\alpha-\gamma)(\alpha-\delta)(\beta-\gamma)(\beta-\delta)$$

$$=\Sigma_6\alpha\beta(\alpha-\beta)^2\{\alpha^2\beta^2-\alpha\beta(\alpha+\beta)(\gamma+\delta)+(\alpha^2+\beta^2)\gamma\delta+\alpha\beta(\gamma+\delta)^2-\gamma\delta(\alpha+\beta)(\gamma+\delta)+\gamma^2\delta^2\}$$

$$\left.\begin{array}{ll}
=\Sigma_6\alpha^3\beta^3(\alpha-\beta)^2,\text{ viz.} & \Sigma_{12}\alpha^5\beta^3\\
& -2\Sigma_6\alpha^4\beta^4\\
-\Sigma_6\alpha^2\beta^2(\alpha-\beta)^2(\alpha+\beta)(\gamma+\delta),\text{ viz.} & -\Sigma_{24}\alpha^5\beta^2\gamma\\
& +\Sigma_{24}\alpha^4\beta^3\gamma\\
+\alpha\beta\gamma\delta\Sigma_6(\alpha-\beta)^2(\alpha^2+\beta^2),\text{ viz} & +\frac{e}{a}3\Sigma_4\alpha^4\\
& -2\frac{e}{a}\Sigma_{12}\alpha^3\beta\\
& +2\frac{e}{a}\Sigma_6\alpha^2\beta^2\\
+\Sigma_6\alpha^2\beta^2(\alpha-\beta)^2(\gamma+\delta)^2,\text{ viz.} & +2\Sigma_{12}\alpha^4\beta^2\gamma^2\\
& -2\Sigma_{12}\alpha^3\beta^3\gamma^2\\
& +2\frac{e}{a}\Sigma_{12}\alpha^3\beta\\
& -4\frac{e}{a}\Sigma_6\alpha^2\beta^2\\
-\alpha\beta\gamma\delta\Sigma_6(\alpha-\beta)^2(\alpha+\beta)(\gamma+\delta),\text{ viz.} & -2\frac{e}{a}\Sigma_{12}\alpha^3\beta\\
& +2\frac{e}{a}\Sigma_{12}\alpha^2\beta\gamma\\
+\alpha\beta\gamma\delta\Sigma_6(\alpha-\beta)^2\gamma\delta,\text{ viz.} & +\frac{e}{a}\Sigma_{12}\alpha^2\beta\gamma\\
& -12\frac{e^2}{a^2}
\end{array}\right\}
=\left.\begin{array}{l}
\Sigma_{12}\alpha^5\beta^3\\
-2\Sigma_6\alpha^4\beta^4\\
-\Sigma_{24}\alpha^2\beta^2\gamma\\
+\Sigma_{24}\alpha^4\beta^3\gamma\\
+2\Sigma_{12}\alpha^4\beta^2\gamma^2\\
-2\Sigma_{12}\alpha^3\beta^3\gamma^2\\
+\frac{e}{a}3\Sigma_4\alpha^4\\
-2\frac{e}{a}\Sigma_{12}\alpha^3\beta\\
-2\frac{e}{a}\Sigma_6\alpha^2\beta^2\\
+3\frac{e}{a}\Sigma_{12}\alpha^2\beta\gamma\\
-12\frac{e^2}{a^2}
\end{array}\right\}
=\begin{array}{l}
(53)\\
-2(4^2)\\
-1(521)\\
+1(431)\\
+2(42^2)\\
-2(3^22)\\
+3e(4)\\
-2e(31)\\
-2e(2^2)\\
+3e(21^2)\\
-12e^2,
\end{array}$$

where for a moment a is put equal to unity.

The value of the last-mentioned expression is then calculated as follows:

e^2	−4	−12	−8	−8	−8	−8	−12	−8	−4	−12	−12	=	−96
bde	+1	+16	+10	+10		+2	+12	+2	+4	+3		=	+60
c^2e	+8	+8	+4		+8	+4	+6	+4	−2			=	+40
b^2ce	−9	−8	−11	−1	−4		−12	−2				=	−47
b^4e	+3		+3				+3					=	+9
cd^2	−7	−8	−5	−1	−4	−2						=	−27
b^2d^2	+3	−4	+1	−2	+2							=	0
bc^2d	+6	+8	+3	+1								=	+18
b^3cd	−3		−1									=	−4
c^4	−2	−2										=	−4
b^2c^2	+1											=	+1

and restoring the powers of a by the principle of homogeneity, and putting

$$M = \begin{array}{|ll|} a^3e^2 & +96 \\ a^2bde & -60 \\ a^2c^2e & -40 \\ a^2cd^2 & +27 \\ ab^2ce & +47 \\ abc^2d & -18 \\ ac^4 & +\ 4 \\ b^4e & -\ 9 \\ b^3cd & +\ 4 \\ b^2c^3 & -\ 1 \end{array}$$

we have

$$\Sigma\theta_1\theta_2 = a^{-6}Me.$$

Next,

$$\begin{aligned}\Sigma_4\theta_1\theta_2\theta_3 &= \Sigma_4\beta\gamma\delta\,\zeta^{\frac{1}{2}}(\beta,\ \gamma,\ \delta) \times -\gamma\delta\alpha\,\zeta^{\frac{1}{2}}(\gamma,\ \delta,\ \alpha) \times \delta\alpha\beta\,\zeta^{\frac{1}{2}}(\delta,\ \alpha,\ \beta) \\ &= \Sigma_4\beta\gamma\delta(\beta-\gamma)(\beta-\delta)(\gamma-\delta) \times -\gamma\delta\alpha(\gamma-\delta)(\gamma-\alpha)(\delta-\alpha) \times \delta\alpha\beta(\delta-\alpha)(\delta-\beta)(\alpha-\beta) \\ &= \alpha^2\beta^2\gamma^2\delta^2(\alpha-\beta)(\alpha-\gamma)(\alpha-\delta)(\beta-\gamma)(\beta-\delta)(\gamma-\delta)\,\Sigma_4\alpha(\alpha-\beta)(\alpha-\gamma)(\alpha-\delta) \\ &= \alpha^2\beta^2\gamma^2\delta^2\,\zeta^{\frac{1}{2}}(\alpha,\ \beta,\ \gamma,\ \delta)\,\Sigma_4\alpha(\alpha-\beta)(\alpha-\gamma)(\alpha-\delta) \\ &= -a^{-5}e^2\sqrt{Z}\,\Sigma_4 a(\alpha-\beta)(\alpha-\gamma)(\alpha-\delta),\end{aligned}$$

and observing that

$$(a,\ b,\ c,\ d,\ e \between v,\ 1)^4 = a(v-\alpha)(v-\beta)(v-\gamma)(v-\delta),$$

and therefore

$$4av^3 + bv^2 + 2cv + d = a(v-\beta)(v-\gamma)(v-\delta) + \&c.,$$

which, putting $v=\alpha$, gives

$$4a\alpha^3 + 3b\alpha^2 + 2c\alpha + d = a(\alpha-\beta)(\alpha-\gamma)(\alpha-\delta),$$

we have

$$\begin{aligned}\Sigma_4\alpha(\alpha-\beta)(\alpha-\gamma)(\alpha-\delta) &= \frac{1}{a}(4a\Sigma\alpha^4 + 3b\Sigma\alpha^3 + 2c\Sigma\alpha^2 + d\Sigma\alpha) \\ &= 4\ (4) + 3b\,(3) + 2c\,(2) + d\,(1),\end{aligned}$$

where for a moment a is put equal to 1. This is calculated by

e	-16				=	-16
bd	$+16$	-9		-1	=	$+\ 6$
c^2	$+\ 8$		-4		=	$+\ 4$
b^2c	-16	$+9$	$+2$		=	$-\ 5$
b^4	$+\ 4$	-3			=	$+\ 1$

or restoring the powers of a, and putting

$$N = \begin{array}{|ll|} \hline a^3e & -16 \\ a^2bd & +\ \ 6 \\ a^2c^2 & +\ \ 4 \\ ab^2c & -\ \ 5 \\ b^4 & +\ \ 1 \\ \hline \end{array}$$

we have

$$\Sigma_4\theta_1\theta_2\theta_3 = a^{-9}Ne^2\sqrt{Z}.$$

Lastly,

$$\begin{aligned} \Sigma_1\theta_1\theta_2\theta_3\theta_4 \text{ or } \theta_1\theta_2\theta_3\theta_4 &= \alpha^2\beta^2\gamma^2\delta^2\,\zeta^{\frac{1}{2}}(\beta,\ \gamma,\ \delta)\,\zeta^{\frac{1}{2}}(\gamma,\ \delta,\ \alpha)\,\zeta^{\frac{1}{2}}(\delta,\ \alpha,\ \beta)\,\zeta^{\frac{1}{2}}(\alpha,\ \beta,\ \gamma) \\ &= \alpha^2\beta^2\gamma^2\delta^2\,\zeta(\alpha,\ \beta,\ \gamma,\ \delta) \\ &= a^{-9}e^3Z, \end{aligned}$$

and the equation $\Pi_4(\theta-\theta_1)=0$ is thus found to be the one given above, No. 10.

267.

ON THE PORISM OF THE IN-AND-CIRCUMSCRIBED POLYGON.

[From the *Philosophical Transactions of the Royal Society of London,* vol. CLI. (for the year 1861), pp. 225—239. Received February 20,—Read March 7, 1861.]

THE Porism referred to is as follows, viz. two conics may be so related to each other, that a polygon may be inscribed in the one, and circumscribed about the other conic, in such manner that any point whatever of the circumscribing conic may be taken for a vertex of the polygon. I gave in the year 1853, in the Philosophical Magazine(1), a general formula for the relation between the two conics, viz. if $U = 0$ is the equation of the inscribed conic, $V=0$ that of the circumscribed conic, and if disct. $(U+\xi V)$, where ξ is an arbitrary multiplier, denotes the discriminant of $U+\xi V$ in regard to the coordinates (x, y, z) (such discriminant being of course a cubic function in regard to ξ, and also in regard to the coefficients of the two conics, U, V, jointly), then if we write

$$\sqrt{\text{disct. } (U+\xi V)} = A + B\xi + C\xi^2 + D\xi^3 + E\xi^4 + F\xi^5 + G\xi^6 + \text{ \&c.},$$

the relations for the cases of the triangle, pentagon, heptagon, &c. are

$$C = 0, \quad \begin{vmatrix} C, & D \\ D, & E \end{vmatrix} = 0, \quad \begin{vmatrix} C, & D, & E \\ D, & E, & F \\ E, & F, & G \end{vmatrix} = 0, \text{ \&c.}$$

1 See the papers—"On the Geometrical Representation of the Integral $\int dx \div \sqrt{(x+a)(x+b)(x+c)}$," *Phil. Mag.*, April 1853, [113].

"Note on the Porism of the in-and-circumscribed Polygon," *Phil. Mag.* August 1853, [115].

"Correction of two Theorems relating to the Porism of the in-and-circumscribed Polygon," *Phil. Mag.* November 1853, [116].

"Developments of the Porism of the in-and-circumscribed Polygon," *Phil. Mag.* May 1854, [128].

respectively, while those in the cases of the quadrangle, hexagon, octagon, &c. are

$$D=0,\quad \begin{vmatrix} D, & E \\ E, & F \end{vmatrix}=0,\quad \begin{vmatrix} D, & E, & F \\ E, & F, & G \\ F, & G, & H \end{vmatrix}=0, \text{ \&c.}$$

respectively. The demonstration of this fundamental theorem is for greater completeness here reproduced; but the chief object of the memoir is to direct attention to a curious analytical theorem which is an easy *à priori* consequence of the Porism, and to obtain the relations for the several polygons up to the enneagon, in a new and simple form which puts in evidence *à posteriori* for these cases the analytical theorem just referred to. The analytical theorem rests upon the following considerations:—the relation for a hexagon ought to include that for a triangle; in fact a triangle with its sides in order twice over is a form of hexagon; the condition for an octagon should in like manner include that for a quadrangle; and so in other cases. Let the cubic function, disct. $(U+\xi V)$, be represented by $1+\beta\xi+\gamma\xi^2+\delta\xi^3$, the coefficients A, B, C, D, E, &c. are functions of β, γ, δ. Write

$$C = (3),$$

$$D = (4),$$

$$\begin{vmatrix} C, & D \\ D, & E \end{vmatrix} = (5),$$

$$\begin{vmatrix} D, & E \\ E, & F \end{vmatrix} = (6),$$

$$\begin{vmatrix} C, & D, & E \\ D, & E, & F \\ E, & F, & G \end{vmatrix} = (7),$$

&c.

then (3), (4), (5) are respectively prime functions of β, γ, δ; that is they cannot be decomposed into factors, rational functions of these quantities; and it is convenient to denote this by writing (3) = [3], (4) = [4], (5) = [5]. But by what precedes, (6) contains the factor (3), that is [3]; and if the other factor, which is prime, is denoted by [6], then we have (6) = [6] [3]. The next term (7) is prime, that is we have (7) = [7]; but the term (8) gives (8) = [8] [4]; the term (9) gives (9) = [9] [3], and so on. Thus we have (12) = [12] [6] [4] [3], the numbers in [ ] being all the factors, the number itself included, and as well composite as prime, of the number in ( ), the factors 2 and 1 being however excluded. To make this clearer, it may be remarked that the last-mentioned equation has the geometrical signification that the relation for a dodecagon is the aggregate of the relations for a proper dodecagon, a proper hexagon, a quadrangle, and a triangle; that is, the relation for a dodecagon implies one or other of the last-mentioned relations. The relations for the several polygons up to the enneagon are in the memoir obtained in a form which puts in evidence the property in question, that is, the series of equations

$$(3) = [3],$$
$$(4) = [4],$$
$$(5) = [5],$$
$$(6) = [6]\,[3],$$
$$(7) = [7],$$
$$(8) = [8]\,[4],$$
$$(9) = [9]\,[3].$$

To do this, the discriminant is represented, not as above in terms of the constants β, γ, δ, but in a somewhat different form, by means of the constants $b\,[=\beta]$, c, d, the last two whereof are such that $c=0$ is the relation for the triangle, $d=0$ the relation for the quadrangle; thus $[3]=c$, $[4]=d$, and for the particular cases considered, the analytical theorem consists herein, that c is a factor of (6), and of (9), and that d is a factor of (8). I have, for the sake of homogeneity, introduced into the formulæ the quantity $a\,(=1)$, but this is a matter of form only.

The functions [3], [4], &c. have been spoken of as *prime;* they are so, in fact, so far as they are calculated; and that they are so in general rests on the assumption that for a polygon of a given number of sides, there is but one form of relation: if, for instance, in the equation $[12]=0$, which is the condition for a proper dodecagon, the function [12] could be decomposed into rational factors, then equating each of these factors to zero, we should have so many distinct forms of relation for a proper dodecagon. I believe that the assumption and reasoning are valid; but without entering further into this, I take it for granted that in the general case the functions [3], [4], &c. are in fact prime. But the coefficients β, γ, δ, or b, c, d instead of being so many independent arbitrary quantities, may be given as rational functions of other quantities (if, for instance, the two conics are circles, radii R, r, and distance between the centres a, then β, γ, δ will be functions of R, r, a): and it is in a case of this kind quite conceivable that the functions [3], [4], &c., considered as functions of these new elements, should cease to be prime functions. In fact, in the case just referred to of the two circles (the original case of the Porism as considered by Füss), the functions [4], [6], &c., which correspond to a polygon of an *even* number of sides, appear to be each of them decomposable into two factors: the memoir contains some remarks tending to show *à priori* that in the case in question this decomposition takes place. I was led to examine the point by the elegant formulæ obtained in an essentially different manner by M. Mention, *Bull. de l'Acad. de St Pét.* t. I. pp. 15, 30, and 507 (1860), in reference to the case of the two circles (it thereby appears that the decomposition takes place for the quadrangle and the hexagon); and these formulae are reproduced in the memoir.

I.

Demonstration of the general Formula of the Relation between the two Conics.

The equation of a conic passing through the points of intersection of the conics

$$U=0,\quad V=0,$$

is of the form

$$U+mV=0,$$

where m is an arbitrary parameter. Suppose that the conic touches a given line, we have for the determination of m a quadratic equation; and conversely, if the roots of this quadratic equation are given, the line is also given; that is, the two roots may be considered as parameters which determine the particular line.

Let k be a given value of m; the parameters of any tangent of the conic $U+kV=0$ are k, p, but as k is common to all the tangents, we may consider the particular tangent as determined by the single parameter p. And a point of the conic $U+kV=0$ may be considered as determined by the same parameter p which determines the tangent at that point.

As regards the conic $V=0$, the common parameter for all its tangents is ∞, and we may consider any other tangent of this conic as determined by the parameter θ, and a point of the conic as determined by the same parameter θ.

Suppose, in the first instance, that the two conics are

$$U=ax^2+by^2+cz^2=0,$$
$$V=\ x^2+\ y^2+\ z^2=0;$$

the equation of the tangent of $U+kV=0$, the parameter whereof is p (in fact a common tangent of the conics $U+kV=0$, $U+pV=0$), is easily found to be

$$x\sqrt{b-c}\sqrt{a+k}\sqrt{a+p}+y\sqrt{c-a}\sqrt{b+k}\sqrt{b+p}+z\sqrt{a-b}\sqrt{c+k}\sqrt{c+p}=0;$$

and if this meet the conic $V=0$ in the points P, P', the parameters whereof are ∞, θ, and ∞, θ', or say θ and θ' respectively, then the coordinates of the point P are given by

$$x:y:z=\sqrt{b-c}\sqrt{a+\theta}:\sqrt{c-a}\sqrt{b+\theta}:\sqrt{a-b}\sqrt{c+\theta};$$

and substituting these values in the foregoing equation, we have

$$(b-c)\sqrt{a+k}\sqrt{a+p}\sqrt{a+\theta}+(c-a)\sqrt{b+k}\sqrt{b+p}\sqrt{b+\theta}+(a-b)\sqrt{c+k}\sqrt{c+p}\sqrt{c+\theta}=0$$

as an equation connecting the parameters p and θ. This equation may be replaced by

$$\sqrt{(a+k)(a+p)(a+\theta)}=\lambda+\mu a,$$
$$\sqrt{(b+k)(b+p)(b+\theta)}=\lambda+\mu b,$$
$$\sqrt{(c+k)(c+p)(c+\theta)}=\lambda+\mu c,$$

from which λ, μ are to be eliminated; and squaring and reducing, we have

$$\lambda^2 = abc + kp\theta,$$
$$-2\lambda\mu = bc + ca + ab - (p\theta + kp - k\theta),$$
$$\mu^2 = a + b + c + k + p + \theta,$$

and thence

$$(bc + ca + ab - p\theta - kp - k\theta)^2 - 4(a + b + c + k + p + \theta)(abc + kp\theta) = 0$$

as the rational form of the original equation. But the same rational equation would, it is clear, be obtained from the system

$$\sqrt{(k + a)(k + b)(k + c)} = L + Mk,$$
$$\sqrt{(p + a)(p + b)(p + c)} = L + Mp,$$
$$\sqrt{(\theta + a)(\theta + b)(\theta + c)} = L + M\theta,$$

by the elimination of L and M. And it follows from Abel's theorem (but the result might be verified by means of Euler's fundamental theorem for the addition of elliptic functions), that if

$$\Pi\xi = \int_\infty \frac{d\xi}{\sqrt{(\xi + a)(\xi + b)(\xi + c)}},$$

then the last-mentioned system is equivalent to the transcendental equation

$$\Pi\theta = \Pi p \pm \Pi k,$$

in which the arbitrary constant which should have been inserted, and the sign of $\Pi\theta$, are determined by the consideration that for $k = \infty$ (which gives $\Pi k = 0$) we ought to have $\theta = p$, and therefore $\Pi\theta = \Pi p$.

There is of course a similar equation in θ', and the terms with Πk must be taken with opposite signs, and we have thence the theorem:

"If θ, θ' are the parameters of the points P, P' in which the conic $V = 0$ is intersected by the tangent, the parameter whereof is p, of the conic $U + kV = 0$, then the equations

$$\Pi\theta = \Pi p - \Pi k,$$
$$\Pi\theta' = \Pi p + \Pi k,$$

determine the parameters θ, θ' of the points in question." And again:

"If the two variable parameters θ, θ' are connected by the equation

$$\Pi\theta' - \Pi\theta = 2\Pi k,$$

then the line PP' will be a tangent of the conic $U + kV = 0$."

The foregoing demonstration relates to the particular forms $U = ax^2 + by^2 + cz^2$, $V = x^2 + y^2 + z^2$; but observing that the function $\sqrt{(\xi+a)(\xi+b)(\xi+c)}$, which enters under the integral sign in the transcendental function $\Pi\xi$, is the square root of the discriminant of $U + \xi V$, the theory of covariants shows at once that the conclusions apply to any forms whatever of U, V, the expression for the transcendental function being

$$\Pi\xi = \int_\infty \frac{d\xi}{\sqrt{\text{disct.}\ (U+\xi V)}}.$$

Consider now a triangle inscribed in the conic $V = 0$, and with its sides touching the conics

$$U + k\ V = 0,$$
$$U + k'\ V = 0,$$
$$U + k'' V = 0;$$

then if θ, θ', θ'' are the parameters of the angles, we have

$$\Pi\theta'' - \Pi\theta' = 2\Pi k\ ,$$
$$\Pi\theta\ - \Pi\theta'' = 2\Pi k',$$
$$\Pi\theta' - \Pi\theta\ = 2\Pi k'',$$

and thence

$$\Pi k + \Pi k' + \Pi k'' = 0$$

as the relation which must subsist between the parameters k, k', k'', of the conics touched by the sides; and similarly for a polygon of n sides, the relation between the parameters is

$$\Pi k_1 + \Pi k_2 + \ldots + \Pi k_n = 0.$$

But by Abel's theorem, this transcendental equation is equivalent to an algebraical one.

In fact, calling the radical $\sqrt{\Box x}$, then if ϕx, χx are rational and integral functions of x with arbitrary coefficients, and if

$$\phi^2 x - \chi^2 x\, \Box x = A\,(x-k_1)(x-k_2)\ldots(x-k_n),$$

(this implies that $\phi^2 x$ is of a degree not exceeding n and $\chi^2 x$ of a degree not exceeding $n-3$; that is, for n even the degrees of ϕx, χx are $\frac{1}{2}n$, $\frac{1}{2}(n-4)$; and for n odd they are $\frac{1}{2}(n-1)$, $\frac{1}{2}(n-3)$), then the algebraical equation is that obtained by the elimination of the arbitrary coefficients from the system of equations

$$\phi k_1 + \chi k_1 \Box k_1 = 0,$$
$$\phi k_2 + \chi k_2 \Box k_2 = 0,$$
$$\vdots$$
$$\phi k_n + \chi k_n \Box k_n = 0;$$

or, what is the same thing, for n odd, $= 2p-1$, it is

$$\{1, \theta, \ldots \theta^{p-1}, \sqrt{\Box\theta}, \theta\sqrt{\Box\theta}, \ldots \theta^{p-2}\sqrt{\Box\theta}\} = 0,$$

and for n even, $= 2p$, it is

$$\{1, \theta, \ldots \theta^{p}, \sqrt{\Box\theta}, \theta\sqrt{\Box\theta}, \ldots \theta^{p-2}\sqrt{\Box\theta}\} = 0,$$

where the expressions in $\{\ \}$ denote respectively the determinants, of $2p-1$ lines, and $2p$ lines, formed by substituting for θ the values $k_1, k_2, \ldots k_n$ respectively. Thus for $n=3$, the equation is

$$\begin{vmatrix} 1, & k_1, & \sqrt{\Box k_1} \\ 1, & k_2, & \sqrt{\Box k_2} \\ 1, & k_3, & \sqrt{\Box k_3} \end{vmatrix} = 0;$$

and for $n=4$, it is

$$\begin{vmatrix} 1, & k_1, & k_1^2, & \sqrt{\Box k_1} \\ 1, & k_2, & k_2^2, & \sqrt{\Box k_2} \\ 1, & k_3, & k_3^2, & \sqrt{\Box k_3} \\ 1, & k_4, & k_4^2, & \sqrt{\Box k_4} \end{vmatrix} = 0,$$

and so on.

Suppose

$$\sqrt{\Box\xi} = A + B\xi + C\xi^2 + D\xi^3 + E\xi^4 + \ldots;$$

then substituting the corresponding expressions for $\sqrt{\Box k_1}$, $\sqrt{\Box k_2}$, &c., the determinant will divide by $\{1, \theta, \theta^2, \ldots \theta^{n-1}\}$, and it may be seen without difficulty that the resulting equation on putting therein $k_1 = k_2 \ldots k_n = 0$, will, according as $n = 3, 4, 5, 6$, &c., be

$$C=0,\quad D=0,\quad \begin{vmatrix} C, & D \\ D, & E \end{vmatrix} = 0,\quad \begin{vmatrix} D, & E \\ E, & F \end{vmatrix} = 0,\quad \begin{vmatrix} C, & D, & E \\ D, & E, & F \\ E, & F, & G \end{vmatrix} = 0, \text{ \&c.},$$

which is the theorem above referred to.

II.

Application to the several Polygons up to the Enneagon.

If the equation of the inscribed conic is $U=0$, and that of the circumscribed conic is $V=0$, and if the discriminant $(U+\xi V)$ is in the first instance represented by $1 + 4\beta\xi + 4\gamma\xi^2 + 4\delta\xi^3$, then the square root of the discriminant is

$$1 + 2\beta\xi + 2(\gamma - \beta^2)\xi^2 + 2(\delta - 2\beta\gamma + 2\beta^3)\xi^2 + \text{\&c.},$$

so that the condition for the triangle is

$$\gamma - \beta^2 = 0,$$

and that for the quadrangle is

$$\delta - 2\beta\gamma + 2\beta^3 = 0.$$

It is obviously convenient to introduce into the formulæ, in the place of γ and δ, the quantities

$$c = \gamma - \beta^2,$$
$$d = \delta - 2\beta\gamma + 2\beta^3;$$

and writing also, for symmetry of notation, b in the place of β, we have

$$\beta = b,$$
$$\gamma = c + b^2,$$
$$\delta = d + 2bc,$$

so that the discriminant will be

$$= 1 + 4b\xi + 4(c + b^2)\xi^2 + 4(d + 2bc)\xi^3,$$

which is

$$= (1 + 2b\xi + 2c\xi^2)^2 + 4(d\xi^3 - c^2\xi^4).$$

But for homogeneity I introduce the quantity $a = 1$, and put the discriminant

$$= (1 + 2b\xi + 2ac\xi^2)^2 + 4a^2(d\xi^3 - c^2\xi^4).$$

The square root, divided by a^2, is

$$= a^{-2}(1 + 2b\xi + 2ac\xi^2)\sqrt{1 + \{4a^2(d\xi^3 - c^2\xi^4) \div (1 + 2b\xi + 2ac\xi^2)^2\}};$$

or developing, this is

$$\begin{aligned} & a^{-2}(1 + 2b\xi + 2ac\xi^2) \\ + \ & 2\ (d\xi^3 - c^2\xi^4) \div (1 + 2b\xi + 2ac\xi^2) \\ - \ & 2a^2(d\xi^3 - c^2\xi^4)^2 \div (1 + 2b\xi + 2ac\xi^2)^3 \\ + \ & 4a^4(d\xi^3 - c^2\xi^4)^3 \div (1 + 2b\xi + 2ac\xi^2)^5 \\ - \ & 10a^6(d\xi^3 - c^2\xi^4)^4 \div (1 + 2b\xi + 2ac\xi^2)^7 \\ + \ & \&c.; \end{aligned}$$

or representing this by $1 + 2B\xi + 2C\xi^2 + 2D\xi^3 + \&c.$, we have

$$C\xi^2 + D\xi^3 + E\xi^4 + F\xi^5 + G\xi^6 + H\xi^7 + I\xi^8 + J\xi^9 + K\xi^{10} + L\xi^{11} + M\xi^{12} + \ldots$$

$$= a^{-1}c\xi^2$$

$$+(d\xi^3-c^2\xi^4)\left(\begin{array}{r|r|r|r|r|r|r|r|l} 1-2b\xi-2ac & \xi^2+8abc & \xi^3+4a^2c^2 & \xi^4-24a^2bc^2 & \xi^5-8a^3c^3 & \xi^6+64a^3bc^3 & \xi^7+16a^4c^4 & \xi^8-160a^4bc^4 & \xi^9+\&c. \\ +b^2 & -8b^3 & -24ab^2c & +64ab^3c & +96a^2b^2c^2 & -320a^2b^3c^2 & -320a^3b^2c^3 & +1280a^3b^3c^3 & \\ & & +16b^4 & -32b^5 & -160ab^4c & +384ab^5c & +960a^2b^4c^2 & -2688a^2b^5c^2 & \\ & & & & +64b^6 & -128b^7 & -896ab^6c & +2048ab^7c & \\ & & & & & & +256b^8 & -512b^9 & \end{array}\right)$$

$$+a^2(-d^2\xi^6+2c^2d\xi^7-c^4\xi^8)\left(\begin{array}{r|r|r|r|r|l} 1-6b\xi-6ac & \xi^2+48abc & \xi^3+24a^2c^2 & \xi^4-240a^2bc^2 & \xi^5-80a^3c^3 & \xi^6+\&c. \\ +24b^2 & -80b^3 & -240ab^2c & +960ab^3c & +1440a^2b^2c^2 & \\ & & +240b^4 & -672b^3 & -3360ab^4c & \\ & & & & +1792b^6 & \end{array}\right)$$

$$+a^4(2d^3\xi^9-6c^2d^2\xi^{10}+6c^4d\xi^{11}-2c^6\xi^{12}\left\{\begin{array}{r|r|l} 1-10b\xi-10ac & \xi^2+120abc & \xi^3-\&c. \\ +60b^2 & -280b^3 & \end{array}\right\}$$

$$+a^6(-5d^4\xi^{12}+\&c.)\{1+\&c.\}+\&c.;$$

and the values of the coefficients C, D, E, &c. thus are

C	D	E	F	G	H	I	J	K	L	M
$a^{-1}c$	d	$bd-2$	$acd-2$	$a^2d^2\ -1$	$a^2bd^2\ +\ 6$	$a^3cd^2\ +\ 6$	$a^4d^3\ +\ 2$	$a^4bd^3\ -\ 20$	$a^5cd^3\ -\ 20$	$a^6d^4\ -$
		$c^2\ -1$	b^2d+4	$abcd+8$	$a^2c^2d^2+\ 6$	$a^2c^4\ -\ 5$	$a^3bcd^2\ -\ 48$	$a^4c^2d^2\ -\ 30$	$a^4b^2d^3\ +\ 120$	$a^5bcd^3\ +\ 24$
			$bc^2\ +2$	$ac^3\ +2$	$ab^2cd-\ 24$	$a^2b^2d^2-24$	$a^3c^3d\ -\ 20$	$a^3b^2cd^2+240$	$a^4bc^2d\ +\ 300$	$a^5c^3d^2\ +\ 14$
				$b^3d\ -8$	$abc^3\ -\ 8$	a^2bc^2d-36	$a^2b^3d^2\ +\ 80$	$a^3bc^3d\ +160$	$a^4c^4d\ +\ 70$	$a^4b^3d^3\ -\ 5$
				$b^2c^2\ -4$	$b^4d\ +\ 16$	$ab^3cd\ +64$	$a^2b^2c^2d\ +144$	$a^3c^5\ +\ 14$	$a^3b^3cd^2-\ 960$	$a^4b^2c^2d^2-18$
					$b^3c^2\ +\ 8$	$ab^2c^3\ +24$	$a^2bc^4\ +\ 30$	$a^2b^4d^2\ -240$	$a^3b^2c^3d-\ 800$	$a^4bc^4d\ -\ 7$
						$b^5d\ -32$	$ab^4cd\ -160$	$a^2b^3c^2d-480$	$a^3bc^5\ -\ 112$	$a^4c^6\ -$
						$b^4c^2\ -16$	$ab^3c^3\ -\ 64$	$a^2b^2c^4\ -120$	$a^2b^5d^2\ +\ 672$	$a^3b^4cd^2+33$
							$b^6d\ +\ 64$	$ab^5cd\ +384$	$a^2b^4c^2d+1440$	$a^3b^3c^3d+32$
							$b^5c^2\ +\ 32$	$ab^4c^3\ +160$	$a^2b^3c^4\ +\ 400$	$a^3b^2c^5\ +\ 5$
								$b^7d\ -128$	$ab^6cd\ -\ 896$	$a^2b^6d^2\ -17$
								$b^6c^2\ -\ 64$	$ab^5c^3\ -\ 384$	$a^2b^5c^2d-40$
									$b^8d\ +\ 256$	$a^2b^4c^4\ -12$
									$b^7c^2\ +\ 128$	$ab^7cd\ +20$
										$ab^6c^3\ +\ 8$
										$b^9d\ -\ 5$
										$b^8c^2\ -\ 2$
+1	+1	−3	+4	−3	+4	−19	+60	−124	+214	−4

But in the sequel only the coefficients up to I are made use of: the expressions for J, K, L, M may however be useful, and they are given accordingly.

The sums of the numerical coefficients are given here and elsewhere, as they are very useful for verifications; thus, putting $a=b=c=d=1$, we have, as should be,

$$\sqrt{1+4\xi+8\xi^2+12\xi^3}=1+2\xi+2\xi^2(1,+1,-3,+4,-3,+4,-19,+60,-124,+214,-455,\&c.\between 1,\xi)^\infty.$$

Proceeding now to form the several terms of the matrix

$$\left(\begin{array}{ccccc} C, & D, & E, & F, & G,\ldots \\ D, & E, & F, & G, & H,\ldots \end{array}\right),$$

which may be represented by **12**, **13**, &c., viz.

$$\mathbf{12} = \begin{vmatrix} C, & D \\ D, & E \end{vmatrix}, \quad \mathbf{13} = \begin{vmatrix} C, & E \\ D, & F \end{vmatrix}, \text{ \&c.,}$$

we have, up to **45**, which is all that is required,

$12 = a^{-1} \times$	$13 = a^{-1} \times$	$14 = a^{-1} \times$	$15 = a^{-1} \times$	$23 =$	$24 =$	$25 =$	$34 =$	$35 =$	$45 =$
$ad^2 - 1$	$abd^2 + 2$	$a^2cd^2 + 1$	$a^3d^3 + 1$	$acd^2 - 2$	$a^2d^3 - 1$	$a^2bd^3 + 4$	$a^2bd^3 + 2$	$a^3cd^3 - 2$	$a^4d^4 - 1$
$bcd - 2$	$ac^2d - 1$	$ab^2d^2 - 4$	$a^2bcd^2 - 2$	$bc^2d - 2$	$abcd^2 + 4$	$a^2c^2d^2 + 5$	$a^2c^2d^2 - 3$	$a^2b^2d^3 - 8$	$a^4bcd^3 + 4$
$c^3 - 1$	$b^2cd + 4$	$abc^2d + 6$	$a^2c^3d + 4$	$c^4 - 1$	$b^2c^2d + 4$	$ab^2cd^2 - 8$	$abc^3d - 4$	$a^2c^4d - 2$	$a^3c^3d^2 - 8$
	$bc^3 + 2$	$ac^4 + 2$	$ab^3d^2 + 8$		$bc^4 + 2$	$abc^3d + 4$	$ac^5 - 2$	$ab^2c^3d + 8$	$a^2b^3d^3 + 8$
		$b^3cd - 8$	$ab^2c^2d - 20$			$ac^5 + 2$		$abc^5 + 4$	$a^2b^2c^2d^2 + 12$
		$b^2c^3 - 4$	$abc^4 - 8$			$b^3c^2d - 8$			$a^2bc^4d - 4$
			$b^4cd + 16$			$b^2c^4 - 4$			$a^2c^6 - 4$
			$b^3c^3 + 8$						
-4	-7	$+7$	-7	-5	$+9$	-5	-7	0	$+7$

Forming in like manner the determinants of the matrix

$$\begin{pmatrix} C, & D, & E, & F, .. \\ D, & E, & F, & G, . \\ E, & F, & G, & H, \end{pmatrix},$$

and representing these by **123**, **124**, &c., viz.

$$\mathbf{123} = \begin{vmatrix} C, & D, & E \\ D, & E, & F \\ E, & F, & G \end{vmatrix}, \text{ \&c.,}$$

we have, up to **234**,

$123 = a^{-1} \times$	$124 = a^{-1} \times$	$134 = a^{-1} \times$	$234 =$
$a^3d^4 + 1$	$a^3bd^4 - 4$	$a^4cd^4 + 1$	$a^4d^5 - 1$
$a^2bcd^3 + 2$	$a^3c^2d^2 - 3$	$a^3b^2d^4 + 4$	$a^3bcd^4 - 4$
$a^2c^3d^2 - 1$	$a^2b^2cd^3 - 8$	$a^3bc^2d^3 + 8$	$a^3c^3d^3 - 4$
$abc^4d - 2$	$a^2bc^3d^2 - 4$	$a^3c^4d^2 - 3$	$a^2b^2c^2d^3 - 4$
$ac^6 - 1$	$a^2c^5d - 2$	$a^2b^3cd^3 + 8$	$a^2bc^4d^2 - 6$
	$ab^2c^4d + 4$	$a^2b^2c^3d^2 + 12$	$a^2c^6d - 2$
	$abc^6 + 2$	$a^2c^7 - 2$	
-1	-15	$+28$	-21

and further, the determinants of the matrix

$$\left(\begin{array}{llll} C, & D, & E, & F, \ldots \\ D, & E, & F, & G \\ E, & F, & G, & H \\ F, & G, & H, & I \end{array}\right),$$

in the present case, the single determinant

$$1234 = \left|\begin{array}{llll} C, & D, & E, & F \\ D, & E, & F, & G \\ E, & F, & G, & H \\ F, & G, & H, & I \end{array}\right|,$$

we have

$$1234 = a^{-1} \times \left(\begin{array}{ll} a^6cd^6 & -\ 3 \\ a^5bc^2d^5 & -\ 12 \\ a^5c^4d^4 & -\ 4 \\ a^4b^2c^3d^4 & -\ 16 \\ a^4bc^5d^3 & -\ 14 \\ a^4c^7d^2 & -\ 3 \\ a^3b^3c^4d^3 & -\ 8 \\ a^3b^2c^6d^2 & -\ 12 \\ a^3bc^8d & -\ 6 \\ a^3c^{10} & -\ 1 \end{array}\right)$$

$$-79$$

The conditions for the triangle and the quadrangle are $c=0$, $d=0$; those for the pentagon and hexagon are $12=0$, $23=0$; for the heptagon and octagon, $123=0$, $234=0$; and that for the enneagon is $1234=0$. The foregoing values show that 23 and 1234 (which belong to the hexagon and the enneagon) divide by c (which belongs to the triangle), and that 234 (which belongs to the octagon) divides by d (which belongs to the quadrangle). But I was not prepared for the destruction which will be observed in the several determinants, of the terms involving the lower powers of a (that is, the terms of the highest orders in b, c, d), and which renders these expressions so much more simple than they would otherwise have been.

Representing the reduced equations for the several polygons, as before, by

$$[3] = 0,$$
$$[4] = 0,$$
$$[5] = 0,$$
$$[6] = 0,$$
$$[7] = 0,$$
$$[8] = 0,$$
$$[9] = 0,$$
$$\&c.;$$

then retaining the quantity $a(=1)$ for homogeneity, but rejecting the powers of a which divide out, and reversing in some cases the signs, the values of the functions [3], [4], &c. are

[3] =	[4] =	[5] =	[6] =	[7] =	[8] =	[9] =
$c+1$	$d+1$	ad^2+1	ad^2+2	$a^2d^4\ +1$	$a^2d^4\ +1$	$a^3d^6\ +\ 3$
		$bcd+2$	$bcd+2$	$abcd^3\ +2$	$abcd^3\ +4$	$a^2bcd^5\ +12$
		$c^3\ +1$	$c^3\ +1$	$ac^3d^2\ -1$	$ac^3d^2\ +4$	$a^2c^3d^4\ +\ 4$
				$bc^4d\ -2$	$b^2c^2d^2\ +4$	$ab^2c^2d^4+16$
				$c^6\ -1$	$bc^4d\ +6$	abc^4d^3-14
					$c^6\ +2$	$ac^6d^2\ +\ 3$
						$b^3c^3d^3\ +\ 8$
						$b^2c^5d^2\ +12$
						$bc^7d\ +\ 6$
						$c^9\ +\ 1$
+1	+1	+4	+5	−1	+21	+79

The similarity of form for the relations corresponding to the pentagon and the hexagon, and for those corresponding to the heptagon and the octagon, is, I am inclined to think, accidental; the functions are homogeneous as regards degree and weight; and the degrees and weights of the two consecutive functions being identical, the literal parts must be similarly constituted.

III.

M. Mention's *Formulæ for the Case of two Circles.*

In the case of two circles, if, as usual, the radii of the inscribed and circumscribed circles are put equal to r and R respectively, and the distance of their centres to a, then the equations of the inscribed and circumscribed circles respectively may be taken to be

$$x^2+y^2-r^2=0,$$
$$(x-a)^2+y^2-R^2=0;$$

and if, in the notation of M. Mention, we put

$$\frac{R^2-r^2}{a^2}=i,$$

$$\frac{1}{r^4}(r^4+R^4+a^4-2r^2R^2-2r^2a^2-2R^2a^2)=-\nu,$$

then the quadratic radical is

$$\sqrt{(1+4\xi)\{[1+(2+2i)\,\xi]^2+4\nu\xi^2\}};$$

and comparing this with

$$\sqrt{1 + 4b\xi + 4(c + b^2)\xi^2 + 4(d + 2bc)\xi^3},$$

we have

$$\begin{aligned} b &= i + 2, \\ c + b^2 &= i^2 + 6i + \nu + 5, \\ d + 2bc &= 4(i^2 + 2i + \nu + 1), \end{aligned}$$

and thence

$$\begin{aligned} b &= i + 2, \\ c &= 2i + \nu + 1, \\ d &= -2i\nu - 2i; \end{aligned}$$

and by means of these values, or by effecting the development in a different manner, we find

$$B = \begin{cases} i \\ +2, \end{cases}$$

$$C = \begin{cases} i \,.\, 2 \\ +\nu + 1, \end{cases}$$

$$D = 2\{\, i(-\nu - 1)\},$$

$$E = \begin{cases} i^2 \,.\, 4\nu \\ + i \,.\, 4(\nu + 1) \\ \quad - (\nu + 1)^2, \end{cases}$$

$$F = 2\begin{cases} i^3 \,.\, -4\nu \\ + i^2 \,.\, -8\nu \\ + i \,.\, (\nu + 1)(3\nu - 5) \\ + \ 2(\nu + 1)^2, \end{cases}$$

$$G = 2\begin{cases} i^4 \,.\, 8\nu \\ + i^3 \,.\, 24\nu \\ + i^2 \,.\, 4\nu(-3\nu + 5) \\ + i \,.\, (\nu + 1)(-18\nu + 14) \\ + \ (\nu + 1)^2(\nu - 7), \end{cases}$$

$$H = 4\begin{cases} i^5 \,.\, -8\nu \\ + i^4 \,.\, -32\nu \\ + i^3 \,.\, 4\nu(5\nu - 11) \\ + i^2 \,.\, 16\nu(3\nu - 1) \\ + i \,.\, (\nu + 1)(\nu - 7)(-5\nu + 3) \\ \quad - 4(\nu + 1)^2(\nu - 3). \end{cases}$$

These values give

$$C = \begin{cases} \;\; 2i \\ + \nu + 1, \end{cases} \qquad = [3],$$

$$D = 2\{\; i(-\nu - 1), \qquad = [4],$$

$$CE - D^2 = \begin{cases} \quad i^3 . 8\nu \\ + i^2 . 4(\nu + 1) \\ + i \;. 2(\nu + 1)^2 \\ \quad - (\nu + 1)^3, \end{cases} \qquad = [5],$$

$$DF - E^2 = \begin{pmatrix} 2i \\ + \nu + 1 \end{pmatrix} \begin{pmatrix} 2i \\ - \nu - 1 \end{pmatrix} \begin{Bmatrix} i^2 . 4\nu \\ + (\nu + 1)^2 \end{Bmatrix} \qquad = [3]\;[6],$$

$$\begin{vmatrix} C, & D, & E \\ D, & E, & F \\ E, & F, & G \end{vmatrix} = \begin{cases} \quad i^6 . \quad 64\nu \\ + i^5 . - 32\nu(\nu + 1)(\nu - 1) \\ + i^4 . - 16\nu(\nu + 1)^2 \\ + i^3 . - 8(3\nu - 1)(\nu + 1)^3 \\ + i^2 . \quad 4(\nu + 1)^4 \\ + i \;. - 4(\nu + 1)^5 \\ \quad - \quad (\nu + 1)^6 \end{cases} \qquad = [7],$$

$$\begin{vmatrix} D, & E, & F \\ E, & F, & G \\ F, & G, & H \end{vmatrix} = i . (-\nu - 1) \begin{cases} 16 i^4 \nu \\ + (\nu + 1)^4. \end{cases} \qquad = [4]\;[8].$$

It will be remarked that $[4], = i(-\nu - 1)$, breaks up into the factors i and $\nu + 1$; and so $[6], = (2i - \nu - 1)\{4\nu i^2 + (\nu + 1)^2\}$, breaks up into the factors $2i - \nu - 1$ and $4\nu i^2 + (\nu + 1)^2$.

It may be added that the developed expression of [6] is

$$= \begin{cases} \quad i^3 . 8\nu \\ + i^2 . - 4\nu(\nu + 1) \\ + i \;. 2(\nu + 1)^2 \\ \quad - (\nu + 1)^3, \end{cases}$$

so that the difference between this and [5] is $i^2 . 4(\nu + 1)^2$, which is $= d^2$; this agrees with a former result.

M. Mention has also given, but not in a developed form, the formulæ for the enneagon and the endecagon, and the following formula for the decagon, viz.

$$[16 i^4 \nu + (\nu + 1)^4]^2 + 16 i^2 \nu (\nu + 1)^2 . \{2 i^2 (1 - \nu) - (\nu + 1)^2\}^2 = 0.$$

IV.

Considerations as to the form of relation, in the case of two circles, for Polygons of an odd and even number of sides respectively.

The relation between the two conics, or condition for the existence of the polygon, is the same whatever point of the circumscribed conic is taken as an angle of the polygon. Take for an angle, a point of intersection of the two conics. Consider first the case of the triangle; if a point of intersection is taken as a vertex A of the triangle, then the sides AB, AC coincide in direction with the tangent at A to the inscribed conic U, hence B and C coincide together at the point where this tangent meets the circumscribed conic V, BC is therefore a tangent of V, but it is by hypothesis a tangent of U; hence for the triangle the relation between the inscribed conic U and the circumscribed conic V is as follows: viz. a tangent to U at a point of intersection with V meets V at a point of contact of a common tangent of U and V.

In like manner for the quadrangle, if A be taken at a point of intersection, the sides AB and AD will coincide in direction with the tangent to U at this point, consequently B and D must coincide at the point where this tangent meets V; hence also CB, CD, the two tangents to U from the point C, must coincide, or C must be a point of intersection of the conics U, V. In other words, the pole, with respect to the inscribed conic U, of a common chord AC of the two conics must lie on the circumscribed conic V; this is therefore the condition for the quadrilateral.

In the ordinary mode of drawing the figures, with two conics which do not intersect, the points and lines employed in the foregoing constructions are imaginary, but the conics may be so drawn that these points and lines are all real.

In general, for a polygon of an odd number, $2n+1$, of sides, then starting from a point of intersection, the sides will coincide in pairs, viz. the first and last, second and last but one, and so on, the middle or $(n+1)$th side being a common tangent of the two conics. But for a figure of an even number, $2n$, of sides, then starting from a point of intersection of the two conics, the nth side will terminate at a second point of intersection, and then the same series of sides will be repeated in the reverse order, so that the sides will coincide in pairs, first and last, second and last but one, nth and $(n+1)$th. For a figure of an odd number of sides, the relation involves only a single point of intersection, but for a figure of an even number of sides, it involves two points of intersection.

Now in the case of two circles, for a polygon of an odd number of sides, the same relation is obtained, whether we take as the point of intersection one of the actual points of intersection, or a circular point at infinity, and the relation $[2n+1]=0$ does not break up into factors. And so for a polygon of an even number of sides, then taking for the two points of intersection, the two actual points of intersection, or the two circular points at infinity, we have one form of result; but taking for them an actual point of intersection and a circular point at infinity, we have a different form of result; and the equation $[2n]=0$ does break up into factors.

This is verified very simply in the case of the quadrangle. Taking for the two points of intersection the circular points at infinity, the line joining them is the line infinity, and its pole (with respect to the inscribed circle) is the centre of this circle; the relation therefore is that the centre of the inscribed circle lies on the circumscribed circle. But when this is the case, it is easy to see that the pole (with respect to the inscribed circle) of the radical axis, lies also on the circumscribed circle; this pole and the centre of the inscribed circle are in fact the extremities of a diameter of the circumscribed circle. The condition thus obtained is $R^2 - a^2 = 0$ (which is M. Mention's condition $i = 0$). We have next to find the analytical relation when the pole (with respect to the inscribed circle), of the line joining one of the actual points of intersection with a circular point at infinity is a point on the circumscribed circle. This I effect as follows:—taking $z = 0$ as the equation of line infinity, if the origin be taken on the middle point of the radical axis, and if $x = 0$ be the radical axis, then the equations of the two circles may be taken to be

$$\text{Inscribed circle,} \quad x^2 + y^2 - 2l\,xz - \nabla z^2 = 0,$$

$$\text{Circumscribed circle,} \quad x^2 + y^2 - 2Lxz - \nabla z^2 = 0,$$

a circular point at infinity is

$$x : y : z = 1 : i : 0, \quad (i = \sqrt{-1}),$$

an actual point of intersection is

$$x : y : z = 0 : \sqrt{\nabla} : 1.$$

The line joining these is

$$xi - y + z\sqrt{\nabla} = 0,$$

its pole, with respect to the inscribed circle, is

$$x : y : z = -i\sqrt{\nabla} : \sqrt{\nabla} - il : 1;$$

and if this be a point of the circumscribed circle

$$-\nabla + (\nabla - 2il - l^2) + 2Li\sqrt{\nabla} - \nabla = 0,$$

that is

$$2(L - l)\,i\sqrt{\nabla} = l^2 + \nabla,$$

or

$$(l^2 + \nabla)^2 + 4(L - l)^2\,\nabla = 0,$$

which is the required relation: but to express it in terms of the ordinary data R, r, a, the equations of the circles, putting therein $z = 1$, become

$$(x - l\,)^2 + y^2 = \nabla + l^2,$$

$$(x - L)^2 + y^2 = \nabla + L^2,$$

and therefore

$$a = L - l,$$
$$r^2 = \nabla + l^2,$$
$$R^2 = \nabla + L^2;$$

these equations give

$$\frac{R^2 - r^2}{a} = L + l,$$
$$a = L - l;$$

and thence

$$L = \frac{R^2 - r^2 + a^2}{2a},$$

and

$$\nabla = R^2 - \left(\frac{R^2 - r^2 - a^2}{2a}\right)^2 = \frac{1}{4a^2}(2a^2R^2 + 2a^2r^2 + 2r^2R^2 - R^4 - r^4 - a^4)$$
$$= \frac{r^4}{4a^2}\nu,$$

if with M. Mention we write

$$\frac{1}{r^4}(r^4 + R^4 + a^4 - 2r^2R^2 - 2r^2a^2 - 2R^2a^2) = -\nu.$$

The equation

$$(l^2 + \nabla)^2 + 4(L - l)^2\nabla = 0$$

thus becomes

$$r^4 + 4a^2 \cdot \frac{r^4}{4a^2}\nu = 0,$$

that is, it becomes $\nu + 1 = 0$, which is the other factor of the complete condition $i(\nu + 1) = 0$.

268.

ON A NEW AUXILIARY EQUATION IN THE THEORY OF EQUATIONS OF THE FIFTH ORDER.

[From the *Philosophical Transactions of the Royal Society of London*, vol. CLI. (for the year 1861), pp. 263—276. Received February 20,—Read March 7, 1861.]

CONSIDERING the equation of the fifth order, or quintic equation,

$$(*\between v,\ 1)^5 = (v - x_1)(v - x_2)(v - x_3)(v - x_4)(v - x_5) = 0,$$

and putting as usual

$$f\omega = x_1 + \omega x_2 + \omega^2 x_3 + \omega^3 x_4 + \omega^4 x_5,$$

where ω is an imaginary fifth root of unity, then, according to Lagrange's general theory for the solution of equations, $f\omega$ is the root of an equation of the order 24, called the Resolvent Equation, but the solution whereof depends ultimately on an equation of the sixth order, viz.

$$(f\omega)^5,\ (f\omega^2)^5,\ (f\omega^3)^5,\ (f\omega^4)^5$$

are the roots of an equation of the fourth order, each coefficient whereof is determined by an equation of the sixth order; and moreover the other coefficients can be all of them rationally expressed in terms of any one coefficient assumed to be known; the solution thus depends on a single equation of the sixth order. In particular the last coefficient, or

$$(f\omega\,.\,f\omega^2\,.\,f\omega^3\,.\,f\omega^4)^5,$$

is determined by an equation of the sixth order; and not only so, but its fifth root, or

$$f\omega\,.\,f\omega^2\,.\,f\omega^3\,.\,f\omega^4,$$

(which is a rational function of the roots, and is the function called by Mr Cockle the Resolvent Product), is also determined by an equation of the sixth order: this equation may be called the Resolvent-Product Equation. But the recent researches of Mr Cockle and Mr Harley [1] show that the solution of an equation of the fifth order may be made to depend on an equation of the sixth order, originating indeed in, and closely connected with, the resolvent-product equation, but of a far more simple form; this is the auxiliary equation referred to in the title of the present memoir. The connexion of the two equations, and the considerations which led to the new one, will be pointed out in the sequel; but I will here state synthetically the construction of the auxiliary equation. Representing for shortness the roots $(x_1, x_2, x_3, x_4, x_5)$ of the given quintic equation by 1, 2, 3, 4, 5, and putting moreover

$$12345 = 12 + 23 + 34 + 45 + 51, \text{ \&c.}$$

(where on the right-hand side 12, 23, &c. stand for x_1x_2, x_2x_3, &c.), then the auxiliary equation, say

$$(*\between\phi,\ 1)^6 = 0,$$

has for its roots

$$\begin{aligned}
\phi_1 &= 12345 - 24135, & \phi_4 &= 21435 - 13245,\\
\phi_2 &= 13425 - 32145, & \phi_5 &= 31245 - 14325,\\
\phi_3 &= 14235 - 43125, & \phi_6 &= 41325 - 12435,
\end{aligned}$$

and, it follows therefrom, is of the form

$$(1,\ 0,\ C,\ 0,\ E,\ F,\ G\between\phi,\ 1)^6 = 0,$$

where C, E, G are rational and integral functions of the coefficients of the given equation, being in fact seminvariants, and F is a mere numerical multiple of the square root of the discriminant.

The roots of the given quintic equation are each of them rational functions of the roots of the auxiliary equation, so that the theory of the solution of an equation of the fifth order appears to be now carried to its extreme limit. We have in fact

$$\begin{aligned}
\phi_1\phi_6 + \phi_2\phi_4 + \phi_3\phi_5 &= (*\between x_1,\ 1)^4,\\
\phi_1\phi_2 + \phi_3\phi_4 + \phi_5\phi_6 &= (*\between x_2,\ 1)^4,\\
\phi_1\phi_5 + \phi_2\phi_3 + \phi_4\phi_6 &= (*\between x_3,\ 1)^4,\\
\phi_1\phi_3 + \phi_2\phi_6 + \phi_3\phi_5 &= (*\between x_4,\ 1)^4,\\
\phi_1\phi_4 + \phi_2\phi_5 + \phi_3\phi_6 &= (*\between x_5,\ 1)^4,
\end{aligned}$$

where $(*\between x_1,\ 1)^4$, &c. are the values, corresponding to the roots x_1, &c. of the given equation, of a given quartic function. And combining these equations respectively with the quintic equations satisfied by the roots x_1, &c. respectively, it follows that, conversely, the roots x_1, x_2, &c. are rational functions of the combinations $\phi_1\phi_6 + \phi_2\phi_4 + \phi_3\phi_5$, $\phi_1\phi_2 + \phi_3\phi_4 + \phi_5\phi_6$, &c. respectively, of the roots of the auxiliary equation.

[1] Cockle, "Researches in the Higher Algebra," *Manchester Memoirs*, t. xv. pp. 131—142 (1858).

Harley, "On the Method of Symmetric Products, and its Application to the Finite Algebraic Solution of Equations," *Manchester Memoirs*, t. xv. pp. 172—219 (1859).

Harley, "On the Theory of Quintics," *Quart. Math. Journ.* t. III. pp. 343—359 (1859).

It is proper to notice that, combining together in every possible manner the six roots of the auxiliary equation, there are in all fifteen combinations of the form $\phi_1\phi_2+\phi_3\phi_4+\phi_5\phi_6$. But the combinations occurring in the above-mentioned equations are a completely determinate set of five combinations: the equation of the order 15, whereon depend the combinations $\phi_1\phi_2+\phi_3\phi_4+\phi_5\phi_6$, is not rationally decomposable into three quintic equations, but only into a quintic equation having for its roots the above-mentioned five combinations, and into an equation of the tenth order, having for its roots the other ten combinations, and being an irreducible equation. Suppose that the auxiliary equation and its roots are known; the method of ascertaining what combinations of roots correspond to the roots of the quintic equation would be to find the rational quintic factor of the equation of the fifth order, and observe what combinations of the roots of the auxiliary equation are also roots of this quintic factor. The direct calculation of the auxiliary equation by the method of symmetric functions would, I imagine, be very laborious. But the coefficients are seminvariants, and the process explained in my memoir on the Equation of Differences, [262], was therefore applicable, and by means of it, the equation, it will be seen, is readily obtained. The auxiliary equation gives rise to a corresponding covariant equation, which is given at the conclusion of the memoir.

1. I will commence by referring to some of the results obtained by Mr Cockle and Mr Harley.

In the paper "Researches on the Higher Algebra," Mr Cockle, dealing with the quintic equation

$$v^5-5Qv+E=0,$$

obtains for the Resolvent Product $\theta\,(=f\omega f\omega^2 f\omega^3 f\omega^4)$ the equation

$$\theta^6+2QE\,5^5\,\theta^4+2Q^4\,5^7\,\theta^3+Q^2E^2\,5^{10}\,\theta^2-(58Q^5-E^3)\,E\theta+5^{14}\,Q^8=0\,;$$

and he remarks that this equation may be written

$$(\theta^3+5^5\,QE\theta+5^7\,Q^4)^2=5^{10}\,(108\,Q^5E-E^4)\,\theta,$$

so that $\sqrt{-\theta}$ is determined by an equation of the sixth order, involving the quadratic radical $\sqrt{E\,(E^3-108Q^5)}$, which is in fact the square root of the discriminant of the quintic equation.

2. Mr Harley, in his paper "On the Symmetric Product &c.," makes use of the functions

$$\tau\ =x_1x_2+x_2x_3+x_3x_4+x_4x_5+x_5x_1\ (=12345),$$

$$\tau'=x_1x_3+x_3x_5+x_5x_2+x_2x_4+x_4x_1\ (=24135),$$

and he obtains for the form $v^5-5Qv^2+E=0$, the relation $\theta=5\tau\tau'$, which, since here $\tau+\tau'=0$, gives $\theta=-5\tau^2$.

Hence $\tau\,(=\sqrt{-\tfrac{1}{5}\theta})$ is the root of an equation of the sixth order involving the radical $\sqrt{E(E^3-108Q^5)}$, and which is in fact $(t=\tau\div\sqrt{5}=\tfrac{1}{5}\sqrt{-\theta})$, the equation

$$t^5+5QE\,t^2+\sqrt{E(E^3-108Q^5)}\,t-5Q^4=0,$$

given in Mr Harley's paper "On the Theory of Quintics."

3. And in the same paper there is given a system of equations

$$t_1t_3+t_2t_5+t_4t_6=x_1(3^2Q-x_1^3),\ \&c.,$$

connecting the five roots of the given quintic equation with the combinations

$$t_1t_3+t_2t_5+t_4t_6,\ \&c.$$

of the roots of the equation in t.

4. I quote also, with a slight change of notation, the following results from the paper "On the Symmetric Product &c.," viz. considering the quintic equation under the form

$$(a,\ b,\ c,\ d,\ e,\ f\between v,\ 1)^5=0,$$

we have

$$f\omega\,f\omega^4=\Sigma x^2+\tau\,(\omega\ +\omega^4)+\tau'\,(\omega^2+\omega^3),$$

$$f\omega^2f\omega^3=\Sigma x^2+\tau\,(\omega^2+\omega^3)+\tau'\,(\omega\ +\omega^4),$$

where

$$\Sigma x^2=x_1^2+x_2^2+x_3^2+x_4^2+x_5^2=\frac{1}{a^2}(b^2-2ac),$$

and thence, observing also that $\tau+\tau'=\dfrac{c}{a}$,

$$a^4\theta\,(=a^4\,f\omega f\omega^2 f\omega^3 f\omega^4)=5a^2c^2-5ab^2c+b^4+5a^4\tau\tau',$$

or, as this equation may also be written,

$$4a^4\theta=(5ac-2b^2)^2-5a^4(\tau-\tau')^2;$$

and hence the Resolvent Product $\theta\,(=f\omega f\omega^2 f\omega^3 f\omega^4)$ being determined by an equation, of the sixth order, this is also the case with the function $(\tau-\tau')^2$.

5. But the twelve functions $\pm(\tau-\tau')$ can be divided into two sets of six functions each, so that each set is determined by an equation of the sixth order involving a single quadratic radical. This was in fact suggested to me by Mr Harley's equation in t; for in the case considered $t+t'$ was $=0$, or $2t=t-t'$, and the equation in t was presumably the particular form of the equation for $\tfrac{1}{2}(t-t')$ in the general case. But it will presently appear in what manner the conclusion should have been arrived at *à priori*.

6. The preceding remarks show the connexion between the function $\phi\,(=\tau-\tau')$ to which belongs the new auxiliary equation, and the Resolvent Product $\theta\,(=f\omega f\omega^2 f\omega^3 f\omega^4)$. The relation was given for the denumerate form of the quintic; but taking, instead, the standard form $(a, b, c, d, e, f \between v, 1)^5=0$, it becomes

$$4a^4\theta = 2500\,(ac-b^2)^2 - 5a^4\phi^2.$$

7. The foregoing equation shows that ϕ is a seminvariantive function of the roots. In fact

$$f\omega, \; = x_1 - x_5 + \omega\,(x_2 - x_5) + \omega^2\,(x_3 - x_5) + \omega^3\,(x_4 - x_5),$$

is seminvariantive, and $f\omega^2$, $f\omega^3$, $f\omega^4$, being in like manner seminvariantive, the product $\theta\,(=f\omega f\omega^2 f\omega^3 f\omega^4)$ is also seminvariantive; $ac-b^2$ and a are seminvariants, and therefore ϕ is a seminvariantive function.

8. But it is easy to show this directly. For representing, as before, the roots by 1, 2, 3, 4, 5, we have

$$(1-5)(2-5)+(2-5)(3-5)+(3-5)(4-5) = 12+23+34-5\,(1+22+23+4)+35^2,$$
$$(2-5)(4-5)+(4-5)(1-5)+(1-5)(3-5) = 24+41+13-5\,(2+24+21+3)+35^2;$$

and the difference of the right-hand sides is

$$\begin{aligned}&12+23+34-5\,(2+3)\\ &-24-41-13+5\,(4+1),\end{aligned}$$

which is $=12345-24135$. So that ϕ,

$$=(1-5)(2-5)+(2-5)(3-5)+(3-5)(4-5)-[(2-5)(4-5)+(4-5)(1-5)+(1-5)(3-5)],$$

is a function of the differences of the roots, that is, it is a seminvariantive function.

9. To account for the division of the twelve values of $\pm(\tau-\tau')$ into two sets as above, and to explain the formation of a set, consider the symbols 1, 2, 3, 4, 5 as belonging to five points. We may with these five points form in all $(\tfrac{1}{2}.1.2.3.4=)\,12$ pentagons, and the symbol 12345 of any pentagon may of course be read backwards or forwards from any point $(12345=23451=\&\text{c.}=15432=\&\text{c.})$ without alteration of its meaning. Now attaching to each arrangement of the five numbers a sign, + or −, according to the ordinary rule of signs, 12345 being as usual positive, the arrangements 12345, 23451, &c. . . 15432, &c., which belong to the same pentagon, have all of them the same sign; and we may consequently connect with each pentagon the sign + or −; there are, in fact, six pentagons with the sign + and six with the sign −; and to each positive pentagon there corresponds a negative pentagon, which is derived from it by *stellation*, viz. to the positive pentagon 12345 there corresponds the negative one 24135, and so for the other positive pentagons. The above-mentioned system of equations

$$\begin{aligned}\phi_1 &= 12345-24135, & \phi_4 &= 21435-13245,\\ \phi_2 &= 13425-32145, & \phi_5 &= 31245-14325,\\ \phi_3 &= 14235-43125, & \phi_6 &= 41325-12435,\end{aligned}$$

in fact exhibits the six positive pentagons, each accompanied by its stellated negative pentagon, and the formation of the system of equations is thus completely explained; the order of arrangement of the pairs *inter se* (or, what comes to the same thing, the order of arrangement of the suffixes of the ϕ's) is wholly immaterial.

10. The six pairs of pentagons, or, what is the same thing, the ϕ's, correspond to each other in pairs in a fivefold manner, *quoad* the numbers 1, 2, 3, 4, 5 respectively; thus, *quoad* 5, the pairs are ϕ_1 and ϕ_4, ϕ_2 and ϕ_5, ϕ_3 and ϕ_6, or say 1 and 4, 2 and 5, 3 and 6. The relation is best seen by means of the positive pentagons; thus, *quoad* 5, in the pentagons 12345 and 21435, the points adjacent to 5 in the one of them are the points 2, 3, and in the other of them the complementary points 1, 4; and so in the other cases. The fivefold correspondence is shown by the symbolical equations

$$1 = 16,\ 24,\ 35,$$
$$2 = 12,\ 34,\ 56,$$
$$3 = 15,\ 23,\ 46,$$
$$4 = 13,\ 26,\ 35,$$
$$5 = 14,\ 25,\ 36,$$

which, in fact, indicate the combinations of the ϕ's which correspond to the several roots of the quintic.

11. It is proper to notice that the right-hand sides of the last-mentioned equations contain all the duads formed with the six numbers 1, 2, 3, 4, 5, 6, each duad once, and once only. There are in all six such synthemes of duads, viz.

12 . 34 . 56 13 . 25 . 46 14 . 26 . 35 15 . 24 . 36 16 . 23 . 45	12 . 35 . 46 13 . 24 . 56 14 . 25 . 36 15 . 26 . 34 16 . 23 . 45	12 . 36 . 45 13 . 25 . 46 14 . 23 . 56 15 . 26 . 34 16 . 24 . 35
12 . 34 . 56 13 . 26 . 45 14 . 25 . 36 15 . 23 . 46 16 . 24 . 35	12 . 35 . 46 13 . 26 . 45 14 . 23 . 56 15 . 24 . 36 16 . 25 . 34	12 . 36 . 45 13 . 24 . 56 14 . 26 . 35 15 . 23 . 46 16 . 25 . 34

which is in fact the theorem whereon depends the existence, for six letters, of a 6-valued function not symmetrical in respect of five letters. There is not any peculiarity in the syntheme of duads which above presented itself; the occurrence of this particular syntheme, instead of any other, arises merely from the arbitrary selection of the suffixes of the ϕ's.

12. It is hardly necessary to remark that if the pentagon 12345 had been assumed negative instead of positive, the only difference would be that the ϕ's would have their signs reversed.

13. I proceed now to the calculation of the Auxiliary Equation. As the working is rather easier for that form, I shall in the first instance take for the given quintic the denumerate form

$$(a, b, c, d, e, f \between v, 1)^5 = 0.$$

Representing, as before, the roots x_1, x_2, x_3, x_4, x_5 of this equation by 1, 2, 3, 4, 5, and writing

$$12345 = 12 + 23 + 34 + 45 + 51, \text{ \&c.}$$

(where on the right-hand side 12, 23, &c. stand for x_1x_2, x_2x_3, &c.), we have to find the equation

$$(* \between \phi, 1)^6 = 0,$$

the roots whereof are

$$\phi_1 = 12345 - 24135, \quad \phi_4 = 21435 - 13245,$$
$$\phi_2 = 13425 - 32145, \quad \phi_5 = 31245 - 14325,$$
$$\phi_3 = 14235 - 43125, \quad \phi_6 = 41325 - 12435.$$

As already remarked, the coefficients are seminvariants, and if the equation is in the first instance calculated for the particular case $f = 0$, the terms in f can be separately determined. But putting $f = 0$, one of the roots, say 5, becomes $= 0$, and the remaining roots 1, 2, 3, 4 are the roots of the quartic equation $(a, b, c, d, e \between v, 1)^4 = 0$.

14. Writing for shortness

$$1234 = 12 + 23 + 34, \text{ \&c.}$$

and putting also

$$A = 12 + 34,$$
$$B = 13 + 42,$$
$$C = 14 + 23,$$

then we have

$$\phi_1 = 1234 - 2413 = 12 + 23 + 34 - 24 - 41 - 13 = A - B + 23 - 14,$$
$$\phi_2 = 1342 - 3214 = 13 + 34 + 42 - 32 - 21 - 14 = B - C + 34 - 12,$$
$$\phi_3 = 1423 - 4312 = 14 + 42 + 23 - 43 - 31 - 12 = C - A + 42 - 13,$$

$$\phi_4 = 2143 - 1324 = 21 + 14 + 43 - 13 - 32 - 24 = A - B - 23 + 14,$$
$$\phi_5 = 3124 - 1423 = 31 + 12 + 24 - 14 - 42 - 23 = B - C - 34 + 12,$$
$$\phi_6 = 4132 - 1243 = 41 + 13 + 32 - 12 - 24 - 43 = C - A - 42 + 13.$$

15. We have then

$$(\phi - \phi_1)(\phi - \phi_4) = (\phi - A + B)^2 - (14 - 23)^2$$
$$= (\phi - A + B)^2 - C^2 + 4 \,.\, 1234,$$

where 1234 denotes the product of the four roots; the functions A, B, C, and the product 1234, are each of the degree zero in the coefficients (a, b, c, d, e); and if we put

$$\begin{aligned} \mathrm{b} &= - c, \\ \mathrm{c} &= -4ae + bd, \\ \mathrm{d} &= 4ace - ad^2 - b^2e, \end{aligned}$$

then we in fact have

$$\begin{aligned} a\,\Sigma A &= -\mathrm{b}, \\ a^2\,\Sigma AB &= \mathrm{c}, \\ a^3\,ABC &= -d, \\ a\,.\,1234 &= e. \end{aligned}$$

But on the understanding that ϕ is ultimately to be changed into $a\phi$, it is allowable, and it will be convenient to write

$$\begin{aligned} \Sigma A &= -\mathrm{b}, \\ \Sigma AB &= \mathrm{c}, \\ ABC &= -\mathrm{d}, \\ 1234 &= ae. \end{aligned}$$

16. I assume also

$$\begin{aligned} B + C - A &= \alpha, \\ C + A - B &= \beta, \\ A + B - C &= \gamma; \end{aligned}$$

we have thus

$$\begin{aligned} (\phi - \phi_1)(\phi - \phi_4) &= (\phi + \alpha)(\phi - \beta) + 4ae, \text{ and therefore also} \\ (\phi - \phi_2)(\phi - \phi_5) &= (\phi + \beta)(\phi - \gamma) + 4ae, \\ (\phi - \phi_3)(\phi - \phi_6) &= (\phi + \gamma)(\phi - \alpha) + 4ae, \end{aligned}$$

so that the equation in ϕ is

$$[(\phi + \beta)(\phi - \gamma) + 4ae]\,[(\phi + \gamma)(\phi - \alpha) + 4ae]\,[(\phi + \alpha)(\phi - \beta) + 4ae] = 0.$$

17. To obtain the symmetrical functions of α, β, γ it is only necessary to remark that if in the identical equation

$$(1,\ \mathrm{b},\ \mathrm{c},\ \mathrm{d}\between\theta,\ 1)^3 = (\theta - A)(\theta - B)(\theta - C),$$

we put $\frac{1}{2}(\chi + A + B + C)$, $= \frac{1}{2}(\chi - \mathrm{b})$, in the place of θ, the equation becomes

$$(1,\ \mathrm{b},\ \mathrm{c},\ \mathrm{d}\between\chi - \mathrm{b},\ 2)^3 = (\chi + \alpha)(\chi + \beta)(\chi + \gamma),$$

so that we have

$$\begin{aligned} \Sigma\alpha &= -\mathrm{b} &&= c, \\ \Sigma\alpha\beta &= -\mathrm{b}^2 + 4\mathrm{c} &&= -16ae + 4bd - c^2, \\ \alpha\beta\gamma &= \mathrm{b}^3 - 4\mathrm{bc} + 8\mathrm{d} &&= 16ace - 8ad^2 - 8b^2e + 4bcd - c^3. \end{aligned}$$

18. The developed expression for the equation in ϕ is easily found to be

$$\left.\begin{array}{l} \phi^6 \\ +\phi^4 . -\Sigma\alpha^2 \quad +12\,ae \\ +\phi^2 . \quad \Sigma\alpha^2\beta^2 - \quad 4\,ae\,(\Sigma\alpha^2+\Sigma\alpha\beta)+48\,a^2e^2 \\ +\phi \;. -4\,ae\,(\alpha-\beta)(\beta-\gamma)(\gamma-\alpha) \\ + \qquad -\alpha^2\beta^2\gamma^2+4\,ae\,\alpha\beta\gamma\,\Sigma\alpha-16\,a^2e^2\,\Sigma\alpha\beta+64\,a^3e^3 \end{array}\right\} = 0.$$

19. In this equation the coefficient of ϕ is

$$\begin{aligned} &-4\;\; ae\,.\,8\,(B-A)\,(C-B)\,(A-C) \\ =\;\; &\;\; 32\,ae \quad (A-B)\,(B-C)\,(C-A)\,; \end{aligned}$$

or, neglecting the multiplier a, it is

$$-32\,.\,1\,.\,2\,.\,3\,.\,4\,(1-2)\,(1-3)\,(1-4)\,(2-3)\,(2-4)\,(3-4),$$

which is the value for $5=0$, of

$$-32\,(1-2)\,(1-3)\,(1-4)\,(1-5)\,(2-3)\,(2-4)\,(2-5)\,(3-4)\,(3-5)\,(4-5),$$

i.e. the coefficient in question is

$$-32\,.\,25\,\sqrt{5}\,\sqrt{a^4f^4+\&c.} = -800\,\sqrt{5}\,\sqrt{a^4f^4+\&c.},$$

where $a^4f^4+\&c.$ denotes the discriminant of the denumerate form

$$(a,\ b,\ c,\ d,\ e,\ f\between v,\ 1)^5.$$

20. The remaining coefficients are rational functions of a, b, c, d, e, which have to be completed by the introduction of the terms in f. We have

Coeff. ϕ^4

$$\begin{array}{ll} =-\quad (\Sigma\alpha)^2 & =-c^2 \\ +\;\; 2\,\Sigma\alpha\beta & \;\; +2\,(-16\,ae+4\,bd-c^2) \\ +12\,ae & \;\; +12\,ae. \end{array}$$

Coeff. ϕ^2

$$\begin{array}{ll} = \qquad (\Sigma\alpha\beta)^2 & = \quad (-16\,ae+4\,bd-c^2)^2 \\ -\quad 2\,\alpha\beta\gamma\Sigma\alpha & \quad -2c\,(16\,ace-8\,ad^2-8\,b^2e+4\,bcd-c^3) \\ -\quad 4\,ae\,(\Sigma\alpha)^2 & \quad -4\,ac^2e \\ +\quad 4\,ae\,(\Sigma\alpha\beta) & \quad +4\,ae\,(-16\,ae+4\,bd-c^2) \\ +48\,a^2e^2 & \quad +48\,a^2e^2. \end{array}$$

Coeff. ϕ^0

$$\begin{array}{ll} =-\qquad \alpha^2\beta^2\gamma^2 & =-(16\,ace-8\,ad^2-8\,b^2e+4\,bcd-c^3)^2 \\ +\quad 4\,ae\,\alpha\beta\gamma\Sigma\alpha & \quad +4\,ace\,(16\,ace-8\,ad^2-8\,b^2e+4\,bcd-c^3) \\ -16\,a^2e^2\,\Sigma\alpha\beta & \quad -16\,a^2e^2\,(-16\,ae+4\,bd-c^2) \\ +64\,a^3e^3. & \quad +64\,a^3e^3. \end{array}$$

21. Effecting the developments, these are

$ae \; - 20$	$a^2e^2 \; + 240$	$a^3e^3 \; + 320$
$bd \; + \;\; 8$	$abde \; - 112$	$a^2bde^2 - \;\; 64$
$c^2 \; - \;\; 3$	$ac^2e \; - \;\;\; 8$	$a^2c^2e^2 \; - 176$
	$acd^2 \; + \;\; 16$	$a^2cd^2e + 224$
	$b^2ce \; + \;\; 16$	$a^2d^4 \; - \;\; 64$
	$b^2d^2 \; + \;\; 16$	$ab^2ce^2 + 224$
	$bc^2d \; - \;\; 16$	$ab^2d^2e - 128$
	$c^4 \; + \;\;\; 3$	$abc^2de - 112$
		$abcd^3 + \;\; 64$
		$ac^4e \; + \;\; 28$
		$ac^3d^2 \; - \;\; 16$
		$b^4e^2 \; - \;\; 64$
		$b^3cde \; + \;\; 64$
		$b^2c^3e \; - \;\; 16$
		$b^2c^2d^2 \; - \;\; 16$
		$bc^4d \; + \;\;\; 8$
		$c^6 \; - \;\;\; 1$

the first of which is in fact complete; the others being completed, we obtain the equation in ϕ, viz.:

22. For the denumerate form $(a, b, c, d, e, f \between v, 1)^5 = 0$, the equation in ϕ is

(

$a^6 \times$		$a^4 \times$		$a^2 \times$	$-800\sqrt{5}a^2\sqrt{\square}\times$	
$+1$	0	$ae - 20$	0	$a^2df \; - 400$	$+1$	$a^3cf^2 \; + 4000$
		$bd + \;\; 8$		$a^2e^2 \; + 240$		$a^3def \; - 1600$
		$c^2 - \;\; 3$		$abcf \; + 240$		$a^3e^3 \; + \;\; 320$
				$abde \; - 112$		$a^2b^2f^2 \; - 1600$
				$ac^2e \; - \;\;\; 8$		$a^2bcef \; - \;\; 640$
				$acd^2 \; + \;\; 16$		$a^2bd^2f + \;\; 640$
				$b^3f \; - \;\; 64$		$a^2bde^2 \; - \;\;\; 64$
				$b^2ce \; + \;\; 16$		$a^2c^2df \; - \;\;\; 80$
				$b^2d^2 \; + \;\; 16$		$a^2c^2e^2 \; - \;\; 176$
				$bc^2d \; - \;\; 16$		$a^2cd^2e \; + \;\; 224$
				$c^4 \; + \;\;\; 3$		$a^2d^4 \; - \;\;\; 64$
						$ab^3ef \; + \;\; 384$
						$ab^2cdf \; - \;\; 192$
						$ab^2ce^2 \; + \;\; 224$
						$ab^2d^2e \; - \;\; 128$
						$abc^3f \; + \;\;\; 48$
						$abc^2de \; - \;\; 112$
						$abcd^3 \; + \;\;\; 64$
						$ac^4e \; + \;\;\; 28$
						$ac^3d^2 \; - \;\;\; 16$
						$b^4e^2 \; - \;\;\; 64$
						$b^3cde \; + \;\;\; 64$
						$b^2c^3e \; - \;\;\; 16$
						$b^2c^2d^2 \; - \;\;\; 16$
						$bc^4d \; + \;\;\;\; 8$
						$c^6 \; - \;\;\;\; 1$

$\between \phi, 1)^6 = 0,$

where $\square$, $= a^4f^4 + \&c.$, denotes the discriminant for the denumerate form.

I proceed now to form the expression for $\phi_1\phi_4 + \phi_2\phi_5 + \phi_3\phi_6$.

23. Writing for convenience x in the place of the root 5, we have

$$\phi_1 = A - B + \{23 - 14 + x(1 + 4 - 2 - 3)\}$$
$$\phi_4 = A - B - \{23 - 14 + x(1 + 4 - 2 - 3)\},$$

or

$$\phi_1\phi_4 = (A - B)^2 - \{23 - 14 + x(1 + 4 - 2 - 3)\}^2.$$

The terms without x are, as before, $(A - B)^2 - C^2 + 4 \,.\, 1234$, or $-\alpha\beta + 4 \,.\, 1234$, and we have

$$\begin{aligned}\phi_1\phi_4 = &-\alpha\beta + 4 \,.\, 1234 \\ &+ 2x(1 + 4 - 2 - 3)(14 - 23) \\ &- x^2(1 + 4 - 2 - 3)^2;\end{aligned}$$

and in like manner

$$\begin{aligned}\phi_2\phi_5 = &-\beta\gamma + 4 \,.\, 1234 \\ &+ 2x(1 + 2 - 3 - 4)(12 - 34) \\ &- x^2(1 + 2 - 3 - 4)^2,\end{aligned}$$

and

$$\begin{aligned}\phi_3\phi_6 = &-\gamma\alpha + 4 \,.\, 1234 \\ &+ 2x(1 + 3 - 4 - 2)(13 - 42) \\ &- x^2(1 + 3 - 4 - 2)^2.\end{aligned}$$

24. The roots 1, 2, 3, 4, contained in these expressions explicitly, and in α, β, γ, are the roots of the equation $\frac{1}{v - x}\ (a, b, c, d, e, f \between v, 1)^5 = 0$, or, what is the same thing,

$$(a', b', c', d', e' \between v, 1)^4 = 0,$$

where

$$\begin{aligned}a' &= a, \\ b' &= ax + b, \\ c' &= ax^2 + bx + c, \\ d' &= ax^3 + bx^2 + cx + d, \\ e' &= ax^4 + bx^3 + cx^2 + dx + e.\end{aligned}$$

Omitting, as before, a power of a, which is ultimately restored, we have

$$\begin{aligned}\phi_1\phi_4 + \phi_2\phi_5 + \phi_3\phi_6 = &-\Sigma\alpha\beta + 12a'e' \\ &+ 2x\Sigma(1 + 4 - 2 - 3)(14 - 23) \\ &- x^2\Sigma(1 + 4 - 2 - 3)^2,\end{aligned}$$

where the Σ's in the second and third lines denote each of them the sum of the three terms obtained by the cyclical permutations of 2, 3, 4.

The first line is

$$(16\,a'e' - 4\,b'd' + c'^2) + 12\,a'e'$$
$$= 28\,a'e' - 4\,b'd' + 1\,c'^2.$$

The second line is $2x$ into $\Sigma\,1^2 2 - 3\,\Sigma\,123$,

$$= (-\,b'c' + 3\,a'd') + 3\,a'd'\,;$$

or it is

$$= 2x\,(6\,a'd' - 1\,b'c')\,;$$

and the third line is $-\,x^2$ into $3\,\Sigma\,1^2 - 2\,\Sigma\,12$,

$$= 3\,(b'^2 - 2\,a'c') - 2\,a'c'\,;$$

or it is

$$= x^2\,(8\,a'c' - 3\,b'^2).$$

Hence, combining the three terms,

$$\begin{aligned}\phi_1\phi_4 + \phi_2\phi_5 + \phi_3\phi_6 = {} & 28\,a'e' - 4\,b'd' + 1\,c'^2 \\ & + x\ \ (12\,a'd' - 2\,b'c') \\ & + x^2\,(8\,a'c' - 3\,b'^2),\end{aligned}$$

or substituting for $(a',\ b',\ c',\ d',\ e')$ their values, the right-hand side is

$$= (40\,a^2,\ 32\,ab,\ 28\,ac - 8\,b^2,\ 44\,ad - 8\,bc,\ 28\,ae - 4\,bd + 1\,c^2 \between x,\ 1)^4,$$

where x stands for x_5, and on the left-hand side the factor a^2 is to be restored.

25. Writing for shortness

$$(* \between x,\ 1)^4 = (40\,a^2,\ 32\,ab,\ 28\,ac - 8\,b^2,\ 44\,ad - 8\,bc,\ 28\,ae - 4\,bd + 1\,c^2 \between x,\ 1)^4,$$

the equation is

$$a^2\,(\phi_1\phi_4 + \phi_2\phi_5 + \phi_3\phi_6) = (* \between x_5,\ 1)^4;$$

and the system of equations to which this belongs is

$$\begin{aligned} a^2\,(\phi_1\phi_6 + \phi_2\phi_4 + \phi_3\phi_5) &= (* \between x_1,\ 1)^4, \\ a^2\,(\phi_1\phi_2 + \phi_3\phi_4 + \phi_5\phi_6) &= (* \between x_2,\ 1)^4, \\ a^2\,(\phi_1\phi_5 + \phi_2\phi_3 + \phi_4\phi_6) &= (* \between x_3,\ 1)^4, \\ a^2\,(\phi_1\phi_3 + \phi_2\phi_6 + \phi_3\phi_5) &= (* \between x_4,\ 1)^4, \\ a^2\,(\phi_1\phi_4 + \phi_2\phi_5 + \phi_3\phi_6) &= (* \between x_5,\ 1)^4; \end{aligned}$$

so that the roots $x_1,\ x_2,\ x_3,\ x_4,\ x_5$ will be rational functions of the combinations $\phi_1\phi_6 + \phi_2\phi_4 + \phi_3\phi_5$, &c. respectively, of the equation in ϕ.

26. Passing now to the standard form $(a, b, c, d, e, f \rangle\!\langle v, 1)^5 = 0$, the equation in ϕ is

(

$a^6 \times$		$-100\,a^4 \times$		$2000\,a^2 \times$	$-800\,a^2\sqrt{5}\sqrt{\square} \times$	$40000 \times$
$+1$	0	$ae\ +1$	0	$a^2df\ -2$	$+1$	$a^3cf^2\ +1$
		$bd\ -4$		$a^2e^2\ +3$		$a^3def\ -2$
		$c^2\ +3$		$abcf\ +6$		$a^3e^3\ +1$
				$abde\ -14$		$a^2b^2f^2\ -1$
				$ac^2e\ -2$		$a^2bcef\ -4$
				$acd^2\ +8$		$a^2bd^2f\ +8$
				$b^3f\ -4$		$a^2bde^2\ -2$
				$b^2ce\ +10$		$a^2c^2df\ -2$
				$b^2d^2\ +20$		$a^2c^2e^2\ -11$
				$bc^2d\ -40$		$a^2cd^2e\ +28$
				$c^4\ +15$		$a^2d^4\ -16$
						$ab^3ef\ +6$
						$ab^2cdf\ -12$
						$ab^2ce^2\ +35$
						$ab^2d^2e\ -40$
						$abc^3f\ +6$
						$abc^2de\ -70$
						$abcd^3\ +80$
						$ac^4e\ +35$
						$ac^3d^2\ -40$
						$b^4e^2\ -25$
						$b^3cde\ +100$
						$b^2c^3e\ -50$
						$b^2c^2d^2\ -100$
						$bc^4d\ +100$
						$c^6\ -25$

$\rangle\!\langle \phi, 1)^6 = 0,$

where $\square$, $= a^4f^4 + \&c.$, denotes the Discriminant for the Standard form.

27. And if we put

$$(* \rangle\!\langle x, 1)^4 = 20\,(2\,a^2,\ 8\,ab,\ 22\,ac - 10\,b^2,\ 18\,ad - 10\,bc,\ 7\,ae - 10\,bd + 5\,c^2 \rangle\!\langle x, 1)^4,$$

then we have

$$a^2(\phi_1\phi_6 + \phi_2\phi_4 + \phi_3\phi_5) = (* \rangle\!\langle x_1, 1)^4,$$

$$a^2(\phi_1\phi_2 + \phi_3\phi_4 + \phi_5\phi_6) = (* \rangle\!\langle x_2, 1)^4,$$

$$a^2(\phi_1\phi_5 + \phi_2\phi_3 + \phi_4\phi_6) = (* \rangle\!\langle x_3, 1)^4,$$

$$a^2(\phi_1\phi_3 + \phi_2\phi_6 + \phi_3\phi_5) = (* \rangle\!\langle x_4, 1)^4,$$

$$a^2(\phi_1\phi_4 + \phi_2\phi_5 + \phi_3\phi_6) = (* \rangle\!\langle x_5, 1)^4,$$

which lead to rational expressions for the roots x_1, x_2, x_3, x_4, x_5 in terms of the combinations $\phi_1\phi_6 + \phi_2\phi_4 + \phi_3\phi_6$, &c. respectively.

28. Consider now the quintic function

$$U=(a,\ b,\ c,\ d,\ e,\ f\between x,\ y)^5=a\,(x-\alpha y)\,(x-\beta y)\,(x-\gamma y)\,(x-\delta y)\,(x-\epsilon y)\,;$$

and treating the numbers 1, 2, 3, 4, 5 as corresponding to α, β, γ, δ, ϵ respectively, write

$$\Phi=\overbrace{12345}-\overbrace{24135},$$

where

$$\overbrace{12345}=\overbrace{12}+\overbrace{23}+\overbrace{34}+\overbrace{45}+\overbrace{51},\ \&c.,$$

in which $\overbrace{12}$, &c. denote respectively

$$\frac{1}{y^2}\frac{1}{x-\alpha y}\cdot\frac{1}{x-\beta y},\ \&c.$$

Then we have

$$\Phi=\frac{a}{U}[\phi x-\chi y],$$

where

$$\phi=12345-24135,$$

and

$$12345=12+23+34+45+51=\alpha\beta+\beta\gamma+\gamma\delta+\delta\epsilon+\epsilon\alpha,$$

$$24135=24+41+13+35+52=\beta\delta+\delta\alpha+\alpha\gamma+\gamma\epsilon+\epsilon\beta,$$

and where

$$\chi=(12345)-(24135),$$

and

$$(12345)=123+234+345+451+512=\alpha\beta\gamma+\beta\gamma\delta+\gamma\delta\epsilon+\delta\epsilon\alpha+\epsilon\alpha\beta,$$

$$(24135)=241+413+135+352+524=\beta\delta\alpha+\delta\alpha\gamma+\alpha\gamma\epsilon+\gamma\epsilon\beta+\epsilon\beta\delta.$$

29. In fact,

$$\Phi-\frac{a}{U}\left\{(\overbrace{12345})-(\overbrace{24135})\right\},$$

where

$$(\overbrace{12345})=\overbrace{123}+\overbrace{234}+\overbrace{345}+\overbrace{451}+\overbrace{512},$$

$$(\overbrace{24135})=\overbrace{241}+\overbrace{413}+\overbrace{135}+\overbrace{352}+\overbrace{524},$$

where $\overbrace{123}$, &c. denote respectively

$$\frac{1}{y^2}(x-\alpha y)\,(x-\beta y)\,(x-\gamma y),\ \&c.,$$

and $(\overbrace{12345})-(\overbrace{24135})$ thus presents itself as a cubic function divided by y^2. But in this cubic function the coefficients of x^3, x^2y vanish. For the coefficient of any power of x will be

$$123+234+345+451+512-241-413-135-352-524,$$

where, first, for x^3, 123, &c. denote respectively unity; the coefficient of x^3 therefore vanishes. Next, for x^2y, 123, &c. denote respectively $-(1+2+3)$, &c. $(=\alpha+\beta+\gamma)$, and the coefficient of x^2y also vanishes. But for xy^2, 123, &c. denote respectively $12+23+31$ $(=\alpha\beta+\beta\gamma+\gamma\alpha)$, &c. respectively; the positive terms are

$$(12+23+31)+(23+34+42)+(34+45+53)+(45+51+14)+(51+12+25),$$

which are

$$=2\,(12+23+34+45+51)+(24+41+13+35+52)$$
$$=2\,.\,12345+24153\,;$$

and the negative terms, taken positively, are

$$(24+41+12)+(41+13+34)+(13+35+51)+(35+52+23)+(52+24+45),$$

which are

$$=(12+23+34+45+51)+2\,(24+41+13+35+52)$$
$$=12345+2\,.\,24135\,;$$

so that the difference, or coefficient of xy^2, is

$$=12345-24135,$$

which is $=\phi$.

And for y^3, 123, &c. denote respectively -123 $(=-\alpha\beta\gamma)$, &c., so that the coefficient of y^3 is $=\chi$.

30. The cubic function is therefore $=\phi xy^2-\chi y^3$; and dividing by y^2, we have

$$\Phi=\frac{a}{U}(\phi x-\chi y).$$

Φ is thus a fractional covariantive function, the leading coefficient whereof is ϕ, and the equation for the determination of Φ is consequently that deduced from the equation for ϕ, by replacing therein the seminvariants by the corresponding covariants. The equation [denoting the covariants as in 141 and 143, A the quintic itself, &c.] is

$$\left(\begin{array}{l} A^6, \\ 0, \\ -100A^4B, \\ 0, \\ +2000\,A^2(6B^2-4H) \\ -800\,A^2\sqrt{5}\,\sqrt{\text{disct.}},=Q, \\ AJ-25D^2, \end{array}\right)(\Phi,\ 1)^6=0,$$

where the coefficients are in regard to $(x,\ y)$ of the orders 30, —, 22, —, 14, 10, 6 respectively. The last coefficient [now given in the form $AJ-25D^2$], being of the degree 6 in the coefficients (a, b, c, d, e, f), is not given in the Tables; it is therefore merely indicated by $(\mathfrak{A}, \mathfrak{B}, \mathfrak{C}, \mathfrak{D}, \mathfrak{E}, \mathfrak{F}, \mathfrak{G} \between x, y)^6$, the leading coefficient $\mathfrak{A}$ being of course the last coefficient in the equation for ϕ, to the standard form.

I refrain from at present entering into the consideration of the values of the expressions $\Phi_1\Phi + \Phi_2\Phi_5 + \Phi_3\Phi_6$, &c.

ADDITION, Nov. 10, 1862. [Originally printed in a later memoir "On Tschirnhausen's Transformation" *post*, 275.]

I take the opportunity of remarking, with reference to [the foregoing] memoir, that I recently discovered that the auxiliary equation there considered is in fact due to Jacobi, who, in his paper, "Observatiunculæ ad theoriam æquationum pertinentes," *Crelle*, t. XIII. (1835), pp. 340—352, under the heading "Observatio de æquatione sexti gradus ad quam æquationes quinti gradus revocari possunt," gives the theory, and observes that the equation is of the form

$$\phi^6 + a_2\phi^4 + a_4\phi^2 + a_6 = 32\sqrt{\square}\,\phi,$$

and mentions that the value of a_2 is easily found to be (I adapt his notation to the denumerate form $(a, b, c, d, e, f \between v, 1)^5 = 0$)

$$= 40ae - 16bd + 6c^2$$

(this ought, however, to be divided by -2), but that the values of a_4, a_6 "paullo ampliores calculos poscunt."

The value of the coefficient in question is correctly obtained (page 270 of my memoir [as printed in the *Philosophical Transactions*]) in the form

$$\begin{aligned} & - c^2 \\ & + 2(-16ae + 4bd - c^2) \\ & + 12\,ae; \end{aligned}$$

but the reduced value is given in two places (page 271) as equal to

$$\begin{array}{lcl} -\,32\,ae, & \text{this should be} & -\,20\,ae, \\ +\;\;8\,db, & ,, & +\;\;8\,bd, \\ -\;\;3\,c^2, & ,, & -\;\;3\,c^2. \end{array}$$

The last-mentioned correct value was used in obtaining the coefficient for the standard form, which coefficient is given correctly, page 274. [The correction here indicated -20 in place of -32, is made *ante* p. 318.]

269.

A SEVENTH MEMOIR ON QUANTICS.

[From the *Philosophical Transactions of the Royal Society of London*, vol. CLI. (for the year 1861), pp. 277—292. Received February 28,—Read March 14, 1861.]

THE present memoir relates chiefly to the theory of ternary cubics. Since the date of my Third Memoir on Quantics, [144], M. Aronhold has published the continuation of his researches on ternary cubics, in the memoir "Theorie der homogenen Functionen dritten Grades von drei Veränderlichen," *Crelle*, t. LV. pp. 97—191 (1858). He there considers two derived contravariants, linear functions of the fundamental ones, and which occupy therein the position which the fundamental contravariants PU, QU do in my Third Memoir; in the notation of the present memoir these derived contravariants are

$$YU = \quad 3T\,.\,PU - 4S\,.\,QU,$$
$$ZU = -48\,S^2\,.\,PU + \ T\,.\,QU;$$

and for the canonical form $x^3 + y^3 + z^3 + 6lxyz$, they acquire respectively the factor $(1+8l^3)^2$, viz. in this case

$$YU = (1+8l^3)^2\,\{\qquad l\ \ (\xi^3+\eta^3+\zeta^3) - \ 3\ \ \xi\eta\zeta\},$$
$$ZU = (1+8l^3)^2\,\{(1+2l^3)\,(\xi^3+\eta^3+\zeta^3) + 18\,l^2\xi\eta\zeta\}.$$

The derived contravariants have with the covariants U, HU, even a more intimate connexion than have the contravariants PU, QU; and the advantage of the employment of YU, ZU fully appears by M. Aronhold's memoir.

But the conclusion is, not that the contravariants PU, QU are to be rejected, but that the system is to be completed by the addition thereto of two derived covariants, linear functions of U, HU; these derived covariants, suggested to me by M. Aronhold's memoir, are in the present memoir called CU, DU; their values are

$$CU = \ \ -T\,.\,U + 24\,S\,.\,HU,$$
$$DU = \ \ 8\,S^2\,.\,U - \ 3\,T\,.\,HU:$$

and for the canonical form $x^3+y^3+z^3+6lxyz$, they acquire respectively, not indeed $(1+8l^3)^2$, but the simple power $(1+8l^3)$, as a factor, viz. in this case

$$\begin{aligned} CU &= (1+8l^3)\{\ \ (-1+4l^3)(x^3+y^3+z^3)+ \qquad\qquad 18lxyz\}, \\ DU &= (1+8l^3)\{l^2(\ \ 5+4l^3)(x^3+y^3+z^3)+3(1-10l^3)\,xyz\}; \end{aligned}$$

it was in fact by means of this condition as to the factor $(1+8l^3)$, that the foregoing expressions for CU, DU were obtained (1).

The formulæ of my Third Memoir and those of M. Aronhold are by this means brought into harmony and made parts of a whole; instead of the two intermediates

$$\alpha U+6\beta HU,\quad 6\alpha PU+\beta QU,$$

in Tables 68 and 69 of my Third Memoir, or of the intermediates

$$\alpha U+6\beta HU,\quad -2\alpha YU+2\beta ZU,$$

of M. Aronhold's theory, we have the four intermediates

$$\alpha U+6\beta HU,\quad -2\alpha YU+2\beta ZU,\quad 2\alpha CU-2\beta DU,\quad 6\alpha PU+\beta QU,$$

in Tables 74, 75, 76, and 77 of the present memoir. These four Tables embrace the former results, and the new ones which relate to the covariants CU, DU; and they are what is most important in the present memoir. I have, however, excluded from the Tables, and I do not in the memoir consider (otherwise than incidentally) the covariant of the sixth order ΘU, or the contravariant (reciprocant) FU.

I have given in the memoir a comparison of my notation with that of M. Aronhold. A short part of the memoir relates to the binary cubic and the binary quartic, viz. each of these quantics has a covariant of its own order, forming with it an intermediate $\alpha U+\beta W$, the covariants whereof contain quantics in (α, β), the coefficients of which are invariants of the original quantic. The formulæ which relate to these cases are in fact given in my Fifth Memoir, [156], but they are reproduced here in order to show the relations between the quantics in (α, β) contained in the formulæ. As regards the binary quartic, these results are required for the discussion of the like question in regard to the ternary cubic, viz. that of finding the relations between the different quantics in (α, β) contained in the formulæ relating to the ternary cubic. Some of these relations have been obtained by M. Hermite in the memoir "Sur les formes cubiques à trois indéterminées," (*Liouville*, t. III. pp. 37—40 (1858), and in that "Sur la Résolution des équations du quatrième degré," *Comptes Rendus*, XLVI. p. 715 (1858), and by M. Aronhold in his memoir already referred to; and in particular I reproduce and demonstrate some of the results in the last-mentioned memoir of M. Hermite. But the relations in question are in the present memoir exhibited in a more complete and systematic form.

1 M. Aronhold, in a letter dated Berlin, 17 June 1861, has pointed out to me that the covariants CU, DU are in his notation P_{S_f}, P_{T_f}, and that they belong to the forms called Conjugate Forms, § 27 of his memoir. But the explicit development of the properties of these covariants is not on this account the less interesting. Added 20 Sept. 1861.—A.C.

The paragraphs and Tables of the present memoir are numbered consecutively with those of my former memoirs on Quantics.

231. For the binary cubic $(a, b, c, d \between x, y)^3$, if U be the cubic itself, HU the Hessian, ΦU the cubicovariant, and $\square$ the discriminant (see Fifth Memoir, Nos. 115, 118), then

Covariant and other Tables, No. 71.

$$\begin{aligned} H(\alpha U+\beta\Phi U) &= (\alpha^2-\beta^2\square)\,HU, \\ \Phi(\alpha U+\beta\Phi U) &= -\tfrac{1}{2}\partial_\beta(\alpha^2-\beta^2\square)\,.\,U \\ &\quad +\tfrac{1}{2}\partial_\alpha(\alpha^2-\beta^2\square)\,.\,U, \\ \square(\alpha U+\beta\Phi U) &= (\alpha^2-\beta^2\square)^2\,\square, \end{aligned}$$

so that the quantics in (α, β) all of them depend on $\alpha^2-\beta^2\square$.

232. For the binary quartic $(a, b, c, d, e \between x, y)^4$, if U be the quartic itself, HU the Hessian, ΦU the cubicovariant, I, J, the quadrinvariant and the cubinvariant, and $\square$ $(=I^3-27J^2)$ the discriminant (see Fifth Memoir, Nos. 128, 134), then

Table No. 72.

$$\begin{aligned} \Phi\,(\alpha U+6\beta HU) &= (1,\ 0,\ -9I,\ -54J \between \alpha,\ \beta)^3\,\Phi U, \\ H\,(\alpha U+6\beta HU) &= -\tfrac{1}{18}\,\partial_\beta\,(1,\ 0,\ -9I,\ -54J \between \alpha,\ \beta)^3\,.\,U \\ &\quad +\tfrac{1}{3}\,\partial_\alpha\,(1,\ 0,\ -9I,\ -54J \between \alpha,\ \beta)^3\,.\,HU, \\ I\,(\alpha U+6\beta HU) &= (I,\ 18J,\ 3I^2 \between \alpha,\ \beta)^2, \\ J\,(\alpha U+6\beta HU) &= (J,\ I^2,\ 9IJ,\ -I^3+54J^2 \between \alpha,\ \beta)^3, \\ \square\,(\alpha U+6\beta HU) &= {}^{(1)}(1,\ 0,\ -18I,\ 108J,\ 81I^2,\ 972IJ,\ 2916J^2 \between \alpha,\ \beta)^6\,\square \\ &= [(1,\ 0,\ -9I,\ -54J \between \alpha,\ \beta)^3]^2\,\square. \end{aligned}$$

233. Writing for the moment

$$G=(1,\ 0,\ -9I,\ -54J \between \alpha,\ \beta)^3,$$

then the Hessian, cubicovariant, and discriminant of this cubic function of (α, β) are respectively

$$\begin{aligned} HG &= -\ 3\,(I,\ 18J,\ 3I^2 \between \alpha,\ \beta)^2, \\ \Phi G &= 54\,(J,\ I^2,\ 9IJ,\ -I^3+54J^2 \between \alpha,\ \beta)^3, \\ \square G &= -108\square\,; \end{aligned}$$

so that the covariants of the intermediate $\alpha U+6\beta HU$ are all of them expressible by means of the cubic function G.

[1] The coefficient $2916J^2$ is in the Fifth Memoir erroneously given as $-2916J^2$. [This correction should have been made, vol. II. p. 549.]

It may be noticed that G is what the left-hand side of the equation

$$4(HU)^3 - 4I \,.\, HU \,.\, U^2 + JU^3 = -(\Phi U)^2$$

(see Fifth Memoir, No. 128) becomes on writing therein α, -6β, for U, HU respectively, and throwing out the factor 4.

234. I take the opportunity of remarking with respect to a binary quartic $U = (a, b, c, d \between x, y)^4$, that the Hessian of the cubicovariant, to fix the numerical factor, say $-\frac{1}{5}\{\partial_x{}^2 \Phi U \,.\, \partial_y{}^2 \Phi U - (\partial_x \partial_y \Phi U)^2\}$, is

$$= I^2 U^2 - 36 J \,.\, U \,.\, HU + 12 I (HU)^2,$$

which is

$$= \left(IU - \frac{18J}{I} HU\right)^2 + \frac{12}{I^2}(I^3 - 27J^2)(HU)^2;$$

or if $I^3 - 27J^2 = 0$, that is if the quartic has a pair of equal factors, the Hessian of the cubicovariant is a perfect square.

235. Coming now to the ternary cubic $U = (a, b, c, f, g, h, i, j, k, l \between x, y, z)^3$, I give in the first place the following comparison of my notation with that of M. Aronhold.

Aronhold.	Cayley.
f	U
Δf	$-6HU$
S	$4S$
T	$-T$
R	$-R$
S_f	$6PU$
T_f	$-2QU$
J	$64S^3 \div T^2$
P_f	$-2YU$
Q_f	$2ZU$
...	CU
...	DU
Θ	...
H	...
F	$-FU$
ψ	$2(\Theta_{,,}U - TU^2 + 4SU \,.\, HU)$,

where the notations YU, ZU (to correspond to M. Aronhold's P_f, Q_f) and the notations CU, DU are first employed in the present memoir. I remark in regard to $P_f (= -2YU)$, where, as already mentioned,

$$YU = (1 + 8l^3)^2 \{l(\xi^3 + \eta^3 + \zeta^3) - 3\xi\eta\zeta\},$$

that in my Memoir on Curves of the Third Order (*Phil. Trans.* t. CXLVII. (1857), see p. 427), [146], I was led incidentally to the curve

$$l(\xi^3+\eta^3+\zeta^3)-3\xi\eta\zeta=0,$$

and that I there gave the equation

$$3T.PU-4S.QU=(1+8l^3)^3\{l(\xi^3+\eta^3+\zeta^3)-3\xi\eta\zeta\}.$$

But the curve

$$(1+2l^3)(\xi^3+\eta^3+\zeta^3)+18l^2\xi\eta\zeta=0,$$

which corresponds to ($Q_f=2ZU$), does not occur in that memoir.

236. I remark, further, in regard to M. Aronhold's Θ, H, that these are what he calls "Zwischenformen," viz. they are covariants of the cubic and of the adjoint linear form $\xi x+\eta y+\zeta z$, or as they might be termed *Contracovariants*. For the canonical form $U=x^3+y^3+z^3+6lxyz$, the value of $\tfrac{1}{2}\Theta$ is

$$(yz-l^2x^2,\ zx-l^2y^2,\ xy-l^2z^2,\ l^2yz-lx^2,\ l^2zx-ly^2,\ l^2xy-lz^2 \between \xi,\ \eta,\ \zeta)^2,$$

which is a form which occurs incidentally in my memoir last referred to (see p. 427). The value of H in the same case is

$$(-2l(1+2l^3)x^2-6lyz,\ldots,-(1+4l^3)x^2+2l(1+2l^3)yz,\ldots \between \xi,\ \eta,\ \zeta)^2,$$

which does not occur in that memoir. In my Third Memoir on Quantics I purposely abstained from the consideration of any such forms.

237. My covariants ΘU and $\Theta_{\prime}U$ involved unsymmetrically the cubic and its Hessian, and it did not occur to me how a similar covariant, such as M. Aronhold's ψ, which involves the two functions symmetrically, was to be formed. Let (A, B, C) be the first derived functions, (a, b, c, f, g, h) the second derived functions of the cubic U, and (A', B', C') the first derived functions, (a', b', c', f', g', h') the second derived functions of the Hessian HU, then disregarding numerical factors, we have

$$\Theta U=(bc-f^2,\ldots,gh-af,\ldots \between A',\ B',\ C')^2,$$

$$\Theta_{\prime}U=(b'c'-f'^2,\ldots,g'h'-a'f',\ldots \between A,\ B,\ C)^2,$$

and

$$\psi=(bc'+b'c-2ff',\ldots,gh'+g'h-af'-a'f,\ldots \between A,\ B,\ C \between A',\ B',\ C');$$

and considering $U=0$ as the equation of a curve of the third order, the equations $\Theta U=0$, $\Theta_{\prime}U=0$, $\psi=0$ have the following significations, viz. $\Theta U=0$ is the locus of a point, such that its second or line polar with respect to the Hessian touches its first or conic polar with respect to the cubic: $\Theta_{\prime}U$ is the locus of a point such that its second or line polar with respect to the cubic touches its first or conic polar with respect to the Hessian: and $\psi=0$ is the locus of a point such that its second or line polar with respect to the cubic, and its second or line polar with respect to the Hessian are reciprocals (that is, each passes through the pole of the other of them) with respect to the conic which is the envelope of a line cutting the first or conic polar of the point with respect to the cubic, and the first or conic polar of the point with respect to the Hessian, in two pairs of points which are harmonically related to each other: such being in fact the immediate interpretation of the analytical formula. But this in passing.

238. The formulæ (Tables 68 and 69 of my Third Memoir) for the discriminants of the intermediates $\alpha U + 6\beta HU$ and $6\alpha PU + \beta QU$ respectively are

$$R(\alpha U + 6\beta HU) = [(1, \; 0, \; -24S, \; -8T, \; -48S^2 \between \alpha, \beta)^4]^3 R,$$
$$R(6\alpha PU + \beta QU) = [(48S, \; 8T, \; -96S^2, \; -24TS, \; -T^2 - 16S^3 \between \alpha, \beta)^4]^3 R.$$

In M. Hermite's paper in the *Comptes Rendus*, already referred to, there are given between these quantics in (α, β) certain relations which (although less simple than the relations that will afterwards be obtained) I now proceed to investigate. Putting in the first formula $\alpha \div \beta = p$, and in the second formula $\alpha \div \beta = \theta$, we have

$$R(pU + 6HU) = 0, \text{ if } (1, \; 0, \; -24S, \; -8T, \; -48S^2 \between p, 1)^4 = 0,$$
$$R(6\theta PU + QU) = 0, \text{ if } (48S, \; 8T, \; -96S^2, \; -24TS, \; -T^2 - 16S^3 \between \theta, 1)^4 = 0,$$

which equations in p, θ, are about to be considered in place of the quantics from which they respectively arise.

239. It is convenient to write [1]

$$A = 4S,$$
$$B = \sqrt[3]{T^2 - 64S^3}$$

(so that $T^2 = A^3 + B^3$ and, for the canonical form,

$$A = -4l + 4l^3, \quad B = 1 + 8l^3).$$

Making this change, and joining to the equation in p that derived from it by writing q for p, and interchanging A, B, we have the three equations

$$(1, \; 0, \; -6A, \; -8T, \; -3A^2 \between p, 1)^4 = 0,$$
$$(1, \; 0, \; -6B, \; -8T, \; -3B^2 \between q, 1)^4 = 0,$$
$$(12A, \; 8T, \; -6A^2, \; -6TA, \; -T^2 - \tfrac{1}{4}A^3 \between \theta, 1)^4 = 0.$$

240. The signification of the equation in q is as follows, viz. if the quantic

$$U = (* \between x, y, z)^3$$

is transformed into the canonical form

$$X^3 + Y^3 + Z^3 + 6lXYZ$$

by means of the linear equations

$$(x, y, z) = (\Lambda \between X, Y, Z),$$

[1] A is (Aronhold's and) Hermite's S, B is Hermite's S_1, and p, q, θ, Λ are Hermite's δ, δ_1, Δ, d: there is a slight inaccuracy in three of his formulæ, which should be

$$\Delta = -\tfrac{1}{2}\frac{1}{S}\left(T + \frac{S_1^2}{\delta_1}\right), \quad \delta_1 = \frac{24S^2}{f'\delta}, \quad \delta = \frac{24S^2}{f_1'\delta},$$

corresponding to formulæ in the present memoir.

where Λ is a matrix, then using the same letter Λ to represent the determinant formed out of this matrix, or determinant of substitution, we have

$$q=\frac{3}{\Lambda^2},$$

so that the equation in q is one that presents itself in the question of the reduction of the cubic to its canonical form.

In fact the linear transformation gives

$$S\Lambda^4=-l+l^4,$$
$$T\Lambda^6=1-20l^3-8l^6,$$

and thence

$$(T^2-64S^3)\,\Lambda'^2=(1+8l^3)^3,$$

which, writing B^3 in the place of T^2-64S^3, becomes

$$B^3\Lambda'^2=(1+8l^3)^3,\text{ or}$$
$$B\,\Lambda^4 = 1+8l^3\ ,\text{ or } 8l^3=B\Lambda^4-1,$$

whence also

$$\begin{aligned}8T\Lambda^6 &= 8-20\,(B\Lambda^4-1)-(B\Lambda^4-1)^2\\ &=27-18\ B\Lambda^4 \qquad -B^2\Lambda^8,\end{aligned}$$

or, as this may be written,

$$\frac{81}{\Lambda^8}-\frac{54B}{\Lambda^4}-\frac{24T}{\Lambda^2}-3B^2=0,$$

which, putting therein $q=\frac{3}{\Lambda^2}$, becomes

$$(1,\ 0,\ -6B,\ -8T,\ -3B^2 \between q,\ 1)^4=0,$$

the above-mentioned equation in q.

241. The relation between θ and q is

$$\theta=-\tfrac{1}{2}\frac{T}{A}+\frac{B^2}{2Aq},$$

as may be verified without difficulty. That between θ and p is

$$\theta=\tfrac{1}{4}\left(p+\frac{A}{p}\right),$$

as appears by the identical equation

$$(12A,\ 8T,\ -6A^2,\ -6TA,\ -T^2-\tfrac{1}{4}A^3 \between \tfrac{1}{4}(p+\frac{A}{p}),\ 1)^4$$

$$=\frac{1}{64p^4}(3A,\ 8T,\ -12A^2,\ -72TA,\ -46A^3-64T^2,\ -72TA^2,\ -12A^4,\ 8TA^2,\ 3A^5 \between p,\ 1)^8$$

$$=\frac{1}{64p^4}(1,\ 0,\ -6A,\ -8T,\ -3A^2 \between p,\ 1)^4\,.\,(3A,\ 8T,\ 6A^2,\ 0,\ -A^3 \between p,\ 1)^4,$$

where the second factor of the product on the right-hand side is

$$-\frac{p^4}{A}(1,\ 0,\ -6A,\ -8T,\ -3A^2 \between \frac{A}{p},\ 1)^4.$$

The relation between p and q is then at once found to be

$$q = \frac{\frac{2B^2}{A}}{p + \frac{A}{p} + \frac{2T}{A}},$$

or (since p, q and A, B may be simultaneously interchanged)

$$p = \frac{\frac{2A^2}{B}}{q + \frac{B}{q} + \frac{2T}{B}}.$$

242. Let the equations in p, q be represented by $\phi p = 0$, $\psi q = 0$ respectively; then we have

$$\phi p = p^4 - 6Ap^2 - 8Tp - 3A^2,$$

and therefore

$$\tfrac{1}{4}\phi' p = p^3 - 3Ap \ - 2T,$$

whence

$$\begin{aligned}\tfrac{1}{4}p\phi' p &= p^4 - 3Ap^2 - 2Tp \\ &= \qquad 3Ap^2 + 6Tp + 3A^2,\end{aligned}$$

and therefore

$$q = \frac{2B^2 p}{\frac{1}{12}p\phi' p} = \frac{24B^2}{\phi' p},$$

with a like formula for p, that is we have

$$q = \frac{24B^2}{\phi' p},\ p = \frac{24A^2}{\psi' q},$$

which with the equation

$$\theta = \tfrac{1}{4}\left(p + \frac{A}{p}\right),$$

are the system of equations connecting θ, p, q.

243. As already remarked, we have to consider the two derived covariants

$$\begin{aligned}CU &= -T\,.\,U + 24S\,.\,HU,\\ DU &= 8S^2\,.\,U - \ 3T\,.\,HU,\end{aligned}$$

and the two derived contravariants

$$\begin{aligned}YU &= \qquad 3T\,.\,PU - 4S\,.\,QU,\\ ZU &= -48S^2\,.\,PU + \ T\,.\,QU,\end{aligned}$$

which for the canonical form $x^3+y^3+z^3+6lxyz$ are as follows:

Table No. 70 (addition to).

$$\begin{aligned}
CU &= (1+8l^3)\ [(-1+4l^3)(x^3+y^3+z^3) + 18l\ xyz],\\
DU &= (1+8l^3)\ [l^2(5+4l^3)(x^3+y^3+z^3) + 3(1-10l^3)\ xyz],\\
YU &= (1+8l^3)^2\ [\ l(\xi^3+\eta^3+\zeta^3) - 3\xi\eta\zeta],\\
ZU &= (1+8l^3)^2\ [(1+2l^3)(\xi^3+\eta^3+\zeta^3) + 18l^2\xi\eta\zeta].
\end{aligned}$$

244. We have conversely

$$\begin{aligned}
3R.\ \ U &= 3T\ .\ CU + 24S\ .\ DU,\\
3R\ .\ HU &= 8S^2\ .\ CU + \ \ T\ .\ DU,
\end{aligned}$$

and

$$\begin{aligned}
-3R\ .\ PU &= \ \ T\ .\ YU + 4S\ .\ QU,\\
-3R\ .\ QU &= 48S^2\ .\ YU + 3T\ .\ PU,
\end{aligned}$$

and also the following formulæ, viz. if

$$2\alpha CU - 2\beta DU = \alpha' U + 6\beta' HU;$$

then

$$\begin{aligned}
\alpha' &= \ -2T\alpha - 16S^2\beta,\\
\beta' &= \ \ \ 8S\alpha + \ \ T\beta,
\end{aligned}$$

which give, conversely,

$$\begin{aligned}
\alpha &= \frac{1}{2R}(\ \ T\alpha' + 16S^2\beta'),\\
\beta &= \frac{1}{2R}(-8S\alpha' - \ 2T\beta');
\end{aligned}$$

and moreover, if

$$-2\alpha YU + 2\beta ZU = 6\alpha' PU + \beta' QU,$$

then

$$\begin{aligned}
\alpha' &= \ -(\ \ T\alpha + 16S^2\beta),\\
\beta' &= \ -(-8S\alpha - \ 2T\beta),
\end{aligned}$$

which give, conversely,

$$\begin{aligned}
\alpha &= -\frac{1}{2R}(-2T\alpha' - 16S^2\beta'),\\
\beta &= -\frac{1}{2R}(\ \ 8S\alpha' + \ \ T\beta');
\end{aligned}$$

so that the relation between (α, β) and (α', β') in the present case is similar to that between (α', β') and (α, β) in the former case. It may be noticed that in all these systems of linear equations, the determinant of transformation is a multiple of $64S^3 - T^2\ (=R)$.

245. It will be convenient, before giving the Tables for the covariants of

$$\alpha U + 6\beta HU,\quad 2\alpha CU - 2\beta DU,\quad 6\alpha PU + \beta QU,\quad 2\alpha YU - 2\beta ZU,$$

which replace Tables 68 and 69 of my Third Memoir, to give the following separate Table of the quantics in (α, β) which enter into the expressions of the invariants in Tables 68 and 69, and in these new Tables.

Table No. 73.

$$(1,\ 0,\ -24S,\ -8T,\ \ -48S^2 \between \alpha, \beta)^4,$$

$$(S,\ T,\ \ 24S^2,\ 4TS,\ T^2 - 48S^3 \between \alpha, \beta)^4,$$

$$(T,\ 96S^2,\ 60TS,\ 20T^2,\ 240TS^2,\ -48T^2S + 4608S^4,\ -8T^3 + 576TS^3 \between \alpha, \beta)^6.$$

$$(48S,\ 8T,\ -96S^2,\ -24TS,\ -T^2 - 16S^3 \between \alpha, \beta)^4.$$

$$\left(\begin{array}{l} T^2 + 192\,S^3\ , \\ 128\,TS^2\ , \\ 18\,T^2S + 384\,S^4\ , \\ T^3 + 64\,TS^3, \\ 5\,T^2S^2 - 64\,S^5\ , \end{array} \between \alpha, \beta\right)^4,$$

$$\left(\begin{array}{l} -8\,T^3 + 4608\,TS^3\ , \\ 1920\,T^2S^2 + 73728\,S^5\ , \\ 360\,T^3S + 38400\,TS^4\ , \\ 20\,T^4 + 8960\,T^2S^3, \\ 840\,T^3S^2 + 7680\,TS^5\ , \\ 36\,T^4S + 384\,T^2S^4 + 24576\,S^7, \\ 1\,T^5 - 40\,T^3S^3 + 2560\,TS^6, \end{array} \between \alpha, \beta\right)^6,$$

where the first part of the Table contains the quantics in (α, β) which relate to the forms $\alpha U + 6\beta HU$ and $2\alpha YU - 2\beta ZU$, and the second part of the Table contains the quantics in (α, β) which relate to the forms $6\alpha PU + \beta QU$ and $-2\alpha CU + 2\beta DU$.

The quantics in (α, β) contained in the foregoing Table are in the sequel indicated by means of their leading coefficients; as thus,

$$(1,\ 0,\ -24S, \ldots \between \alpha, \beta)^4,\quad (S,\ T, \ldots \between \alpha, \beta)^4,\quad (T^2 + 192S^3, \ldots \between \alpha, \beta)^4,\ \&c.$$

246. It is easy to see what transformations must be performed on the results in Tables 68 and 69, in order to obtain the new Tables. Thus, in the formation of Table 74, Table 68 gives $\alpha U + 6\beta HU$ and $H(\alpha U + 6\beta HU)$, and from these $C(\alpha U + 6\beta HU)$,

$D(\alpha U+6\beta HU)$ have to be found: the same Table gives also $P(\alpha U+6\beta HU)$, $Q(\alpha U+6\beta HU)$, but the expressions of these quantities YU, ZU have to be introduced in the place of PU, QU; and from the expressions so transformed are deduced also the expressions for $Y(\alpha U+6\beta HU)$, $Z(\alpha U+6\beta HU)$. Table 75 is to be deduced from Table 69 by writing therein (α', β'), for (α, β), and then putting $6\alpha' PU+\beta' QU = 2\alpha YU - 2\beta ZU$, which, as is seen above, gives α', β' as functions of α, β and of the invariants S and T; but in some of the formulæ YU, ZU, have to be introduced in the place of PU, QU. And so for the Tables 76 and 77. The actual effectuation of the transformations would, it is almost needless to remark, be very laborious, but the forms of the results are easily foreseen, and the results can then be verified by means of one or two coefficients only. The new Tables are

Table No. 74.

$$R(\alpha U+6\beta HU) = R\times[(1,\ 0,\ -24S,\ ..\between\alpha,\ \beta)^4]^3,$$

$$S(\alpha U+6\beta HU) = (S,\ T\ ,\ ..\between\alpha,\ \beta)^4,$$

$$T(\alpha U+6\beta HU) = [(T,\ 96S^2,\ ..\between\alpha,\ \beta)]^6.$$

$$(\alpha U+6\beta HU) = \alpha U+6\beta HU,$$

$$H(\alpha U+6\beta HU) = -\tfrac{1}{24}\times\left\{\begin{array}{l}\partial_\beta(1,\ 0,\ -24S,\ ..\between\alpha,\ \beta)^4\,.\,U\\ -6\partial_\alpha(1,\ 0,\ -24S,\ ..\between\alpha,\ \beta)^4\,.\,HU,\end{array}\right.$$

$$C(\alpha U+6\beta HU) = (1,\ 0,\ -24S,\ ..\between\alpha,\ \beta)^4\times\left\{\begin{array}{l}\partial_\beta(S,\ T,\ ..\between\alpha,\ \beta)^4\,.\,U\\ -6\partial_\alpha(S,\ T,\ ..\between\alpha,\ \beta)^4\,.\,HU,\end{array}\right.$$

$$D(\alpha U+6\beta HU) = \tfrac{1}{12}(1,\ 0,\ -24S,\ ..\between\alpha,\ \beta)^4\times\left\{\begin{array}{l}\partial_\beta(T,\ 96S^2,\ ..\between\alpha,\ \beta)^6\,.\,U\\ -6\partial_\alpha(T,\ 96S^2,\ ..\between\alpha,\ \beta)^6\,.\,HU,\end{array}\right.$$

$$P(\alpha U+6\beta HU) = -\frac{1}{3R}\times\left\{\begin{array}{l}\partial_\beta(S,\ T,\ ..\between\alpha,\ \beta)^4\,.\,YU\\ +\partial_\alpha(S,\ T,\ ..\between\alpha,\ \beta)^4\,.\,ZU,\end{array}\right.$$

$$Q(\alpha U+6\beta HU) = -\frac{1}{6R}\times\left\{\begin{array}{l}\partial_\beta(T,\ 96S^2,\ ..\between\alpha,\ \beta)^6\,.\,YU\\ +\partial_\alpha(T,\ 96S^2,\ ..\between\alpha,\ \beta)^6\,.\,ZU,\end{array}\right.$$

$$Y(\alpha U+6\beta HU) = [(1,\ 0,\ -24S,\ ..\between\alpha,\ \beta)^4]^2\times(-2\alpha YU+2\beta ZU),$$

$$Z(\alpha U+6\beta HU) = \tfrac{1}{4}[(1,\ 0,\ -24S,\ ..\between\alpha,\ \beta)^4]^2\times\left\{\begin{array}{l}\partial_\beta(1,\ 0,\ -24S,\ ..\between\alpha,\ \beta)^4\,.\,YU\\ +\partial_\alpha(1,\ 0,\ -24S,\ ..\between\alpha,\ \beta)^4\,.\,ZU.\end{array}\right.$$

No. 75.

$$R\,(-2\alpha YU+2\beta ZU)=-4096R^8\times[(S,\ T,\ ..\between\alpha,\ \beta)^4]^3,$$

$$S\,(-2\alpha YU+2\beta ZU)=\quad -\ R^3\times(1,\ 0,\ -24S,\ ..\between\alpha,\ \beta)^4,$$

$$T\,(-2\alpha YU+2\beta ZU)=\quad -8R^4\times(T,\ 96S^2,\ ..\between\alpha,\ \beta)^6.$$

$$-2\alpha YU+2\beta ZU=\qquad -2\alpha YU+2\beta ZU,$$

$$H\,(-2\alpha YU+2\beta ZU)=-\tfrac{2}{3}\,R\times\left\{\begin{array}{l}\quad\partial_\beta(S,\ T,\ ..\between\alpha,\ \beta)^4.\,YU\\ +\ \partial_\alpha(S,\ T,\ ..\between\alpha,\ \beta)^4.\,ZU,\end{array}\right.$$

$$C\,(-2\alpha YU+2\beta ZU)=-16\,R^4(S,\ T,\ ..\between\alpha,\ \beta)^4\times$$

$$\left\{\begin{array}{l}\quad\partial_\beta(1,\ 0,\ -24S,\ ..\between\alpha,\ \beta)^4.\,YU\\ +\ \partial_\alpha(1,\ 0,\ -24S,\ ..\between\alpha,\ \beta)^4.\,ZU\end{array}\right.$$

$$D\,(-2\alpha YU+2\beta ZU)=-\tfrac{32}{3}\,R^5(S,\ T,\ ..\between\alpha,\ \beta)^4\times$$

$$\left\{\begin{array}{l}\quad\partial_\beta(T,\ 96S^2,\ ..\between\alpha,\ \beta)^6.\,YU\\ +\ \partial_\alpha(T,\ 96S^2,\ ..\between\alpha,\ \beta)^6.\,ZU,\end{array}\right.$$

$$P\,(-2\alpha YU+2\beta ZU)=\ \tfrac{1}{6}\,R^2\times\left\{\begin{array}{l}\quad\partial_\beta(1,\ 0,\ -24S,\ ..\between\alpha,\ \beta)^4.\quad U\\ -6\partial_\alpha(1,\ 0,\ -24S,\ ..\between\alpha,\ \beta)^4.\,HU,\end{array}\right.$$

$$Q\,(-2\alpha YU+2\beta ZU)=-\tfrac{2}{3}\,R^3\times\left\{\begin{array}{l}\quad\partial_\beta(T,\ 96S^2,\ ..\between\alpha,\ \beta)^6.\quad U\\ -6\partial_\alpha(T,\ 96S^2,\ ..\between\alpha,\ \beta)^6.\,HU,\end{array}\right.$$

$$Y\,(-2\alpha YU+2\beta ZU)=\ 256R^6[(S,\ T,\ ..\between\alpha,\ \beta)^4]^2\times$$

$$(\alpha U+6\beta HU),$$

$$Z\,(-2\alpha YU+2\beta ZU)=-512R^7[(S,\ T,\ ..\between\alpha,\ \beta)^4]^2\times$$

$$\left\{\begin{array}{l}\quad\partial_\beta(S,\ T,\ ..\between\alpha,\ \beta)^4.\quad U\\ -6\partial_\alpha(S,\ T,\ ..\between\alpha,\ \beta)^4.\,HU.\end{array}\right.$$

No. 76.

$$R\,(2\alpha CU-2\beta DU)=-4096R^4\times[(T^2+192S^3\quad,\ldots\between\alpha,\ \beta)^4]^3,$$
$$S\,(2\alpha CU-2\beta DU)=\quad-\ R^2\times\ (48S,\ 8T\quad,\ldots\between\alpha,\ \beta)^4,$$
$$T\,(2\alpha CU-2\beta DU)=\quad-8R^2\times\ (-8T^3+4608TS^3,\ldots\between\alpha,\ \beta)^6.$$

$$2\alpha CU\quad-2\beta DU\ =\qquad 2\alpha CU-2\beta DU,$$

$$H\,(2\alpha CU-2\beta DU)=\qquad\tfrac{2}{3}\times\begin{cases}\ \ \partial_\beta\,(T^2+192S^3,\ldots\between\alpha,\ \beta)^4.\,CU\\ +\partial_\alpha\,(T^2+192S^3,\ldots\between\alpha,\ \beta)^4.\,DU,\end{cases}$$

$$C\ (2\alpha CU-2\beta DU)=\qquad -16R^2\,(T^2+192S^3,\ldots\between\alpha,\ \beta)^4\times$$
$$\begin{cases}\ \ \partial_\beta\,(48S,\ 8T\quad,\ldots\between\alpha,\ \beta)^4.\,CU\\ +\partial_\alpha\,(48S,\ 8T\quad,\ldots\between\alpha,\ \beta)^4.\,DU,\end{cases}$$

$$D\,(2\alpha CU-2\beta DU)=\qquad \tfrac{32}{3}R^2\,(T^2+192S^3\quad,\ldots\between\alpha,\ \beta)^4\times$$
$$\begin{cases}\ \ \partial_\beta\,(-8T^3+4608TS^3,\ldots\between\alpha,\ \beta)^6.\,CU\\ +\partial_\alpha\,(-8T^3+4608TS^3,\ldots\between\alpha,\ \beta)^6.\,DU,\end{cases}$$

$$P\,(2\alpha CU-2\beta DU)=\quad\tfrac{1}{6}R\quad\times\begin{cases}\ \ 6\partial_\beta\,(48S,\ 8T,\ldots\between\alpha,\ \beta)^4.\,PU\\ -\ \ \partial_\alpha\,(48S,\ 8T,\ldots\between\alpha,\ \beta)^4.\,QU,\end{cases}$$

$$Q\,(2\alpha CU-2\beta DU)=\quad\tfrac{2}{3}R\quad\times\begin{cases}\ \ 6\partial_\beta\,(-8T^3+4608TS^3,\ldots\between\alpha,\ \beta)^6.\,PU\\ -\ \ \partial_\alpha\,(-8T^3+4608TS^3,\ldots\between\alpha,\ \beta)^6.\,QU,\end{cases}$$

$$Y\,(2\alpha CU-2\beta DU)=\ 216R^3\quad[(T^2+192S^3,\ldots\between\alpha,\ \beta)^4]^2\times$$
$$(6\alpha PU+\beta QU),$$

$$Z\ (2\alpha CU-2\beta DU)=-512R^3\,[(T^2+192S^3,\ldots\between\alpha,\ \beta)^4]^2\times$$
$$\begin{cases}\ \ 6\partial_\beta\,(T^2+192S^3,\ldots\between\alpha,\ \beta)^4.\,PU\\ -\ \ \partial_\alpha\,(T^2+192S^3,\ldots\between\alpha,\ \beta)^4.\,QU.\end{cases}$$

No. 77.

$$R\ (6\alpha PU+\beta QU)=R^2\times[(48S,\ 8T\quad ,\ldots\between\alpha,\ \beta)^4]^3,$$

$$S\ (6\alpha PU+\beta QU)=\quad (T^2+192S^3\quad ,\ldots\between\alpha,\ \beta)^4,$$

$$T\ (6\alpha PU+\beta QU)=\quad (-8T^3+4608TS^3,\ldots\between\alpha,\ \beta)^6.$$

$$6\alpha PU+\beta QU)=\quad 6\alpha PU+\beta QU,$$

$$H(6\alpha PU+\beta QU)=-\tfrac{1}{24}\times\begin{cases}\ \ 6\partial_\beta(48S,\ 8T,\ldots\between\alpha,\ \beta)^4.PU\\ -\ \partial_\alpha(48S,\ 8T,\ldots\between\alpha,\ \beta)^4.QU,\end{cases}$$

$$C\ (6\alpha PU+\beta QU)=\quad -(48S,\ 8T,\ldots\between\alpha,\ \beta)^4\times$$
$$\begin{cases}\ \ 6\partial_\beta(T^2+192S^3,\ldots\between\alpha,\ \beta)^4.PU\\ -\ \partial_\alpha(T^2+192S^3,\ldots\between\alpha,\ \beta)^4.QU,\end{cases}$$

$$D\ (6\alpha PU+\beta QU)=\quad \tfrac{1}{12}(48S,\ 8T,\ldots\between\alpha,\ \beta)^4\times$$
$$\begin{cases}\ \ 6\partial_\beta(-8T^3+4608TS^3,\ldots\between\alpha,\ \beta)^6.PU\\ -\ \partial_\alpha(-8T^3+4608TS^3,\ldots\between\alpha,\ \beta)^6.QU,\end{cases}$$

$$P\ (6\alpha PU+\beta QU)=\frac{1}{3R}\times\begin{cases}\ \ \partial_\beta(T^2+192S^3,\ldots\between\alpha,\ \beta)^4.CU\\ +\partial_\alpha(T^2+192S^3,\ldots\between\alpha,\ \beta)^4.DU,\end{cases}$$

$$Q\ (6\alpha PU+\beta QU)=\frac{1}{6R}\times\begin{cases}\ \ \partial_\beta(-8T^3+4608TS^3,\ldots\between\alpha,\ \beta)^6.CU\\ +\partial_\alpha(-8T^3+4608TS^3,\ldots\between\alpha,\ \beta)^6.DU,\end{cases}$$

$$Y\ (6\alpha PU+\beta QU)=\ -\tfrac{1}{2}R.[(48S,\ 8T,\ldots\between\alpha,\ \beta)^4]^2\times$$
$$(-2\alpha CU+2\beta DU),$$

$$Z\ (6\alpha PU+\beta QU)=\ -\tfrac{1}{4}R[(48S,\ 8T,\ldots\between\alpha,\ \beta)^4]^2\times$$
$$\begin{cases}\ \ \partial_\beta(48S,\ 8T,\ldots\between\alpha,\ \beta)^4.CU\\ +\partial_\alpha(48S,\ 8T,\ldots\between\alpha,\ \beta)^4.DU.\end{cases}$$

247. It will be noticed how Tables 74 and 75 form a system involving only the quantics in (α, β) contained in the first part of Table 73, and how, in like manner, Tables 76 and 77 form a system involving only the quantics in (α, β) contained in the second part of Table 73; and, moreover, how in each pair of Tables the covariants, &c. correspond to each other as follows, viz.

$$\begin{array}{l} R,\ S,\ T,\ 1,\ H,\ C,\ D,\ P,\ Q,\ Y,\ Z \text{ to} \\ S,\ R,\ T,\ Y,\ P,\ Z,\ Q,\ H,\ D,\ 1,\ C. \end{array}$$

Thus in Table 74,—the formula for $H(\alpha U \quad + 6\beta HU)$,

and in Table 75,—the formula for $P(2\alpha YU - 2\beta ZU)$,

each of them involve the same factor

$$\left\{\begin{array}{l} \partial_\beta (1,\ 0,\ -24S, \ldots \between \alpha,\ \beta)^4 \,.\, U \\ -6\partial_\alpha (1,\ 0,\ -24S, \ldots \between \alpha,\ \beta)^4 \,.\, HU, \end{array}\right.$$

and so in all the other cases.

248. The quantics in (α, β) in each part of the foregoing Table 73 are covariantively connected together. In fact, considering the function $(1, 0, -24S, \ldots \between \alpha, \beta)^4$, which for shortness I call G, we have

$$\begin{aligned} G &= (1,\ 0,\ -24S, \ldots \between \alpha,\ \beta)^4, \\ IG &= 0, \\ JG &= 4(64S^3 - T^2) = 4R, \\ \square G &= (IG)^3 - 27(JG)^2 = -432R^2, \\ HG &= -4(S,\ T, \ldots \between \alpha,\ \beta)^4, \\ \Phi G &= 2(T,\ 96S^2, \ldots \between \alpha,\ \beta)^6. \end{aligned}$$

The last-mentioned formulæ, by the aid of Table 72, give rise to the following more general system in which they are themselves included.

Table No. 78.

$$\begin{aligned} \lambda G + 6\mu HG &= \lambda G + 6\mu HG, \\ H(\lambda G + 6\mu HG) &= 36\mu^2 G + \lambda^2 HG, \\ \Phi(\lambda G + 6\mu HG) &= (\lambda^3 - 216R\mu^3)\, 2(T,\ 96S^2, \ldots \between \alpha,\ \beta)^6, \\ I(\lambda G + 6\mu HG) &= 72R\lambda\mu, \\ J(\lambda G + 6\mu HG) &= 4R(\lambda^3 + 216R\mu^3), \\ \square(\lambda G + 6\mu HG) &= -432R^2(\lambda^3 - 216R\mu^3)^2. \end{aligned}$$

The expression for $H(\lambda G+6\mu HG)$, putting therein $\lambda=0$, shows that, to a numerical factor *près*, $H.HG$ is equal to G, and hence, disregarding numerical factors, we may say that each of the quartics $(1, 0, 24S, ..\between\alpha, \beta)^4$, $(S, T, ..\between\alpha, \beta)^4$, is the Hessian of the other of them, and that the sextic $(T, 96S^2, ..\between\alpha, \beta)^6$ is the cubicovariant of each of them.

249. Similarly, if the function $(48S, 8T, ..\between\alpha, \beta)^4$ is for shortness called G, then we have

$$
\begin{aligned}
G &= (48S,\ 8T, ..\between\alpha, \beta)^4, \\
IG &= 0, \\
JG &= 4(64S^3 - T^2)^2 = 4R^2, \\
\square G &= (IG)^3 - 27(JG)^2 = -432R^4, \\
HG &= -4(T^2+192S^3, ..\between\alpha, \beta)^4, \\
\Phi G &= -2(-8T^3+4608TS^3, ..\between\alpha, \beta)^4.
\end{aligned}
$$

The last-mentioned formulæ, by the aid of the same Table 72, give rise to the more general system in which they are themselves included.

Table No. 79.

$$
\begin{aligned}
\lambda G+6\mu HG &= \lambda G+6\mu HG, \\
H(\lambda G+6\mu HG) &= 36R^2\mu^2G+\lambda^2 HG, \\
\Phi(\lambda G+6\mu HG) &= (\lambda^3-216R^2\mu^3)\times -2(-8T^3+4608TS^3, ..\between\alpha, \beta)^6, \\
I(\lambda G+6\mu HG) &= 72R^2\lambda\mu, \\
J(\lambda G+6\mu HG) &= 4R^2(\lambda^3+216R^2\mu^3), \\
\square(\lambda G+6\mu HG) &= -432R^4(\lambda^3-216R^2\mu^3)^2.
\end{aligned}
$$

The expression for $H(\lambda G+6\mu HG)$, putting therein $\lambda=0$, shows that, to a numerical factor *près*, $H.HG$ is equal to G; so that, disregarding numerical factors, we may say that each of the quartics $(48S, T, ..\between\alpha, \beta)^4$, $(T^2+192S^3, ..\between\alpha, \beta)^4$, is the Hessian of the other of them, and that the sextic $(-8T^3+4608TS^3, ..\between\alpha, \beta)^6$ is the cubicovariant of each of them.

250. But besides this, the quantics in (α, β) in the two parts of the Table 73 are linearly connected together: the linear relations in question are in fact the equations whereon depend the expressions for the invariants in Tables 76 and 77 as deduced from those in Tables 74 and 75; and in the order of proof, they precede the formulæ in these four Tables. The linear relations are

Table No. 80.

$$(1,\ 0,\ -24S,\ldots \between -2T\alpha-16S^2\beta,\ \ 8S\alpha+\ T\beta)^4=-16R\ (T^2+192S^3,\ldots \between \alpha,\ \beta)^4,$$

$$(S,\ T,\ldots \between -2T\alpha-16S^2\beta,\ \ 8S\alpha+\ T\beta)^4=\ R^2\,(48S,\ 8T,\ldots \between \alpha,\ \beta)^4,$$

$$(T,\ 96S^2,\ldots \between -2T\alpha-16S^2\beta,\ \ 8S\alpha+\ T\beta)^6=-\ 8R^2\,(-8T^3+4608TS^3,\ldots \between \alpha,\ \beta)^6.$$

$$(48S,\ 8T,\ldots \between\ T\alpha+16S^2\beta,\ -8S\alpha-2T\beta)^4=-16R^2\,(S,\ T,\ldots \between \alpha,\ \beta)^4,$$

$$(T^2+192S^3,\ldots \between\ T\alpha+16S^2\beta,\ -8S\alpha-2T\beta)^4=-\ R^3\,(1,\ 0,\ -24S,\ldots \between \alpha,\ \beta)^4,$$

$$(-8T^3+4608TS^3,\ldots \between\ T\alpha+16S^2\beta,\ -8S\alpha-2T\beta)^6=-\ 8R^4\,(T,\ 96S^2,\ldots \between \alpha,\ \beta)^6.$$

Hence, attending to the remarks on the Tables 78 and 79, we may say that the quartics

$$(1,\ 0,\ -24S,\ldots \between \alpha,\ \beta)^4,$$

$$(48S,\ T,\ldots \between \alpha,\ \beta)^4,$$

which belong to the two parts respectively of Table 73, and which are related, the first of them to the discriminants of $\alpha U+6\beta HU$ and $2\alpha YU-2\beta ZU$, and the second to the discriminants of $6\alpha PU+\beta QU$, $-2\alpha CU+2\beta DU$, have these relations to each other, viz. each is a linear transformation of the Hessian of the other of them, and the cubicovariant of each is a linear transformation of the cubicovariant of the other.

270.

ON THE DOUBLE TANGENTS OF A CURVE OF THE FOURTH ORDER.

[From the *Philosophical Transactions of the Royal Society of London*, vol. CLI. (for the year 1861), pp. 357—362. Received May 30,—Read June 20, 1861.]

THE present memoir is intended to be supplementary to that "On the Double Tangents of a Plane Curve (*Phil. Trans.*, vol. CXLIX. (1859), pp. 193—212) [260]." I take the opportunity of correcting an error which I have there fallen into, and which is rather a misleading one, viz. the emanants U_1, U_2, .. were numerically determined in such manner as to become equal to U on putting (x_1, y_1, z_1) equal to (x, y, z); the numerical determination should have been (and in the latter part of the memoir is assumed to be) such as to render H_1, H_2, &c. equal to H, on making the substitution in question; that is, in the place of the formulæ

$$U_1 = \frac{1}{n}(x_1\partial_x + y_1\partial_y + z_1\partial_z)\,U,$$

$$U_2 = \frac{1}{n(n-1)}(x_1\partial_x + y_1\partial_y + z_1\partial_z)^2\,U, \text{ \&c.,}$$

$$\vdots$$

there ought to have been

$$U_1 = \frac{1}{(n-2)}(x_1\partial_x + y_1\partial_y + z_1\partial_z)\,U,$$

$$U_2 = \frac{1}{(n-2)(n-3)}(x_1\partial_x + y_1\partial_y + z_1\partial_z)^2\,U, \text{ \&c.}$$

$$\vdots$$

[this error is corrected *ante* p. 189].

The points of contact of the double tangents of the curve of the fourth order or quartic $U=0$, are given as the intersections of the curve with a curve of the fourteenth order $\Pi=0$; the last-mentioned curve is not absolutely determinate, since instead of $\Pi=0$, we may, it is clear, write $\Pi + MU = 0$, where M is an arbitrary function of the tenth order. I have in the memoir spoken of Hesse's original form (say $\Pi_1 = 0$) of the curve of the fourteenth order obtained by him in 1850, and of his transformed form (say $\Pi_2 = 0$) obtained in 1856. The method in the memoir itself (Mr Salmon's method) gives, in the case in question of a quartic curve, a third form, say $\Pi_3 = 0$. It appears by his paper "On the Determination of the Points of Contact of Double Tangents to an Algebraic Curve (*Quart. Math. Journ.* vol. III. p. 317 (1859))," that Mr Salmon has verified by algebraic transformations the equivalence of the last-mentioned form with those of Hesse; but the process is not given. The object of the present memoir is to demonstrate the equivalence in question, viz. that of the equation $\Pi_3 = 0$ with the one or other of the equations $\Pi_1 = 0$, $\Pi_2 = 0$, in virtue of the equation $U = 0$. The transformation depends, 1st, on a theorem used by Hesse for the deduction of his second form $\Pi_2 = 0$ from the original form $\Pi_1 = 0$, which theorem is given in his paper "Transformation der Gleichung der Curven 14ten Grades welche eine gegebene Curve 4ten Grades in den Berührungspuncten ihrer Doppeltangenten schneiden," *Crelle*, t. LII. pp. 97—103 (1856), containing the transformation in question; I prove this theorem in a different and (as it appears to me) more simple manner; 2nd, on a theorem relating to a cubic curve proved incidentally in my memoir "On the Conic of Five-pointic Contact at any point of a Plane Curve (*Phil. Trans.*, vol. CXLIX. (1859), see p. 385 [261])," the cubic curve being in the present case any first emanent of the given quartic curve: the demonstration occupies only a single paragraph, and it is here reproduced; and I reproduce also Hesse's demonstration of the equivalence of the two forms $\Pi_1 = 0$ and $\Pi_2 = 0$.

Let $U = (*\between x, y, z)^4$ be a quartic function of (x, y, z); (a, b, c, f, g, h) its second differential coefficients; $(\text{A}, \text{B}, \text{C}, \text{F}, \text{G}, \text{H})$ the reciprocal system

$$(bc - f^2,\ ca - g^2,\ ab - h^2,\ gh - af,\ hf - bg,\ fg - ch);$$

and let H be the Hessian of U, or determinant $abc - af^2 - bg^2 - ch^2 + 2fgh$ (H is of course a sextic function of x, y, z); (a', b', c', f', g', h') the second differential coefficients of H; $(\text{A}', \text{B}', \text{C}', \text{F}', \text{G}', \text{H}')$ the reciprocal system

$$(b'c' - f'^2,\ c'a' - g'^2,\ a'b' - h'^2,\ g'h' - a'f',\ h'f' - b'g',\ f'g' - c'h').$$

Then $U = 0$ being the equation of a quartic curve, the equation of the curve of the fourteenth order which by its intersections determines the points of contact of the double tangents of the quartic curve, may be taken to be (Hesse's original form)

$$\Pi_1 = (\text{A}, \text{B}, \text{C}, \text{F}, \text{G}, \text{H}\between \partial_x H, \partial_y H, \partial_z H)^2 - 3H(\text{A}, \text{B}, \text{C}, \text{F}, \text{G}, \text{H}\between \partial_x, \partial_y, \partial_z)^2 H = 0\ (^1).$$

Or it may be taken to be (Hesse's transformed form)

$$\Pi_2 = 5(\text{A}, \text{B}, \text{C}, \text{F}, \text{G}, \text{H}\between \partial_x H, \partial_y H, \partial_z H)^2 - 3(\text{A}', \text{B}', \text{C}', \text{F}', \text{G}', \text{H}'\between \partial_x U, \partial_y U, \partial_z U)^2 = 0.$$

And moreover, if $U_1 = \frac{1}{2}(x_1\partial_x + y_1\partial_y + z_1\partial_z)\,U$, and if H_1 be the Hessian of U_1, and $(a'', b'', c'', f'', g'', h'')$ the second differential coefficients of $H - 3H_1$, where in the differentiations (x_1, y_1, z_1) are treated as constants but after the differentiations are

[1] In quoting this formula in my former memoir, the numerical factor 3 is by mistake omitted. [This correction should have been made *ante* p. 187.]

effected they are replaced by (x, y, z), and if $(\text{A}'', \text{B}'', \text{C}'', \text{F}'', \text{G}'', \text{H}'')$ be the reciprocal system

$$(b''c''-f''^2,\ c''a''-g''^2,\ a''b''-h''^2,\ g''h''-a''f'',\ h''f''-b''g'',\ f''g''-c''h''),$$

then the equations of the curve of the fourteenth order may be taken to be (Salmon's form)

$$\Pi_3 = (\text{A}'', \text{B}'', \text{C}'', \text{F}'', \text{G}'', \text{H}''\between \partial_x U, \partial_y U, \partial_z U)^2 = 0.$$

I have preferred to write the three equations in the foregoing forms; but it is clear that the terms

$$(\text{A}, \text{B}, \text{C}, \text{F}, \text{G}, \text{H}\between \partial_x, \partial_y, \partial_z)^2 H \qquad ; \ (\text{A}', \text{B}', \text{C}', \text{F}', \text{G}', \text{H}'\between \partial_x, \partial_y, \partial_z)^2 U$$

might also have been written

$$(\text{A}, \text{B}, \text{C}, \text{F}, \text{G}, \text{H}\between a', b', c', 2f', 2g', 2h');\ (\text{A}', \text{B}', \text{C}', \text{F}', \text{G}', \text{H}'\between a, b, c, 2f, 2g, 2h).$$

As already noticed, it has been shown by Hesse (and his demonstration is to be here reproduced) that the two forms $\Pi_1 = 0$ and $\Pi_2 = 0$ are equivalent to each other. And the object of the memoir is to show that the third form $\Pi_3 = 0$ is equivalent to the other two. The equivalences in question subsist in virtue of the equation $U = 0$, that is, the functions Π_1, Π_2, Π_3 are not identical, but differ from each other by multiples of U.

Demonstration of Hesse's Theorem.

Let (a, b, c, f, g, h), (a', b', c', f', g', h') be any systems of coefficients of a ternary quadratic function; $(\text{A}, \text{B}, \text{C}, \text{F}, \text{G}, \text{H})$, $(\text{A}', \text{B}', \text{C}', \text{F}', \text{G}', \text{H}')$ the reciprocal systems as above, (x, y, z) arbitrary quantities. Consider the function

$$\begin{aligned}\square = (a, b, c, f, g, h\between x, y, z)^2 . (\text{A}', \text{B}', \text{C}', \text{F}', \text{G}', \text{H}'\between a, b, c, 2f, 2g, 2h)\\ -(\text{A}', \text{B}', \text{C}', \text{F}', \text{G}', \text{H}'\between ax+hy+gz,\ hx+by+fz,\ gx+fy+cz)^2.\end{aligned}$$

The term involving A' is

$$a(a, b, c, f, g, h\between x, y, z)^2 - (ax+hy+gz)^2,$$

which is

$$\begin{aligned}&= (ab-h^2)y^2 + (ac-g^2)z^2 + 2(af-gh)yz,\\ &= Cy^2 + Bz^2 - 2Fyz;\end{aligned}$$

and the term involving $2F'$ is

$$f(a, b, c, f, g, h\between x, y, z)^2 - (hx+by+fz)(gx+fy+cz),$$

which is

$$\begin{aligned}&= (af-gh)x^2 + (f^2-bc)yz + (fg-ch)zx + (hf-bg)xy,\\ &= -Fx^2 - Ayz + Hzx + Gxy;\end{aligned}$$

and the entire expression for $\square$ is thus

$$\begin{aligned} & A'(Cy^2 + Bz^2 - 2Fyz) \\ + \ & B'(Az^2 + Cx^2 - 2Gzx) \\ + \ & C'(Bx^2 + Ay^2 - 2Hxy) \\ + \ & 2F'(-Fx^2 - Ayz + Hzx + Gxy) \\ + \ & 2G'(-Gy^2 - Bzx + Fxy + Hyz) \\ + \ & 2H'(-Hz^2 - Cxy + Gyz + Fzx); \end{aligned}$$

or, what is the same thing,

$$\square = (BC' + B'C - 2FF',\ CA' + C'A - 2GG',\ AB' + A'B - 2HH',$$
$$GH' + G'H - AF' - A'F,\ HF' + H'F - BG' - B'G,\ FG' + F'G - CH' - C'H \between x,\ y,\ z)^2,$$

which is really the fundamental theorem. It is however used as follows; viz. the right-hand side being symmetrical in regard to the two systems

$$(a,\ b,\ c,\ f,\ g,\ h),\quad (a',\ b',\ c',\ f',\ g',\ h'),$$

the left-hand side, which is not in form symmetrical as regards the two systems, must be so in reality; or if $\square'$ is what $\square$ becomes by interchanging the two systems, then $\square' = \square$; or substituting for $\square$ and $\square'$ their values, we have

$$\begin{aligned} & (a,\ b,\ c,\ f,\ g,\ h \between x,\ y,\ z)^2 . (\text{A}',\ \text{B}',\ \text{C}',\ \text{F}',\ \text{G}',\ \text{H}' \between a,\ b,\ c,\ 2f,\ 2g,\ 2h) \\ & \qquad - (\text{A}',\ \text{B}',\ \text{C}',\ \text{F}',\ \text{G}',\ \text{H}' \between ax + hy + gz,\ hx + by + fz,\ gx + fy + cz)^2 \\ = \ & (a',\ b',\ c',\ f',\ g',\ h' \between x,\ y,\ z)^2 . (\text{A},\ \text{B},\ \text{C},\ \text{F},\ \text{G},\ \text{H} \between a',\ b',\ c',\ 2f',\ 2g',\ 2h') \\ & \qquad - (\text{A},\ \text{B},\ \text{C},\ \text{F},\ \text{G},\ \text{H} \between a'x + h'y + g'z,\ h'x + b'y + f'z,\ g'x + f'y + c'z)^2, \end{aligned}$$

which is Hesse's theorem.

If in particular $(a,\ b,\ c,\ f,\ g,\ h)$ are the second differential coefficients of a function $u = (* \between x,\ y,\ z)^p$, and $(a',\ b',\ c',\ f',\ g',\ h')$ the second differential coefficients of a function $u' = (* \between x,\ y,\ z)^{p'}$, then the equation becomes

$$p(p-1)u . (\text{A}',\ \text{B}',\ \text{C}',\ \text{F}',\ \text{G}',\ \text{H}' \between \partial_x,\ \partial_y,\ \partial_z)^2 u - (p-1)^2 (\text{A}',\ \text{B}',\ \text{C}',\ \text{F}',\ \text{G}',\ \text{H}' \between \partial_x u,\ \partial_y u,\ \partial_z u)^2$$
$$= p'(p'-1)u' . (\text{A},\ \text{B},\ \text{C},\ \text{F},\ \text{G},\ \text{H} \between \partial_x,\ \partial_y,\ \partial_z)^2 u' - (p'-1)^2 (\text{A},\ \text{B},\ \text{C},\ \text{F},\ \text{G},\ \text{H} \between \partial_x u',\ \partial_y u',\ \partial_z u')^2;$$

and if for u, u' we take the quartic function U and the sextic function H, its Hessian, we have

$$12U . (\text{A}',\ \text{B}',\ \text{C}',\ \text{F}',\ \text{G}',\ \text{H}' \between \partial_x,\ \partial_y,\ \partial_z)^2 U - 9(\text{A}',\ \text{B}',\ \text{C}',\ \text{F}',\ \text{G}',\ \text{H}' \between \partial_x U,\ \partial_y U,\ \partial_z U)^2$$
$$= 30H . (\text{A},\ \text{B},\ \text{C},\ \text{F},\ \text{G},\ \text{H} \between \partial_x,\ \partial_y,\ \partial_z)^2 H - 25(\text{A},\ \text{B},\ \text{C},\ \text{F},\ \text{G},\ \text{H} \between \partial_x H,\ \partial_y H,\ \partial_z H)^2;$$

and if in this identical equation we write $U = 0$, then from the resulting equation and the equation

$$\Pi_1 = -3H(\text{A},\ \text{B},\ \text{C},\ \text{F},\ \text{G},\ \text{H} \between \partial_x,\ \partial_y,\ \partial_z)^2 H + (\text{A},\ \text{B},\ \text{C},\ \text{F},\ \text{G},\ \text{H} \between \partial_x H,\ \partial_y H,\ \partial_z H)^2$$

we may eliminate any one of the three terms

$$(\text{A}', \text{B}', \text{C}', \text{F}', \text{G}', \text{H}' \between \partial_x U, \partial_y U, \partial_z U)^2,$$

$$H.(\text{A}, \text{B}, \text{C}, \text{F}, \text{G}, \text{H} \between \partial_x, \partial_y \quad \partial_z)^2 H,$$

$$(\text{A}, \text{B}, \text{C}, \text{F}, \text{G}, \text{H} \between \partial_x H, \partial_y H, \partial_z H)^2;$$

and in particular if the second term be eliminated, we obtain the equation

$$\Pi_2 = 5(\text{A}, \text{B}, \text{C}, \text{F}, \text{G}, \text{H} \between \partial_x H, \partial_y H, \partial_z H)^2 - 3(\text{A}', \text{B}', \text{C}', \text{F}', \text{G}', \text{H}' \between \partial_x U, \partial_y U, \partial_z U)^2,$$

and the equivalence of the two forms $\Pi_1 = 0$ and $\Pi_2 = 0$ is thus established.

But Hesse's theorem leads also to the demonstration of the equivalence of the third form $\Pi_3 = 0$. To use it for this purpose, I remark that if $(a'', b'', c'', f'', g'', h'')$ are the second differential coefficients of $H - 3H_1$, where after the differentiations (x_1, y_1, z_1) are to be replaced by (x, y, z), then the theorem gives

$$\begin{aligned} 12U.(\text{A}'', \text{B}'', \text{C}'', \text{F}'', \text{G}'', \text{H}'' \between \partial_x, \partial_y, \partial_z)^2 U - 9(\text{A}'', \text{B}'', \text{C}'', \text{F}'', \text{G}'', \text{H}'' \between \partial_x U, \partial_y U, \partial_z U)^2 \\ = (a'', b'', c'', f'', g'', h'' \between x, y, z)^2.(\text{A}, \text{B}, \text{C}, \text{F}, \text{G}, \text{H} \between \partial_x, \partial_y, \partial_z)^2 (H - 3H_1) \\ - (\text{A}, \text{B}, \text{C}, \text{F}, \text{G}, \text{H} \between a''x + h''y + g''z, h''x + b''y + f''z, g''x + f''y + c''z)^2. \end{aligned}$$

But on putting (x, y, z) for (x_1, y_1, z_1) we have (since H is a homogeneous function of the order 6, and H_1 before the change is a homogeneous function of the order 3 in (x, y, z)) $a''x + h''y + g''z = 5\partial_x H - 3.2\partial_x H_1 = 5\partial_x H - 3\partial_x H$ (since, on making the substitution, $H_1 = H$, but $\partial_x H_1 = \frac{1}{2}\partial_x H) = 2\partial_x H$; and thus

$$(a''x + h''y + g''z, h''x + b''y + f''z, g''x + f''y + c''z) = (2\partial_x H, 2\partial_y H, 2\partial_x H);$$

and similarly, on making the substitution,

$$(a'', b'', c'', f'', g'', h'' \between x, y, z)^2 = 6.5H - 3.3.2H_1 = (30 - 18)H = 12H.$$

Hence writing therein $U = 0$, the foregoing equation becomes

$$\begin{aligned} -9(\text{A}'', \text{B}'', \text{C}'', \text{F}'', \text{G}'', \text{H}'' \between \partial_x U, \partial_y U, \partial_z U)^2 \\ = 12H.(\text{A}, \text{B}, \text{C}, \text{F}, \text{G}, \text{H} \between \partial_x, \partial_y, \partial_z)^2 (H - 3H_1) \\ - 4 (\text{A}, \text{B}, \text{C}, \text{F}, \text{G}, \text{H} \between \partial_x U, \partial_y U, \partial_z U)^2, \end{aligned}$$

which may also be written

$$\begin{aligned} -9(\text{A}'', \text{B}'', \text{C}'', \text{F}'', \text{G}'', \text{H}'' \between \partial_x U, \partial_y U, \partial_z U)^2 \\ = 12H.(\text{A}, \text{B}, \text{C}, \text{F}, \text{G}, \text{H} \between \partial_x, \partial_y, \partial_z)^2 H \\ - 36H.(\text{A}, \text{B}, \text{C}, \text{F}, \text{G}, \text{H} \between \partial_x, \partial_y, \partial_z)^2 H_1 \\ - 4 (\text{A}, \text{B}, \text{C}, \text{F}, \text{G}, \text{H} \between \partial_x U, \partial_y U, \partial_z U)^2, \end{aligned}$$

where (x_1, y_1, z_1) are ultimately to be replaced by (x, y, z). The second line in fact vanishes, which I show as follows:

Demonstration of my Theorem for a Cubic Curve.

Let $U = (*\between x, y, z)^3$ be a cubic function; it may by a linear transformation of the coordinates be reduced to the canonical form $x^3 + y^3 + z^3 + 6lxyz$, and we then have

$$
\begin{aligned}
&(\text{A, B, C, F, G, H}\between \partial_x, \partial_y, \partial_z)^2\, H \div 6^5 \\
&\quad = (yz - l^2x^2) \,.\, -6l^2x \\
&\quad\quad + (zx - l^2y^2) \,.\, -6l^2y \\
&\quad\quad + (xy - l^2z^2) \,.\, -6l^2z \\
&\quad\quad + 2\,(l^2yz - lx^2)\,.\,(1 + 2l^3)\,x \\
&\quad\quad + 2\,(l^2zx - ly^2)\,.\,(1 + 2l^3)\,y \\
&\quad\quad + 2\,(l^2xy - lz^2)\,.\,(1 + 2l^3)\,z \\
&\quad = -18l^2\,xyz \qquad + 6l^4\ (x^3 + y^3 + z^3) \\
&\quad\quad + 6l^2\,(1 + 2l^3)\,xyz - 2l\,(1 + 2l^3)\,(x^3 + y^3 + z^3) \\
&\quad = (-12l^2 + 12l^5)\,xyz + (-2l + 2l^4)\,(x^3 + y^3 + z^3) \\
&\quad = 2\,(-l + l^4)\,(x^3 + y^3 + z^3 + 6lxyz);
\end{aligned}
$$

or since $-l + l^4$ is equal to the quartinvariant S, and the equation is an invariantive one, we have for any cubic function whatever

$$(\text{A, B, C, F, G, H}\between \partial_x, \partial_y, \partial_z)^2\, H \div 6^5 = 2S\,.\,U,$$

which is the theorem in question. There is a difference of notation, and consequently a different numerical factor, in the theorem as stated in the memoir on the conic of five-pointic contact, referred to above.

If, as above, U is a quartic function $(*\between x, y, z)^4$, and $U_1 = \tfrac{1}{2}\,(x_1\partial_x + y_1\partial_y + z_1\partial_z)\,U$, then U_1 is a cubic function, and we have

$$(\text{A}_1, \text{B}_1, \text{C}_1, \text{F}_1, \text{G}_1, \text{H}_1 \between \partial_x, \partial_y, \partial_z)^2\, H_1 \div 6^5 = 2S_1\,.\,U_1,$$

where it is to be noticed that S_1 denotes a quartic function in the coefficients of U_1, and consequently a quartic function in (x_1, y_1, z_1), the coefficients being quartic functions of the coefficients of U. On writing (x, y, z) in the place of (x_1, y_1, z_1), S_1 becomes a quartic function of (x, y, z), which is in fact a quarticovariant quartic of U.

If in the foregoing equation we write (x, y, z) in the place of (x_1, y_1, z_1), then U_1 becomes equal to $2U$; and consequently, if $U = 0$, the right-hand side of the equation vanishes. Moreover $(a_1, b_1, c_1, f_1, g_1, h_1)$ (the second differential coefficients of U_1) become equal to (a, b, c, f, g, h), and consequently the coefficients $(\text{A}_1, \text{B}_1, \text{C}_1, \text{F}_1, \text{G}_1, \text{H}_1)$ become equal to (A, B, C, F, G, H). Hence, assuming always that $U=0$, the equation becomes

$$(\text{A}, \text{B}, \text{C}, \text{F}, \text{G}, \text{H} \between \partial_x, \partial_y, \partial_z)^2 H_1 = 0,$$

where after the differentiations (x_1, y_1, z_1) are replaced by (x, y, z). This is the form which is required for the present purpose.

Returning to the foregoing expression of $-9(\text{A}'', \text{B}'', \text{C}'', \text{F}'', \text{G}'', \text{H}'' \between \partial_x U, \partial_y U, \partial_z U)^2$, this now becomes

$$\begin{aligned} -9\Pi_3 &= -9(\text{A}'', \text{B}'', \text{C}'', \text{F}'', \text{G}'', \text{H}'' \between \partial_x U, \partial_y U, \partial_z U)^2 \\ &= 4\{3H.(\text{A}, \text{B}, \text{C}, \text{F}, \text{G}, \text{H} \between \partial_x, \partial_y, \partial_z)^2 H - (\text{A}, \text{B}, \text{C}, \text{F}, \text{G}, \text{H} \between \partial_x U, \partial_y U, \partial_z U)^2\}, \end{aligned}$$

so that the equation $\Pi_3 = 0$ gives

$$\Pi_1 = (\text{A}, \text{B}, \text{C}, \text{F}, \text{G}, \text{H} \between \partial_x U, \partial_y U, \partial_z U)^2 - 3H.(\text{A}, \text{B}, \text{C}, \text{F}, \text{G}, \text{H} \between \partial_x, \partial_y, \partial_z)^2 H = 0,$$

and the equivalence of the equations $\Pi_1 = 0$ and $\Pi_3 = 0$ is thus established.

271.

SUR L'INVARIANT LE PLUS SIMPLE D'UNE FONCTION QUADRATIQUE BI-TERNAIRE, ET SUR LE RÉSULTANT DE TROIS FONCTIONS QUADRATIQUES TERNAIRES.

[From the *Journal für die reine und angewandte Mathematik* (Crelle), tom. LVII. (1860), pp. 139—148.]

M. SYLVESTER a trouvé depuis longtemps, pour le résultant des trois fonctions quadratiques ternaires

$$(a\ ,\ b\ ,\ c\ ,\ f\ ,\ g\ ,\ h\ )(x,\ y,\ z)^2,$$
$$(a',\ b',\ c',\ f',\ g',\ h')(x,\ y,\ z)^2,$$
$$(a'',\ b'',\ c'',\ f'',\ g'',\ h'')(x,\ y,\ z)^2,$$

une expression remarquable

$$12R = 16C_{12} - C_6^2,$$

où C_{12}, C_6 représentent des fonctions données des coefficients qui sont non seulement des invariants mais aussi des *combinants* des trois fonctions quadratiques (Voir le mémoire de M. Sylvester, "On the Calculus of forms otherwise the theory of Invariants," § VII. *Camb. et Dub. Math. Journ.*, t. VIII. pp. 256—269 année 1853, équation (A.), p. 267). Pour expliquer la formation de la fonction C_6, il convient de considérer la fonction quadratique bi-ternaire

$$U = \left(\begin{array}{cccccc} a, & b, & c, & f, & g, & h \\ a', & b', & c', & f', & g', & h' \\ a'', & b'', & c'', & f'', & g'', & h'' \\ A, & B, & C, & F, & G, & H \\ A', & B', & C', & F', & G', & H' \\ A'', & B'', & C'', & F'', & G'', & H'' \end{array}\right)(x,\ y,\ z)^2(\xi,\ \eta,\ \zeta)^2$$

cette notation représentant la fonction

$$\begin{aligned}&(a\ ,b\ ,c\ ,f\ ,g\ ,h\ )(x,\ y,\ z)^2.\xi^2\\+&(a'\ ,b'\ ,c'\ ,f'\ ,g'\ ,h'\ )(x,\ y,\ z)^2.\eta^2\\+&(a''\ ,b''\ ,c''\ ,f''\ ,g''\ ,h''\ )(x,\ y,\ z)^2.\zeta^2\\+&(A\ ,B\ ,C\ ,F\ ,G\ ,H\ )(x,\ y,\ z)^2.2\eta\zeta\\+&(A',B',C',F',G',H')(x,\ y,\ z)^2.2\zeta\xi\\+&(A'',B'',C'',F'',G'',H'')(x,\ y,\ z)^2.2\xi\eta,\end{aligned}$$

ou, ce qui est la même chose,

$$\begin{aligned}&(a\ x^2+b\ y^2+c\ z^2+2f\ yz+2g\ zx+2h\ xy)\,\xi^2\\+&(a'\ x^2+b'\ y^2+c'\ z^2+2f'\ yz+2g'\ zx+2h'\ xy)\,\eta^2\\+&(a''x^2+b''\ y^2+c''\ z^2+2f''yz+2g''zx+2h''xy)\,\zeta^2\\+&(A\ x^2+B\ y^2+C\ z^2+2F\ yz+2G\ zx+2H\ xy)\,2\eta\zeta\\+&(A'x^2+B'y^2+C'z^2+2F'\ yz+2G'zx+2H'xy)\,2\zeta\xi\\+&(A''x^2+B''y^2+C''z^2+2F''yz+2G''zx+2H''xy)\,2\xi\eta.\end{aligned}$$

Soit $\overline{123}$ le déterminant

$$\begin{vmatrix}\partial_{x_1}, & \partial_{y_1}, & \partial_{z_1}\\ \partial_{x_2}, & \partial_{y_2}, & \partial_{z_2}\\ \partial_{x_3}, & \partial_{y_3}, & \partial_{z_3}\end{vmatrix},$$

et $\overline{1'2'3'}$ le déterminant

$$\begin{vmatrix}\partial_{\xi_1}, & \partial_{\eta_1}, & \partial_{\zeta_1}\\ \partial_{\xi_2}, & \partial_{\eta_2}, & \partial_{\zeta_2}\\ \partial_{\xi_3}, & \partial_{\eta_3}, & \partial_{\zeta_3}\end{vmatrix},$$

et soient U_1, U_2, U_3 ce que devient U lorsqu'on y écrit x_1, y_1, z_1, ξ_1, η_1, ζ_1; x_2, y_2, z_2, ξ_2, η_2, ζ_2; x_3, y_3, z_3, ξ_3, η_3, ζ_3 au lieu de x, y, z, ξ, η, ζ; on a alors cet invariant (analogue à l'invariant quadratique d'une fonction binaire d'ordre pair), savoir:

$$\frac{1}{384}\overline{123}^2.\overline{1'2'3'}^2\,U_1U_2U_3$$

où comme à l'ordinaire les valeurs x, y, z, ξ, η, ζ sont rétablies après les différentiations au lieu de x_1, y_1, z_1, ξ_1, η_1, ζ_1, etc.

Je représente cet invariant par

$$\left\{\begin{matrix}a\ , & b\ , & c\ , & f\ , & g\ , & h\\ a'\ , & b'\ , & c'\ , & f'\ , & g'\ , & h'\\ a''\ , & b''\ , & c''\ , & f''\ , & g''\ , & h''\\ A\ , & B\ , & C\ , & F\ , & G\ , & H\\ A'\ , & B'\ , & C'\ , & F'\ , & G'\ , & H'\\ A''\ , & B''\ , & C''\ , & F''\ , & G''\ , & H''\end{matrix}\right\};$$

on voit sans peine que l'invariant peut être développé, et qu'alors il prend la forme

$$\left\{\begin{matrix} a, & b, & c, & f, & g, & h \\ a', & b', & c', & f', & g', & h' \\ a'', & b'', & c'', & f'', & g'', & h'' \end{matrix}\right\}$$

$$-\left\{\begin{matrix} a, & b, & c, & f, & g, & h \\ A, & B, & C, & F, & G, & H \\ A, & B, & C, & F, & G, & H \end{matrix}\right\}$$

$$-\left\{\begin{matrix} a', & b', & c', & f', & g', & h' \\ A', & B', & C', & F', & G', & H' \\ A', & B', & C', & F', & G', & H' \end{matrix}\right\}$$

$$-\left\{\begin{matrix} a'', & b'', & c'', & f'', & g'', & h'' \\ A'', & B'', & C'', & F'', & G'', & H'' \\ A'', & B'', & C'', & F'', & G'', & H'' \end{matrix}\right\}$$

$$+2\left\{\begin{matrix} A, & B, & C, & F, & G, & H \\ A', & B', & C', & F', & G', & H' \\ A'', & B'', & C'', & F'', & G'', & H'' \end{matrix}\right\}$$

où le premier terme

$$\left\{\begin{matrix} a, & b, & c, & f, & g, & h \\ a', & b', & c', & f', & g', & h' \\ a'', & b'', & c'', & f'', & g'', & h'' \end{matrix}\right\}$$

dénote la fonction

$$\begin{aligned} & ab'c'' + ab''c' + a'b''c + a'bc'' + a''bc' + a''b'c \\ & - 2af'f'' - 2a'f''f - 2a''ff' \\ & - 2bg'g'' - 2b'g''g - 2b''gg' \\ & - 2ch'h'' - 2c'h''h - 2c''hh' \\ & + 2fg'h'' + 2fg''h' + 2f'g''h + 2f'gh'' + 2f''gh' + 2f''g'h \end{aligned}$$

et de même pour les autres termes. L'invariant contient les termes

$$\begin{aligned} & ab'c'' + 2AB'C'' + 2fg'h'' + 4FG'H'' \\ & - 2af'f'' - 2aBC + 2aF^2 \\ & - 4AF'F'' + 4fAF - 4fGH, \end{aligned}$$

et on déduit de là son expression complète en y écrivant

$$ab'c'' + ab''c' + a'b''c + a'bc'' + a''bc' + a''b'c$$

au lieu de $ab'c''$;

$$af'f'' + a'f''f + a''ff' + bg'g'' + b'g''g + c''gg' + ch'h'' + c'h''h + c''hh'$$

au lieu de $af'f''$, et ainsi de suite; la somme des coefficients est $6 + 2.6 + 2.6 + 4.6 + 2.9 + 2.9 + 2.9 + 4.9 + 4.9 + 4.9 = 216$, ce qui est exacte, car il n'y a pas de termes qui se détruisent, et dans l'expression $\frac{1}{384}\overline{123}^2\,\overline{1'2'3'}^2\,U_1U_2U_3$, le nombre des termes qui composent le produit des deux déterminants carrés est $36 \times 36 = 1296$, chacun de ces termes est multiplié par le facteur $8 \times 8 = 64$ introduit par la différentiation, et l'on a

$$\frac{1}{384} \times 64 \times 1296 = 216.$$

L'invariant qui vient d'être donné est lié avec l'invariant T de M. Aronhold (lequel se rapporte à une fonction ternaire cubique). C'est pour cela que je le désigne par la même lettre T, et pour fixer sa valeur j'écris

$$2T = ab'c'' + \text{etc.}$$

Cela posé, je forme les deux combinants C_6 et C_{12} de la manière suivante.

Combinant du sixième ordre C_6. Soient U, V, W les trois fonctions quadratiques ternaires; considérez la fonction syzygétique $\lambda U + \mu V + \nu W$, où λ, μ, ν sont des quantités arbitraires; formez avec les variables ξ, η, ζ le reciprocant (¹) (multiplié par 2) de cette fonction, le résultat sera de l'ordre 2 par rapport aux coefficients de U, V, W, de l'ordre 2 par rapport à λ, μ, ν, et de l'ordre 2 par rapport à ξ, η, ζ; c. à d. ce résultat sera une fonction quadratique bi-ternaire de λ, μ, ν et ξ, η, ζ dont l'invariant $2T$ (lequel sera par conséquent de l'ordre 6 par rapport aux coefficients de U, V, W) est le combinant C_6 qu'il s'agissait de trouver.

Combinant du douzième ordre C_{12}. Considérez comme auparavant la fonction syzygétique $\lambda U + \mu V + \nu W$, formez le discriminant (multiplié par 6) de cette fonction; le résultat sera de l'ordre 3 par rapport aux coefficients de U, V, W, et de l'ordre 3 par rapport à λ, μ, ν; c. à d. il sera une fonction cubique ternaire de λ, μ, ν, dont

[1] Le réciprocant de $(a, b, c, f, g, h)(x, y, z)^2$ est

$$= -\begin{vmatrix} & \xi, & \eta, & \zeta \\ \xi, & a, & h, & g \\ \eta, & h, & b, & f \\ \zeta, & g, & f, & c \end{vmatrix}$$

$$= -(\mathfrak{A}, \mathfrak{B}, \mathfrak{C}, \mathfrak{F}, \mathfrak{G}, \mathfrak{H})(\xi, \eta, \zeta)^2.$$

l'invariant S de M. Aronhold [1] (lequel sera de l'ordre 12 par rapport aux coefficients de U, V, W) est le combinant C_{12} qu'il s'agissait de trouver.

On peut exprimer ces résultats d'une manière abrégée en disant que C_6 *est l'invariant* $2T$ *de la fonction bi-ternaire* $2R$ et que C_{12} *est l'invariant* S *de la fonction cubique ternaire* $6D$, en désignant par R et D le réciprocant et le discriminant de la fonction $\lambda U + \mu V + \nu W$.

Pour démontrer l'équation

$$12R = 16C_{12} - C_6^2$$

il suffit de faire voir que cette équation est vérifiée lorsqu'au lieu des trois fonctions données U, V, W on substitue trois fonctions de la forme $lU + mV + nW$, les coefficients l, m, n ayant été choisis de manière à simplifier autant que possible les expressions des trois fonctions. Il est facile de voir qu'il est permis de prendre pour les trois fonctions les formes dont s'est servi M. Sylvester dans son mémoire, savoir

$$\begin{gathered} \rho\,(x^2 - y^2), \\ \sigma\,(y^2 - z^2), \\ y^2 + 2fyz + 2gzx + 2hxy. \end{gathered}$$

Mais il suit des recherches de M. Hesse, et M. Sylvester aussi a depuis reconnu, qu'on peut prendre pour les trois fonctions les formes encore plus simples

$$\begin{gathered} x^2 + 2lyz, \\ y^2 + 2lzx, \\ z^2 + 2lxy, \end{gathered}$$

qui se réduisent aux fonctions dérivées de la fonction cubique $x^3 + y^3 + z^3 + 6lxyz$. Or, en prenant ces valeurs de U, V, W, on obtient

$$\lambda U + \mu V + \nu W = (\lambda,\ \mu,\ \nu,\ \lambda l,\ \mu l,\ \nu l)(x,\ y,\ z)^2.$$

[1] Pour la fonction cubique ternaire

$$(a,\ b,\ c,\ i,\ j,\ k,\ i_1,\ j_1,\ k_1,\ l)(x,\ y,\ z)^3$$

l'invariant S est

$$\begin{aligned} = &\ l^4 \\ &- 2l^2\,(jk_1 + ki_1 + ij_1) \\ &+ 3l\,(ijk + i_1j_1k_1) \\ &+ l\,(aii_1 + bjj_1 + ckk_1) \\ &- l\,abc \\ &- (ii_1j_1k + jj_1k_1i + kk_1i_1j) \\ &+ (j^2k_1^2 + k^2i_1^2 + i^2j_1^2) \\ &+ (bcj_1k + cak_1i + abi_1j) \\ &- (ai^2j + bj^2k + ck^2i + ai_1^2k_1 + bj_1^2i_1 + ck_1^2j_1). \end{aligned}$$

Le réciprocant (multiplié par 2) est

$$= -2 \begin{vmatrix} & \xi, & \eta, & \zeta \\ \xi, & \lambda, & \nu l, & \mu l \\ \eta, & \nu l, & \mu, & \lambda l \\ \zeta, & \mu l, & \lambda l, & \nu \end{vmatrix},$$

lequel, en forme développée, donne le résultat

$$\begin{array}{lcccccccl} & (-2l^2\lambda^2 & . & . & +2\mu\nu & . & . &) & \xi^2 \\ +(& . & -2l^2\mu^2 & . & . & +2\nu\lambda & . &) & \eta^2 \\ +(& . & . & -2l^2\nu^2 & . & . & +2\lambda\mu &) & \zeta^2 \\ +(& -2l\lambda^2 & . & . & +2l^2\mu\nu & . & . &) & 2\zeta\eta \\ +(& . & -2l\mu^2 & . & . & +2l^2\nu\lambda & . &) & 2\xi\zeta \\ +(& . & . & -2l\nu^2 & . & . & +2l^2\lambda\mu &) & 2\eta\xi, \end{array}$$

que l'on peut aussi exprimer comme fonction quadratique bi-ternaire, savoir:

$$\begin{pmatrix} -2l^2 & . & . & 1 & . & . \\ . & -2l^2 & . & . & 1 & . \\ . & . & -2l^2 & . & . & 1 \\ -2l & . & . & l^2 & . & . \\ . & -2l & . & . & l^2 & . \\ . & . & -2l & . & . & l^2 \end{pmatrix} (\lambda, \mu, \nu)^2 (\xi, \eta, \zeta)^2.$$

En représentant l'invariant $2T$ par le développement incomplet donné ci-dessus, on obtient dans le cas dont il s'agit

$$2T = \left\{ \begin{matrix} -2l^2 & . & . & 1 & . & . \\ . & -2l^2 & . & . & 1 & . \\ . & . & -2l^2 & . & . & 1 \end{matrix} \right\}$$

$$- \left\{ \begin{matrix} -2l^2 & . & . & 1 & . & . \\ -2l & . & . & l^2 & . & . \\ -2l & . & . & l^2 & . & . \end{matrix} \right\} - \text{etc. etc.}$$

$$+ 2 \left\{ \begin{matrix} -2l & . & . & l^2 & . & . \\ . & -2l & . & . & l^2 & . \\ . & . & -2l & . & . & l^2 \end{matrix} \right\},$$

ce qui se réduit à la valeur finale

$$\begin{aligned}
& - \; 8l^6 + 2 \\
& - (4l^6 + 4l^3 + 4l^3) \\
& - (4l^6 + 4l^3 + 4l^3) \\
& - (4l^6 + 4l^3 + 4l^3) \\
& + 2\,(-8l^3 + 2l^6) \\
= & \;\; 2\,(1 - 20l^3 - 8l^6).
\end{aligned}$$

Les invariants S et T de la fonction cubique $x^3 + y^3 + z^3 + 6lxyz$ sont

$$S = -l + l^4,$$
$$T = 1 - 20l^3 - 8l^6;$$

l'invariant $2T$ de la fonction quadratique bi-ternaire est donc précisément égal à $2T$, où T est l'invariant qui vient d'être donné, et on a

$$C_6 = 2T.$$

Le discriminant (multiplié par 6) de la fonction $\lambda U + \mu V + \nu W$ est

$$= 6 \begin{vmatrix} \lambda\,, & l\nu, & l\mu \\ l\lambda, & \mu\,, & l\lambda \\ l\mu, & l\nu, & \nu \end{vmatrix}$$

$$= 6\,\{-l^2\lambda^3 - l^2\mu^3 - l^2\nu^3 + (1 + 2l^3)\,\lambda\mu\nu\},$$

où la fonction en parenthèses est ce que devient le Hessien de la fonction cubique $x^3 + y^3 + z^3 + 6lxyz$, lorsqu'on y substitue λ, μ, ν au lieu de x, y, z. Son invariant S est

$$= (1 + 2l^3)\,216l^6 + (1 + 2l^3)^4$$
$$= 1 + 8l^3 + 240l^6 + 464l^9 + 16l^4.$$

Cette quantité est en même temps un invariant de la fonction cubique

$$x^3 + y^3 + z^3 + 6lxyz,$$

et comme tel elle doit s'exprimer en fonction des deux invariants S et T de cette dernière: en effet elle se réduit à

$$(1 - 20l^3 - 8l^6)^2 - 48\,(-l + l^4)^3 = T^2 - 48S^3.$$

Nous avons donc

$$C_{12} = T^2 - 48S^3,$$

et comme nous venons de trouver

$$C_6 = 2T,$$

l'équation $12R = 16C_{12} - C_6^2$ se réduit à

$$12R = 16\,(T^2 - 48S^3) - 4T^2$$
$$= 12T^2 - 768S^3,$$

ou enfin à

$$R = T^2 - 64S^3,$$

équation qui fait voir que le résultant R dont il s'agit est effectivement égal à $R=(1+8l^3)^3$, c. à d. au discriminant de la fonction cubique

$$x^3 + y^3 + z^3 + 6lxyz,$$

ce qui donne la vérification du théorème général. Je m'étais servi d'abord d'une analyse moins simple, en considérant le cas dans lequel on prend pour les trois fonctions les formes

$$\begin{aligned}&x^2 - y^2,\\ &z^2 - y^2,\\ &x^2 + y^2 + z^2 + 2lyz + 2mzx + 2nxy.\end{aligned}$$

Il vaut la peine, je crois, de donner les expressions correspondantes de C_6, C_{12}, R. On a ici à considérer la fonction

$$\lambda(x^2 - y^2) + \mu(x^2 + y^2 + z^2 + 2lyz + 2mzx + 2nxy) + \nu(z^2 - y^2),$$

laquelle peut s'écrire comme suit,

$$(\mu + \lambda,\ \mu - \lambda - \nu,\ \mu + \nu,\ l\mu,\ m\mu,\ n\mu)(x,\ y,\ z)^2.$$

Le réciprocant (multiplié par 2) est

$$= -2 \begin{vmatrix} & \xi, & \eta, & \zeta \\ \xi, & \mu+\lambda, & n\mu, & m\mu \\ \eta, & n\mu, & \mu-\lambda-\nu, & l\mu \\ \zeta, & m\mu, & l\mu, & \mu+\nu \end{vmatrix}$$

et en écrivant a, b, c, f, g, h au lieu de $1-l^2$, $1-m^2$, $1-n^2$, $mn-l$, $nl-m$, $lm-n$, cette fonction est

$$\begin{aligned} = \quad & (2a\mu^2 - 2\lambda\mu - 2\nu\lambda - 2\nu^2)\,\xi^2 \\ & + (2b\mu^2 + 2\lambda\mu + 2\mu\nu + 2\nu\lambda)\,\eta^2 \\ & + (2c\mu^2 - 2\nu\mu - 2\nu\lambda - 2\lambda^2)\,\zeta^2 \\ & + (2f\mu^2 - 2l\lambda\mu)\,2\eta\zeta \\ & + (2g\mu^2 + 2m\lambda\mu + 2m\mu\nu)\,2\zeta\xi \\ & + (2h\mu^2 - 2n\nu\mu)\,2\xi\eta, \end{aligned}$$

ou, écrite en forme de fonction quadratique bi-ternaire,

$$= \left(\begin{array}{cccccc} 0, & 2a, & -2, & 0, & 1, & -1 \\ 0, & 2b, & 0, & 1, & 1, & 1 \\ -2, & 2c, & 0, & -1, & -1, & 0 \\ 0, & 2f, & 0, & 0, & 0, & -l \\ 0, & 2g, & 0, & m, & 0, & m \\ 0, & 2h, & 0, & -n, & 0, & 0 \end{array}\right)(\lambda,\ \mu,\ \nu)^2(\xi,\ \eta,\ \zeta)^2.$$

En exprimant comme auparavant l'invariant $2T$ par une somme de cinq termes, on trouve sans peine que ces termes sont respectivement $18-4l^2-4m^2-4n^2$, $-4l^2$, $-4m^2$, $-4n^2$, 0; l'invariant sera donc $-8(l^2+m^2+n^2)+18$; et l'on a pour le combinant du sixième ordre

$$C_6=-8(l^2+m^2+n^2)+18.$$

Le discriminant (multiplié par 6) de la fonction $\lambda U+\mu V+\nu W$ est

$$=6\begin{vmatrix} \mu+\lambda, & n\mu & , & m\mu \\ n\mu & , & \mu-\lambda-\nu, & l\mu \\ m\mu & , & l\mu & , & \mu+\nu \end{vmatrix}$$

$$=0\lambda^3+6(1-l^2-m^2-n^2+2lmn)\mu^3+0\nu^3$$

$$+6(m^2-n^2)\mu^2\nu-6\nu^2\lambda-6\lambda^2\mu-6\mu\nu^2-6\nu\lambda^2+6(m^2-l^2)\lambda\mu^2-6\lambda\mu\nu$$

$$=\big(0,\ 6(1-l^2-m^2-n^2+2lmn),\ 0,\ 2(m^2-n^2),\ -2,\ -2,\ -2,\ -2,\ 2(m^2-l^2),\ -1\big)(\lambda,\ \mu,\ \nu)^3$$

$$=(a,\ b,\ c,\ i,\ j,\ k,\ i_1,\ j_1,\ k_1,\ l)(\lambda,\ \mu,\ \nu)^3,$$

où l'on a posé $a=0$, $b=6(1-l^2-m^2-n^2+2lmn)$, etc. L'invariant S de cette dernière fonction est

$$\begin{aligned} = \quad & 1 \\ & -2\big(-4(m^2-l^2)+4-4(m^2-n^2)\big) \\ & -3\big(\ 8(m^2-n^2)\quad +8(m^2-l^2)\big) \\ & -4b \\ & +16(m^2-n^2)-16(m^2-n^2)(m^2-l^2)+16(m^2-l^2) \\ & +16(m^2-n^2)^2+16+16(m^2-l^2)^2 \\ & +16b \end{aligned}$$

$$=9+12b+16(l^4+m^4+n^4-m^2n^2-n^2l^2-l^2m^2),$$

ou, si l'on substitue la valeur de b,

$$\begin{aligned} C_{12}=\ & 16(l^4+m^4+n^4-m^2n^2-n^2l^2-l^2m^2) \\ & +144lmn-72(l^2+m^2+n^2)+81. \end{aligned}$$

On vient de trouver

$$C_6=-8(l^2+m^2+n^2)+18,$$

équation qui donne

$$\begin{aligned} C_6^2=\ & 64(l^4+m^4+n^4+2m^2n^2+2n^2l^2+2l^2m^2) \\ & -288(l^2+m^2+n^2)+324. \end{aligned}$$

De ces résultats combinés on conclue

$$16C_{12}-C_6^2=192\,(l^4+m^4+n^4-2m^2n^2-2n^2l^2-2l^2m^2)+2304lmn-864\,(l^2+m^2+n^2)+972.$$

Évidemment l'expression du résultant est, à un facteur numérique près,

$$R=(3+2l+2m+2n)(3+2l-2m-2n)(3-2l+2m-2n)(3-2l-2m+2n),$$

ou en réduisant

$$R=16\,(l^4+m^4+n^4-2m^2n^2-2n^2l^2-2l^2m^2)+192lmn-72\,(l^2+m^2+n^2)+81.$$

On retrouve donc

$$12R=16C_{12}-C_6^2,$$

équation qu'il s'agissait de vérifier.

En terminant je remarque que la plus grande partie de ce mémoire est tirée d'un manuscrit daté "Machynlleth, 13 Août 1853."

Londres, 22 *Mars* 1859.

272.

DÉMONSTRATION D'UN THÉORÈME DE JACOBI PAR RAPPORT AU PROBLÈME DE PFAFF.

[From the *Journal für die reine und angewandte Mathematik* (Crelle), tom. LVII. (1860), pp. 273—277.]

DANS le mémoire de Jacobi "Theoria novi multiplicatoris etc." (t. XXIX. de ce Journal, 1845) on trouve p. 253 le passage que voici: "Methodum ad solvendum problema Pfaffianum ab ipso auctore adhibitam, data occasione observo per plures et altiores procedere integrationes quam methodus vera et genuina poscat. Quam novam methodum pro exemplo simplice explicabo. Ad aequationem differentialem

$$X_1dx_1 + X_2dx_2 + X_3dx_3 + X_4dx_4 = 0$$

per duas aequationes integrandam poscit Pfaffiana methodus...integrationem completam systematis trium aequationum differentialium primi ordinis inter quatuor variabiles ac deinde unius aequationis differentialis primi ordinis inter duas variabiles. Illius igitur systematis integrali uno invento, secundum illam methodum restat integratio completa duarum aequationum differentialium primi ordinis inter tres variabiles sive unius aequationis differentialis secundi ordinis inter duas variabiles ac deinde aequationis differentialis primi ordinis inter duas variabiles. At observo si integrali illo invento exprimatur x_4 per x_1, x_2, x_3, aequationem differentialem propositam abire in aliam linearem primi ordinis inter tres variabiles *conditioni integrabilitatis satisfacientem;* cujus integrationem vidimus (voir p. 246) absolvi posse per integrationes separatas duarum aequationum differentialium primi ordinis inter duas variabiles. Unde in locum aequationis differentialis secundi ordinis, tantum integrandae sunt duae aequationes differentiales separatae primi ordinis, quae est reductio maxime insignis; integrationi autem aequationis differentialis primi ordinis postremo praestandae omnino supersedetur. Tractatio hujus rei gravissimae completa ac generalis alii commentationi reservanda est." Le théorème dont il s'agit est énoncé dans les mots "conditioni integrabilitatis satisfacientem," mais j'ai cité le passage en entier pour montrer l'importance qu'à bon droit Jacobi attachait à la théorie dont ce théorème fait partie. On sait que le problème de la solution d'une équation à différences partielles du premier ordre n'est qu'un cas particulier du problème de Pfaff et que la solution des équations différentielles de la Dynamique et celle d'autres systèmes est intimement liée avec la question de la solution de ce cas particulier: le problème de Pfaff est ainsi de la plus haute importance dans l'analyse. Jacobi a parlé autre part d'un mémoire sur la mécanique analytique dont il s'est beaucoup occupé, lequel aurait naturellement des rapports avec celui-là; malheureusement ni l'un ni l'autre de ces mémoires n'ont paru.

Pour démontrer le théorème, considerons l'expression différentielle

$$\nabla = X_1\,dx_1 + X_2\,dx_2 + X_3\,dx_3 + X_4\,dx_4.$$

En supposant que $a = \phi(x_1, x_2, x_3, x_4)$ soit une intégrale des équations différentielles de Pfaff, si au moyen de cette équation on exprime x_4 en fonction de (x_1, x_2, x_3, a) l'expression ∇ prendra la forme

$$\nabla_1 = Y_1\,dx_1 + Y_2\,dx_2 + Y_3\,dx_3 + A\,da$$

et il s'agit de faire voir que l'équation

$$Y_1\,dx_1 + Y_2\,dx_2 + Y_3\,dx_3 = 0$$

à laquelle $\nabla_1 = 0$ se réduit, lorsqu'on fait $a =$ constante, sera intégrable au moyen d'un facteur, ou autrement dit, que l'expression $Y_1dx_1 + Y_2dx_2 + Y_3dx_3$ sera de la forme Udu, c. à d. qu'il existe une quantité u, fonction de x_1, x_2, x_3, a, telle qu'en traitant a comme une constante on a

$$Y_1\,dx_1 + Y_2\,dx_2 + Y_3\,dx_3 = U\,du.$$

Ainsi dans cette formule du dénote $\frac{du}{dx_1}dx_1 + \frac{du}{dx_2}dx_2 + \frac{du}{dx_3}dx_3$, et si l'on veut regarder a comme variable il faudrait y écrire $du - \frac{du}{da}da$ au lieu de du; on aurait ainsi $\nabla_1 = U\left(du - \frac{du}{da}da\right) + Ada$, ou en changeant la signification de A,

$$\nabla = X_1\,dx_1 + X_2\,dx_2 + X_3\,dx_3 + X_4\,dx_4 = U\,du + A\,da,$$

où u, a, et de même U, A, seront des fonctions données de x_1, x_2, x_3, x_4; c'est en effet la forme à laquelle dans le problème de Pfaff il s'agit de réduire l'expression différentielle ∇.

Or les équations différentielles de Pfaff sont le système que voici, savoir en écrivant

$$12 = \frac{dX_1}{dx_2} - \frac{dX_2}{dx_1},\ \text{etc.},$$

ce qui implique $12 = -21$, $11 = 0$, etc., et en introduisant la variable auxiliaire t, qui doit être éliminée, le système est

$$\begin{aligned}
0 &= X_1dx_1 + X_2dx_2 + X_3dx_3 + X_4dx_4,\\
X_1dt &= \quad * \quad 12dx_2 + 13dx_3 + 14dx_4,\\
X_2dt &= 21dx_1 \quad * \quad + 23dx_3 + 24dx_4,\\
X_3dt &= 31dx_1 + 32dx_2 \quad * \quad + 34dx_4,\\
X_4dt &= 41dx_1 + 42dx_2 + 43dx_3 \quad * \quad ,
\end{aligned}$$

lequel ne contient que quatre équations indépendantes. On donne à ce système une forme plus symétrique en écrivant x_0 au lieu de t, et en y mettant de plus

$$X_1 = 01 = -10, \text{ etc.}$$

les équations deviennent par ce moyen

$$\begin{aligned}
0 &= \quad * \quad 01dx_1 + 02dx_2 + 03dx_3 + 04dx_4,\\
0 &= 10dx_0 \quad * \quad + 12dx_2 + 13dx_3 + 14dx_4,\\
0 &= 20dx_0 + 21dx_1 \quad * \quad + 23dx_3 + 24dx_4,\\
0 &= 30dx_0 + 31dx_1 + 32dx_2 \quad * \quad + 34dx_4,\\
0 &= 40dx_0 + 41dx_1 + 42dx_2 + 43dx \quad * \quad ,
\end{aligned}$$

et on déduit delà

$$dx_0 : dx_1 : dx_2 : dx_3 : dx_4 = 1234 : 2340 : 3401 : 4012 : 0123,$$

où 1234, etc. sont les fonctions de Jacobi que j'ai nommées "Pfaffiens," savoir on a

$$\begin{aligned}
1234 &= 12\,.\,34 + 13\,.\,42 + 14\,.\,23,\\
2340 &= 23\,.\,40 + 24\,.\,03 + 20\,.\,34, = \quad * \quad - X_2 34 - X_3 42 - X_4 23,\\
3401 &= 34\,.\,01 + 30\,.\,14 + 31\,.\,40, = \quad X_1 34 \quad * \quad + X_3 41 + X_4 13,\\
4012 &= 40\,.\,12 + 41\,.\,20 + 42\,.\,01, = -X_1 24 - X_2 41 \quad * \quad - X_4 12,\\
0123 &= 01\,.\,23 + 02\,.\,31 + 03\,.\,12, = \quad X_1 23 + X_2 31 + X_3 12 \quad *
\end{aligned}$$

Cette transformation se trouve en effet dans le mémoire de Jacobi "Ueber die Pfaffsche Methode etc." (t. II. de ce Journal p. 347 et suivantes, 1827), où cependant Jacobi ne s'est pas servi de l'algorithme $X_1 = 01 = -10$ etc. au moyen duquel la forme de la solution devient si simple.

Faisons attention à présent à ce que l'équation

$$a = \phi(x_1,\ x_2,\ x_3,\ x_4)$$

est supposée être une intégrale des équations différentielles. On en déduit d'abord $x_4 = \text{fonc.}\ (x_1, x_2, x_3, a)$, et de là en traitant a comme constante

$$dx_4 = \frac{dx_4}{dx_1}dx_1 + \frac{dx_4}{dx_2}dx_2 + \frac{dx_4}{dx_3}dx_3,$$

équation qui doit être satisfaite par les valeurs qu'on vient de trouver pour $dx_1 : dx_2 : dx_3 : dx_4$, et qui par conséquent donne:

$$0 = 0123 - \frac{dx_4}{dx_1}\, 2340 - \frac{dx_4}{dx_2}\, 3401 - \frac{dx_4}{dx_3}\, 4012.$$

En y substituant pour 0123, etc. leurs valeurs, modifiées en sorte que tous les termes deviennent positifs (ce qui se fait à l'aide des relations $12 = -21$ etc.), on trouve

$$\left.\begin{aligned} &\frac{dx_4}{dx_1}(\quad * \quad X_2\,34 + X_3\,42 + X_4\,23) \\ +&\frac{dx_4}{dx_2}(X_1\,43 \quad * \quad + X_3\,14 + X_4\,31) \\ +&\frac{dx_4}{dx_3}(X_1\,24 + X_2\,41 \quad * \quad + X_4\,12) \\ +&\quad\;(X_1\,32 + X_2\,13 + X_3\,21 \quad * \quad) \end{aligned}\right\} = 0,$$

équation aux différences partielles à laquelle satisfait la variable x_4, lorsqu'au moyen de l'équation $a = \phi(x_1, x_2, x_3, x_4)$ (qui est une intégrale des équations différentielles de Pfaff) elle est donnée en fonction de x_1, x_2, x_3. Nous allons voir que c'est cette dernière équation, dans laquelle est contenu le théorème dont la démonstration fait l'objet de cette note. En effet l'expression différentielle

$$\nabla = X_1\,dx_1 + X_2\,dx_2 + X_3\,dx_3 + X_4\,dx_4$$

ayant été transformée en

$$\nabla_1 = Y_1\,dx_1 + Y_2\,dx_2 + Y_3\,dx_3 + A\,da$$

par la substitution de a au lieu de x_4, il s'en suit qu'on a les relations:

$$\begin{aligned} Y_1 &= X_1 + X_4\frac{dx_4}{dx_1}, \\ Y_2 &= X_2 + X_4\frac{dx_4}{dx_2}, \\ Y_3 &= X_3 + X_4\frac{dx_4}{dx_3}, \\ A &= \qquad X_4\frac{dx_4}{da}; \end{aligned}$$

la variable x_4 contenue dans les quantités X_1, X_2, X_3, X_4 doit être remplacée par son expression en fonction de x_1, x_2, x_3, a.

Cela posé, la condition pour que l'équation

$$Y_1\,dx_1 + Y_2\,dx_2 + Y_3\,dx_3 = 0$$

soit intégrable par un facteur étant désignée par

$$Z = 0,$$

on sait que

$$Z = Y_1\left(\frac{dY_2}{dx_3} - \frac{dY_3}{dx_2}\right) + Y_2\left(\frac{dY_3}{dx_1} - \frac{dY_1}{dx_3}\right) + Y_3\left(\frac{dY_2}{dx_1} - \frac{dY_1}{dx_2}\right).$$

Mais d'après les relations que l'on vient d'établir on a

$$\frac{dY_2}{dx_3} = \frac{dX_2}{dx_3} + \frac{dX_2}{dx_4}\,\frac{dx_4}{dx_3} + \left(\frac{dX_4}{dx_3} + \frac{dX_4}{dx_4}\,\frac{dx_4}{dx_3}\right)\frac{dx_4}{dx_2} + X_4\frac{d^2x_4}{dx_2dx_3},$$

$$\frac{dY_3}{dx_2} = \frac{dX_3}{dx_2} + \frac{dX_3}{dx_4}\,\frac{dx_4}{dx_2} + \left(\frac{dX_4}{dx_2} + \frac{dX_4}{dx_4}\,\frac{dx_4}{dx_2}\right)\frac{dx_4}{dx_3} + X_4\frac{d^2x_4}{dx_2dx_3};$$

d'où l'on tire

$$\frac{dY_2}{dx_3} - \frac{dY_3}{dx_2} = 23 + 24\,\frac{dx_4}{dx_3} + 43\,\frac{dx_4}{dx_2},$$

et on trouve de même

$$\frac{dY_3}{dx_1} - \frac{dY_1}{dx_3} = 31 + 34\,\frac{dx_4}{dx_1} + 41\,\frac{dx_4}{dx_3},$$

$$\frac{dY_1}{dx_2} - \frac{dY_2}{dx_1} = 12 + 14\,\frac{dx_4}{dx_2} + 42\,\frac{dx_4}{dx_1};$$

donc en ajoutant ces équations après les avoir multipliées par $X_1 + X_4\frac{dx_4}{dx_1}$, $X_2 + X_4\frac{dx_4}{dx_2}$, $X_3 + X_4\frac{dx_4}{dx_3}$, les termes qui contiennent les produits $\frac{dx_4}{dx_1}\,\frac{dx_4}{dx_2}$, etc. se détruisent, et l'on obtient l'équation finale

$$Z = \left\{\begin{array}{l} \dfrac{dx_4}{dx_1}(\quad * \qquad X_2\,34 + X_3\,42 + X_4\,23) \\ + \dfrac{dx_4}{dx_2}(X_1\,43 \qquad * \quad + X_3\,14 + X_4\,31) \\ + \dfrac{dx_4}{dx_3}(X_1\,24 + X_2\,41 \qquad * \quad + X_4\,12) \\ + \qquad (X_1\,32 + X_2\,13 + X_3\,21 \qquad * \quad) \end{array}\right\},$$

dont la seconde partie s'évanouit comme il a été démontré ci-dessus. Par conséquent la condition d'intégrabilité $Z = 0$ se trouve remplie, ce qu'il s'agissait de prouver.

Londres, 3 *Septembre* 1859

273.

NOTE SUR LA TRANSFORMATION DE TSCHIRNHAUSEN.

[From the *Journal für die reine und angewandte Mathematik* (Crelle), tom. LVIII. (1861), pp. 259—262.]

On trouve dans le mémoire de M. Hermite "Sur quelques théorèmes d'algèbre et la résolution de l'équation du quatrième degré" (*Comptes Rendus*, t. XLVI. p. 961, 1858) un théorème très-important relatif à la transformation de Tschirnhausen, à l'aide de laquelle une équation $f(x) = 0$ est transformée en une autre du même degré en y par la substitution $y = \phi x$, où ϕx désigne une fonction rationnelle. En considérant pour fixer les idées une équation du quatrième degré

$$(a,\ b,\ c,\ d,\ e \between x,\ 1)^4 = 0,$$

M. Hermite donne à l'équation $y = \phi x$ la forme que voici,

$$y = aT + (ax + 4b)\,T_0 + (ax^2 + 4bx + 6c)\,T_1 + (ax^3 + 4bx^2 + 6cx + 4d)\,T_2,$$

et il fait voir que les coefficients de la transformée en y ont la propriété suivante: toute fonction de ces coefficients, laquelle exprimée en fonction de $a, b, c, d, e, T, T_0, T_1, T_2$ ne contient pas T, est un *invariant*, c. à d. un invariant des deux fonctions

$$(a,\ b,\ c,\ d,\ e \between \xi,\ \eta)^4, \quad (T_0,\ T_1,\ T_2 \between \eta,\ -\xi)^2.$$

Cela revient à dire qu'en supposant la valeur de T déterminée de manière à faire évanouir dans l'équation en y le coefficient du second terme (de y^3), les autres coefficients seront des invariants, de sorte que, si dans le polynome en y qui est égalé à zéro on considère y comme une constante absolue, le polynome tout entier sera un invariant des deux fonctions ci-dessus mentionnées. On trouve sans peine la valeur que doit avoir T, elle est donnée par l'équation

$$0 = aT + 3bT_0 + 3cT_1 + dT_2,$$

ce qui conduit pour y à la valeur

$$y = (ax + b)\, T_0 + (ax^2 + 4bx + 3c)\, T_1 + (ax^3 + 4bx^2 + 6cx + 3d)\, T_2,$$

et en même temps la transformée en y aura la propriété dont il s'agit.

Par rapport à la forme de l'expression que l'on vient de trouver pour y il est bon de remarquer que les coefficients numériques qu'on y rencontre, hormis ceux du terme en x^0, ou de $bT_0 + 3cT_1 + 3dT_2$, sont les coefficients de la puissance $(1, 1)^4$, tandis que les coefficients qui ont été désignés comme faisant exception sont ceux de la puissance $(1, 1)^3$. Une remarque pareille s'applique au cas général. Pour démontrer le théorème énoncé, je représente l'équation qui vient d'être écrite par $y = V$, la transformée en y sera donc

$$(y - V_1)(y - V_2)(y - V_3)(y - V_4) = 0,$$

où V_1, V_2, V_3, V_4 sont ce que devient V lorsqu'on y substitue successivement pour x les quatre racines de l'équation $(a, b, c, d, e \between x, 1)^4 = 0$. Or, en considérant y comme une constante, pour que l'expression qui forme la première partie de l'équation que l'on vient d'écrire soit un invariant, les conditions à remplir consistent en ce que cette expression se réduise à zéro au moyen de l'un et l'autre des opérateurs

$$a\partial_b + 2b\partial_c + 3c\partial_d + 4d\partial_e - (\;T_2\,\partial_{T_1} + 2T_1\,\partial_{T_0}),$$
$$4b\partial_a + 3c\partial_b + 2d\partial_c + \;\;e\partial_d - (2T_1\,\partial_{T_2} + \;\;T_0\,\partial_{T_1}).$$

Ces conditions seront satisfaites si chacun des facteurs $y - V_1$, etc. est doué de cette même propriété, c. à d. si $y - V$, ou plus simplement si V, en y faisant x égal à l'une des racines de l'équation en x, se réduit à zéro au moyen de l'un et l'autre des deux opérateurs ci-dessus écrits. Je considère le premier des deux opérateurs et pour abréger je le désigne par

$$\Delta - (T_2\,\partial_{T_1} + 2T_1\,\partial_{T_0}).$$

Pour avoir ΔV, il faut d'abord former la valeur de Δx. On l'obtient en opérant avec Δ sur l'équation $(a, b, c, d, e \between x, 1)^4 = 0$, ce qui donne

$$(a, b, c, d \between x, 1)^3\, \Delta x + (a, b, c, d \between x, 1)^3 = 0, \text{ ou } \Delta x = -1.$$

La partie de ΔV qui dépend de la variation de x est par conséquent

$$-aT_0 + (-2ax - 4b)\, T_1 + (-3ax^2 - 8bx - 6c)\, T_2.$$

Pour l'autre partie de ΔV on trouve aisément

$$aT_0 + (4ax + 6b)\, T_1 + (4ax^2 + 12bx + 9c)\, T_2,$$

et de là en ajoutant

$$\Delta V = 2\,(ax + b)\, T_1 + (ax^2 + 4bx + 3c)\, T_2$$

ce qui est précisément égal à

$$(T_2\partial_{T_1} + 2T_1\partial_{T_0})\, V.$$

Donc V se réduit à zéro par l'opérateur

$$\Delta - (T_2 \partial_{T_1} + 2T_1 \partial_{T_0}).$$

De même en représentant le second opérateur par

$$\nabla - (2T_1 \partial_{T_2} + T_0 \partial_{T_1})$$

on trouve d'abord

$$(a, b, c, d \between x, 1)^3 \nabla x + x . (b, c, d, e \between x, 1)^3 = 0,$$

mais en ayant égard à l'équation $(a, b, c, d, e \between x, 1)^4 = 0$ la valeur de ∇x se réduit à $\Delta x = x^2$. La partie de ∇V due à la variation de x est par conséquent

$$ax^2 T_0 + (2ax^3 + 4bx^2) T_1 + (3ax^4 + 8bx^3 + 6cx^2) T_2.$$

L'autre partie de ∇V est

$$(4bx + 3c) T_0 + (4bx^2 + 12cx + 6d) T_1 + (4bx^3 + 12cx^2 + 12dx + 3c) T_2.$$

En les ajoutant, le coefficient de T_2 s'évanouit en vertu de l'équation en x, et l'on trouve

$$\nabla V = (ax^2 + 4bx + 3c) T_0 + 2 (ax^3 + 4bx^2 + 6cx + 3d) T_1,$$

ce qui est précisément égal à

$$(2T_1 \partial_{T_2} + T_0 \partial_{T_1}) V.$$

V se réduit donc à zéro au moyen de l'opérateur

$$\nabla - (2T_1 \partial_{T_2} + T_0 \partial_{T_1}),$$

ce qui achève la démonstration dont il s'agissait. Il va sans dire que la démonstration serait conduite d'une manière semblable pour une équation de degré quelconque. Pour avoir l'exemple le plus simple, je prends les équations

$$(a, b, c, d \between x, 1)^3 = 0,$$
$$y = (ax + b) T_0 + (ax^2 + 3bx + 2c) T_1,$$

et pour effectuer l'élimination j'écris

$$yx = (ax^2 + bx) T_0 + (-cx - d) T_1,$$
$$yx^2 = (-2bx^2 - 3cx - d) T_0 + (-cx^2 - dx) T_1.$$

Maintenant on a les trois équations

$$\begin{aligned}
0 &= bT_0 + 2cT_1 - y + x (aT_0 + 3bT_1) + x^2 . aT_1, \\
0 &= \quad - dT_1 + x (bT_0 - cT_1 - y) + x^2 . aT_0, \\
0 &= \quad - dT_0 + x (-3cT_0 - dT_1) + x^2 (-2bT_0 - cT_1 - y).
\end{aligned}$$

donc l'équation en y est

$$\begin{vmatrix} y - bT_0 - 2cT_1, & -aT_0 - 3bT_1, & -aT_1 \\ dT_1 \quad , & y - bT_0 + cT_1, & -aT_0 \\ dT_0 \quad , & 3cT_0 + dT_1, & y + 2bT_0 + cT_1 \end{vmatrix} = 0.$$

En ordonnant ce déterminant suivant les puissances de y, les coefficients de y^3, y^2, y, y^0 deviennent des formes binaires en T_0, T_1, des ordres 0, 1, 2, 3. En calculant les valeurs de ces quatre coefficients on les trouve respectivement

$$= 1,$$

$$= 0,$$

$$= 3\,(ac - b^2,\ ad - bc,\ bd - c^2 \between T_0,\ T_1)^2,$$

$$= (a^2d - 3abc + 2b^3,\ 3abd - 6ac^2 + 3b^2c,\ -3acd + 6b^2d - 3bc^2,\ -ad^2 + 3bcd - 2c^3 \between T_0,\ T_1)^3,$$

c'est-à-dire que l'équation en y est celle-ci:

$$y^3 + 3y\,(ac - b^2,\ ad - bc,\ bd - c^2 \between T_0,\ T_1)^2$$
$$+ (a^2d - 3abc + 2b^3,\ 3abd - 6ac^2 + 3b^2c,\ -3acd + 6b^2d - 3bc^2,\ -ad^2 + 3bcd - 2c^3 \between T_0,\ T_1)^3 = 0,$$

équation qui remplit en effet la condition ci-dessus mentionnée, d'avoir pour coefficients des invariants des deux formes:

$$(a,\ b,\ c,\ d \between \xi,\ \eta)^3, \quad (T_0,\ T_1 \between \eta,\ -\xi).$$

Dans le cas particulier dont il s'agit, la fonction $(T_0,\ T_1 \between \eta,\ -\xi)$ est linéaire, et l'on peut même dire que les coefficients sont des covariants de la seule fonction $(a,\ b,\ c,\ d \between T_0,\ T_1)^3$ en y considérant T_0, T_1 comme les variables.

J'ai cru qu'il y avait de l'intérêt de donner cette vérification. D'ailleurs je remarque qu'au moyen du théorème même on aurait pu trouver tout de suite l'équation en y, en écrivant d'abord $T_1 = 0$, ce qui donne le système

$$(a,\ b,\ c,\ d \between x,\ 1)^3 = 0,$$
$$y = (ax + b)\,T_0,$$

et de là

$$\frac{1}{a}(a,\ b,\ c,\ d \between y - bT_0,\ aT_0)^3 = 0$$

ou enfin

$$y^3 + 3y\,(b^2 - ac)\,T_0^2 + (a^2d - 3abc + 2b^3)\,T_1^3 = 0.$$

Les valeurs des coefficients peuvent être complétées, eu égard à ce qu'ils doivent être des invariants de $(a,\ b,\ c,\ d \between \xi,\ \eta)^3$, $(T_0,\ T_1 \between \eta,\ -\xi)$ (ou, comme on vient de le dire, des covariants de $(a,\ b,\ c,\ d \between T_0,\ T_1)^3$). Mais ce n'est que dans le cas particulier, où les coefficients T_0, T_1 sont au nombre de deux, que l'on peut réduire la seconde équation à une équation linéaire.

Londres, 18 *Avril* 1860.

274.

DEUXIÈME NOTE SUR LA TRANSFORMATION DE TSCHIRNHAUSEN.

[From the *Journal für die reine und angewandte Mathematik* (Crelle), tom. LVIII. (1861), pp. 263—269.]

À LA fin de ma première note sur ce sujet j'ai appliqué la transformation de Tschirnhausen à l'équation du troisième degré mise sous la forme

$$ax^3+3bx^2+3cx+d=(a,\ b,\ c,\ d\between x,\ 1)^3=0.$$

En y substituant $\frac{1}{3}b$, $\frac{1}{3}c$ au lieu de b, c, cette équation se change en

$$ax^3+bx^2+cx+d=(a,\ b,\ c,\ d\between x,\ 1)^3=0,$$

et en même temps le résultat obtenu dans ma première note s'énonce de la manière suivante:

En calculant pour l'équation

$$(a,\ b,\ c,\ d\between x,\ 1)^3=0,$$

la transformée en

$$y=(ax+\tfrac{1}{3}b)\,T_0+(ax^2+bx+\tfrac{2}{3}c)\,T_1,$$

on obtient

$$\left.\begin{array}{l} y^3 \\ +\frac{1}{3}y\left\{\begin{array}{ll} T_0^2 & (3ac-b^2) \\ +T_0T_1 & (9ad-bc) \\ +T_1^2 & (3bd-c^2) \end{array}\right\} \\ +\frac{1}{27}\left\{\begin{array}{ll} T_0^3 & (\ \ 27a^2d-\ \ 9abc+2b^3) \\ +T_0^2T_1 & (\ \ 27abd-18ac^2+3b^2c) \\ +T_0T_1^2 & (-27acd+18b^2d-3bc^2) \\ +T_1^3 & (-27ad^2+\ \ 9bcd-2c^3) \end{array}\right\} \end{array}\right\}=0.$$

Je vais me servir de cette formule, pour en déduire l'équation qui d'une manière analogue est la transformée de l'équation du quatrième ordre

$$(a,\ b,\ c,\ d,\ e \between x,\ 1)^4 = 0.$$

J'écris d'abord

$$(a,\ b,\ c,\ d,\ e \between x,\ 1)^4 = 0,$$

$$y = (ax + \tfrac{1}{4}b)\,T_0 + (ax^2 + bx + \tfrac{1}{2}c)\,T_1 + (ax^3 + bx^2 + cx + \tfrac{3}{4}d)\,T_2,$$

et je remarque qu'en faisant $e = 0$, le système proposé se partage en deux, dont le premier est:

$$x = 0,$$

$$y = \tfrac{1}{4}(bT_0 + 2cT_1 + 3dT_2),$$

et le second:

$$(a,\ b,\ c,\ d \between x,\ 1)^3 = 0,$$

$$y = (ax + \tfrac{1}{4}b)\,T_0 + (ax^2 + bx + \tfrac{1}{2}c)\,T_1 - \tfrac{1}{4}dT_2;$$

ou, ce qui est la même chose,

$$(a,\ b,\ c,\ d \between x,\ 1)^3 = 0,$$

$$y + \tfrac{1}{12}(bT_0 + 2cT_1 + 3dT_2) = (ax + \tfrac{1}{3}b)\,T_0 + (ax^2 + bx + \tfrac{2}{3}c)\,T_1.$$

Une circonstance analogue a lieu dans l'équation en y, résultat de l'élimination du système proposé. Pour $e = 0$ son premier membre se résout de même en deux facteurs qui égalés à zéro sont les résultats de l'élimination du premier et du second système ci-dessus écrits. Le premier de ces deux facteurs est donc

$$y - \tfrac{1}{4}(bT_0 + 2cT_1 + 3dT_2);$$

et le second (en vertu de la formule donnée antérieurement)

$$\begin{aligned} &\{y + \tfrac{1}{12}(bT_0 + 2cT_1 + 3dT_2)\}^3 \\ &+ \tfrac{1}{3}\,[(3ac - b^2)\,T_0^2 + \text{etc.}]\ \{y + \tfrac{1}{12}(bT_0 + 2cT_1 + 3dT_2)\} \\ &+ \tfrac{1}{27}[(27a^2d - 9abc + 2b^3)\,T_0^3 + \text{etc.}]\,; \end{aligned}$$

donc en multipliant les deux facteurs, et en égalant à zéro leur produit, on a la transformée en y de la forme $(a,\ b,\ c,\ d,\ 0 \between x,\ 1)^4$. Or dans le cas général, où e est différent de zéro, les coefficients de la transformée en y sont des invariants des deux formes

$$(a,\ b,\ c,\ d,\ e \between \xi,\ \eta)^4,\quad (T_0,\ T_1,\ T_2 \between \eta,\ -\xi)^2.$$

Cette propriété permet de déduire leurs valeurs générales des valeurs particulières qu'ils ont pour $e = 0$. Je formerai de cette manière la transformée en y pour la forme $(a,\ b,\ c,\ d,\ 0 \between x,\ 1)^4$, je passerai de là à la forme $(a,\ b,\ c,\ d,\ 0 \between x,\ 1)^4$ (ce qui se fait en écrivant $4b$, $6c$, $4d$ au lieu de b, c, d), et enfin je complèterai les valeurs des coefficients en y introduisant e au moyen de la propriété que doivent posséder les coefficients d'être des invariants des deux formes

$$(a,\ b,\ c,\ d,\ e \between \xi,\ \eta)^4,\quad (T_0,\ T_1,\ T_2 \between \eta,\ -\xi)^2.$$

On obtient d'abord l'équation en y sous la forme

$$(1,\ 0,\ \mathfrak{C},\ \mathfrak{D},\ \mathfrak{E} \between y,\ 1)^4 = 0,$$

où

$$\begin{aligned}
8\mathfrak{C} = {} & T_0^2 (\ 8ac - 3b^2) \\
& + T_0T_1 (\ 24ad - 4bc) \\
& + T_1^2 (\ 8bd - 4c^2) \\
& + T_0T_2 (-\ 2bd) \\
& + T_1T_2 (-\ 4cd) \\
& + T_2^2 (-\ 3d^2),
\end{aligned}$$

$$\begin{aligned}
8\mathfrak{D} = {} & T_0^3 (\ 8a^2d - 4abc + 1b^3) \\
& + T_0^2T_1 (\ 4abd - 8ac^2 + 2b^2c) \\
& + T_0T_1^2 (-16acd + 4b^2d) \\
& + T_1^3 (-\ 8ad^2) \\
& + T_0^2T_2 (-\ 4acd + 1b^2d) \\
& + T_0T_1T_2 (-12ad^2) \\
& + T_1^2T_2 (-\ 4bd^2) \\
& + T_0T_2^2 (-\ 1bd^2) \\
& + T_1T_2^2 (-\ 2cd^2) \\
& + T_2^3 (-\ 1d^3),
\end{aligned}$$

$$\begin{aligned}
256\mathfrak{E} = {} & T_0^4 (\ 64a^2bd + 16ab^2c - 3b^4) \\
& + T_0^3T_1 (-128a^2cd - 80ab^2d + 64abc^2 - 8b^3c) \\
& + T_0^2T_1^2 (-128abcd + 64ac^3 - 48b^3d + 8b^2c^2) \\
& + T_0T_1^3 (\ 64abd^2 + 64ac^2d - 128b^2cd + 32bc^3) \\
& + T_1^4 (\ 128acd^2 - 64bc^2d + 16c^4) \\
& + T_0^3T_2 (-192a^2d^2 + 32abcd - 4b^3d) \\
& + T_0^2T_1T_2 (-288abd^2 + 64ac^2d + 8b^2cd) \\
& + T_0T_1^2T_2 (-160b^2d^2 + 48bc^2d) \\
& + T_1^3T_2 (+192ad^3 - 128bcd^2 + 32c^3d) \\
& + T_0^2T_2^2 (-\ 48acd^2 + 14b^2d^2) \\
& + T_0T_1T_2^2 (-144ad^3 + 8bcd^2) \\
& + T_1^2T_2^2 (-\ 48bd^3 + 8c^2d^2) \\
& + T_0T_2^3 (-\ 4bd^3) \\
& + T_1T_2^3 (-\ 1cd^3) \\
& + T_2^4 (-\ 8d^4).
\end{aligned}$$

Ce calcul achevé et substituant la forme $(a, b, c, d, 0 \mathfrak{X} \xi, \eta)^4$ au lieu de $(a, b, c, d, 0 \mathfrak{X} \xi, \eta)^4$ (ou $4b$, $6c$, $4d$ au lieu de b, c, d) on obtient tous les termes de l'équation cherchée, hormis ceux qui contiennent e: et ces derniers s'obtiennent au moyen de ce que les coefficients des différentes puissances de y se réduisent à zéro par l'opérateur

$$a\partial_b + 2b\partial_c + 3c\partial_d + 4d\partial_e - T_2\partial_{T_1} - 2T_1\partial_{T_0}.$$

Cela ne présente pas de difficulté, je supprime donc les calculs intermédiaires et je donne le résultat final que voici: les équations

$$(a,\ b,\ c,\ d,\ e \between x,\ 1)^4 = 0,$$

$$y = (ax + b)\, T_0 + (ax^2 + 4bx + 3c)\, T_1 + (ax^3 + 4bx^2 + 6cx + 3d)\, T_2,$$

conduisent à la transformée:

$$(1,\ 0,\ \mathfrak{C},\ \mathfrak{D},\ \mathfrak{E} \between y,\ 1)^4 = 0,$$

où l'on a

	T_0^2	T_0T_1	T_0T_2	T_1^2	T_1T_2	T_2^2
$\mathfrak{C} = 2$	$ac + 3$	$ad + 6$	$ae + 2$	$ae + 1$	$be + 6$	$ce + 3$
	$b^2 - 3$	$bc - 6$	$bd - 2$	$bd + 8$	$cd - 6$	$d^2 - 3$
				$c^2 - 9$		

	T_0^3	$T_0^2T_1$	$T_0^2T_2$	$T_0T_1^2$	$T_0T_1T_2$	T_1^3	$T_0T_2^2$	$T_1^2T_2$	$T_1T_2^2$	T_2^3
$\mathfrak{D} = 4$	$a^2d + 1$	$a^2e + 1$	$abe + 1$	$abe + 4$	$ad^2 - 6$	$ad^2 - 4$	$ade - 1$	$ade - 4$	$ae^2 - 1$	$be^2 - 1$
	$abc - 3$	$abd + 2$	$acd - 3$	$acd - 12$	$b^2e + 6$	$b^2e + 4$	$bce + 3$	$bce + 12$	$bde - 2$	$cde + 3$
	$b^3 + 2$	$ac^2 - 9$	$b^2d + 2$	$b^2d + 8$			$bd^2 - 2$	$bd^2 - 8$	$c^2e + 9$	$d^3 - 2$
		$b^2c + 6$							$cd^2 - 6$	

	T_0^4	$T_0^3T_1$	$T_0^3T_2$	$T_0^2T_1^2$	$T_0^2T_1T_2$	$T_0T_1^3$	$T_0^2T_2^2$	$T_0T_1^2T_2$
$\mathfrak{E} =$	$a^3e + 1$	$a^2be + 8$	$a^2ce + 12$	$a^2ce - 6$	$abce + 60$	$a^2de - 4$	$a^2e^2 + 2$	$a^2e^2 - 4$
	$a^2bd - 4$	$a^2cd - 12$	$a^2d^2 - 12$	$ab^2e + 30$	$abd^2 - 72$	$abce - 12$	$abde - 16$	$abde + 20$
	$ab^2c + 6$	$ab^2d - 20$	$ab^2e - 8$	$abcd - 48$	$ac^2d + 36$	$abd^2 + 16$	$ac^2e + 36$	$ac^2e + 36$
	$b^4 - 3$	$abc^2 + 36$	$abcd + 12$	$ac^3 + 54$	$b^3e - 36$	$ac^2d + 36$	$acd^2 - 18$	$b^2d^2 - 160$
		$b^3c - 12$	$b^3d + 4$	$b^3d - 48$	$b^2cd + 12$	$b^3e + 48$	$b^2ce - 18$	$bc^2d + 108$
				$b^2c^2 + 18$		$b^2cd - 192$	$b^2d^2 + 14$	
						$bc^3 + 108$		

T_1^4	$T_0T_1T_2^2$	$T_1^3T_2$	$T_0T_2^3$	$T_1^2T_2^2$	$T_1T_2^3$	T_2^4
$a^2e^2 + 1$	$acde + 60$	$abe^2 - 4$	$ace^2 + 12$	$ace^2 - 6$	$ade^2 + 8$	$ae^3 + 1$
$abde - 16$	$ad^3 - 36$	$acde - 12$	$ad^2e - 8$	$ad^2e + 30$	$bce^2 - 12$	$bde^2 - 4$
$ac^2e - 18$	$b^2de - 72$	$ad^3 + 48$	$b^2e^2 - 12$	$bcde - 48$	$bd^2e - 20$	$cd^2e + 6$
$acd^2 + 48$	$bc^2e + 36$	$b^2de + 16$	$bcde + 12$	$bd^3 - 48$	$c^2de + 36$	$d^4 - 3$
$b^2ce + 48$	$bcd^2 + 12$	$bc^2e + 36$	$bd^3 - 4$	$c^3e + 54$	$cd^3 - 12$	
$bc^2d - 144$		$bcd^2 - 192$		$c^2d^2 + 18$		
$c^4 + 81$		$c^3d + 108$				

J'écris

$$U' = aT_0^2 + 4bT_0T_1 + c\,(2T_0T_2 + 4T_1^2) + 4dT_1T_2 + eT_2^2,$$

$$H' = (ac - b^2)\, T_0^2 + 2\,(ad - bc)\, T_0T_1 + (ae - 2bd + c^2)\, T_0T_2 + 4\,(bd - c^2)\, T_1^2$$
$$+ 2\,(be - cd)\, T_1T_2 + (ce - d^2)\, T_2^2,$$

et je représente par $4\Phi'$ la valeur qui vient d'être trouvée pour $\mathfrak{D}$. Ces expressions U', H', Φ' sont des invariants des deux formes $(a, b, c, d, e \between \xi, \eta)^4$, $(T_0, T_1, T_2 \between \eta, -\xi)^2$, on a de plus les invariants

$$ae - 4bd + 3c^2, \quad ace - ad^2 - b^2e + 2bcd - c^3,$$

que je représente comme à l'ordinaire par I, J, et l'invariant $T_0T_2 - T_1^2$ que je représente par Θ'. Cela posé on a

$$\begin{aligned}
\mathfrak{C} &= 6H' - 2I\Theta', \\
\mathfrak{D} &= 4\Phi', \\
\mathfrak{E} &= IU'^2 - 3H'^2 + I^2\Theta'^2 + 12J\Theta'U' + 2I\Theta'H'.
\end{aligned}$$

La dernière de ces équations peut être vérifiée aisément, pour cela on a seulement besoin de remarquer qu'en posant $a = e = 1$, $b = d = 0$, $c = \theta$, elle devient

$$\begin{aligned}
&(1 + 3\theta^2)\left(T_0^2 + \theta\,(2T_0T_2 + 4T_1^2) + T_2^2\right)^2 \\
&- 3\left(\theta T_0^2 + (1+\theta^2)\,T_0T_2 - 4\theta^2T_1^2 + \theta T_2^2\right)^2 \\
&+ (1 + 3\theta^2)^2\,(T_0T_2 - T_1^2)^2 \\
&+ 12\,(\theta - \theta^3)\,(T_0T_2 - T_1^2)\left(T_0^2 + \theta\,(2T_0T_2 + 4T_1^2) + T_2^2\right) \\
&+ 2\,(1 + 3\theta^2)\,(T_0T_2 - T_1^2)\left(\theta T_0^2 + (1+\theta^2)\,T_0T_2 - 4\theta^2T_1^2 + \theta T_2^2\right)
\end{aligned}$$

$$\begin{aligned}
= \quad & T_0^4 \\
&+ T_0^3T_2\,(\;12\theta) \\
&+ T_0^2T_1^2\,(-6\theta + 54\theta^3) \\
&+ T_0^2T_2^2\,(2 + 36\theta^2) \\
&+ T_0T_1^2T_2\,(-4 + 36\theta^2) \\
&+ T_1^4\,(1 - 18\theta^2 + 81\theta^4) \\
&+ T_1^2T_2^2\,(-6\theta + 54\theta^3) \\
&+ T_0T_2^3\,(\;12\theta) \\
&+ T_2^4,
\end{aligned}$$

équation qui est identique. L'expression de l'invariant I (quadrinvariant) de la fonction $(1, 0, \mathfrak{C}, \mathfrak{D}, \mathfrak{E} \between y, 1)^4$ est $\mathfrak{E} + 3\,(\frac{1}{6}\mathfrak{C})^2$, ou $\mathfrak{E} + 3\,(H' - \frac{1}{3}I\Theta')^2$, c'est-à-dire

$$\begin{aligned}
IU'^2 - 3H'^2 + {} & I^2\Theta'^2 + 12J\Theta'U' + 2I\Theta'H' \\
+ 3H'^2 + {} & \tfrac{1}{3}I^2\Theta'^2 \qquad\qquad - 2I\Theta'H',
\end{aligned}$$

ou enfin

$$IU'^2 + \tfrac{4}{3}I^2\Theta'^2 + 12J\Theta'U',$$

ce qui est égal à

$$\frac{1}{I}[(IU' + 6J\Theta')^2 + \tfrac{4}{3}(I^3 - 27J^2)\,\Theta'^2].$$

La condition à remplir pour que cet invariant se réduise à zéro peut donc être présentée sous la forme

$$IU' + [6J \pm 2\sqrt{-\tfrac{1}{3}(I^3 - 27J^2)}]\,\Theta' = 0,$$

ce qui s'accorde avec un résultat trouvé par M. Hermite.

Il doit y avoir, ce me semble, une équation identique de la forme

$$JU'^2 - IU'^2H' + 4H'^3 + M\Theta' = -\Phi'^2$$

qui servirait à exprimer le carré de Φ' au moyen des autres invariants U', H', Θ', I, J, mais en supposant que cette équation existe, la forme du facteur M, que je n'ai pas encore cherchée, reste à déterminer; l'invariant J (cubinvariant) de la forme $(1, 0, \mathfrak{C}, \mathfrak{D}, \mathfrak{E} \between y, 1)^4$ contient Φ'^2, et il faudrait employer l'identité dont je viens de parler pour réduire à sa forme la plus simple cet invariant; dans l'état actuel de la question je ne m'occupe donc pas de l'expression du cubinvariant de $(1, 0, \mathfrak{C}, \mathfrak{D}, \mathfrak{E} \between y, 1)^4$.

Pour passer au cas d'une équation du cinquième ordre, on devra faire usage de la formule qui se rapporte à la forme $(a, b, c, d, e \between x, 1)^4$. En faisant la substitution nécessaire on arrive à ce résultat que pour l'équation

$$(a, b, c, d, e \between x, 1)^4 = 0$$

la transformée en

$$y = (ax + \tfrac{1}{4}b)\,T_0 + (ax^2 + bx + \tfrac{1}{2}c)\,T_1 + (ax^3 + bx^2 + cx + \tfrac{3}{4}d)\,T_2$$

est la suivante

$$(1, 0, \mathfrak{C}, \mathfrak{D}, \mathfrak{E} \between y, 1)^4 = 0,$$

où

	T_0^2	T_0T_1	T_0T_2	T_1^2	T_1T_2	T_2^2
$\mathfrak{C} = \frac{1}{8}$	ac + 8 b^2 − 3	ad + 24 bc − 4	ae + 32 bd − 2	ae + 16 bd + 8 c^2 − 4	be + 24 cd − 4	ce + 8 d^2 − 3

	T_0^3	$T_0^2T_1$	$T_0^2T_2$	$T_0T_1^2$	$T_0T_1T_2$	T_1^3	$T_0T_2^2$	$T_1^2T_2$	$T_1T_2^2$	T_2^3
$\mathfrak{D} = \frac{1}{8}$	a^2d + 8 abc − 4 b^3 + 1	a^2e + 32 abd + 4 ac^2 − 8 b^2c + 2	abe + 8 acd − 4 b^2d + 1	abe + 32 acd − 16 b^2d + 4	ad^2 − 12 b^2e + 12	ad^2 − 8 b^2e + 8	ade − 8 bce + 4 bd^2 − 1	ade − 32 bce + 16 bd^2 − 4	ae^2 − 32 bde − 4 c^2e + 8 cd^2 − 2	be^2 − 8 cde + 4 d^3 − 1

$\mathfrak{E} = \frac{1}{256}$

T_0^4	$T_0^3T_1$	$T_0^3T_2$	$T_0^2T_1^2$	$T_0^2T_1T_2$	$T_0T_1^3$	$T_0^2T_2^2$	$T_0T_1^2T_2$
a^3e + 256	a^2be + 512	a^2ce + 512	a^2ce − 256	$abce$ + 640	a^2de − 256	a^2e^2 + 512	a^2e^2 − 1024
a^2bd − 64	a^2cd − 128	a^2d^2 − 192	ab^2e + 480	abd^2 − 288	$abce$ − 128	$abde$ − 256	$abde$ + 320
ab^2c + 16	ab^2d − 80	ab^2e − 128	$abcd$ − 128	ac^2d + 64	abd^2 + 64	ac^2e + 256	ac^2e + 256
b^4 − 3	abc^2 + 64	$abcd$ + 32	ac^3 + 64	b^3e − 144	ac^2d + 64	acd^2 − 48	b^2d^2 − 160
	b^3c − 8	b^3d − 4	b^3d − 48	b^2cd + 8	b^3e + 192	b^2ce − 48	bc^2d + 48
			b^2c^2 + 8		b^2cd − 128	b^2d^2 + 14	
					bc^3 + 32		

T_1^4	$T_0T_1T_2^2$	$T_1^3T_2$	$T_0T_2^3$	$T_1^2T_2^2$	$T_1T_2^3$	T_2^4
a^2e^2 + 256	$acde$ + 640	abe^2 − 256	ace^2 + 512	ace^2 − 256	ade^2 + 512	ae^3 + 256
$abde$ − 256	ad^3 − 144	$acde$ − 128	ad^2e − 128	ad^2e + 480	bce^2 − 128	bde^2 − 64
ac^2e − 128	b^2de − 288	ad^3 + 192	b^2e^2 − 192	$bcde$ − 128	bd^2e − 80	cd^2e + 16
acd^2 + 128	bc^2e + 64	b^2de + 64	$bcde$ + 32	bd^3 − 48	c^2de + 64	d^4 − 3
b^2ce + 128	bcd^2 + 8	bc^2e + 64	bd^3 − 4	c^3e + 64	cd^3 − 8	
bc^2d − 64		bcd^2 − 128		c^2d^2 + 8		
c^4 + 16		c^3d + 32				

En m'appuyant sur ce résultat j'espère être à même de trouver la formule pour l'équation du cinquième ordre.

Londres, 11 *Mai* 1860.

275.

ON TSCHIRNHAUSEN'S TRANSFORMATION.

[From the *Philosophical Transactions of the Royal Society of London*, vol. CLI. (for the year 1861), pp. 561—578. Received November 7,—Read December 5, 1861.]

THE memoir of M. Hermite, "Sur quelques théorèmes d'algèbre et la résolution de l'équation du quatrième degré," *Comptes Rendus*, t. XLVI. p. 961 (1858), contains a very important theorem in relation to Tschirnhausen's Transformation of an equation $f(x)=0$ into another of the same degree in y, by means of the substitution $y=\phi x$, where ϕx is a rational and integral function of x. In fact, considering for greater simplicity a quartic equation,

$$(a, b, c, d, e \between x, 1)^4=0,$$

M. Hermite gives to the equation $y=\phi x$ the following form,

$$y=aT+(ax+4b)B+(ax^2+4bx+6c)C+(ax^3+4bx^2+6cx+4d)D,$$

(I write B, C, D in the place of his T_0, T_1, T_2), and he shows that the transformed equation in y has the following property: viz., every function of the coefficients which, expressed as a function of a, b, c, d, e, T, B, C, D, does not contain T, is an *invariant*, that is, an invariant of the two quantics

$$(a, b, c, d, e \between X, Y)^4, \ (B, C, D \between Y, -X)^2.$$

This comes to saying that if T be so determined that in the equation for y the coefficient of the second term (y^3) shall vanish, the other coefficients will be invariants; or if, in the function of y which is equated to zero, we consider y as an absolute constant, the function of y will be an invariant of the two quantics. It is easy to find the value of T; this is in fact given by the equation

$$0=aT+3bB+3cC+dD;$$

and we have thence for the value of y,

$$y = (ax+b)B + (ax^2+4bx+3c)C + (ax^3+4bx^2+6cx+3d)D;$$

so that for this value of y the function of y which equated to zero gives the transformed equation will be an invariant of the two quantics. It is proper to notice that in the last-mentioned expression for y, all the coefficients except those of the term in x^0, or $bB+3cC+3dD$, are those of the binomial $(1, 1)^4$, whereas the excepted coefficients are those of the binomial $(1, 1)^3$; this suffices to show what the expression for y is in the general case.

I have in the two papers, "Note sur la Transformation de Tschirnhausen," and "Deuxième Note sur la Transformation de Tschirnhausen," *Crelle*, t. LVIII. pp. 259 and 263 (1861), [273 and 274], obtained the transformed equations for the cubic and quartic equations; and by means of a grant from the Government Grant Fund, I have been enabled to procure the calculation, by Messrs Davis and Otter, under my superintendence, of the transformed equation for the quintic equation. The several results are given in the present memoir; and for greater completeness, I reproduce the demonstration which I have given in the former of the above-mentioned two Notes, of the general property, that the function of y is an invariant. At the end of the memoir I consider the problem of the reduction of the general quintic equation to Mr Jerrard's form $x^5+ax+b=0$.

Considering for simplicity the foregoing two equations

$$(a, b, c, d, e\between x, 1)^4 = 0,$$
$$y = (ax+b)B + (ax^2+4bx+3c)C + (ax^3+4bx^2+6cx+3d)D;$$

let the second of these be represented by $y=V$, the transformed equation in y is

$$(y-V_1)(y-V_2)(y-V_3)(y-V_4)=0,$$

where V_1, V_2, V_3, V_4 are what V becomes upon substituting therein for x the roots x_1, x_2, x_3, x_4 of the quartic equation respectively. Considering y as a constant, the conditions to be satisfied in order that the function in y may be an invariant are that this function shall be reduced to zero by each of the two operators

$$a\partial_b + 2b\partial_c + 3c\partial_d + 4d\partial_e - (\ D\partial_C + 2C\partial_B),$$
$$4b\partial_a + 3c\partial_b + 2d\partial_c + \ e\partial_d \ - (2C\partial_D + \ B\partial_C).$$

These conditions will be satisfied if each of the factors $y-V_1$, &c. has the property in question; that is, if $y-V$, or (what is the same thing) if V, supposing that x denotes therein a root of the quartic equation, is reduced to zero by each of the two operators. Considering the first operator, which for shortness I represent by

$$\Delta - (D\partial_C + 2C\partial_B),$$

in order to obtain ΔV we require the value of Δx. To find it, operating with Δ on the quartic equation, we have

$$(a, b, c, d\between x, 1)^3 \Delta x + (a, b, c, d\between x, 1)^3 = 0,$$

or $\Delta x = -1$. In ΔV, the part which depends on the variation of Δx then is

$$-aB + (-2ax - 4b)\,C + (-3ax^2 - 8bx - 6c)\,D,$$

and the other part of ΔV is at once found to be

$$+aB + (4ax + 6b)\,C + (4ax^2 + 12bx + 9c)\,D;$$

whence, adding,

$$\Delta V = 2\,(ax + b)\,C + (ax^2 + 4bx + 3c)\,D,$$

and this is precisely equal to

$$(D\partial_C + 2C\partial_B)\,V;$$

so that V is reduced to zero by the operator $\Delta - (D\partial_C + 2C\partial_B)$.

Similarly, if the second operator is represented by

$$\nabla - (2C\partial_D + B\partial_C),$$

then we have

$$(a,\ b,\ c,\ d \between x,\ 1)^3\,\nabla x + x\,(b,\ c,\ d,\ e \between x,\ 1)^3 = 0,$$

which by means of the equation

$$(a,\ b,\ c,\ d,\ e \between x,\ 1)^4 = 0$$

is reduced to $\nabla x = x^2$. Hence in ∇V the part depending on the variation of x is

$$(ax^2\qquad)\,B + (2ax^3 + 4bx^2\qquad)\,C + (3ax^4 + 8bx^3 + 6cx^2\qquad)\,D,$$

and the other part of ∇V is at once found to be

$$(4bx + 3c)\,B + (4bx^2 + 12cx + 6d)\,C + (4bx^3 + 12cx^2 + 12dx + 3e)\,D;$$

and, adding, the coefficient of D vanishes on account of the quartic equation, and we have

$$\nabla V = (ax^2 + 4bx + 3c)\,B + 2\,(ax^3 + 4bx^2 + 6cx + 3d)\,C,$$

which is precisely equal to

$$(2C\partial_D + B\partial_C)\,V,$$

so that V is reduced to zero by the operator

$$\nabla - (2C\partial_D + B\partial_C),$$

which completes the demonstration; and the demonstration in the general case is precisely similar.

In the case of the cubic equation we have

$$(a,\ b,\ c,\ d \between x,\ 1)^3 = 0,$$
$$y = (ax + b)\,B + (ax^2 + 3bx + 2c)\,C;$$

and writing the second equation in the form

$$(y - bB - 2cC) + x\,(\;\; - aB - 3bC) + x^2\,(\qquad\quad - aC) = 0,$$

multiplying by x and reducing by the cubic equation, we have

$$dC \qquad\qquad + x\,(y - bB + \;\; cC) + x^2\,(\;-\;\; aB) \qquad = 0,$$

and repeating the process,

$$dB \qquad\qquad + x\,(\;\; 3cB + \;\; dC) + x^2\,(y + 2bB + cC) = 0\,;$$

or, what is the same thing, we have the system of equations

$$\left(\begin{array}{ccc} y - bB - 2cC, & -\;aB - 3bC, & -cC \\ dC, & y - \;bB + \;cC, & -\;aB \\ dB\qquad , & 3cB + \;dC, & y + 2b\,B + cC \end{array}\right)\!\!\left(1,\; x,\; x^2\right) = 0,$$

and the resulting equation in y is of course that formed by equating to zero the determinant formed out of the matrix in this equation. The developed expression is

$$(1,\; 0,\; \mathfrak{C},\; \mathfrak{D}\,\big)\!\!\big(y,\; 1)^3 = 0,$$

where

$\tfrac{1}{3}\mathfrak{C} =$

	B^2		BC		C^2
ac	$+1$	ad	$+1$	bd	$+1$
b^2	-1	bc	-1	c^2	-1
	± 1		± 1		± 1

$\mathfrak{D} =$

	B^3		B^2C		BC^2		C^3
a^2d	$+1$	abd	$+3$	acd	-3	ad^2	-1
abc	-3	ac^2	-6	b^2d	$+6$	bcd	$+3$
b^3	$+2$	b^2c	$+3$	bc^2	-3	c^3	-2
	± 3		± 6		± 6		± 3

The sum of the coefficients in each column should here and elsewhere in the present memoir be equal to zero, and I have by way of verification annexed to each column the sums ($\pm$ a number) of the positive and negative coefficients. The coefficients $\mathfrak{C}$, $\mathfrak{D}$, and therefore the function in y, are invariants of the two forms,

$$(a,\; b,\; c,\; d\,\big)\!\!\big(X,\; Y)^3, \quad (B,\; C\,\big)\!\!\big(Y,\; -X);$$

or in the present case, where there are only two coefficients B, C, the coefficients $\mathfrak{C}$, $\mathfrak{D}$, and therefore also the function in y, are covariants of the single form $(a, b, c, d\,\big)\!\!\big(B, C)^3$, considering therein (B, C) as the facients.

It may be remarked that in the present case, assuming the invariance of the function in y, we may obtain the transformed equation in a very simple manner by writing in the first instance $C=0$, this gives

$$(a,\ b,\ c,\ d\,\between\, x,\ 1)^3=0,$$

$$y=(ax+b)\,B,$$

and thence

$$\frac{1}{a}\,(a,\ b,\ c,\ d\,\between\, y-bB,\ aB)^3=0\,,$$

or developing,

$$y^3+3y\,(ac-b^2)\,B^2+(a^2d-3abc+2b^3)\,B^3=0,$$

where the expressions for the coefficients are to be completed by the consideration that these coefficients are covariants of the form $(a,\ b,\ c,\ d\,\between\, B,\ C)^3$. But it is only in the case in hand of a cubic equation that the transformed equation can be obtained in this manner.

In the case of a quartic equation, we have

$$(a,\ b,\ c,\ d,\ e\,\between\, x,\ 1)^4=0,$$

$$y=(ax+b)\,B+(ax^2+4bx+3c)\,C+(ax^3+4bx^2+6cx+3d)\,D,$$

and these give the system of equations

$$\left(\begin{array}{llll} y-bB-3cC-3dD, & -\ aB-4bC-6cD, & -\ aC-4bD, & -aD \\ eD, & y-\ bB-3cC+\ dD, & -\ aB-4bC\quad, & -\ aC \\ eC\quad, & 4dC+\ eD, & y-\ bB+3cC+\ dD, & -\ aB \\ eB\quad, & 4dB+\ eC\quad, & 6cB+4dC+\ eD, & y+3bB+3cC+dD \end{array}\right|\between\, 1,\ x,\ x^2,\ x^3)=0,$$

and the transformed equation is therefore found by equating to zero the determinant formed out of the matrix contained in this equation.

The developed result, which was obtained by a different process in the 'Deuxième Note' above referred to, is

$$(1,\ 0,\ \mathfrak{C},\ \mathfrak{D},\ \mathfrak{E}\,\between\, y,\ 1)^4=0,$$

where

$\frac{1}{2}\,\mathfrak{C}=$

	B^2		BC		BD^2	C^2		CD		D^2
ac b^2	$+3$ -3	ad bc	$+6$ -6	ae bd c^2	$+2$ 2	$+1$ $+8$ -9	be cd	$+6$ -6	ce d^2	$+3$ -3
	± 3		± 6		± 2	± 9		± 6		± 3

$\frac{1}{4}\mathfrak{D} =$

	B^3		B^2C		B^2D	BC^2		BCD	C^3		BD^2	C^2D		CD^2		D^3
							ac^2									
a^2d	+1	a^2e	+1	abe	+1	+4	ad^2	−6	−4	ade	−1	−4	ae^2	−1	be^2	−1
abc	−3	abd	+2	acd	−3	−12	b^2e	+6	+4	bcd	+3	+12	bde	−2	cde	+3
b^3	+2	ac^2	−9	b^2d	+2	+8	bcd			bd^2	−2	−8	c^2e	+9	d^3	−2
		b^2c	+6	bc^2			c^3			c^2d			cd^2	−6		
	±3		±9		±3	±12		±6	±4		±3	±12		±9		±3

and

$\mathfrak{E} =$

	B^4		B^3C		B^3D	B^2C^2		B^2CD	BC^3		B^2D^2	BC^2D	C^4
a^3e	+1	a^2be	+8	a^2ce	+12	−6	a^2de	...	−4	a^2e^2	+2	−4	+1
a^2bd	−4	a^2cd	−12	a^2d^2	−12	...	$abce$	+60	−12	$abde$	−16	+20	−16
a^2c^2	...	ab^2d	−20	ab^2e	−8	+30	abd^2	−72	+16	ac^2e	+36	+36	−18
ab^2c	+6	abc^2	+36	$abcd$	+12	−48	ac^2d	+36	+36	acd^2	−18	...	+48
b^4	−3	b^3c	−12	ac^3	...	+54	b^3e	−36	+48	b^2ce	−18	...	+48
				b^3d	−4	−48	b^2cd	+12	−192	b^2d^2	+14	−160	...
				b^2c^2		+18	bc^3		+108	bc^2d		+108	−144
										c^4			+81
	±7		±44		±24	±102		±108	±208		±52	±164	±178

	BCD^2	C^3D		BD^3	C^2D^2		CD^3		D^4
abe^2	...	−4	ace^2	+12	−6	ade^2	+8	ae^3	+1
$acde$	+60	−12	ad^2e	−8	+30	bce^2	−12	bde^2	−4
ad^3	−36	+48	b^2e^2	−12	...	bd^2e	−20	c^2e^2	...
b^2de	−72	+16	$bcde$	+12	−48	c^2de	+36	cd^2e	+6
bc^2e	+36	+36	bd^3	−4	−48	cd^3	−12	d^4	−3
bcd^2	+12	−192	c^3e		+54				
c^3d		+108	c^2d^2		+18				
	±108	±208		±24	±102		±44		±7

I write

$$U' = aB^2 + 4bBC + c(2BD + 4C^2) + 4dCD + eD^2,$$

$$H' = (ac - b^2)B^2 + 2(ad - bc)BC + (ae - 2bd + c^2)BD + 4(bd - c^2)C^2 + 2(be - cd)CD + (ce - d^2)D^2;$$

and I represent by Φ' the expression which has just been found for $\frac{1}{4}\mathfrak{D}$. These functions, U', H', Φ', are invariants of the two forms

$$(a, b, c, d, e \between X, Y)^4, \quad (B, C, D \between Y, -X)^2;$$

we have, moreover, the invariants

$$ae - 4bd + 3c^2, \; ace - ad^2 - b^2e + 2bcd - c^3,$$

which I represent as usual by I, J, and the invariant $BD-C^2$, which I represent by Θ'. This being so, we have

$$\mathfrak{C} = 6H' - 2I\Theta',$$

$$\mathfrak{D} = 4\Phi',$$

$$\mathfrak{E} = IU'^2 - 3H'^2 + I^2\Theta'^2 + 12J\Theta' U' + 2I\Theta' H',$$

the last of which may be verified as follows:—viz. writing $a=e=1$, $b=d=0$, $c=\theta$, it becomes

$$\begin{aligned}
&(1+3\theta^2)\{B^2+\theta(2BD+4C^2)+D^2\}^2\\
&-3\{\theta B^2+(1+\theta^2)BD-4\theta^2C^2+\theta D^2\}^2\\
&+(1+3\theta^2)^2(BD-C^2)^2\\
&+12(\theta-\theta^3)(BD-C^2)\{B^2+\theta(2BD+4C^2)+D^2\}\\
&+2(1+3\theta^2)(BD-C^2)\{\theta B^2+(1+\theta^2)BD-4\theta^2C^2+\theta D^2\}
\end{aligned}$$

$$\begin{aligned}
=\ &B^4\\
&+B^3D\,(12\theta)\\
&+B^2C^2\,(-6\theta+54\theta^3)\\
&+B^2D^2\,(2+36\theta^2)\\
&+BC^2D\,(-4+36\theta^2)\\
&+C^4\,(1-18\theta^2+81\theta^4)\\
&+C^2D^2\,(-6\theta+54\theta^3)\\
&+BD^3\,(12\theta)\\
&+D^4,
\end{aligned}$$

which is an identical equation. [This is in effect the identity of the "Deuxième Note," see *ante* p. 372, but the left-hand side was in the original printed incorrectly, with four lines only instead of five.]

The expression for the invariant I (quadrinvariant) of the function $(1, 0, \mathfrak{C}, \mathfrak{D}, \mathfrak{E}\between y, 1)^4$ is $\mathfrak{E}+3(\tfrac{1}{6}\mathfrak{C})^2$, or $\mathfrak{E}+3(H'-\tfrac{1}{3}I\Theta')^2$, viz. it is

$$\begin{aligned}
IU'^2 - 3H'^2 + I^2\Theta'^2 + 12J\Theta'U' + 2I\Theta'H'&\\
+3H'^2+\tfrac{1}{3}I^2\Theta'^2 \qquad\qquad - 2I\Theta'H'&,
\end{aligned}$$

or, finally, it is

$$IU'^2 + \tfrac{4}{3}I^2\Theta'^2 + 12J\Theta'U',$$

which is equal to

$$\frac{1}{I}\left[(IU'+6J\Theta')^2 + \tfrac{4}{3}(I^3-27J^2)\,\Theta'^2\right];$$

so that the condition in order that this invariant may be equal to zero is

$$IU' + \left[6J \pm 2\sqrt{-\tfrac{1}{3}(I^3-27J^2)}\right]\Theta' = 0,$$

which agrees with a result of M. Hermite's.

There should, I think, be an identical equation of the form

$$JU'^2 - IU'^2H' + 4H'^3 + M\Theta' = -\Phi'^2,$$

which would serve to express the square of the invariant Φ' in terms of the other invariants U', H', Θ', I, J; but assuming that such an equation exists, the form of the factor M remains to be ascertained. The invariant J (cubinvariant) of the form $(1, 0, \mathfrak{C}, \mathfrak{D}, \mathfrak{E} \between y, 1)^4$ contains Φ'^2, and it would be necessary to make use of the identical equation just referred to in order to reduce it to its simplest form; and (this being so) I have not sought for the expression of the cubinvariant of $(1, 0, \mathfrak{C}, \mathfrak{D}, \mathfrak{E} \between y, 1)^4$.

For the quintic we have the equations

$$(a, b, c, d, e, f \between x, 1)^5 = 0,$$

$$\begin{aligned} y = \;& (ax + b)\,B \\ &+ (ax^2 + 5bx + 4c)\,C \\ &+ (ax^3 + 5bx^2 + 10cx + 6d)\,D \\ &+ (ax^4 + 5bx^3 + 10cx^2 + 10dx + 4e)\,E, \end{aligned}$$

and this leads to the system of equations

$$\begin{pmatrix} y - bB - 4cC - 6dD - 4eE, & -aB - 5bC - 10cD - 10dE, & -aC - 5bD - 10cE, & -aD - 5bE, & -aE \\ fE, & y - bB - 4cC - 6cD + eE, & -aB - 5bC - 10cD, & -aC - 5bD, & -aD \\ fD, & 5eD + fE, & y - bB - 4cC + 4dD + eE, & -aB - 5bC, & -aC \\ fC, & 5eC + fD, & 10dC + 5eD + fE, & y - bB + 6cC + 4dD + eE, & -aB \\ fB, & 5eB + fC, & 10dB + 5eC + fD, & 10cB + 10dC + 5eD + fE, & y + 4bB + 6cC + 4dD + eE \end{pmatrix} \between 1, x, x^2, x^3, x^4) = 0,$$

and the transformed equation is obtained by equating to zero the determinant formed out of the matrix contained in this equation.

The determinant in question was calculated by the formula

	Div.
$\square = -12.345$	1
$+13.245$	2
-14.235	3
$+15.234$	4
-23.145	5
$+24.135$	6
-25.134	7
-34.125	8
$+35.124$	9
-45.123	10,

where the duadic symbols refer to the first and fifth columns, viz. 12 is the determinant formed out of the lines 1 and 2 of these columns, and so for the other like symbols; and the triadic symbols refer to the second, third, and fourth columns, viz. 345 is the determinant formed out of the lines 3, 4, 5 of these columns, and so for the other like symbols.

The ten divisions were separately calculated. It is to be noticed that these divisions other than 4 and 6 correspond to each other in pairs, while each of the divisions 4 and 6 corresponds to itself, as thus:

Div. 1,	-10
2,	-9
3,	-7
5,	-8
4,	-4
6,	-6,

viz. if in the place of

$$y;\ a,\ b,\ c,\ d,\ e,\ f;\ B,\ C,\ D,\ E,$$

we write

$$-y;\ f,\ e,\ d,\ c,\ b,\ a;\ E,\ D,\ C,\ B,$$

then division 1 becomes division 10 with its sign reversed, and so for divisions 2 and 9, 3 and 7, 5 and 8; while each of the divisions 4 and 6 is unaltered, except that the sign is reversed. But the corresponding divisions were each of them calculated, and the property in question was used as a verification. Another very convenient verification, which was employed for the several divisions, was obtained by putting

$$a=b=c=d=e=f=B=C=D=E=1,$$

upon which supposition the determinant becomes

$$\begin{vmatrix} y-15, & -26, & -16, & -6, & -1 \\ 1, & y-10, & -16, & -6, & -1 \\ 1, & 6, & y\ , & -6, & -1 \\ 1, & 6, & 16, & y+10, & -1 \\ 1, & 6, & 16, & 16, & y+15 \end{vmatrix}$$

and the values of the ten divisions respectively are

y^5,	y^4,	y^3,	y^2,	y,	1	
		6,	− 288,	+ 4608,	− 24576	1
		16,	− 576,	+ 6144,	− 16384	2
		26,	− 544,	+ 3584,	− 24576	3
1,	0,	− 96,	0 ,	− 28672,	0	4
				0 ,	0	5
				0 ,	0	6
		26,	+ 544,	+ 3584,	+ 24576	7
				0 ,	0	8
		16,	+ 576,	+ 6144,	+ 16384	9
		6,	+ 288,	+ 4608,	+ 24576	10
1,	0,	0,	0 ,	0 ,	0	

A verification similar to this was in fact employed at each step of the calculation of a division: viz. in forming a product such as $(\lambda X+\mu Y+\&c.)(\lambda' X+\mu' Y+\&c.)$, where λ, μ, &c., λ', μ', .. &c. are numerical coefficients, and X, Y, &c. are monomial products of a, b, c, d, e, f and B, C, D, E, the sum of the numerical coefficients of the product is $(\lambda+\mu+\&c.)(\lambda'+\mu'+\&c.)$.

It was of course necessary to employ such verifications, as a test of the correctness of the several divisions, before proceeding to collect them together, but the collection itself affords an exceedingly good ultimate verification. The following is an exemplification: the terms in y which involve the product $BCDE$ are obtained by the collection of the corresponding terms in the ten divisions, as follows:

		1	2	3	4	5	6	7	8	9	10	
y	$BCDE$. − 5 a^2f^2	+ 1	− 2	− 2	+ 1	+ 1	− 2	− 2	+ 1	− 2	+ 1	
	+ 30 $abef$	+ 49	− 16	− 20	+ 4	− 25	+ 50	− 20	− 25	− 16	+ 49	
	+ 980 $acdf$	+ 80	+ 200	+ 184	− 148	+ 100		+ 184	+ 100	+ 200	+ 80	
	− 280 ace^2	+ 80		+ 80	− 440							
	− 180 ad^2e		− 60	− 60	− 60							
	− 280 b^2df				− 440			+ 80			+ 80	
	− 825 b^2e^2				− 825							
	− 180 bc^2f				− 60			− 60		− 60		
	+ 740 $bcde$				+ 740							
	bd^3											
	c^3e											
	c^2d^2											
	± 1750 = 0	+ 210	+ 122	+ 182	− 1228	+ 76	+ 48	+ 182	+ 76	+ 122	+ 210	= ± 1228

where it may be remarked that the greater part, but not all, of the component coefficients are divisible by 5. I soon observed in the process of summing the ten divisions that all the resulting coefficients should be divisible by 5 (the only exception is as to the terms in y^0 which contain B^5, C^5, D^5, and E^5 respectively), and the circumstance that they are so in each particular instance is as far as it goes a verification, which, however, only applies to those of the component coefficients which are not themselves divisible by 5. But it was known *à priori* (I will presently show how this is so) that the sum of the resulting coefficients should be equal to zero, and that they are so in fact is a verification as to *all* the coefficients. The foregoing specimen term $BCDE$ is one which remains unaltered when B, C, D, E are changed into E, D, C, B; and on making the further change a, b, c, d, e, f into f, e, d, c, b, a, the coefficient of $BCDE$ remains, as it should do, unaltered; this is a verification of the coefficients of the terms ace^2, b^2df; ad^2e, bc^2f, which are respectively interchanged by the substitution in question, but not of the other terms a^2f^2, $abef$, $acdf$, b^2e^2, $bcdf$, which are respectively unaltered by the substitution. I did *not* employ what would have been another convenient verification of the several divisions, viz. the comparison of their values on putting therein $a=b=c=d=e=f=1$, with the corresponding values as calculated independently from the determinant

$$\left|\begin{array}{ccccc} y-B-4C-6D-E, & -B-5C-10D-10E, & -C-5D-10E, & D-5E, & -E \\ E, & y-B-4C-6D+E, & -B-5C-10D\;\;, & -C-5D\;\;, & -D \\ D\;\;, & 5D+E, & y-B-4C+4D+E, & -B-5C\;\;, & -C \\ C\;\;, & 5C+D\;\;, & 10C+5D+E, & y-B+6C+4D+E, & -B \\ B\;\;, & 5B+C\;\;, & 10B+5C+D\;\;, & 10B+10C+5D+E, & y+4B+6C+4D+E \end{array}\right|;$$

the calculation of the ten divisions of this determinant would however itself have been a work of some labour.

The last-mentioned determinant is $=y^5$; in fact, equating it to zero, we have the transformed equation corresponding to the system of equations

$$(1,\ 1,\ 1,\ 1,\ 1,\ 1 \between x,\ 1)^5=0,$$

$$y=(x+1)\,B+(x^2+5x+4)\,C+(x^3+5x^2+10x+6)\,D+(x^4+5x^3+10x^2+10x+4)\,E.$$

But the first of these equations is $(x+1)^5=0$, and the second is

$$y=(x+1)\,\{B+(x+4)\,C+(x^2+4x+6)\,D+(x^3+4x^2+6x+4)\,E\},$$

so that for each of the five equal roots $x=-1$, we have $y=0$, or the transformed equation in y is $y^5=0$.

And since upon writing $a=b=c=d=e=f=1$ the transformed equation becomes $y^5=0$, it is clear that in the coefficient of any monomial product of B, C, D, E, the sum of the numerical coefficients of the several monomial products of a, b, c, d, e, f must be $=0$, which is the property above referred to as affording a verification of the calculated expression of the transformed equation.

The final result is that the equations

$$(a,\ b,\ c,\ d,\ e,\ f\between x,\ 1)^5=0,$$

$$\begin{aligned} y = &\ \ \ (ax + b)\,B \\ &+ (ax^2 + 5bx + 4c)\,C \\ &+ (ax^3 + 5bx^2 + 10cx + 6d)\,D \\ &+ (ax^4 + 5bx^3 + 10cx^2 + 10dx + 4e)\,E \end{aligned}$$

give for the transformed equation in y

$$(1,\ 0,\ \mathfrak{C},\ \mathfrak{D},\ \mathfrak{E},\ \mathfrak{F}\between y,\ 1)^5=0,$$

where

		B^2		BC		BD	C^2		BE	CD		CE	D^2		DE		E^2
	ac	$+2$	ab	$+6$	ae	$+4$	$+2$	af	$+1$	$+1$	bf	$+4$	$+2$	cf	$+6$	df	$+2$
$\frac{1}{5}\mathfrak{C}=$	b^2	-2	bc	-6	bd	-4	$+10$	be	-1	$+15$	ce	-4	$+10$	de	-6	e^2	-2
					c^2		-12	cd		-16	d^2		-12				
		± 2		± 6		± 4	± 12		± 1	± 16		± 4	± 12		± 6		± 2

		B^3		B^2C		B^2D	BC^2		B^2E	BCD	C^3		BCE	BD^2	C^2D
	a^2d	$+2$	a^2e	$+4$	a^2f	$+1$	$+1$	abf	$+1$	$+10$	$+3$	acf	...	$+6$	$+4$
	abc	-6	abd	$+2$	abe	$+3$	$+19$	ace	-4	-8	-4	ade	-12	-22	-46
$\frac{1}{5}\mathfrak{D}=$	b^3	$+4$	ac^2	-24	acd	...	-52	ad^2	...	-48	-20	b^2f	$+8$	$+4$	$+20$
			b^2c	$+18$	b^2d	-4	$+20$	b^2e	$+3$	$+30$	$+25$	bce	$+4$	$+20$	$+70$
					bc^2		$+12$	bcd		$+16$	-20	bd^2		-8	-80
								c^3			$+16$	c^2d			$+32$
		± 6		± 24		± 4	± 52		± 4	± 56	± 44		± 12	± 30	± 126

	BDE	C^2E	CD^2		BE^2	CDE	D^3		CE^2	D^2E		DE^2		E^3
adf	...	-6	-4	aef	-1	-10	-3	af^2	-1	-1	bf^2	-4	cf^2	-2
ae^2	-8	-4	-20	bdf	$+4$	$+8$	$+4$	bef	-3	-19	cef	-2	def	$+6$
bcf	$+12$	$+22$	$+46$	be^2	-3	-30	-25	cdf	...	$+52$	d^2f	$+24$	e^3	-4
bde	-4	-20	-70	c^2f		$+48$	$+20$	ce^2	$+4$	-20	de^2	-18		
c^2e		$+8$	$+80$	cde		-16	$+20$	d^2e		-12				
cd^2			-32	d^3			-16							
	± 12	± 30	± 126		± 4	± 56	± 44		± 4	± 52		± 24		± 6

$\frac{1}{5}\mathfrak{E}=$

	B^4
a^3e	+1
a^2bd	−4
a^2c^2	...
ab^2c	+6
b^4	−3
	±7

	B^3C
a^3f	+1
a^2be	+7
a^2cd	−16
ab^2d	−22
abc^2	+48
b^3c	−18
	±56

	B^3D	B^2C^2
a^2bf	+2	+9
a^2ce	+20	−22
a^2d^2	−24	...
ab^2e	−18	+31
$abcd$	+32	−56
ac^3	...	+96
b^3d	−12	−70
b^2c^2		+12
	±54	±148

	B^3E	B^2CD	BC^3
a^2cf	+6	+10	−2
a^2de	−6	−18	−10
ab^2f	−5	+17	+29
$abce$	+8	+96	−92
abd^2	...	−144	+40
ac^2d	...	+128	+128
b^3e	−3	−105	+75
b^2cd		+16	−360
bc^3			+192
	±14	±267	±464

	B^2CE	B^2D^2	BC^2D	C^4
a^2df	+12	−6	...	−4
a^2e^2	−12	+10	−20	+5
$abcf$	+28	+38	+48	+6
$abde$	−36	−86	−8	−50
ac^2e	+32	+100	+72	−64
acd^2	...	−24	+144	+160
b^3f	−28	−14	+60	+25
b^2ce	+4	−70	−240	+50
b^2d^2		+52	−440	...
bc^2d			+384	−320
c^4				+192
	±76	±200	±708	±438

	B^2DE	BC^2E	BCD^2	C^3D
a^2ef	+5	−5	−7	+1
$abdf$	−8	+92	−8	−36
abe^2	−29	−67	+15	−25
ac^2f	+60	+12	+72	−20
$acde$	−12	+36	+196	+28
ad^3	...	...	−72	+240
b^2cf	−42	−54	+78	+190
b^2de	+26	−110	−510	−50
bc^2e		+96	−60	−280
bcd^2			+296	−560
c^3d				+512
	±91	±236	±657	±971

	B^2E^2	$BCDE$	BD^3	C^3E	C^2D^2
a^2f^2	+1	−1	−1	−1	+1
$abef$	−5	+6	−4	−4	−25
$acdf$	+20	+196	+28	+28	−80
ace^2	−14	−56	+100	−28	+10
ad^2e	...	−36	−68	+60	+336
b^2df	−14	−56	−28	+100	+10
b^2e^2	+12	−165	−75	−75	...
bc^2f		−36	+60	−68	+336
$bcde$		+148	−140	−140	−940
bd^3			+128	...	−120
c^3e				+128	−120
c^2d^2					+592
	±33	±350	±316	±316	±1285

	BCE^2	BD^2E	C^2DE	CD^3
abf^2	+5	−5	−7	+1
$acef$	−8	+92	−8	−36
ad^2f	+60	+12	+72	−20
ade^2	−42	−54	+78	+190
b^2ef	−29	−67	+15	−25
$bcdf$	−12	+36	+196	+28
bce^2	+26	−110	−510	−50
bd^2e		+96	−60	−280
c^3f			−72	+240
c^2de			+296	−560
cd^3				+512
	±91	±236	±657	±971

	BDE^2	C^2E^2	CD^2E	D^4
acf^2	+12	−6	...	−4
$adef$	+28	+38	+48	+6
ae^3	−28	−14	+60	+25
b^2f^2	−12	+10	−20	+5
$bcef$	−36	−86	−8	−50
bd^2f	+32	+100	+72	−64
bde^2	+4	−70	−240	+50
c^2df		−24	+144	+160
c^2e^2		+52	−440	...
cd^2e			+384	−320
d^4				+192
	±76	±200	±708	±438

	BE^3	CDE^2	D^3E
adf^2	+6	+10	−2
ae^2f	−5	+17	+29
bcf^2	−6	−18	−10
$bdef$	+8	+96	−92
be^3	−3	−105	+75
c^2ef		−144	+40
cd^2f		+128	+128
cde^2		+16	−360
d^3e			+192
	±14	±267	±464

	CE^3	D^2E^2
aef^2	+2	+9
bdf^2	+20	−22
be^2f	−18	+31
c^2f^2	−24	...
$cdef$	+32	−56
ce^3	−12	−70
d^3f		+96
d^2e^2		+12
	±54	±148

	DE^3
af^3	+1
bef^2	+7
cdf^2	−16
ce^2f	−22
d^2ef	+48
de^3	−18
	±56

	E^4
bf^3	+1
cef^2	−4
d^2f^2	...
de^2f	+6
e^4	−3
	±7

$\mathfrak{F} =$

	B^5
a^4f	+ 1
a^3be	− 5
a^3cd	...
a^2b^2d	+ 10
a^2bc^2	...
ab^3c	− 10
b^5	+ 4
	± 15

	B^4C
a^3bf	+ 15
a^3ce	− 20
a^3d^2	...
a^2b^2e	− 55
a^2bcd	+ 80
a^2c^3	...
ab^3d	+ 70
ab^2c^2	− 120
b^4c	+ 30
	± 195

	B^4D	B^3C^2
a^3cf	+ 30	− 10
a^3de	− 30	...
a^2b^2f	− 15	+ 100
a^2bce	− 100	− 190
a^2bd^2	+ 120	...
a^2c^2d	...	+ 160
ab^3e	+ 55	− 260
ab^2cd	− 80	+ 540
abc^3	...	− 480
b^4d	+ 20	+ 200
b^3c^2		− 60
	± 225	± 1000

	B^4E	B^3CD	B^2C^3
a^3df	+ 20	...	− 10
a^3e^2	− 20	...	...
a^2bcf	− 30	+ 320	− 60
a^2bde	+ 30	− 360	+ 50
a^2c^2e	...	− 400	+ 120
a^2cd^2	...	+ 480	...
ab^3f	+ 15	− 140	+ 340
ab^2ce	− 20	− 440	− 1020
ab^2d^2	...	+ 960	− 100
abc^2d	...	− 640	+ 960
ac^4	...	...	− 640
b^4e	+ 5	+ 300	− 500
b^3cd		− 80	+ 1900
b^2c^3			− 1040
	± 70	± 2060	± 3370

	B^3CE	B^3D^2	B^2C^2D	BC^4
a^3ef	...	+ 10	− 20	+ 5
a^2bdf	+ 240	− 40	+ 20	− 80
a^2be^2	− 240	− 50	+ 100	− 25
a^2c^2f	− 120	+ 300	+ 20	− 40
a^2cde	+ 120	− 600	+ 60	+ 200
a^2d^3	...	+ 360	...	...
ab^2cf	− 140	− 220	+ 1190	+ 20
ab^2de	+ 240	+ 540	− 1980	+ 250
abc^2e	− 160	− 500	− 3160	+ 320
$abcd^2$	...	+ 120	+ 4080	− 800
ac^3d	...	...	− 1280	− 320
b^4f	+ 80	+ 40	− 400	+ 500
b^3ce	− 20	+ 200	+ 850	− 2750
b^3d^2		− 160	+ 2600	...
b^2c^2d			− 2080	+ 5600
bc^4				− 2800
	± 680	± 1570	± 8920	± 6895

	B^3DE	B^2C^2E	B^2CD^2	BC^3D	C^5
a^3f^2	+ 5	− 5	− 5	+ 5	− 1
a^2bef	− 30	+ 30	+ 60	− 130	+ 25
a^2cdf	+ 220	− 40	+ 380	− 380	+ 80
a^2ce^2	− 400	+ 40	− 200	+ 400	− 100
a^2d^2e	+ 180	...	− 180	+ 300	...
ab^2df	− 40	+ 1070	− 640	+ 380	− 250
ab^2e^2	+ 185	− 1145	+ 25	+ 125	...
abc^2f	− 300	− 940	+ 1640	+ 940	− 360
$abcde$	+ 60	+ 1020	− 4380	− 1140	+ 1000
abd^3	...	...	+ 3960	− 1200	...
ac^3e	...	− 320	− 2000	− 2080	+ 960
ac^2d^2	...	...	+ 480	+ 960	− 1600
b^3cf	+ 120	+ 160	− 620	+ 800	+ 750
b^3de	− 80	+ 650	+ 2900	− 3500	...
b^2c^2e		− 520	− 100	− 2600	− 300
b^2cd^2			− 1320	+ 14800	...
bc^3d				− 7680	+ 4800
c^5					− 2304
	± 810	± 2970	± 9445	± 18710	± 7615

	B^3E^2	B^2CDE	B^2D^3	BC^3E	BC^2D	C^4D		B^2CE^2	B^2D^2E	BC^2DE	BCD^3	C^4E	C^3D^2
a^2bf^2	+ 5	+ 15	− 10	− 25	...	+ 5	a^2cf^2	− 50	+ 70	+ 10	− 50	− 10	+ 20
a^2cef	− 70	− 60	+ 200	+ 80	− 210	+ 30	a^2def	+ 110	+ 80	− 340	+ 160	+ 130	− 30
a^2d^2f	+ 100	+ 240	+ 100	− 200	− 180	+ 120	a^2e^3	− 80	− 200	+ 400	...	− 100	...
a^2de^2	− 40	− 240	− 300	+ 200	+ 600	− 150	ab^2f^2	+ 80	− 70	− 40	+ 20	− 25	...
ab^2ef	+ 30	− 30	− 90	+ 70	...	− 125	$abcef$	− 520	− 460	+ 620	+ 380	− 20	− 500
$abcdf$	− 100	+ 1380	− 640	+ 260	+ 1920	− 1180	abd^2f	+ 700	+ 460	+ 1240	− 760	− 1000	+ 540
$abce^2$	+ 70	− 2820	− 500	− 360	+ 450	+ 500	$abde^2$	− 190	+ 620	− 2140	− 200	+ 1000	+ 750
abd^2e	...	+ 1980	+ 1540	− 300	− 3180	+ 1500	ac^2df	− 400	+ 200	− 1720	+ 2840	− 80	− 880
ac^3f	...	− 1200	+ 1000	− 560	+ 1080	− 240	ac^2e^2	+ 280	− 2000	− 2240	− 2000	+ 880	+ 1400
ac^2de	...	+ 240	− 3000	+ 240	− 8160	+ 3120	acd^2e	...	+ 1080	+ 2520	− 6440	− 1200	− 120
acd^3	...	...	+ 1440	...	+ 5040	− 4800	ad^4	...	...	...	+ 4320	...	− 3600
b^3df	+ 40	+ 240	+ 120	+ 2000	− 1800	+ 500	b^3ef	+ 160	+ 430	− 200	− 500	+ 125	...
b^3e^2	− 35	+ 975	+ 500	− 2125	...	...	b^2cdf	− 20	− 440	+ 5320	− 2240	+ 3000	− 700
b^2c^2f		− 60	− 400	− 1060	+ 1620	+ 2100	b^2ce^2	− 70	+ 650	− 2950	+ 2750	− 3500	...
b^2cde		− 660	+ 600	+ 3700	+ 2700	− 6500	b^2d^2e		− 420	+ 4800	+ 7400	...	− 3000
b^2d^3			− 560	...	+ 9600	...	bc^3f			− 3240	+ 1800	− 1680	+ 3960
bc^3e				− 1920	− 5400	− 4800	bc^2de			− 2040	− 10200	+ 4400	− 5600
bc^2d^2					− 4080	+ 17600	bcd^3				+ 2720	...	+ 20400
c^4d						− 7680	c^4e					− 1920	− 7200
							c^3d^2						− 5440
	± 245	± 5070	± 5500	± 6550	± 23010	± 25475		± 1330	± 3590	± 14910	± 22390	± 9535	± 27070

	B^2DE^2	BC^2E^2	BCD^2E	BD^4	C^3DE	C^2D^3		B^2E^3	$BCDE^2$	BD^3E	C^3E^2	C^2D^2E	CD^4
a^2df^2	+ 50	− 70	− 10	+ 10	+ 50	− 20	a^2ef^2	− 5	− 15	+ 25	+ 10	...	− 5
a^2e^2f	− 80	+ 70	+ 40	+ 25	− 20	...	$abdf^2$	+ 70	+ 60	− 80	− 200	+ 210	− 30
$abcf^2$	− 110	− 80	+ 340	− 130	− 160	+ 30	abe^2f	− 30	+ 30	− 70	+ 90	...	+ 125
$abdef$	+ 520	+ 460	− 620	+ 20	− 380	+ 500	ac^2f^2	− 100	− 240	+ 200	− 100	+ 180	− 120
abe^3	− 160	− 430	+ 200	− 125	+ 500	...	$acdef$	+ 100	− 1380	− 260	+ 640	− 1920	+ 1180
ac^2ef	− 700	− 460	− 1240	+ 1000	+ 760	− 540	ace^3	− 40	− 240	− 2000	− 120	+ 1800	− 500
acd^2f	+ 400	− 200	+ 1720	+ 80	− 2840	+ 880	ad^3f	...	+ 1200	+ 560	− 1000	− 1080	+ 240
$acde^2$	+ 20	+ 440	− 5320	− 3000	+ 2240	+ 700	ad^2e^2	...	+ 60	+ 1060	+ 400	− 1620	− 2100
ad^3e	...	...	+ 3240	+ 1680	− 1800	− 3960	b^2cf^2	+ 40	+ 240	− 200	+ 300	− 600	− 150
b^3f^2	+ 80	+ 200	− 400	+ 100	...	...	b^2def	− 70	+ 2820	+ 360	+ 500	− 450	− 500
b^2cef	+ 190	− 620	+ 2140	− 1000	+ 200	− 750	b^2e^3	+ 35	− 975	+ 2125	− 500	...	...
b^2d^2f	− 280	+ 2000	+ 2240	− 880	+ 2000	− 1400	bc^2ef		− 1980	+ 300	− 1540	+ 3180	+ 1500
b^2de^2	+ 70	− 650	+ 2950	+ 3500	− 2750	...	bcd^2f		− 240	− 240	+ 3000	+ 8160	− 3120
bc^2df		− 1080	− 2520	+ 1200	+ 6440	+ 120	$bcde^2$		+ 660	− 3700	− 600	− 2700	+ 6500
bc^2e^2		+ 420	− 4800	...	− 7400	+ 3000	bd^3e			+ 1920	...	+ 5400	+ 4800
bcd^2e			+ 2040	− 4400	+ 10200	+ 5600	c^3df				− 1440	− 5040	+ 4800
bd^4				+ 1920	...	+ 7200	c^3e^2				+ 560	− 9600	...
c^4f					− 4320	+ 3600	c^2d^2e					+ 4080	− 17600
c^3de					− 2720	− 20400	cd^4						+ 7680
c^2d^3						+ 5440							
	± 1330	± 3590	± 14910	± 9535	± 22390	± 27070		± 245	± 5070	± 6550	± 5500	± 23010	± 25475

	BCE^3	BD^2E^2	CD^2E^2	CD^3E	D^5
a^2f^3	− 5	+ 5	+ 5	− 5	+ 1
$abef^2$	+ 30	− 30	− 60	+ 130	− 25
$acdf^2$	− 220	+ 40	− 380	+ 380	− 80
ace^2f	− 40	− 1070	+ 640	− 380	+ 250
ad^2ef	+ 300	+ 940	− 1640	− 940	+ 360
ade^3	− 120	− 160	+ 620	− 800	− 750
b^2df^2	+ 400	− 40	+ 200	− 400	+ 100
b^2e^2f	− 185	+ 1145	− 25	− 125	...
bc^2f^2	− 180	...	+ 180	− 300	...
$bcdef$	− 60	− 1020	+ 4380	+ 1140	− 1000
bce^3	+ 80	− 650	− 2900	+ 3500	...
bd^3f		+ 320	+ 2000	+ 2080	− 960
bd^2e^2		+ 520	+ 100	+ 2600	+ 3000
c^3ef			− 3960	+ 1200	...
c^2d^2f			− 480	− 960	+ 1600
c^2de^2			+ 1320	− 14800	...
cd^3e				+ 7680	− 4800
d^5					+ 2304
	± 810	± 2970	± 9445	± 18710	± 7615

	BDE^3	C^2E^3	CD^2E^2	D^4E
abf^3	...	− 10	+ 20	− 5
$acef^2$	− 240	+ 40	− 20	+ 80
ad^2f^2	+ 120	− 300	− 20	+ 40
ade^2f	+ 140	+ 220	− 1190	− 20
ae^4	− 80	− 40	+ 400	− 500
b^2ef^2	+ 240	+ 50	− 100	+ 25
$bcdf^2$	− 120	+ 600	− 60	− 200
bce^2f	− 240	− 540	+ 1980	− 250
bd^2ef	+ 160	+ 500	+ 3160	− 320
bde^3	+ 20	− 200	− 850	+ 2750
c^3f^2		− 360	...	...
c^2def		− 120	− 4080	+ 800
c^2e^3		+ 160	− 2600	...
cd^3f			+ 1280	+ 320
cd^2e^2			+ 2080	− 5600
d^4e				+ 2880
	± 680	± 1570	± 8920	± 6895

	BE^4	CDE^3	D^3E^2
acf^3	− 20	...	+ 10
$adef^2$	+ 30	− 320	+ 60
ae^3f	− 15	+ 140	− 340
b^2f^3	+ 20	...	...
$bcef^2$	− 30	+ 360	− 50
bd^2f^2	...	+ 400	− 120
bde^2f	+ 20	+ 440	+ 1020
be^4	− 5	− 300	+ 500
c^2df^2		− 480	...
c^2e^2f		− 960	+ 100
cd^2ef		+ 640	− 960
cde^3		+ 80	− 1900
d^4f			+ 640
d^3e^2			+ 1040
	± 70	± 2060	± 3370

	CE^4	D^2E^3
adf^3	− 30	+ 10
ae^2f^2	+ 15	− 100
bcf^3	+ 30	...
$bdef^2$	+ 100	+ 190
be^3f	− 55	+ 260
c^2ef^2	− 120	...
cd^2f^2	...	− 160
cde^2f	+ 80	− 540
ce^4	− 20	− 200
d^3ef		+ 480
d^2e^3		+ 60
	± 225	± 1000

	DE^4
aef^3	− 15
bdf^3	+ 20
be^2f^2	+ 55
c^2f^3	...
$cdef^2$	− 80
ce^3f	− 70
d^3f^2	...
d^2e^2f	+ 120
de^4	− 30
	± 195

	E^5
af^4	− 1
bef^3	+ 5
cdf^3	...
ce^2f^2	− 10
d^2ef^2	...
de^3f	+ 10
e^5	− 4
	± 15

Upon writing $(B, C, D, E) = (x^3, xy^2, xy^2, y^3)$, the foregoing values of $\mathfrak{C}$, $\mathfrak{D}$, $\mathfrak{E}$, $\mathfrak{F}$ become covariants of the quintic $(a, b, c, d, e, f \between x, y)^5$. In fact [using for the covariants of the quintic the notation of 141 and 143, A the quintic itself, &c.], we have

$$\tfrac{1}{5}\mathfrak{C} = 2C,$$

$$\tfrac{1}{5}\mathfrak{D} = 2F,$$

$$\tfrac{1}{5}\mathfrak{E} = A^2B - 3C^2,$$

$$\mathfrak{F} = A^2E - 2CF.$$

This [the circumstance that the values thus become covariants of the quintic] is a further verification of the foregoing result.

I will conclude by showing how the formula may be applied to the reduction of the general quintic equation to Mr Jerrard's form $x^5+ax+b=0$. It was long ago remarked by Professor Sylvester that Tschirnhausen's Transformation, in its original form, gave the means of effecting this reduction. In fact, if the transforming equation be

$$y=\alpha+\beta x+\gamma x^2+\delta x^3+\epsilon x^4,$$

then the equation in y is of the form

$$(\mathfrak{A},\ \mathfrak{B},\ \mathfrak{C},\ \mathfrak{D},\ \mathfrak{E},\ \mathfrak{F}\between y,\ 1)^5=0,$$

where $\mathfrak{B}$, $\mathfrak{C}$, $\mathfrak{D}$, $\mathfrak{E}$, $\mathfrak{F}$ are, in regard to α, β, γ, δ, ϵ, of the degrees 1, 2, 3, 4, 5 respectively. And by assuming, say α a linear function of β, γ, δ, ϵ, we may make $B=0$, and we have then $\mathfrak{C}$, $\mathfrak{D}$, $\mathfrak{E}$, $\mathfrak{F}$ functions of the degrees 2, 3, 4, 5 respectively of the quantities β, γ, δ, ϵ: and these can be determined by means of a quadric equation and a cubic equation in such manner that $\mathfrak{C}=0$, $\mathfrak{D}=0$, in which case the equation in y will be of the required form. For considering β, γ, δ, ϵ as the coordinates of a point in space, the equations $\mathfrak{C}=0$, $\mathfrak{D}=0$ will be the equations of a quadric surface and a cubic surface respectively, and if β, γ, δ, ϵ be the coordinates of a point on the curve of intersection, the required conditions will be satisfied. And by combining with the equation of the quadric surface, the equation of any tangent plane thereto (or by the different process which is made use of in the sequel), we may, by means of a quadric equation, find a generating line of the quadric surface, and then, by means of a cubic equation, a point of intersection of this line with the cubic surface, i.e. a point the coordinates whereof give the required values β, γ, δ, ϵ. And similarly for the new form of Tschirnhausen's Transformation; the only difference being that, starting with an equation in y which contains the four arbitrary quantities B, C, D, E, we obtain in the first instance an equation

$$(1,\ 0,\ \mathfrak{C},\ \mathfrak{D},\ \mathfrak{E},\ \mathfrak{F}\between y,\ 1)^5=0$$

wanting the second term. And then B, C, D, E are to be so determined that $\mathfrak{C}=0$, $\mathfrak{D}=0$.

To proceed with the reduction, I write the foregoing value of $\mathfrak{C}$ in the form

$$\tfrac{2}{5}\mathfrak{C}=\left(\begin{array}{cccc} 4ac-4b^2, & 6ad-6bc, & 4ae-4bd, & af-be \\ 6ad-6bc, & 4ae+20bd-24c^2, & af+15be-16cd, & 4bf-4ce \\ 4ae-4bd, & af+15be-16cd, & 4bf+20ce-24d^2, & 6cf-6de \\ af-be, & 4bf-4ce, & 6cf-6de, & 4df-4e^2 \end{array}\between B,\ C,\ D,\ E\right)^2,$$

which for shortness may be represented by

$$\tfrac{2}{5}\mathfrak{C}=\left(\begin{array}{cccc} \mathrm{a}, & \mathrm{h}, & \mathrm{g}, & \mathrm{l} \\ \mathrm{h}, & \mathrm{b}, & \mathrm{f}, & \mathrm{m} \\ \mathrm{g}, & \mathrm{f}, & \mathrm{c}, & \mathrm{n} \\ \mathrm{l}, & \mathrm{m}, & \mathrm{n}, & \mathrm{p} \end{array}\between B,\ C,\ D,\ E\right)^2,$$

or say

$$\tfrac{2}{5}\mathfrak{C} = (\Omega \between B,\ C,\ D,\ E)^2.$$

Now, by a formula in my memoir "On the Automorphic Linear Transformation of a Bipartite Quadric Function," *Phil. Trans.*, vol. CXLVIII. (1858), see p. 44, [153], if Υ denote any skew symmetric matrix of the order 4, then if

$$(B,\ C,\ D,\ E) = (\Omega^{-1}(\Omega - \Upsilon)(\Omega + \Upsilon)^{-1}\,\Omega \between B',\ C',\ D',\ E'),$$

in which formula Ω^{-1}, $\Omega - \Upsilon$, $(\Omega + \Upsilon)^{-1}$, Ω are all matrices which are to be compounded together into a single matrix, we have identically

$$(\Omega \between B,\ C,\ D,\ E)^2 = (\Omega \between B',\ C',\ D',\ E')^2.$$

Let Q denote the determinant $|\,\Omega + \Upsilon\,|$, then the terms of the matrix $(\Omega + \Upsilon)^{-1}$ are respectively divided by Q, and we may write

$$(\Omega + \Upsilon)^{-1} = \frac{1}{Q} \cdot Q(\Omega + \Upsilon)^{-1},$$

where $Q(\Omega + \Upsilon)^{-1}$ is the matrix obtained from the matrix $(\Omega + \Upsilon)^{-1}$ by multiplying each term by Q, the terms of $Q(\Omega + \Upsilon)^{-1}$ being thus rational and integral functions of the terms of the matrix $(\Omega + \Upsilon)$. Hence if, instead of the before-mentioned relation between $(B,\ C,\ D,\ E)$ and $(B',\ C',\ D',\ E')$, we assume

$$(B,\ C,\ D,\ E) = (\Omega^{-1}(\Omega - \Upsilon)\,Q(\Omega + \Upsilon)^{-1}\,\Omega \between B',\ C',\ D',\ E'),$$

we find

$$(\Omega \between B,\ C,\ D,\ E)^2 = Q^2(\Omega \between B',\ C',\ D',\ E')^2.$$

If here the matrix Υ is such that we have $Q = 0$, i.e. Det. $(\Omega + \Upsilon) = 0$ (which is a quadric relation between the terms of the skew matrix, that is, each term is contained therein in the first and second powers only), then the equation becomes

$$(\Omega \between B,\ C,\ D,\ E)^2 = 0.$$

It is clear that this can only be the case in consequence of the coefficients of transformation in the equation

$$(B,\ C,\ D,\ E) = (\Omega^{-1}(\Omega - \Upsilon)\,Q(\Omega + \Upsilon)^{-1}\,\Omega \between B',\ C',\ D',\ E')$$

being such that there shall exist at least two linear relations between the quantities $(B,\ C,\ D,\ E)$, and I assume (without stopping to prove it) that they are such that the number of such linear relations is in fact two. That is, the last-mentioned equation establishes between the quantities $(B,\ C,\ D,\ E)$ two linear relations, in virtue whereof $\mathfrak{C} = 0$. And this being so, we may, without loss of generality, write $D' = 0$, $E' = 0$, or put

$$(B,\ C,\ D,\ E) = (\Omega^{-1}(\Omega - \Upsilon)\,Q(\Omega + \Upsilon)^{-1}\,\Omega \between B',\ C',\ 0,\ 0);$$

so that B, C, D, E are linear functions of B', C', such that $\mathfrak{C} = 0$. And then substituting these values for (B, C, D, E), we find $\mathfrak{D}$ a cubic function of B', C'; so that, putting $\mathfrak{D} = 0$, we have a cubic equation to determine the ratio $B' : C'$.

The foregoing reasoning presents no real difficulty, but it is expressed by means of a very condensed notation, and it may be proper to illustrate it by the case of the quadric function $x^2 + y^2 + z^2$. Considering the equations

$$\begin{aligned}
x &= (1 + \lambda^2 - \mu^2 - \nu^2)\, x' + 2(\lambda\mu - \nu)\, y' + 2(\lambda\nu + \mu)\, z', \\
y &= 2(\lambda\mu + \nu)\, x' + (1 - \lambda^2 + \mu^2 - \nu^2)\, y' + 2(\mu\nu - \lambda)\, z', \\
z &= 2(\nu\lambda - \mu)\, x' + 2(\mu\nu + \lambda)\, y' + (1 - \lambda^2 - \mu^2 + \nu^2)\, z',
\end{aligned}$$

these equations, if the expressions for x, y, z had been divided by $1 + \lambda^2 + \mu^2 + \nu^2$, would have given

$$x^2 + y^2 + z^2 = x'^2 + y'^2 + z'^2.$$

Hence they actually do give

$$x^2 + y^2 + z^2 = (1 + \lambda^2 + \mu^2 + \nu^2)^2\, (x'^2 + y'^2 + z'^2);$$

or if

$$1 + \lambda^2 + \mu^2 + \nu^2 = 0,$$

they give

$$x^2 + y^2 + z^2 = 0.$$

But if

$$1 + \lambda^2 + \mu^2 + \nu^2 = 0,$$

then

$$\begin{aligned}
& 1 + \lambda^2 - \mu^2 - \nu^2 : 2(\lambda\mu - \nu) : 2(\lambda\nu + \mu) \\
= {} & 2(\lambda\mu + \nu) : 1 - \lambda^2 + \mu^2 - \nu^2 : 2(\mu\nu - \lambda) \\
= {} & 2(\nu\lambda - \mu) : 2(\mu\nu + \lambda) : 1 - \lambda^2 - \mu^2 + \nu^2;
\end{aligned}$$

so that we have

$$x : y : z = 1 + \lambda^2 - \mu^2 - \nu^2 : 2(\lambda\mu + \nu) : 2(\nu\lambda - \mu),$$

which is the same result as would have been found by writing $y' = z' = 0$, and which comes to saying that x, y, z are not independent, but are connected by two linear relations.

The equation Det. $(\Omega + \Upsilon) = 0$, written at length, will be

$$\begin{vmatrix}
\mathrm{a}, & \mathrm{h} - \tau, & \mathrm{g} + \sigma, & \mathrm{l} + \lambda \\
\mathrm{h} + \tau, & \mathrm{b}, & \mathrm{f} - \rho, & \mathrm{m} + \mu \\
\mathrm{g} - \sigma, & \mathrm{f} + \rho, & \mathrm{c}, & \mathrm{n} + \nu \\
\mathrm{l} - \lambda, & \mathrm{m} - \mu, & \mathrm{n} - \nu, & \mathrm{p}
\end{vmatrix} = 0,$$

where $\lambda, \mu, \nu, \rho, \sigma, \tau$ are the arbitrary constituents of the skew matrix; or developing, this is

$$\begin{vmatrix} a, & h, & g, & l \\ h, & b, & f, & m \\ g, & f, & c, & n \\ l, & m, & n, & p \end{vmatrix}$$

$$+\left(\begin{array}{llllll} bc - f^2, & fg - ch, & hf - bg, & mg - nh, & fm - bn, & -fn + cm \\ fg - ch, & ca - g^2, & gh - af, & -gl + an, & nh - lf, & gn - cl \\ hf - bg, & gh - af, & ab - h^2, & hl - am, & -hm + bl, & lf - mg \\ mg - nh, & fm - bn, & -fn + cm, & ap - l^2, & ph - lm, & pg - ln \\ -gl + an, & nh - lf, & gn - cl, & ph - lm, & bp - m^2, & pf - mn \\ hl - am, & -hm + bl, & lf - mg, & pg - ln, & pf - mn, & cp - n^2 \end{array}\right)\!\!\!\!\between\!\!\lambda, \mu, \nu, \rho, \sigma, \tau)^2$$

$$+ (\lambda\rho + \mu\sigma + \nu\tau)^2 = 0,$$

the first term whereof, substituting for (a, b, c, f, g, h, l, m, n, p) their values, is in fact equal to the discriminant $a^4 f^4 + \&c.$ of the quintic $(a, b, c, d, e, f \between X, Y)^5$. There is no loss of generality in putting all but two of the quantities $(\lambda, \mu, \nu, \rho, \sigma, \tau)$ equal to zero; in fact this leaves in the formulæ a single arbitrary quantity, which is the right number, since the ratios $B : C : D : E$ have to satisfy only the two conditions $\mathfrak{C} = 0$, $\mathfrak{D} = 0$.

[An Addition of a half page, dated Nov. 10, 1862, and referring to the Memoir "On a New Auxiliary Equation in the Theory of Equations of the Fifth Order", 268, is printed at the end of that paper, *ante* p. 324.]

276.

ON THE ANALYTICAL THEORY OF THE CONIC.

[From the *Philosophical Transactions of the Royal Society of London*, vol. CLII. (for the year 1862), pp. 639—662. Received May 8,—Read May 15, 1862.]

THE decomposition into its linear factors of a decomposable quadric function cannot be effected in a symmetrical manner otherwise than by formulæ containing supernumerary arbitrary quantities; thus, for a binary quadric (which of course is always decomposable) we have

$$(a,\ b,\ c\between x,\ y)^2 = \frac{1}{(a,\ b,\ c\between x',\ y')^2}\ \text{Prod.}\ \{(a,\ b,\ c\between x,\ y\between x',\ y') \pm \sqrt{ac-b^2}\,(xy'-x'y)\}\ ;$$

or the expression for a linear factor is

$$\frac{1}{\sqrt{(a,\ b,\ c\between x',\ y')^2}}\{(a,\ b,\ c\between x,\ y\between x',\ y') \pm \sqrt{ac-b^2}\,(xy'-x'y)\},$$

which involves the arbitrary quantities (x', y'). And this appears to be the reason why, in the analytical theory of the conic, the questions which involve the decomposition of a decomposable ternary quadric have been little or scarcely at all considered: thus, for instance, the expressions for the coordinates of the points of intersection of a conic by a line (or say the line-equations of the two ineunts), and the equations for the tangents (separate each from the other) drawn from a given point not on the conic, do not appear to have been obtained. These questions depend on the decomposition of a decomposable ternary quadric, which decomposition itself depends on that for the simplest case, when the quadric is a perfect square. Or we may say that in the first instance they depend on the transformation of a given quadric function $U = (*\between x,\ y,\ z)^2$ into the form $W^2 + V$, where W is a linear function, given save as to a constant factor (that is, $W = 0$ is the equation of a given line), and V is a decomposable quadric function, which is ultimately decomposed into its linear factors, $= QR$, so that we have $U = W^2 + QR$.

The formula for this purpose, which is exhibited in the eight different forms I, II, III, IV, I (bis), II (bis), III (bis), IV (bis), is the analytical basis of the whole theory; and the greater part of the memoir relates to the establishment of these forms.

The solution of the geometrical questions above referred to is (as shown in the memoir) involved in and given immediately by these forms. It is also shown that the formulæ are greatly simplified in the case e.g. of tangents drawn to a conic from a point in a conic having double contact with the first-mentioned conic, and that in this case they lead to the linear Automorphic Transformation of the ternary quadric. The memoir concludes with some formulæ relating to the case of two conics, which however is treated of in only a cursory manner.

Article Nos. 1—17, *relating to a single conic.*

1. The point-equation of the conic is

$$(a,\ b,\ c,\ f,\ g,\ h \between x,\ y,\ z)^2 = 0,$$

which expresses that the point $(x,\ y,\ z)$ is an ineunt of the conic.

The line-equation of the same conic is

$$\begin{vmatrix} & \xi, & \eta, & \zeta \\ \xi, & a, & h, & g \\ \eta, & h, & b, & f \\ \zeta, & g, & f, & c \end{vmatrix} = 0,$$

or putting

$$(A,\ B,\ C,\ F,\ G,\ H) = (bc - f^2,\ ca - g^2,\ ab - h^2,\ gh - af,\ hf - bg,\ fg - ch),$$

the line-equation is

$$(A,\ B,\ C,\ F,\ G,\ H \between \xi,\ \eta,\ \zeta)^2 = 0,$$

which expresses that the line $(\xi,\ \eta,\ \zeta)$ (that is, the line the point-equation whereof is $\xi x + \eta y + \zeta z = 0$) is a tangent of the conic. We are thus in the analytical theory of the conic concerned with the quadrics $(a,\ b,\ c,\ f,\ g,\ h \between x,\ y,\ z)^2$ and $(A,\ B,\ C,\ F,\ G,\ H \between \xi,\ \eta,\ \zeta)^2$, which are the characteristics or *nilfactums* of these equations respectively.

2. I put also

$$K = \begin{vmatrix} a, & h, & g \\ h, & b, & f \\ g, & f, & c \end{vmatrix},$$

or, what is the same thing,

$$K = abc - af^2 - bg^2 - ch^2 + 2fgh.$$

3. It may be convenient to notice that when $(a,..\between x, y, z)^2$ breaks up into factors, the conic the equation whereof is $(a,..\between x, y, z)^2=0$, becomes a pair of lines; and that when $(a,..\between x, y, z)^2$ is a perfect square, the conic becomes a pair of coincident lines, or say a *twofold* line. But a pair of lines, distinct or coincident, cannot be represented by a line-equation. The analytical formulæ presently given show that in the former case $(A,..\between \xi, \eta, \zeta)^2$ is the square of a linear function, which equated to zero gives the line-equation of the point of intersection of the two lines, or node of the conic; and the equation $(A,..\between \xi, \eta, \zeta)^2=0$ accordingly represents such point considered as a pair of coincident points, or say a *twofold* point. But in the latter case, where the conic is a twofold line, $(A,..\between \xi, \eta, \zeta)^2$ is identically equal to zero, and the line-equation $(A,..\between \xi, \eta, \zeta)^2=0$ is a mere identity $0=0$, thus ceasing to have any signification at all. And the like remarks apply to the conic as represented by the line-equation $(A,..\between \xi, \eta, \zeta)^2=0$, the conic here breaking up into a pair of distinct or coincident points, &c.

4. It is proper to remark also that

$$(a, ...\between x', y', z'\between x, y, z)=0$$

is the equation of the polar of the point (x', y', z') in regard to the conic, and that

$$(A,..\between \xi', \eta', \zeta'\between \xi, \eta, \zeta)=0$$

is the line-equation of the pole of the line (ξ', η', ζ'); or, what is the same thing, the point-coordinates of the pole are

$$A\xi'+H\eta'+G\zeta' : H\xi'+B\eta'+F\zeta' : G\xi'+F\eta'+C\zeta'.$$

5. The inverse matrix is

$$\left(\begin{array}{ccc} a, & h, & g \\ h, & b, & f \\ g, & f, & c \end{array}\right)^{-1}, = \frac{1}{K}\left(\begin{array}{ccc} A, & H, & G \\ H, & B, & F \\ G, & F, & C \end{array}\right);$$

but it is convenient to disregard the factor $\frac{1}{K}$, and speak of (A, B, C, F, G, H) as the inverse or reciprocal coefficients. The equation just written down implies the relations $Aa+Hh+Gg=K$, $Ah+Hb+Gf=0$, &c., which may be arranged in two different ways as a system of nine equations.

6. We have also

$$(BC-F^2, CA-G^2, AB-H^2, GH-AF, HF-BG, FG-CH)=K(a, b, c, f, g, h),$$

and

$$ABC-AF^2-BG^2-CH^2+2FGH=K^2,$$

which are well-known theorems.

7. I notice also the theorem

$$(a,..\between x, y, z)^2 . (a,..\between x', y', z')^2 - [(a,..\between x, y, z\between x', y', z')]^2$$
$$= (A,..\between yz' - y'z, zx' - z'x, xy' - x'y)^2,$$

which is much used in the sequel: it may be mentioned, in passing, that this is included in the more general theorem

$$\begin{vmatrix} (a,..\between x, y, z\between l, m, n), & (a,..\between x', y', z'\between l, m, n) \\ (a,..\between x, y, z\between l', m', n'), & (a,..\between x', y', z'\between l, m, n) \end{vmatrix}$$
$$= (A,..\between yz' - y'z, zx' - z'x, xy' - x'y\between mn' - m'n, nl' - n'l, lm' - l'm),$$

which is at once deducible from

$$\begin{vmatrix} Ll + Mm + Nn, & L'l + M'm + N'n \\ Ll' + Mm' + Nn', & L'l' + M'm' + N'n' \end{vmatrix}$$
$$= (MN' - M'N)(mn' - m'n) + (NL' - N'L)(nl' - n'l) + (LM' - L'M)(lm' - l'm),$$

by writing therein

$$(L, M, N) = (ax + hy + gz, hx + by + fz, gx + fy + cz),$$
$$(L', M', N') = (ax' + hy' + gz', hx' + by' + fz', gx' + fy' + cz').$$

8. Suppose now that

$$(a, b, c, f, g, h\between x, y, z)^2$$

breaks up into factors, or say that we have

$$(a, b, c, f, g, h\between x, y, z)^2 = 2(\alpha x + \beta y + \gamma z)(\alpha' x + \beta' y + \gamma' z),$$

the values of the coefficients $(a,..)$ then are

$$(a, b, c, f, g, h) = (2\alpha\alpha', 2\beta\beta', 2\gamma\gamma', \beta\gamma' + \beta'\gamma, \gamma\alpha' + \gamma'\alpha, \alpha\beta' + \alpha'\beta),$$

and forming from these the inverse coefficients $(A,..)$ and the discriminant K, we find

$$(A, B, C, F, G, H) = -(\beta\gamma' - \beta'\gamma, \gamma\alpha' - \gamma'\alpha, \alpha\beta' - \alpha'\beta)^2.$$
$$K = 0.$$

9. The last-mentioned equation, $K = 0$, is the condition in order that $(a,..\between x, y, z)^2$ may break up into factors; and when it does so, we have

$$(A,..\between \xi, \eta, \zeta)^2 = -[(\beta\gamma' - \beta'\gamma, \gamma\alpha' - \gamma'\alpha, \alpha\beta' - \alpha'\beta\between \xi, \eta, \zeta)]^2,$$

that is, $(a,..\between x, y, z)^2$ breaking up into factors, $(A,..\between \xi, \eta, \zeta)^2$ is a perfect square; and equating it to zero, we have

$$[(\beta\gamma' - \beta'\gamma, \gamma\alpha' - \gamma'\alpha, \alpha\beta' - \alpha'\beta\between \xi, \eta, \zeta)]^2 = 0;$$

which, (ξ, η, ζ) being line-coordinates, gives (as a two-fold point) the point of intersection of the lines (α, β, γ), $(\alpha', \beta', \gamma')$, that is, the lines $\alpha x + \beta y + \gamma z = 0$, $\alpha' x + \beta' y + \gamma' z = 0$.

10. If $(a,..\between x, y, z)^2$ is a perfect square, then $\alpha' : \beta' : \gamma' = \alpha : \beta : \gamma$; whence not only, as before, $K=0$, but the coefficients (A, B, C, F, G, H) all vanish (this implies the first-mentioned condition, $K=0$); and the line-equation $(A,..\between \xi, \eta, \zeta)^2=0$ becomes the mere identity $0=0$.

11. Conversely if $K=0$, then $(a,..\between x, y, z)^2$ breaks up into factors; and if (A, B, C, F, G, H) all vanish, then $(a,..\between x, y, z)^2$ is a perfect square. The conclusions stated *ante*, No. 3, are thus sustained.

12. I assume, first, that $(a,..\between x, y, z)^2$ is a perfect square (No. 13); and secondly, that it breaks up into factors (No. 14); and I proceed to inquire how in the one case the root, and in the other case the factors, can be determined in a symmetrical form.

13. Considering the before-mentioned identical equation

$$(a,..\between x, y, z)^2.(a,..\between x', y', z')^2-[(a,..\between x, y, z\between x', y', z')]^2=(A,..\between yz'-y'z, zx'-z'x, xy'-x'y)^2,$$

if $(a,..\between x, y, z)^2$ is a perfect square, then by what precedes, the right-hand side of the equation vanishes, and we have

$$(a,..\between x, y, z)^2=\frac{[(a,..\between x, y, z\between x', y', z')]^2}{(a,..\between x', y', z')^2};$$

and the root of $(a,..\between x, y, z)^2$ is thus seen to be

$$=\pm\frac{(a,..\between x, y, z\between x', y', z')}{\sqrt{(a,..\between x', y', z')^2}},$$

an expression which involves the quantities (x', y', z'), the values whereof may be assumed at pleasure without altering the value of the expression. For instance, assuming for (x', y', z') the values $(1, 0, 0)$, $(0, 1, 0)$, $(0, 0, 1)$ successively, the different values of the expression are

$$\frac{ax+hy+gz}{\sqrt{a}}, \quad \frac{hx+by+fz}{\sqrt{b}}, \quad \frac{gx+fy+cz}{\sqrt{c}}.$$

But if, as assumed, $(a,..\between x, y, z)^2$ be a perfect square $=(\alpha x+\beta y+\gamma z)^2$, then

$$(a, b, c, f, g, h)=(\alpha^2, \beta^2, \gamma^2, \beta\gamma, \gamma\alpha, \alpha\beta),$$

and each of the foregoing values becomes equal to the root $\alpha x+\beta y+\gamma z$. It is somewhat singular that it is not possible to obtain symmetrical formulæ without employing in this manner supernumerary arbitrary quantities such as (x', y', z').

14. Next, if $(a, ...\between x, y, z)^2$, instead of being a perfect square, only breaks up into factors, then in the foregoing identical equation the right-hand side is a perfect square, and by the formula just obtained its value is

$$\frac{[(A,..\between X, Y, Z\between yz'-y'z, zx'-z'x, xy'-x'y)]^2}{(A,..\between X, Y, Z)^2},$$

where (X, Y, Z) are supernumerary arbitrary quantities. The identical equation then gives

$$(a,..\between x,y,z)^2=\frac{1}{(a,...\between x',y',z')^2}\left\{[(a,..\between x,y,z\between x',y',z')]^2+\frac{[(A,..\between X,Y,Z\between yz'-y'z,zx'-z'x,xy'-x'y)]^2}{(A,..\between X,\ Y,\ Z)^2}\right\},$$

and consequently

$$(a,..\between x,\ y,\ z)^2=\frac{1}{(a,..\between x',\ y',\ z')^2}\text{ Product of}$$

$$\left\{(a,..\between x,\ y,\ z\between x',\ y',\ z')\pm\frac{(A,..\between X,\ Y,\ Z\between yz'-y'z,\ zx'-z'x,\ xy'-x'y)}{\sqrt{-(A,..\between X,\ Y,\ Z)^2}}\right\},$$

a formula which exhibits the decomposition of $(a,..\between x,\ y,\ z)^2$ assumed to be a function which breaks up into factors; the formula contains the two sets of supernumerary arbitrary quantities $(x',\ y',\ z')$ and $(X,\ Y,\ Z)$. It will be remembered that $(A,..)$ denotes the system of inverse or reciprocal coefficients $(bc-f^2,..)$.

15. Consider the formula

$$(a,\ b,\ c,\ f,\ g,\ h\between \eta\zeta'-\eta'\zeta,\ \zeta\xi'-\zeta'\xi,\ \xi\eta'-\xi'\eta)^2=(\mathrm{a},\ \mathrm{b},\ \mathrm{c},\ \mathrm{f},\ \mathrm{g},\ \mathrm{h}\between \xi',\ \eta',\ \zeta')^2,$$

which gives

$$\begin{aligned}
\mathrm{a} &= c\eta^2+b\zeta^2-2f\eta\zeta,\\
\mathrm{b} &= a\zeta^2+c\xi^2-2g\zeta\xi,\\
\mathrm{c} &= b\xi^2+a\eta^2-2h\xi\eta,\\
\mathrm{f} &= -a\eta\zeta-f\xi^2+g\xi\eta+h\xi\zeta,\\
\mathrm{g} &= -b\zeta\xi+f\xi\eta-g\eta^2+h\eta\zeta,\\
\mathrm{h} &= -c\xi\eta+f\xi\zeta+g\eta\zeta-h\zeta^2;
\end{aligned}$$

and from these we deduce

$$\left(\begin{array}{ccc}\mathrm{a}, & \mathrm{h}, & \mathrm{g}\\ \mathrm{h}, & \mathrm{b}, & \mathrm{f}\\ \mathrm{g}, & \mathrm{f}, & \mathrm{c}\end{array}\right\between \xi,\ \eta,\ \zeta)=(0,\ 0,\ 0),$$

viz. $\mathrm{a}\xi+\mathrm{h}\eta+\mathrm{g}\zeta=0$, &c.

Also

$$(\mathrm{bc}-\mathrm{f}^2,\ \mathrm{ca}-\mathrm{g}^2,\ \mathrm{ab}-\mathrm{h}^2,\ \mathrm{gh}-\mathrm{af},\ \mathrm{hf}-\mathrm{bg},\ \mathrm{fg}-\mathrm{ch})=(\xi,\ \eta,\ \zeta)^2.(A,\ B,\ C,\ F,\ G,\ H\between \xi,\ \eta,\ \zeta)^2,$$

that is

$$\mathrm{bc}-\mathrm{f}^2=\xi^2(A,\ B,\ C,\ F,\ G,\ H\between \xi,\ \eta,\ \zeta)^2,\ \&c.$$

Hence also

$$(\mathrm{bc}-\mathrm{f}^2,..\between l,\ m,\ n)^2 = (l\xi+m\eta+n\zeta)^2(A,..\between \xi,\ \eta,\ \zeta)^2,$$

and

$$(\mathrm{bc}-\mathrm{f}^2,..\between l,\ m,\ n\between l',\ m',\ n')=(l\xi+m\eta+n\zeta)(l'\xi+m'\eta+n'\zeta)(A,..\between \xi,\ \eta,\ \zeta)^2;$$

and moreover

$$\mathrm{abc}-\mathrm{af}^2-\mathrm{bg}^2-\mathrm{ch}^2+2\mathrm{fgh}=0.$$

16. The last equation shows that $(a, \ldots \between \eta\zeta' - \eta'\zeta,\ \zeta\xi' - \zeta'\xi,\ \xi\eta' - \xi'\eta)^2$, considered as a function of (ξ', η', ζ'), breaks up into factors. Or since the expression is not altered by interchanging (ξ', η', ζ') and (ξ, η, ζ), the same expression, considered as a function of (ξ, η, ζ), breaks up into factors. It is in fact easy to see that any quantic whatever, $(* \between \eta\zeta' - \eta'\zeta,\ \zeta\xi' - \zeta'\xi,\ \xi\eta' - \xi'\eta)^m$, considered as a function of (ξ, η, ζ), breaks up into linear factors; for in virtue of the equation $\xi'(\eta\zeta' - \eta'\zeta) + \eta'(\zeta\xi' - \zeta'\xi) + \zeta'(\xi\eta' - \xi'\eta) = 0$, any one of the quantities $\eta\zeta' - \eta'\zeta,\ \zeta\xi' - \zeta'\xi,\ \xi\eta' - \xi'\eta$ can be expressed as a linear function of the other two; so that the quantic can be expressed as a linear function of any two of the three quantities; and *quà* homogeneous function of two quantities, it of course breaks up into factors, linear functions of these two quantities.

We may in all the formulæ interchange (x', y', z') and (x, y, z), writing $(\mathrm{a}', \mathrm{b}', \mathrm{c}', \mathrm{f}', \mathrm{g}', \mathrm{h}')$ in the place of $(\mathrm{a}, \mathrm{b}, \mathrm{c}, \mathrm{f}, \mathrm{g}, \mathrm{h})$.

17. Putting, in like manner,

$$\begin{aligned}(A, B, C, F, G, H \between yz' - y'z,\ zx' - z'x,\ xy' - x'y)^2 \\ = (\mathfrak{A}, \mathfrak{B}, \mathfrak{C}, \mathfrak{F}, \mathfrak{G}, \mathfrak{H} \between x', y', z')^2,\end{aligned}$$

so that

$$\begin{aligned}\mathfrak{A} &= Cy^2 + Bz^2 - 2Fyz, \\ \mathfrak{B} &= Az^2 + Cx^2 - 2Gzx, \\ \mathfrak{C} &= Bx^2 + Ay^2 - 2Hxy, \\ \mathfrak{F} &= -Ayz - Fx^2 + Gxy + Hzx, \\ \mathfrak{G} &= -Bzx + Fxy - Gy^2 + Hyz, \\ \mathfrak{H} &= -Cxy + Fzx + Gyz - Hz^2,\end{aligned}$$

we obtain

$$\left(\begin{array}{ccc} \mathfrak{A}, & \mathfrak{H}, & \mathfrak{G} \\ \mathfrak{H}, & \mathfrak{B}, & \mathfrak{F} \\ \mathfrak{G}, & \mathfrak{F}, & \mathfrak{C} \end{array} \between x, y, z\right) = (0, 0, 0),$$

viz.

$$\mathfrak{A}x + \mathfrak{H}y + \mathfrak{G}z = 0, \text{ \&c.}$$

Also

$$\begin{aligned}(\mathfrak{BC} - \mathfrak{F}^2,\ \mathfrak{CA} - \mathfrak{G}^2,\ \mathfrak{AB} - \mathfrak{H}^2,\ \mathfrak{GH} - \mathfrak{AF},\ \mathfrak{HF} - \mathfrak{BG},\ \mathfrak{HG} - \mathfrak{CG}) \\ = (x, y, z)^2 . K(a, b, c, f, g, h \between x, y, z)^2;\end{aligned}$$

that is

$$\mathfrak{BC} - \mathfrak{F}^2 = x^2 K(a, b, c, f, g, h \between x, y, z)^2, \text{ \&c.};$$

whence also

$$\begin{aligned}(\mathfrak{BC} - \mathfrak{F}^2, \ldots \between \lambda, \mu, \nu)^2 &= (\lambda x + \mu y + \nu z)^2 . K(a, \ldots \between x, y, z)^2, \\ (\mathfrak{BC} - \mathfrak{F}^2, \ldots \between \lambda, \mu, \nu \between \lambda', \mu', \nu') &= (\lambda x + \mu y + \nu z)(\lambda' x + \mu' y + \nu' z) . K(a, \ldots \between x, y, z)^2;\end{aligned}$$

and moreover

$$\mathfrak{ABC} - \mathfrak{AF}^2 - \mathfrak{BG}^2 - \mathfrak{CH}^2 + 2\mathfrak{FGH} = 0.$$

The last equation shows that $(A, \ldots \between yz' - y'z,\ zx' - z'x,\ xy' - x'y)^2$, considered as a function of (x', y', z'), breaks up into factors, or (what is the same thing) this expression, considered as a function of (x, y, z), breaks up into factors; we may in all the formulæ interchange (x, y, z) and (x', y', z'), writing $(\mathfrak{A}', \mathfrak{B}', \mathfrak{C}', \mathfrak{F}', \mathfrak{G}', \mathfrak{H}')$ in the place of $(\mathfrak{A}, \mathfrak{B}, \mathfrak{C}, \mathfrak{F}, \mathfrak{G}, \mathfrak{H})$.

Article Nos. 18 *to* 28, *relating to a single conic in connexion with a point or line.*

18. I apply the decomposition formula to the function $(A, \ldots \between yz' - y'z, \ldots)^2$, which, considered as a function of (x, y, z), breaks up into factors. We have

$$(A, \ldots \between yz' - y'z, \ldots)^2 = (\mathfrak{A}', \ldots \between x, y, z)^2$$

$$= \frac{1}{(\mathfrak{A}', \ldots \between l, m, n)^2} \text{ Product of}$$

$$(\mathfrak{A}', \ldots \between l, m, n \between x, y, z) \pm \frac{(\mathfrak{B}'\mathfrak{C}' - \mathfrak{F}'^2, \ldots \between mz - ny, \ldots \between \lambda, \mu, \nu)}{\sqrt{-(\mathfrak{B}'\mathfrak{C}' - \mathfrak{F}'^2, \ldots \between \lambda, \mu, \nu)^2}}.$$

But we have

$$(\mathfrak{A}', \ldots \between l, m, n)^2 = (A, \ldots \between mz' - ny', \ldots)^2,$$

$$(\mathfrak{A}', \ldots \between l, m, n \between x, y, z) = (A, \ldots \between mz' - ny', \ldots \between yz' - y'z, \ldots),$$

$$(\mathfrak{B}'\mathfrak{C}' - \mathfrak{F}'^2, \ldots \between mz - ny, \ldots \between \lambda, \mu, \nu)$$
$$= [x'(mz - ny) + y'(nx - lz) + z'(ly - mx)]\,(\lambda x' + \mu y' + \nu z')\ K(a, \ldots \between x', y', z')^2,$$

$$(\mathfrak{B}'\mathfrak{C}' - \mathfrak{F}'^2, \ldots \between \lambda, \mu, \nu)^2 = (\lambda x' + \mu y' + \nu z')^2 K(a, \ldots \between x', y', z')^2,$$

and thence

$$\frac{(\mathfrak{B}'\mathfrak{C}' - \mathfrak{F}'^2, \ldots \between mz - ny, \ldots \between \lambda, \mu, \nu)}{\sqrt{-(\mathfrak{B}'\mathfrak{C}' - \mathfrak{F}'^2, \ldots \between \lambda, \mu, \nu)^2}} = \begin{vmatrix} x, & y, & z \\ x', & y', & z' \\ l, & m, & n \end{vmatrix} \sqrt{-K(a, \ldots \between x', y', z')^2},$$

whence we have

$$(A, \ldots \between yz' - y'z, \ldots)^2 = \frac{1}{(A, \ldots \between mz' - ny')^2} \text{ Product of}$$

$$(A, \ldots \between mz' - ny', \ldots \between yz' - y'z, \ldots) \pm \begin{vmatrix} x, & y, & z \\ x', & y', & z' \\ l, & m, & n \end{vmatrix} \sqrt{-K(a, \ldots \between x', y', z)^2};$$

and the identical equation

$$(a, \ldots \between x, y, z)^2 . (a, \ldots \between x', y', z')^2 - [(a, \ldots \between x, y, z \between x', y', z')]^2 = (A, \ldots \between yz' - y'z, \ldots)^2$$

now gives

$$\text{I.}\quad (a,..\between x, y, z)^2 = \text{Quotient by } (a,..\between x', y', z')^2 \text{ of } \left\{\begin{array}{l} [(a,..\between x, y, z\between x', y', z')]^2 \\ +\text{Quotient by } (A,..\between mz'-ny',..)^2 \text{ of Product} \\ \{(A,..\between mz'-ny',..\between yz'-y'z,..) \pm \begin{vmatrix} x, & y, & z \\ x', & y', & z' \\ l, & m, & n \end{vmatrix} \sqrt{-K(a,..\between x', y', z')^2}\}, \end{array}\right.$$

where the Product part may also be written

$$\begin{array}{ll} (a,...\between l, m, n\between x', y', z') & .\,(a,..\between x, y, z\between x', y', z') \\ -(a,...\between x', y', z')^2 & .\,(a,..\between x, y, z\between l, m, n) \\ & \pm\sqrt{-K(a,..\between x', y', z')^2} \begin{vmatrix} x, & y, & z \\ x', & y', & z' \\ l, & m, & n \end{vmatrix}. \end{array}$$

19. Writing in the formula I.

$$\left(\begin{array}{ccc} a, & h, & g \\ h, & b, & f \\ g, & f, & c \end{array}\between x', y', z'\right) = (\xi', \eta', \zeta'),$$

we have

$$(x', y', z') = \frac{1}{K}\left(\begin{array}{ccc} A, & H, & G \\ H, & B, & F \\ G, & F, & C \end{array}\between \xi', \eta', \zeta'\right),$$

and thence

$$\begin{aligned} K(a,..\between x', y', z') &= (A,..\between \xi', \eta', \zeta')^2 \\ (a,..\between x, y, z\between x', y', z') &= \xi' x + \eta' y + \zeta' z. \end{aligned}$$

Assume

$$(l, m, n) = (\nu\eta' - \mu\zeta',\ \lambda\zeta' - \nu\xi',\ \mu\xi' - \lambda\eta'),$$

then from the foregoing values of (x', y', z')

$$\begin{aligned} mz' - ny' &= \frac{1}{K}\left\{(\lambda\zeta' - \nu\xi')(G\xi' + F\eta' + C\zeta') - (\mu\xi' - \lambda\eta')(H\xi' + B\eta' + F\zeta')\right\} \\ &= \frac{1}{K}\left\{\lambda\,[\xi'(A\xi' + H\eta' + G\zeta') + \eta'(H\xi' + B\eta' + F\zeta') + \zeta'(G\xi' + F\eta' + C\zeta')]\right. \\ &\qquad \left.-\lambda\xi'(A\xi' + H\eta' + G\zeta') - \mu\xi'(H\xi' + B\eta' + F\zeta') - \nu\xi'(G\xi' + F\eta' + C\zeta')\right\}, \end{aligned}$$

that is

$$mz' - ny' = \frac{1}{K}\left\{\lambda\,(A, \ldots \between \xi', \eta', \zeta')^2 - \xi'\,(A, \ldots \between \xi', \eta', \zeta' \between \lambda, \mu, \nu)\right\},$$

and similarly

$$nx' - lz' = \frac{1}{K}\left\{\mu\,(A, \ldots \between \xi', \eta', \zeta')^2 - \eta'\,(A, \ldots \between \xi', \eta', \zeta' \between \lambda, \mu, \nu)\right\},$$

$$ly' - mx' = \frac{1}{K}\left\{\nu\,(A, \ldots \between \xi', \eta', \zeta')^2 - \zeta'\,(A, \ldots \between \xi', \eta', \zeta' \between \lambda, \mu, \nu)\right\};$$

and thence

$$\begin{aligned}(A, H, G \between mz' - ny', \ldots) \\ = \frac{1}{K}\Big\{ & (A, H, G \between \lambda, \mu, \nu)\,.\,(A, \ldots \between \xi', \eta', \zeta')^2 \\ & - (A, H, G \between \xi', \eta', \zeta')\,.\,(A, \ldots \between \xi', \eta', \zeta' \between \lambda, \mu, \nu)\Big\}\end{aligned}$$

with the like equations, writing H, B, F and G, F, C in the place of A, H, G successively: and we then have

$$\begin{aligned}(A, \ldots \between mz' - ny', \ldots)^2 \\ = \frac{1}{K}\Big\{ & (A, \ldots \between \lambda, \mu, \nu \between mz' - ny', \ldots)\,.\,(A, \ldots \between \xi', \eta', \zeta')^2 \\ & - (A, \ldots \between \xi', \eta', \zeta' \between mz' - ny, \ldots)\,.\,(A, \ldots \between \xi', \eta', \zeta' \between \lambda, \mu, \nu)\Big\}.\end{aligned}$$

But the foregoing values of $mz' - ny'$, $nx' - lz'$, $ly' - mx'$ give also

$$\begin{aligned}(A, \ldots \between \lambda, \mu, \nu \between mz' - ny', \ldots) \\ = \frac{1}{K}\left\{(A, \ldots \between \lambda, \mu, \nu)^2\,.\,(A, \ldots \between \xi', \eta', \zeta')^2 - [(A, \ldots \between \lambda, \mu, \nu \between \xi', \eta', \zeta')]^2\right\},\end{aligned}$$

$$\begin{aligned}(A, \ldots \between \xi', \eta', \zeta, \between mz' - ny', \ldots) \\ = \frac{1}{K}\left\{(A, \ldots \between \lambda, \mu, \nu \between \xi', \eta', \zeta')\,.\,(A, \ldots \between \xi', \eta', \zeta')^2 - (A, \ldots \between \xi', \eta', \zeta')^2\,.\,(A, \ldots \between \lambda, \mu, \nu \between \xi', \eta', \zeta')\right\} = 0;\end{aligned}$$

so that

$$\begin{aligned}(A, \ldots \between mz' - ny', \ldots)^2 \\ = \frac{1}{K^2}(A, \ldots \between \xi', \eta', \zeta')^2\,.\,\{(A, \ldots \between \lambda, \mu, \nu)^2\,.\,(A, \ldots \between \xi', \eta', \zeta')^2 - [(A, \ldots \between \lambda, \mu, \nu \between \xi', \eta', \zeta')]^2\} \\ = \frac{1}{K}(A, \ldots \between \xi', \eta', \zeta')^2\,.\,(a, \ldots \between \nu\eta' - \mu\zeta', \ldots)^2.\end{aligned}$$

Similarly,

$$(A,..\between mz'-ny',..\between yz'-y'z,..)$$

$$=\frac{1}{K}\left\{\begin{array}{l}(A,..\between \lambda,\ \mu,\ \nu\between yz'-y'z,..).(A,..\between \xi',\ \eta',\ \zeta')^2\\ \quad -(A,..\between \xi',\ \eta',\ \zeta'\between yz'-y'z,..).(A,..\between \xi',\ \eta',\ \zeta'\between \lambda,\ \mu,\ \nu)\end{array}\right\}.$$

But

$$\begin{aligned}(A,..\between \lambda,\ \mu,\ \nu\between yz'-y'z,..)\\ =\ &(A\lambda+H\mu+G\nu)(yz'-y'z)\\ &+(H\lambda+B\mu+F\nu)(zx'-z'x)\\ &+(G\lambda+F\mu+C\nu)(xy'-x'y),\\ =\ &x\,[y'(G\lambda+F\mu+C\nu)-z'(H\lambda+B\mu+F\nu)]\\ &+y\,[z'(A\lambda+H\mu+G\nu)-x'(G\lambda+F\mu+C\nu)]\\ &+z\,[x'(H\lambda+B\mu+F\nu)-y'(A\lambda+H\mu+G\nu)],\end{aligned}$$

which, substituting for x', y', z' their values

$$(x',\ y',\ z')=\frac{1}{K}\left(\begin{array}{ccc} A, & H, & G\\ H, & B, & F\\ G, & F, & C\end{array}\right|\!\between \xi',\ \eta',\ \zeta'),$$

becomes

$$=\frac{1}{K}\left\{\begin{array}{l}x\,[(BC-F^2)(\nu\eta'-\mu\zeta')+(FG-CH)(\lambda\zeta'-\nu\xi')+(HF-BG)(\mu\xi'-\lambda\eta')]\\ +y\,[(FG-CH)(\nu\eta'-\mu\zeta')+(CA-G^2)(\lambda\zeta'-\nu\xi')+(GH-AF)(\mu\xi'-\lambda\eta')]\\ +z\,[(HF-BG)(\nu\eta'-\mu\zeta')+(GH-AF)(\lambda\zeta'-\nu\xi')+(AB-H^2)(\mu\xi'-\lambda\eta')]\end{array}\right\},$$

which is

$$=(a,..\between x,\ y,\ z\between \nu\eta'-\mu\zeta',..);$$

and by merely writing $(\xi',\ \eta',\ \zeta')$ in the place of $(\lambda,\ \mu,\ \nu)$, we have

$$(A,..\between \xi',\ \eta',\ \zeta'\between yz'-y'z,..)=0;$$

so that we find

$$(A,..\between mz'-ny',..\between yz'-y'z,..)$$

$$=\frac{1}{K}(A,..\between \xi',\ \eta',\ \zeta')^2.(a,..\between x,\ y,\ z\between \nu\eta'-\mu\zeta',\ \lambda\zeta'-\nu\xi',\ \mu\xi'-\lambda\eta').$$

Now, writing the formula I. in the form

$$(a,..\between x, y, z)^2 = \text{Quotient by } K(a,..\between x', y', z')^2 \text{ of } \left\{ \begin{array}{l} K[(a,..\between x, y, z\between x', y', z')]^2 \\ + \text{Quotient by } K(A,..\between mz'-ny',..)^2 \text{ of} \\ K^2\{[(A,..\between mz'-ny',..\between yz'-y'z,..)]^2 + \begin{vmatrix} x, & y, & z \\ x', & y', & z' \\ l, & m, & n \end{vmatrix}^2 K(a,..\between x, y, z)^2\}, \end{array} \right.$$

the right-hand side is

$$= \text{Quotient by } (A,..\between \xi', \eta', \zeta')^2 \text{ of } \left\{ \begin{array}{l} K(\xi' x + \eta' y + \zeta' z)^2 \\ + \text{Quotient by } (a,..\between \nu\eta' - \mu\zeta',..)^2 (A,..\between \xi', \eta', \zeta')^2 \text{ of} \\ \{[(a,..\between \nu\eta' - \mu\zeta',..\between x, y, z).(A,..\between \xi', \eta', \zeta')^2]^2 + \Pi^2 (A,..\between \xi', \eta', \zeta')^2\}, \end{array} \right.$$

where

$$\Pi = K \begin{vmatrix} x, & y, & z \\ x', & y', & z' \\ l, & m & n \end{vmatrix},$$

or, what is the same thing,

$$\Pi = \begin{vmatrix} x, & y, & z \\ Kx', & Ky', & Kz' \\ l, & m, & n \end{vmatrix} = \begin{vmatrix} x & y & z \\ A\xi' + H\eta' + G\zeta', & H\xi' + B\eta' + F\zeta', & G\xi' + F\eta' + C\zeta' \\ \nu\eta' - \mu\zeta', & \lambda\zeta' - \nu\xi', & \mu\xi' - \lambda\eta'. \end{vmatrix}.$$

More simply, the right-hand side is

$$= \text{Quotient by } (A,..\between \xi', \eta', \zeta')^2 \text{ of } \left\{ \begin{array}{l} K(\xi' x + \eta' y + \zeta' z)^2 \\ + \text{Quotient by } (a,..\between \nu\eta' - \mu\zeta',..)^2 \text{ of} \\ \{[(a,..\between \nu\eta' - \mu\zeta',..\between x, y, z)]^2.(A,..\between \xi', \eta', \zeta')^2 + \Pi^2\}; \end{array} \right.$$

or restoring the left-hand side, and resolving into its linear factors the function in { }, we have

$$\text{II.} \quad (a,..\between x, y, z)^2 = \text{Quotient by } (A,..\between \xi', \eta', \zeta')^2 \text{ of } \left\{ \begin{array}{l} K(\xi' x + \eta' y + \zeta' z)^2 \\ + \text{Quotient by } (a,..\between \nu\eta' - \mu\zeta',..)^2 \text{ of Product} \\ \Pi \pm \sqrt{-(A,..\between \xi', \eta', \zeta')^2}.(a,..\between \nu\eta' - \mu\zeta',..\between x, y, z), \end{array} \right.$$

where Π has the value given above, which may also be written

$$\begin{aligned}\Pi = \ &(A,..\between\xi', \eta', \zeta'\between\lambda, \mu, \nu)(\xi' x + \eta' y + \zeta' z)\\ &-(A,..\between\xi', \eta', \zeta')^2(\lambda x + \mu y + \nu z).\end{aligned}$$

20. We deduce at once the inverse or reciprocal formulæ

$$\text{III.}\quad \left.\begin{array}{l}(A,..\between\xi, \eta, \zeta)^2 = \text{Quotient by } (A,..\between\xi', \eta', \zeta')^2 \text{ of}\end{array}\right\}$$

$$\left\{\begin{array}{l}[(A,..\between\xi, \eta, \zeta\between\xi', \eta', \zeta')]^2\\ + \text{Quotient by } (a,..\between\nu\eta' - \mu\zeta', ..)^2 \text{ of } K \text{ into Product}\\ (a,..\between\nu\eta' - \mu\zeta', ..\between\eta\zeta' - \eta'\zeta, ..) \pm\sqrt{-(A,..\between\xi', \eta', \zeta')^2}\begin{vmatrix}\xi, & \eta, & \zeta\\ \xi', & \eta', & \zeta'\\ \lambda, & \mu, & \nu\end{vmatrix},\end{array}\right.$$

where the Product part may also be written

$$\begin{aligned}\text{Product}\Bigg[&\frac{1}{K}(A,..\between\xi', \eta', \zeta'\between\lambda, \mu, \nu).(A,..\between\xi', \eta', \zeta'\between\xi, \eta, \zeta)\\ &-\frac{1}{K}(A,..\between\xi', \eta', \zeta')^2 \quad .(A,..\between\lambda, \mu, \nu\between\xi, \eta, \zeta)\\ &\pm\sqrt{-(A,..\between\xi', \eta', \zeta')^2}\begin{vmatrix}\xi, & \eta, & \zeta\\ \xi', & \eta', & \zeta'\\ \lambda, & \mu, & \nu\end{vmatrix}\Bigg].\end{aligned}$$

21. And also

$$(A,..\between\xi, \eta, \zeta)^2 = \text{Quotient by } (a,..\between x', y', z')^2 \text{ of}$$

$$\text{IV.}\quad\left\{\begin{array}{l}K(\xi x' + \eta y' + \zeta z')^2\\ + \text{Quotient by } K(A,..\between ny' - mz', ..)^2 \text{ of Product}\\ K\Phi \pm\sqrt{-K(a,..\between x', y', z')^2}(A,..\between ny' - mz', ..\between\xi, \eta, \zeta),\end{array}\right.$$

where

$$\Phi = \begin{vmatrix}\xi & \eta & \zeta\\ ax' + hy' + gz', & hx' + by' + fz', & gx' + fy' + cz'\\ ny' - mz', & lz' - nx', & mx' - ly'\end{vmatrix},$$

which may also be written

$$\begin{aligned}= \ &(a,..\between x', y', z'\between l, m, n)(x'\xi + y'\eta + z'\zeta)\\ &-(a,..\between x', y', z')^2 \quad . \quad (l\xi + m\eta + n\zeta).\end{aligned}$$

22. The geometrical signification is obvious. The formulæ I. and II. each of them show that the equation

$$(a,..\between x, y, z)^2 = 0$$

of the conic may be written in the form

$$W^2 + \frac{1}{M} QR = 0,$$

where $Q=0$, $R=0$ are any two tangents of the conic, and $W=0$ is the line joining the points of contact, or chord of contact corresponding to the two tangents; viz., in the formula I. we have

$$W = (a, .. \between x', y', z' \between x, y, z),$$

$$\left.\begin{matrix} Q \\ R \end{matrix}\right\} = (A, .. \between mz' - ny', .. \between yz' - y'z, ..) \pm \sqrt{-K(a, .. \between x, y, z)^2} \begin{vmatrix} x, & y, & z \\ x', & y', & z' \\ l, & m, & n \end{vmatrix},$$

(or for a different form of Q, R see the formula). The quantities (x', y', z') are the coordinates of the point of intersection of the two tangents, or pole of the chord of contact: (l, m, n) are supernumerary arbitrary quantities, the values whereof do not affect the result [1]. And in the formula II. we have

$$W = \xi' x + \eta' y + \zeta' z,$$

$$\left.\begin{matrix} Q \\ R \end{matrix}\right\} = \Pi \pm \sqrt{-(A, .. \between \xi', \eta', \zeta')^2}\ (a, .. \between \nu\eta' - \mu\zeta', .. \between x, y, z),$$

(for the value of Π see the formula). The quantities (ξ', η', ζ') are the line-coordinates of the chord of contact (viz. the point-equation of this line is $\xi' x + \eta' y + \zeta' z = 0$); (λ, μ, ν) are supernumerary arbitrary quantities.

23. In the like manner the formulæ III. and IV. each of them show that the line-equation

$$(A, .. \between \xi, \eta, \zeta)^2 = 0$$

of the conic may be written in the form

$$W^2 + \frac{1}{M} QR = 0,$$

where $Q=0$, $R=0$ are any two ineunts of the conic, and $W=0$ is the point of intersection of the corresponding tangents; viz. in the formula III. we have

$$W = (A, ... \between \xi', \eta', \zeta' \between \xi, \eta, \zeta),$$

$$\left.\begin{matrix} Q \\ R \end{matrix}\right\} = (a, ... \between \nu\eta' - \mu\zeta', ..) \pm \sqrt{-(A, .. \between \xi', \eta', \zeta')^2} \begin{vmatrix} \xi, & \eta, & \zeta \\ \xi', & \eta', & \zeta' \\ \lambda, & \mu, & \nu \end{vmatrix},$$

(for another form of Q, R see the formula).

[1] In a different point of view, viz. if we consider the formula I. as a transformation of the function $(a, ... \between x, y, z)^2$, then (x', y', z') and (l, m, n) would be each of them supernumerary arbitrary quantities: and so in the other like cases.

The quantities ξ', η', ζ' are the line-coordinates of the line through the two ineunts, or chord of contact; (λ, μ, ν) are supernumerary arbitrary quantities; and so in the formula IV. we have

$$W = x'\xi + y'\eta + z'\zeta,$$

$$\left.\begin{matrix} Q \\ R \end{matrix}\right\} = K\Phi \pm \sqrt{-K(a,..\between x', y', z')^2}\,(A,..\between ny'-mz,..\between \xi, \eta, \zeta)$$

(for the value of Φ see the formula), where x', y', z' are the point-coordinates of the intersection of tangents at the two ineunts, or pole of the chord of contact; (l, m, n) are supernumerary arbitrary quantities.

24. We may, instead of the supernumerary arbitrary quantities (l, m, n) of the formula I., introduce the quantities (λ, μ, ν), where

$$(l, m, n) = \frac{1}{K}\left(\begin{array}{ccc} A, & H, & G \\ H, & B, & F \\ G, & F, & C \end{array}\right|\!\between \lambda, \mu, \nu).$$

This gives

$$\begin{aligned}
&(A, H, G\between mz'-ny', ..)\\
&\quad = A(mz'-ny') + H(nx'-lz') + G(ly'-mx')\\
&\quad = x'(Hn-Gm) + y'(Gl-An) + z'(Am-Hl)\\
&\quad = \frac{1}{K}\cdot\ x'[H(G\lambda+F\mu+C\nu) - G(H\lambda+B\mu+F\nu)]\\
&\qquad\quad + y'[G(A\lambda+H\mu+G\nu) - A(G\lambda+F\mu+C\nu)]\\
&\qquad\quad + z'[A(H\lambda+B\mu+F\nu) - H(H\lambda+B\mu+F\nu)]\\
&\quad = x'(g\mu-h\nu) + y'(f\mu-b\nu) + z'(c\mu-f\nu)\\
&\quad = \mu(gx'+fy'+cz') - \nu(hx'+by'+fz');
\end{aligned}$$

we have thus the system

$$\begin{aligned}
(A, H, G\between mz'-ny', ..) &= \mu(gx'+fy'+cz') - \nu(hx'+by'+fz'),\\
(H, B, F\between mz'-ny', ..) &= \nu(ax'+hy'+gz') - \lambda(gx'+fy'+cz'),\\
(G, F, C\between mz'-ny', ..) &= \lambda(hx'+by'+fz') - \mu(ax'+hy'+gz'),
\end{aligned}$$

and thence

$$\begin{aligned}
&(A,..\between mz'-ny',..\between yz'-y'z,..)\\
&\quad = -\left|\begin{array}{ccc} yz'-y'z, & zx'-z'x, & xy'-x'y \\ ax'+hy'+gz', & hx'+by'+fz', & gx'+fy'+cz' \\ \lambda, & \mu, & \nu \end{array}\right|;
\end{aligned}$$

or observing that the term in λ is

$$-(zx'-z'x)(gx'+fy'+cz') + (xy'-x'y)(hx'+by'+fz'),$$

which is

$$\begin{aligned}
&= x\,(x'(ax'+hy'+gz')+y'(hx'+fy'+fz')+z'(gx'+fy'+cz'))\\
&\quad - x\,.\,x'(ax'+hy'+gz')\\
&\quad - y\,.\,x'(hx'+by'+fz')\\
&\quad - z\,.\,x'(gx'+fy'+cz')\\
&= -x'(a,..\between x,\ y,\ z\between x',\ y',\ z') + x\,(a,..\between x',\ y',\ z')^2,
\end{aligned}$$

with similar expressions for the terms in μ, ν, we have

$$(A,..\between mz'-ny',..\between yz'-y'z,..)$$
$$= -(\lambda x'+\mu y'+\nu z').(a,..\between x,\ y,\ z\between x',\ y',\ z') + (\lambda x+\mu y+\nu z).(a,..\between x',\ y',\ z')^2;$$

and so also

$$(A,..\between mz'-ny',..)^2$$
$$= -(\lambda x'+\mu y'+\nu z').(a,..\between l,\ m,\ n\between x',\ y',\ z') + (\lambda l+\mu m+\nu n).(a,..\between x',\ y',\ z')^2,$$

where

$$\begin{aligned}
(a,..\between l,\ m,\ n\between x',\ y',\ z') &= \lambda x'+\mu y'+\nu z',\\
\lambda l+\mu m+\nu n &= \frac{1}{K}(A,..\between \lambda,\ \mu,\ \nu)^2,
\end{aligned}$$

so that

$$(A,..\between mz'-ny',..)^2 = -(\lambda x'+\mu y'+\nu z')^2 + \frac{1}{K}(A,..\between \lambda,\ \mu,\ \nu)^2.(a,..\between x',\ y',\ z')^2.$$

Moreover,

$$\begin{vmatrix} x, & y, & z \\ x', & y', & z' \\ l, & m, & n \end{vmatrix} = \frac{1}{K}\left\{(A\lambda+H\mu+G\nu)(yz'-y'z)+(H\lambda+B\mu+F\nu)(zx'-z'x)+(G\lambda+F\mu+C\nu)(xy'-x'y)\right\},$$

which is

$$= \frac{1}{K}(A,..\between \lambda,\ \mu,\ \nu\between yz'-y'z,..);$$

and hence instead of the formula I. we have

$$(a,..\between x,\ y,\ z)^2 = \text{Quotient by } (a,..\between x',\ y',\ z')^2 \text{ of}$$

$$\text{I. (bis)}\left\{\begin{array}{l}
[(a,..\between x,\ y,\ z\between x',\ y',\ z')]^2\\
+\text{ Quotient by } (A,..\between \lambda,\mu,\nu)^2(a,..\between x',y',z')^2 - K(\lambda x'+\mu y'+\nu z')^2 \text{ of } K \text{ into Product}\\
\left\{\begin{array}{l}(\lambda x'+\mu y'+\nu z').(a,..\between x,\ y,\ z\between x',\ y',\ z') - (\lambda x+\mu y+\nu z).(a,..\between x',\ y',\ z')^2\\
\pm\frac{1}{K}\sqrt{-K(a,..\between x',\ y',\ z')^2}\,(A,..\between \lambda,\ \mu,\ \nu\between yz'-y'z,..).\end{array}\right\}^{\!\cdot}
\end{array}\right.$$

25. If, in like manner, in the formula II. we introduce, instead of $(\lambda,\ \mu,\ \nu)$, the new quantities $(l,\ m,\ n)$, where

$$(\lambda,\ \mu,\ \nu) = \left(\begin{array}{ccc} a, & h, & g \\ h, & b, & f \\ g, & f, & c \end{array}\right\between l,\ m,\ n),$$

or, what is the same thing,

$$(l,\ m,\ n) = \frac{1}{K}\left(\begin{array}{ccc} A, & H, & G \\ H, & B, & F \\ G, & F, & C \end{array}\right)\!\!\left(\lambda,\ \mu,\ \nu\right),$$

then we have

$$\begin{aligned}
(a,\ h,\ g\between \nu\eta' - \mu\zeta', ..) &= n\ (H\xi' + B\eta' + F\zeta') - m\,(G\xi' + F\eta' + C\zeta'),\\
(h,\ b,\ f\between \nu\eta' - \mu\zeta', ..) &= l\ \ (G\xi' + F\eta' + C\zeta') - n\ (A\xi' + H\eta' + G\zeta'),\\
(g,\ f,\ c\between \nu\eta' - \mu\zeta', ..) &= m\,(A\xi' + H\eta' + G\zeta') - l\ \ (H\xi' + B\eta' + F\zeta');
\end{aligned}$$

and thence

$$(a, ..\between \nu\eta' - \mu\zeta', ..\between x,\ y,\ z) = \left|\begin{array}{ccc} x & , \quad y & , \quad z \\ A\xi' + H\eta' + G\zeta', & H\xi' + B\eta' + F\zeta', & G\xi' + F\eta' + C\zeta' \\ l & , \quad m & , \quad n \end{array}\right|$$

$$= \ (A, ..\between mz - ny, ..\between \xi',\ \eta',\ \zeta'),$$

$$(a, ..\between \nu\eta' - \mu\zeta', ..)^2 = \frac{1}{K}\left\{(A, ..\between \lambda,\ \mu,\ \nu)^2 . (A, ..\between \xi',\ \eta',\ \zeta')^2 - [(A, ..\between \lambda,\ \mu,\ \nu\between \xi',\ \eta',\ \zeta')]^2\right\}$$

$$= (a, ..\between l,\ m,\ n)^2 . (A, ..\between \xi',\ \eta',\ \zeta')^2 - K\,(l\xi' + m\eta' + n\zeta')^2;$$

$$V = \left|\begin{array}{ccc} x & , \quad y & , \quad z \\ A\xi' + H\eta' + G\zeta', & H\xi' + B\eta' + F\zeta', & G\xi' + F\eta' + C\zeta' \\ \nu\eta' - \mu\zeta' & , \quad \lambda\zeta' - \nu\xi' & , \quad \mu\xi' - \lambda\eta' \end{array}\right|$$

$$\begin{aligned}
&= (\nu\eta' - \mu\zeta') . y\,(G\xi' + F\eta' + C\zeta') - z\,(H\xi' + B\eta' + F\zeta')\\
&\quad (\lambda\zeta' - \nu\xi') . z\,(A\xi' + H\eta' + G\zeta') - x\,(G\xi' + F\eta' + C\zeta')\\
&\quad (\mu\xi' - \lambda\eta') . x\,(H\xi' + B\eta' + F\zeta') - y\,(A\xi' + H\eta' + G\zeta')\\
&= \ \ \lambda\,\{(\xi' x + \eta' y + \zeta' z)(A\xi' + H\eta' + G\zeta') - x\,(A, ..\between \xi',\ \eta',\ \zeta')^2\}\\
&\quad + \mu\,\{(\xi' x + \eta' y + \zeta' z)(H\xi' + B\eta' + F\zeta') - y\,(A, ..\between \xi',\ \eta',\ \zeta')^2\}\\
&\quad + \nu\,\{(\xi' x + \eta' y + \zeta' z)(G\xi' + F\eta' + C\zeta') - z\,(A, ..\between \xi',\ \eta',\ \zeta')^2\}\\
&= (\xi' x + \eta' y + \zeta' z) . (A, ..\between \lambda,\ \mu,\ \nu\between \xi',\ \eta',\ \zeta') - (\lambda x + \mu y + \nu z) . (A, ..\between \xi',\ \eta',\ \zeta')^2\\
&= K\,(l\xi' + m\eta' + n\zeta') . (\xi' x + \eta' y + \zeta' z) - (a, ..\between l,\ m,\ n\between x,\ y,\ z) . (A, ..\between \xi',\ \eta'\ \zeta')^2;
\end{aligned}$$

and the formula II. thus becomes

$(a, ..\between x,\ y,\ z)^2 =$ Quotient by $(A, ..\between \xi',\ \eta',\ \zeta')^2$ of

II. (bis)
$$\left\{\begin{array}{l} K\,(\xi' x + \eta' y + \zeta' z)^2 \\ + \text{Quotient by } (a, ..\between l,\ m,\ n)^2 . (A, ..\between \xi',\ \eta',\ \zeta')^2 - K(l\xi' + m\eta' + n\zeta')^2 \text{ of Product} \\ \left\{\begin{array}{l} K\,(l\xi' + m\eta' + n\zeta') . (\xi' x + \eta' y + \zeta' z) - (a, ..\between l,\ m,\ n\between x,\ y,\ z) . (A, ..\between \xi',\ \eta',\ \zeta')^2 \\ \pm \sqrt{-(A, ..\between \xi',\ \eta',\ \zeta')^2}\,(A, ..\between mz - ny, ..\between \xi',\ \eta',\ \zeta'). \end{array}\right\} \end{array}\right.$$

26. And from these we at once deduce the inverse or reciprocal formulæ

$$\text{III. (bis)}\quad (A,\ldots)(\xi,\ \eta,\ \zeta)^2 = \text{Quotient by } (A,\ldots)(\xi',\ \eta',\ \zeta')^2 \text{ of } \left\{\begin{array}{l} [(A,\ldots)(\xi,\ \eta,\ \zeta)(\xi',\ \eta',\ \zeta')]^2 \\ +\text{Quotient by}\,(a,\ldots)(l,m,n)^2\,.\,(A,\ldots)(\xi',\eta',\zeta')^2 - K(l\xi'+m\eta'+n\zeta')^2 \text{ of } K \text{ into Product} \\ \left\{\begin{array}{l}(l\xi'+m\eta'+n\zeta')(A,\ldots)(\xi,\ \eta,\ \zeta)(\xi',\ \eta',\ \zeta') - (l\xi+m\eta+n\zeta)\,.\,(A,\ldots)(\xi',\ \eta',\ \zeta')^2 \\ \pm\sqrt{-(A,\ldots)(\xi',\ \eta',\ \zeta')^2}\,(a,\ldots)(l,\ m,\ n)(\eta\zeta'-\eta'\zeta,\ldots),\end{array}\right\}.\end{array}\right.$$

27. And

$$\text{IV. (bis)}\quad (A,\ldots)(\xi,\ \eta,\ \zeta)^2 = \text{Quotient by } (a,\ldots)(x',\ y',\ z')^2 \text{ of } \left\{\begin{array}{l} K(x'\xi+y'\eta+z'\zeta)^2 \\ +\text{Quotient by}\,(A,\ldots)(\lambda,\ \mu,\ \nu)^2\,.\,(a,\ldots)(x',\ y',\ z')^2 - K(\lambda x'+\mu y'+\nu z')^2 \text{ of Product} \\ \left\{\begin{array}{l}K(\lambda x'+\mu y'+\nu z')\,.\,(x'\xi+y'\eta+z'\zeta) - (A,\ldots)(\xi,\ \eta,\ \zeta)(l,\ m,\ n)\,.\,(a,\ldots)(x',\ y',\ z')^2 \\ \pm\sqrt{-K(a,\ldots)(x',\ y',\ z')^2}\,(a,\ldots)(x',\ y',\ z')(\mu\zeta-\nu\eta,\ldots),\end{array}\right\}.\end{array}\right.$$

These four formulæ have the same geometrical significations with the original four formulæ to which they correspond respectively.

28. The eight formulæ become all of them the same or very similar for the quadric form $(a,\ldots)(x,\ y,\ z)^2 = x^2+y^2+z^2$, which of course implies $(A,\ldots)(\xi,\ \eta,\ \zeta)^2 = \xi^2+\eta^2+\zeta^2$. Thus selecting any one of them at pleasure, e.g. the formula II. (bis), this becomes

$$\begin{aligned} &\{(x^2+y^2+z^2)(\xi'^2+\eta'^2+\zeta'^2) - (\xi'x+\eta'y+\zeta'z)^2\} \\ &\times\{(l^2+m^2+n^2)(\xi'^2+\eta'^2+\zeta'^2) - (l\xi'+m\eta'+n\zeta')^2\} \\ = \;&\{(l\xi'+m\eta'+n\zeta')(x\xi'+m\eta'+n\zeta') - (lx+my+nz)(\xi'^2+\eta'^2+\zeta'^2)\}^2 \\ &+(\xi'^2+\eta'^2+\zeta'^2)\begin{vmatrix}\xi', & \eta', & \zeta' \\ x, & y, & z \\ l, & m, & n\end{vmatrix}^2, \end{aligned}$$

where the terms independent of $\xi'^2+\eta'^2+\zeta'^2$ destroy each other. Omitting these terms, and dividing by $\xi'^2+\eta'^2+\zeta'^2$, the resulting equation is found to be

$$\begin{vmatrix}\xi', & \eta', & \zeta' \\ x, & y, & z \\ l, & m, & n\end{vmatrix}^2 = \begin{vmatrix}\xi'^2+\eta'^2+\zeta'^2, & \xi'x+\eta'y+\zeta'z, & \xi'l+\eta'm+\zeta'n \\ x\xi'+y\eta'+z\zeta', & x^2+y^2+z^2, & xl+ym+zn \\ l\xi'+m\eta'+n\zeta', & lx+my+nz, & l^2+m^2+n^2,\end{vmatrix}$$

which is a well-known identical equation.

Article Nos. 29 *to* 33, *relating to a single conic in connexion with an ineunt or a tangent of a conic of double contact.*

29. The formulæ assume a very simple form when the point of intersection of the two tangents, or the line of junction of the two ineunts of the conic, is an ineunt or a tangent of a conic having double contact with the first-mentioned conic. Thus, if to the conic

$$(a,..\between x,\ y,\ z)^2=0$$

tangents are drawn from a point $(x',\ y',\ z')$ of the conic

$$(a,..\between x,\ y,\ z)^2+(\xi'x+\eta'y+\zeta'z)^2=0,$$

then we have

$$(a,..\between x',\ y',\ z')^2=-(\xi'x'+\eta'y'+\zeta'z')^2;$$

and using the form I. (bis), and putting therein $(\xi',\ \eta',\ \zeta')$ in the place of the arbitrary quantities $(\lambda,\ \mu,\ \nu)$, the equation of the tangent divides out by $\xi'x'+\eta'y'+\zeta'z'$, and omitting this factor it becomes

$$(a,..\between x',\ y',\ z'\between x,\ y,\ z)+(\xi'x'+\eta'y'+\zeta'z')(\xi'x+\eta'y+\zeta'z)$$
$$\pm\frac{1}{\sqrt{K}}(A,..\between \xi',\ \eta',\ \zeta'\between yz'-y'z,\ zx'-z'x,\ xy'-x'y)=0,$$

which is of the form

$$\left(\begin{array}{ccc}\alpha, & \beta, & \gamma\\ \alpha', & \beta', & \gamma'\\ \alpha'', & \beta'', & \gamma''\end{array}\between x',\ y',\ z'\between x,\ y,\ z\right)=0,$$

where the matrix

$$\left(\begin{array}{ccc}\alpha, & \beta, & \gamma\\ \alpha', & \beta', & \gamma'\\ \alpha'', & \beta'', & \gamma''\end{array}\right) \text{ is }=$$

$$\begin{array}{lll}
a+\xi'^2 & ,\ h+\xi'\eta'+\frac{1}{\sqrt{K}}(G\xi'+F\eta'+C\zeta'), & g+\xi'\zeta'-\frac{1}{\sqrt{K}}(H\xi'+B\eta'+F\zeta')\\
h+\xi'\eta'-\frac{1}{\sqrt{K}}(G\xi'+F\eta'+C\zeta'), & b+\eta'^2 & ,\ f+\eta'\zeta'+\frac{1}{\sqrt{K}}(A\xi'+H\eta'+G\zeta')\\
g+\xi'\zeta'+\frac{1}{\sqrt{K}}(H\xi'+B\eta'+F\zeta'), & f+\eta'\zeta'-\frac{1}{\sqrt{K}}(A\xi'+H\eta'+G\zeta'), & c+\zeta'^2.
\end{array}$$

30. But instead of further developing these formulæ, I prefer to consider the formulæ which give the points of contact of the tangents in question, viz. the ineunts of the conic $(a,..\between x,\ y,\ z)^2=0$, or the tangents through the point $(x',\ y',\ z')$ of the conic $(a,..\between x,\ y,\ z)^2+(\xi'x+\eta'y+\zeta'z)^2=0$.

We have as before

$$(a,\ldots\between x',\ y',\ z')^2 = -(\xi' x' + \eta' y' + \zeta' z')^2,$$

and using the formula IV. (bis) and writing therein $(\xi',\ \eta',\ \zeta')$ in the place of the arbitrary quantities $(\lambda,\ \mu,\ \nu)$, the equation contains the factor $\xi' x' + \eta' y' + \zeta' z'$, and dividing by this factor, and by K, the line-equation of the ineunt is

$$x'\xi + y'\eta + z'\zeta + \frac{1}{K}(x'\xi' + y'\eta' + z'\zeta')\,.\,(A,\ldots\between \xi',\ \eta',\ \zeta'\between \xi,\ \eta,\ \zeta)$$

$$\pm\frac{1}{\sqrt{K}}(a,\ldots\between x',\ y',\ z'\between \eta\zeta' - \eta'\zeta,\ldots) = 0.$$

Selecting the positive sign, the coordinates of the corresponding ineunt are

$$x' + \frac{1}{K}(x'\xi' + y'\eta' + z'\zeta')(A\xi' + H\eta' + G\zeta') + \frac{1}{\sqrt{K}}\left\{\eta'(gx' + fy' + cz') - \zeta'(hx' + by' + fz')\right\},$$

$$y' + \frac{1}{K}(x'\xi' + y'\eta' + z'\zeta')(H\xi' + B\eta' + F\zeta') + \frac{1}{\sqrt{K}}\left\{\zeta'(ax' + hy' + gz') - \xi'(gx' + fy' + cz')\right\},$$

$$z' + \frac{1}{K}(x'\xi' + y'\eta' + z'\zeta')(G\xi' + F\eta' + C\zeta') + \frac{1}{\sqrt{K}}\left\{\xi'(hx' + by' + fz') - \eta'(ax' + hy' + gz')\right\};$$

and taking $(X,\ Y,\ Z)$ for the coordinates of the ineunt in question, and putting for shortness

$$\alpha = 1 - \frac{1}{\sqrt{K}}(g\eta' - h\zeta'),\quad \beta = -\frac{1}{\sqrt{K}}(f\eta' - b\zeta'),\quad \gamma = -\frac{1}{\sqrt{K}}(c\eta' - f\zeta'),$$

$$\alpha' = -\frac{1}{\sqrt{K}}(a\zeta' - g\xi'),\quad \beta' = 1 - \frac{1}{\sqrt{K}}(h\zeta' - f\xi'),\quad \gamma' = -\frac{1}{\sqrt{K}}(g\zeta' - c\xi'),$$

$$\alpha'' = -\frac{1}{\sqrt{K}}(h\xi' - a\eta'),\quad \beta'' = -\frac{1}{\sqrt{K}}(b\xi' - h\eta'),\quad \gamma'' = 1 - \frac{1}{\sqrt{K}}(f\xi' - g\eta'),$$

we may write

$$(1+P)X = (2-\alpha)x' - \beta\, y' - \gamma\, z' + \frac{1}{K}(A\xi' + H\eta' + G\zeta')(\xi' x' + \eta' y' + \zeta' z'),$$

$$(1+P)Y = -\alpha'\, x' + (2-\beta')y' - \gamma'\, z' + \frac{1}{K}(H\xi' + B\eta' + F\zeta')(\xi' x' + \eta' y' + \zeta' z'),$$

$$(1+P)Z = \alpha''\, x'' - \beta''\, y' + (2-\gamma'')z' + \frac{1}{K}(G\xi' + F\eta' + C\zeta')(\xi' x' + \eta' y' + \zeta' z'),$$

where P, which is arbitrary, may be put

$$= \frac{1}{K}(A,\ldots\between \xi',\ \eta',\ \zeta')^2.$$

31. These equations then give

$$(x', y', z') = \left(\begin{array}{ccc} \alpha, & \beta, & \gamma \\ \alpha', & \beta', & \gamma' \\ \alpha'', & \beta'', & \gamma'' \end{array}\right)\!\!\left(X, Y, Z\right),$$

which can be verified without difficulty by reversing the process; and we have thus the coordinates (X, Y, Z) in terms of (x', y', z'), and reciprocally.

32. If $(X_{\prime}, Y_{\prime}, Z_{\prime})$ are the coordinates of the other ineunt, we have, it is clear,

$$(x', y', z') = \left(\begin{array}{ccc} 2-\alpha, & -\beta, & -\gamma \\ -\alpha', & 2-\beta', & -\gamma' \\ -\alpha'', & -\beta'', & 2-\gamma'' \end{array}\right)\!\!\left(X_{\prime}, Y_{\prime}, Z_{\prime}\right);$$

or substituting for (x', y', z') their values in terms of (X, Y, Z),

$$(2X_{\prime}, 2Y_{\prime}, 2Z_{\prime}) = \left(\begin{array}{ccc} \alpha, & \beta, & \gamma \\ \alpha', & \beta', & \gamma' \\ \alpha'', & \beta'', & \gamma'' \end{array}\right)\!\!\left(X + X_{\prime}, Y + Y_{\prime}, Z + Z_{\prime}\right),$$

so that $(X + X_{\prime}, Y + Y_{\prime}, Z + Z_{\prime})$ are the same linear functions of $2X_{\prime}, 2Y_{\prime}, 2Z_{\prime}$ that (X, Y, Z) are of (x', y', z'); that is, we have

$$\tfrac{1}{2}(1+P)(X+X_{\prime}) = (2-\alpha)X_{\prime} - \beta\, Y_{\prime} - \gamma\, Z_{\prime} + \frac{1}{K}(A\xi' + H\eta' + G\zeta')(\xi' x' + \eta' y' + \zeta' z'),$$

$$\tfrac{1}{2}(1+P)(X+X_{\prime}) = -\alpha'\, X_{\prime} + (2-\beta')\, Y_{\prime} - \gamma'\, Z_{\prime} + \frac{1}{K}(H\xi' + B\eta' + F\zeta')(\xi' x' + \eta' y' + \zeta' z'),$$

$$\tfrac{1}{2}(1+P)(X+X_{\prime}) = -\alpha''\, X_{\prime} - \beta''\, Y_{\prime} + (2-\gamma'')\, Z_{\prime} + \frac{1}{K}(G\xi' + F\eta' + C\zeta')(\xi' x' + \eta' y' + \zeta' z'),$$

which equations may be written

$$(1+P)(X, Y, Z) = \left(\begin{array}{ccc} \mathrm{a}, & \mathrm{b}, & \mathrm{c} \\ \mathrm{a}', & \mathrm{b}', & \mathrm{c}' \\ \mathrm{a}'', & \mathrm{b}'', & \mathrm{c}'' \end{array}\right)\!\!\left(X_{\prime}, Y_{\prime}, Z_{\prime}\right),$$

where the values of the coefficients are

$$h\zeta') + \frac{1}{K}\,(A\xi'^2 - B\eta'^2 - 2F\eta'\zeta' - C\zeta'^2), \quad -\frac{2}{\sqrt{K}}(f\eta' - b\zeta') + \frac{2}{K}\eta'(A\xi' + H\eta' + G\zeta'), \quad -\frac{2}{\sqrt{K}}(c\eta' - f\zeta') + \frac{2}{K}\zeta'(A\xi' + H\eta' + G\zeta')$$

$$g\xi') + \frac{2}{K}\xi'(H\xi' + B\eta' + F\zeta'), \quad 1 - \frac{2}{\sqrt{K}}(h\zeta' - f\eta') + \frac{1}{K}\,(B\eta'^2 - C\zeta'^2 - 2G\zeta'\xi' - A\xi'^2), \quad -\frac{2}{\sqrt{K}}(g\zeta' - c\xi') + \frac{2}{K}\zeta'(H\xi' + B\eta' + F\zeta')$$

$$a\eta') + \frac{2}{K}\xi'(G\xi' + F\eta' + C\zeta'), \quad -\frac{2}{\sqrt{K}}(b\xi' - h\eta') + \frac{2}{K}\eta'(G\xi' + F\eta' + C\zeta'), \quad 1 - \frac{2}{\sqrt{K}}(f\xi' - g\eta') + \frac{1}{K}\,(C\zeta'^2 - A\xi'^2 - 2H\xi'$$

and considering (X, Y, Z) and $(X_{,}, Y_{,}, Z_{,})$ as quantities connected by the foregoing linear relations, we have identically

$$(a, .. \between X, Y, Z)^2 = (a, .. \between X_{,}, Y_{,}, Z_{,})^2.$$

The investigation leads thus to the automorphic transformation of the quadric function, a transformation first effected by M. Hermite[1].

33. It is to be remarked that the foregoing formulæ show that (x', y', z') being the coordinates of a point on the conic $(a, .. \between x, y, z)^2 + (\xi' x + \eta' y + \zeta' z)^2 = 0$, from which point tangents are drawn to the conic $(a, .. \between x, y, z)^2 = 0$, then the coordinates (x', y', z') enter *linearly* into the equations of the tangents, the ineunts (or points of contact), and the polar. And it may be added that the equation of the conic enveloped by the polar (that is, the polar conic of $(a, .. \between x, y, z)^2 + (\xi' x + \eta' y + \zeta' z)^2 = 0$) has for its equation

$$\{K + (A, .. \between \xi', \eta', \zeta')^2\} (a, .. \between x, y, z)^2 - K (\xi' x + \eta' y + \zeta' z)^2 = 0,$$

and that the coordinates of the point of contact of the polar with this conic are

$$x' + \frac{1}{K} (A\xi' + H\eta' + G\zeta') (\xi' x' + \eta' y' + \zeta' z'),$$

$$y' + \frac{1}{K} (H\xi' + B\eta' + F\zeta') (\xi' x' + \eta' y' + \zeta' z'),$$

$$z' + \frac{1}{K} (G\xi' + F\eta' + C\zeta') (\xi' x' + \eta' y' + \zeta' z');$$

so that (x', y', z') also enter linearly into the expressions for the coordinates of the last-mentioned point.

Article Nos. 34 *to* 37, *relating to two conics.*

34. Considering now the two conics

$$U = (a, b, c, f, g, h \between x, y, z)^2 = 0,$$
$$U' = (a', b', c', f', g', h' \between x, y, z)^2 = 0;$$

suppose that the conic

$$\theta U + \theta' U' = (\theta a + \theta' a', .. \qquad \between x, y, z)^2 = 0$$

represents a pair of lines.

The condition for this is

$$\text{Disct. } (\theta a + \theta' a', .. \between x, y, z)^2 = 0,$$

which is

$$(\mathfrak{A}, \mathfrak{B}, \mathfrak{C}, \mathfrak{D} \between \theta, \theta')^3 = 0,$$

where

$$\mathfrak{A} = K,$$
$$\mathfrak{B} = Aa' + Bb' + Cc' + 2Ff' + 2Gg' + 2Hh',$$
$$\mathfrak{C} = A'a + B'b + C'c + 2F'f + 2G'g + 2H'h,$$
$$\mathfrak{D} = K'$$

[1] See my "Memoir on the Automorphic Transformation of a Bipartite Quadric Function," *Phil. Trans.* vol. CXLVIII. (1858), pp. 39—46, [153].

(the significations of K', A', B', C', F', G', H' being of course analogous to those of K, A, B, C, F, G, H). The three roots $\theta : \theta'$ correspond, it is clear, to the three pairs of lines which can be drawn through the intersections of the two conics.

35. The equation

$$\text{Disct. } (\mathfrak{A}, \mathfrak{B}, \mathfrak{C}, \mathfrak{D} \between \theta, \theta')^3 = 0,$$

which is of the fourth order in $\mathfrak{A}$, $\mathfrak{B}$, $\mathfrak{C}$, $\mathfrak{D}$, and of the sixth order as regards (a, b, c, f, g, h) and (a', b', c', f', g', h') respectively, is the condition in order that the two conics may touch each other. Assuming that it is satisfied, the cubic equation in $\theta : \theta'$ has a pair of equal roots; or say there is a twofold root and a onefold root; the twofold root gives the pair of lines drawn from the point of contact to the other two points of intersection, the onefold root gives the pair made up of the common tangent and the line joining the other two points of intersection.

36. In particular, suppose that the two conics are

$$2(\rho x + \sigma y + \tau z)(\rho' x + \sigma' y + \tau' z) = 0,$$
$$2(\lambda x + \mu y + \nu z)(\lambda' x + \mu' y + \nu' z) = 0;$$

so that

$$(a, b, c, f, g, h) = (2\rho\rho', 2\sigma\sigma', 2\tau\tau', \sigma\tau' + \sigma'\tau, \tau\rho' + \tau'\rho, \rho\sigma' + \rho'\sigma),$$
$$(a', b', c', f', g', h') = (2\lambda\lambda', 2\mu\mu', 2\nu\nu', \mu\nu' + \mu'\nu, \nu\lambda' + \nu'\lambda, \lambda\mu' + \lambda'\mu),$$
$$(A, B, C, F, G, H) = -(\sigma\tau' - \sigma'\tau, \tau\rho' - \tau'\rho, \rho\sigma' - \rho'\sigma)^2,$$
$$(A', B', C', F', G', H') = -(\mu\nu' - \mu'\nu, \nu\lambda' - \nu'\lambda, \lambda\mu' - \lambda'\mu)^2;$$

and thence also

$$\mathfrak{A} = K = 0,$$

$$\mathfrak{B} = Aa' + \&\text{c.} = -2 \begin{vmatrix} \lambda, & \mu, & \nu \\ \rho, & \sigma, & \tau \\ \rho', & \sigma', & \tau' \end{vmatrix} \begin{vmatrix} \lambda', & \mu', & \nu' \\ \rho, & \sigma, & \tau \\ \rho', & \sigma', & \tau' \end{vmatrix},$$

$$\mathfrak{C} = A'a + \&\text{c.} = -2 \begin{vmatrix} \rho, & \sigma, & \tau \\ \lambda, & \mu, & \nu \\ \lambda', & \mu', & \nu' \end{vmatrix} \begin{vmatrix} \rho', & \sigma', & \tau' \\ \lambda, & \mu, & \nu \\ \lambda', & \mu, & \nu' \end{vmatrix}$$

$$\mathfrak{D} = K' = 0;$$

and the equation in (θ, θ') is

$$\mathfrak{B}\theta + \mathfrak{C}\theta' = 0;$$

hence writing $\theta = \mathfrak{C}$, $\theta' = -\mathfrak{B}$, the equation of the pair of lines is

$$\begin{vmatrix} \sigma, & \tau \\ \mu, & \nu \\ \mu', & \nu' \end{vmatrix} \begin{vmatrix} \rho', & \sigma', & \tau' \\ \lambda, & \mu, & \nu \\ \lambda', & \mu', & \nu' \end{vmatrix} (\rho x + \sigma y + \tau z)(\rho' x + \sigma' y + \tau' z) - \begin{vmatrix} \lambda, & \mu, & \nu \\ \rho, & \sigma, & \tau \\ \rho', & \sigma', & \tau' \end{vmatrix} \begin{vmatrix} \lambda', & \mu', & \nu' \\ \rho, & \sigma, & \tau \\ \rho', & \sigma', & \tau' \end{vmatrix} (\lambda x + \mu y + \nu z)(\lambda' x + \mu' y + \nu' z) = 0;$$

and it is easy to see that the left-hand side does in fact break up into factors, and that the equation is

$$\begin{vmatrix} x, & y, & z \\ \mu\tau' - \nu\sigma', & \nu\rho' - \lambda\tau', & \lambda\sigma' - \mu\rho' \\ \sigma\nu' - \tau\mu', & \tau\lambda' - \rho\nu', & \rho\mu' - \sigma\lambda' \end{vmatrix} \begin{vmatrix} x, & y, & z \\ \mu\tau - \nu\sigma, & \nu\rho - \lambda\tau, & \lambda\sigma - \mu\rho \\ \sigma'\nu' - \tau'\mu', & \tau'\lambda' - \rho'\nu', & \rho'\mu' - \sigma'\lambda' \end{vmatrix} = 0,$$

which of course might have been obtained at once by means of the four points which are the intersection of each component line of the first conic by each component line of the second conic.

37. Suppose that the first conic is

$$(a,\ b,\ c,\ f,\ g,\ h \between x,\ y,\ z)^2 = 0,$$

while the second conic is the pair of lines

$$2(\lambda x + \mu y + \nu z)(\lambda' x + \mu' y + \nu' z) = 0;$$

then, putting as before,

$$(\theta a + \theta' . 2\lambda\lambda', .. \between x,\ y,\ z)^2 = 0,$$

we have

$$(\mathfrak{A},\ \mathfrak{B},\ \mathfrak{C},\ \mathfrak{D} \between \theta,\ \theta')^3 = 0,$$

where

$$\begin{aligned} \mathfrak{A} &= K, \\ \mathfrak{B} &= 2(A,\ B,\ C,\ F,\ G,\ H \between \lambda,\ \mu,\ \nu \between \lambda',\ \mu',\ \nu'), \\ \mathfrak{C} &= -(a,\ b,\ c,\ f,\ g,\ h \between \mu\nu' - \mu'\nu,\ \nu\lambda' - \nu'\lambda,\ \lambda\mu' - \lambda'\mu)^2, \\ \mathfrak{D} &= 0; \end{aligned}$$

and the equation in $(\theta,\ \theta')$ is

$$K\theta^2 + 2(A, .. \between \lambda,\ \mu,\ \nu \between \lambda',\ \mu',\ \nu')\,\theta\theta' - (a, .. \between \mu\nu - \mu'\nu, ..)^2\,\theta'^2 = 0,$$

which may be written

$$\begin{aligned} \{K\theta + (A, .. \between \lambda,\ \mu,\ \nu \between \lambda',\ \mu',\ \nu')\,\theta'\}^2 &= \{[(A, .. \between \lambda,\ \mu,\ \nu \between \lambda',\ \mu',\ \nu')]^2 + K(a, .. \between \mu\nu' - \mu'\nu, ..)^2\}\theta'^2, \\ &= (A, .. \between \lambda,\ \mu,\ \nu)^2 . (A, .. \between \lambda',\ \mu',\ \nu')^2 . \theta'^2, \end{aligned}$$

that is

$$K\theta = [\pm\sqrt{(A, .. \between \lambda,\ \mu,\ \nu)^2}\sqrt{(A, .. \between \lambda',\ \mu',\ \nu')^2} - (A, .. \between \lambda,\ \mu,\ \nu \between \lambda',\ \mu',\ \nu')]\theta';$$

we may assume

$$\theta = \pm\sqrt{(A, .. \between \lambda,\ \mu,\ \nu)^2}\sqrt{(A, .. \between \lambda',\ \mu',\ \nu')^2} - (A, .. \between \lambda,\ \mu,\ \nu \between \lambda',\ \mu',\ \nu'), \quad \theta' = K,$$

so that the conic

$$\begin{aligned} \{\pm\sqrt{(A, .. \between \lambda,\ \mu,\ \nu)^2}\sqrt{(A, .. \between \lambda',\ \mu',\ \nu')^2} - (A, .. \between \lambda,\ \mu,\ \nu \between \lambda',\ \mu',\ \nu')\}\,(a, ... \between x,\ y,\ z)^2 \\ + 2K(\lambda x + \mu y + \nu z)(\lambda' x + \mu' y + \nu' z) = 0 \end{aligned}$$

breaks up into a pair of lines.

Putting for shortness

$$\pm\sqrt{(A, .. \between \lambda,\ \mu,\ \nu)^2}\sqrt{(A, .. \between \lambda',\ \mu',\ \nu')^2} - (A, .. \between \lambda,\ \mu,\ \nu \between \lambda',\ \mu',\ \nu') = \Omega$$

the coefficients on the left-hand side of the equation are

$$(\Omega a + 2K\lambda\lambda', \ldots \Omega f + K(\mu\nu' + \mu'\nu), \ldots),$$

whence, after all reductions, the inverse function is

$$\{(A, \ldots \between \lambda, \mu, \nu \between \xi, \eta, \zeta) \sqrt{(A, \ldots \between \lambda', \mu', \nu')^2} \mp (A, \ldots \between \lambda', \mu', \nu' \between \xi, \eta, \zeta) \sqrt{(A, \ldots \between \lambda, \mu, \nu)^2}\}^2,$$

and the remainder of the process of decomposition is effected without difficulty.

ADDITION, 18 December, 1862.

The formulæ II. and II. (bis) each of them give the tangents of the conic $(a, \ldots \between x, y, z)^2 = 0$ at the ineunts of intersection with the line $\xi' x + \eta' y + \zeta' z = 0$. A very elegant formula for these ineunts themselves was communicated to me by Mr Spottiswoode, and I have since found that the same or an equivalent formula is made use of by M. Aronhold in his recent valuable memoir, "Ueber eine neue algebraische Behandlungsweise der Integrale irrationaler Differentiale, &c.," *Crelle*, t. LXII. pp. 95—145 (1862). The formula is as follows, viz. for the conic and line,

$$(a, b, c, f, g, h \between x, y, z)^2 = 0,$$

$$\xi' x + \eta' y + \zeta' z = 0,$$

then

$$x : y : z =$$

$$(l\xi' + m\eta' + n\zeta') \frac{1}{2\sqrt{\phi}} \frac{d\phi}{d\xi'} + \eta'(gl + fm + c\eta) - \zeta'(hl + bm + f\eta) + l\sqrt{\phi},$$

$$: (l\xi' + m\eta' + n\zeta') \frac{1}{2\sqrt{\phi}} \frac{d\phi}{d\eta'} + \zeta'(al + hm + g\eta) - \xi'(gl + fm + c\eta) + m\sqrt{\phi},$$

$$: (l\xi' + m\eta' + n\zeta') \frac{1}{2\sqrt{\phi}} \frac{d\phi}{d\zeta'} + \xi'(hl + bm + f\eta) - \eta'(al + hm + g\eta) + n\sqrt{\phi};$$

where

$$\phi = \begin{vmatrix} & \xi', & \eta', & \zeta' \\ \xi', & a, & h, & g \\ \eta', & h, & b, & f \\ \zeta', & g, & f, & c \end{vmatrix} = -(A, B, C, F, G, H \between \xi', \eta', \zeta')^2,$$

so that $\frac{1}{2}\frac{d\phi}{d\xi'}, \frac{1}{2}\frac{d\phi}{d\eta'}, \frac{1}{2}\frac{d\phi}{d\zeta'}$ are respectively

$$= -(A\xi' + H\eta' + G\zeta'), \; -(H\xi' + B\eta' + F\zeta'), -(G\xi' + F\eta' + C\zeta'),$$

and where l, m, n are supernumerary arbitrary quantities.

277.

ON THE WAVE SURFACE.

[From the *Quarterly Journal of Pure and Applied Mathematics*, vol. III. (1860), pp. 16—22.]

SOME very beautiful results in relation to the Wave Surface have been recently obtained by Herr Zech, in the Memoirs "Die Eigenschaften der Wellenflächen der zweiaxigen Krystalle mittels der höhern Geometrie abgeleitet," *Crelle*, t. LII. pp. 243—254 (1856), and "Die Krümmungslinien der Wellenfläche zweiaxiger Krystalle," *Crelle*, t. LIV. pp. 72—77 (1857). According to the former of Fresnel's two modes of generation, the wave surface is the envelope of a plane whose perpendicular distance v is a certain given function of the direction cosines l, m, n. For the same system of direction cosines, there are in fact two values of the perpendicular distance: call these v and w, and let the corresponding planes (parallel of course to each other) be called P and Q. Then the entire system of the planes P and Q envelope the wave surface, viz. the planes P may be considered as enveloping one sheet and the planes Q the other sheet of the surface. But if instead of considering the entire system of planes we consider only the planes P and the parallel planes Q, for which the perpendicular on the plane P has a given constant value v, then the planes P will envelope a developable F, and the planes Q will envelope a developable G, these two developables being, it is to be observed, distinct surfaces, not sheets of one and the same surface. Or, what is the same thing, the planes P for which the perpendicular distance has a given constant value v will envelope a developable F, and the planes P for which the perpendicular distance of the parallel planes Q has a given constant value w, will envelope a developable G. And it is obvious that the developables F and G touch the wave surface along curves. The equation of the developable F contains of course the arbitrary parameter v, and the equation of the developable G contains in like manner the arbitrary parameter w, so that we in fact have two series of developables

F and G respectively touching the wave surface along two series of curves. And it is shown in the second of the Memoirs above referred to that these curves are the *curves of curvature* of the wave surface.

The developable F is obtained as the envelope of the plane P, whose equation is

$$lx + my + nz = v,$$

where v has a given constant value and l, m, n are parameters which vary, subject to the two conditions

$$l^2 + m^2 + n^2 = 1,$$

$$\frac{l^2}{a^2 - v^2} + \frac{m^2}{b^2 - v^2} + \frac{n^2}{c^2 - v^2} = 0.$$

And in like manner the developable G is obtained as the envelope of the plane Q, whose equation is

$$lx + my + nz = \frac{abc}{w} \sqrt{\left(\frac{l^2}{a^2} + \frac{m^2}{b^2} + \frac{n^2}{c^2}\right)},$$

where w has a given constant value and l, m, n are parameters which vary, subject to the two conditions

$$l^2 + m^2 + n^2 = 1,$$

$$\frac{l^2}{a^2 - w^2} + \frac{m^2}{b^2 - w^2} + \frac{n^2}{c^2 - w^2} = 0.$$

{It is hardly necessary to remark that if in the last-mentioned system of equations, the parameter w is also treated as variable, we obtain the wave surface: in fact v^2, w^2 being the roots of the equation

$$\frac{l^2}{a^2 - \theta} + \frac{m^2}{b^2 - \theta} + \frac{n^2}{c^2 - \theta} = 0,$$

we have, attending to the condition $l^2 + m^2 + n^2 = 1$, the identical equation

$$l^2 (b^2 - \theta)(c^2 - \theta) + m^2 (c^2 - \theta)(a^2 - \theta) + n^2 (a^2 - \theta)(b^2 - \theta) = (v^2 - \theta)(w^2 - \theta),$$

and thence

$$vw = abc \sqrt{\left(\frac{l^2}{a^2} + \frac{m^2}{b^2} + \frac{n^2}{c^2}\right)},$$

which shows that the system of equations for the plane Q (w being treated as variable) is in fact identical with the system of equations for the plane P.}

The form of the equations of the planes P and Q respectively shows that each of these planes is parallel to a tangent plane of the cone

$$\frac{x^2}{a^2 - v^2} + \frac{y^2}{b^2 - v^2} + \frac{z^2}{c^2 - v^2} = 0,$$

or in other words, that the planes P and Q are respectively tangents to the conic or infinitely thin surface of the second order, which is the second of the last-mentioned cone by the plane at infinity. Moreover it is obvious from the same equations that the plane P is a tangent plane of the sphere whose equation is

$$x^2 + y^2 + z^2 = v^2,$$

and that the plane Q is a tangent plane of the ellipsoid whose equation is

$$a^2x^2 + b^2y^2 + c^2z^2 = \frac{a^2b^2c^2}{v^2}.$$

Hence each of the developables F and G is the envelope of a plane which is the common tangent plane of two surfaces of the second order; such developables are in general of the eighth order, see my paper "On the Developable Surfaces which arise from Two Surfaces of the Second Order," *Camb. and Dubl. Math. Journ.*, t. v. pp. 46—57 (1850), [84], and it will be presently seen that this is in fact the order of the developables F and G respectively.

The before-mentioned cone

$$\frac{x^2}{a^2 - v^2} + \frac{y^2}{b^2 - v^2} + \frac{z^2}{c^2 - v^2} = 0,$$

(Zech's cone K) is a cone having for its focal lines the optic axes (or normals to the circular sections) of the ellipsoid $a^2x^2 + b^2y^2 + c^2z^2 = 1$ (Zech's ellipsoid E) which is used in the theory of the Wave Surface in the place of Fresnel's Surface of Elasticity. The complementary cone

$$(a^2 - v^2)\, x^2 + (b^2 - v^2)\, y^2 + (c^2 - v^2)\, z^2 = 0,$$

(Zech's cone C) meets the last-mentioned ellipsoid in a curve lying on the sphere whose equation is $x^2 + y^2 + z^2 = \frac{1}{v^2}$, a property which may be considered as affording the geometrical construction of the magnitude v, by means of the cone K. The ellipsoid $a^2x^2 + b^2y^2 + c^2z^2 = \frac{a^2b^2c^2}{v^2}$ is obviously an ellipsoid similar to the ellipsoid E, and the value of v being determined as above, the sphere $x^2+y^2+z^2=v^2$ and the ellipsoid $a^2x^2+b^2y^2+c^2z^2=\frac{a^2b^2c^2}{v^2}$ (which are used in the preceding geometrical construction of the developables F and G respectively) may be considered as given by construction.

To find the equation of the developable F, we have, from the equations

$$\begin{aligned} lx + my + nz &= v, \\ l^2 + m^2 + n^2 &= 1, \\ \frac{l^2}{a^2 - v^2} + \frac{m^2}{b^2 - v^2} + \frac{n^2}{c^2 - v^2} &= 0, \end{aligned}$$

treating them by the method of arbitrary multipliers in the usual manner, we obtain

$$x+\left(\rho+\frac{\theta}{a^2-v^2}\right) l=0,$$

$$y+\left(\rho+\frac{\theta}{b^2-v^2}\right) m=0,$$

$$z+\left(\rho+\frac{\theta}{c^2-v^2}\right) n=0,$$

equations which give in the first instance $v+\rho=0$ or $\rho=-v$, and then substituting this value for ρ,

$$l=\frac{x\,(v^2-a^2)}{v\,(v^2-a^2)+\theta},\qquad m=\frac{y\,(v^2-b^2)}{v\,(v^2-b^2)+\theta},\qquad z=\frac{z\,(v^2-c^2)}{v\,(v^2-c^2)+\theta},$$

and thence

$$\frac{x^2\,(v^2-a^2)}{v\,(v^2-a^2)+\theta}+\frac{y^2\,(v^2-b^2)}{v\,(v^2-b^2)+\theta}+\frac{z^2\,(v^2-c^2)}{v\,(v^2-c^2)+\theta}=v,$$

$$\frac{x^2\,(v^2-a^2)}{[v\,(v^2-a^2)+\theta]^2}+\frac{y^2\,(v^2-b^2)}{[v\,(v-b^2)+\theta]^2}+\frac{z^2\,(v^2-c^2)}{[v\,(v^2-c^2)+\theta]^2}=0,$$

the latter of which is the derived equation of the former with respect to the parameter θ, hence writing the former equation under the form

$$\begin{aligned}&\{\theta+v\,(v^2-a^2)\}\,\{\theta+v\,(v^2-b^2)\}\,\{\theta+v\,(v^2-c^2)\}\\&-\frac{x^2}{v^2}\,v\,(v^2-a^2)\;\{\theta+v\,(v^2-b^2)\}\,\{\theta+v\,(v^2-c^2)\}\\&-\text{\&c.}=0,\end{aligned}$$

or what is the same thing

$$(A,\ B,\ C,\ D)(\theta,\ 1)^3=0,$$

where

$A=3v^2$,
$B=(v^2-x^2)(v^2-a^2)+(v^2-y^2)(v^2-b^2)+(v^2-z^2)(v^2-c^2)$,
$C=(v^2-y^2-z^2)(v^2-b^2)(v^2-c^2)+(v^2-z^2-x^2)(v^2-c^2)(v^2-a^2)+(v^2-x^2-y^2)(v^2-a^2)(v^2-b^2)$,
$D=3\,(v^2-x^2-y^2-z^2)(v^2-a^2)(v^2-b^2)(v^2-c^2)$,

the equation of the developable F is

$$(AD-BC)^2-4\,(AC-B^2)(BD-C^2)=0.$$

The investigation for the developable G is very similar to the preceding. Write for shortness

$$\Lambda^2=\frac{l^2}{a^2}+\frac{m^2}{b^2}+\frac{n^2}{c^2},$$

then the system of equations is

$$\begin{aligned}lx+my+nz-\frac{abc}{w}\Lambda&=0,\\ l^2+m^2+n^2&=1,\\ \frac{l^2}{a^2-w^2}+\frac{m^2}{b^2-w^2}+\frac{n^2}{c^2-w^2}&=0,\end{aligned}$$

and we then have

$$x - \frac{abc}{wa^2\Lambda}\, l + \left(\rho + \frac{\theta}{a^2 - w^2}\right) l = 0,$$

$$y - \frac{abc}{wb^2\Lambda}\, m + \left(\rho + \frac{\theta}{b^2 - w^2}\right) m = 0,$$

$$z - \frac{abc}{wc^2\Lambda}\, n + \left(\rho + \frac{\theta}{c^2 - w^2}\right) n = 0,$$

equations which give $\rho = 0$, and substituting this value, we obtain

$$l = \frac{x\,(w^2 - a^2)}{\dfrac{abc}{wa^2\Lambda}(w^2 - a^2) + \theta},$$

$$m = \frac{y\,(w^2 - b^2)}{\dfrac{abc}{wb^2\Lambda}(w^2 - b^2) + \theta},$$

$$n = \frac{z\,(w^2 - c^2)}{\dfrac{abc}{wc^2\Lambda}(w^2 - c^2) + \theta}.$$

Substituting these values in the equations

$$\frac{l^2}{a^2} + \frac{m^2}{b^2} + \frac{n^2}{c^2} = \Lambda^2,$$

$$\frac{l^2}{w^2 - a^2} + \frac{m^2}{w^2 - b^2} + \frac{n^2}{w^2 - c^2} = 0,$$

we find

$$\frac{\dfrac{x^2}{a^2}(w^2 - a^2)^2}{\left\{\dfrac{abc}{wa^2}(w^2 - a^2) + \Lambda\theta\right\}^2} + \frac{\dfrac{y^2}{b^2}(w^2 - b^2)^2}{\left\{\dfrac{abc}{wb^2}(w^2 - b^2) + \Lambda\theta\right\}^2} + \frac{\dfrac{z^2}{c^2}(w^2 - c^2)^2}{\left\{\dfrac{abc}{wc^2}(w^2 - c^2) + \Lambda\theta\right\}^2} = 1,$$

$$\frac{x^2\,(w^2 - a^2)}{\left\{\dfrac{abc}{wa^2}(w^2 - a^2) + \Lambda\theta\right\}^2} + \frac{y^2\,(w^2 - b^2)}{\left\{\dfrac{abc}{wb^2}(w^2 - b^2) + \Lambda\theta\right\}^2} + \frac{z^2\,(w^2 - c^2)}{\left\{\dfrac{abc}{wc^2}(w^2 - c^2) + \Lambda\theta\right\}^2} = 0,$$

and multiplying the first equation by $\dfrac{abc}{w}$ and the second equation by $\Lambda\theta$ and adding, we obtain

$$\frac{x^2\,(w^2 - a^2)}{\dfrac{abc}{wa^2}(w^2 - a^2) + \Lambda\theta} + \frac{y^2\,(w^2 - b^2)}{\dfrac{abc}{wb^2}(w^2 - b^2) + \Lambda\theta} + \frac{z^2\,(w^2 - c^2)}{\dfrac{abc}{wc^2}(w^2 - c^2) + \Lambda\theta} = \frac{abc}{w},$$

of which equation the latter of the foregoing two equations is the derived equation with respect to the parameter $\Lambda\theta$.

Comparing this with the foregoing equation

$$\frac{x^2(v^2-a^2)}{v(v^2-a^2)+\theta}+\frac{y^2(v^2-b^2)}{v(v^2-c^2)+\theta}+\frac{z^2(v^2-c^2)}{v(v^2-c^2)+\theta}=v,$$

we see that it is deduced from it by writing

$$a^2x^2,\ b^2y^2,\ c^2z^2,\ \frac{w^2}{a^2}-1,\ \frac{w^2}{b^2}-1,\ \frac{w^2}{c^2}-1,\ \frac{abc}{w^2},\ \Lambda\theta$$

in the place of

$$x^2,\quad y^2,\quad z^2,\ v^2-a^2,\ v^2-b^2,\ v^2-c^2,\ v\ ,\quad \theta$$

respectively, and the equation of the developable G is therefore

$$(A'D'-B'C')^2-4(A'C'-B'^2)(B'D'-C'^2)=0,$$

where A', B', C', D' are what A, B, C, D become when

$$x^2,\quad y^2,\quad z^2,\ v^2-a^2,\ v^2-b^2,\ v^2-c^2,\ v\ ,$$

are replaced by

$$a^2x^2,\ b^2y^2,\ c^2z^2,\ \frac{w^2}{a^2}-1,\ \frac{w^2}{b^2}-1,\ \frac{w^2}{c^2}-1,\ \frac{abc}{w}.$$

The equations of the developables F and G, although radically distinct from each other, are consequently similar in form, and each is at once deducible from the other.

2, *Stone Buildings, W.C., 9th March,* 1858.

278.

NOTE ON THE SINGULAR SOLUTIONS OF DIFFERENTIAL EQUATIONS.

[From the *Quarterly Journal of Pure and Applied Mathematics,* vol. III. (1860), pp. 36, 37.]

THE following investigation (which has been in my possession for a good many years) affords I think a simple explanation of the theory of the singular solutions of differential equations.

Let the primitive equation be

$$c^n + Pc^{n-1} + Qc^{n-2} + \dots = 0,$$

where c is the arbitrary constant and $P, Q \dots$ are any functions of x, y; then the differential equation is obtained by eliminating c from the foregoing equation and the derived equation

$$P'c^{n-1} + Q'c^{n-2} + \dots = 0,$$

and the result may be represented by

$$F(P, Q, \dots, P', Q', \dots) = 0.$$

Assume now

$$c^n + Pc^{n-1} + Qc^{n-2} + \dots = (c+X)(c+Y)(c+Z)\dots,$$

then we have

$$\begin{aligned} P &= X + Y + Z + \text{\&c.}, \\ Q &= XY + XZ + YZ + \text{\&c.}, \\ &\text{\&c.}, \end{aligned}$$

and consequently

$$\begin{aligned} P' &= X' + Y' + Z' + \text{\&c.}, \\ Q' &= (Y + Z + \text{\&c.})\,X' + \text{\&c.}, \\ &\text{\&c.}, \end{aligned}$$

and substituting these values in the function $F(P, Q, \ldots P', Q', \ldots)$ it is clear that we shall have $F(P, Q, \ldots, P', Q', \ldots) = UX'Y'Z'\ldots$ where U is a symmetrical function of X, Y, Z, &c., and therefore a function of $P, Q, \ldots$; and this equation will be identically true whatever values we attribute to $X', Y', Z', \ldots$, hence putting these quantities respectively equal to unity, we have

$$\begin{aligned} P' &= n, \\ Q' &= (n-1)P, \\ R' &= (n-2)Q, \\ &\text{\&c.,} \end{aligned}$$

and with these values

$$U = F(P, Q, \ldots, P', Q', \ldots),$$

that is $U = 0$ is the result obtained by eliminating c from the primitive equation and the equation

$$nc^{n-1} + (n-1)Pc^{n-2} + \ldots = 0,$$

which is the equation obtained by differentiating the primitive equation with respect to the arbitrary constant c: that is, $U = 0$ being the singular solution, the differential equation is

$$UX'Y'Z'\ldots = 0.$$

It is to be remarked that (P, Q, &c. being rational and integral functions) then if the roots X, Y, Z, &c. are also rational and integral functions, the differential equation contains U as a separable rational and integral factor, but if the roots are irrational then the differential equation does not really contain the rational and integral factor U, but $X'Y'Z'\ldots$ is here a rational fraction containing U in the denominator and $UX'Y'Z'\ldots$ is an indecomposable rational and integral function. This is easily verified *à posteriori* for a quadratic equation.

2, *Stone Buildings, W.C.*, 28*th January*, 1858.

279.

ON A THEOREM RELATING TO SPHERICAL CONICS.

[From the *Quarterly Journal of Pure and Applied Mathematics*, vol. III. (1860), p. 53.]

THE following theorem was given by Prof. Maccullagh: "If three lines at right angles to each other pass through a fixed point O so that two of them are confined to given planes: the third line traces out a cone of the second order whose sections parallel to the given planes are circles, and the plane containing the other two lines envelopes a cone of the second order whose sections by planes parallel to the given planes are parabolas."

Referring the figure to the sphere we have a trirectangular triangle XYZ, of which two angles X, Y lie on fixed arcs A, B. The angle Z generates a spherical conic U' having A, B for its cyclic arcs. The side XY envelopes a spherical conic U touched by the arcs A, B. The conic U' is evidently the supplementary conic of U, hence the poles of A, B are the foci of U. We may drop altogether the consideration of the triangle XYZ and consider only the side XY, we have then the theorem:

If a quadrantal arc XY slides between the two fixed arcs A, B, the envelope of XY is a spherical conic U touched by the fixed arcs A, B, and which has for its foci the poles of these same arcs A, B.

It is worth while to notice the great reduction of order which takes place in consequence of the arc XY being a quadrant. If XY had been an arc of a given magnitude θ, the envelope would have been a spherical curve of an order certainly higher than 6. For considering the corresponding problem in plano, the envelope in the particular case where the fixed lines A, B are at right angles to each other is a curve of the sixth order, and in the general case where the two fixed lines are not at right angles the order is higher: the problem in plano corresponds of course, not to the general problem on the sphere, but to that in which θ is indefinitely small.

280.

ON THE CONICS WHICH TOUCH FOUR GIVEN LINES.

[From the *Quarterly Journal of Pure and Applied Mathematics,* vol. III. (1860), pp. 94—96.

THERE are considerable practical difficulties in drawing a figure of the system of conics which touch four given lines, but a notion of the figure of the system may be obtained as follows:

Figure 20 may be taken to represent any quadrilateral whatever, having all its sides real; and if we attend only to the unshaded spaces, it will be seen that there are five regions which are called the inner, upper, lower, right-hand, and left-hand regions respectively. The inner diagonals are AC the vertical diagonal and BD the horizontal diagonal; the former of these traverses the inner, upper, and lower regions;

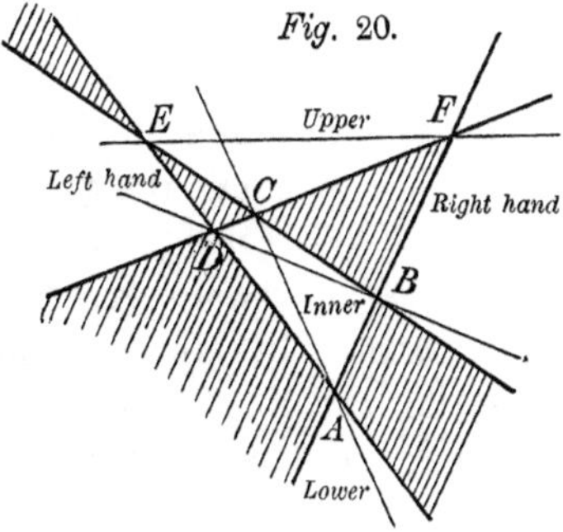

Fig. 20.

the latter the inner, right-hand, and left-hand regions; the outer diagonal is EF which traverses the upper region and the right-hand and left-hand regions. The inner or vertical and horizontal diagonals meet in the inner region, the vertical and outer diagonals meet in the upper region, the horizontal and outer diagonals meet in the left-hand region. No conic touching the four lines lies wholly or in part in the shaded regions, and every conic touching the four lines lies wholly in the inner region or wholly in the upper region, or partly in the upper and partly in the lower region, or partly in the right-hand and partly in the left-hand region. It will be convenient to consider, 1°. the conics which lie in the inner region; 2°. the conics which lie in the upper and lower regions, or in the upper region only; 3°. the conics which lie in the right-hand and left-hand regions.

1°. The conics in the inner region are obviously ellipses; an extreme term is the finite right line BD, considered as an indefinitely thin ellipse; this gradually broadens out and there is (as a mean term) an ellipse which touches the four lines in the points in which they are intersected two and two by the lines joining the points E, F with the point of intersection of the diagonals AC and BD, the ellipse then narrows in the transverse direction and at length reduces itself to the finite line AC considered as an indefinitely thin ellipse.

2°. Considering first the conics which lie in the upper and lower regions, these are of course hyperbolas, an extreme term is the infinite portions $A\infty$ and $C\infty$ of the line AC, considered as an indefinitely thin hyperbola: we have then hyperbolas such that for each of them, one branch lies in the lower region and touches the two lines through A, while the other branch lies in the upper region and touches the two lines through C; the points of contact of the lower branch with the lines through A gradually recede from A, but the point of contact on the line AD, which adjoins the left-hand region, recedes with the greater rapidity, and it at last becomes infinite while the point of contact with the line AB which adjoins the right-hand region remains finite: we have thus a hyperbola having the line AD for its asymptote; the point at infinity of this line belongs of course indifferently to the upper and lower regions; and we may therefore consider one branch as lying in the lower region and touching AB and (at infinity) AD; the other branch as lying in the upper region and touching the two lines through C, and besides (at infinity) the line AD. We have next a series of hyperbolas such that for each of them, one branch lies in the lower region and touches only the line AB, the other branch lies in the upper region and touches the two lines through C and besides the line AD. We arrive again at a limiting case when the point of contact on the line AB passes off to infinity, or AB becomes an asymptote; we have here in the lower region a branch touching (at infinity) AB and in the upper region a branch touching the two lines through C, the line AD, and besides (at infinity) the line AB. To this succeeds a series of hyperbolas such that for each of them, one branch lies in the lower region but does not touch either of the lines through A, the other branch lies in the upper region and touches as well the two lines through C as also the two lines through A. Finally the branch in the lower region passes off to infinity, or the conic becomes a parabola lying wholly in the upper region.

We have thus arrived at the conics which lie wholly in the upper region, the extreme term being a parabola: this passes into an ellipse touching of course the four lines, and gradually reducing itself to the finite line EF considered as an indefinitely thin ellipse.

3°. We come now to the conics which lie in the right-hand and left-hand regions; an extreme term is the conic composed of the infinite portions $E\infty$ and $F\infty$ of the line EF, considered as an indefinitely thin hyperbola; we have then a series of hyperbolas such that for each of them the branch in the right-hand region touches the two lines through E, and the branch in the left-hand region the two lines through F; we then arrive at a limiting case where the branch in the right-hand region touches the line EC the upper boundary of the right-hand region at ∞, or where EC is an asymptote; we may say that the branch in the right-hand region touches the line EB and (at infinity) the line EC, while the branch in the left-hand region touches not only the two lines through F but also (at infinity) the line EC; we have next a series of hyperbolas such that for each of them the branch in the right-hand region touches only the line EB while the branch in the left-hand region touches as well the two lines through F as the line EC; we have then again a limiting case, viz. the branch in the right-hand region touches the line BC at infinity, or BC is an asymptote; we may here say that the branch in the right-hand region touches the line BE and (at infinity) the line BC, while the branch in the left-hand region touches the lines through D and (at infinity) the line BC; and to this succeeds a series of hyperbolas such that for each of them the branch in the right-hand region touches the two lines through B while the branch in the left-hand region touches the two lines through D; the hyperbola finally becomes the infinite portions $B\infty$ and $D\infty$ of the line BD, considered as an indefinitely thin hyperbola.

It is to be noticed that the entire system commences with the finite line BD considered as an indefinitely thin ellipse and passes on continuously to the infinite portions $B\infty$ and $D\infty$ of the same line, considered as an indefinitely thin hyperbola: it is therefore a cyclical series, and any term whatever might have been taken for the commencement of the series.

281.

NOTE ON THE WAVE SURFACE.

[From the *Quarterly Journal of Pure and Applied Mathematics*, vol. III. (1860), pp. 142—144.]

IN the paper "On the Wave Surface" in the last Number of the *Journal*, [277], I stated that it was shown in the second of Zech's Memoirs that the curves of contact of the developables F and G with the wave surface were *the curves of curvature* of the wave surface. I had not examined the demonstration of this theorem, and I had overlooked the author's note "Die Krümmungslinien der Wellenfläche zweiaxiger Krystalle, &c.," *Crelle*, t. LIV. p. 94 (Feb. 1858), where he points out that in a phrase which he quotes, he had assumed without demonstration a theorem which was in fact erroneous, and that he retracted all that he had said in No. 11 of his Memoir (the portion which contains the theorem as to the curves of curvature of the wave surface). M. Bertrand in a note in the *Comptes Rendus*, t. XLVII. pp. 817—819 (Nov. 1858), after referring to my paper, remarks that the theorem as to the curves of curvature of the wave surface appeared to him so remarkable that he hastened to investigate a proof of it, but that he very soon discovered that the theorem was unfortunately erroneous; and he proceeds to show why the theorem cannot be true. M. Bertrand's demonstration is as follows:

THEOREM I. If from a point O, we let fall perpendiculars on the tangent planes of a surface, the locus of their feet is a new surface. Let P be a point of this surface corresponding to the point M of the first surface, the normal at P passes through the middle point of OM.

THEOREM II. If the curve of curvature of a surface is such that the tangent planes at the several points of the curve are equidistant from a point O, the curve of curvature is situate on a sphere having the point O for its centre.

For suppose that we let fall from the point O perpendiculars on the tangent planes to the surface at the several points of the curve of curvature in question. The locus of the feet will be a spherical curve, a normal at any point P of the curve will it is clear pass through the point O; besides, in virtue of the first theorem, another normal will pass through the middle point of the radius OM drawn to the corresponding point M of the surface; the tangent to the curve which is the locus of the points P is therefore perpendicular to the plane MOP, and consequently to the line MP. But when a developable surface is circumscribed about a sphere, the perpendiculars let fall from the centre of the sphere on the tangent planes of the developable surface have their feet on the generating lines; and consequently the curve which is the locus of the points P is situate on the developable surface (viz., the developable surface enveloped by the tangent planes at the several points of the curve of curvature of the given surface) and cuts the generating lines at right angles. But the curve, the locus of the point M, being by hypothesis a curve of curvature of the given surface, also cuts at right angles the generating lines of the developable surface, the two curves are therefore equidistant curves on the developable surface, viz. MP is constant, and since by hypothesis OP is constant, OM is also constant, which proves the second theorem.

This being premised, it is to be recollected that the wave surface may be generated in two different ways; 1°. it is the locus of the extremities of the central perpendiculars to the diametral sections of an ellipsoid E, equal to the axes of these sections; 2°. it is the envelope of planes parallel to the diametral sections of a second ellipsoid E' at distances inversely proportional to the axes of the sections. To obtain all the tangent planes of the wave surface situate at a distance h from the centre, it is necessary to find in the ellipsoid the diametral sections which have an axis equal to $\frac{1}{h}$; for this, we may cut the ellipsoid by a concentric sphere of the radius $\frac{1}{h}$, and draw tangent planes to the cone having for its vertex the centre of the ellipsoid and passing through the curve of intersection with the sphere. The tangent planes of the wave surface respectively parallel to the tangent planes of the cone will be at the distance h from the centre, and, if they touched the wave surface along a curve of curvature, it would follow from the foregoing theorem II. that their points of contact would be all at the same distance from the centre; but the distance of the centre of the wave surface from the point of contact of a tangent plane is inversely proportional to the perpendicular let fall from the centre of the ellipsoid E' on the tangent plane at the corresponding point of this ellipsoid: it follows that the curve of intersection of an ellipsoid with a concentric sphere is not such that the tangent planes of the ellipsoid at the several points of the curve are at the same distance from the centre, and consequently the tangent planes which were under consideration do *not* determine by their points of contact a curve of curvature of the wave surface.

I remark in relation to M. Bertrand's theorem I. that it is immediately connected with a well-known theorem which occurs in optics—viz., if rays proceeding from a point are reflected at a surface, and if from the radiant point to the surface and thence along the reflected ray, we measure off a constant distance, the surface which is the locus of the point so obtained (the secondary focal surface of Dandelin and Quetelet) is an orthogonal trajectory of the reflected rays. In fact, if O be the radiant point, and OM' the incident ray, and if from M' we measure off on the reflected ray a distance $= - OM'$, that is, on the reflected ray produced backwards, a distance $M'P = OM'$, then the whole distance from O is $OM' - OM' = 0$, and the surface which is the locus of the point P is consequently an orthogonal trajectory to the reflected rays. But the point P may, it is clear, be constructed as follows: viz., on the tangent plane at M' let fall the perpendicular OP', and produce it to a point P such that $OP' = P'P$. And if we produce OM' to M so that $OM' = M'M$, then it is clear that the locus of M is a surface similar and similarly situated with the original surface, but of double the magnitude, and that OP is the perpendicular from O upon the tangent plane at M of the last-mentioned surface. And, by what precedes, the line PM' from P to the middle point of OM is a normal of the surface which is the locus of P: the corresponding theorem *in plano* is in fact actually given by Dandelin.

31*st Dec.*, 1858.

282.

ON A PARTICULAR CASE OF CASTILLON'S PROBLEM.

[From the *Quarterly Journal of Pure and Applied Mathematics*, vol. III. (1860), pp. 157—164.

THE problem referred to as Castillon's problem, being itself a particular case of the problem of the in-and-circumscribed triangle, is as follows: viz. "In a circle to inscribe a triangle the sides of which pass through three given points," and the reciprocal problem of course presents itself "about a circle to circumscribe a triangle, the angles of which lie in given lines." If in Castillon's problem the three given points are the angles of a triangle circumscribed about the circle, or if in the reciprocal problem the three given lines are the sides of a triangle circumscribed about the circle, we have a circle and an inscribed and circumscribed triangle, such that the sides of the inscribed triangle pass through the angles of the circumscribed triangle, and the problem arises "given one of the triangles to determine the other triangle." The problem, so far as I am aware, was first proposed by Clausen, who has given, *Crelle*, t. IV. (1829), p. 391, a very elegant solution, which I propose to reproduce here.

Let the angle of a point be defined as the inclination to a fixed radius, of the line from the centre through the given point.

Let α, β, γ, be the angles of the points of contact of the sides of the circumscribed triangle.

α', β', γ', the angles of the angular points of the circumscribed triangle.

a, b, c, the angles of the angular points of the inscribed triangle.

a', b', c', the angles of the intersections of the perpendiculars from the centre on the sides of the inscribed triangle.

Therefore

$$\begin{aligned} 2a' &= b + c, & 2\alpha' &= \beta + \gamma, \\ 2b' &= c + a, & 2\beta' &= \gamma + \alpha, \\ 2c' &= a + b, & 2\gamma' &= \alpha + \beta. \end{aligned}$$

Hence observing that the distance of one of the angles of the circumscribed triangle is $\sec\frac{1}{2}(\beta-\gamma)$, and that the projection of this line perpendicular to the corresponding side of the inscribed triangle is equal to $\sec\frac{1}{2}(\beta-\gamma)\cos(\alpha'-a')$, which is equal to the projection of the radius perpendicular to the same side, or to $\cos\frac{1}{2}(b-c)$, we have

$$\cos(\alpha'-a')=\cos\tfrac{1}{2}(\beta-\gamma)\cos\tfrac{1}{2}(b-c),$$
$$\cos(\beta'-b')=\cos\tfrac{1}{2}(\gamma-\alpha)\cos\tfrac{1}{2}(c-a),$$
$$\cos(\gamma'-c')=\cos\tfrac{1}{2}(\alpha-\beta)\cos\tfrac{1}{2}(a-b).$$

Write

$$b-c=4f,\quad b+c-2\alpha=4x,$$
$$c-a=4g,\quad c+a-2\beta=4y,$$
$$a-b=4h,\quad a+b-2\gamma=4z;$$

therefore

$$\cos(y+z+g-h)=\cos 2f\cos(f+y-z),$$
$$\cos(z+x+h-f)=\cos 2g\cos(g+z-x),$$
$$\cos(x+y+f-g)=\cos 2h\cos(h+x-y),$$

or, since $f+g+h=0$,

$$\cos\{(y-h)+(z+g)\}=\cos 2f\cos\{(y-h)-(z+g)\},$$
$$\cos\{(z-f)+(x+h)\}=\cos 2g\cos\{(z-f)-(x+h)\},$$
$$\cos\{(x-g)+(y+f)\}=\cos 2h\cos\{(x-g)-(y+f)\},$$

or

$$\tan^2 f=\tan(y-h)\tan(z+g),$$
$$\tan^2 g=\tan(z-f)\tan(x+h),$$
$$\tan^2 h=\tan(x-g)\tan(y+f).$$

Put

$$x+h=\eta,\quad x-g=\zeta_1,$$
$$y+f=\zeta,\quad y-h=\xi_1,$$
$$z+g=\xi,\quad z-f=\eta_1,$$

$$\tan f=l,\quad \tan\xi=\mathrm{x},\quad \tan\xi_1=\mathrm{x}_1,$$
$$\tan g=m,\quad \tan\eta=\mathrm{y},\quad \tan\eta_1=\mathrm{y}_1,$$
$$\tan h=n,\quad \tan\zeta=\mathrm{z},\quad \tan\zeta_1=\mathrm{z}_1.$$

We have then

$$\tan^2 f=\tan\xi\tan\xi_1,$$
$$\tan^2 g=\tan\eta\tan\eta_1,$$
$$\tan^2 h=\tan\zeta\tan\zeta_1,$$

or, what is the same thing,

$$l^2=\mathrm{xx}_1,\quad m^2=\mathrm{yy}_1,\quad n^2=\mathrm{zz}_1.$$

But to obtain equations involving only one of the sets (x, y, z), (x_1, y_1, z_1), it is proper to write

$$\tan^2 f = \tan(\zeta + g)\tan\xi = \tan\xi_1 \tan(\zeta_1 - h),$$
$$\tan^2 g = \tan(\xi + h)\tan\eta = \tan\eta_1 \tan(\xi_1 - f),$$
$$\tan^2 h = \tan(\eta + f)\tan\zeta = \tan\zeta_1 \tan(\eta_1 - g);$$

taking the first set of equations, we have

$$l^2 = \frac{x(z+m)}{1-mz},$$

$$m^2 = \frac{y(x+n)}{1-nx},$$

$$n^2 = \frac{z(y+l)}{1-ly};$$

therefore

$$z = \frac{l^2 - mx}{x + l^2 m}, \qquad y = \frac{m^2(1-nx)}{x+n};$$

therefore

$$y + l : 1 - ly = m^2 + ln + (l - m^2 n)x : n - lm^2 + (1 + lm^2 n)x;$$

therefore

$$n^2\{n - lm^2 + (1 + lm^2 n)x\}(l^2 m + x) = (l^2 - mx)\{m^2 + ln + (l - m^2 n)x\};$$

therefore

$$\begin{aligned}&(l^2m^2 + l^3 n - l^2 mn^3 + l^3 m^3 n^2)\\ &+ x\,(-n^3 + lm^2n^2 - l^2mn^2 - l^3m^3n^3 - m^3 - lmn + l^3 - l^2m^2n)\\ &+ x^2(-lm + m^3 n - n^2 - lm^2 n^3) = 0.\end{aligned}$$

Now $l + m + n = lmn$, and by means of this relation we find

$$\begin{aligned}\text{Coef. } x^\circ &= l\{lm^2 + l^2 n - n^2(l+m+n) + m(l+m+n)^2\},\\ \text{Coef. } x\ &= -n^3 + mn(l+m+n) - ln(l+m+n) - (l+m+n)^3 - m^3\\ &\qquad\qquad - lmn + l^3 - lm(l+m+n),\end{aligned}$$

$$\text{Coef. } x^2 = \frac{1}{l}\{-l^2 m + m^2(l+m+n) - ln^2 - n(l+m+n)^2\};$$

or, reducing,

$$\begin{aligned}\frac{1}{l}\text{ coef. } x^\circ &= (m^3 + 2m^2 n - n^3) + l(3m^2 + 2mn - n^2) + l^2(m+n)\\ &= (m+n)\{(m^2 + mn - n^2) + l(3m - n) + l^2\}\\ &= (m+n)\{(l+m)^2 + (l+m)(l+n) - (l+n)^2\},\end{aligned}$$

$$\begin{aligned}-\tfrac{1}{2}\text{ coef. } x &= m^3 + m^2 n + mn^2 + n^3 + 2l(m^2 + 2mn + n^2) + 2l^2(m+n)\\ &= (m+n)\{m^2 + n^2 + 2l(m+n) + 2l^2\}\\ &= (m+n)\{(l+m)^2 + (l+n)^2\},\end{aligned}$$

$$\begin{aligned}l\text{ coef. } x^2 &= (m^3 - 2mn^2 - n^3) + l(m^2 - 2mn - 3n^2) - l^2(m+n)\\ &= (m+n)\{(m^2 - mn - n^2) + l(m - 3n) - l^2\}\\ &= (m+n)\{(l+m)^2 - (l+m)(l+n) - (l+n)^2\};\end{aligned}$$

and the equation becomes

$$\begin{aligned} & l^2 \{(l+m)^2 + (l+m)(l+n) - (l+n)^2\} \\ & - 2l\mathrm{x} \{(l+m)^2 + (l+n)^2\} \\ & + \mathrm{x}^2 \{(l+m)^2 - (l+m)(l+n) - (l+n)^2\} = 0, \end{aligned}$$

which may be written

$$\left[\frac{\mathrm{x}}{l}\{(l+m)^2 - (l+m)(l+n) - (l+n)^2\} - \{(l+m)^2 + (l+n)^2\}\right]^2 = 5(l+m)^2(l+n)^2,$$

and we have therefore

$$\frac{\mathrm{x}}{l} = \frac{(l+m)^2 + \sqrt{(5)}(l+m)(l+n) + (l+n)^2}{(l+m)^2 - (l+m)(l+n) - (l+n)^2},$$

or, reducing and observing also that $l^2 = \mathrm{x}\mathrm{x}_1$, we have

$$\frac{\mathrm{x}}{l} = \frac{l}{\mathrm{x}_1} = \frac{l+m + \frac{1}{2}\{1+\sqrt{(5)}\}(l+n)}{(l+m) - \frac{1}{2}\{1+\sqrt{(5)}\}(l+n)}.$$

The values of (l, m, n) are

$$l = \tan\tfrac{1}{4}(b-c), \quad m = \tan\tfrac{1}{4}(c-a), \quad n = \tan\tfrac{1}{4}(a-b),$$

and those of $(\mathrm{x}, \mathrm{y}, \mathrm{z})$, $(\mathrm{x}_1, \mathrm{y}_1, \mathrm{z}_1)$ are

$$\begin{aligned} \mathrm{x} &= \tan\tfrac{1}{4}(b+c-2\gamma), \quad \mathrm{y} = \tan\tfrac{1}{4}(c+a-2\alpha), \quad \mathrm{z} = \tan\tfrac{1}{4}(a+b-2\beta), \\ \mathrm{x}_1 &= \tan\tfrac{1}{4}(b+c-2\beta), \quad \mathrm{y}_1 = \tan\tfrac{1}{4}(c+a-2\beta), \quad \mathrm{z}_1 = \tan\tfrac{1}{4}(a+b-2\alpha), \end{aligned}$$

and the foregoing result shows that the values of

$$\frac{\mathrm{x}}{l} \text{ or } \frac{l}{\mathrm{x}_1}; \quad \frac{\mathrm{y}}{m} \text{ or } \frac{m}{\mathrm{y}_1}; \quad \frac{\mathrm{z}}{n} \text{ or } \frac{n}{\mathrm{z}_1}$$

are

$$\frac{\sin\frac{1}{2}(a-b) + \frac{1}{2}\{1+\sqrt{(5)}\}\sin\frac{1}{2}(c-a)}{\sin\frac{1}{2}(a-b) - \frac{1}{2}\{1+\sqrt{(5)}\}\sin\frac{1}{2}(c-a)},$$

$$\frac{\sin\frac{1}{2}(b-c) + \frac{1}{2}\{1+\sqrt{(5)}\}\sin\frac{1}{2}(a-b)}{\sin\frac{1}{2}(b-c) - \frac{1}{2}\{1+\sqrt{(5)}\}\sin\frac{1}{2}(a-b)},$$

$$\frac{\sin\frac{1}{2}(c-a) + \frac{1}{2}\{1+\sqrt{(5)}\}\sin\frac{1}{2}(b-c)}{\sin\frac{1}{2}(c-a) - \frac{1}{2}\{1+\sqrt{(5)}\}\sin\frac{1}{2}(b-c)}.$$

Write for a moment $\frac{\mathrm{x}}{l} = \frac{l}{\mathrm{x}_1} = k$; therefore

$$\frac{\mathrm{x}-\mathrm{x}_1}{1+\mathrm{x}\mathrm{x}_1} = \frac{lk - \frac{l}{k}}{1+l^2} = \frac{l\left(k - \frac{1}{k}\right)}{1+l^2} = \frac{2l}{1+l^2} \div \frac{2k}{k^2-1} = \sin\tfrac{1}{2}(b-c) \cdot \frac{k^2-1}{2k}$$

$$= \sin\tfrac{1}{2}(b-c) \cdot \frac{P^2-Q^2}{2PQ} \text{ if } k = \frac{P}{Q};$$

and taking P, Q for the numerator and the denominator respectively of the foregoing fractional expression for k, we find

$$P^2 - Q^2 = 2\{1+\surd(5)\}\sin\tfrac{1}{2}(a-b)\sin\tfrac{1}{2}(c-a),$$

$$2PQ = 2\left[\sin^2\tfrac{1}{2}(a-b) - \tfrac{1}{4}\{1+\surd(5)\}^2\sin^2\tfrac{1}{2}(c-a)\right]$$

$$= -\{1+\surd(5)\}\left[\tfrac{1}{2}\{1-\surd(5)\}\sin^2\tfrac{1}{2}(a-b) + \tfrac{1}{2}\{1+\surd(5)\}\sin^2\tfrac{1}{2}(c-a)\right];$$

also

$$\frac{\mathrm{x}-\mathrm{x}_1}{1+\mathrm{x}\mathrm{x}_1} = \tan\tfrac{1}{2}(\beta-\gamma),$$

we have therefore

$$\tan\tfrac{1}{2}(\beta-\gamma) = \frac{-2\sin\tfrac{1}{2}(b-c)\sin\tfrac{1}{2}(c-a)\sin\tfrac{1}{2}(a-b)}{\tfrac{1}{2}\{1-\surd(5)\}\sin^2\tfrac{1}{2}(a-b) + \tfrac{1}{2}\{1+\surd(5)\}\sin^2\tfrac{1}{2}(c-a)},$$

$$\tan\tfrac{1}{2}(\gamma-\alpha) = \frac{-2\sin\tfrac{1}{2}(b-c)\sin\tfrac{1}{2}(c-a)\sin\tfrac{1}{2}(a-b)}{\tfrac{1}{2}\{1-\surd(5)\}\sin^2\tfrac{1}{2}(b-c) + \tfrac{1}{2}\{1+\surd(5)\}\sin^2\tfrac{1}{2}(a-b)},$$

$$\tan\tfrac{1}{2}(\alpha-\beta) = \frac{-2\sin\tfrac{1}{2}(b-c)\sin\tfrac{1}{2}(c-a)\sin\tfrac{1}{2}(a-b)}{\tfrac{1}{2}\{1-\surd(5)\}\sin^2\tfrac{1}{2}(c-a) + \tfrac{1}{2}\{1+\surd(5)\}\sin^2\tfrac{1}{2}(b-c)},$$

equations which determine the circumscribed triangle when the inscribed triangle is given.

The more general problem, for a conic and an inscribed and circumscribed triangle such that the sides of the inscribed triangle pass through the angles of the circumscribed triangle, "given one of the triangles to determine the other" is solved by Möbius, *Crelle*, t. v. (1830), p. 103, by means of his Barycentric Calculus, which is in fact the method of trilinear coordinates. The solution is in effect as follows:

Let $\xi=0$, $\eta=0$, $\zeta=0$ be the equations of the sides of the inscribed triangle, ξ, η, ζ may be considered as containing each of them an implicit constant factor, and the equation of the conic may be taken to be

$$\eta\zeta+\zeta\xi+\xi\eta=0,$$

moreover, if $x=0$, $y=0$, $z=0$ be the equations of the sides of the circumscribed triangle, then considering x, y, z as also containing each of them an implicit constant factor, the equation of the conic may be taken to be

$$x^2+y^2+z^2-2yz-2zx-2xy=0.$$

Suppose now the sides of the inscribed triangle pass through the angles of the circumscribed triangle, we have equations such as

$$\xi=b'y+cz,\quad \eta=c'z+ax,\quad \zeta=a'x+by;$$

substituting these values we must have identically

$$\eta\zeta+\zeta\xi+\xi\eta+m(x^2+y^2+z^2-2yz-2zx-2xy)=0,$$

or

$$aa'+m=0,\quad bc+bc'+b'c=2m,$$

$$bb'+m=0,\quad ca+ca'+c'a=2m,$$

$$cc'+m=0,\quad ab+ab'+a'b=2m,$$

and, substituting for a', b', c' their values,

$$(m-bc)^2=mb^2,\quad (m-ca)^2=mc^2,\quad (m-ab)^2=ma^2,$$

or, if $m=n^2$,

$$n^2-bc=nb,\quad n^2-ca=nc,\quad n^2-ab=na,$$

{the signs on the second side must be all + or all − and the − sign may be omitted without loss of generality}.

Hence

$$a=b=c=-\tfrac{1}{2}\{1+\surd(5)\}\,n,\quad \surd(5) \text{ being written for } \pm\surd(5),$$

$$a'=b'=c'=-\frac{n^2}{a}=-\tfrac{1}{2}\{1-\surd(5)\}\,n=-\nu a,$$

if for shortness

$$-\nu=\frac{1-\surd(5)}{1+\surd(5)},\ \text{ i.e. } \nu=\frac{\surd(5)-1}{\surd(5)+1},\ \text{ or } \nu^2-3\nu+1=0,$$

and ν having this value, the equations give

$$\xi=\nu y-z,\quad \eta=\nu z-x,\quad \zeta=\nu x-y,$$

whence also

$$\begin{aligned}4(2\nu-1)\,x&=\nu\xi+\eta+(3\nu-1)\,\zeta,\\ 4(2\nu-1)\,y&=(3\nu-1)\,\xi+\nu\eta+\zeta,\\ 4(2\nu-1)\,z&=\xi+(3\nu-1)\,\eta+\nu\zeta.\end{aligned}$$

Each side of the circumscribed triangle has on it four points, viz. two angles of the circumscribed triangle, a point of contact with the conic, and an intersection with the corresponding side of the inscribed triangle. Thus for the side $x=0$, the four points are given as the intersections of $x=0$, with

$$y=0,\quad z=0,\quad y-z=0,\quad \nu y+z=0,$$

and the anharmonic ratio of the four points is therefore a given quantity.(*)

Again, each side of the inscribed triangle has on it four points, viz. two angles of the inscribed triangle, a point of intersection with the tangent at the opposite angle of the inscribed conic, and a point of intersection with the corresponding side of the circumscribed triangle.

Thus for the side $\xi=0$, the four points are given as the points of intersection of $\xi=0$ with the lines

$$\eta=0,\quad \zeta=0,\quad \eta+\zeta=0,\quad \eta+(3\nu-1)\,\zeta=0,$$

and the anharmonic ratio of the four points is therefore a given quantity.(*)

If we draw tangents at the angles of the inscribed triangle, we have a new triangle, the sides of which are $\eta+\zeta=0$, $\zeta+\xi=0$, $\eta+\xi=0$, and joining the angles of this triangle with the points of contact of the opposite sides (i.e. the angles of the inscribed triangle), we have three lines $\eta - \zeta = 0$, $\zeta - \xi = 0$, $\xi - \eta = 0$ meeting in a point $\xi = \eta = \zeta$, which is obviously the same as the point $x = y = z$, which is the point of intersection of the lines joining the angles of the circumscribed triangle with the points of contact of the opposite sides.(*)

The coordinates of the points of contact of the sides of the circumscribed triangle are given by ($x = 0$, $y - z = 0$), ($y = 0$, $z - x = 0$), ($z = 0$, $x - y = 0$), these points form therefore an inscribed triangle the sides of which are

$$y + z - x = 0, \quad z + x - y = 0, \quad x + y - z = 0.$$

Again, the tangents at the angles of the inscribed triangle form a circumscribed triangle the sides of which are

$$\eta + \zeta = 0, \quad \zeta + \xi = 0, \quad \xi + \eta = 0;$$

therefore

$$(\zeta + \xi) - \nu(\xi + \eta) = (1 - \nu)\,\xi - \nu\eta + \zeta = 2\nu x\,(1 - \nu + \nu^2)\,y - (1 - \nu + \nu^2)\,z = -\,2\nu\,(y + z - x),$$

and we thus have

$$\begin{aligned} \zeta + \xi - \nu(\xi + \eta) &= -\,2\nu\,(y + z - x),\\ \xi + \eta - \nu(\eta + \zeta) &= -\,2\nu\,(z + x - y),\\ \eta + \zeta - \nu(\zeta + \xi) &= -\,2\nu\,(x + y - z), \end{aligned}$$

equations which show that the sides of the second inscribed triangle pass through the angles of the second circumscribed triangle, and that the two systems are consequently reciprocal.(*)

The four theorems marked (*) are all of them contained in the paper by Möbius.

283.

ON A THEOREM RELATING TO HOMOGRAPHIC FIGURES.

[From the *Quarterly Journal of Pure and Applied Mathematics*, vol. III. (1860), pp. 177—180.]

THE following theorem is not new, but I do not remember where to find it:

Given any two homographic figures; there exists in the first figure a point S, which is the focus of an (ellipse or hyperbola, say an) ellipse Σ; and in the second figure a point S', which is the focus of a (hyperbola or ellipse, say a) hyperbola Σ', such that the points S, S' and the conics Σ, Σ' correspond to each other, and the conics Σ, Σ' are so related that the foci and vertices of the one may be superimposed upon the vertices and foci of the other. Moreover the perpendiculars from S, S' upon corresponding tangents of Σ, Σ' will be equal.

Write

$$x : y : 1 = \alpha\xi + \beta\eta + \gamma : \alpha'\xi + \beta'\eta + \gamma' : \alpha''\xi + \beta''\eta + \gamma'',$$

then l, m, λ, μ may be so determined that $(x-l)+i(y-m)$ shall vanish with $(\xi-\lambda)+i(\eta-\mu)$, for if we write

$$\begin{aligned} J &= \alpha - l\alpha'', & J' &= \alpha' - m\alpha'', \\ K &= \beta - l\beta'', & K' &= \beta' - m\beta'', \\ L &= \gamma - l\gamma'', & L' &= \gamma' - m\gamma'', \end{aligned}$$

we have

$$(x-l)+i(y-m) = \frac{(J+iJ')\{(\xi-\lambda)+i(\eta-\mu)\}}{\alpha''\xi+\beta''\eta+\gamma''};$$

and therefore

$$\begin{aligned} K + iK' &= i(J+iJ'), \\ L + iL' &= (J+iJ')(\lambda+i\mu), \end{aligned}$$

that is

$$K = -J, \quad K' = J,$$

or

$$l\beta'' + m\alpha'' = \alpha' + \beta,$$
$$-l\alpha'' + m\beta'' = -\alpha + \beta',$$

whence also

$$l(\alpha''^2 + \beta''^2) = \alpha'\beta'' - \alpha''\beta' + \alpha\alpha'' + \beta\beta'',$$
$$m(\alpha''^2 + \beta''^2) = \alpha''\beta - \alpha\beta'' + \alpha\alpha' + \beta\beta',$$

which determine l, m; and then λ, μ are given by

$$\lambda + i\mu = \frac{L + iL'}{J + iJ'},$$

but more simply as follows, viz. remarking that if $c = \beta'\gamma'' - \beta''\gamma'$, &c., so that

$$\xi : \eta : 1 = ax + a'y + a'' : bx + b'y + b'' : cx + c'y + c'',$$

then we have

$$\lambda(c^2 + c'^2) = bc' - b'c + ac + a'c',$$
$$\mu(c^2 + c'^2) = ca' - c'a + bc + b'c';$$

the equations $x = l$, $y = m$ give the point S and $\xi = \lambda$, $\eta = \mu$ the point S'. The values of J, J' are

$$J = \frac{-c\alpha'' - c'\beta''}{\alpha''^2 + \beta''^2}, \quad J' = \frac{-c'\alpha'' + c\beta''}{\alpha''^2 + \beta''^2};$$

and therefore

$$J^2 + J'^2 = \frac{c^2 + c'^2}{\alpha''^2 + \beta''^2} = \frac{k^2}{K^2},$$

if

$$k = \sqrt{(c^2 + c'^2)}, \quad K = \sqrt{(\alpha''^2 + \beta''^2)}:$$

we have therefore

$$\sqrt{\{(x - l)^2 + (y - m)^2\}} = \frac{k}{K} \frac{\sqrt{\{(\xi - \lambda)^2 + (\eta - \mu)^2\}}}{\alpha''\xi + \beta''\eta + \gamma''}. \qquad (*)$$

Consider now the expression

$$(x - l)\cos\vartheta + (y - m)\sin\vartheta - \varpi,$$

which made equal to 0 would be the equation of a line the perpendicular distance of which from S is ϖ, and which distance is inclined at an angle ϑ to the axis of x; then

$$(x - l)\cos\vartheta + (y - m)\sin\vartheta = \tfrac{1}{2}[(\cos\vartheta - i\sin\vartheta)\{(x - l) + i(y - m)\}$$
$$+ (\cos\vartheta + i\sin\vartheta)\{(x - l) - i(y - m)\}],$$

$$= \frac{1}{2(\alpha''\xi + \beta''\eta + \gamma'')}[(\cos\vartheta - i\sin\vartheta)\{(\xi - \lambda) + i(\eta - \mu)\}(J + iJ')$$
$$+ (\cos\vartheta + i\sin\vartheta)\{(\xi - \lambda) - i(\eta - \mu)\}(J - iJ')]$$

$$= \frac{k}{2(\alpha''\xi + \beta''\eta + \gamma'')K} [\{\cos(\vartheta - \vartheta_0) - i\sin(\vartheta - \vartheta_0)\}\{(\xi - \lambda) + i(\eta - \mu)\}$$

$$+ \{\cos(\vartheta - \vartheta_0) + i\sin(\vartheta - \vartheta_0)\}\{(\xi - \lambda) - i(\eta - \mu)\}]$$

$$= \frac{k}{(\alpha''\xi + \beta''\eta + \gamma'')K} \{(\xi - \lambda)\cos(\vartheta - \vartheta_0) + (\eta - \mu)\sin(\vartheta - \vartheta_0)\},$$

where J, J' have been replaced by

$$J = \frac{k}{K}\cos\vartheta_0, \quad J' = \frac{k}{K}\sin\vartheta_0.$$

And putting besides

$$\vartheta - \vartheta_0 = \theta,$$

we have more simply

$$(x - l)\cos\vartheta + (y - m)\sin\vartheta = \frac{k}{(\alpha''\xi + \beta''\eta + \gamma'')K}\{(\xi - \lambda)\cos\theta + (\eta - \mu)\sin\theta\},$$

whence

$$(x - l)\cos\vartheta + (y - m)\sin\vartheta \quad - \varpi$$

$$= \frac{k}{(\alpha''\xi + \beta''\eta + \gamma'')K}\left\{(\xi - \lambda)\cos\theta + (\eta - \mu)\sin\theta - \frac{\varpi K}{k}(\alpha''\xi + \beta''\eta + \gamma'')\right\}$$

$$= \frac{k}{(\alpha''\xi + \beta''\eta + \gamma'')K}\left\{(\xi - \lambda)\left(\cos\theta - \frac{\varpi K\alpha''}{k}\right) + (\eta - \mu)\left(\sin\theta - \frac{\varpi K\beta''}{k}\right) - \frac{\varpi K}{k}(\alpha''\lambda + \beta''\mu + \gamma'')\right\}.$$

Write now

$$\alpha'' = K\cos j, \ \beta'' = K\sin j; \ \mathrm{a} = \frac{k}{K^2}, \ \mathrm{b} = \frac{1}{K}(\alpha''\lambda + \beta''\mu + \gamma'');$$

we have

$$(x - l)\cos\vartheta + (y - m)\sin\vartheta \quad - \varpi$$

$$= \frac{K\mathrm{b}}{\alpha''\xi + \beta''\eta + \gamma''}\left\{\frac{\mathrm{a}}{\mathrm{b}}\left(\cos\theta - \frac{\varpi\cos j}{\mathrm{a}}\right)(\xi - \lambda) + \frac{\mathrm{a}}{\mathrm{b}}\left(\sin\theta - \frac{\varpi\sin j}{\mathrm{a}}\right)(\eta - \mu) - \varpi\right\}$$

$$= \frac{K\mathrm{b}}{\alpha''\xi + \beta''\eta + \gamma''}\{(\xi - \lambda)\cos\phi + (\eta - \mu)\sin\phi - \varpi\},$$

that is

$$(x - l)\cos\vartheta + (y - m)\sin\vartheta \quad - \varpi \qquad (*)$$

$$= \frac{K\mathrm{b}}{\alpha''\xi + \beta''\eta + \gamma''}\{(\xi - \lambda)\cos\phi + (\eta - \mu)\sin\phi - \varpi\},$$

where it will be noticed that

$$(x - l)\cos\vartheta + (y - m)\sin\vartheta - \varpi = 0,$$

and

$$(\xi - \lambda)\cos\phi + (\eta - \mu)\sin\phi - \varpi = 0,$$

are lines in the first figure and in the second figure, corresponding to each other and at the same distance ϖ from the points S, S' respectively. Call these lines T and T'.

But the relations between θ, ϕ, ϖ are given by

$$\varpi \cos j = \mathrm{a} \cos \theta - \mathrm{b} \cos \phi,$$

$$\varpi \sin j = \mathrm{a} \sin \theta - \mathrm{b} \sin \phi,$$

or, by changing the fixed axes from which θ, ϕ are respectively measured,

$$\varpi = \mathrm{a} \cos \theta - \mathrm{b} \cos \phi,$$

$$0 = \mathrm{a} \sin \theta - \mathrm{b} \sin \phi.$$

Now in these equations ϖ is the perpendicular distance from S upon the line $(x-l)\cos\vartheta + (y-m)\sin\vartheta - \varpi = 0$ and θ (which only differs from ϑ by a constant angle) is the inclination of this perpendicular to a certain fixed line; ϖ is also the perpendicular distance of the line $(\xi-\lambda)\cos\phi + (\eta-\mu)\sin\phi - \varpi = 0$ from the point S', and ϕ is the inclination of this perpendicular to a fixed line. Eliminating successively ϕ and θ, we have

$$\varpi = \mathrm{a} \cos \theta - \sqrt{(\mathrm{b}^2 - \mathrm{a}^2 \sin^2 \theta)},$$

$$\varpi = -\mathrm{b} \cos \phi + \sqrt{(\mathrm{a}^2 - \mathrm{b}^2 \sin^2 \phi)},$$

or, as these equations may also be written,

$$\varpi^2 - 2\varpi \mathrm{a} \cos \theta + \mathrm{a}^2 - \mathrm{b}^2 = 0,$$

$$\varpi^2 + 2\varpi \mathrm{b} \cos \phi + \mathrm{b}^2 - \mathrm{a}^2 = 0.$$

Suppose $\mathrm{a} > \mathrm{b}$, the former equation shows that the line T is a tangent to a certain hyperbola, and the latter equation shows that the line T' is a tangent to a certain ellipse, and it is easily seen that, taking for the transverse axes the lines from which the angles θ and ϕ are respectively measured, the equation of the hyperbola is

$$\frac{x^2}{\mathrm{b}^2} - \frac{y^2}{\mathrm{a}^2 - \mathrm{b}^2} = 1,$$

and that of the ellipse is

$$\frac{\xi^2}{\mathrm{a}^2} + \frac{y^2}{\mathrm{a}^2 - \mathrm{b}^2} = 1,$$

which are the conics Σ, Σ' referred to in the enunciation. And if the second conic is superimposed upon the first in such manner that the coordinates ξ, η may belong to the same axes with x, y; then the two conics will have the assumed relation, viz. the foci of either conic will coincide with the vertices of the other conic.

284.

ON A NEW ANALYTICAL REPRESENTATION OF CURVES IN SPACE.

[From the *Quarterly Journal of Pure and Applied Mathematics*, vol. III. (1860), pp. 225—236.]

THE ordinary analytical representation of a curve in space by the equations of two surfaces passing through the curve is, even in the case where the curve is the complete intersection of the two surfaces, inappropriate as involving the consideration of surfaces which are extraneous to the curve; and the objection becomes more serious when the curve is not the complete intersection of any two surfaces; for in this case the curve can only be represented in conjunction with a curve or curves extraneous to itself. A curve in space can be in a mode represented by means of the equation of the developable surface having the curve for its edge of regression; but this corresponds to the representation of a plane curve by means of its equation in line coordinates; a representation which is very useful in addition to, but which is not to be substituted for, the equation in point coordinates. It occurred to me some years ago that it might be advantageous to represent a curve in space by means of the cone passing through the curve and having for its vertex an arbitrary point[1]; and although I have not advanced beyond the first steps of the theory, the results which I have obtained may, I think, be interesting to geometers. The conclusion is that a curve in space may be represented by a homogeneous equation $V=0$ between six coordinates (p, q, r, s, t, u) such that $ps+qt+ru=0$; the function V is moreover such that in virtue of this relation between the coordinates, and of the equation $V=0$ itself, we have

$$d_pV\,.\,d_sV+d_qV\,.\,d_tV+d_rV\,.\,d_uV=0,$$

[1] See my paper "On the cones which pass through a curve of the third order in space," *Phil. Mag.*, t. XII. (1856), p. 20, [200], where a curve of the third order in space is in effect so represented.

or what is the same thing we have identically

$$d_p V . d_s V + d_q V . d_t V + d_r V . d_u V = LV + M(ps + qt + ru),$$

where L and M are functions of p, q, r, s, t, u. But the converse proposition, viz. any equation whatever $V = 0$, where V satisfies the condition just referred to represents a curve in space, is not true; it would obviously be an important point in the theory to ascertain what further conditions must be satisfied by the function V.

The establishment of the foregoing results is very easy; in fact if x, y, z, w are current coordinates of the ordinary kind (point-coordinates) and $\alpha, \beta, \gamma, \delta$ the coordinates of an arbitrary point; the equation of any cone whatever having for its vertex the point $(\alpha, \beta, \gamma, \delta)$ may be represented by a homogeneous equation between the six determinants of the matrix

$$\left(\begin{array}{cccc} x, & y, & z, & w \\ \alpha, & \beta, & \gamma, & \delta \end{array}\right)$$

or if we write

$$\begin{aligned} p &= \gamma y - \beta z, & s &= \delta x - \alpha w, \\ q &= \alpha z - \gamma x, & t &= \delta y - \beta w, \\ r &= \beta x - \alpha y, & u &= \delta z - \gamma w, \end{aligned}$$

values which it is well known give identically

$$ps + qt + ru = 0,$$

then the cone will be represented by a homogeneous equation

$$V = 0$$

between the six coordinates (p, q, r, s, t, u). It remains to find the conditions in order that all the cones so represented, viz. the cones obtained by giving any values whatever to the arbitrary quantities $\alpha, \beta, \gamma, \delta$ which enter implicitly into the coordinates p, q, r, s, t, u, pass through one and the same curve; for when this is the case, the equation $V = 0$ may be properly considered as the equation of the curve.

Assume then that all the cones pass through the same curve; if we give to one of the arbitrary quantities $\alpha, \beta, \gamma, \delta$, say α, the infinitesimal variation $d\alpha$, then the function V becomes $V + d_\alpha V . d\alpha$, and each of the equations $V = 0$, $V + d_\alpha V . d\alpha = 0$ belongs to a cone passing through the curve; the equation $d_\alpha V = 0$ is therefore the equation of a surface passing through the curve; and in like manner the four equations

$$d_\alpha V = 0, \quad d_\beta V = 0, \quad d_\gamma V = 0, \quad d_\delta V = 0$$

are each of them the equation of a surface passing through the curve, or these equations must be simultaneously satisfied for all the points of the curve, they must consequently reduce themselves to two independent relations. But V is given as a function of p, q, r, s, t, u, through which quantities it is a function of $\alpha, \beta, \gamma, \delta$; the last-mentioned four equations become therefore

$$\begin{array}{rl} 0 = & \quad . \qquad - d_r V . y + d_q V . z - d_s V . w, \\ 0 = & d_r V . x \qquad . \qquad - d_p V . z - d_t V . w, \\ 0 = & - d_q V . x + d_p V . y \qquad . \qquad - d_u V . w, \\ 0 = & d_s V . x + d_t V . y + d_u V . z \qquad . \quad , \end{array}$$

and from the first, second, and third equation, or any other combination of three equations, we obtain at once the condition

$$d_p V . d_s V + d_q V . d_t V + d_r V . d_u V = 0,$$

so that if this condition is satisfied, the four equations do in fact reduce themselves to two independent equations; the condition in question is thus shown to be necessary. But the condition only implies that all the cones having their vertices in the neighbourhood of the point $(\alpha, \beta, \gamma, \delta)$ pass through one and the same curve; this will be the case for instance for a series of cones all of them circumscribed about one and the same surface, those having their vertices in the neighbourhood of the point $(\alpha, \beta, \gamma, \delta)$ will all pass through the curve of contact with the surface of the cone having for its vertex the point in question. But the curve of contact is not a fixed curve for all positions of the vertex, and the condition before referred to is consequently insufficient.

It may be noticed that the systems p, q, r and s, t, u are not similar to each other and that the six coordinates cannot be in any way divided into two systems which are similar to each other: the symmetry of the coordinates is in fact that of the vertices (or sides) of a complete quadrilateral (or quadrangle); thus we may divide the coordinates into two sets in a fourfold manner as follows:

$$\begin{array}{l} u, t, p; \; r, q, s, \\ s, q, u; \; p, t, r, \\ r, t, s; \; u, q, p, \\ p, q, r; \; s, t, u, \end{array}$$

where each left-hand set corresponds to three vertices forming a triangle and each right-hand set to the remaining three vertices *in lineo*. It may be noticed also that if in the equation $V = 0$ of any curve in space we substitute for p, q, r, s, t, u, their values, and equate to zero the coefficients of the different powers and products of $\alpha, \beta, \gamma, \delta$, each of the equations so obtained will belong to a surface passing through the curve, and the entire system of these equations will be equivalent to two relations only between the coordinates x, y, z, w. But any two of these surfaces will not in general intersect only in the curve, i.e. the curve will not be the complete intersection of any two of the surfaces. It may be added that the equation of any other surface whatever through the curve will be obtained by equating to zero a syzygetic function of the functions which equated to zero give the surfaces first referred to.

As an example of the theory, the equation, in the new coordinates, of a line in space will be

$$Ap + Bq + Cr + Fs + Gt + Hu = 0,$$

where the constants are such that

$$AF + BG + CH = 0.$$

This may be verified as follows: suppose that the line in question is given as the line of junction of the points $(\alpha', \beta', \gamma', \delta')$ and $(\alpha'', \beta'', \gamma'', \delta'')$; then the cone through the line and the arbitrary point $(\alpha, \beta, \gamma, \delta)$ is nothing else than the plane through the three points; its equation is therefore

$$\begin{vmatrix} x, & y, & z, & w \\ \alpha, & \beta, & \gamma, & \delta \\ \alpha', & \beta', & \gamma', & \delta' \\ \alpha'', & \beta'', & \gamma'', & \delta'' \end{vmatrix} = 0,$$

which may be written

$$Ap + Bq + Cr + Fs + Gt + Hu = 0,$$

the values of A, &c. being

$$\begin{aligned} A &= \alpha'\delta'' - \alpha''\delta', & F &= \beta'\gamma'' - \beta''\gamma', \\ B &= \beta'\delta'' - \beta''\delta', & G &= \gamma'\alpha'' - \gamma''\alpha', \\ C &= \gamma'\delta'' - \gamma''\delta', & H &= \alpha'\beta'' - \alpha''\beta', \end{aligned}$$

which in fact satisfy the relation

$$AF + BG + CH = 0.$$

So again if the line is given as the intersection of the planes

$$\begin{aligned} ax + by + cz + dw &= 0, \\ a'x + b'y + c'z + d'w &= 0, \end{aligned}$$

then the equation of the cone (plane) through this line and the arbitrary point $(\alpha, \beta, \gamma, \delta)$ is

$$(ax + by + cz + dw)(a'\alpha + b'\beta + c'\gamma + d'\delta) - (a'x + b'y + c'z + d'w)(a\alpha + b\beta + c\gamma + d\delta) = 0,$$

which is

$$Ap + Bq + Cr + Fs + Gt + Hu = 0,$$

the values of A, &c. being

$$\begin{aligned} A &= bc' - b'c, & F &= ad' - a'd, \\ B &= ca' - c'a, & G &= bd' - b'd, \\ C &= ab' - a'b, & H &= cd' - c'd, \end{aligned}$$

which also satisfy the relation

$$AF + BG + CH = 0.$$

I annex the following further investigation: let p', q', r', s', t', u' and p'', q'', r'', s'', t'', u'' be what p, q, r, s, t, u become when α, β, γ, δ are changed into α', β', γ', δ' and α'', β'', γ'', δ'' respectively: the line represented by the equation

$$Ap + Bq + Cr + Fs + Gt + Hu = 0,$$

is also represented by the equations

$$\begin{aligned} Ap' + Bq' + Cr' + Fs' + Gt' + Hu' &= 0, \\ Ap'' + Bq'' + Cr'' + Fs'' + Gt'' + Hu'' &= 0, \end{aligned}$$

which, reverting to the coordinates x, y, z, w, may be written

$$\begin{aligned} \mathfrak{A}'x + \mathfrak{B}'y + \mathfrak{C}'z + \mathfrak{D}'w &= 0, \\ \mathfrak{A}''x + \mathfrak{B}''y + \mathfrak{C}''z + \mathfrak{D}''w &= 0, \end{aligned}$$

where

$$\left.\begin{array}{llll} \mathfrak{A}' = & . & CB' - B\gamma' + F\delta', \\ \mathfrak{B}' = -C\alpha' & . & + A\gamma' + G\delta', \\ \mathfrak{C}' = & B\alpha' - A\beta' & . \; + H\delta', \\ \mathfrak{D}' = -F\alpha' - G\beta' - H\gamma' & . & , \end{array}\right| \quad \begin{array}{llll} \mathfrak{A}'' = & . & C\beta'' - B\gamma'' + F\delta'', \\ \mathfrak{B}'' = -C\alpha'' & . & + A\gamma'' + G\delta'', \\ \mathfrak{C}'' = & B\alpha'' - A\beta'' & . \; + H\delta'', \\ \mathfrak{D}'' = -F\alpha'' - G\beta'' - H\gamma'' & . & , \end{array}$$

in which form the equations represent two planes each of them through the given line: the equation of the cone (plane) through the given line and the arbitrary point $(\alpha, \beta, \gamma, \delta)$ is

$$\begin{aligned} &(\mathfrak{A}'x + \mathfrak{B}'y + \mathfrak{C}'z + \mathfrak{D}'w)(\mathfrak{A}''\alpha + \mathfrak{B}''\beta + \mathfrak{C}''\gamma + \mathfrak{D}''\delta) \\ &\qquad - (\mathfrak{A}'\alpha + \mathfrak{B}'\beta + \mathfrak{C}'\gamma + \mathfrak{D}'\delta)(\mathfrak{A}''x + \mathfrak{B}''y + \mathfrak{C}''z + \mathfrak{D}''w) = 0, \end{aligned}$$

or, developing,

$$\begin{aligned} &(\mathfrak{B}'\mathfrak{C}'' - \mathfrak{B}''\mathfrak{C}')\,p + (\mathfrak{C}'\mathfrak{A}'' - \mathfrak{C}''\mathfrak{A}')\,q + (\mathfrak{A}'\mathfrak{B}'' - \mathfrak{A}''\mathfrak{B}')\,r \\ &\qquad + (\mathfrak{A}'\mathfrak{D}'' - \mathfrak{A}''\mathfrak{D}')\,s + (\mathfrak{B}'\mathfrak{D}'' - \mathfrak{B}''\mathfrak{D}')\,t + (\mathfrak{C}'\mathfrak{D}'' - \mathfrak{C}''\mathfrak{D}')\,u = 0. \end{aligned}$$

Now

$$\begin{aligned} \mathfrak{B}'\mathfrak{C}'' - \mathfrak{B}''\mathfrak{C}' = &(-C\alpha' + A\gamma' + G\delta')(B\alpha'' - A\beta'' + H\delta'') \\ &- (-C\alpha'' + A\gamma'' + G\delta'')(B\alpha' - A\beta' + H\delta'), \end{aligned}$$

which putting

$$\begin{aligned} \beta'\gamma'' - \beta''\gamma' &= p, & \alpha'\delta'' - \alpha''\delta' &= s, \\ \gamma'\alpha'' - \gamma''\alpha' &= q, & \beta'\delta'' - \beta''\delta' &= t, \\ \alpha'\beta'' - \alpha''\beta' &= r, & \gamma'\delta'' - \gamma''\delta' &= u, \end{aligned}$$

become

$$A^2p_1 + ABq_1 + ACr_1 - (BG + CH)\,s_1 + AGt_1 + AHu_1,$$

or as it may be written

$$A\,(Ap_1 + Bq_1 + Cr_1 + Fs_1 + Gt_1 + Hu_1) - (AF + BG + CH)\,s_1.$$

Instead of writing at once $AF + BG + CH = 0$, I denote it by ∇ and I write also

$$Ap_1 + Bq_1 + Cr_1 + Fs_1 + Gt_1 + Hu_1 = \Omega.$$

The series of coefficients $\mathfrak{B}'\mathfrak{C}'' - \mathfrak{B}''\mathfrak{C}''$, &c. is

$$A\Omega - \nabla s_1,\quad B\Omega - \nabla t_1,\quad C\Omega - \nabla u_1,$$
$$F\Omega - \nabla p_1,\quad G\Omega - \nabla q_1,\quad H\Omega - \nabla r_1,$$

and the required equation, restoring for Ω, ∇ their values, is

$$(Ap + Bq + Cr + Fs + Gt + Hu)(Ap_1 + Bq_1 + Cr_1 + Fs_1 + Gt_1 + H_1)$$
$$- (AF + BG + CH)(ps_1 + p_1 s + qt_1 + q_1 t + ru_1 + r_1 u) = 0,$$

which in virtue of $AF + BG + CH = 0$ becomes

$$(Ap + Bq + Cr + Fs + Gt + Hu)(Ap_1 + Bq_1 + Cr_1 + Fs_1 + Gt_1 + Hu_1) = 0.$$

The equation

$$Ap_1 + Bq_1 + Cr_1 + Fs_1 + Gt_1 + Hu_1 = 0$$

would imply that the two points $(\alpha', \beta', \gamma', \delta')$, $(\alpha'', \beta'', \gamma'', \delta'')$ are in the same plane with the line

$$Ap + Bq + Cr + Fs + Gt + Hu = 0,$$

(or, what is the same thing, that this line is intersected by the line through the two points); in this exceptional case, the planes determined by the given line and the two points respectively are one and the same plane, and they do not by their intersection determine the given line. But in every other case the factor $Ap_1 + Bq_1 + Cr_1 + Fs_1 + Gt_1 + Hu_1$ is not equal to zero, and the foregoing equation becomes

$$Ap + Bq + Cr + Fs + Gt + Hu = 0,$$

which is as it should be.

In what precedes it has been shown that the equation of the line through the points $(\alpha', \beta', \gamma', \delta')$ and $(\alpha'', \beta'', \gamma'', \delta'')$ is

$$Ap + Bq + Cr + Fs + Gt + Hu = 0,$$

where

$$A = \alpha'\delta'' - \alpha''\delta',\quad F = \beta'\gamma'' - \beta''\gamma',$$
$$B = \beta'\delta'' - \beta''\delta',\quad G = \gamma'\alpha'' - \gamma''\alpha',$$
$$C = \gamma'\delta'' - \gamma''\delta',\quad H = \alpha'\beta'' - \alpha''\beta',$$

and that the equation of the line of intersection of the planes $ax + by + cz + dw = 0$, $a'x + b'y + c'z + d'w = 0$ is

$$Ap + Bq + Cr + Fs + Gt + Hu = 0,$$

where

$$A = bc' - b'c,\quad F = ad' - a'd,$$
$$B = ca' - c'a,\quad G = bd' - b'd,$$
$$C = ab' - a'b,\quad H = cd' - c'd.$$

It may be added that the condition of intersection of the two lines

$$Ap + Bq + Cr + Fs + Gt + Hu = 0,$$

$$A'p + B'q + C'r + F's + G't + H'u = 0,$$

is

$$AF' + A'F + BG' + B'G + CH' + C'H = 0,$$

and any other problems in relation to the line, for instance to find the condition that a line may pass through a given point or lie in a given plane, &c. may also be solved by means of the new coordinates.

A curve of the second order in space is either a pair of lines or a plane conic, each of which species may degenerate into a pair of intersecting lines. If we write

$$W = Ap + Bq + Cr + Fs + Gt + Hu,$$

$$W' = A'p + B'q + C'r + F's + G't + H'u,$$

then the equation of the pair of lines will be $V = WW' = 0$, and it is worth while to show that this value of V satisfies the fundamental equation $d_pV . d_sV + d_qV . d_tV + d_rV . d_uV = 0$; we have in fact $d_pV = AW' + A'W$, &c. and thence the left-hand side is

$$W'^2(AF + BG + CH) + WW'(AF' + A'F + BG' + B'G + CH' + C'H) + W^2(A'F' + B'G' + G'H'),$$

which in fact vanishes in virtue of the relations

$$AF + BG + CH = 0, \quad A'F' + B'G' + C'H' = 0, \quad WW' = 0.$$

The like proof applies to any curve made up of two or more curves.

Consider in the next place the plane conic given by the equations

$$ax + by + cz + dw = 0,$$

$$x^2 + y^2 + z^2 + w^2 = 0.$$

To find the equation of the cone having for its vertex the point $(\alpha, \beta, \gamma, \delta)$ and passing through the conic we must, according to Joachimsthal's method, substitute in the two equations for x, y, z, w the values $\lambda x + \mu\alpha$, $\lambda y + \mu\beta$, $\lambda z + \mu\gamma$, $\lambda w + \mu\delta$, and from the resulting equations eliminate λ, μ. The two equations become

$$\lambda^2(x^2 + y^2 + z^2 + w^2) + 2\lambda\mu(\alpha x + \beta y + \gamma z + \delta w) + \mu^2(\alpha^2 + \beta^2 + \gamma^2 + \delta^2) = 0,$$

$$\lambda\ (ax + by + cz + dw) + \mu\ (\alpha x + \beta y + \gamma z + \delta w) = 0,$$

and thence eliminating λ, μ we find

$$\begin{aligned}&(a\alpha + b\beta + c\gamma + d\delta)^2(x^2 + y^2 + z^2 + w^2)\\ -\,&2(a\alpha + b\beta + c\gamma + d\delta)(ax + by + cz + dw)(\alpha x + \beta y + \gamma z + \delta w)\\ +\,&(ax + by + cz + dw)^2(\alpha^2 + \beta^2 + \gamma^2 + \delta^2) = 0,\end{aligned}$$

where the left-hand side is as it should be a function of p, q, r, s, t, u; the equation may in fact be written

$$\left.\begin{array}{l} a^2(\,.\quad q^2+r^2+s^2) \\ +\ b^2(p^2\quad .\ +r^2+t^2) \\ +\ c^2(p^2+q^2\quad .\ +u^2) \\ +\ d^2(s^2+t^2+u^2\quad .\,) \\ +2bc\,(-qr+tu) \\ +2ca\,(-rp+us) \\ +2ab\,(-pq+st) \\ +2ad\,(\ qu-rt) \\ +2bd\,(\ rs-pu) \\ +2cd\,(\ pt-qs) \end{array}\right\}=0,$$

which treated as a quadric in (a, b, c, d) may be written

$$\left(\begin{array}{c|c|c|c} q^2+r^2+s^2 & -pq+st & -rp+us & qu-rt \\ \hline -pq+st & p^2+r^2+t^2 & -qr+tu & rs-pu \\ \hline -rp+us & -qr+tu & p^2+q^2+u^2 & pt-qs \\ \hline qu-rt & rs-pu & pt-qs & s^2+t^2+u^2 \end{array}\right)(a, b, c, d)^2=0,$$

or treated as a quadric in (p, q, r, s, t, u), in the form

$$\left(\begin{array}{c|c|c|c|c|c} b^2+c^2 & -ab & -ac & . & cd & -bd \\ \hline -ba & c^2+a^2 & -bc & -cd & . & ad \\ \hline -ca & -cb & a^2+b^2 & bd & -ad & . \\ \hline . & -cd & bd & a^2+d^2 & ab & ac \\ \hline cd & . & -ad & ba & b^2+d^2 & bc \\ \hline -bd & ad & . & ca & cb & c^2+d^2 \end{array}\right)(p, q, r, s, t, u)^2=0.$$

Or again, in a form which is one of a system of four forms,

$$(b^2+c^2,\ c^2+a^2,\ a^2+b^2,\ -bc,\ -ca,\ -ab)(p,\ q,\ r)^2+(a^2+d^2,\ b^2+d^2,\ c^2+d^2,\ bc,\ ca,\ ab)(s,\ t,\ u)^2$$

$$+2\left\{\begin{array}{c|c|c|c} & s & t & u \\ \hline p & . & cd & -bd \\ \hline q & -cd & . & ad \\ \hline r & bd & -ad & . \end{array}\right\}=0.$$

Representing the equation by $V=0$, the function V should verify the fundamental equation $d_pV \,.\, d_sV + d_qV \,.\, d_tV + d_rV \,.\, d_sV = 0$, and we in fact have

$$d_pV = (b^2+c^2)\,p - abq - acr + \qquad cdt - bdu,$$

$$d_sV = \qquad - cdq + bdr + (a^2+d^2)\,s + abt + acu,$$

and thence forming the product $d_pV \,.\, d_sV$, and from it the two analogous products, we find

	$d_pV \,.\, d_sV =$	$d_qV \,.\, d_tV =$	$d_rV \,.\, d_uV =$
p^2	0	$- abcd$	$+ abcd$
q^2	$+ abcd$	0	$- abcd$
r^2	$- abcd$	$+ abcd$	0
s^2	0	$- abcd$	$+ abcd$
t^2	$+ abcd$	0	$- abcd$
u^2	$- abcd$	$+ abcd$	0
pq	$- cd\,(b^2+c^2)$	$+ cd\,(c^2+a^2)$	$- cd\,(a^2-b^2)$
qr	$- ad\,(b^2-c^2)$	$- ad\,(c^2+a^2)$	$+ ad\,(a^2+b^2)$
rp	$+ bd\,(b^2+c^2)$	$- bd\,(c^2-a^2)$	$- bd\,(a^2+b^2)$
su	$- bd\,(a^2+d^2)$	$+ bd\,(c^2+d^2)$	$- bd\,(c^2-a^2)$
ut	$- ad\,(b^2-c^2)$	$- ad\,(c^2+d^2)$	$+ ad\,(b^2+d^2)$
ts	$+ cd\,(a^2+d^2)$	$- cd\,(a^2-d^2)$	$- cd\,(b^2+d^2)$
ps	$+ (a^2+d^2)\,(b^2+c^2)$	$- a^2b^2 - c^2d^2$	$- a^2c^2 - b^2d^2$
pt	$+ ab\,(b^2+c^2)$	$- ab\,(b^2+d^2)$	$- ab\,(c^2-d^2)$
pu	$+ ac\,(b^2+c^2)$	$- ac\,(b^2-d^2)$	$- ac\,(c^2+d^2)$
qs	$- ab\,(a^2+d^2)$	$+ ab\,(c^2+a^2)$	$- ab\,(c^2-d^2)$
qt	$- a^2b^2 - c^2d^2$	$+ (b^2+d^2)\,(c^2+a^2)$	$- b^2c^2 - a^2d^2$
qu	$- bc\,(a^2-d^2)$	$+ bc\,(c^2+a^2)$	$- bc\,(c^2+d^2)$
rs	$- ac\,(a^2+d^2)$	$- ac\,(b^2-d^2)$	$+ ac\,(a^2+b^2)$
rt	$- bc\,(a^2-d^2)$	$- bc\,(b^2+d^2)$	$+ bc\,(a^2+b^2)$
ru	$- a^2c^2 - b^2d^2$	$- b^2c^2 - a^2d^2$	$+ (c^2+d^2)(a^2+b^2)$

by which the equation is verified.

I conclude with some remarks relating to a different part of the subject. It is shown by M. E. de Jonquières in his "Essai sur la generation des courbes geometriques, &c." *Mem. Prés. à l'Acad. de Paris*, t. XVI. (1858), that the equation $U=0$ of any geometrical plane curve can be presented in a variety of different ways in the form

$$\begin{vmatrix} P, & Q \\ P', & Q' \end{vmatrix} = 0,$$

this being in fact the fundamental theorem of his very beautiful investigations. I am not aware that it has ever been considered whether the equation $U=0$ of a plane curve can in general be represented in the form

$$\begin{vmatrix} P, & Q, & R \\ P', & Q', & R' \\ P'', & Q'', & R'' \end{vmatrix} = 0,$$

or in the analogous forms where the left-hand side is a determinant of a higher order. As regards surfaces, the equation $U=0$ of a geometrical surface cannot in general be presented in the form

$$\begin{vmatrix} P, & Q \\ P', & Q' \end{vmatrix} = 0,$$

nor in the similar forms where the left-hand side is a determinant of a higher order. We may consequently classify surfaces of a given order according to the forms of this nature by which they can be represented, or as I propose to term it, according to their "frangibility." It is obvious that this question is immediately connected with that of the representation of curves in space by means of the ordinary coordinates of analytical geometry; for instance if we have a surface $U=0$, which can be represented in the form

$$\begin{vmatrix} P, & Q, & R \\ P', & Q', & R' \\ P'', & Q'', & R'' \end{vmatrix} = 0,$$

then we can describe upon the surface curves such as

$$\begin{Vmatrix} P', & Q', & R' \\ P'', & Q'', & R'' \end{Vmatrix} = 0,$$

viz. the curve so represented is the curve which in conjunction with the curve $(R'=0,\ R''=0)$ makes up the complete intersection of the two surfaces $P'R''-P''R'=0$, $Q'R''-Q''R'=0$: the curve in question is not in general the complete intersection of any two surfaces. If the surface $U=0$ can be represented in the form $\begin{vmatrix} P, & Q \\ P', & Q' \end{vmatrix} = 0$, then we can describe upon the surface, curves such as the curve $(P=0,\ Q=0)$ which, although it is the complete intersection of two surfaces, is not the complete intersection of the given surface $U=0$ by any other surface. But if the surface $U=0$ cannot be represented in any such form, or as we may express it, if the surface is infrangible, then it would appear that the only curves which can be described upon the surface are those which are the complete intersection of the given surface by some other surface. The question is, I think, an interesting one in the theory of surfaces, but I doubt whether much will be done in this manner as regards the theory of curves in space, and it appears to me that there is more to be hoped for from the theory previously explained in the present paper.

2, *Stone Buildings, W.C., June* 2, 1859.

285.

ON THE SYSTEM OF CONICS HAVING DOUBLE CONTACT WITH EACH OTHER.

[From the *Quarterly Journal of Pure and Applied Mathematics*, vol. III. (1860), pp. 246—250.]

CONSIDER the conics which pass through four points coinciding two and two together; the two points of each pair of coincident points are to be regarded as lying in a line which will be a tangent to the conic, and the system is thus a system of conics touching the same two lines at the same two points: or if we replace the two lines by a conic of the system, it is the system of conics having double contact with a given conic at the same two points. The lines may be spoken of as the tangents, and the points as the ineunts of the system; the line joining the two points is the axis, and the point of intersection of the two lines, the pole. It is assumed that the system contains real conics; the pole and axis must consequently be real, but the tangents and ineunts may be either all of them real or else all of them imaginary—we may for shortness say that in the former case the double contact is real, and that in the latter case it is imaginary. Consider first the case of a real double contact, the system of conics commences with an ellipse differing only slightly from the finite portion of the axis included between the ineunts, this ellipse being the first of a series of ellipses the last of which only slightly differs from a parabola, we have then a parabola, then a hyperbola differing only slightly from the parabola, and which is the first of a series of hyperbolas of which the last differs only slightly from the two tangents: all the before-mentioned conics are included within the angles at the pole or vertex of the triangle, viz. the ellipses and parabola within the angle towards the axis or base, and the hyperbolas, one branch of each of them within this angle and the other branch within the opposite angle. And it will be convenient to term these conics the "Multiform Series." Passing over the intermediate case, we come to a hyperbola differing only slightly from the two tangents, the first of a series of hyperbolas of which the last differs only slightly from the infinite portions, not included between the ineunts, of the axis; all these hyperbolas lying without the angles at the pole or vertex of the triangle. The last-mentioned hyperbolas may be termed the "Uniform Series." Passing over another intermediate case we return to the multiform series, the entire system of conics forming in fact a complete cycle. The intermediate case forming the transition from the multiform to the uniform series is a conic which considered as generated by a point is the pair of tangents, but which considered as enveloped by a line is a pair of points coincident with the pole. The intermediate case forming the transition from the uniform series to the multiform series is a conic which considered as generated by a point is a pair of lines coincident with the axis, but considered as enveloped by a line is the pair of ineunts.

When the double contact is imaginary, the conics of the system may still be considered as forming a multiform and a uniform series, but the uniform series consists entirely of imaginary conics. The multiform series commences with an ellipse differing only slightly from the pole; this is the first of a series of ellipses of which the last differs only slightly from a parabola, we have next the parabola, and then a series of hyperbolas, the first of which differs only slightly from the parabola, and the last of which differs only slightly from the axis. The ellipses and parabolas lie all of them on the same side of the axis; the hyperbolas lie, one branch of each on the one side of the axis, and the other branch on the other side. The uniform series (being imaginary) of course does not admit of description. The intermediate or transition cases are the same as for a real double contact, with only the variation occasioned by the tangents and ineunts being imaginary. The conics of the entire system are considered (as in the case of a real double contact) to form a cycle, but here part of the cycle, viz. the uniform series, is imaginary.

Suppose that M and U stand for the multiform and uniform series, MU for the transition form from the multiform to the uniform series, and UM for the transition form from the uniform to the multiform series. The cycle may be represented as in the figure.

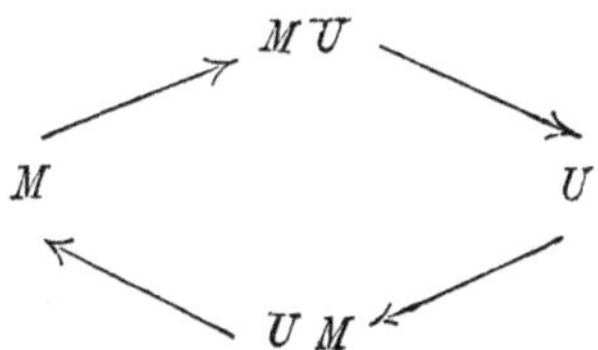

Call either of the conics MU, UM the centre of the cycle, and say that the cycle is arranged line-wise when MU is considered as the centre, and point-wise when UM is considered as the centre. We may pass from one conic of the cycle to another incentrically or excentrically, i.e. by passing towards the centre or away from the centre. The intermediates of two conics are the conics passed through in going from the first to the second incentrically, the extramediates are those passed through in going from the first to the second excentrically. In speaking of the extramediates or the intermediates of a single conic, such conic is considered as the first conic and the second conic is understood to be the centre. The nearer extramediates are those passed through previous to reaching an extremity of the cycle, the further extramediates are those passed through subsequent to reaching an extremity of the cycle. It is clear that the intermediates and the nearer extramediates make up the series multiform or uniform to which the conic belongs, and that the further extramediates are the other series; and moreover that if the cycle be arranged line-wise and point-wise successively, the intermediates of the one arrangement are the nearer extramediates of the other arrangement and *vice versâ*; the further extramediates of each arrangement being the same.

Two points of a conic may be said to be conjunctive with respect to a given line when it is possible to pass from the one to the other along the curve without crossing the line and disjunctive in the contrary case—it being understood that both the parabola and the hyperbola are to be treated as closed curves, viz. the points at infinity of the parabola are to be considered as one and the same point, and so the points at each extremity of either asymptote of the hyperbola are to be considered as one and the same point—and in like manner two tangents of a conic are said to be conjunctive with respect to a given point, when it is possible by the revolution of the one tangent to arrive at the other tangent without sweeping through the point in question, and disjunctive in the contrary case.

Consider now any conic of the series, and call this simply the *conic*. Take upon the conic any two points a, a' and joining these with a variable point in the axis, let the lines so obtained meet the conic in the points b', b (so that ab' and $a'b$ meet on the axis). We have thus on the conic a series of points $a, b, c, \ldots$ and a second series $a', b', c', \ldots$ which are homographically related to each other, and which possess the property that taking any two points b, c of the first series and the corresponding points b', c' of the second series, the lines bc', $b'c$ meet on the axis. It is immaterial how the point a is chosen, but a being chosen at pleasure, the system will obviously depend on the way in which the point a' is chosen; so that the points of a conic may be considered as homographically related to each other in an infinity of different ways. The reciprocal construction of course applies to the tangents of a conic, so that the tangents of a conic may be considered as homographically related to each other in an infinity of different ways: and not only this, but it is clear that the tangents at points homographically related to each other are also homographically related to each other, and *vice versâ*.

The line joining corresponding points envelopes a conic, one of the conics of the system, and which may for shortness be spoken of simply as the *envelope*. The point of intersection of corresponding tangents generates a conic, one of the conics of a system, and which may in like manner be spoken of simply as the *locus*.

We may now enunciate the following theorem: Let the system be arranged line-wise, the envelope is an extramediate of the conic; viz. if a pair of corresponding points of the conic (all pairs have in this respect the same property) be conjunctive with respect to the axis, a nearer extramediate, but if disjunctive, then a further extramediate.

And again:

Let the system be arranged point-wise, the locus is an extramediate of the conic; viz. if a pair of corresponding tangents of the conic (all pairs have in this respect the same property) be conjunctive with respect to the pole, a nearer extramediate, but if disjunctive, then a further extramediate.

It is however proper to remark that when the double contact is imaginary, the pair of corresponding points or tangents (in fact any pair of points or tangents) is necessarily conjunctive, so that in this case the second alternatives in the two theorems have no application, and the envelope or locus is always a nearer extramediate. When the double contact is real, the pair of corresponding points or tangents may be either conjunctive or disjunctive, and the character (nearer or further) of the extramediate is determined accordingly.

The double contact being either real or imaginary, we may as a limiting case assume that the corresponding points or tangents coincide, the envelope or locus is in this case the conic itself. Suppose that the double contact is real, we have here two other limiting cases, viz. first, one of a pair (and therefore of each pair) of corresponding points or tangents may be situate in the axis or pass through the pole; this is the limit between the two cases of the pair of corresponding points or tangents being conjunctive and disjunctive, and therefore the envelope or locus must be the conic which is the limit between the nearer extramediates and the further extramediates, i.e. the envelope is the pair of ineunts and the locus is the pair of tangents.

Secondly, the line through a pair of corresponding points may pass through the pole, or the point of intersection of a pair of corresponding tangents may lie in the axis; the envelope or locus is here the furthest of the extramediates, viz. the envelope is the pair of points coincident with the pole, the locus is the pair of lines coincident with the axis. If considering one of the pair of corresponding points or lines as fixed, the other point or line passes through the last-mentioned limiting position, the envelope or locus returns back through the series of further extramediates.

In the statement of the preceding theorems, the envelope and locus have been considered separately, but we may if we please consider the locus as generated by the point of intersection of the tangents at the corresponding points: the locus and envelope are in this case reciprocal polars with respect to the conic; it should be noticed that the pair of corresponding points and the pair of corresponding tangents are either both conjunctive, or else both disjunctive.

2, *Stone Buildings, W.C.*, 3*rd June*, 1859.

286.

NOTE ON THE VALUE OF CERTAIN DETERMINANTS, THE TERMS OF WHICH ARE THE SQUARED DISTANCES OF POINTS IN A PLANE OR IN SPACE.

[From the *Quarterly Journal of Pure and Applied Mathematics*, vol. III. (1860), pp. 275—277.]

THE values of the several determinants mentioned in my paper "On a Certain Theorem in the Geometry of Position," *Cambridge Mathematical Journal*, Old Series, t. II. (1842), p. 267, [1], are as follows:

(1)
$$\begin{vmatrix} 0, & \overline{12}^2, & \overline{13}^2, & 1 \\ \overline{21}^2, & 0, & \overline{23}^2, & 1 \\ \overline{31}^2, & \overline{32}^2, & 0, & 1 \\ 1, & 1, & 1, & 0 \end{vmatrix} = \Sigma\, \overline{12}^2\, \overline{21}^2 - \Sigma\, \overline{12}^2\, \overline{23}^2,$$

where the Σ, Σ contain 3 and 6 terms respectively.

(2)
$$\begin{vmatrix} 0, & \overline{12}^2, & \overline{13}^2, & \overline{14}^2, & 1 \\ \overline{21}^2, & 0, & \overline{23}^2, & \overline{24}^2, & 1 \\ \overline{31}^2, & \overline{32}^2, & 0, & \overline{34}^2, & 1 \\ \overline{41}^2, & \overline{42}^2, & \overline{43}^2, & 0, & 1 \\ 1, & 1, & 1, & 1, & 0 \end{vmatrix} = \Sigma\, \overline{12}^2\, \overline{23}^2\, \overline{34}^2 - \Sigma\, \overline{12}^2\, \overline{34}^2\, \overline{43}^2 - \Sigma\, \overline{12}^2\, \overline{23}^2\, \overline{31}^2,$$

where the Σ, Σ, Σ contain 24, 12 and 8 terms respectively.

(3)
$$\begin{vmatrix} 0, & \overline{12}^2, & \overline{13}^2, & \overline{14}^2, & \overline{15}^2, & 1 \\ \overline{21}^2, & 0, & \overline{23}^2, & \overline{24}^2, & \overline{25}^2, & 1 \\ \overline{31}^2, & \overline{32}^2, & 0, & \overline{34}^2, & \overline{35}^2, & 1 \\ \overline{41}^2, & \overline{42}^2, & \overline{43}^2, & 0, & \overline{45}^2, & 1 \\ \overline{31}^2, & \overline{52}^2, & \overline{53}^2, & \overline{54}^2, & 0, & 1 \\ 1, & 1, & 1, & 1, & 1, & 0 \end{vmatrix} = \begin{array}{l} -\Sigma\, \overline{12}^2\, \overline{23}^2\, \overline{34}^2\, \overline{45}^2 \\ -\Sigma\, \overline{12}^2\, \overline{21}^2\, \overline{34}^2\, \overline{43}^2 \\ +\Sigma\, \overline{12}^2\, \overline{23}^2\, \overline{45}^2\, \overline{54}^2 \\ +\Sigma\, \overline{12}^2\, \overline{23}^2\, \overline{34}^2\, \overline{41}^2 \\ +\Sigma\, \overline{12}^2\, \overline{23}^2\, \overline{31}^2\, \overline{45}^2 \end{array}$$

where the Σ, Σ, Σ, Σ, Σ contain 120, 15, 60, 30, and 40 terms respectively.

(4)
$$\begin{vmatrix} 0, & \overline{12}^2, & \overline{13}^2, & \overline{14}^2 \\ \overline{21}^2, & 0, & \overline{23}^2, & \overline{24}^2 \\ \overline{31}^2, & \overline{32}^2, & 0, & \overline{34}^2 \\ \overline{41}^2, & \overline{42}^2, & \overline{43}^2, & 0 \end{vmatrix} = \Sigma\, \overline{12}^2\, \overline{21}^2\, \overline{34}^2\, \overline{43}^2 - \Sigma\, \overline{12}^2\, \overline{23}^2\, \overline{34}^2\, \overline{41}^2,$$

where the Σ, Σ contain 3 and 6 terms respectively.

(5)
$$\begin{vmatrix} 0, & \overline{12}^2, & \overline{13}^2, & \overline{14}^2, & \overline{15}^2 \\ \overline{21}^2, & 0, & \overline{23}^2, & \overline{24}^2, & \overline{25}^2 \\ \overline{31}^2, & \overline{32}^2, & 0, & \overline{34}^2, & \overline{35}^2 \\ \overline{41}^2, & \overline{42}^2, & \overline{43}^2, & 0, & \overline{45}^2 \\ \overline{51}^2, & \overline{52}^2, & \overline{53}^2, & \overline{54}^2, & 0 \end{vmatrix} = \begin{array}{l} \Sigma\, \overline{12}^2\, \overline{23}^2\, \overline{34}^2\, \overline{45}^2\, \overline{51}^2 \\ -\Sigma\, \overline{12}^2\, \overline{23}^2\, \overline{31}^2\, \overline{45}^2\, \overline{54}^2 \end{array}$$

where the Σ, Σ contain 24 and 20 terms respectively.

And it is proper to remark that it is not in the preceding formulæ (as in the memoir above referred to in which $\overline{12}$ denotes a distance between two points 1 and 2) assumed that $\overline{12}$ and $\overline{21}$ are equal.

The formulæ (1) gives, if a, b, c represent the sides, the well-known expression for the area of a triangle

$$(\text{area})^2 = \tfrac{1}{16}(2b^2c^2 + 2c^2a^2 + 2a^2b^2 - a^4 - b^4 - c^4).$$

Similarly the formula (2), if a, b, c, f, g, h represent the edges, viz. $\overline{23} = a$, $\overline{31} = b$, $\overline{12} = c$, $\overline{14} = f$, $\overline{24} = g$, $\overline{34} = h$, gives for the volume of a tetrahedron

$$\begin{aligned}(\text{volume})^2 = \tfrac{1}{144}\{\; & b^2c^2(g^2+h^2) + c^2a^2(h^2+f^2) + a^2b^2(f^2+g^2) \\ & + g^2h^2(b^2+c^2) + h^2f^2(c^2+a^2) + f^2g^2(a^2+b^2) \\ & - a^2f^2(a^2+f^2) - b^2g^2(b^2+g^2) - c^2h^2(c^2+h^2) \\ & - a^2g^2h^2 \quad - b^2h^2f^2 \quad - c^2f^2g^2 \quad - a^2b^2c^2\}, \\ = \tfrac{1}{144}\, & W \text{ suppose.}\end{aligned}$$

Now

$$\begin{aligned}
4 \times \text{surface} \\
&= \sqrt{2h^2g^2 + 2g^2a^2 + 2a^2h^2 - a^4 - h^4 - g^4} \\
&+ \sqrt{2f^2h^2 + 2h^2b^2 + 2b^2f^2 - h^4 - b^4 - f^4} \\
&+ \sqrt{2g^2f^2 + 2f^2c^2 + 2c^2g^2 - g^4 - f^4 - c^4} \\
&+ \sqrt{2b^2c^2 + 2c^2a^2 + 2a^2b^2 - a^4 - b^4 - c^4}, \\
&= x + y + z + w \text{ suppose,}
\end{aligned}$$

and the norm of this is

$$(x^4 + y^4 + z^4 + w^4 - 2y^2z^2 - 2z^2x^2 - 2x^2y^2 - 2x^2w^2 - 2y^2w^2 - 2z^2w^2)^2 - 64x^2y^2z^2w^2,$$

which is of the sixteenth order, and must be of the form WQ where Q is a function of a, b, c, f, g, h of the tenth order. The expression of this function is given by Prof. Sylvester in his paper "On the Relation between the Volume of a Tetrahedron &c.," *Camb. and Dubl. Math. Jour.*, t. VIII. (1853), pp. 171—178, viz. the value is

$$\begin{aligned}
Q = \ & a^2b^2c^2\,\{f^4 + g^4 + h^4 + g^2h^2 + h^2f^2 + f^2g^2 + b^2c^2 + c^2a^2 + a^2b^2 - (f^2 + g^2 + h^2)(a^2 + b^2 + c^2)\} \\
& + a^2g^2h^2\,\{f^4 + b^4 + c^4 + b^2c^2 + c^2f^2 + f^2b^2 + g^2h^2 + h^2a^2 + a^2g^2 - (f^2 + b^2 + c^2)(a^2 + g^2 + h^2)\} \\
& + b^2h^2f^2\,\{g^4 + c^4 + a^4 + c^2a^2 + a^2g^2 + g^2c^2 + h^2f^2 + f^2b^2 + b^2h^2 - (g^2 + c^2 + a^2)(b^2 + h^2 + f^2)\} \\
& + c^2f^2g^2\,\{h^4 + a^4 + b^4 + a^2b^2 + b^2h^2 + h^2a^2 + f^2g^2 + g^2c^2 + c^2f^2 - (h^2 + a^2 + b^2)(c^2 + f^2 + g^2)\},
\end{aligned}$$

and, as there remarked, the equation $Q = 0$ expresses the condition that the radius of the inscribed sphere may be infinite.

2, *Stone Buildings, W.C., June* 10*th*, 1859.

287.

NOTE ON THE EQUATION FOR THE SQUARED DIFFERENCES OF THE ROOTS OF A CUBIC EQUATION.

[From the *Quarterly Journal of Pure and Applied Mathematics*, vol. III. (1860), pp. 307—309.]

THE question of finding the equation for the squared differences of the roots, presents, in the case of a cubic equation, a peculiarity which does not occur for equations of a higher order, viz. we may in the first instance form the equation for the differences of the roots taken in a given cyclical order, and thence deduce the equation for the squared differences of the roots. Let the cubic equation be

$$U = (a, b, c, d \between x, 1)^3 = a(x-\alpha)(x-\beta)(x-\gamma) = 0,$$

the function

$$\Pi \{\theta - (\beta - \gamma)\},$$

which equated to zero gives for θ the values $\beta-\gamma$, $\gamma-\alpha$, $\alpha-\beta$, which are the differences of the roots taken in a given circular order, has for any interchanges whatever of the roots, two values only, viz. that just written down, and the value $\Pi\{\theta-(\gamma-\beta)\}$, which may be deduced therefrom by changing first the sign of θ and then the sign of the entire expression (or what is the same thing, by changing the signs of the terms containing the even powers of θ); we may consequently write

$$\Pi\{\theta-(\beta-\gamma)\} = P - Q\,\zeta^{\frac{1}{2}}(\alpha, \beta, \gamma),$$

which P, Q are symmetrical functions of the roots, and $\zeta^{\frac{1}{2}}(\alpha, \beta, \gamma)$ or $(\alpha-\beta)(\beta-\gamma)(\gamma-\alpha)$ is a function the square of which is a symmetrical function of the roots, and such symmetrical functions of the roots can of course be expressed as functions of the coefficients. We have in fact

$$P = \theta^3 + \theta(\Sigma\beta\gamma - \Sigma\alpha^2) = a^{-2}\{a^2\theta^3 + 9(ac - b^2)\theta\},$$
$$Q = 1,$$

and

$$\zeta^{\frac{1}{2}}(\alpha, \beta, \gamma) = a^{-2}\sqrt{-27\Box},$$

where $\Box$ is the discriminant of the cubic function,

$$= a^2d^2 - 6abcd + 4ac^3 + 4b^3d - 3b^2c^2.$$

Consequently

$$\Pi\{\theta - (\beta - \gamma)\} = a^{-2}\{a^2\theta^3 + 9(ac - b^2)\theta - \sqrt{-27\Box}\},$$

and forming the similar equation

$$\Pi\{\theta + (\beta - \gamma)\} = a^{-2}\{a^2\theta^3 + 9(ac - b^2)\theta + \sqrt{-27\Box}\},$$

multiplying the two equations together and writing u in the place of θ^2, we find

$$\Pi\{u - (\beta - \gamma)^2\} = a^{-4}\{[a^2u + 9(ac - b^2)]^2 u + 27\Box\},$$

and the equation for the squared differences of the roots is thus seen to be

$$[a^2u + 9(ac - b^2)]^2 u + 27\Box = 0,$$

or what is the same thing

$$a^4u^3 + 18a^2(ac - b^2)u^2 + 81(ac - b^2)^2 u + 27\Box = 0.$$

I remark that if ω is an imaginary cube root of unity (so that $(\omega - \omega^2)^2 = -3$, $\omega - \omega^2$ being thus only another form of $\sqrt{-3}$) then if in the expression for $\Pi\{\theta - (\beta - \gamma)\}$ we write $\dfrac{3\theta}{(\omega - \omega^2)a}$ in the place of θ, the equation assumes the more simple form

$$\Pi\{\theta - \tfrac{1}{3}a(\omega - \omega^2)(\beta - \gamma)\} = \theta^3 - 3(ac - b^2)\theta - a\sqrt{\Box},$$

which if U be the cubic function, H the Hessian $= (ac - b^2,\ ad - bc,\ bd - c^2 \between x, y)^2$, and $\Box$ the discriminant as before, is a particular case (obtained by writing $x = 1$, $y = 0$) of the equation

$$\Pi\{\theta - \tfrac{1}{3}a(\omega - \omega^2)(x - \alpha y)\} = \theta^3 - 3H\theta - U\sqrt{\Box},$$

which equation can be at once obtained from the equation (where Φ is the cubicovariant of the cubic function)

$$\sqrt[3]{\tfrac{1}{2}(\Phi + U\sqrt{\Box})} - \sqrt[3]{\tfrac{1}{2}(\phi - U\sqrt{\Box})} = \tfrac{1}{3}a(\omega - \omega^2)(\beta - \gamma)(x - \alpha y),$$

given in my Fifth Memoir on Quantics, *Phil. Trans.*, t. CXLVIII. (1858), [156]. For writing for a moment

$$\theta = \sqrt[3]{X} - \sqrt[3]{Y},$$

we find

$$\theta^3 = X - Y - 3\sqrt[3]{XY\theta},$$

or

$$\theta^3 + 3\sqrt[3]{XY}\theta - (X - Y) = 0,$$

where $\sqrt[3]{XY} = \sqrt[3]{\tfrac{1}{4}(\Phi^2 - U^2\Box)}$, which by the equation

$$\phi^2 - U^2\Box = -4H^3$$

(given in the Memoir) is $= -H$, and $(X - Y)$ is $= U\sqrt{\Box}$, so that the equation in θ is, as above, $\theta^3 - 3H\theta - U\sqrt{\Box} = 0$, an equation which is satisfied by $\theta = \tfrac{1}{3}a(\omega - \omega^2)(\beta - \gamma)(x - \alpha y)$; and the other two roots being of course of the like form, the cubic function in θ is equal to $\Pi\{\theta - \tfrac{1}{3}a(\omega - \omega^2)(\beta - \gamma)(x - \alpha y)\}$ which proves the theorem.

2, *Stone Buildings, W.C., Nov. 3rd*, 1859.

288.

NOTE ON THE CURVATURE OF A PLANE CURVE AT A DOUBLE POINT, AND ON THE CURVATURE OF SURFACES.

[From the *Quarterly Journal of Pure and Applied Mathematics*, vol. III. (1860), pp. 322—326.]

THE radius of curvature of a plane curve at a double point is most readily determined as follows: viz. if x, y be the coordinates of the double point, θ the inclination to the axis of x of the tangent to the branch under consideration, and u, v the coordinates measured from the double point as origin, parallel and perpendicular to the tangent, of the consecutive point on such branch, then the radius of curvature for the branch in question is

$$=\frac{u^2}{2v};$$

the coordinates of the consecutive point are

$$x+u\cos\theta-v\sin\theta,$$
$$y+u\sin\theta+v\cos\theta,$$

and the value of $u^2\div 2v$ is to be found by substituting these expressions for x, y in the equation of the curve. Let $U=0$ be the equation of the curve; then at the double point, U and the differential coefficients of the first order vanish, let those of the second order be $(a,\ b,\ c)$ and those of the third order $(\mathrm{a},\ \mathrm{b},\ \mathrm{c},\ \mathrm{d})$, then substituting we have

$$\tfrac{1}{2}(a,\ b,\ c)(u\cos\theta-v\sin\theta,\ u\sin\theta+v\cos\theta)^2$$
$$+\tfrac{1}{6}(\mathrm{a},\ \mathrm{b},\ \mathrm{c},\ \mathrm{d})(u\cos\theta-v\sin\theta,\ u\sin\theta+v\cos\theta)^3=0.$$

The coefficient of u^2 is

$$\tfrac{1}{2}(a,\ b,\ c)(\cos\theta,\ \sin\theta)^2,$$

which vanishes, since it is by putting this function equal to zero that we find the direction of the tangents at the double point. For the present purpose we are concerned with one of the two branches only, and in all that follows the ratio $\cos\theta : \sin\theta$ will denote a determinate root of the quadratic equation; viz. the root which corresponds to the branch in question.

The equation takes therefore the form

$$Buv + \tfrac{1}{2}Cv^2 + \tfrac{1}{6}Du^3 + \&c. = 0,$$

which might be satisfied by assuming $Buv + \tfrac{1}{2}Cv^2 = 0$, but the values so obtained belong to the branch which does not touch the tangent; the proper solution is

$$Buv + \tfrac{1}{6}Du^3 = 0,$$

and thence

$$\frac{u^2}{2v} = -\frac{3B}{D},$$

or substituting for B, D their values, the radius of curvature is

$$= \frac{3(a,\ b,\ c)(\cos\theta,\ \sin\theta)(-\sin\theta,\ \cos\theta)}{(\mathrm{a},\ \mathrm{b},\ \mathrm{c},\ \mathrm{d})(\cos\theta,\ \sin\theta)^3},$$

where the expression in the numerator is

$$= (-a+c)\sin\theta\cos\theta + b(\cos^2\theta - \sin^2\theta),$$

$$= \tfrac{1}{2}\{(c-a)\sin 2\theta + 2b\cos 2\theta\}.$$

The curvature, at the point of contact, of the curve in which any surface is intersected by a tangent plane cannot be found by the ordinary theory of the curvature of surfaces, being (by reason that the point of contact is a double point on the curve) a case of exception from that theory. The foregoing method may be applied to the solution of the question as follows: Let $U=0$ be the equation of the surface, and suppose that (x, y, z) refer to the point of contact of the tangent plane. Let L, M, N be the first differential coefficients of U at this point, (a, b, c, f, g, h) the second differential coefficients, (a, b, c, f, g, h, i, j, k, l) the third differential coefficients. Let λ, μ, ν be proportional to the cosines of the inclinations to the axis of the tangent to one of the two branches through the double point; the ratios $\lambda : \mu : \nu$ are determined by

$$(L,\ M,\ N)(\lambda,\ \mu,\ \nu) = 0,$$

$$(a,\ b,\ c,\ f,\ g,\ h)(\lambda,\ \mu,\ \nu)^2 = 0.$$

Let u, v be proportional to the coordinates of a consecutive point measured, from the point of contact as origin, in the direction of the tangent and in the perpendicular direction in the tangent plane. The cosines of the inclinations of u to the axes are as $\lambda : \mu : \nu$; those for the inclinations of the normal to the axes are as $L : M : N$; hence for the coordinate v which is perpendicular to the plane of the last-mentioned two lines, the cosines of the inclinations to the axes are as $N\mu - M\nu : L\nu - N\lambda : M\lambda - L\mu$, and the coordinates of the consecutive point may be taken to be

$$\begin{aligned} &x + \lambda u + (N\mu - M\nu)\, v, \\ &y + \mu u + (L\nu \ \ - N\lambda)\, v, \\ &z + \nu u \ + (M\lambda - L\mu)\, v. \end{aligned}$$

Substituting these values in the equation of the surface, the terms involving the first powers of u, v vanish, and the term involving u^2 also vanishes in virtue of the relation $(a, b, c, f, g, h)(\lambda, \mu, \nu)^2 = 0$. The equation consequently becomes

$$Buv + \tfrac{1}{2}Cv^2 + \tfrac{1}{6}Du^3 + \ldots = 0,$$

and we have as before for the branch in question,

$$\frac{u^2}{2v} = -\frac{3B}{D}.$$

In the present case u, v have been taken, not as before equal, but only proportional, to the coordinates of the consecutive point measured from the point of contact parallel and perpendicular to the tangent, the values of the coordinates are in fact

$$\sqrt{\lambda^2 + \mu^2 + \nu^2}\, u, \ \ \sqrt{L^2 + M^2 + N^2}\ \sqrt{\lambda^2 + \mu^2 + \nu^2}\ v,$$

and the expression for the radius of curvature is

$$= \frac{\sqrt{\lambda^2 + \mu^2 + \nu^2}}{\sqrt{L^2 + M^2 + N^2}} \frac{u^2}{2v},$$

or substituting for $\dfrac{u^2}{2v}$ the value $-\dfrac{3B}{D}$ and for B, D their values, the radius of curvature is

$$= -\frac{3\sqrt{\lambda^2 + \mu^2 + \nu^2}}{\sqrt{L^2 + M^2 + N^2}} \times \frac{(a, b, c, f, g, h)(\lambda, \mu, \nu)(N\mu - M\nu,\ L\nu - N\lambda,\ M\lambda - L\mu)}{(\mathrm{a, b, c, f, g, h, i, j, k, l})(\lambda, \mu, \nu)^3},$$

where, as already noticed, the ratios $\lambda : \mu : \nu$ are determined by the equations

$$\begin{aligned} (L, M, N)(\lambda, \mu, \nu) &= 0, \\ (a, b, c, f, g, h)(\lambda, \mu, \nu)^2 &= 0, \end{aligned}$$

the system of roots selected being that which corresponds to the branch under consideration. It may be noticed that these two equations give

$$(\mathfrak{A}, \mathfrak{B}, \mathfrak{C}, \mathfrak{F}, \mathfrak{G}, \mathfrak{H})(L, M, N)^2 . (a, b, c, f, g, h)(\lambda, \mu, \nu)^2 - K[(L, M, N)(\lambda, \mu, \nu)]^2 = 0,$$

where as usual

$$\begin{aligned} (\mathfrak{A}, \mathfrak{B}, \mathfrak{C}, \mathfrak{F}, \mathfrak{G}, \mathfrak{H}) &= (bc - f^2,\ ca - g^2,\ ab - h^2,\ gh - af,\ hf - bg,\ fg - ch), \\ K &= abc - af^2 - bg^2 - ch^2 + 2fgh, \end{aligned}$$

and the expression on the left-hand side considered as a function of (λ, μ, ν) is decomposable into a pair of factors. Selecting the proper factor and equating it to zero, we have in addition to the linear equation

$$(L, M, N)(\lambda, \mu, \nu) = 0,$$

a new linear equation; these two equations determine the ratios $\lambda : \mu : \nu$.

P.S.—I remark that the process adopted for finding the radius of curvature at a double point of a plane curve is the most simple one for the case of an ordinary point on the curve. In fact let $U = 0$ be the curve, (x, y) the coordinates of the point in question, L, M the corresponding values of the first differential coefficients, (a, b, c) those of the second differential coefficients. The coordinates of the consecutive point are

$$x + u \cos\theta - v \sin\theta,$$

$$y + u \sin\theta + v \cos\theta.$$

Substituting these in the equation of the curve, we have

$$L(u\cos\theta - v\sin\theta) + M(u\sin\theta + v\cos\theta) + \tfrac{1}{2}(a, b, c)(u\cos\theta - v\sin\theta,\ u\sin\theta + v\cos\theta)^2 = 0,$$

the coefficient of u must be zero, or we have

$$L\cos\theta + M\sin\theta = 0,$$

giving

$$\sin\theta = -\frac{L}{\sqrt{L^2+M^2}}, \quad \cos\theta = \frac{M}{\sqrt{L^2+M^2}},$$

and the equation may be reduced to

$$\sqrt{L^2+M^2}\, v + \tfrac{1}{2}(a, b, c)(\cos\theta, \sin\theta)^2 u^2 = 0,$$

or what is the same thing

$$(L^2+M^2)^{\frac{3}{2}} v + \tfrac{1}{2}(a, b, c)(M, -L)^2 u^2 = 0,$$

whence the radius of curvature

$$= \frac{u^2}{2v} = -\frac{(a, b, c)(M, -L)^2}{(L^2+M^2)^{\frac{3}{2}}},$$

which is the ordinary form for the radius of curvature at any point of a curve represented by the equation $U = 0$.

2, *Stone Buildings, W.C., October 2nd,* 1859.

289.

ON SOME NUMERICAL EXPANSIONS.

[From the *Quarterly Journal of Pure and Applied Mathematics*, vol. III. (1860), pp. 366—369.]

THERE are in the theory of series and the calculus of finite differences and the differential calculus, various sets of numbers, such for instance as Bernoulli's Numbers, or the successive differences of 0^m, which from their more or less frequent occurrence and the complexity of their law it is desirable to tabulate. To these belong, I think, the coefficients of the successive powers of x in the expansion of a given power of $\frac{\log(1+x)}{x}$, or when the index of the power of the function is an indeterminate quantity r, then the coefficients of the several terms of the rational and integral functions of r which form the coefficients of the successive powers of x in the expansion in question. We have

$$\log\frac{\log(1+x)}{x} = -\tfrac{1}{2}x + \tfrac{5}{24}x^2 - \tfrac{1}{8}x^3 + \tfrac{251}{2880}x^4 - \tfrac{19}{288}x^5 + \tfrac{19087}{362880}x^6 - \&c.,$$

which is most easily verified by differentiating and multiplying by $\log(1+x)$, which gives on the left-hand side $\frac{1}{1+x}$, and on the right-hand side the expansion $1 - x + x^2 - \&c.$ And from the above, multiplying each side by r, and taking the exponential, we find

$$\begin{aligned}\left\{\frac{(\log 1+x)}{x}\right\}^r = 1 & \\ & - (\tfrac{1}{2}r)\,.\,x \\ & + (\tfrac{1}{8}r^2 + \tfrac{5}{24}r)\,x^2 \\ & - (\tfrac{1}{48}r^3 + \tfrac{5}{48}r^2 + \tfrac{1}{8}r)\,x^3 \\ & + (\tfrac{1}{384}r^4 + \tfrac{5}{192}r^3 + \tfrac{97}{1152}r^2 + \tfrac{251}{2880}r)\,x^4 \\ & - (\tfrac{1}{3840}r^5 + \tfrac{5}{1152}r^5 + \tfrac{61}{2304}r^3 + \tfrac{401}{5760}r^2 + \tfrac{19}{288}r)\,x^5 \\ & + (\tfrac{1}{46080}r^6 + \tfrac{5}{9216}r^5 + \tfrac{49}{9216}r^4 + \tfrac{10543}{414720}r^3 + \tfrac{4075}{69120}r^2 + \tfrac{19087}{362880}r)\,x^6 \\ & \mp \&c.\end{aligned}$$

which may be verified by writing $r = 1$.

As an instance of the use of the preceding table, let it be required to find expressions for the combinations without repetitions of the series of natural numbers 1, 2, 3, ... $(n-1)$, or what is the same thing for the coefficients of the powers of k in $k(k-1)(k-2)\ldots(k-n+1)$. We have

$$\begin{aligned} &k(k-1)(k-2)\ldots(k-n+1)\\ &= k^n - A_1{}^n k^{n-1} + A_2{}^n k^{n-2} - \&c.\\ &= \Pi n \text{ coefficient } x^n \text{ in } (1+x)^k; \end{aligned}$$

whence

$$\begin{aligned} (-)^r A_r{}^n &= \Pi n \text{ coefficient } x^n k^{n-r} \text{ in } (1+x)^k\\ &= \Pi n \text{ coefficient } x^n k^{n-r} \text{ in } e^{k\log(1+x)}\\ &= \frac{\Pi n}{\Pi(n-r)} \text{ coefficient } x^n \text{ in } \{\log(1+x)\}^{n-r}\\ &= \frac{\Pi n}{\Pi(n-r)} \text{ coefficient } x^r \text{ in } \left\{\frac{\log(1+x)}{x}\right\}^{n-r}. \end{aligned}$$

Thus

$$\begin{aligned} &A_1{}^n = n\{\tfrac{1}{2}(n-1)\},\\ &A_2{}^n = n(n-1)\{\tfrac{1}{8}(n-2)^2 + \tfrac{5}{24}(n-2)\},\\ &A_3{}^n = n(n-1)(n-2)\{\tfrac{1}{48}(n-3)^3 + \tfrac{5}{48}(n-3)^2 + \tfrac{1}{8}(n-3)\},\\ &\vdots \end{aligned}$$

and so on, as far as the expansion has been effected. It may be remarked that the general expression for the algebraical transcendent $A_r{}^n$ is given in Dr Schläfli's paper, "Sur les coefficients du développement du product $(1+x)(1+2x)\ldots\{1+(n-1)x\}$ suivant les puissances ascendantes de x." *Crelle*, t. XLIII. [1852], pp. 1—22, but the law is a very complicated one.

We have

$$\log(1+x) = x - \tfrac{1}{2}x^2 + \tfrac{1}{3}x^3 - \tfrac{1}{4}x^4 + \tfrac{1}{5}x^5 - \tfrac{1}{6}x^6 + \tfrac{1}{7}x^7 - \&c.$$

and dividing by x, and taking the logarithm, we have as before

$$\log\left(\frac{1}{x}\log(1+x)\right) = -\tfrac{1}{2}x + \tfrac{5}{24}x^2 - \tfrac{1}{8}x^3 + \tfrac{251}{2880}x^4 - \tfrac{19}{288}x^5 + \tfrac{19087}{362880}x^6 - \&c.$$

which may be considered as the first of a series of logarithmic derivatives, viz., dividing by $-\frac{1}{2}x$ and taking the logarithm we have

$$\log\left(-\frac{2}{x}\log\left(\frac{1}{x}\log(1+x)\right)\right) = -\tfrac{5}{12}x + \tfrac{47}{288}x^2 - \tfrac{2443}{25920}x^3 + \tfrac{5303}{82944}x^4 - \tfrac{19631}{580608}x^5 + \&c.,$$

and by the like process

$$\log\left(-\frac{12}{5x}\log\left(-\frac{2}{x}\log\left(\frac{1}{x}\log(1+x)\right)\right)\right) =$$
$$-\tfrac{47}{120}x + \tfrac{11317}{86400}x^2 - \tfrac{439989}{5184000}x^3 + \tfrac{9390960319}{156764160000}x^4 - \&c.,$$

and so on.

Suppose in general that

$$\phi x = x + B_1x^2 + C_1x^3 + \ldots,$$

and let it be required to find the rth function $\phi^r x$. It is easy to see that the successive coefficients are rational and integral functions of r of the degrees 1, 2, 3, &c. respectively; we have in fact

$$\begin{aligned}
\phi^r x = & \qquad \phi^0 x \\
& + \frac{r}{1} \qquad (\phi^1 x - \phi^0 x) \\
& + \frac{r \,.\, r-1}{1 \,.\, 2} \qquad (\phi^2 x - 2\phi^1 x + \phi^0 x) \\
& + \frac{r \,.\, r-1 \,.\, r-2}{1 \,.\, 2 \,.\, 3} (\phi^3 x - 3\phi^2 x + 3\phi^1 x - \phi^0 x), \\
& \qquad \&\text{c.},
\end{aligned}$$

and by successive substitutions,

$$\begin{aligned}
\phi^0 x &= x, \\
\phi^1 x &= x + B_1x^2 + C_1x^3 + D_1x^4 + \ldots, \\
\phi^2 x &= x + 2B_1x^2 + (2B_1^2 + 2C_1)\,x^3 + (B_1^3 + 5B_1C_1 + 2D_1)\,x^4 + \ldots, \\
\phi^3 x &= x + 3B_1x^3 + (6B_1^2 + 3C_1)\,x^3 + (9B_1^3 + 15B_1C_1 + 3D_1)\,x^4 + \ldots. \\
&\vdots
\end{aligned}$$

Whence

$$\begin{aligned}
\phi^r x = x & \\
& + rB_1\,x^2 \\
& + \{(r^2 - r)\,B_1^2 + rC_1\}\,x^3 \\
& + \{(r^3 - \tfrac{5}{2}r^2 + \tfrac{3}{2}r)\,B_1^3 + (\tfrac{5}{2}r^2 - \tfrac{5}{2}r)\,B_1C_1 + rD_1\}\,x^4 \\
& + \&\text{c.}
\end{aligned}$$

It would, I think, be worth while to continue the expansion some steps further.

2, *Stone Buildings, W.C., Oct. 2nd,* 1859.

290.

A DISCUSSION OF THE STURMIAN CONSTANTS FOR CUBIC AND QUARTIC EQUATIONS.

[From the *Quarterly Journal of Pure and Applied Mathematics*, vol. IV. (1861), pp. 7—12.]

FOR the cubic equation

$$(a,\ b,\ c,\ d)(x,\ 1)^3 = 0,$$

the Sturmian Constants (or leading coefficients of the Sturmian functions) are

$$a,\ a,\ b^2 - ac,\ -a^2d^2 + 6abcd - 4ac^3 - 4b^3d + 3b^2c^2.$$

If the signs of the constants, that is, of the functions for $+\infty$, are	then the signs of the functions for $-\infty$ are	
$+ + + +$	$- + - +$	three real roots.
$+ + - +$	$- + + +$	case cannot occur.
$+ + + -$	$- + - -$	one real root.
$+ + - -$	$- + + -$	

The second case would give a loss of variations of sign in passing from ∞ to $-\infty$, which is inconsistent with Sturm's theorem. To show *à posteriori* that the case cannot occur, we may form the identical equation

$$(a^2d - 3abc + 2b^3)^2 = -a^2(-a^2d^2 + 6abcd - 4ac^3 - 4b^3d + 3b^2c^2) + 4(b^2 - ac)^3,$$

and, this being so, then in the case in question, the right-hand side would consist of two terms, each of them negative, while the left-hand side is essentially positive.

In the particular case where the third constant vanishes, or

$$b^2 - ac = 0,$$

we have

$$\begin{aligned} &- a^2d^2 + 6abcd - 4ac^3 - 4b^3d + 3b^2c^2 \\ &= -(ad-bc)^2 + 4(b^2-ac)(c^2-bd) \\ &= -(ad-bc)^2, \text{ is negative;} \end{aligned}$$

hence, regarding the evanescent term as being at pleasure positive or negative, we have in each case a combination of signs corresponding to one real root.

The general result (which is well known) is, that there are three real roots or one real root according as

$$- a^2d^2 + 6abcd - 4ac^3 - 4b^3d + 3b^2c^2$$

is positive or negative.

For the quartic equation

$$(a,\ b,\ c,\ d,\ e)(x,\ 1)^4 = 0,$$

the Sturmian constants are

$$a,\ a,\ b^2 - ac,\ 3aJ + 2(b^2 - ac)I,\ I^3 - 27J^2,$$

if, as usual,

$$\begin{aligned} I &= ae - 4bd + 3c^2, \\ J &= ace - ad^2 - b^2e + 2bcd - c^3. \end{aligned}$$

If the signs of the constants, that is, of the functions for $+\infty$, are		then the signs of the functions for $-\infty$ are		
+ + + + +	0	+ − + − +	4	4 real roots.
+ + − + +	2	+ − − − +	2	no real root.
+ + + − +	2	+ − + + +	2	no real root.
+ + − − +	2	+ − − + +	2	no real root.
+ + + + −	1	+ − + − −	3	2 real roots.
+ + − + −	3	+ − − − −	1	− 2, cannot occur.
+ + + − −	1	+ − + + −	3	2 real roots.
+ + − − −	1	+ − − + −	3	2 real roots.

The non-existing combination of signs is

$$\begin{aligned} I^3 - 27J^2 \quad &= -, \\ 3aJ + 2(b^2 - ac)I &= +, \\ b^2 - ac \quad &= -. \end{aligned}$$

To show *à posteriori* that this case cannot occur, write

$$\begin{aligned} \vartheta &= a^2d - 3abc + 2b^3, \\ X &= 3aJ + 2(b^2 - ac)I, \end{aligned}$$

then we have identically

$$9(3a^2J^2+X^2)\vartheta^2=-4a^2X^3+36(b^2-ac)^3X^2-4a^2(b^2-ac)^3(I^3-27J^2),$$

which is impossible under the given combination of signs, since the left-hand side would be positive, and the right-hand side negative.

To prove the above identity—the relation $JU^3-IU^2H+4H^3+\Phi^2=0$, between the covariants of the quartic, gives

$$a^3J+a^2(b^2-ac)I-4(b^2-ac)^3+\vartheta^2=0,$$

or, what is the same thing,

$$\vartheta^2=-a^3J-a^2(b^2-ac)I+4(b^2-ac)^3.$$

But

$$X=3aJ+2(b^2-ac)I,$$

and thence

$$3\vartheta^2+a^2X=-a^2(b^2-ac)I+12(b^2-ac)^3,$$

or

$$3\vartheta^2=-a^2X-a^2(b^2-ac)I+12(b^2-ac)^3,$$

and the identity will be true, if

$$(3X^2+9a^2J^2)\left\{-X-(b^2-ac)I+12\frac{(b^2-ac)^3}{a^2}\right\}$$

$$=-4X^3+36\frac{(b^2-ac)^3}{a^2}X^2-4(b^2-ac)^3(I^3-27J^2).$$

This gives

$$(3X^2+9a^2J^2)\{-X-(b^2-ac)I\}=-4X^3-4(b^2-ac)^3I^3,$$

or, what is the same thing,

$$(3X^2+9a^2J^2)\{\ X+(b^2-ac)I\}=4\{\ X^3+\ (b^2-ac)^3I^3\},$$

or, dividing by $X+(b^2-ac)I$,

$$3X^2+9a^2J^2=4\{X^2-X(b^2-ac)I+(b^2-ac)^2I^2\},$$

and reducing

$$X^2-4X(b^2-ac)I-9a^2J^2+4(b^2-ac)I^2=0,$$

or finally

$$\{X-3aJ-2(b^2-ac)I\}\{X+3aJ-2(b^2-ac)I\}=0,$$

which is true in virtue of

$$X=3aJ+2(b^2-ac)I,$$

and the identity is thus proved.

The general conclusion is,

if $I^3 - 27J^2$ is positive, the four roots are all real or all imaginary,

viz., all real if $b^2 - ac$ and $3aJ + 2(b^2 - ac)\,I$ are both positive, imaginary if otherwise. But if $I^3 - 27J^2$ is negative, the roots are two of them real, and the other two imaginary.

The following special cases may be noticed,

1°. $b^2 - ac = 0$,

here

$$9(3a^2J^2 + X^2)\,\vartheta^2 = -4a^2X^3, \text{ or } X = 3aJ + 2(b^2 - ac)\,I = 3aJ, \text{ is negative,}$$

so that,

if $I^3 - 27J^2$ is $+$, the roots are all imaginary;

if $I^3 - 27J^2$ is $-$, the roots are two real and two imaginary.

2°. $X = 3aJ + 2(b^2 - ac)\;I = 0$,

here

$$27a^2J^2\vartheta^2 = -4a^2(b^2 - ac)^3(I^3 - 27J^2),$$

or $b^2 - ac$, $I^3 - 27J^2$ are of opposite signs, and if

$b^2 - ac = -$, $I^3 - 27J^2 = +$, the roots are all imaginary,

$b^2 - ac = +$, $I^3 - 27J^2 = -$, the roots are two real and two imaginary.

3°. $b^2 - ac = 0$, $X = 3aJ + 2(b^2 - ac)\,I = 0$,

here $J = 0$, that is,

$$2bcd - ad^2 - c^3 = 0, \text{ or } (ad - bc)^2 + c^2(ac - b^2) = 0, \text{ or } ad - bc = 0, \text{ and } I^3 - 27J^2 = I^3,$$

$$I = ae - 4bd + 3c^2 = ae - 4\frac{b^4}{a^2} + 3\frac{b^4}{a^2} = ae - \frac{b^4}{a^2} = \frac{1}{a^2}(a^3e - b^4),$$

whence $I = +$, the roots are all imaginary.

$I = -$, the roots are two real and two imaginary.

This is easily verified, in fact $ac - b^2 = 0$, $ad - bc = 0$, give $c = \frac{b^2}{a}$, $d = \frac{bc}{a} = \frac{b^3}{a^2}$, and the equation becomes

$$ax^4 + 4bx^3 + 6\frac{b^2}{a}x^2 + 4\frac{b^3}{a^2}x + e = 0,$$

or, which is the same thing,

$$(ax + b)^4 + (a^3e - b^4) = 0,$$

so that the roots are all imaginary, or two real and two imaginary, according to the sign of $a^3e - b^4$ as above.

It may be noticed that for a quintic equation

$$(a, b, c, d, e, f)(x, 1)^5,$$

if the Sturmian Constants are

$$a, a, C, D, E, F,$$

where as before a is positive, then the roots are real or imaginary as follows: viz.,

C	D	E	F	
+	+	+	+,	5 real roots.
−	+	+		
+	−	+		
−	−	+	+,	1 real root, 4 imaginary roots.
+	+	−		
+	−	−		
−	−	−		
+	+	+		
+	+	−	−,	3 real roots, 2 imaginary roots.
+	−	−		
−	−	−		
−	+	−	+,	case which does not occur.
−	+	+		
+	−	+	−,	cases which do not occur.
−	−	+		
−	+	−		

The values of C, D, E, and F are given in my "Tables of the Sturmian Functions for Equations of the Second, Third, Fourth, and Fifth Degrees," *Phil. Trans.*, t. 147 (1857), pp. 733—736, [151], but I have not further examined this case.

2, *Stone Buildings, W.C., September* 29, 1859.

291.

ON THE DEMONSTRATION OF A THEOREM RELATING TO THE MOMENTS OF INERTIA OF A SOLID BODY.

[From the *Quarterly Journal of Pure and Applied Mathematics*, vol. IV. (1861), pp. 25—27.]

CONSIDERING in the first instance the analogous question *in plano*, let $a = \int x^2 dm$, $b = \int y^2 dm$, $h = \int xy dm$, where the integration extends over any closed figure whatever, then it is to be shown that the equations

$$(a,\ h,\ b)(p,\ q)^2 = 1, \text{ and } (a+b)(p^2+q^2) - (a,\ h,\ b)(p,\ q)^2 = 1,$$

represent respectively ellipses.

If in the first case, by a transformation of coordinates,

$$(a,\ h,\ b)(p,\ q)^2 = a_1 p_1^2 + b_1 q_1^2,$$

then a_1, b_1 are the roots of the quadratic equations,

$$\begin{vmatrix} a-\rho, & h \\ h\ , & b-\rho \end{vmatrix} = 0,$$

and if in the second case,

$$(a+b)(p^2+q^2) - (a,\ h,\ b)(p,\ q)^2 = a_1 p_1^2 + b_1 q_1^2,$$

then a_1, b_1 are the roots of

$$\begin{vmatrix} b-\rho, & -h \\ -h, & a-\rho \end{vmatrix} = 0,$$

the two equations being in fact the same equation,

$$\rho^2 - (a+b)\rho + ab - h^2 = 0,$$

and the conditions that the curve may be an ellipse, are

$$a+b=+,$$
$$ab-h^2=+,$$

the former of which requires no demonstration; to prove the latter, changing merely the variables under the integral sign, I write

$$a' = \int x'^2 dm', \quad b' = \int y'^2 dm', \quad h' = \int x'y' dm',$$

these quantities being of course respectively equal to a, b, h, we have then

$$ab' + a'b - 2hh' = \iint (xy' - x'y)^2 \, dm dm' = 2(ab - h^2),$$

or since the quantity under the integral sign is a square, $ab-h^2$ is positive.

For the analogous problem *in solido*, we have

$$a = \int x^2 dm, \ b = \int y^2 dm, \ c = \int z^2 dm, \ f = \int yz dm, \ g = \int zx dm, \ h = \int xy dm,$$

and it is to be shown that the equations

$$(a, b, c, f, g, h)(p, q, r)^2 = 1,$$
$$(a+b+c)(p^2+q^2+r^2) - (a, b, c, f, g, h)(p, q, r)^2 = 1,$$

represent respectively ellipsoids.

The conditions in the first problem are

$$a+b+c = +,$$
$$bc+ca+ab-f^2-g^2-h^2 = +,$$
$$abc-af^2-bg^2-ch^2+2fgh = +,$$

the first of which is obviously true: as regards the second, the theorem *in plano* shows that each of the quantities $bc-f^2$, $ca-g^2$, $ab-h^2$ is positive, or merely reproducing the investigation, we find

$$2(bc+ca+ab-f^2-g^2-h^2) = \iint [(yz'-y'z)^2 + (zx'-z'x)^2 + (xy'-x'y)^2] \, dm dm',$$

which proves the theorem, and where it is to be observed that the integral may also be written

$$\iint [(x^2+y^2+z^2)(x'^2+y'^2+z'^2) - (xx'+yy'+zz')^2] \, dm dm';$$

and for the third, we find in a precisely similar manner,

$$6(abc-af^2-bg^2-ch^2+2fgh) = \iint \begin{vmatrix} x, & y, & z \\ x', & y', & z' \\ x'', & y'', & z'' \end{vmatrix}^2 dm dm',$$

which proves the theorem. The integral may also be written

$$\iint \begin{vmatrix} x^2 + y^2 + z^2, & xx' + yy' + zz', & xx'' + yy'' + zz'' \\ x'x + y'y + z'z, & x'^2 + y'^2 + z'^2, & x'x'' + y'y'' + z'z'' \\ x''x + y''y + z''z, & x''x' + y''y' + z''z', & x''^2 + y''^2 + z''^2 \end{vmatrix} dm\,dm'.$$

The conditions in the second problem are

$$\begin{aligned} &(b+c)+(c+a)+(a+b)=+, \\ &(c+a)(a+b)+(a+b)(b+c)+(b+c)(c+a)-f^2-g^2-h^2 =+, \\ &(a+b)(b+c)(c+a)-(b+c)f^2-(c+a)g^2-(a+b)h^2-2fgh=+, \end{aligned}$$

the first and second of which are respectively equivalent to

$$\begin{aligned} &a+b+c=+, \\ &(a+b+c)^2+bc+ca+ab-f^2-g^2-h^2=+, \end{aligned}$$

which are already proved. The last may be written

$$(a+b+c)(bc+ca+ab-f^2-g^2-h^2)-(abc-af^2-bg^2-ch^2+2fgh)=+,$$

which, putting for shortness,

$$\begin{aligned} &A = x^2 + y^2 + z^2, \quad B = x'^2 + y'^2 + z'^2, \quad C = x''^2 + y''^2 + z''^2, \\ &F = x'x'' + y'y'' + z'z'', \quad G = x''x + y''y + z''z, \quad H = xx' + yy' + zz', \end{aligned}$$

is by what precedes expressible in the form

$$\tfrac{1}{6}\iint \{A(BC-F^2)+B(CA-G^2)+C(AB-H^2)-(ABC-AF^2-BG^2-CH^2+2FGH)\}\, dm\,dm'$$

$$= \tfrac{1}{3}\iint (ABC - FGH)\, dm\,dm',$$

or, since $\sqrt{BC} > F$, $\sqrt{CA} > G$, $\sqrt{AB} > H$, we have $ABC > FGH$, or $ABC - FGH = +$, and therefore the value of the integral is also positive.

2, *Stone Buildings, W.C., 6th March*, 1860.

292.

A THEOREM IN CONICS.

[From the *Quarterly Journal of Pure and Applied Mathematics,* vol. IV. (1861), pp. 131—133.]

THE following theorem is given in Todhunter's *Conic Sections,* [Ed. 7, p. 304], "If ellipses be inscribed in a triangle, each with one focus in a fixed straight line, the locus of the other focus is a conic section through the angular points of the triangle." A focus is the intersection of tangents to the conic from the circular points at infinity; and instead of the circular points at infinity we may substitute any two points whatever. This being so, let the equations of the sides of the triangle be $x=0$, $y=0$, $z=0$, and let a pair of tangents to the curve from the points (α, β, γ), $(\alpha', \beta', \gamma')$ meet in the point (ξ, η, ζ), and the other pair of tangents from the same two points meet in the point (X, Y, Z). I find that we have the very simple relation

$$X\xi : Y\eta : Z\zeta = \alpha\alpha' : \beta\beta' : \gamma\gamma',$$

and consequently, when the locus of the point (ξ, η, ζ) is given, that of the point (X, Y, Z) is at once determined by substituting in the equation of the first-mentioned locus, in the place of ξ, η, ζ, the values $\frac{\alpha\alpha'}{\xi}, \frac{\beta\beta'}{\eta}, \frac{\gamma\gamma'}{\zeta}$, or as we may express it, the second locus is derived from the first by the method of reciprocal trilinear substitutions. And, in particular, when the first locus is a line, the second locus is a conic through the angular points of the triangle, which is Mr Todhunter's theorem. I have considered some of the properties of this substitution in a Memoir "Sur quelques transmutations des Courbes," *Liouville,* t. XIV. (1849), pp. 40—46 and t. XV. (1850), pp. 351—356, [80 and 81].

To demonstrate the theorem, I take for the equation of the conic

$$\sqrt{(lx)} + \sqrt{(my)} + \sqrt{(nz)} = 0,$$

and I write for shortness

$$\beta\zeta-\gamma\eta,\ \gamma\xi-\alpha\zeta,\ \alpha\eta-\beta\xi=A,\ B,\ C,$$

$$\beta'\zeta-\gamma'\eta,\ \gamma'\xi-\alpha'\zeta,\ \alpha'\eta-\beta'\xi=A',\ B',\ C',$$

$$\xi(\beta\gamma'-\beta'\gamma)+\eta(\gamma\alpha'-\gamma'\alpha)+\zeta(\alpha\beta'-\alpha'\beta)=\Delta,$$

$$\eta\zeta\alpha\alpha'(\beta\gamma'-\beta'\gamma)+\zeta\xi\beta\beta'(\gamma\alpha'-\gamma'\alpha)+\xi\eta\gamma\gamma'(\alpha\beta'-\alpha'\beta)=\square,$$

so that in fact

$$\Delta=\begin{vmatrix}\xi, & \eta, & \zeta\\ \alpha, & \beta, & \gamma\\ \alpha', & \beta', & \gamma'\end{vmatrix},\quad \square=\alpha\beta\gamma\alpha'\beta'\gamma'\xi\eta\zeta\begin{vmatrix}\frac{1}{\xi}, & \frac{1}{\eta}, & \frac{1}{\zeta}\\ \frac{1}{\alpha}, & \frac{1}{\beta}, & \frac{1}{\gamma}\\ \frac{1}{\alpha'}, & \frac{1}{\beta'}, & \frac{1}{\gamma'}\end{vmatrix},$$

and we have

$$BC'-B'C=\xi\Delta,\quad CA'-C'A=\eta\Delta,\quad AB'-A'B=\zeta\Delta,$$

$$\beta\gamma'BC'-\beta'\gamma B'C=\gamma\alpha'CA'-\gamma'\alpha C'A=\alpha\beta'AB'-\alpha'\beta A'B=\square.$$

The conditions in order that the conic

$$\sqrt{(lx)}+\sqrt{(my)}+\sqrt{(nz)}=0$$

may touch the line through (α, β, γ) and (ξ, η, ζ) is

$$\frac{l}{A}+\frac{m}{B}+\frac{n}{C}=0,$$

and the condition in order that it may touch the line through $(\alpha', \beta', \gamma')$ and (ξ, η, ζ) is

$$\frac{l}{A'}+\frac{m}{B'}+\frac{n}{C'}=0,$$

and we thus have

$$l:m:n=\frac{1}{BC'}-\frac{1}{B'C}:\frac{1}{CA'}-\frac{1}{C'A}:\frac{1}{AB'}-\frac{1}{A'B},$$

or, what is the same thing,

$$l:m:n=AA'\xi:BB'\eta:CC'\zeta,$$

which determine the constants in the equation of the conic.

Consider now the tangents to the conic from the point (α, β, γ); if the equation of the tangent is assumed to be

$$px+qy+rz=0,$$

then we have

$$p\alpha+q\beta+r\gamma=0,$$

$$\frac{l}{p}+\frac{m}{q}+\frac{n}{r}=0,$$

and these equations are of course satisfied by $p : q : r = A : B : C$, since the line through (ξ, η, ζ) is a tangent. They are also satisfied by

$$p : q : r = \frac{l}{A\alpha} : \frac{m}{B\beta} : \frac{n}{C\gamma},$$

as is obvious by substitution, we have therefore

$$\frac{l}{A\alpha}x + \frac{m}{B\beta}y + \frac{n}{C\gamma}z = 0,$$

or more simply

$$\frac{A'\xi}{\alpha}x + \frac{B'\eta}{\beta}y + \frac{C'\zeta}{\gamma}z = 0,$$

for the equation of the other tangent through (α, β, γ), and we have in like manner

$$\frac{A\xi}{\alpha'}x + \frac{B\eta}{\beta'}y + \frac{C\zeta}{\gamma'}z = 0,$$

for the equation of the other tangent through $(\alpha', \beta', \gamma')$; the last-mentioned two lines intersect in the point X, Y, Z, that is we have

$$X\xi : Y\eta : Z\zeta = \frac{B'C}{\beta\gamma'} - \frac{BC'}{\beta'\gamma} : \frac{C'A}{\gamma\alpha'} - \frac{CA'}{\gamma'\alpha} : \frac{A'B}{\alpha\beta'} - \frac{AB'}{\alpha'\beta},$$

or attending to an above-mentioned equation, we have

$$X\xi : Y\eta : Z\zeta = \alpha\alpha' : \beta\beta' : \gamma\gamma',$$

which is the property in question. In the particular case, where the points (α, β, γ), $(\alpha', \beta', \gamma')$ are the foci, the theorem is an immediate consequence of the well-known proposition that the product of the perpendiculars let fall from the two foci upon any tangent of the conic is a constant.

2, *Stone Buildings, W.C.*, 17*th March*, 1860.

293.

ON A CERTAIN SYSTEM OF FUNCTIONAL SYMBOLS.

[From the *Quarterly Journal of Pure and Applied Mathematics*, vol. IV. (1861), pp. 225—230.]

M. HERMITE, in the Memoir "Sur la Résolution de l'équation du cinquième degré," *Comptes Rendus*, t. XLVI., (1858), p. 508, has given for equations of the fifth degree a solution by means of elliptic functions, analogous to the well-known trigonometrical solution of (the irreducible case of) the cubic equation. He has for this purpose to consider the effect of transforming certain functions $\phi\omega$, $\psi\omega$, connected with the elliptic functions, by replacing therein the argument ω by a linear fraction of the form $\frac{c + d\omega}{a + b\omega}$, where a, b, c, d are integer numbers satisfying the condition

$$ad - bc = 1,$$

and he finds for $\phi\left(\frac{c + d\omega}{a + b\omega}\right)$ (and the like would apply to $\psi\left(\frac{c + d\omega}{a + b\omega}\right)$), six different expressions according to the evenness, or values in relation to the modulus 2, of the coefficients a, b, c, d, viz., the congruence

$$ad - bc \equiv 1 \text{ (mod. 2)}$$

admits of the six solutions shown by the table

	a	b	c	d
I.	1	0	0	1
II.	0	1	1	0
III.	1	1	0	1
IV.	1	1	1	0
V.	1	0	1	1
VI.	0	1	1	1

(where 0 or 1 denotes that the coefficient to which it refers is even or odd), and to each solution there corresponds a distinct value of $\phi\left(\frac{c+d\omega}{a+b\omega}\right)$ or $\psi\left(\frac{c+d\omega}{a+b\omega}\right)$.

It appeared to me that the formulæ in question were not only of the highest importance in the theory of elliptic functions, but that they were very remarkable in and for themselves, and that the question might be looked at in the following manner.

Suppose that A, B are functional symbols, or, so to term them, substitutions, the laws of operation being

$$A(X,\ Y) = \left(\theta\frac{X}{Y},\ \frac{1}{Y}\right),$$

$$B(X,\ Y) = \quad (Y,\ X),$$

where θ is a constant. The equations mean that the effect of A is to change X into $\theta\frac{X}{Y}$, and Y into $\frac{1}{Y}$, and the effect of B to change X into Y, and Y into X. We have

$$A^2X = A\,.\,\theta\frac{X}{Y} = \theta\frac{AX}{AY} = \theta\,.\,\theta\frac{X}{Y} \div \frac{1}{Y} = \theta^2 X,$$

$$A^2Y = A\quad\frac{1}{Y} = \frac{1}{AY} = \quad 1\div\frac{1}{Y} = Y, \qquad \text{or } A^2(X,\ Y) = (\theta^2 X,\ Y),$$

and so

$$B^2X = X, \quad B^2Y = Y, \qquad \text{or } B^2(X,\ Y) = (X,\ Y);$$

thus B is periodic of the second order; and in the particular case $\theta = 1$, A is also periodic of the second order.

This being so, let K denote any one of the compound symbols A^p, BA^p, A^pB, A^qBA^p, &c.; I say that the *species* of K is determined by connecting with each of the symbols in question, a certain continued fraction as follows:

with

$$A,\ p+\omega,$$

$$BA^p,\ p-\frac{1}{\omega},$$

$$A^qBA^p,\ p-\frac{1}{q+\omega},$$

$$BA^qBA^p,\ p-\cfrac{1}{q-\cfrac{1}{\omega}},$$

$$A^rBA^qBA^p,\ p-\cfrac{1}{q-\cfrac{1}{r+\omega}},$$

&c.,

$$A^pB,\ -\frac{1}{p+\omega},$$

$$BA^pB,\ -\cfrac{1}{p-\cfrac{1}{\omega}},$$

$$A^qBA^pB,\ -\cfrac{1}{p-\cfrac{1}{q+\omega}},$$

&c.,

and when the continued fraction is reduced to the form $\dfrac{c+d\omega}{a+b\omega}$, where, of course, $ad-bc=1$, then, that the corresponding symbol K is of the species I. if $a, b, c, d \equiv 1, 0, 0, 1$ (mod. 2), and in like manner, that the species is II., III., IV., V., or VI., according as a, b, c, d are in regard to the modulus 2 of the form specified in the second, third, fourth, fifth, or sixth line of the foregoing table.

We have then a system of equations which I write in the form

$$K(X,\ Y) = \left\{ \begin{array}{ll|c} \theta^x & \theta^y & \text{for } K \text{ of the species} \\ \hline X, & Y & \text{I.} \\ Y, & X & \text{II.} \\ \dfrac{1}{X}, & \dfrac{Y}{X} & \text{III.} \\ \dfrac{1}{Y}, & \dfrac{X}{Y} & \text{IV.} \\ \dfrac{X}{Y}, & \dfrac{1}{Y} & \text{V.} \\ \dfrac{Y}{X}, & \dfrac{1}{X} & \text{VI.} \end{array} \right.$$

viz., the first line denotes that $K(X,\ Y)=(\theta^x X,\ \theta^y Y)$, and so for the other lines, the values of the indices x, y, being different in the several cases, but I have written the system in the above form, to put in evidence more distinctly the theorem as applied to the most simple case, where $\theta=1$.

It will be sufficient to indicate how the theorem is proved in this particular case: I will suppose that for

$$K = BA^q BA^p \ldots,$$

we have

$$K(X,\ Y) = \left\{ \begin{array}{llc} & & \text{for } K \text{ of the species} \\ X, & Y & \text{I.} \\ Y, & X & \text{II.} \\ \dfrac{1}{X}, & \dfrac{Y}{X} & \text{III.} \\ \dfrac{1}{Y}, & \dfrac{X}{Y} & \text{IV.} \\ \dfrac{X}{Y}, & \dfrac{1}{X} & \text{V.} \\ \dfrac{Y}{X}, & \dfrac{1}{X} & \text{VI.} \end{array} \right.$$

and let it be inquired what is the effect of the symbol $K' = A^r BA^q BA^p \ldots$, which, since $A^2(X,\ Y) = (X,\ Y)$, is equivalent to AK or K, according as r is odd or even.

Let $\Theta = \dfrac{c + d\omega}{a + b\omega}$ be the continued fraction which corresponds to K, and $\Theta' = \dfrac{c' + d'\omega}{a' + b'\omega}$ the continued fraction which corresponds to K', then Θ' is deduced from Θ by putting therein $r + \omega$ for ω, or we have

$$\frac{c + d(\omega + r)}{a + b(\omega + r)} = \frac{c' + d'\omega}{a' + b'\omega},$$

or

$$c' = c + dr,\ d' = d,$$
$$a' = a + br,\ b' = b,$$

and therefore when r is even, or $\equiv 0$ (mod. 2), then

$$c' \equiv c,\ d' \equiv d, \quad (\text{mod. } 2),$$
$$a' \equiv a,\ b' \equiv b,$$

or K' is of the *same* species as K. But in this case since $A^2(X, Y) = (X, Y)$ and r is even, we ought to have K' equivalent to K; and hence r being even, the theorem, assumed to be true for K, is also true for K'.

But if r be odd, or $\equiv 1$ (mod. 2), then

$$c' \equiv c + 1,\ d' \equiv d, \quad (\text{mod. } 2),$$
$$a' \equiv a + 1,\ b' \equiv b,$$

whence, according as

K is of species I., II., III., IV., V., VI., so

K' . . . . V., IV., VI., II., I., III.

But if from $K(X, Y)$ we derive $AK(X, Y)$, we find that of whatever species K may be, the species of AK is identical with the corresponding species of K', and in the case in question, where r is odd, it is clear that we ought to have K' equivalent to AK. Hence in this case also, and therefore generally, if the theorem be true for K it will be also true for K'.

In a very similar manner it may be shown that if the theorem is true for $K' = A^rBA^qBA^p\ldots$, it will be true for $BA^rBA^qBA^p\ldots$, and the foregoing cases include the case where the symbol is of the form BA^rBA^qB or A^rBA^qB (terminating in a B), for it is only necessary to assume that the last index of A is zero; hence the theorem, if true in any case, as it obviously is for $A^0(X, Y) = (X, Y)$, is true universally.

When θ is not equal 1, the only difference is that the terms of $K(X, Y)$ are respectively multiplied by certain powers θ^x, θ^y of θ, and it is not difficult to see that for $K = A^rBA^qBA^p\ldots$ the indices x, y are of the forms

$$x = \mathfrak{c}r + \mathfrak{b}q + \mathfrak{a}p + \ldots,$$
$$y = \mathfrak{c}'r + \mathfrak{b}'q + \mathfrak{a}'p + \ldots,$$

where c, b, a, ..., c′, b′, a′, ... have only the values 0, $+1$, -1, the values of

c, c′ depending on the species of $BA^qBA^p...$,
b, b′ . . . . $BA^p...$,
a, a′ . . . .,
&c.,

in a manner which can easily be determined. In the particular case, where $\theta^{16}=1$ (as it is in M. Hermite's problem), it is only necessary to know the values of x, y in regard to the modulus 16, and these values can be expressed in terms of the numbers (a, b, c, d) which correspond to the symbol K (this is in itself a remarkable theorem), and I find that (to modulus 16)

$$x \equiv \begin{cases} d(c+d)-1 \\ c(c-d)-1 \\ d(d-c)-1 \\ c(d-c)+1 \\ cd \\ -cd \end{cases}, \quad y = \begin{cases} a(a-b)-1 & \text{I.} \\ b(a+b)-1 & \text{II.} \\ ab & \text{III.} \\ -ab & \text{IV.} \\ a(b-a)+1 & \text{V.} \\ b(b-a)-1 & \text{VI.} \end{cases} \quad \text{for } K \text{ of the species}$$

This may be shown by induction in like manner as the theorem in regard to the parts independent of θ. Thus K, K' as before, let K be of the species I., and assume that $K(X, Y)=(\theta^x X, \theta^y Y)$ where

$$x \equiv d(c+d)-1,\ y \equiv a(a-b)-1 \text{ (mod. 16)},$$

then if r be even, since $A^2(X, Y)=(\theta^2 X, Y)$, we ought to have $K'(X, Y)=(\theta^{x+2r} X, \theta^y Y)$. But in this case K' is of the same species with K, viz., of the species I., and by the theorem, $K'(X, Y)=(\theta^{x'} X, \theta^{y'} Y)$ where

$$x' \equiv d'(c'+d')-1,\ y' \equiv a'(a'-b')-1 \text{ (mod. 16)};$$

and we ought to have

$$x' \equiv x+r,\ y' \equiv y \text{ (mod. 16)},$$

that is

$$\begin{aligned} d'(c'+d')-1 &\equiv d(c+d)-1+r, \text{ (mod. 16)}, \\ a'(a'-b')-1 &\equiv a(a-b)-1 \end{aligned}$$

or substituting for a', b', c', d' their values $a+br$, b, $c+dr$, d, these become

$$\begin{aligned} d(c+d+dr) &\equiv d(c+d)+r, \text{ (mod. 16)}, \\ (a+br)(a-b+br) &\equiv a(a-b) \end{aligned}$$

or

$$\begin{aligned} (d^2-1)r &\equiv 0, \qquad \text{(mod. 16)}. \\ (2ab-b^2+b^2r)r &\equiv 0 \end{aligned}$$

But r being even, these become

$$d^2-1\equiv 0,\ 2ab-b^2+b^2r\equiv 0 \text{ (mod. 8)};$$

or, since in the case under consideration, $a, b, c, d\equiv 1, 0, 0, 1$ (mod. 2), and therefore $b^2r\equiv 0$ (mod. 8), the two congruences become

$$d^2-1\equiv 0,\ a^2-(a-b)^2\equiv 0 \text{ (mod. 8)},$$

which are satisfied since d, a, and $a-b$ are all odd. And in a similar way the theorem is proved for all the other cases.

In M. Hermite's problem, X, Y are replaced by $\phi\omega$, $\psi\omega$, AX, AY are $\phi(\omega+1)$, $\psi(\omega+1)$, and BX, BY are $\phi\left(-\frac{1}{\omega}\right)$, $\psi\left(-\frac{1}{\omega}\right)$, the properties of the functions ϕ, ψ being such that, as in the preceding investigation,

$$A(X,\ Y)=\left(\theta\frac{X}{Y},\ \frac{1}{Y}\right),\quad B(X,\ Y)=(Y,\ X),$$

where $\theta\,(=e^{\frac{1}{8}i\pi})$ is a sixteenth root of unity, and KX, KY are respectively equal to $\phi\left(\frac{c+d\omega}{a+b\omega}\right)$, $\psi\left(\frac{c+d\omega}{a+b\omega}\right)$, and the formulæ are precisely similar to those of the present paper.

2, *Stone Buildings, W.C.*, 5 *May*, 1860.

294.

ON A NEW ANALYTICAL REPRESENTATION OF CURVES IN SPACE.

[From the *Quarterly Journal of Pure and Applied Mathematics*, vol. v. (1862), pp. 81—86.]

THE employment of a new kind of coordinates for the analytical representation of curves in space is suggested in my former paper under the same title *Journal*, t. III., pp. 225—236 (1859). The idea was as follows: viz. if (x, y, z, w) are current coordinates of a point in space (ordinary point coordinates), and $(\alpha, \beta, \gamma, \delta)$ the coordinates of a particular point, then taking (p, q, r, s, t, u) to represent the minor determinants formed out of the matrix

$$\begin{pmatrix} x, & y, & r, & w \\ \alpha, & \beta, & \gamma, & \delta \end{pmatrix},$$

viz.

$$\begin{aligned} p &= \gamma y - \beta z, & s &= \delta x - \alpha w, \\ q &= \alpha z - \gamma x, & t &= \delta y - \beta w, \\ r &= \beta x - \alpha y, & u &= \delta z - \gamma w, \end{aligned}$$

values which satisfy identically

$$ps + qt + ru = 0,$$

then the equation of a cone passing through a given curve and having for its vertex the arbitrary point $(\alpha, \beta, \gamma, \delta)$, is of the form

$$V = 0,$$

V being a homogeneous function of the six new coordinates (p, q, r, s, t, u). And it was proposed to consider $V = 0$ as the equation of the curve.

But as remarked in the paper, it is not every function V of the coordinates (p, q, r, s, t, u) which equated to zero, does in fact represent a curve. In order that the equation $V=0$ may represent a curve, it is necessary, that when any infinitesimal variations whatever are given to the constants $(\alpha, \beta, \gamma, \delta)$, thus converting the equation into $V+\delta V=0$, the two equations $V=0$, $\delta V=0$ (considered as equations in ordinary point coordinates) shall represent one and the same curve, whatever the system of infinitesimal variations attributed to $\alpha, \beta, \gamma, \delta$ may be. Let P, Q, R, S, T, U denote the differential coefficients of V in regard to p, q, r, s, t, u respectively, then the equation $\delta V=0$, breaks up into the equations

$$\begin{array}{llll} . & -Ry+Qz & -Sw & =0, \\ Rx & . & -Pz-Tw & =0, \\ -Qx+Py & . & -Uw & =0, \\ Sx+Ty+Uz & & . & =0, \end{array}$$

and the system composed of these four equations and the equation $V=0$ (considered as equations in ordinary point coordinates) must belong to one and the same curve.

The four equations gave

$$PS+QT+RU=0,$$

a relation between the differential coefficients of V which must be satisfied either identically or in virtue of the equation $V=0$. And this relation existing, any two of the four equations lead to the other two. Attending exclusively to the coordinates (p, q, r, s, t, u) and considering (x, y, z, w) as mere arbitrary multipliers, the above equation

$$PS+QT+RU=0$$

is the only relation between the differential coefficients of V which is deducible from the four equations.

But it was noticed that the equation $V=0$, even when V is a function such that we have (identically or in virtue of the equation $V=0$) the equation $PS+QT+RU=0$, does not of necessity represent a curve. Some further relation or relations between the differential coefficients of V must therefore exist, either identically or in virtue of the equation $V=0$; and such relations can be found by resorting to the second differential $\delta^2 V$ of the function V. In fact not only the equation $\delta V=0$ but the entire series of relations $\delta^2 V=0$, $\delta^3 V=0$,... should be satisfied by the coordinates of any point of the curve. I find by means of the equation $\delta^2 V=0$ a plexus of equations, which are consequently necessary, and I am inclined to believe sufficient, in order that the equation $V=0$ may in fact represent a curve; the equations of the plexus are, it will be seen, very numerous, and certainly only a small number of them are independent, but this is a question which I have not as yet investigated.

Attending to the expressions for p, q, r, s, t, u, we have

$$\begin{array}{rl} d_\alpha = & . \quad -yd_r+zd_q-wd_s=(1), \\ d_\beta = & xd_r \quad . \quad -zd_p-wd_t=(2), \\ d_\gamma = & -xd_q+yd_p \quad . \quad -wd_u=(3); \\ d_\delta = & xd_s+yd_t+zd_u \quad . \quad =(4), \end{array}$$

and writing for convenience a, b, c, d instead of d_α, d_β, d_γ, d_δ, we have

$$d = (1)\,\mathrm{a} + (2)\,\mathrm{b} + (3)\,\mathrm{c} + (4)\,\mathrm{d}.$$

It was in effect by operating on V with this symbol and equating to zero the coefficients of a, b, c, d, that the before-mentioned equations

$$\begin{array}{rrrrl} . & -Ry & +Qz & -Sw & =0, \\ Rx & . & -Pz & -Tw & =0, \\ -Qx & +Py & . & -Uw & =0, \\ Sx & +Ty & +Uz & . & =0, \end{array}$$

were found.

If to these equations we join the equation

$$Ax + By + Cz + Dw = 0,$$

where A, B, C, D are arbitrary multipliers, we can express x, y, z, w in terms of A, B, C, D in such manner as to satisfy the four equations, viz. we have

$$\begin{array}{rrrrl} x = & . & BU & -CT & +DP, \\ y = & -AU & . & +CS & +DQ, \\ z = & AT & -BS & . & +DR, \\ w = & -AP & -BQ & -CR & . \;, \end{array}$$

and if in the expressions for (1), (2), (3), (4) we substitute for x, y, z, w these values, and form therewith the value of d, which value I will for distinction call $\mathfrak{D}$, we have

$$\mathfrak{D} = \left(\begin{array}{llll} Ud_r + Td_q + Pd_s, & Qd_s - Sd_q, & Rd_s - Sd_r, & Rd_q - Qd_r \\ Pd_t - Td_p, & Ud_r + Qd_t + Sd_p, & Rd_t - Td_r, & Pd_r - Rd_p \\ Pd_u - Ud_p, & Qd_u - Ud_q, & Rd_u + Sd_p + Td_q, & Qd_p - Pd_q \\ Td_u - Ud_t, & Ud_s - Sd_u, & Sd_t - Td_s, & Pd_s + Qd_t + Rd_u \end{array}\right)(A, B, C, D)(\mathrm{a}, \mathrm{b}, \mathrm{c}, \mathrm{d}),$$

viz. $\mathfrak{D}$ is a lineo-linear function of the two sets of indeterminate quantities (A, B, C, D), (a, b, c, d), the coefficients thereof being the operators

$$Ud_r + Td_q + Pd_s,\ Qd_s - Sd_q,\ \&\text{c.}$$

It may be remarked that we have identically

$$\mathfrak{D}V = (PS + QT + RU)(A\mathrm{a} + B\mathrm{b} + C\mathrm{c} + D\mathrm{d}),$$

since obviously each term such as $(Qd_s - Sd_q)\,V$, which is equal to $QS - SQ$, vanishes identically. The equation $\mathfrak{D}V = 0$ gives therefore only the before-mentioned equation $PS + QT + RU = 0$, which is as it should be.

The equation $\mathfrak{D}^2 V=0$, is then to be satisfied independently of the values of (A, B, C, D) and (a, b, c, d), and as $\mathfrak{D}$ contains 16 distinct terms, $\mathfrak{D}^2$ will contain in all $\frac{1}{2}16.17$ or 136 distinct terms. The equation $\mathfrak{D}^2 V=0$ gives therefore a plexus of 136 equations, and the equations in each succeeding plexus, involved in $\mathfrak{D}^3 V=0$, $\mathfrak{D}^4 V=0$, &c. will, of course, be still be more numerous.

If $V=0$ be the plane conic which is the intersection of the surfaces

$$\begin{aligned} x^2+y^2+z^2+w^2&=0,\\ ax+by+cz+dw&=0, \end{aligned}$$

then we have

$$V=\left(\begin{array}{cccccc} b^2+c^2, & -ab, & -ac, & ., & cd, & -bd \\ -ba, & c^2+a^2, & -bc, & -cd, & ., & ad \\ -ca, & -cb, & a^2+b^2, & bd, & -ad, & . \\ ., & -cd, & bd, & a^2+d^2, & ab, & ac \\ cd, & ., & -ad, & ba, & b^2+d^2, & bc \\ -bd, & ad, & ., & ca, & cb, & c^2+d^2 \end{array}\right)(p, q, r, s, t, u)^2.$$

The values of P, Q, R, S, T, U (omitting a common factor 2) are

$$P=(b^2+c^2, -ab, -ac, ., +cd, -bd)(p, q, r, s, t, u),$$
&c.,

and if we proceed to form a term in $\mathfrak{D}^2 V$, say the coefficient of $A^2\mathrm{a}^2$, this is $(Ud_r+Td_q+Pd_s)^2\ V$, or

$$(a^2+b^2, c^2+a^2, a^2+d^2, -cd, +bd, -cb)(U, T, P)^2.$$

The coefficient therein of p^2 is

$$(a^2+b^2, c^2+a^2, a^2+d^2, -cd, +bd, -cb)(-bd, cd, b^2+c^2)^2,$$

that is, it is

$$\begin{aligned} &(a^2+b^2)\,b^2d^2 && -2cd\,.\ cd\,(b^2+c^2)\\ +&(c^2+a^2)\,c^2d^2 && +2bd\,.-bd\,(b^2+c^2)\\ +&(a^2+d^2)\,(b^2+c^2)^2 && -2cb\,.-bd\,.\,cd, \end{aligned}$$

where the terms in which (b^2+c^2) does not appear as a factor are together equal to

$$a^2d^2(b^2+c^2)+d^2(b^2+c^2)^2,$$

the entire expression thus divides by b^2+c^2, the quotient being

$$(a^2+d^2)(b^2+c^2)-2c^2d^2-2b^2d^2+a^2d^2+d^2(b^2+c^2),$$

which is equal to $a^2(b^2+c^2+d^2)$, or restoring the factor b^2+c^2, we see that in $\mathfrak{D}^2 V$ the coefficient of $A^2\mathrm{a}^2$ is

$$a^2(b^2+c^2+d^2)(b^2+c^2)\,p^2+\text{\&c.}$$

The complete value must, it is clear, be of the form

$$a^2(b^2+c^2+d^2)\,V + k\,(ps+qt+ru),$$

vanishing in virtue of the equations $V=0$, $ps+qt+ru=0$, and this being so, observing that V contains no term in ps, we have $k=$ coefficient ps in

$$(a^2+b^2,\ c^2+a^2,\ a^2+d^2,\ -cd,\ +bd,\ -cb)\,(U,\ T,\ P)^2,$$

that is

$$k = 2\,(a^2+b^2,\ c^2+a^2,\ a^2+d^2,\ -cd,\ +bd,\ -cb)\,(-bd,\ cd,\ b^2+c^2)\,(ca,\ ba,\ 0),$$

or

$$\begin{aligned}\tfrac{1}{2}k = {} & (a^2+b^2).-bd\,.\,ca \ - \ cd\,\{cd\,.\,0+(b^2+c^2)\,ba\}\\ & +(c^2+a^2).\ \ cd\,.\,ba \ + \ bd\,\{(b^2+c^2)\,ca-bd\,.\,0\}\\ & +(a^2+d^2).\qquad 0 \ - \ cb\,\{-bd\,.\,ba+cd\,.\,ca\},\end{aligned}$$

which is

$$= abcd\left\{\begin{matrix} -(a^2+b^2) & -(b^2+c^2) \\ +(c^2+a^2) & +(b^2+c^2) \\ & +(b^2-c^2)\end{matrix}\right\},\ =0.$$

The coefficient k consequently vanishes, and therefore in $\mathfrak{D}^2V$ the coefficient of $A^2\mathfrak{a}^2$ is $a^2(b^2+c^2+d^2)\,V$, but I have not worked out the coefficients of the other terms.

2, *Stone Buildings, W.C.*, 30*th October*, 1860.

295.

ON THE CONSTRUCTION OF THE NINTH POINT OF INTERSECTION OF THE CUBICS WHICH PASS THROUGH EIGHT GIVEN POINTS.

[From the *Quarterly Journal of Pure and Applied Mathematics*, vol. v. (1862), pp. 222—233.]

I REPRODUCE with additional developments the solution which has been given of this interesting problem.

The equation of a given cubic may be written

$$qU - pV = 0,$$

where $U=0$, $V=0$ are any two conics meeting the cubic in the same four points; $p=0$ is the line joining the remaining two points of intersection of the cubic with the conic $U=0$; and $q=0$ is the line joining the remaining two points of intersection of the cubic with the conic $V=0$; the relation between the arbitrary constant factors implicitly contained in the functions qU and pV is assumed to be properly determined.

The form is employed by Plücker, *Theorie der algebraischen Curven*, p. 56 (1839), and in connexion therewith he gives some geometrical considerations which, he remarks, contain implicitly the solution of the above-mentioned problem.

The form is also the analytical basis of the investigations of M. Chasles, "Construction de la courbe du troisième ordre par neuf points," *Comptes Rendus*, t. XXXVI. (1853), pp. 942—952. In fact calling to mind the theorem that in the pencil of conics $U - \alpha V = 0$ (where α is an arbitrary multiplier) the anharmonic ratio of the multipliers α is equal to the anharmonic ratio of the tangents at any one of the points of intersection, or (what is the same thing) to the anharmonic ratio of the polars of any point whatever in regard to the conics, and recollecting that the anharmonic ratio in question is said to be the anharmonic ratio of the conics themselves; then since the equation $qU-pV=0$ is satisfied by the system

$$U-\alpha V=0,\ p-\alpha q=0,$$

(where α is arbitrary) we have at once the theorem, p. 949, viz. that if there be a pencil of lines through a *point*, and corresponding anharmonically thereto, a pencil of conics through the same *four points;* the locus of the intersections of a line by the corresponding conic is a cubic through the five points: and, conversely, that a given cubic may be so generated, the point of the pencil of lines, and the four points of the pencil of conics being any five points whatever of the cubic.

This gives at once the construction (M. Chasles' *first* construction) for the cubic through nine given points. In fact if the points are called 1, 2, 3, 4, 5, 6, 7, 8, 9; then grouping the points in any manner, it is only necessary to find a point x such that

$$x\,(1,\ 2,\ 3,\ 4,\ 5)=6789\ (1,\ 2,\ 3,\ 4,\ 5),$$

that is, such that the pencil of lines $x1$, $x2$, $x3$, $x4$, $x5$ shall correspond anharmonically to the pencil of conics 67891, 67892, 67893, 67894, 67895. The foregoing notation is that employed in M. de Jonquières' "Essai sur la génération des Courbes géométriques, &c.," *Mém. Sav. Etrang.*, t. XVI. (1858), which I take the opportunity of referring to. In fact, if x satisfies the foregoing condition, then taking through the point x any other line, and corresponding anharmonically thereto a conic through the points 6, 7, 8, 9, the locus of the intersections of the line and conic will be a cubic through the nine points. But the condition in question gives

$$x\,(1,\ 2,\ 3,\ 4)=6789\,(1,\ 2,\ 3,\ 4),$$
$$x\,(1,\ 2,\ 3,\ 5)=6789\,(1,\ 2,\ 3,\ 5),$$

which (by the anharmonic property of the points of a conic) show, the first that x is in a certain conic passing through 1, 2, 3, 4, and the second that x is in a certain conic passing through 1, 2, 3, 5; the two conics intersect in the points 1, 2, 3, and in a fourth point which is the required point x. Or we may say that x is given by the condition that the pencils $x\,(1,\ 2,\ 3,\ 4)$ and $x\,(1,\ 2,\ 3,\ 5)$ shall have given anharmonic ratios. It will presently be seen how x can be determined by the ruler alone.

Suppose now that the points 1, 2, 3, 4, 5, 6, 7, 8, 9 are the points of intersection of two cubics; the construction should become indeterminate; this is only the case when the two conics which by their intersection should determine x become one and the same conic. This implies that the conic $x\,(1,\ 2,\ 3,\ 4)$ passes through 5, or that x, 1, 2, 3, 4, 5 are points of the same conic. And then since by the anharmonic property of the points of a conic $x\,(1,\ 2,\ 3,\ 4)=5\,(1,\ 2,\ 3,\ 4)$, we have

$$5\,(1,\ 2,\ 3,\ 4)=6789\,(1,\ 2,\ 3,\ 4).$$

The grouping of the nine points is altogether arbitrary, hence there are in all ($9 \times 70 =$) 630 such equations, which are really equivalent to only two equations, and which when eight of the points are given, determine the ninth point. Supposing that the given points are 1, 2, 3, 4, 5, 6, 7, 8, the equations for the determination of the remaining point 9 may be taken to be

$$9\,(5,\ 6,\ 7,\ 8) = 1234\,(5,\ 6,\ 7,\ 8),$$
$$9\,(4,\ 6,\ 7,\ 8) = 1235\,(4,\ 6,\ 7,\ 8),$$

which (it is to be remarked) determine 9 in a similar way to that in which x is given in the construction of the cubic through nine points; viz. 9 is the fourth intersection of two conics which pass through the points 5, 6, 7, 8 and the points 4, 6, 7, 8 respectively. Or we may say that 9 is given by the conditions that the pencils 9 (5, 6, 7, 8) and 9 (4, 6, 7, 8) shall have given anharmonic ratios.

The foregoing equations

$$9\,(5,\ 6,\ 7,\ 8) = 1234\,(5,\ 6,\ 7,\ 8),$$
$$9\,(4,\ 6,\ 7,\ 8) = 1235\,(4,\ 6,\ 7,\ 8),$$

are equivalent to and constitute the geometrical interpretation of the equations obtained (previous to M. Chasles' Memoir) by Weddle in the paper "On the construction of the ninth point of intersection of two curves of the third degree when the other eight points are given," *Cambridge and Dublin Mathematical Journal*, t. VI., pp. 83—86 (1851). In fact, reproducing his analysis with only a slight change of notation, let 012 denote the determinant

$$\begin{vmatrix} x, & y, & z \\ x_1, & y_1, & z_1 \\ x_2, & y_2, & z_2 \end{vmatrix},$$

so that $012 = 0$ is the equation of the line through the points 1 and 2; and in like manner let 012345 denote the determinant

$$\begin{vmatrix} x^2, & y^2, & z^2, & yz, & zx, & xy \\ x_1^2, & \&c. & & & & \\ \vdots & & & & & \end{vmatrix},$$

so that $012345 = 0$ is the equation of the conic through the points 1, 2, 3, 4, 5. Of course 123, 123456, &c. will denote given functions of the coordinates of the points 1, 2, 3, the points 1, 2, 3, 4, 5, 6, &c. This being so

$$012345\,.\,078 = \lambda\,.\,012347\,.\,058$$

is the equation of a particular cubic passing through the points 1, 2, 3, 4, 5, 7, 8, and which if we properly determine λ, viz. if we write

$$\lambda = \frac{612345\,.\,678}{612347\,.\,658}$$

will also pass through the point 6.

And similarly

$$012345 \,.\, 076 = \mu \,.\, 012347 \,.\, 056$$

is the equation of a particular cubic curve passing through the points 1, 2, 3, 4, 5, 6, 7 and which if we properly determine μ, viz. if we write

$$\mu = \frac{812345 \,.\, 876}{812347 \,.\, 856}$$

will also pass through the point 8. Hence the two curves, each of them passing through the points 1, 2, 3, 4, 5, 6, 7, 8 will intersect in the remaining point 9; and writing 9 for 0, and combining the two equations, we have

$$\frac{978 \,.\, 956}{958 \,.\, 976} = \frac{\lambda}{\mu} = \frac{612345}{612347} \frac{812347}{812345},$$

or, what is the same thing,

$$\frac{956 \,.\, 978}{958 \,.\, 967} = \frac{123456 \,.\, 123478}{123458 \,.\, 123467},$$

which is Weddle's equation, and is equivalent to the above-mentioned equation

$$9\ (5,\ 6,\ 7,\ 8) = 1234\ (5,\ 6,\ 7,\ 8).$$

To prove this I remark that we have identically

$$012 \,.\, 034 \,.\, 514 \,.\, 523 - 014 \,.\, 023 \,.\, 512 \,.\, 534 = 012345.$$

In fact the left-hand side equated to zero is the equation of the conic through 1, 2, 3, 4, 5, and such left-hand side must therefore, save to a mere numerical factor, be equal to 012345. And to determine this factor it is to be observed that 012345 contains the term

$$+ x_0{}^2 \,.\, y_1{}^2 \,.\, z_2{}^2 \,.\, y_3 z_3 \,.\, z_4 x_4 \,.\, x_5 y_5,$$

but that there is no such term in 012.034.514.523, and that there is in $-$ 014.023.512.534 the equivalent term

$$- x_0 y_1 z_4 \,.\, - x_0 y_3 z_2 \,.\, x_5 y_1 z_2 \,.\, y_5 z_3 x_4,$$

so that the numerical factor is rightly determined.

The foregoing identity written under the form

$$\frac{012 \,.\, 034}{014 \,.\, 023} - \frac{512 \,.\, 534}{514 \,.\, 523} = \frac{012345}{014 \,.\, 023 \,.\, 514 \,.\, 523}$$

shows that, when $012345 = 0$, i.e. if 0 be a point of the conic through 1, 2, 3, 4, 5, then we have

$$0\ (1,\ 2,\ 3,\ 4) = 5\ (1,\ 2,\ 3,\ 4),$$

which is in fact the anharmonic property of the points of a conic. And observing that $012345 = 051234$, and substituting 5, 6, for 0, 5 respectively, the identity becomes

$$\frac{512 \,.\, 534}{514 \,.\, 523} - \frac{612 \,.\, 634}{614 \,.\, 623} = \frac{561234}{514 \,.\, 523 \,.\, 614 \,.\, 623}.$$

The equation $012345 = 0$ may be written

$$012 \,.\, 034 - \frac{512 \,.\, 534}{514 \,.\, 523} 014 \,.\, 023 = 0,$$

and hence the anharmonic ratio of the conics

$$012345 = 0,\ 012346 = 0,\ 012347 = 0,\ 012348 = 0$$

is equal to that of the quantities

$$\frac{512 \,.\, 534}{514 \,.\, 523},\ \frac{612 \,.\, 634}{614 \,.\, 623},\ \frac{712 \,.\, 734}{714 \,.\, 723},\ \frac{812 \,.\, 834}{814 \,.\, 823},$$

or calling these quantities for a moment α, β, γ, δ, it is

$$= \frac{(\alpha - \beta)(\gamma - \delta)}{(\alpha - \delta)(\beta - \gamma)},$$

where

$$\alpha - \beta = \frac{512 \,.\, 534}{514 \,.\, 523} - \frac{612 \,.\, 634}{614 \,.\, 623} = \frac{561234}{514 \,.\, 523 \,.\, 614 \,.\, 623};$$

and forming in this manner the expressions of each of the four factors $\alpha - \beta$, $\gamma - \delta$, $\alpha - \delta$, $\beta - \gamma$, we have

$$\frac{(\alpha - \beta)(\gamma - \delta)}{(\alpha - \delta)(\beta - \gamma)} = \frac{561234 \,.\, 781234}{581234 \,.\, 671234},$$

so that in the equation

$$\frac{956 \,.\, 978}{958 \,.\, 967} = \frac{561234 \,.\, 781234}{581234 \,.\, 671234},$$

the right-hand side is

$$= 1234\,(5,\ 6,\ 7,\ 8),$$

and since by what precedes the left-hand side is $= 9\,(5,\ 6,\ 7,\ 8)$, the equation is

$$9\,(5,\ 6,\ 7,\ 8) = 1234\,(5,\ 6,\ 7,\ 8),$$

which is the transformation in question.

Now resuming the two equations

$$9\,(5,\ 6,\ 7,\ 8) = 1234\,(5,\ 6,\ 7,\ 8),$$
$$9\,(4,\ 6,\ 7,\ 8) = 1235\,(4,\ 6,\ 7,\ 8),$$

the right-hand sides are given anharmonic ratios, and as we have seen the question is to find 9 so that the anharmonic ratios $9\,(5,\ 6,\ 7,\ 8)$, $9\,(4,\ 6,\ 7,\ 8)$ shall have given values. But for the geometrical solution by the ruler alone, we have the preliminary question, from the given eight points, without the assistance of the before-mentioned conics, to construct the given anharmonic ratios $1234\,(5,\ 6,\ 7,\ 8)$ and $1235\,(4,\ 6,\ 7,\ 8)$. The solution of both questions is given in Dr Hart's paper, "Construction by the ruler alone to determine the ninth point of intersection of two curves of the third degree," *Cambridge and Dublin Mathematical Journal*, t. VI., pp. 181, 182 (1851).

The Preliminary Question. The anharmonic ratio 1234 (5, 6, 7, 8) is equal to that of the polars of an arbitrary point X in regard to the conics 12345, 12346, 12347, 12348 respectively (these polars all pass through one and the same point). Now to construct the polars of X in regard to these conics, and first in regard to the conic 12345: the fourth harmonics of X in regard to the lines 12, 34, in regard to the lines 13, 42, and in regard to the lines 14, 23, meet in a point; and considering the several combinations 1234, 1235, 1245, 1345, 2345 we have thus five points; these lie on a line which is the required polar of X in regard to the conic 12345. The polars in regard to the other conics are obtained in the same manner; and it is clear that the first above-mentioned point (viz. that deduced from the points 1, 2, 3, 4) is in fact the point of intersection of the four polars, or point of the pencil formed by the polars.

The Principal Question then is, given the points 4, 5, 6, 7, 8 to find the point 9, such that

$$9\,(5,\ 6,\ 7,\ 8) = 1234\,(5,\ 6,\ 7,\ 8),$$

$$9\,(4,\ 6,\ 7,\ 8) = 1235\,(4,\ 6,\ 7,\ 8),$$

where the right-hand sides represent given anharmonic ratios.

For this, let 65, 74 (see the figure) meet in M, and on 74 find a point Q such

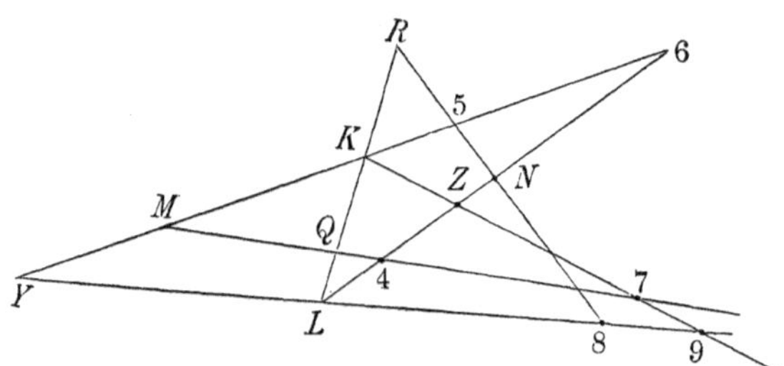

that the anharmonic ratio of the points (4, M, 7, Q) may be equal to the given ratio 1235 (4, 6, 7, 8), say

$$(4,\ M,\ 7,\ Q) = 1235\,(4,\ 6,\ 7,\ 8),$$

and let 64, 85 meet in N and on 85 find a point R, such that the anharmonic ratio of the points 5, N, R, 8 is equal to the given anharmonic ratio 1234 (5, 6, 7, 8), say

$$(5,\ N,\ R,\ 8) = 1234\,(5,\ 6,\ 7,\ 8).$$

Join QR meeting 65 in K and 64 in L; then $7K$ and $8L$ will meet in the required point 9.

In fact taking Y as the intersection of the lines $65KM$ and $8L$, and Z as the intersection of the lines $64NL$ and $7K$; then, as is clear from the figure, first, the anharmonic ratio $9(4, 6, 7, 8)$ is equal to that of the points $(4, 6, Z, L)$ on the line $46ZL$, that is

$$9(4, 6, 7, 8) = (4, 6, Z, L),$$

which is

$$= (4, M, 7, Q),$$

since the lines $6M$, $Z7$, LQ meet in the point K; but by the construction $(4, M, 7, Q) = 1235\,(4, 6, 7, 8)$, that is, we have

$$9(4, 6, 7, 8) = 1235\,(4, 6, 7, 8);$$

and, secondly, the anharmonic ratio $9(5, 6, 7, 8)$ is equal to that of the points $(5, 6, K, Y)$ on the line $56KY$, that is

$$9(5, 6, 7, 8) = (5, 6, K, Y),$$

which is

$$= (5, N, R, 8),$$

since the lines $6N$, KR, $Y8$ meet in the same point L; but by the construction $(5, N, R, 8) = 1234\,(5, 6, 7, 8)$, that is, we have

$$9(5, 6, 7, 8) = 1234\,(5, 6, 7, 8),$$

so that the point 9 satisfies the required conditions.

It has been already remarked that the point x in M. Chasles' theorem for the construction of the cubic through nine points is determined by precisely similar conditions to those which determine the ninth intersection of the two cubics; that is, the foregoing construction by the ruler alone is applicable to the determination of the point x; and when this is once obtained, the remainder of the construction, giving the points of the cubic through the nine given points, can obviously be performed by the ruler alone. The construction for the cubic through nine points gives implicitly the relation between ten points of the cubic and such relation is accordingly expressed by the equation

$$x(1, 2, 3, 4, 5, 10) = 6789\,(1, 2, 3, 4, 5, 10),$$

which is one out of 210 similar forms. But it is possible that some more convenient form of the relation between the ten points may yet be found.

I proceed to further develope the analytical theory. Writing for convenience ω in the place of 10, we have

$$x(1, 2, 3, 4, 5, 6) = 789\omega\,(1, 2, 3, 4, 5, 6),$$

or, what is the same thing,

$$x(1, 2, 3, 4) = 789\omega\,(1, 2, 3, 4),$$
$$x(1, 2, 3, 5) = 789\omega\,(1, 2, 3, 5),$$
$$x(1, 2, 3, 6) = 789\omega\,(1, 2, 3, 6),$$

which belong to three conics each of them passing through 1, 2, 3, and which must have a remaining fourth point of intersection.

The equation of the first conic is

$$012\,.\,034 = \frac{\lambda}{\mu}\,014\,.\,023,$$

where

$$\lambda = 789\omega12\,.\,789\omega34,$$

$$\mu = 789\omega14\,.\,789\omega23,$$

and thence, in virtue of an identical equation already referred to,

$$-\lambda - \mu = 789\omega13\,.\,789\omega24.$$

But we have identically

$$123\,.\,0pq = 1pq\,.\,023 + 2pq\,.\,031 + 3pq\,.\,012,$$

and thence in particular

$$123\,.\,034 = 134\,.\,023 + 234\,.\,031,$$

$$123\,.\,014 = 214\,.\,031 + 314\,.\,012,$$

and the equation of the conic may therefore be written

$$012\,(134\,.\,023 + 234\,.\,031)\,\mu - 023\,(214\,.\,031 + 314\,.\,012)\,\lambda = 0,$$

that is

$$031\,.\,012\,.\,234\,.\,\mu + 012\,.\,023\,.\,314\,(-\lambda - \mu) + 023\,.\,031\,.\,124\,.\,\lambda = 0,$$

or, substituting for μ, $-\lambda - \mu$, λ, their values, this is

$$\begin{aligned} &031\,.\,012\,.\,234\,.\,789\omega14\,.\,789\omega23 \\ +\,&012\,.\,023\,.\,314\,.\,789\omega13\,.\,789\omega42 \\ +\,&023\,.\,031\,.\,124\,.\,789\omega12\,.\,789\omega34 = 0, \end{aligned}$$

or, what is the same thing,

$$\frac{234\,.\,789\omega14\,.\,789\omega23}{023} + \frac{314\,.\,789\omega13\,.\,789\omega24}{031} + \frac{124\,.\,789\omega12\,.\,789\omega34}{012} = 0,$$

or, making a slight change of form, the equation of the conic is

$$\frac{423\,.\,789\omega41\,.\,789\omega23}{023} + \frac{431\,.\,789\omega42\,.\,789\omega31}{031} + \frac{412\,.\,789\omega43\,.\,789\omega12}{012} = 0.$$

The equations of the other two conics are deduced by writing successively 5 and 6 in the place of 4; and the condition in order that the conics may have a remaining fourth point of intersection is

$$\begin{vmatrix} 423\,.\,789\omega41, & 431\,.\,789\omega42, & 412\,.\,789\omega43 \\ 523\,.\,789\omega51, & 531\,.\,789\omega52, & 512\,.\,789\omega53 \\ 623\,.\,789\omega61, & 631\,.\,789\omega62, & 612\,.\,789\omega63 \end{vmatrix} = 0.$$

This equation, say $\square = 0$, expresses the relation between the coordinates of the ten points 1, 2, 3, 4, 5, 6, 7, 8, 9, ω, of the cubic. Hence if 123456789ω denote the determinant

$$\left|\begin{matrix} x_1^3, & y_1^3, & z_1^3, & y_1^2z_1, & z_1^2x_1, & x_1^2y_1, & y_1z_1^2, & z_1x_1^2, & x_1y_1^2, & x_1y_1z_1 \\ x_2^3, & \&c. & & & & & & & & \\ \vdots & & & & & & & & & \end{matrix}\right|,$$

123456789ω must be a factor of $\square$, and a little consideration shows that the other factor which is of the order one as regards the coordinates of each of the points 1, 2, 3, and of the order three as regards the coordinates of each of the points 7, 8, 9, ω, must be of the form $123 . 789 . 78\omega . 79\omega . 89\omega$. We must therefore have

$$\square = \epsilon . 123 . 789 . 78\omega . 79\omega . 89\omega . 123456789\omega,$$

where the merely numerical factor ϵ is, I believe, equal to $+1$ or else to -1.

In order to verify the factor $123 . 789 . 78\omega . 79\omega . 89\omega$, observing that the points 7, 8, 9, ω enter symmetrically, it will be sufficient to show that 123, 789 are each of them factors of $\square$, or, what is the same thing, that if $123 = 0$, or if $789 = 0$, then in either case $\square = 0$.

First, if $123 = 0$, we may write

$$\begin{aligned} x_3 &= \lambda x_1 + \mu x_2, \\ y_3 &= \lambda y_1 + \mu y_2, \\ z_3 &= \lambda z_1 + \mu z_2 \end{aligned}$$

equations which give

$$423 = \lambda . 421, \quad 431 = \lambda . 421, \quad \&c.,$$

and the equation $\square = 0$, thus becomes

$$\left|\begin{matrix} 789\omega41, & 789\omega42, & 789\omega43 \\ 789\omega51, & 789\omega52, & 789\omega53 \\ 789\omega61, & 789\omega62, & 789\omega63 \end{matrix}\right| = 0,$$

or, what is the same thing,

$$\left|\begin{matrix} 789\omega14, & 789\omega24, & 789\omega34 \\ 789\omega15, & 789\omega25, & 789\omega35 \\ 789\omega16, & 789\omega27, & 789\omega36 \end{matrix}\right| = 0.$$

Now if the terms in the same column are multiplied by $789\omega56$, $789\omega64$, $789\omega45$ respectively and added, then for the first column the sum is

$$789\omega14 . 789\omega56 + 789\omega15 . 789\omega64 + 789\omega16 . 789\omega45,$$

which is $= 0$, and the sums for the second and third columns are each $= 0$ in the same manner: wherefore the determinant vanishes as it should do.

Next, if $789=0$, we have identically

$$789\omega 41 = 789 \,.\, 741 \,.\, \omega 81 \,.\, \omega 94 + 781 \,.\, 794 \,.\, \omega 89 \,.\, \omega 41,$$

which when $789=0$ gives

$$789\omega 41 = 781 \,.\, 794 \,.\, \omega 89 \,.\, \omega 41,$$

and in like manner,

$$789\omega 42 = 782 \,.\, 794 \,.\, \omega 89 \,.\, \omega 42,$$

$$789\omega 43 = 783 \,.\, 794 \,.\, \omega 89 \,.\, \omega 43,$$

in which three equations 4 may be changed into 5 and 6 successively. The equation $\square=0$ thus becomes

$$\begin{vmatrix} 423 \,.\, \omega 41, & 431 \,.\, \omega 42, & 412 \,.\, \omega 43 \\ 523 \,.\, \omega 51, & 531 \,.\, \omega 52, & 512 \,.\, \omega 53 \\ 623 \,.\, \omega 61, & 631 \,.\, \omega 62, & 612 \,.\, \omega 63 \end{vmatrix} = 0,$$

or, what is the same thing,

$$\begin{vmatrix} 423 \,.\, 41\omega, & 421 \,.\, 4\omega 3, & 42\omega \,.\, 431 \\ 523 \,.\, 51\omega, & 521 \,.\, 5\omega 3, & 52\omega \,.\, 531 \\ 623 \,.\, 61\omega, & 621 \,.\, 6\omega 3, & 69\omega \,.\, 631 \end{vmatrix} = 0,$$

and since the sum of the terms in each line of the determinant is $=0$, the determinant is as it should be $=0$.

The foregoing equation

$$\begin{vmatrix} 412 \,.\, 789\omega 43, & 423 \,.\, 789\omega 41, & 431 \,.\, 789\omega 42 \\ 512 \,.\, 789\omega 53, & 523 \,.\, 789\omega 51, & 531 \,.\, 789\omega 52 \\ 612 \,.\, 789\omega 63, & 623 \,.\, 789\omega 61, & 631 \,.\, 789\omega 62 \end{vmatrix}$$

$$= \epsilon \,.\, 123 \,.\, 789 \,.\, 78\omega \,.\, 79\omega \,.\, 89\omega \,.\, 123456789\omega,$$

(since 412, $789\omega 43$, &c. are interpretable functions of the coordinates) affords a geometrical interpretation of the equation

$$123456789\omega = 0$$

between the coordinates of the ten points of the cubic; but it would be more satisfactory if a similar identical equation could be found, having on the right-hand side the function 123456789ω without the irrelevant factor

$$123 \,.\, 789 \,.\, 78\omega \,.\, 79\omega \,.\, 89\omega.$$

2, *Stone Buildings, W.C., 6th March,* 1862.

296.

ON THE CONICS WHICH PASS THROUGH THE FOUR FOCI OF A GIVEN CONIC.

[From the *Quarterly Journal of Pure and Applied Mathematics*, vol. v. (1862), pp. 275—280.]

THE foci of a conic are the points of intersection of the tangents through the circular points at infinity; the pair of tangents through each of the circular points at infinity is a conic through the four foci; and we have thus two conics $P=0$, $Q=0$ passing through the four foci; the equation of any other conic through the four foci is of course $P+\lambda Q=0$; and in particular if λ be suitably determined this equation gives the axes of the conic.

I was led to develope the solution, in seeking to obtain the elegant formulæ given in Mr P. J. Hensley's paper "Determination of the foci of the conic section expressed by trilinear coordinates," *Journal*, t. v., pp. 177—183, (March, 1862).

I take the coordinates to be proportionate to the perpendicular distances of the point from the sides of the fundamental triangle, each distance divided by the perpendicular distance of the side from the opposite angle. This being so, the equation of the line infinity is

$$x+y+z=0,$$

and, α, β, γ denoting the sides of the fundamental triangle, the equation of the circle circumscribed about the triangle is

$$\frac{\alpha^2}{x}+\frac{\beta^2}{y}+\frac{\gamma^2}{z}=0.$$

The foregoing two equations determine the circular points at infinity; and if (x_1, y_1, z_1) are the coordinates, there is no difficulty in obtaining the system of values

$$\begin{aligned} x_1 : y_1 : z_1 &= -\alpha : \beta(\cos C+i\sin C) : \gamma(\cos B+i\sin B) \\ &= \alpha(\cos C-i\sin C) : -\beta : \gamma(\cos A+i\sin A) \\ &= \alpha(\cos B+i\sin B) : \beta(\cos A-i\sin A) : -\gamma \end{aligned},$$

where as usual $i=\sqrt{(-1)}$, and where A, B, C denote the angles of the triangle, so that the cosines and sines of these angles denote given functions of α, β, γ. The coordinates (x_2, y_2, z_2) of the other circular point at infinity are of course obtained by merely writing $-i$ for i. We find also

$$x_1x_2 : y_1y_2 : z_1z_2 : y_1z_2+y_2z_1 : z_1x_2+z_2x_1 : x_1y_2+x_2y_1$$
$$=-\alpha^2 : -\beta^2 : -\gamma^2 : \beta^2+\gamma^2-\alpha^2 : \gamma^2+\alpha^2-\beta^2 : \alpha^2+\beta^2-\gamma^2,$$

which are the formula chiefly made use of in the sequel.

Suppose now that the equation of the conic is

$$U=(a, b, c, f, g, h)(x, y, z)^2=0;$$

then putting for a moment

$$U_1=(a, b, c, f, g, h)(x_1, y_1, z_1)^2,$$
$$W_1=(a, b, c, f, g, h)(x, y, z)(x_1, y_1, z_1),$$

and the like as regards U_2 and W_2; the tangents from the points (x_1, y_1, z_1) and (x_2, y_2, z_2) respectively are

$$UU_1-W_1^2=0,$$
$$UU_2-W_2^2=0,$$

which are respectively pairs of lines intersecting in the four foci. And it is moreover clear that the equation of the axes is

$$U_2W_1^2-U_1W_2^2=0.$$

The foregoing equations may be written

$$(\mathfrak{A}, \mathfrak{B}, \mathfrak{C}, \mathfrak{F}, \mathfrak{G}, \mathfrak{H})(yz_1-y_1z, zx_1-z_1x, xy_1-x_1y)^2=0,$$
$$(\mathfrak{A}, \mathfrak{B}, \mathfrak{C}, \mathfrak{F}, \mathfrak{G}, \mathfrak{H})(yz_2-y_2z, zx_2-z_2x, xy_2-x_2y)^2=0,$$

where $(\mathfrak{A}, \mathfrak{B}, \mathfrak{C}, \mathfrak{F}, \mathfrak{G}, \mathfrak{H})$ are the inverse system of coefficients.

These may be written

$$(\mathrm{a}, \mathrm{b}, \mathrm{c}, \mathrm{f}, \mathrm{g}, \mathrm{h})(x_1, y_1, z_1)^2=0,$$
$$(\mathrm{a}, \mathrm{b}, \mathrm{c}, \mathrm{f}, \mathrm{g}, \mathrm{h})(x_2, y_2, z_2)^2=0,$$

where a, b, c, f, g, h are quadric functions of x, y, z, viz.

$$\begin{aligned}
\mathrm{a} &= \mathfrak{B}z^2+\mathfrak{C}y^2-2\mathfrak{F}yz,\\
\mathrm{b} &= \mathfrak{C}x^2+\mathfrak{A}z^2-2\mathfrak{G}zx,\\
\mathrm{c} &= \mathfrak{A}y^2+\mathfrak{B}x^2-2\mathfrak{H}xy,\\
\mathrm{f} &= -\mathfrak{A}yz-\mathfrak{F}x^2+\mathfrak{G}xy+\mathfrak{H}xz,\\
\mathrm{g} &= -\mathfrak{B}zx-\mathfrak{G}y^2+\mathfrak{H}yz+\mathfrak{F}yx,\\
\mathrm{h} &= -\mathfrak{C}xy-\mathfrak{H}z^2+\mathfrak{F}zx+\mathfrak{G}xy,
\end{aligned}$$

and this being so, I combine the two equations as follows:

$$\begin{array}{llll}
x_2^2 & (\mathrm{a}, \ldots)(x_1, y_1, z_1)^2 + x_1^2 & (\mathrm{a}, \ldots)(x_2, y_2, z_2)^2 & = 0, \\
y_2^2 & \quad " \qquad + y_1^2 & \quad " & = 0, \\
z_2^2 & \quad " \qquad + z_1^2 & \quad " & = 0, \\
y_2 z_2 & (\mathrm{a}, \ldots)(x_1, y_1, z_1)^2 + y_1 z_1 & (\mathrm{a}, \ldots)(x_2, y_2, z_2)^2 & = 0, \\
z_2 x_2 & \quad " \qquad + z_1 x_1 & \quad " & = 0, \\
x_2 y_2 & \quad " \qquad + x_1 y_1 & \quad " & = 0,
\end{array}$$

any one of which is the equation of a conic passing through the four foci; the current coordinates being always (x, y, z).

The first of these equations is

$$\begin{aligned}
&\mathrm{a}\,(-2x_1^2x_2^2) + \mathrm{b}\,(x_2^2y_1^2 + x_1^2y_2^2) + \mathrm{c}\,(x_2^2z_1^2 + x_1^2z_2^2) \\
&\qquad + 2\mathrm{f}\,(x_2^2y_1z_1 + x_1^2y_2z_2) + 2\mathrm{g}\,(x_2^2z_1x_1 + x_1^2z_2x_2) + 2\mathrm{h}\,(x_2^2x_1y_1 + x_1^2x_2y_2) = 0,
\end{aligned}$$

where the quantities multiplied by a, b, &c. are all of them easily expressible in terms of x_1x_2, y_1y_2, z_1z_2, $y_1z_2 + y_2z_1$, $z_1x_2 + z_2x_1$, $x_1y_2 + x_2y_1$, which are respectively proportional to given functions of (α, β, γ); and replacing for a, b, &c. their values, the equation is

$$\begin{array}{ll}
\;\;\;(\;\mathfrak{B}z^2 + \mathfrak{C}x^2 - 2\mathfrak{F}yz\;) & .\;\; 2\alpha^4 \\
+\;(\;\mathfrak{C}x^2 + \mathfrak{A}y^2 - 2\mathfrak{G}zx) & .\;\; (\alpha^2 + \beta^2 - \gamma^2)^2 - 2\alpha^2\beta^2 \\
+\;(\;\mathfrak{A}y^2 + \mathfrak{B}x^2 - 2\mathfrak{H}xy) & .\;\; (\gamma^2 + \alpha^2 - \beta^2)^2 - 2\gamma^2\alpha^2 \\
+\,2\,(-\mathfrak{A}yz - \mathfrak{F}x^2 + \mathfrak{G}xy + \mathfrak{H}xz) & .\;\; \alpha^2(\beta^2 + \gamma^2) - (\beta^2 - \gamma^2)^2 \\
+\,2\,(-\mathfrak{B}zx - \mathfrak{G}y^2 + \mathfrak{H}yz + \mathfrak{F}yz\;) & .- \alpha^2(\gamma^2 + \alpha^2 - \beta^2) \\
+\,2\,(-\mathfrak{C}xy - \mathfrak{H}z^2 + \mathfrak{F}zx + \mathfrak{G}zy) & .- \alpha^2(\alpha^2 + \beta^2 - \gamma^2) = 0.
\end{array}$$

Now putting for shortness

$$\square = \alpha^4 + \beta^4 + \gamma^4 - 2\beta^2\gamma^2 - 2\gamma^2\alpha^2 - 2\alpha^2\beta^2,$$

so that $-\square$ is equal to sixteen times the square of the area of the fundamental triangle, the coefficient of x^2 is

$$= \mathfrak{B}(\square + 2\alpha^2\gamma^2) + \mathfrak{C}(\square + 2\alpha^2\beta^2) - 2\mathfrak{F}\{-\square - \alpha^2(\beta^2 + \gamma^2 - \alpha^2)\},$$

which is

$$= (\mathfrak{B} + \mathfrak{C} + 2\mathfrak{F})\,\square + 2\alpha^2\{\mathfrak{B}\gamma^2 + \mathfrak{C}\beta^2 + \mathfrak{F}(\beta^2 + \gamma^2 - \alpha^2)\}.$$

And reducing in a similar manner the other coefficients, the equation is

$$\square\,\{(\mathfrak{B} + \mathfrak{C} + 2\mathfrak{F})\,x^2 + \mathfrak{A}(y+z)^2 - 2(\mathfrak{H} + \mathfrak{G})\,x(y+z)\} + 2\alpha^2\Theta = 0,$$

where for shortness

$$\begin{array}{rl}
\Theta = \;\; x^2\,. & \mathfrak{B}\gamma^3 + \mathfrak{C}\,\beta^2 + \mathfrak{F}\,(\beta^2 + \gamma^2 - \alpha^2) \\
+\,y^2\,. & \mathfrak{C}\alpha^2 + \mathfrak{A}\,\gamma^2 + \mathfrak{G}\,(\gamma^2 + \alpha^2 - \beta^2) \\
+\,z^2\,. & \mathfrak{A}\beta^2 + \mathfrak{B}\,\alpha^2 + \mathfrak{H}\,(\alpha^2 + \beta^2 - \gamma^2) \\
+\,yz\,. & -2\mathfrak{F}\alpha^2 + \mathfrak{A}(\beta^2 + \gamma^2 - \alpha^2) - \mathfrak{H}(\gamma^2 + \alpha^2 - \beta^2) - \mathfrak{G}(\alpha^2 + \beta^2 - \gamma^2) \\
+\,zx\,. & -2\mathfrak{G}\beta^2 - \mathfrak{H}(\beta^2 + \gamma^2 - \alpha^2) + \mathfrak{B}(\gamma^2 + \alpha^2 - \beta^2) - \mathfrak{F}(\alpha^2 + \beta^2 - \gamma^2) \\
+\,xy\,. & -2\mathfrak{H}\gamma^2 - \mathfrak{G}(\beta^2 + \gamma^2 - \alpha^2) - \mathfrak{F}(\gamma^2 + \alpha^2 - \beta^2) + \mathfrak{C}(\alpha^2 + \beta^2 - \gamma^2),
\end{array}$$

or, what is the same thing, the equation is

$$\square\,\{(\mathfrak{A}+\mathfrak{B}+\mathfrak{C}+2\mathfrak{F}+2\mathfrak{G}+2\mathfrak{H})\,x^2-2\,(\mathfrak{A}+\mathfrak{H}+\mathfrak{G})\,x\,(x+y+z)+\mathfrak{A}\,(x+y+z)^2\}+2\alpha^2\Theta=0,$$

or putting for shortness

$$\begin{aligned}\mathfrak{A}+\mathfrak{B}+\mathfrak{C}+2\mathfrak{F}+2\mathfrak{G}+2\mathfrak{H}&=\mathfrak{P},\\ \mathfrak{A}+\mathfrak{H}+\mathfrak{G}&=\mathfrak{L},\\ \mathfrak{H}+\mathfrak{B}+\mathfrak{F}&=\mathfrak{M},\\ \mathfrak{G}+\mathfrak{F}+\mathfrak{C}&=\mathfrak{N},\end{aligned}$$

so that in fact

$$\mathfrak{P}=\mathfrak{L}+\mathfrak{M}+\mathfrak{N},$$

the equation is

$$\square\,\{\mathfrak{P}x^2-2\mathfrak{L}x\,(x+y+z)+\mathfrak{A}\,(x+y+z)^2\}+2\alpha^2\Theta=0.$$

The equation with y_2z_2, y_1z_1 is in a similar manner found to be

$$\square\,\{\mathfrak{F}x^2-(\mathfrak{C}+\mathfrak{G})\,y^2-(\mathfrak{B}+\mathfrak{H})\,z^2+(\mathfrak{A}+2\mathfrak{F}+\mathfrak{G}+\mathfrak{H})\,yz+(-\mathfrak{B}-\mathfrak{H}+\mathfrak{F})\,zx+(-\mathfrak{C}-\mathfrak{G}+\mathfrak{F})\,xy\}$$
$$-(\beta^2+\gamma^2-\alpha^2)\,\Theta=0,$$

or, what is the same thing,

$$\square\,[(\mathfrak{A}+\mathfrak{B}+\mathfrak{C}+2\mathfrak{F}+2\mathfrak{G}+2\mathfrak{H})\,yz-\{(\mathfrak{G}+\mathfrak{F}+\mathfrak{C})\,y+(\mathfrak{H}+\mathfrak{B}+\mathfrak{F})\,z\}(x+y+z)+\mathfrak{F}\,(x+y+z)^2]$$
$$-(\beta^2+\gamma^2-\alpha^2)\,\Theta=0,$$

or, finally,

$$\square\,\{\mathfrak{P}yz-(\mathfrak{N}y+\mathfrak{M}z)\,(x+y+z)+\mathfrak{F}\,(x+y+z)^2\}-(\beta^2+\gamma^2-\alpha^2)\,\Theta=0.$$

Hence the entire system of equations is

$$\begin{aligned}&\square\,\{\mathfrak{P}x^2-2\mathfrak{L}x\,(x+y+z)+\mathfrak{A}\,(x+y+z)^2\}+2\alpha^2\Theta=0,\\ &\square\,\{\mathfrak{P}y^2-2\mathfrak{M}y\,(x+y+z)+\mathfrak{B}\,(x+y+z)^2\}+2\beta^2\Theta=0,\\ &\square\,\{\mathfrak{P}z^2-2\mathfrak{N}z\,(x+y+z)+\mathfrak{C}\,(x+y+z)^2\}+2\gamma^2\Theta=0,\\ &\square\,\{\mathfrak{P}yz-(\mathfrak{N}y+\mathfrak{M}z)\,(x+y+z)+\mathfrak{F}\,(x+y+z)^2\}-(\beta^2+\gamma^2-\alpha^2)\,\Theta=0,\\ &\square\,\{\mathfrak{P}zx-(\mathfrak{L}z+\mathfrak{N}x)\,(x+y+z)+\mathfrak{G}\,(x+y+z)^2\}-(\gamma^2+\alpha^2-\beta^2)\,\Theta=0,\\ &\square\,\{\mathfrak{P}xy-(\mathfrak{M}x+\mathfrak{L}y)\,(x+y+z)+\mathfrak{H}\,(x+y+z)^2\}-(\alpha^2+\beta^2-\gamma^2)\,\Theta=0,\end{aligned}$$

which are the equations of six conics, each of them passing through the four foci.

From the first three of these, we have

$$\frac{1}{\alpha^2}\{\mathfrak{P}x^2-2\mathfrak{L}x\,(x+y+z)+\mathfrak{A}\,(x+y+z)^2\}$$
$$=\frac{1}{\beta^2}\{\mathfrak{P}y^2-2\mathfrak{M}y\,(x+y+z)+\mathfrak{B}\,(x+y+z)^2\}$$
$$=\frac{1}{\gamma^2}\{\mathfrak{P}z^2-2\mathfrak{N}z\,(x+y+z)+\mathfrak{C}\,(x+y+z)^2\},$$

which, allowing for the difference of notation, are Mr Hensley's equations: it appears by his investigation that their geometrical signification is as follows; viz. if for shortness we denote the equations by

$$A = B = C,$$

then if we consider the tangents parallel to the x-side of the fundamental triangle, and the tangents parallel to the y-side of the fundamental triangle, the equation $A = B$ is the locus of a point such that the feet of the perpendiculars let fall from it on the four tangents lie in a circle. And similarly for the equations $A = C$, $B = C$.

If we multiply the six equations by 1, 1, 1, 2, 2, 2 and add, we obtain the identical equation $0 = 0$; if we multiply them by a, b, c, $2f$, $2g$, $2h$ and add, then after some easy reductions, we obtain for the equation of a new conic passing through the four foci

$$\square \{\mathfrak{P} U + K(x + y + z)^2\} + 2S\Theta = 0,$$

where

$$U = (a,\ b,\ c,\ f,\ g,\ h)(x,\ y,\ z)^2,$$

K is the discriminant $abc - af^2 - bg^2 - ch^2 + 2fgh$, and

$$S = a\alpha^2 + b\beta^2 + c\gamma^2 - f(\beta^2 + \gamma^2 - \alpha^2) - g(\gamma^2 + \alpha^2 - \beta^2) - h(\alpha^2 + \beta^2 - \gamma^2),$$

or, what is the same thing,

$$S = (f + a - h - g)\,\alpha^2 + (g - h + b - f)\,\beta^2 + (h - g - f + c)\,\gamma^2.$$

It would be interesting to ascertain the geometrical signification of the six conics and of the last-mentioned new conic.

2, *Stone Buildings, W.C., March* 13*th*, 1862.

297.

ON SOME FORMULÆ RELATING TO THE DISTANCES OF A POINT FROM THE VERTICES OF A TRIANGLE, AND TO THE PROBLEM OF TACTIONS.

[From the *Quarterly Journal of Pure and Applied Mathematics*, vol. v. (1862), pp. 381—384.]

THE relation between the distances of four points 1, 2, 3, 4 in a plane is

$$\begin{vmatrix} 0, & \overline{12}^2, & \overline{13}^2, & \overline{14}^2, & 1 \\ \overline{21}^2, & 0, & \overline{23}^2, & \overline{24}^2, & 1 \\ \overline{31}^2, & \overline{32}^2, & 0, & \overline{34}^2, & 1 \\ \overline{42}^2, & \overline{42}^2, & \overline{43}^2, & 0, & 1 \\ 1, & 1, & 1, & 1, & 0 \end{vmatrix} = 0,$$

where, see my paper "Note on the value of certain Determinants the terms of which are the squared distances of Points in a plane or in space," *Quarterly Journal of Mathematics*, t. III., p. 275 (1859), [286], the determinant is

$$= \Sigma\overline{12}^2 . \overline{23}^2 . \overline{34}^2 - \Sigma\overline{12}^2 . \overline{34}^2 . \overline{43}^2 - \Sigma\overline{12}^2 . \overline{23}^2 . \overline{31}^2,$$

an identity which subsists without the aid of the relations $12 = 21$, &c., and in which the Σ, Σ, Σ contain 24, 12, and 8 terms respectively.

Writing $23 = f$, $31 = g$, $12 = h$, $14 = a$, $24 = b$, $34 = c$, the determinant is

$$\begin{aligned} = 2 \{ & g^2h^2(b^2+c^2) + h^2f^2(c^2+a^2) + f^2g^2(a^2+b^2) \\ & + b^2c^2(g^2+h^2) + c^2a^2(a^2+f^2) + a^2b^2(f^2+g^2) \\ & - a^2f^2(a^2+f^2) - b^2g^2(b^2+g^2) - c^2h^2(c^2+h^2) \\ & - b^2c^2f^2 - c^2a^2g^2 - a^2b^2h^2 - f^2g^2h^2 \quad \} \\ = \; & -2\square, \end{aligned}$$

if $\square$ denote the function in { } with the signs reversed. The function $\square$ may be expressed in the form

$$\begin{aligned}\square = \;& a^4f^2+b^4g^2+c^4h^2+f^2h^2g^2\\ &+(a^2f^2+b^2c^2)(f^2-g^2-h^2)\\ &+(b^2g^2+c^2a^2)(g^2-h^2-f^2)\\ &+(c^2h^2+a^2b^2)(h^2-f^2-g^2),\end{aligned}$$

and also in the form

$$\square = U^2+(f+g+h)\,V,$$

if for shortness

$$\begin{aligned}U &= a^2f+b^2g+c^2h+fgh,\\ V &= (a^2f^2+b^2c^2)(f-g-h)\\ &\quad+(b^2g^2+c^2a^2)(g-h-f)\\ &\quad+(c^2h^2+a^2b^2)(h-f-g);\end{aligned}$$

and it may be remarked that since $\square$ is an even function of f, g, h, we may in this last formula change at pleasure the signs of these quantities; we thus obtain in all four similar forms of the function $\square$.

It is clear that considering a triangle, and any point in the plane of the triangle, f, g, h may be taken to denote the sides of the triangle, and a, b, c the distances of the point from the vertices: and the equation $\square=0$ is the relation connecting the sides and distances.

The equation $f+g+h=0$ denotes that the vertices are *in lineâ*, and when this equation is satisfied we have

$$U=a^2f+b^2g+c^2h+fgh=0,$$

which is in fact, as it is easy to see, the relation connecting the distances of a point from any three points *in lineâ*.

For a, b, c write $a+x, b+x, c+x$; x will be the radius of a circle touching the circles, radii a, b, c, described about the vertices as centres. The equation $\square=0$ becomes after all reductions

$$\begin{aligned}&U^2-(f+g+h)\,V\\ &+x\,[\;4U(af+bg+ch)\\ &\qquad-2(f+g+h)\{(af^2+bc(b+c))(f-g-h)\\ &\qquad\qquad+(bg^2+ca(c+a))(g-h-f)\\ &\qquad\qquad+(ch^2+ab(a+b))(h-f-g)\}]\\ &+x^2\,[\;f^2\{-4a^2+6a(b+c)-6bc\}\\ &\qquad+g^2\{-4b^2+6b(c+a)-6ca\}\\ &\qquad+h^2\{-4c^2+6c(a+b)-6ab\}]=0,\end{aligned}$$

which is a quadratic equation only: the two circles thus obtained are those which touch the given circles all three externally or all three internally. But by changing in every possible manner the signs of a, b, c we obtain in all four equations giving the eight tangent circles. It may be noticed that if as before $f+g+h=0$, $U=0$, then not only the constant term vanishes, but the coefficient of x also vanishes or the equation becomes simply $x^2 = 0$.

In particular, suppose $f = b + c$, $g = c + a$, $h = a + b$; developing this *de novo*, and putting for shortness

$$\begin{aligned} a + b + c &= p, \\ bc + ca + ab &= q, \\ abc &= r, \end{aligned}$$

we find

$$U = 2\{px^2 + 2qx + pq - 2r\},$$
$$V = 2\{px^4 + 4qx^3 + (2pq + 12r)x^2 + 4q^2x + pq^2 - 4qr\},$$

and then the equation $\square = U^2 - 2pV = 0$ gives

$$\tfrac{1}{4}\square = (px^2 + 2qx + pq - 2r)^2 - p\{px^4 + 4qx^3 + (2pq + 12r)x^2 + 4q^2x + pq^2 - 4qr\}$$
$$= 4\{(q^2 - 4pr)x^2 - 2qrx + r^2\},$$

so that we have

$$\tfrac{1}{16}\square = (q^2 - 4pr)x^2 - 2qrx + r^2 = (qx - r)^2 - 4prx = 0,$$

and thence

$$qx - r = \pm x\sqrt{pr}, \text{ or } x = \frac{r}{q \pm},$$

which gives the radii of the circles inscribed in and circumscribed about the three circles radii a, b, c, whereof each touches the two others: a formula given by Descartes, *Epistolæ* (Ed. 2, Franc. 1792), Pars III., p. 261, in a letter to the Princess Elizabeth, viz. Descartes has

$$(d^2e^2 + d^2f^2 + e^2f^2 - 2def^2 - 2d^2ef - 2de^2f)x^2 - 2(de^2f^2 + d^2ef^2 + d^2e^2f)x + d^2e^2f^2 = 0,$$

which putting a, b, c for his d, e, f, becomes *ut suprà*

$$x^2(q^2 - 4pr) - 2qrx + r^2 = 0.$$

In conclusion I notice the following formula which is obtained without difficulty, viz. if as before we have a triangle the sides whereof are f, g, h, and if a, b, c are the distances of a point from the vertices (so that as before $\square = 0$) then the perpendicular distances of the point from the sides, each perpendicular distance divided by the perpendicular distance of the opposite vertex from the same side, are as follows: viz. the quotient for the side f is

$$= \frac{1}{16\Delta^2}[(b^2 - c^2)(g^2 - h^2) + f^2(b^2 + c^2 + g^2 + h^2 - 2a^2) - f^4],$$

where Δ is the area of the triangle. It is clear that we ought to have

$$\Sigma\{(b^2 - c^2)(g^2 - h^2) + f^2(b^2 + c^2 + g^2 + h^2 - 2a^2) - f^4\} = 16\Delta^2,$$

and this equation in fact reduces itself to

$$2g^2h^2 + 2h^2f^2 + 2f^2g^2 - f^4 - g^4 - h^4 = 16\Delta^2,$$

which is right.

2, *Stone Buildings, W.C.*, 17*th September*, 1862.

298.

REPORT ON THE PROGRESS OF THE SOLUTION OF CERTAIN SPECIAL PROBLEMS OF DYNAMICS.

[From the *Report of the British Association for the Advancement of Science*, 1862, pp. 184—252.]

MY "Report on the Recent Progress of Theoretical Dynamics" was published in the Report of the British Association for the year 1857, [195]. The present Report (which is in some measure supplemental thereto) relates to the *Special Problems* of Dynamics: to give a general idea of the contents, I will at once mention the heads under which these problems are considered; viz., relating to the motion of a particle or system of particles, we have

Rectilinear Motion;
Central Forces, and in particular
Elliptic Motion;
The Problem of two Centres;
The Spherical Pendulum;
Motion as affected by the Rotation of the Earth, and Relative Motion in general;
Miscellaneous Problems;
The Problem of three bodies.

And relating to the motion of a solid body, we have

The Transformation of Coordinates;
Principal Axes, and Moments of Inertia;
Rotation of a Solid Body;
Kinematics of a Solid Body;
Miscellaneous Problems.

As regards the first division of the subject, I remark that the lunar and planetary theories, as usually treated, do not (properly speaking) relate to the problem of three bodies, but to that of disturbed elliptic motion—a problem which is not considered in the present Report. The problem of the spherical pendulum is that of a particle moving on a spherical surface; but, with this exception, I do not much consider the motion of a particle on a given curve or surface, nor the motion in a resisting medium; what is said on these subjects is included under the head Miscellaneous Problems. The first six heads relate exclusively, and the head Miscellaneous Problems relates principally, to the motion of a single particle. As regards the second division of the subject, I will only remark that, from its intimate connexion with the theory of the motion of a solid body, I have been induced to make a separate head of the geometrical subject, "Transformation of Coordinates," and to treat of it in considerable detail.

I have inserted at the end of the present Report a list of the memoirs and works referred to, arranged (not, as in the former Report, in chronological order, but) alphabetically according to the authors' names: those referred to in the former Report formed for the purpose thereof a single series, which is not here the case. The dates specified are for the most part those on the title-page of the volume, being intended to show approximately the date of the researches to which they refer, but in some instances a more particular specification is made.

I take the opportunity of noticing a serious omission in my former Report, viz., I have not made mention of the elaborate memoir, Ostrogradsky "Mémoire sur les équations différentielles relatives au problème des Isopérimètres," *Mém. de St Pét.*, t. IV. (6 sér.) pp. 385—517, 1850, which among other researches contains, and that *in the most general form*, the transformation of the equations of motion from the Lagrangian to the Hamiltonian form, and indeed the transformation of the general isoperimetric system (that is, the system arising from any problem in the calculus of variations) to the Hamiltonian form. I remark also, as regards the memoir of Cauchy referred to in the note p. 12 as an *unpublished* memoir of 1831, there is an "Extrait du Mémoire présenté à l'Académie de Turin le 11 Oct. 1831," published in lithograph under the date Turin, 1832, with an addition dated 6 Mar. 1833. The Extract begins thus:—"§ I. Variation des Constantes Arbitraires. Soient données entre la variable $t, \ldots$ n fonctions de t désignées par $x, y, z \ldots$ et n autres fonctions de t désignées par $u, v, w, \ldots$ $2n$ équations différentielles du premier ordre et de la forme

$$\frac{dx}{dt}=\frac{dQ}{du},\quad \frac{dy}{dt}=\frac{dQ}{dv},\quad \frac{dz}{dt}=\frac{dQ}{dw},$$

$$\frac{du}{dt}=-\frac{dQ}{dx},\quad \frac{dv}{dt}=-\frac{dQ}{dy},\quad \frac{dw}{dt}=-\frac{dQ}{dz},\ \text{\&c.''}$$

without any explanation as to the origin of these equations; and the formulæ are then given for the variations of the constants in the integrals of the foregoing system; this seems sufficient to establish that Cauchy in the year 1831 was familiar with the Hamiltonian form of the equations of motion.

Bour's "Mémoire sur l'intégration des équations différentielles de la Mécanique," as published, *Mém. prés. à l'Inst.*, t. XIV. pp. 792—821, is substantially the same as the extract thereof in *Liouville's Journal*, referred to in my former Report; but since the date of that Report there have been published in the *Comptes Rendus*, 1861 and 1862, several short papers by the same author; also Jacobi's great memoir, see list, Jacobi, Nova Methodus &c., 1862, edited after his decease by Clebsch; some valuable memoirs by Natani and Clebsch (*Crelle*, 1861 and 1862) relating to the Pfaffian system of equations (which includes those of Dynamics), and Boole "On Simultaneous Differential Equations of the First Order, in which the number of the Variables exceeds by more than one the number of the Equations," *Phil. Trans.*, t. CLII. (1862), pp. 437—454.

Rectilinear Motion. Article Nos. 1 to 5.

1. The determination of the motion of a falling body, which is the case of a constant force, is due to Galileo.

2. A variable force, assumed to be a force depending only on the position of the particle, may be considered as a function of the distance from any point in the line, selected at pleasure as a centre of force; but if, as usual, the force is given as a function of the distance from a certain point, it is natural to take that point for the centre of force. The problem thus becomes a particular case of that of central forces; and it is so treated in the *Principia*, Book I. § 7; the method has the advantage of explaining the paradoxical result which presents itself in the case Force $\propto$ (Dist.)$^{-2}$, and in some other cases where the force becomes infinite. According to theory, the velocity becomes infinite at the centre, but the direction of the motion is there abruptly reversed; so that the body in its motion does not pass through the centre, but on arriving there, forthwith returns towards its original position; of course such a motion cannot occur in nature, where neither a force nor a velocity ever is actually infinite.

3. Analytically the problem may be treated separately by means of the equation $\frac{d^2x}{dt^2} = X$, which is at once integrable in the form $\left(\frac{dx}{dt}\right)^2 = C + 2\int X dx$.

4. The following cases may be mentioned:

Force $\propto$ Dist. The law of motion is well known, being in fact the same as for the cycloidal pendulum.

Force $\propto$ (Dist.)$^{-2}$, $= \frac{\mu}{x^2}$, which is the case above alluded to.

Assuming that the body falls from rest at a distance a, we have

$$x = a(1 - \cos\phi),$$

where, if $n = \frac{a^{\frac{3}{2}}}{\sqrt{\mu}}$, ϕ is given in terms of the time by means of the equation

$$nt = \phi - \sin\phi.$$

If the body had initially a small transverse velocity, the motion would be in a very eccentric ellipse, and the formulæ are in fact the limiting form of those for elliptic motion.

5. There are various laws of force for which the motion may be determined. In particular it can be determined by means of Elliptic Integrals, in the case of a body attracted to two centres, force $\propto$ (dist.)$^{-2}$: see Legendre, *Exercices de Cal. Intég.*, t. II. pp. 502—512, and *Théorie des Fonct. Ellip.*, t. I. pp. 531—538.

Central Forces, Article Nos. 6 to 26.

6. The theory of the motion of a body under the action of a given central force was first established in the *Principia*, Book I. §§ 2 and 3: viz. Prop. I. the areas are proportional to the times, that is (using the ordinary analytical notation), $r^2 d\theta = h dt$, and Prop. VI. Cor. 3, $P \propto \frac{1}{SY^2 . PV}, = h^2u^2 \left(\frac{d^2u}{d\theta^2} + u\right)$, so that

$$\frac{d^2u}{d\theta^2} + u - \frac{P}{h^2u^2} = 0.$$

7. It is to be noticed that, given the orbit, the law of force is at once determined; and § 2 contains several instances of such determination; thus,

Prop. VII. If a body revolve in a circle, the law of force to a point S is force $\propto \frac{1}{SP^2 . PV^3}$ (P the body, PV the chord through S).

Prop. IX. If a body move in a logarithmic spiral, force $\propto$ (dist.)$^{-3}$.

Prop. X. If a body move in an ellipse, force to centre $\propto$ dist., and as a particular case, if the body move in a parabola under the action of a force parallel to the axis, the force is constant. The particular case of motion in a parabola had been obtained by Galileo.

And § 3, Props. XI. XII. XIII. If a body move in an ellipse, hyperbola, or parabola under the action of a force tending to the focus, force $\propto$ (dist.)$^{-2}$.

8. But Newton had no direct method of solving the inverse problem (which depends on the solution of the differential equation), "Given the force to find the orbit." Thus force $\propto$ (dist.)$^{-2}$, after it has been shown that an ellipse, a hyperbola, and a parabola may each of them be described under the action of such a force, the remainder of the solution consists in showing that, given the initial circumstances of the motion, a conic section (ellipse, parabola, or hyperbola, as the case may be) can be constructed, passing through the point of projection, having its tangent in the direction of the initial motion, and such that the velocity of the body describing the conic section under the action of the given central force is equal to the velocity of projection; which being so, the orbit will be the conic section so constructed. This is what is done, Prop. XVII.; it may be observed that the latus rectum is constructed not very elegantly by means of the latus rectum of an auxiliary orbit.

9. A more elegant construction was obtained by Cotes (see the *Harmonia Mensurarum*, pp. 103—105, and demonstration from the author's papers in the Notes by R. Smith, pp. 124, 125), depending on the position of a point C, such that the velocity acquired in falling under the action of the central force from C *directly or through infinity* [1] to P the point of projection, is equal to the given velocity of projection.

10. But Newton's original construction is now usually replaced by a construction which employs the space due to the velocity of projection considered as produced by a constant force equal to the central force at the point of projection.

11. § 9 of Book I. relates to revolving orbits, viz., it is shown that a body may be made to move in an orbit revolving round the centre of force, by adding to the central force required to make the body move in the same orbit at rest, a force $\propto (\text{dist.})^{-3}$. This appears very readily by means of the differential equation (*antè*, No. 6), viz. writing therein $P+cu^3$ for P, and then θ', h' in the place of $\theta\sqrt{1-\frac{c}{h^2}}$, $h\sqrt{1-\frac{c}{h^2}}$ respectively, the equation retains its original form, with θ', h' in the place of θ, h respectively.

12. It may be remarked that when the original central force vanishes, the fixed orbit is a right line (not passing through the centre of force). It thus appears by § 9 that the curve $u=A\cos(n\theta+B)$ may be described under the action of a force $\propto (\text{dist.})^{-3}$. A proposition in § 2, already referred to, shows that a logarithmic spiral may be described under the action of such a force.

13. But the case of a force $\propto (\text{dist.})^{-3}$ was first completely discussed by Cotes in

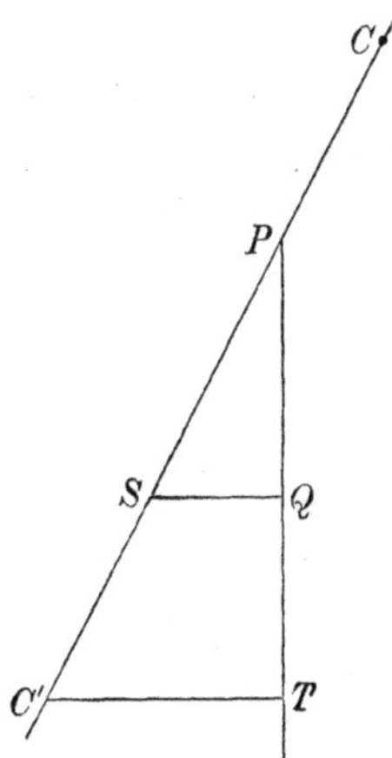

the *Harmonia Mensurarum*, pp. 31—35, 98—104, and Notes, pp. 117—173. There are in all five cases, according as the velocity of projection is

1. Less than that acquired in falling from infinity, or say equal to that acquired in falling from a point C to P, the point of projection.

[1] In the second case C lies on the radius vector produced beyond the centre, and the body is supposed to fall from rest at C (under the action of the central force considered as repulsive) to infinity, and then from the opposite infinity (with an initial velocity equal to the velocity so acquired) to P.

2. Equal to that acquired in falling from infinity.

3, 4, 5. Greater than that acquired in falling from infinity, or say equal to that acquired in falling from a point C', *through infinity*, to P; viz. PQ being the direction of projection, and SQ, $C'T$ perpendiculars thereon from S and C' respectively,

$$3. \quad SQ < TQ;$$
$$4. \quad SQ = TQ;$$
$$5. \quad SQ > TQ;$$

the equations of the orbits being

1. $u = Ae^{m\theta} + Be^{-m\theta}$, A and B same sign, so that rad. vector is never infinite.

2. $u = Ae^{m\theta}$ or $Be^{-m\theta}$, logarithmic spiral.

3. $u = Ae^{m\theta} + Be^{-m\theta}$, A and B opposite signs, so that rad. vector becomes infinite.

4. $u = A\theta + B$, $m = 0$, reciprocal spiral.

5. $u = A\cos(n\theta + B)$, $m = n\sqrt{-1}$.

14. The before-mentioned equation,

$$\frac{d^2u}{d\theta^2} + u - \frac{P}{h^2u^2} = 0,$$

is in effect given (but the equation is encumbered with a tangential force) in Clairaut's *Théorie de la Lune*, 1765. It is given in its actual form, and extensively used (in particular for obtaining the above-mentioned equations for Cotes' spirals) in Whewell's *Dynamics*, 1823. The equation appears to be but little known to continental writers, and (under the form $u'' + u - \alpha^2 r^2 R = 0$) it is given *as new* by Schellbach as late as 1853. The formulæ used in place of it are those which give t and θ each of them in terms of r; viz.

$$dt = \frac{rdr}{\{-h^2 + r^2(C - 2\int Pdr)\}^{\frac{1}{2}}},$$

$$d\theta = \frac{hdr}{r\{-h^2 + r^2(C - 2\int Pdr)\}^{\frac{1}{2}}},$$

which, however, assume that P is a function of r only.

15. Force $\propto$ (dist.)$^{-2}$. The law of motion in the conic sections is implicitly given by Newton's theorem for the equable description of the areas. For the parabola, if α denote the pericentric distance, and f the angle from pericentre or true anomaly, we have

$$t = \frac{\alpha^{\frac{3}{2}}\sqrt{2}}{\sqrt{\mu}}\left(\tan\tfrac{1}{2}f + \tfrac{1}{3}\tan^3\tfrac{1}{2}f\right).$$

For the ellipse we have an angle g, the mean anomaly varying directly as the time $\left(g=nt \text{ if } n=\frac{\sqrt{\mu}}{a^{\frac{3}{2}}}\right)$; an auxiliary angle u, the eccentric anomaly, connected with g by the equation

$$g = u - e \sin u;$$

and then the radius vector r and the true anomaly f are given in terms of u by the equations $r = a(1 - e\cos u)$, and

$$\cos f = \frac{\cos u - e}{1 - e\cos u}, \quad \sin f = \frac{\sqrt{1-e^2}\sin u}{1 - e\cos u}, \text{ and therefore } \tan\tfrac{1}{2}f = \sqrt{\frac{1+e}{1-e}}\tan\tfrac{1}{2}u.$$

16. It is very convenient to have a notation for $\frac{r}{a}$ and f considered as functions of e, g, and I have elsewhere proposed to write

$$r = a \text{ elqr } (e,\ g),\quad f = \text{elta } (e,\ g),$$

read elqr elliptic quotient radius, and elta elliptic true anomaly.

17. The formulæ for the hyperbola correspond to those for the ellipse, but they contain exponential in the place of circular functions (see *post*, Elliptic Motion).

18. Euler, in the memoir "Determinatio Orbitæ Cometæ Anni 1742," (1743), p. 16 *et seq.*, obtained an expression for the time of describing a parabolic arc in terms of the radius vectors and the chord; viz. these being f, g, and k, the expression is

$$\text{Time} = \frac{1}{6\sqrt{\mu}}\left\{(f+g+k)^{\frac{3}{2}} - (f+g-k)^{\frac{3}{2}}\right\},$$

which, however, as remarked by Lagrange, *Méc. Anal.*, t. II. (3rd edit. p. 28), is deducible from Lemma X. of the third book of the *Principia*. But the theorem in its actual form is due to Euler.

19. Lambert, in the *Proprietates Insigniores &c.* (1761), Theorem VII. Cor. 2, obtained the same theorem, and in section 4 he obtained the corresponding theorem for elliptic motion; viz. the expression for the time is

$$= \frac{a^{\frac{3}{2}}}{\sqrt{\mu}}\left\{\phi - \phi' - (\sin\phi - \sin\phi')\right\},$$

if

$$\sin\tfrac{1}{2}\phi = \tfrac{1}{2}\sqrt{\frac{f+g+k}{a}}, \quad \sin\tfrac{1}{2}\phi' = \tfrac{1}{2}\sqrt{\frac{f+g-k}{a}}.$$

The form of the formula is, it will be observed, similar to that for motion in a straight line (*antè*, No. 4), and in fact the motion in the ellipse is, by an ingenious geometrical transformation, made to depend upon that in the straight line. The geometrical theorems upon which the transformation depends are stated, Cayley "On Lambert's Theorem &c." (1861).

20. The theorem was also obtained by Lagrange in the memoir "Recherches &c." (1767) as a corollary to his solution of the problem of two centres; viz. upon making the attractive force of one of the centres equal to zero, and assuming that such centre is situate on the curve, the expression for the time presents itself in the form given by Lambert's theorem.

21. Two other demonstrations of the theorem are given by Lagrange in the memoir "Sur une manière particulière d'exprimer le temps &c." (1778), reproduced in Note V. of the second volume of the last edition (Bertrand's) of the *Mécanique Analytique*. As M. Bertrand remarks, these demonstrations are very complete, very elegant, and very natural, assuming that the theorem is known beforehand.

Demonstrations were also given by Gauss, "Theoria Motus" (1809), p. 119 *et seq.*; Pagani, "Démonstration d'un théorème &c." (1834); and (in connexion with Hamilton's Principal Function) by Sir W. R. Hamilton, "On a General Method &c." (1834), p. 282; Jacobi, "Zur Theorie &c." (1837), p. 122; Cayley, "Note on the Theory of Elliptic Motion" (1856).

22. Connected with the problem of central forces, we have Sir W. R. Hamilton's "Hodograph," which in the paper (*Proc. R. Irish Acad.* 1847) is defined, and the fundamental properties thereof are stated; viz. if in an orbit round a centre of force there be taken on the perpendicular from the centre on the tangent at each point, a length equal to the velocity at that point of the orbit, the extremities of these lengths will trace out a curve which is the hodograph. As the product of the velocity into the perpendicular on the tangent is equal to twice the area swept out in a unit of time ($vp = h$), the hodograph is the reciprocal polar of the orbit with respect to a circle described about the centre of force, radius $= \sqrt{h}$. Whence also the tangent at any point of the hodograph is perpendicular to the radius vector through the corresponding point of the orbit, and the product of the perpendicular on the tangent into the corresponding radius vector is $= h$.

If force $\propto (\text{dist.})^{-2}$, the hodograph, *quà* reciprocal polar of a conic section with respect to a circle described about the focus, is a circle.

23. The following theorem is also given without demonstration; viz. if two circular hodographs, which have a common chord passing or tending through a common centre of force, be both cut at right angles by a third circle, the times of hodographically describing the intercepted arcs (that is, the times of describing the corresponding elliptic arcs) will be equal.

24. Droop, "On the Isochronism &c." (1856), shows geometrically that the last-mentioned property is equivalent to Lambert's theorem; and an analytical demonstration is also given, Cayley, "A demonstration of Sir W. R. Hamilton's Theorem &c." (1857). See also Sir W. R. Hamilton's *Lectures on Quaternions* (1853), p. 614.

25. The laws of central force which have been thus far referred to are force $\propto r$, $\propto \frac{1}{r^2}$, $\propto \frac{1}{r^3}$; and it has been seen that the case of a force $P + \frac{C}{r^3}$ depends upon that

of a force P, so that the motions for the forces $Ar+\frac{C}{r^3}$ and $\frac{B}{r^2}+\frac{C}{r^3}$ are deducible from those for the forces Ar and $\frac{B}{r^2}$ respectively. Some other laws of force, e.g. $\frac{A}{r^2}\pm Br$, $\frac{A}{r^2}+\frac{B}{r^3}+\frac{C}{r^4}+\frac{D}{r^5}$, are considered by Legendre, "Théorie des Fonctions Elliptiques" (1825), being such as lead to results expressible by elliptic integrals, and also the law $\frac{M}{r}$, for which the result involves a peculiar logarithmic integral. But the most elaborate examination of the different cases in which the solution can be worked out by elliptic integrals or otherwise is given in Stader's memoir "De Orbitis &c." (1852), where the investigation is conducted by means of the formulæ which give t and θ in terms of r (*antè*, No. 14).

26. In speaking of a central force, it is for the most part implied that the force is a function of the distance: for some problems in which this is not the case, see *post*, Miscellaneous Problems, Nos. 86 and 87.

It is to be noticed that, although the problem of central forces may be (as it has so far been) considered as a problem *in plano* (viz. the plane of the motion has been made the plane of reference), yet that it is also interesting to consider it as a problem in space; in fact, in this case the integrals, though of course involved in those which belong to the plane problem, present themselves under very distinct forms, and afford interesting applications of the theory of canonical integrals, of the derivation of the successive integrals by Poisson's method, and of other general dynamical theories. Moreover, in the lunar and planetary theories, the problem must of necessity be so treated. Without going into any details on this point, I will refer to Bertrand's memoir, "Sur les Équations différentielles de la Mécanique" (1852), Donkin's memoir "On a Class of Differential Equations &c." (1855), and Jacobi's posthumous memoir, "Nova Methodus &c." (1862).

Elliptic Motion, Article Nos. 27—40.

27. The question of the development of the true anomaly in terms of the mean anomaly (Kepler's problem), and of the other developments which present themselves in the theory of elliptic motion, is one that has very much occupied the attention of geometers. The formulæ on which it depends are mentioned *antè*, No. 15; they involve as an auxiliary quantity the eccentric anomaly u.

28. Consider first the equation

$$g=u-e\sin u,$$

which connects the mean anomaly g with the eccentric anomaly u.

Any function of u, and in particular u itself, and the functions $\begin{matrix}\cos\\ \sin\end{matrix}\, nu$ may be expanded in terms of g by means of Lagrange's theorem (Lagrange, *Mém. de Berlin*, 1768—1769, "Théorie des Fonctions," chap. 16, and "Traité de la Résolution des Équations Numériques," Note 11).

29. Considering next the equation

$$\tan \tfrac{1}{2}f = \sqrt{\frac{1+e}{1-e}} \tan \tfrac{1}{2}u,$$

which gives the true anomaly in terms of the eccentric anomaly, then, by replacing the circular functions by their exponential values (a process employed by Lagrange, *Mém. de Berlin*, 1776), f can be expressed in terms of u; viz. the result is

$$f = u + 2\lambda \sin u + 2\lambda^2 . \tfrac{1}{2} \sin 2u + 2\lambda^3 . \tfrac{1}{3} \sin 3u + \&c.,$$

where $\lambda = \dfrac{1-\sqrt{1-e^2}}{e} \left(= \dfrac{e}{1+\sqrt{1-e^2}} \right)$. Hence if u, $\sin u$, $\sin 2u$, &c. are expressed in terms of the mean anomaly, f will be obtained in the form $f = g +$ a series of multiple sines of g, the coefficients of the different terms being given in the first instance as functions of e and λ; and to complete the development λ and its powers have to be developed in powers of e. The solution is carried thus far in the *Mécanique Analytique* (1788), and in the *Mécanique Celeste* (1799).

30. We have next Bessel's investigations in the Berlin Memoirs for 1816, 1818, and 1824, and which are carried on mainly by means of the integral

$$2\pi I_k^h = \int_0^{2\pi} \cos(hu - k \sin u)\, du,$$

and various properties are there obtained and applications made of this important transcendant.

31. Relating to this integral we have Jacobi's memoir, "Formulæ transformationis &c." (1836), Liouville, "Sur l'intégrale $\int_0^{\pi} \cos i\,(u - x \sin u)\, du$" (1841), and Hansen's "Ermittelung der absoluten Störungen" (1843); the researches of Poisson in the *Connaissance des Temps* for 1825 and 1836 are closely connected with those of Bessel.

32. A very elegant formula, giving the actual expression of the coefficients considered as functions of e and λ, is given by Greatheed in the paper "Investigation of the General Term &c." (1838); viz. this is

$$f = g + 2\Sigma\lambda^r \left\{ \epsilon^{\frac{1}{2}re\,(\lambda+\lambda^{-1})} + \lambda^{-r}\, \epsilon^{-\frac{1}{2}(\lambda+\lambda^{-1})} \right\} \frac{\sin rg}{r},$$

where, after developing in powers of λ, the negative powers of λ must be rejected, and the term independent of λ divided by 2. This result is extended to other functions of f, Cayley "On certain Expansions &c." (1842).

33. An expression for the coefficient of the general term as a function of e only is obtained, Lefort, "Expression Numérique &c." (1846). The expression, which, from the nature of the case, is a very complicated one, is obtained by means of Bessel's integral. This is an indirect process which really comes to the combination of the developments of f in terms of u, and u in terms of g; and an equivalent result is obtained directly in this manner, Creedy, "General and Practical Solution &c." (1855).

34. We have also on the subject of these developments the very valuable and interesting researches of Hansen, contained in his *Fundamenta Nova &c.* (1838), in the memoir "Ermittelung der absoluten Störungen &c." (1845), and in particular in the memoir "Entwickelung des Products &c." (1853).

35. But the expression for the coefficient of the general term $\begin{smallmatrix}\cos\\ \cdot\\ \sin\end{smallmatrix} rg$ in any of these expansions is so complicated that it was desirable to have for the coefficients corresponding to the values $r = 0, 1, 2, 3, \ldots$ the finally reduced expressions in which the coefficient of each power of e is given as a numerical fraction. Such formulæ for the development of $\left(\frac{r}{a} - 1\right)^m \begin{smallmatrix}\cos\\ \cdot\\ \sin\end{smallmatrix} jf$, where j is a general symbol, the expansion being carried as far as e^7, were given, Leverrier, *Annales de l'Observatoire de Paris*, t. I. (1855).

36. And starting from these I deduced the results given in my "Tables of the Developments &c." (1861); viz. these tables give $\left(x = \frac{r}{a} - 1\right)$,

$$(x^1, \ldots. x^7),$$

$$(x^0, x^1, \ldots x^7) \begin{smallmatrix}\cos\\ \cdot\\ \sin\end{smallmatrix} jf,\ j = 1 \text{ to } j = 7,$$

$$\left(\left(\frac{r}{a}\right)^4, \left(\frac{r}{a}\right)^3 \ldots \log\frac{r}{a}, \left(\frac{r}{a}\right)^{-1}, \ldots \left(\frac{r}{a}\right)^{-5}\right),$$

$$\left(\left(\frac{r}{a}\right)^4, \ldots \left(\frac{r}{a}\right)^1, \left(\frac{r}{a}\right)^{-1}, \ldots \left(\frac{r}{a}\right)^{-5}\right) \begin{smallmatrix}\cos\\ \cdot\\ \sin\end{smallmatrix} jf,\ j = 1 \text{ to } j = 7,$$

all carried to e^7.

37. The true anomaly f has been repeatedly calculated to a much greater extent, in particular by Schubert (*Ast. Théorique*, St Pét. 1822), as far as e^{20}. The expression for $\frac{r}{a}$ as far as e^{13} is given in the same work, and that for $\log\frac{r}{a}$ as far as e^9 was calculated by Oriani, see Introd. to Delambre's *Tables du Soleil*, Paris (1806).

38. It may be remarked that when the motion of a body is referred to a plane which is not the plane of the elliptic orbit, then we have questions of development similar in some measure to those which regard the motion in the orbit; if, for instance, z be the distance from node, ϕ the inclination, and x the reduced distance from node, then $\cos z = \cos\phi \cos x$, from which we may derive $z = x +$ series of multiple sines of x. And there are, moreover, the questions connected with the development of the reciprocal distance of two particles—say $(a^2 + a'^2 - 2aa' \cos\theta)^{-\frac{1}{2}}$—which present themselves in the planetary theory; but this last is a wide subject, which I do not here enter upon. I will, however, just refer to Hansen's memoir, "Ueber die Entwickelung der negativen und ungeraden Potenzen &c." (1854).

39. The question of the convergence of the series is treated in Laplace's memoir of 1823, where he shows that in the series which express r and f in multiple cosines or sines of g, the coefficient of a term $\begin{smallmatrix}\cos\\ \cdot\\ \sin\end{smallmatrix} ig$, where i is very great, is at most equal in absolute value to a quantity of the form $\frac{A}{i\sqrt{i}}\left(\frac{e}{\lambda}\right)^i$, A and λ being finite quantities independent of i, whence he concludes that, in order to the convergency of the series, the limiting value of the eccentricity is $e=\lambda$, the numerical value being $e=0{\cdot}66195$.

40. The following important theorem was established by Cauchy, as part of a theory of the convergence of series in general; viz. so long as e is less than $0{\cdot}6627432$, which is the least modulus of e for which the equations

$$\frac{\pi}{2}=u-e\sin u,\quad 1-e\cos u=0,$$

can be satisfied, the development of the true anomaly and other developments in the theory of elliptic motion will be convergent. This was first given in the "Mémoire sur la Mécanique Céleste," read at Turin in 1831, but it is reproduced in the memoir "Considérations nouvelles sur les suites &c.," *Mém. d'Anal. et de Phys. Math.* t. I. (1840); and see also the memoirs in *Liouville's Journal* by Puiseux, and his Note i. to vol. II. of the 3rd ed. of the *Mécanique Analytique* (1855). There are on this subject, and on subjects connected with it, several papers by Cauchy in the *Comptes Rendus*, 1840 *et seq.*, which need not be particularly referred to.

The Problem of two Centres, Article Nos. 41 to 64.

41. The original problem is that of the motion of a body acted upon by forces tending to two centres, and varying inversely as the squares of the distances; but, as will be noticed, the solutions apply with but little variation to more general laws of force.

42. It may be convenient to notice that the coordinates made use of (in the several solutions) for determining the position of the body, are either the sum and difference of the two radius vectors, or else quantities which are respectively functions of the sum and the difference of these radius vectors [1]. If the plane of the motion is not given, then there is a third coordinate, which is the inclination of the plane through the body and the two centres to a fixed plane through the two centres, or say the azimuth of the axial plane, or simply the azimuth.

[1] If v, u are the distances of the body P from the centres A and B, a the distance AB, ζ, η the angles at A and B respectively, and $p=\tan\frac{1}{2}\zeta\tan\frac{1}{2}\eta$, $q=\tan\frac{1}{2}\zeta\div\tan\frac{1}{2}\eta$, then, as may be shown without difficulty, $v+u=a\frac{1+p}{1-p}$, $v-u=a\frac{1-q}{1+q}$, so that p and q are functions of $v+u$ and $v-u$ respectively; these quantities p and q are Euler's original coordinates.

43. Calling the first-mentioned two coordinates r and s, and the azimuth ψ, the solution of the problem leads ultimately to equations of the form

$$\frac{dr}{\sqrt{R}}=\frac{ds}{\sqrt{S}},\qquad dt=\frac{\lambda dr}{\sqrt{R}}+\frac{\mu ds}{\sqrt{S}},\qquad \delta\psi=\frac{\rho dr}{\sqrt{R}}+\frac{\sigma ds}{\sqrt{S}},$$

where R and S are rational and integral functions (of the third or fourth degree, in the case of forces varying as (dist.)$^{-2}$) of r, s respectively (but they are not in general the same functions of r, s respectively); λ and ρ are simple rational functions of r, and μ and σ simple rational functions of s; so that the equations give by quadratures, the first of them the curve described in the axial plane, the second the position of the body in this curve at a given time, and the third of them the position of the axial plane. In the ordinary case, where R and S are each of them of the third or the fourth order, the quadratures depend on elliptic integrals ([1]); but on account of the presence in the formulæ of the two distinct radicals $\sqrt{R}$, $\sqrt{S}$, it would appear that the solution is not susceptible of an ulterior development by means of elliptic and Jacobian functions ([1]) similar to those obtained in the problems of Rotation and the Spherical Pendulum.

44. It has just been noticed that when R, S are each of them of the fourth order, the quadratures depend on elliptic integrals; in the particular cases in which the relation between r, s is of the form $\frac{mdr}{\sqrt{R}}=\frac{nds}{\sqrt{S}}$, R and S being the same functions of r, s respectively, and m and n being integers (or more generally for other relations between the forms of R, S given by the theory of elliptic integrals), the equation admits of algebraical integration; but as the relations in question do not in general hold good, the theory of the algebraical integration of the equations plays only a secondary part in the solution of the problem. It is, however, proper to remark that Euler, when he wrote his first two memoirs "On the Problem of the two Centres" (*post*, Nos. 45 and 46), had already discovered and was acquainted with the theory of the algebraic integration of the equation $\frac{mdr}{\sqrt{R}}=\frac{nds}{\sqrt{S}}$ (R, S, m, n, *ut suprà*), although his memoir, "Integratio æquationis

$$\frac{dx}{\sqrt{A+Bx+Cx^2+Dx^3+Ex^4}}=\frac{dy}{\sqrt{A+By+Cy^2+Dy^3+Ey^4}},"$$

N. Comm. Petrop. t. XII. 1766—1767 ?, bears in fact a somewhat later date.

45. Having made these preliminary remarks, I come to the history of the problem.

It is I think clear that Euler's *earliest* memoir is the one "De Motu Corporis &c." in the Petersburg Memoirs for 1764 (printed 1766). In this memoir the forces vary

[1] The elliptic integrals are Legendre's functions F, E, Π; the elliptic and Jacobian functions are sinam, cosam, Δam, and the higher transcendants, Θ, H.

as (dist.)$^{-2}$, and the body moves in a given plane. The equations of motion are taken to be

$$\frac{d^2x}{dt^2} = 2g\left(-\frac{Ax}{v^3} + \frac{B(a-x)}{u^3}\right),$$

$$\frac{d^2y}{dt^2} = 2g\left(-\frac{Ay}{v^3} - \frac{By}{u^3}\right),$$

which, if ζ, η are the inclinations of the distances v, u to the axis respectively (see foot-note to No. 42), lead to

$$dv^2 + v^2dt^2 = 4gdt^2\left(\frac{A}{v} + \frac{B}{u} + \frac{D+E}{a}\right),$$

$$v^2u^2\, d\zeta\, d\eta = 2gadt^2\,(A\cos\zeta + B\cos\eta + D),$$

where D, E are constants of integration. Substituting for v, u their values in terms of η, ζ and eliminating dt, Euler obtains

$$\frac{d\zeta\sin\eta}{d\eta\sin\zeta} = \frac{P + \sqrt{P^2 - Q^2}}{Q},$$

where

$$\begin{aligned} A\cos\eta + B\cos\zeta + D\cos\zeta\cos\eta + E\sin\zeta\sin\eta &= P, \\ A\cos\zeta + B\cos\eta + D &= Q. \end{aligned}$$

And he then enters into a very interesting discussion of the particular case $A = 0$ or $B = 0$ (viz. the case where one of the attracting masses vanishes, which was of course known to be integrable); and after arriving at some paradoxical conclusions which he does not completely explain, although he remarks that the explanation depends on the circumstance that the integral found is a *singular solution* of a derivative equation, and as such does not satisfy the original equations of motion,—he proceeds to notice that an inquiry into the cause of the difficulty led him to a substitution by which the variables were separated.

46. But in the memoir "Problème, un Corps &c." in the Berlin Memoirs for 1760 (printed 1767), after obtaining the last-mentioned formulæ, he gives at once, without explaining how he was led to it, the analytical investigation of the substitution in question, viz. in *each* of the two memoirs he in fact writes

$$\frac{d\zeta\sin\eta + d\eta\sin\zeta}{d\zeta\sin\eta - d\eta\sin\zeta} = \sqrt{\frac{P+Q}{P-Q}},$$

$$\tan\tfrac{1}{2}\zeta = f, \quad \tan\tfrac{1}{2}\eta = g, \quad fg = p, \quad \frac{f}{g} = q,$$

that is

$$p = \tan\tfrac{1}{2}\zeta\tan\tfrac{1}{2}\eta\,; \quad q = \tan\tfrac{1}{2}\zeta \div \tan\tfrac{1}{2}\eta\,;$$

and in terms of these quantities p, q, the equation becomes

$$\frac{dp}{\sqrt{P}} = \frac{dq}{\sqrt{Q}},$$

where

$$P = (\ \ A + B + D)\,p + 2Ep^2 + (-A - B + D)\,p^3,$$
$$Q = (-A + B - D)\,q + 2Eq^2 + (\ \ A - B - D)\,q^3,$$

so that P and Q are cubic functions (not the same functions) of p and q respectively and the equation for the time is found to be

$$\frac{dt\sqrt{2g}}{a\sqrt{a}} = \frac{pdp}{(1-p)^2\sqrt{P}} + \frac{qdq}{(1+q)^2\sqrt{Q}},$$

which are the formulæ for the solution of the problem, as obtained in Euler's first and second memoirs.

47. In his third memoir, viz. that "De Motu Corporis &c." in the Petersburg Memoirs for 1765 (printed 1767), Euler considers the body as moving in space, the forces being as before as (dist.)$^{-2}$. Assuming that the coordinates y, z are in the plane perpendicular to the axis, there is in this case the equation of areas $y\dfrac{dz}{dt} - z\dfrac{dy}{dt} = \text{const.}$; and writing $y = y' \sin\psi$, $z = y' \cos\psi$, that is, $y' = \sqrt{y^2 + z^2}$, and ψ the azimuth, the integral equations for the motion in the variable plane (coordinates x, y') are not materially different in form from those which belong to the motion in a fixed plane, coordinates x, y (see *post*, No. 56, Jacobi); and the last-mentioned equation, which reduces itself to the form $y'^2\dfrac{d\psi}{dt} = \text{const.}$, gives at once $d\psi$ in a form such as that above alluded to (*antè*, No. 43), and therefore ψ by quadratures. The variables employed by Euler in the memoir in question are

$$v + u, \quad v - u, \ (\text{say } r,\ s), \text{ and } \psi,$$

v, u being, as above, the distances from the two centres, and ψ the azimuth of the axial plane. The functions of r, s under the radical signs are of the fourth order; this is so, with these variables, even if the motion is in a fixed plane; but this is no disadvantage, since, as is well known, the case of a quartic radical is not really more complicated than that of a cubic radical, the two forms being immediately convertible the one into the other.

48. Lagrange's first memoir (Turin Memoirs, 1766—1769) refers to Euler's three memoirs, but the author mentions that it was composed in 1767 without the knowledge of Euler's third memoir. The coordinates ultimately made use of are $v + u$, $v - u$, (say r, s), and ψ, the same as in Euler's third memoir, and the results consequently present themselves in the like form.

49. If the attractive force of one of the centres is taken equal to zero, then the position of such centre is arbitrary, and it may be assumed that the centre lies on the curve, which is in this case an ellipse (conic section); the expression of the time presents itself as a function of the focal radius vectors and the chord of the arc described; which, as remarked, *antè*, No. 20, leads to Lambert's theorem for elliptic motion.

50. The case presents itself of an ellipse or hyperbola described under the action of the two forces, viz. the equation $\frac{dr}{\sqrt{R}} = \frac{ds}{\sqrt{S}}$ will be satisfied by $r - \alpha = 0$, if $r - \alpha$ be a double factor of R, or by $s - \beta = 0$, if $s - \beta$ be a double factor of S, a case which is also considered in the *Mécanique Analytique*; and see in regard to the analytical theory, t. II. 3rd ed. Note III. by M. Serret, and "Thèse," *Liouv.* 1848. It is remarked by M. Bonnet, Note IV. and *Liouv.* t. IX. p. 113, (1844), that the result is a mere corollary of a general theorem, which is in effect as follows, viz. if a particle under the separate actions of the forces F, F',... starting in each case from the same point in the same direction but with the initial velocities v, v', &c. respectively, describe the same curve, then such curve will also be described under the conjoint action of all the forces, provided the body start from the same point in the same direction, with the initial velocity $V = \sqrt{v^2 + v'^2 + \ldots}$

51. Lagrange's second memoir (same volume of the Turin Memoirs) contains an exceedingly interesting discussion as to the laws of force for which the problem can be solved. Writing U, V, u, v in the place of Lagrange's P, Q, p, q, the equations of motion are

$$\frac{d^2x}{dt^2} + \frac{(x-a)\,U}{u} + \frac{(x-\alpha)\,V}{v} = 0,$$

$$\frac{d^2y}{dt^2} + \frac{(y-b)\,U}{u} + \frac{(y-\beta)\,V}{v} = 0,$$

$$\frac{d^2z}{dt^2} + \frac{(z-c)\,U}{u} + \frac{(z-\gamma)\,V}{v} = 0,$$

where

$$u = \sqrt{(x-a)^2 + (y-b)^2 + (z-c)^2},$$

$$v = \sqrt{(x-\alpha)^2 + (y-\beta)^2 + (z-\gamma)^2},$$

and putting also $f(= \sqrt{(a-\alpha)^2 + (b-\beta)^2 + (z-\gamma)^2})$ the distance of the centres, and then $u^2 = f^2x$, $v^2 = f^2y$, $\frac{U}{u} = X$, $\frac{V}{v} = Y$ (x, y are of course not to be confounded with the coordinates originally so represented), Lagrange obtains the equations

$$\tfrac{1}{2}\frac{d^2x}{dt^2} + Xx + \frac{(x+y-1)\,Y}{2} + \int(Xdx + Ydy) = 0,$$

$$\tfrac{1}{2}\frac{d^2y}{dt^2} + Yy + \frac{(x+y-1)\,X}{2} + \int(Xdx + Ydy) = 0,$$

which he represents by

$$\tfrac{1}{2}\frac{d^2x}{dt^2} + M = 0,$$

$$\tfrac{1}{2}\frac{d^2y}{dt^2} + N = 0;$$

and he then inquires as to the conditions of integrability of these equations, for which purpose he assumes that the equations multiplied by $mdx + ndy$ and $\mu dx + \nu dy$ respectively and added, give an integrable equation.

52. A case satisfying the required conditions is found to be

$$X = 2\alpha + \frac{\beta}{x\sqrt{x}}, \qquad Y = 2\alpha + \frac{\gamma}{y\sqrt{y}},$$

or, what is the same thing,

$$U = 2\alpha u + \frac{\beta f^3}{u^2}, \qquad V = 2\alpha v + \frac{\gamma f^3}{v^2};$$

that is, besides the forces $\frac{\beta f^3}{u^2}$, $\frac{\gamma f^3}{v^2}$, which vary as (dist.)$^{-2}$, there are the forces $2\alpha u$, $2\alpha v$, varying directly as the distance, and of the same amount at equal distances; or, what is the same thing, there is, besides the forces varying as (dist.)$^{-2}$, a force varying directly as the distance, tending to a third centre midway between the other two, a case which is specially considered in the memoir; it is found that the functions in r, s under the radicals (instead of rising only to the order 4) rise in this case to the order 6.

53. Among other cases are found the following, viz.:

1°.
$$U = \alpha u + \frac{7\lambda}{f^2} u^3 + \frac{5\lambda}{f^4} u^5,$$

$$V = \alpha v + \frac{7\lambda}{f^2} v^3 + \frac{5\lambda}{f^4} v^5;$$

2°.
$$U = \alpha u + \frac{\beta}{f^2} u^3,$$

$$V = \delta v + \frac{\epsilon}{f^2} v^3,$$

where $\beta = \epsilon$, or else $\alpha\epsilon = \beta\delta = 2\beta\epsilon$.

In regard to the subject of this second memoir of Lagrange, see *post*, Miscellaneous Problems, Liouville's Memoirs, Nos. 100 to 105.

54. In the *Mécanique Analytique* (1st ed. 1788, and 2nd ed. t. II. 1813), Lagrange in effect reproduces his solution for the above-mentioned law of force (say $U = \frac{\alpha}{u^2} + 2\gamma u$, $V = \frac{\beta}{v^2} + 2\gamma v$).(1) There are even in the third edition a few trifling errors of work to be corrected. The remarks above referred to, as made by Lagrange in his first memoir, are also reproduced (see *antè*, Nos. 49 and 50).

55. Legendre, *Exercices de Calcul Intégral*, t. II. (1817), and *Théorie des Fonctions Elliptiques*, t. I. (1825), uses p^2 and q^2 in the place of Euler's p, q; the forces are assumed to vary as (dist.)$^{-2}$, and in consequence of the change Euler's cubic radicals are replaced by quartic radicals involving only even powers of p and q

[1] In the *Mécanique Analytique*, Lagrange's letters are r, q for the distances $r+q=s$, $r-q=u$: the change in the present Report was occasioned by the retention of Euler's variables p, q.

respectively; that is, the radicals are in a form adapted for the transformation to elliptic integrals; in certain cases, however, it becomes necessary to attribute to Legendre's variables p and q imaginary values. The various cases of the motion are elaborately discussed by means of the elliptic integrals; in particular Legendre notices certain cases in which the motion is oscillatory, and which, as he remarks, seem to furnish the first instance of the description by a free particle of only a finite portion of the curve which is analytically the orbit of the particle; there is, however, nothing surprising in this kind of motion, although its existence might easily not have been anticipated.

56. § 26 of Jacobi's memoir "Theoria Novi Multiplicatoris &c." (1845) is entitled "Motus puncti versus duo centra secundum legem Neutonianum attracti." The equations for the motion in space are by a general theorem given in the memoir "De Motu puncti singularis" (1842), reduced to the case of motion in a plane: viz. if x, y are the coordinates, the centre point of the axis being the origin, and y being at right angles to the axis, and if the distance of the centres is $2a$; then the only difference is that to the expression for $\frac{d^2y}{dt^2}$ there is added a term $\frac{\alpha^2}{y^3}$, which arises from the rotation about the axis. Two integrals are obtained, one the integral of *Vis Viva*, and the other of them an integral similar to one of those of Euler's or Lagrange's. And then x', y' being the differential coefficients of x, y with regard to the time, the remaining equation may be taken to be $y'dx - x'dy = 0$, where x', y' are to be expressed as functions of x, y by means of the two given integrals. This being so, the principle of the Ultimate Multiplier[1] furnishes a multiplier of this differential equation, and the integral is found to be

$$\int \frac{y'dx - x'dy}{xy\,(x'^2 - y'^2) + (a^2 - x^2 + y^2)\,x'y'} = \epsilon,$$

the quantity under the integral sign being a complete differential. To verify *à posteriori* that this is so, Jacobi introduces the auxiliary quantities λ', λ'' defined as the roots of the equation $\lambda^2 + \lambda\,(x^2 + y^2 - a^2) - a^2y^2 = 0$, which in fact, if as before u, v are the distances from the centres, leads to

$$u + v = 2\sqrt{a^2 - \lambda'}, \quad u - v = 2\sqrt{a^2 - \lambda''},$$

so that λ', λ'' are functions of $u + v$, $u - v$ respectively; and the formulæ, as ultimately expressed in terms of λ', λ'', are substantially of the same form with those of Euler and Lagrange.

57. The investigations contained in Liouville's three memoirs "Sur quelques cas particuliers &c." (1846), find their chief application in the problem of two centres, and by leading in the most direct and natural manner to the general law of force for which the integration is possible, they not only give some important extension of the problem, but they in fact exhibit the problem itself and the preceding solutions of it in their true light. But as they do not relate to this problem exclusively, it will be convenient to consider them separately under the head Miscellaneous Problems.

[1] Explained in Jacobi's memoir "Theoria Novi Multiplicatoris &c.," *Crelle*, tt. XXVII., XXVIII., XXIX. 1844–45.

58. In Serret's "Thèse sur le Mouvement &c." (1848), the problem is very elegantly worked out according to the principles of Liouville's memoirs as follows: viz. assuming that the expression of the distance between two consecutive positions of the body is

$$ds^2 = \lambda\,(md\mu^2 + ndv^2) + \lambda'' d\gamma^2,$$

where m, n are functions of μ, ν respectively, and if the forces can be represented by means of a force-function U, then the motion can be determined, provided only λ, λU, $\dfrac{\lambda''}{\lambda}$ are of the forms

$$\lambda = \phi\mu - \Phi\nu,$$

$$\lambda U = \psi\mu - \Psi\nu,$$

$$\frac{\lambda}{\lambda''} = \varpi\mu - \Pi\nu,$$

where the functional symbols ϕ, Φ, &c. denote any arbitrary functions whatever.

59. It is then assumed that μ, ν are the parameters of the confocal ellipses and hyperbolas situate in the moveable plane through the axis, viz. that we have

$$\frac{x^2}{\mu^2} + \frac{y^2}{\mu^2 - b^2} = 1,$$

$$\frac{x^2}{\nu^2} - \frac{y^2}{b^2 - \nu^2} = 1,$$

(the origin is midway between the two centres, $2b$ being their distance; $\tfrac{1}{2}\mu$, $\tfrac{1}{2}\nu$ are in fact equal to the sum and difference $u+v$, $u-v$ of the two centres respectively); and that the position of the moveable plane is determined by means of γ, the inclination to a fixed plane through the axis, or say, as before, its azimuth. In fact, with these values of the coordinates, the expression of ds^2 is

$$ds^2 = (\mu^2 - \nu^2)\left(\frac{d\mu^2}{\mu^2 - b^2} + \frac{d\nu^2}{b^2 - \nu^2}\right) + \frac{(\mu^2 - b^2)(b^2 - \nu^2)}{b^2}\,d\gamma^2,$$

which is of the required form. And moreover if the forces to the two centres vary as $(\text{dist.})^{-2}$, and there is besides a force to the middle point varying as the distance, then

$$U = \frac{g}{\mu+\nu} + \frac{g'}{\mu-\nu} + K\,(\mu^2 + \nu^2 - b^2),$$

whence (observing that $\lambda = \mu^2 - \nu^2$) λU is of the required form. The equations obtained by substituting for U the above value give the ordinary solution of the problem.

60. Liouville's note to the last-mentioned memoir (1848) contains the demonstration of a theorem obtained by a different process in his *second* memoir, but which is in the present note, starting from Serret's formulæ, demonstrated by the more simple method of the *first* memoir, viz., it is shown that the motion can be obtained if the two centres, instead of being fixed, revolve about the point midway between them in

a circle in such manner that the diameter through the two centres always passes through the projection of the body on the plane of the circle. It will be observed that the circular motion of the two centres is neither a uniform nor a given motion, but that they are, as it were, carried along with the moving body.

61. In Desboves's memoir "Sur le Mouvement d'un point matériel &c." (1848), the author developes the solution of the foregoing problem of moving centres, chiefly by the aid of the method employed in Liouville's *second* memoir. And he shows also that the methods of Euler and Lagrange for the case of two fixed centres apply with modification to the more complicated problem of the moving centres.

62. The problem of two centres is considered in Bertrand's "Mémoire sur les équations différentielles &c." (1852), by means of Jacobi's form of the equations of motion, viz., the problem is reduced to a plane one by means of the addition of a force $\propto \frac{1}{y^3}$ (*antè*, No. 56).

63. Cayley's "Note on Lagrange's Solution &c." (1857) is merely a reproduction of the investigation in the *Mécanique Analytique;* the object was partly to correct some slight errors of work, and partly to show what were the combinations of the differential equations, which give at once the integrals of the problem.

64. In § II. of Bertrand's "Mémoire sur quelques unes des formes &c." (1857), the following question is considered, viz., assuming that the dynamical equations

$$\frac{d^2x}{dt^2}=\frac{dU}{dx}, \quad \frac{d^2y}{dt^2}=\frac{dU}{dy},$$

have an integral of the form

$$\alpha = Px'^2 + Qx'y' + Ry'^2 + Sy' + Tx' + K$$

(where α is the arbitrary constant, and $P, Q, \ldots K$ are functions of x and y), it is required to find the form of the force-function U. It is found that U must satisfy a certain partial differential equation of the second order, the general solution of which is not known; but taking U to be a function of the distance from any fixed point (or rather the sum of any number of such functions), it is shown that the *only* case in which the differential equations for the motion of a point attracted to a fixed centre of forces have an integral of the form in question is the above-mentioned one of two centres, each attracting according to the inverse square of the distance, and a third centre midway between them, attracting as the distance.

The Spherical Pendulum, Article Nos. 65 to 73.

65. The problem is obviously the same as that of a heavy particle on the surface of a sphere.

I have not ascertained whether the problem was considered by Euler. Lagrange refers to a solution by Clairaut, *Mém. de l'Acad.*, 1735.

The question was considered by Lagrange, *Méc. Anal.*, 1st edit. p. 283. The angles which determine the position are ψ the inclination of the string to the horizon, ϕ the inclination of the vertical plane through the string to a fixed vertical plane, or say the azimuth. And then forming the equations of motion, two integrals are at once obtained; these are the integrals of *Vis Viva*, and an integral of areas. And these give equations of the form $dt = \text{funct.}\,(\psi)\,d\psi$, $d\phi = \text{funct.}\,(\psi)\,d\psi$; so that t, ϕ are each of them given by a quadrature in terms of ψ, which is the point to which the solution is carried. It is noticed that ψ may have a constant value, which is the case of the conical pendulum.

66. In the second edition, t. XI. p. 197 (1815), the solution is reproduced; only, what is obviously more convenient, the angles are taken to be

ψ, the inclination to the vertical,

ϕ, the azimuth.

It is remarked that ψ will always lie between a greatest value α and a least value β, and the integrals are transformed by introducing therein instead of ψ the angle σ, which is such that

$$\cos\psi = \cos\alpha\sin^2\sigma + \cos\beta\cos^2\sigma,$$

by which substitution they assume a more elegant form, involving only the radical

$$\sqrt{1 + k^2(\cos\beta - \cos\alpha)\cos 2\sigma},$$

where k is a constant depending on $\cos\alpha$, $\cos\beta$; and the integration is effected approximately in the case where $\cos\beta - \cos\alpha$ is small.

M. Bravais has noticed, however, that by reason of some errors in the working out, Lagrange has arrived at an incorrect value for the angle Φ, which is the apsidal angle, or difference of the azimuths for the inclinations α and β: see the 3rd edition (1855), Note VII., where M. Bravais resumes the calculation, and he arrives at the value $\Phi = \frac{1}{2}\pi(1 + \frac{3}{8}\alpha\beta)$, α and β being small.

Lagrange considers also the case where the motion takes place in a resisting medium, the resistance varying as velocity squared.

67. A similar solution to Lagrange's, not carried quite so far, is given in Poisson's *Mécanique*, t. I. pp. 385 *et seq.* (2nd ed. 1833).

A short paper by Puiseux, "Note sur le Mouvement d'un point matériel sur une sphère" (1842), shows merely that the angle Φ is $> \frac{1}{2}\pi$.

68. The ulterior development of the solution consists in the effectuation of the integrations by the elliptic and Jacobian functions. It is proper to remark that the dynamical problem the solution whereof by such functions was first fairly worked out,

is the more difficult one of the rotation of a solid body, as solved by Jacobi (1839), in completion of Rueb's solution (1834), *post*, Nos. 186 and 197.

69. In relation to the present problem we have Gudermann's memoir "De pendulis sphæricis &c." (1849), who, however, does not arrive at the actual expressions of the coordinates in terms of the time; and the perusal of the memoir is rendered difficult by the author's peculiar notations for the elliptic functions [1].

70. It would appear that a solution involving the Jacobian functions was obtained by Durège, in a memoir completed in 1849, but which is still unpublished; see § XX. of his *Theorie der elliptischen Functionen* (1861), where the memoir is in part reproduced. It is referred to by Richelot in the Note presently mentioned.

71. We have next Tissot's *Thèse de Mécanique*, 1852, where the expressions for the variables in terms of the time are completely obtained by means of the Jacobian functions H, Θ, and which appears to be the earliest published one containing a complete solution and discussion of the problem.

72. Richelot, in the Note "Bemerkungen zur Theorie des Raumpendels" (1853), gives also, but without demonstration, the final expressions for the coordinates in terms of the time.

Donkin's memoir "On a Class of Differential Equations &c." (1855) contains (No. 59) an application to the case of the spherical pendulum.

73. The first part of the memoir by Dumas, "Ueber die Bewegung des Raumpendels, &c." (1855), comprises a very elegant solution of the problem of the spherical pendulum based upon Jacobi's theorem of the Principal Function (1837), and which is completely developed by the elliptic and Jacobian functions. The latter part of the memoir relates to the effect of the rotation of the Earth; and we thus arrive at the next division of the general subject.

Motion as affected by the Rotation of the Earth, and Relative Motion in general. Article Nos. 74 to 85.

74. Laplace (*Méc. Céleste*, Book X. c. 5) investigates the equations for the motion of a terrestrial body, taking account of the rotation of the Earth (and also of the resistance of the air), and he applies them to the determination of the deviations of falling bodies, &c. He does not, however, apply them to the case of the pendulum.

75. We have also the memoir of Gauss, "Fundamental-gleichungen &c." (1804): the equations ultimately obtained are similar to those of Poisson. I have not had the opportunity of consulting this memoir.

[1] The mere use of sn, cn, dn as an abbreviation of the somewhat cumbrous sinam, cosam, Δam of the *Fundamenta Nova* is decidedly convenient.

76. Poisson, in the "Mémoire sur le mouvement des Projectiles &c." (1838), also obtains the general equations of motion, viz. (omitting terms involving n^2), these may be taken to be

$$\frac{d^2x}{dt^2} = \quad X + 2n\left(\frac{dy}{dt}\sin\beta + \frac{dz}{dt}\cos\beta\right),$$

$$\frac{d^2y}{dt^2} = \quad Y + 2n \quad \frac{dx}{dt}\sin\beta,$$

$$\frac{d^2z}{dt^2} = g + Z + 2n \quad \frac{dx}{dt}\cos\beta$$

(see p. 20), where the axes of x, y, z are fixed on the Earth and moveable with it: viz., z is in the direction of gravity; x, y in the directions perpendicular to gravity, viz., y in the plane of the meridian northwards, x westwards; g is the actual force of gravity as affected by the resolved part of the centrifugal force; β is the latitude. There are some niceties of definition which are carefully given by Poisson, but which need not be noticed here.

77. Poisson applies his formulæ incidentally to the motion of a pendulum, which he considers as vibrating in a plane; and after showing that the time of oscillation is not sensibly affected, he remarks that upon calculating the force perpendicular to the plane of oscillation, arising from the rotation of the Earth, it is found to be too small sensibly to displace the plane of oscillation or to have any appreciable influence on the motion—a conclusion which, as is well known, is erroneous. He considers also the motion of falling bodies, but the memoir relates principally to the theory of projectiles.

78. That the motion of the spherical pendulum is sensibly affected by the rotation of the Earth is the well-known discovery of Foucault; it appears by his paper, "Démonstration Physique &c.," *Comptes Rendus*, t. XXXII. 1851, that he was led to it by considering the case of a pendulum oscillating at the pole; the plane of oscillation, if actually fixed in space, will by the rotation of the Earth appear to rotate with the same velocity in the contrary direction; and he remarks that although the case of a different latitude is more complicated, yet the result of an apparent rotation of the plane of oscillation, diminishing to zero at the equator, may be obtained either from analytical or from mechanical and geometrical considerations. Some other Notes by Foucault on the subject are given, *Comptes Rendus*, t. XXXV. (1853).

79. An analytical demonstration of the theorem was given by Binet, *Comptes Rendus*, t. XXXII. (1851), and by Baehr (1853). Various short papers on the subject will be found in the *Philosophical Magazine*, and elsewhere.

80. In regard to the above-mentioned problem of falling bodies, we have a Note by W. S., *Camb. and Dubl. Math. Journ.* t. III. (1848), containing some errors which are rectified in a subsequent paper, "Remarks on the Deviation of Falling bodies, &c." t. IV. (1849), by Dr Hart and Professor W. Thomson.

81. The theory of relative motion is considered in a very general manner in M. Quet's memoir, "Des Mouvements relatifs en général &c." (1853). Suppose that x, y, z are the coordinates of a particle in relation to a set of moveable axes; let ξ', η', ζ' be the coordinates of the moveable origin in reference to a fixed set of axes, and treating the accelerations $\frac{d^2\xi'}{dt^2}$, $\frac{d^2\eta'}{dt^2}$, $\frac{d^2\zeta'}{dt^2}$ as if they were coordinates, let these, when resolved along the moveable axes, give u', v', w': suppose, moreover, that p, q, r denote the angular velocities of the system of the moveable axes (or axes of x, y, z) round the axes of x, y, and z respectively; u', v', w', p, q, r are considered as given functions of the time, and then, if

$$u = \frac{d^2x}{dt^2} + 2\left(q\frac{dz}{dt} - r\frac{dy}{dt}\right) + z\frac{dq}{dt} - y\frac{dr}{dt} + q(py - qx) - r(rx - pr) + u',$$

$$v = \frac{d^2y}{dt^2} + 2\left(r\frac{dx}{dt} - p\frac{dz}{dt}\right) + x\frac{dr}{dt} - z\frac{dp}{dt} + r(qz - ry) - p(py - qx) + v',$$

$$w = \frac{d^2z}{dt^2} + 2\left(p\frac{dy}{dt} - q\frac{dx}{dt}\right) + y\frac{dp}{dt} - x\frac{dq}{dt} + p(rx - pz) - q(qz - ry) + w',$$

it is shown that the equations of motion are to be obtained from the equation

$$\Sigma m\,[(u - X)\,\delta x + (v - Y)\,\delta y + (w - Z)\,\delta z] = 0,$$

where δx, δy, δz are the virtual velocities of the particle m in the directions of the moveable axes. This equation is in fact obtained as a transformation of the equation

$$\Sigma m\left[\left(\frac{d^2\xi}{dt^2} - X\right)\delta\xi + \left(\frac{d^2\eta}{dt^2} - Y\right)\delta\eta + \left(\frac{d^2\zeta}{dt^2} - Z\right)\delta\zeta\right] = 0,$$

which belongs to a set of fixed axes of ξ, η, ζ.

82. The equations for the motion of a free particle are of course $u = X$, $v = Y$, $w = Z$. In the case where the moveable axes are fixed on the Earth, and moveable with it (the diurnal motion being alone attended to), these lead to equations for the motion of a particle in reference to the Earth, similar to those obtained by Gauss and Poisson. The formulæ are applied to the case of the spherical pendulum, which is developed with some care; and Foucault's theorem of the rotation of the plane of oscillation very readily presents itself. The general formulæ are applied to the relative motion of a solid body, and in particular to the question of the gyroscope; the memoir contains other interesting results.

83. The principal memoirs on the motion of the spherical pendulum, as affected by the rotation of the Earth, are those of Hansen, "Theorie der Pendelbewegung &c." (1853), which contains an elaborate investigation of all the physical circumstances (resistance of the air, torsion of the string, &c.) which can affect the actual motion, and the before-mentioned memoir by Dumas, "Ueber der Bewegung des Raumpendels &c." (1855). The investigation is conducted by means of the variation of the constants;

the integrals for the undisturbed problem were, as already noticed, obtained by means of Jacobi's Principal Function, that is, in a form which leads at once to the expressions for the variation of the constants; and the investigation appears to be carried out in a most elaborate and complete manner.

84. In concluding this part of the subject I refer to Mr Worm's work, *The Rotation of the Earth* (1862), where the last-mentioned questions (falling bodies, the pendulum, and the gyroscope) are, in reference to the proofs they afford of the rotation of the Earth, considered as well in an experimental as in a mathematical point of view. The second part of the volume contains the theory (after Laplace and Gauss) of falling bodies, that of the pendulum (after Hansen), and that of the gyroscope (after Yvon Villarceau); and the whole appears to be a complete and satisfactory *résumé* of the experimental and mathematical theories to which it relates.

85. We have also Cohen "On the Differential Coefficients and Determinants of Lines &c." (1862), where the equations for relative motion are obtained in a very elegant manner. The fundamental notion of the memoir may be considered to be the dealing *directly* with lines, velocities, &c., which are variable in direction as well as in magnitude, instead of referring them, as in the ordinary analytical method, to axes fixed in space. The memoir is a highly interesting and valuable one, and the results are brought out with great facility; but I cannot but think that the great care required to apply the method correctly is an objection to it, if used otherwise than by way of interpretation of previously obtained results, and that the ordinary method is preferable.

I may remark that the theory of relative motion connects itself with the lunar and planetary theories as regards the reference of the plane of the orbit to the variable ecliptic, and as regards the variations of the position of the orbit; but this is a subject which I have abstained from entering upon.

Miscellaneous Problems. Articles Nos. 86 to 111 (several subheadings).

Motion of a single particle.

86. Jacobi, in the memoir "De Motu puncti singularis" (1842), notices (§ 5) the case of a body acted on by a central force which is any homogeneous function of the degree -2 of the coordinates; or representing these by $r\cos\phi$, $r\sin\phi$, then the force is $=\frac{\Phi}{r^2}$, where Φ is any function of the angle ϕ. In fact, after integrating by a process different from the ordinary one the case of a central force $\propto\frac{1}{r^2}$, he remarks that the method in fact applies to the more general law of force just mentioned.

87. Jacobi, in the memoir "Theoria Novi Multiplicatoris &c." (1845), considers (§ 25) the case of a body acted on by a central force P a function of the distance, and

besides by forces X and Y, which are homogeneous functions of the degree -3 of the coordinates (x, y); viz. the equations of motion are in this case

$$\frac{d^2x}{dt^2} = -\frac{Px}{r} + X,$$

$$\frac{d^2y}{dt^2} = -\frac{Py}{r} + Y,$$

and there is an integral

$$\tfrac{1}{2}(xy' - x'y)^2 - \int x^2 (xY - yX)\, d\frac{y}{x} = \text{const.}$$

(the function under the integral sign is obviously a function of the degree 0 in (x, y), that is, it is a function of $\frac{y}{x}$). If X, Y are the derived functions of a force-function U of the degree -2 in (x, y), then there is, besides, the integral of *Vis Viva*, and thence a third integral is obtained by means of the theorem of the Ultimate Multiplier. It may be noticed that in the last-mentioned case the force-function is of the form $\frac{\Phi}{r^2}$, so that if we represent also the central force by means of a force-function R (= function of r), then the entire force-function is $R + \frac{\Phi}{r^2}$. The case is a very interesting one; it includes that considered § iv. of Bertrand's "Mémoire sur les équations différentielles de la Mécanique" (1852), where the force-function is of the form $= \frac{\Phi}{r^2}$.

Motion of three mutually attracting bodies in a right line.

88. The problem is considered by Euler in the memoir "De Motu rectilineo &c." (1765), the forces being as the inverse square of the distance; and a solution is obtained for an interesting particular case. Let A, B, C be the masses, and suppose that at the commencement of the motion the distances CB, BA are in the ratio $\alpha : 1$, and that the velocities (assumed to be in the same sense) are proportional to the distances from a fixed point. Then, if α be the real root (there is only one) of the equation of the fifth order

$$C(1 + 3\alpha + 3\alpha^2) = A\alpha^3(\alpha^2 + 3\alpha + 3) + B(\alpha + 1^2)(\alpha^3 - 1),$$

the distances CB, BA will always continue in the ratio $\alpha : 1$. It may be added that the distances CB, BA each of them vary as $r^2 - a^2$, where a is a constant, and r is, according to the initial circumstances, a function of t defined by one or the other of the two equations

$$t = n^3 r\sqrt{r^2 - a^2} - n^3a^2 \log\frac{r + \sqrt{r^2 - a^2}}{a},$$

$$t = n^3 r\sqrt{a^2 - r^2} + n^3a^2 \sin^{-1}\frac{r}{a}.$$

89. The bodies are considered as restricted to move in a given line; but it is clear that if the bodies, considered as free points in space, are initially in a line, and the initial velocities are also in this line, then the bodies will always continue in this line, which will be a fixed line in space. But if the distances and velocities are as above, except only that the velocities, instead of being along the line, are parallel to each other in any direction whatever, then the bodies will always continue in a line, which is in this case a moveable line in space (see *post*, No. 93).

90. Euler resumes the problem in the memoir of 1776 in the *Nova Acta Petrop.* The distances AB, BC being p and q, then

$$\frac{d^2p}{dt^2} = -\frac{A+B}{p^2} + \frac{C}{q^2} - \frac{C}{(p+q)^2},$$

$$\frac{d^2q}{dt^2} = \frac{A}{p^2} - \frac{A}{(p+q)^2} - \frac{B+C}{q^2};$$

and in particular he considers the before-mentioned case of a solution of the form $p = nq$; and also the particular problem where one of the masses vanishes, $C=0$; in this case, introducing (instead of p, q), the new variables u, s, where $q = up$, $dq = sdp$ (a transformation suggested by the homogeneity of the equations), and making, moreover, the particular supposition that the integral of the first equation is $\left(\frac{dp}{dt}\right)^2 = \frac{2(A+B)}{p}$ (viz. making the constant of integration to vanish), he obtains between s and u the equation of the first order

$$2(A+B)\frac{ds}{du}(s-u) = (A+B)s + A - \frac{A}{(1+u)^2} - \frac{B}{u^2},$$

which, however, he is not able to integrate.

91. Jacobi has given in the memoir "Theoria Novi Multiplicatoris &c." (1845) (§ 28, entitled "De Problemate trium corporum in eâdem rectâ motorum. Substitutio Euleriana. Theoremata de viribus homogeneis") a very symmetrical and elegant investigation of the same problem. The centre of gravity being assumed to be at rest, the coordinates x, x_1, x_2 of the three bodies are in the first instance expressed as linear functions of the two variables u, v (being, as Jacobi remarks, the transformation employed in his memoir "Sur l'élimination des Nœuds" (1843), *post*, No. 114), $\frac{d^2u}{dt^2}$ and $\frac{d^2v}{dt^2}$ come out respectively equal to homogeneous functions of the degree -2 of these variables u and v, and the integral of *Vis Viva* exists. The subsequent transformation consists in the introduction of the variables r, ϕ, s, η, where $u = r\cos\phi$, $v = r\sin\phi$, $s = \sqrt{r}\frac{dr}{dt}$, $\eta = \sqrt{r^3}\frac{d\phi}{dt}$; this gives a system of equations independent of r; viz.,

$$d\phi : ds : d\eta = \eta : \tfrac{1}{2}s^2 + \eta^2 - \Phi : -\tfrac{1}{2}s\eta + \Phi',$$

where Φ is a given function of ϕ, and Φ' is the derived function. If these equations were integrated, the equation of *Vis Viva* gives at once $r = \frac{1}{h}\left(\Phi - \tfrac{1}{2}(s^2+\eta^2)\right)$; and finally the time t would be given by a quadrature. The system of three equations has the multiplier $M=\frac{1}{\sqrt{\phi-\frac{1}{2}(s^2+\eta^2)}}$, hence if one integral were known the other would be at once furnished by the general theory. There is a simplification in the form of the solution if h (the constant of *Vis Viva*) $=0$. It is remarked that the method is equally applicable when the force varies as *any* power of the distance; and moreover that when the force varies as (dist.)$^{-3}$, then the solution depends upon one quadrature only.

92. The concluding part of the section relates to the very general problem of a system of n particles acted on by any forces homogeneous functions of the coordinates (this includes the case of n particles mutually attracting each other according to a power of the distance), and this more general investigation illustrates the method employed in regard to the three bodies in a line. It may be remarked that in the general theorem for the n particles "sint vires &c.," the constant of *Vis Viva* is supposed to vanish.

Particular cases of the motion of three bodies.

93. In the case of three bodies attracting each other according to the inverse square of the distance, the bodies may move in such manner as to be constantly in a line (a moveable line in space); this appears by the memoir, Euler, "Considérations générales &c." (1764), in which memoir, however (which it will be observed precedes the memoir "De Motu rectilineo &c." (1765), referred to No. 88), Euler assumes that the mass of one of the bodies is so small as not to affect the relative motion of the other two. Calling the bodies the Sun, Earth, and Moon, and taking the masses to be 1, m, 0, then a result obtained is, that in order that the Moon may be perpetually in conjunction, its distance must be to that of the Sun as $\alpha : 1$, where $m(1-\alpha)^2=3\alpha^3-3\alpha^4+\alpha^5$, or $\alpha=\sqrt[3]{\frac{1}{3}m}$ nearly. It appears, however (*antè*, No. 88), that the foregoing restriction as to the masses is unnecessary, and, as will be mentioned, the problem has since been treated without such restriction. Euler investigates the motion in the case where the initial circumstances are nearly but not exactly as originally supposed; this assumes, however, that the motion is stable—i.e. that the bodies will continue to move nearly, but not exactly as originally supposed, which is at variance with the conclusions of Liouville's memoir, *post*, No. 95. I have not examined the cause of this discrepancy.

94. In the *Mécanique Céleste* (1799), Book X. c. 6, Laplace considers two cases where the motion can be exactly determined.

1°. Force varies as any function of the distance. It is shown that the motion may be such that the bodies form always an equilateral triangle of variable magnitude—the motion of each body about the centre of gravity being the same as if that point were a centre of force attracting the body according to a similar law.

2°. Force $\propto$ (dist.)^n. The motion may be such that the three bodies are always in a right line (moveable in space), the relative distances being in fixed ratios to each other. In particular, if force $\propto$ (dist.)$^{-2}$, then m, m', m'' being the masses, the quantity z which determines the ratio of the distances $m''m'$, $m'm$ is given by

$$0 = mz^2 [(1+z)^3 - 1] - m' (1+z)^2 (1-z^3) - m'' [(1+z)^3 - z^3] = 0,$$

which is, in fact, the formula in Euler's memoir "De Motu rectilineo &c."

95. Liouville's memoir "Sur un cas particulier &c." (1842) has for its object to show that if the initial circumstances are not precisely as supposed in the second of the two cases considered by Laplace, or, what is the same thing, in Euler's memoir "Considérations générales &c.," then the motion is unstable; the instability manifests itself in the usual manner, viz. the expressions for the deviations from the normal positions are found to contain real exponentials which increase indefinitely with the time.

96. It may be proper to refer here to Jacobi's theorem, *Comptes Rendus*, t. III. p. 61 (1836), quoted in the foot-note to No. 31 of my Report of 1857, [195], which relates to the motion of a point *without mass* revolving round the Sun, and disturbed by a planet moving in a circular orbit, and properly belongs (as I have there remarked) to the problem of two centres, one of them moveable and the other revolving round it in a circle with uniform velocity. The theorem (given without demonstration by Jacobi) is proved by Liouville in his last-mentioned memoir, and he remarks that the theorem follows very simply as a corollary of the theorem by Coriolis, "Mémoire sur le principe des forces vives dans les mouvemens relatifs des Machines," *Journ. de l'Ecole Polyt.* t. XIII. pp. 268—302 (1832). There is, however, no difficulty in proving the theorem; another proof is given, Cayley, "Note on a Theorem of Jacobi's &c." (1862).

Motion in a resisting medium.

97. I do not consider the various integrable cases of the motion of a particle in a resisting medium, the resistance varying with the velocity according to some assumed law, the particle being either not acted on by any force, or acted upon by gravity only. Some interesting cases are considered in Jacobi's memoir "De Motu puncti singularis" (1842), §§ 6 and 7 (see *post*, No. 108).

98. In the case of a central force varying as (dist.)$^{-2}$, the effect of a resisting medium ($R \propto v^2$) is considered in reference to the lunar theory, in the *Mécanique Céleste*, Book VII. c. 6. Formulæ for the variations of the elliptic elements are given in the *Mécanique Analytique*, t. II. (2nd edition). But the variations of the elliptic elements are fully worked out by means of elliptic and Jacobian functions in Sohncke's valuable memoir "Motus Corporum &c." (1833).

99. The effect of the resistance of the air on a pendulum has been elaborately considered by Poisson, Bessel, Stokes, and others; as the dimensions of the ball are attended to, the problem is in fact a hydrodynamical one.

The effect on the spherical pendulum is considered in Hansen's memoir "Theorie der Pendelbewegung &c. (1853).

The effect on the motion of a projectile is considered in Poisson's memoirs "Sur le Mouvement des Projectiles &c." (1838).

Liouville's *memoirs "Sur quelques Cas particuliers où les équations du mouvement d'un point matériel peuvent s'intégrer"* (1846—49).

100. In the *first* memoir (§ 1) the author considers a point moving in a plane or on a given surface, where the principle of *Vis Viva* holds good (or say where there is a force-function U). The coordinates of the point, and the function U, may be expressed in terms of two variables α, β, and it is assumed that these are such that

$$ds^2 = \lambda\,(d\alpha^2 + d\beta^2),$$

where λ is a function of α and β. That is, we have $T = \tfrac{1}{2}\lambda\,(\alpha'^2 + \beta'^2)$; and the equations of motion are

$$\frac{d\,.\,\lambda\alpha'}{dt} = \tfrac{1}{2}\frac{d\lambda}{d\alpha}(\alpha'^2 + \beta'^2) + \frac{dU}{d\alpha},$$

$$\frac{d\,.\,\lambda\beta'}{dt} = \tfrac{1}{2}\frac{d\lambda}{d\beta}(\alpha'^2 + \beta'^2) + \frac{dU}{d\beta}.$$

One integral of these is

$$\lambda\,(\alpha'^2 + \beta'^2) = 2U + C\,;$$

and by means of it the equations take the form

$$\frac{d\,.\,\lambda\alpha'}{dt} = \frac{1}{2\lambda}\frac{d\lambda}{d\alpha}(2U + C) + \frac{dU}{d\alpha},$$

$$\frac{d\,.\,\lambda\beta'}{dt} = \frac{1}{2\lambda}\frac{d\lambda}{d\beta}(2U + C) + \frac{dU}{d\beta}.$$

These equations, it is easy to show, may be integrated if

$$(2U + C)\,\lambda = f\alpha - F\beta,$$

and they then in fact give

$$\lambda^2\alpha'^2 = f\alpha - A\;,$$

$$\lambda^2\beta'^2 = A - F\beta,$$

where A is an arbitrary constant. And we then have

$$\frac{d\alpha}{\sqrt{f\alpha - A}} = \frac{d\beta}{\sqrt{A - F\beta}},$$

which gives the path, and the expression for the time is easily obtained by means of a quadrature.

It is not more general, but it is frequently convenient to employ instead of α, β, two variables μ and ν, such that

$$ds^2 = \lambda\,(m\,d\mu^2 + n\,d\nu^2),$$

where m is a function of μ only and n of ν only, while λ contains μ and ν. The geometrical signification of the equation $ds^2 = \lambda(d\alpha^2 + d\beta^2)$, or of the last-mentioned equivalent form, is that the curves

$$\alpha \text{ or } \lambda = \text{const.}, \qquad \beta \text{ or } \mu = \text{const.}$$

intersect at right angles.

The foregoing differential equation of the path, writing $f\mu$, $F\nu$ in the place of $f\alpha$, $F\beta$ respectively, may be expressed in the form

$$f\mu \cos^2 i + F\nu \sin^2 i = A,$$

where i, $90° - i$ are the inclinations of the path at the point (λ, μ) to the two orthotomic curves through this point.

101. The before-mentioned equation

$$(2U + C)\lambda = f\alpha - F\beta$$

may be satisfied independently of C, or else only for a particular value of C. In the former case the law of force is much more restricted, but on the other hand there is no restriction as regards the initial circumstances of the motion; it is the more important one, and is alone attended to in the sequel of the memoir. In the case in question (changing the functional symbols) we must have

$$\lambda = \phi\alpha - \varpi\beta, \quad \lambda U = f\alpha - F\beta;$$

so that the functions denoted above by $f\alpha$, $F\beta$ now are $2f\alpha + C\phi\alpha$, $2F\beta + C\varpi\beta$; the equation of the trajectory is

$$\frac{d\alpha}{\sqrt{2f\alpha + C\phi\alpha - A}} = \frac{d\beta}{\sqrt{A - 2F\beta + C\varpi\beta}},$$

and for the *time* the formula is

$$dt = \frac{\phi\alpha\, d\alpha}{\sqrt{2f\alpha + C\phi\alpha - A}} - \frac{\varpi\beta\, d\beta}{\sqrt{A - 2F\beta + C\varpi\beta}}.$$

It is noticed also that taking B, ϵ to denote two new arbitrary constants, and writing

$$\Theta \int d\alpha \sqrt{2f\alpha + C\phi\alpha - A} + \int d\beta \sqrt{A - 2F\beta + C\varpi\beta},$$

the equation of the trajectory and the expression for the time assume the forms

$$\frac{d\Theta}{dA} = B, \quad t = 2\frac{d\Theta}{dC} + \epsilon,$$

as is known *à priori* by a theorem of Jacobi's.

If the forces vanish, the path is a geodesic line; and denoting by a the ratio of the constants A, C, we have

$$\frac{d\alpha}{\sqrt{\phi\alpha - a}} = \frac{d\beta}{\sqrt{a - \varpi\beta}};$$

and moreover

$$ds = d\alpha \sqrt{\phi\alpha - a} + d\beta \sqrt{a - \phi\beta},$$

which are geometrical properties relating to the geodesic line.

102. Passing to the applications: in the first place, if α, β are rectangular coordinates of a point *in plano*, then writing instead of them x, y, we have $ds^2 = dx^2 + dy^2$, which is of the required form; but the result obtained is the self-evident one, that the equations may be integrated by quadratures when U is of the form funct. x − funct. y.

But taking instead the elliptic coordinates μ, ν, of a point *in plano*—viz., as employed by the author, these are the semiaxes of the confocal ellipse and hyperbola represented by the equations

$$\frac{x^2}{\mu^2} + \frac{y^2}{\mu^2 - b^2} = 1, \quad \frac{x^2}{\nu^2} - \frac{y^2}{b^2 - \nu^2} = 1,$$

—very interesting results are obtained. The equations give

$$b^2x^2 = \mu^2\nu^2, \quad b^2y^2 = (\mu^2 - b^2)(b^2 - \nu^2),$$

and thence

$$ds^2 = (\mu^2 - \nu^2)\left(\frac{d\mu^2}{\mu^2 - b^2} + \frac{d\nu^2}{\nu^2 - b^2}\right),$$

which is of the proper form, and the corresponding expression of U is

$$U = \frac{f\mu - F\nu}{\mu^2 - \nu^2};$$

so that the force-function having this value ($f\mu$, $F\nu$ being arbitrary functions of μ and ν respectively), the equations of motion may be integrated by quadratures.

103. In particular, if

$$\begin{aligned} f\mu &= g\mu + g'\mu + k(\mu^4 - b^2\mu^2), \\ F\nu &= g\nu - g'\nu + k(\nu^4 - b^2\nu^2), \end{aligned}$$

then

$$U = \frac{g}{\mu + \nu} + \frac{g'}{\mu - \nu} + k(\mu^2 + \nu^2 - b^2).$$

But $\mu+\nu$, $\mu-\nu$ are the distances of the point from the two foci, and $\mu^2+\nu^2-b^2 (=x^2+y^2)$ is the square of the distance from the centre, so that the expression for U is

$$U = \frac{g}{r} + \frac{g'}{r'} + kR^2;$$

and the case is that of forces to the foci varying inversely as the squares of the distances, and a force to the centre varying directly as the distance—the case considered by Lagrange in the problem of two centres. But this is merely one particular case of those given by the general formula.

The cases $g = 0$, $g' = 0$, $k = 0$ (no forces), and $g = 0$, $g' = 0$ (a force to the centre) lead to some interesting results; it is noticed also that the expression for the force-function may be written $U = \dfrac{\text{funct.}(r + r') - \text{funct.}(r - r')}{rr'}$, and that it may be thereby ascertained (without transforming to elliptic coordinates) whether a given value of the force-function is of the form considered in the theory.

In § 3 the author considers the expression $dx^2+dy^2=\lambda(d\alpha^2+d\beta^2)$, λ being in the first instance any function whatever of α and β; and he shows that the expressions of x, y are given by the equation

$$x+y\sqrt{-1}=\psi(\alpha+\beta\sqrt{-1}),$$

ψ being any real function. If, however, it is besides assumed that λ is of the required form $=f\alpha-F\beta$, then he shows that the system of elliptic coordinates is the only one for which the conditions are satisfied. §§ 4, 5, 6 and 7 relate to the motion of a point on a sphere, an ellipsoid, a surface of revolution, and the skew helicoid respectively; and the concluding § 8 contains only a brief reference to the author's second memoir.

104. Liouville's second and third memoirs may be more briefly noticed. In the *second* memoir the author starts from Jacobi's theorem of the V function, viz., assuming that there is a force-function U independent of the time, then in order to integrate the equations of motion $\left(\frac{d^2x}{dt^2}=\frac{dU}{dx},\ \frac{d^2y}{dt^2}=\frac{dU}{dy},\ \frac{d^2z}{dt^2}=\frac{dU}{dz}\right)$, all that is required is to find a function Θ of x, y, z containing three arbitrary constants A, B, C (distinct from the constant attached to Θ by mere addition) satisfying the differential equation

$$\left(\frac{d\Theta}{dx}\right)^2+\left(\frac{d\Theta}{dy}\right)^2+\left(\frac{d\Theta}{dz}\right)^2=2(U+C);$$

for then the required integrals of the equations of motion are

$$\frac{d\Theta}{dA}=A',\quad \frac{d\Theta}{dB}=B',\quad \frac{d\Theta}{dC}=t+C',$$

A', B', C' being new arbitrary constants. Liouville introduces in place of x, y, z, the elliptic coordinates ρ, μ, ν, which are such that

$$\frac{x^2}{\rho^2}+\frac{y^2}{\rho^2-b^2}+\frac{z^2}{\rho^2-c^2}=1,$$

$$\frac{x^2}{\mu^2}+\frac{y^2}{\mu^2-b^2}-\frac{z^2}{c^2-\mu^2}=1,$$

$$\frac{x^2}{\nu^2}-\frac{y^2}{b^2-\nu^2}-\frac{z^2}{c^2-\nu^2}=1,$$

or, what is the same thing,

$$x=\frac{\rho\mu\nu}{bc},$$

$$y=\frac{\sqrt{\rho^2-b^2}\sqrt{\mu^2-b^2}\sqrt{b^2-\nu^2}}{b\sqrt{c^2-b^2}},$$

$$z=\frac{\sqrt{\rho^2-c^2}\sqrt{c^2-\mu^2}\sqrt{c^2-\nu^2}}{c\sqrt{c^2-b^2}};$$

and he then finds that the resulting partial differential equation in ρ, μ, ν may be integrated provided that U is of the form

$$U = \frac{(\mu^2 - \nu^2) f\rho + (\rho^2 - \nu^2) F\mu + (\rho^2 - \nu^2) \varpi\nu}{(\rho^2 - \mu^2)(\rho^2 - \nu^2)(\mu^2 - \nu^2)},$$

f, F, ϖ being any functional symbols whatever: viz., the expression for Θ is

$$\begin{aligned}\Theta = &\int d\rho \sqrt{\frac{2f\rho + A + B\rho^2 + 2C\rho^4}{(\rho^2 - b^2)(\rho^2 - c^2)}}, \\ &+ \int d\mu \sqrt{\frac{2F\mu + A + B\mu^2 + 2C\mu^4}{(\mu^2 - b^2)(c^2 - \mu)^2}}, \\ &+ \int d\nu \sqrt{\frac{2\varpi\nu + A + B\nu^2 + 2C\nu^4}{(b^2 - \nu^2)(c^2 - \nu^2)}}.\end{aligned}$$

In the case where $U = 0$ we have a particle not acted on by any forces, and the path is of course a straight line. The peculiar form in which these equations are obtained leads to very interesting results in regard to the theory of Abelian integrals, and to that of the geodesic lines of an ellipsoid.

The formulæ require to be modified in certain cases, such as $c = b$ or $b = 0$. The case $b = 0$ leads to the theory developed in the first memoir in relation to the problem of two centres. The case is indicated where $b = 0$, $c = 0$, the ratio $b : c$ remaining finite.

The case is briefly considered of a particle moving on a given surface.

105. The *third* memoir purports to relate to a system of particles, but the formulæ are exhibited under a purely analytical point of view; so much so, that the coordinates of the points (3 for each point) are considered as forming a single system of variables $x_1, x_2, \ldots x_i$. The partial differential equation is

$$\left(\frac{d\Theta}{dx_1}\right)^2 + \left(\frac{d\Theta}{dx_2}\right)^2 \ldots + \left(\frac{d\Theta}{dx_i}\right)^2 = 2(U + h),$$

which is transformed by introducing therein the new variables $\rho_1, \rho_2, \ldots \rho_i$ analogous to the elliptic coordinates of the second memoir. The memoir really belongs rather to the theory of the Abelian integrals (in regard to which it appears to be a very valuable one) than to dynamics.

Memoirs by Jacobi, Bertrand, *and* Donkin, *relating to various Special Problems.*

106. I have inserted this heading for the sake of showing at a single view what are the special problems incidentally considered in the undermentioned memoirs which are referred to in several places in the present Report.

107. Jacobi, "De Motu puncti singularis" (1842).—I call to mind that the memoir chiefly depends on the theorem of the Ultimate Multiplier (the theory in its generality being developed in the later memoir "Theoria Novi Multiplicatoris &c.," 1844—45). § 4 is entitled "The motion of a point on the surface of revolution," which, the principle of the conservation of areas holding good, is reduced to the problem of the motion on the meridian curve, and thus depends upon quadratures only. § 5 is entitled "On the motion of a point about a fixed centre attracted according to a certain law more general than the Newtonian one" (*antè*, No. 85). § 6, "On the motion of a point on a given curve and in a resisting medium" (resistance $=a+be^{cv^2}$, or $=a+bv^2$); and § 7, "On the Ballistic Curve," viz., the forces are gravity and a resistance $=a+bv^n$.

108. In Jacobi's memoir "Theoria Novi Multiplicatoris &c." (1845), § 25 is entitled "On the motion of a point attracted towards a fixed centre" (see *antè*, No 87); § 26, "On the motion of a point attracted towards two fixed centres according to the Newtonian law" (*antè*, No. 56); § 27, "On the rotation of a solid body about a fixed point" (*post*, No. 193); § 28, "On the problem of three bodies moving in a right line; the Eulerian substitution; theorems on homogeneous forces" (*antè*, No. 91); and § 29, "The principle of the ultimate multiplier applied to a free system of material points moving in a resisting medium; on the motion of a comet in a resisting medium about the sun."

109. And in Jacobi's memoir "Nova Methodus &c." (1862), besides § 64 and § 65, which are applications of the method to general dynamical theorems, we have § 66, containing a simultaneous solution of the problem of the motion of a point attracted to a fixed centre and of that of the rotation of a solid body (*post*, No. 206) and § 67, relating to the motion of a point attracted to a fixed centre according to the Newtonian law.

110. Bertrand's "Mémoire sur les intégrales différentielles de la Mécanique" (1852).—§ III. relates to the motion of a point attracted to a fixed centre by a force varying as a function of the distance; § IV. to the case where the forces arise from a force-function $U=\frac{1}{x^2+y^2}\phi\left(\frac{x}{y}\right)$ $\left(\text{or, what is the same thing, } =\frac{\Phi}{r^2}\right)$ (*antè*, No. 87); § V. to the problem of two centres (*antè*, No. 62), and § VI. to the problem of three bodies (*post*, No. 117).

111. Donkin's memoir "On a Class of Differential Equations &c." (1855). Part I. Nos. 27 to 30 relate to the problem of central forces (in space), No. 31 to the rotation of a solid body, and § III. to the same subject, viz. Nos. 40 and 41 to the general case, Nos. 42 to 44 to the particular case $A=B$; and Nos. 45 to 48 to the reduction thereto of the general case by treating the forces which arise from the inequality of A and B as disturbing forces. Part II. Nos. 59 and 60 relate to the spherical pendulum; Nos. 72 and 73 to "Transformation from fixed to moving axes of coordinates," say to Relative Motion; and Nos. 84 to 96 to the problem of three bodies (*post*, No. 120).

The Problem of Three Bodies, Article Nos. 112 to 123.

112. A system of differential equations, such as

$$\frac{dx_1}{X_1}=\frac{dx_2}{X_2}\cdots=\frac{dx_{n+1}}{X_{n+1}}$$

(n equations between $n+1$ variables), may be termed a system of the nth order, or more simply a system of n equations. Let (u_1, $u_2 \ldots . u_{n+1}$) be any functions of the original variables (x_1, $x_2 \ldots . x_{n+1}$), the system may be transformed into the similar system

$$\frac{du_1}{U_1}=\frac{du_2}{U_2}\cdots=\frac{du_{n+1}}{U_{n+1}},$$

and if it happens that we have e.g. U_1 identically equal to zero, then the system becomes

$$du_1=0, \quad \left(\frac{du_2}{U_2}=\frac{du_3}{U_3}\cdots=\frac{du_{n+1}}{U_{n+1}}\right),$$

so that we have an integral $u_1=c$, and then in the remaining equations substituting this value, or treating u_1 as constant, the system is reduced to one of $m-1$ equations. Or again, if it happen that we have in the transformed system m equations ($m < n$), say

$$\frac{du_1}{U_1}=\frac{du_2}{U_2}\cdots=\frac{du_{m+1}}{U_{m+1}},$$

which are such that U_1, $U_2 \ldots U_{m+1}$ are functions of only the $m+1$ variables $u_1, u_2 \ldots u_{m+1}$, then the integration of the proposed system of n equations depends on the integration in the first instance of a system of m equations. It is to be observed that if the system of m equations can be integrated, then the completion of the integration of the original system depends on the integration of a system of $n-m$ equations, and in this sense the original system of n equations may be said to be broken up into two systems of m equations and $n-m$ equations respectively: but *non constat* that the system of m equations admits of integration; and it is therefore more correct to say that, from the original system of the n equations, there has been *separated off* a system of m equations.

113. The bearing of the foregoing remarks on the problem of three bodies will presently appear. It will be seen that whereas the problem as it stood before Jacobi depends on a system of seven equations, it has been shown by him that there may be separated off from this a system of *six* equations.

114. Jacobi's memoir "Sur l'élimination des Nœuds &c." (1843).—The problem of the motion of three mutually attracting bodies is in the first instance reduced to that of the motion of two fictitious bodies (which may be considered as mutually attracting bodies, attracted by a fixed centre of force) (1). In fact, in the original problem the

1 This is the effect of Jacobi's reduction; but the explicit statement of the theorem, and actual replacement of the problem of the three bodies by that of the two bodies attracted to a fixed centre, is due to Bertrand (*post*, No. 117).

centre of gravity of the three bodies moves uniformly in a right line, and it may without any real loss of generality be taken to be at rest; that is, if the x-coordinates of the three bodies are ξ_1, ξ_2, ξ_3, then $m_1\xi_1+m_2\xi_2+m_3\xi_3=0$, or ξ_1, ξ_2, ξ_3 may be taken to be linear functions of two quantities x_1 and x_2. And similarly for the y-coordinates and the z-coordinates respectively. And (x_1, y_1, z_1), (x_2, y_2, z_2) may be regarded as the coordinates of two bodies revolving about a fixed centre of force. Hence representing the differential coefficients in regard to the time by x_1', &c., and treating these as new variables, the equations of motion will assume the form

$$\frac{dx_1}{x_1'}=\frac{dy_1}{y_1'}=\frac{dz_1}{z_1'}=\frac{dx_2}{x_2'}=\frac{dy_2}{y_2'}=\frac{dz_2}{z_2'}=\frac{dx_1'}{X_1}=\frac{dy_1'}{Y_1}=\frac{dz_1'}{Z_1}=\frac{dx_2'}{X_2}=\frac{dy_2'}{Y_2}=\frac{dz_2'}{Z_2}\,(=dt),$$

where $X_1, Y_1, Z_1, X_2, Y_2, Z_2$ are forces capable of representation by means of a force-function U. This is a system of twelve equations; but since $X_1, Y_1, Z_1, X_2, Y_2, Z_2$ are independent of the time, we may omit the equation $(=dt)$, and treat the system as one of eleven equations between the variables $x_1, y_1, z_1, x_2, y_2, z_2, x_1', y_1', z_1', x_2', y_2', z_2'$: if this system were integrated, the determination of the time would then depend on a quadrature only. But for the system of eleven equations we have four integrals, viz., the integral of *Vis Viva* and the three integrals of areas, and the system is thus reducible to one of $(11-4=)$ seven equations. This preliminary transformation in Jacobi's memoir explains the remark that the problem, as it stood before him, depended on a system of seven equations.

115. Jacobi remarks that in the transformed problem the three integrals of areas show (1) that the intersection of the planes of the orbits of the two bodies lie in a fixed plane, the invariable plane of the system; (2) that the inclinations of the planes of the two orbits to this fixed plane, and the parameters of the two orbits considered as variable ellipses, are four elements any two of which rigorously determine the two others.

And then choosing for variables the inclinations of the two orbits to the invariable plane, the two radius vectors, the angles which they form with the intersection of the planes of the two orbits, and lastly the angle between this intersection (being as already mentioned a line in the invariable plane) with a fixed line in the invariable plane, he finds *that the last-mentioned angle entirely disappears from the system of differential equations, and is determined after their integration by a quadrature.* In this new form of the differential equations there is no trace of the nodes. The differential equations which determine the relative motion of the three bodies are reduced to five equations of the first order and one of the second order. The equations in question are the equations I. to VI. given at the end of the memoir. It is to be remarked that the differential dt is not eliminated from these equations; the last of them is $\frac{d^2}{dt^2}(\mu r^2+\mu_1 r_1^2)=2U-2h$; and if to reduce them to a system of equations of the first order we write $\frac{d}{dt}(\mu r^2+\mu_1 r_1^2)=\theta$, and therefore $\frac{d\theta}{dt}=2U-2h$, the system may be presented in the form

$$\frac{du}{U}=\frac{du_1}{U_1}=\frac{di}{I}=\frac{di_1}{I_1}=\frac{dr}{R}=\frac{dr_1}{R_1}=\frac{d\theta}{\Theta}\,(=dt),$$

which if we do, and then omit the equation $(=dt)$, we have a system of six equations between the seven quantities $u, u_1, i, i_1, r, r_1, \theta$; when this is integrated, the equation $(=dt)$ gives the time by a quadrature; and finally, Jacobi's equation VII. $\left(d\Omega = \tan u \dfrac{di}{\sin i}\right)$ gives by a quadrature the angle before referred to as disappearing from the system of equations I. to VI.

116. But when Jacobi says, "Par suite on a fait cinq intégrations. Les intégrales connues n'étant qu'au nombre de quatre, on pourra donc dire que l'on a fait une intégration de plus dans le système du monde. Je dis dans le système du monde puisque la même méthode s'applique à un nombre quelconque de corps," the language used is not, I think, quite accurate. It in fact appears from the memoir that it is only on the assumption of the integration of the system of six equations that, besides the integral of *Vis Viva* and the integrals of areas, the remaining two integrals are known; in fact, after, but not before, the system of the order six has been integrated, the time t and the angle Ω are each of them given by a quadrature.

117. Bertrand's "Mémoire sur l'intégration des équations différentielles de la Mécanique" (1852).—I have spoken of this memoir in No. 56 of my former Report. The course of investigation is the inquiry as to the integrals, which, combined according to Poisson's theorem with the integral of *Vis Viva* or any other given integral, give rise to an illusory result. But as regards the application made to the problem of three bodies, it will be more convenient to state from a different point of view the conclusions arrived at: and I may mention that when the author says "Je parviens.. à réduire la question à l'intégration de six équations toutes du premier ordre, c'est-à-dire que j'effectue une intégration de plus que ne l'avait fait Jacobi," he seems to have overlooked that Jacobi's system of five equations of the first order and one of the second order really is, as above noticed, a system of the six equations with another equation which then gives the time by a quadrature, and that, at least as appears to me, he has not advanced the solution beyond the point to which it had been carried by Jacobi [1].

118. Presenting Bertrand's results in the slightly different notation in which they are reproduced in Bour's memoir (*post.* No. 122), then if (x, y, z), (x_1, y_1, z_1) are the coordinates of the two bodies (the problem actually considered being, as with Jacobi, that of the motion of two bodies about a fixed centre of force), and representing the functions $x^2+y^2+z^2$, $x_1^2+y_1^2+z_1^2$, $m^2(x'^2+y'^2+z'^2)$, $m_1^2(x_1'^2+y_1'^2+z_1'^2)$, $m(xx'+yy'+zz')$, $m_1(x_1x_1'+y_1y_1'+z_1z_1')$, $m(x_1x'+y_1y'+z_1z')$, $m_1(xx_1'+yy_1'+zz_1')$, $(xx_1+yy_1+zz_1)$, $mm_1(x'x_1'+y'y_1'+z'z_1')$ by $u, u_1, v, v_1, w, w_1, r, r_1, q, s$ respectively, then the last-mentioned quantities are connected by a single geometrical relation, so that any one of them, say s, may be considered as a given function of the remaining nine. And the author *in effect* shows that the equations of motion give a system

$$\frac{du}{U} = \frac{du_1}{U_1} = \frac{dv}{V} = \frac{dv_1}{V_1} = \frac{dw}{W} = \frac{dw_1}{W_1} = \frac{dr}{R} = \frac{dr_1}{R_1} = \frac{dq}{Q} = (dt),$$

[1] These remarks were communicated by me to M. Bertrand—see my letter "Sur l'intégration des équations différentielles de la Mécanique," *Comptes Rendus* (1863)—and, in the answer he kindly sent me, he agrees that they are correct.

where U, U_1, &c. are functions of the quantities u, u_1, v, &c. Omitting from the system the equation $(=dt)$, there are eight equations between nine quantities; but there are two known integrals, viz., the integral of *Vis Viva* and the integral of principal moment (or sum of the squares of the integrals of areas); that is to say, the system is really a system of *six* equations.

119. Painvin, "Recherche du dernier Multiplicateur &c." (1854).—The author investigates the ultimate multiplier for two systems of differential equations:

1°. The system of the equations I. to VI. in Jacobi's memoir "Sur l'élimination des Nœuds &c." (*antè*, No. 114). Writing in the equations $\frac{dr}{dt}=r'$, $\frac{dr_1}{dt}=r_1'$, and treating r', r_1' as new variables, the system may be written in the form

$$\frac{du}{U}=\frac{du_1}{U_1}=\frac{di}{I}=\frac{di_1}{I_1}=\frac{dr}{R}=\frac{dr_1}{R_1}=\frac{dr'}{R'}=\frac{dr_1'}{R_1'}\,(=dt),$$

which, omitting the equation $(=dt)$, is a system of seven equations between eight variables; and it is for this form of the system that the value of M is determined, the result obtained being the simple and elegant one, $M=\frac{\sin i\sin i_1}{\sin^2 I}$. The system of seven equations has an integral which is in fact the equation V. of the system in Jacobi's form, so that it is really a system of *six* equations (*antè*, No. 115).

2°. The system secondly discussed is Bertrand's system of nine equations (*antè*, No. 118). The multiplier M is obtained under four different forms, $M=\frac{1}{\sqrt{B^2-AC}}=\frac{1}{\sqrt{\alpha\alpha_1}}=\frac{1}{AZ+B}=\frac{1}{mn}$ (I do not stop to explain the notation), the last of them being referred to as a result announced by M. Bertrand in his course. But it is shown by M. Bour in the memoir next referred to (*post*, No. 122), that the multiplier for the system in question can be obtained in a very much more simple manner, almost without calculation.

120. In connexion with Jacobi's theory of the elimination of the Nodes, I may refer to the investigations "Application to the Problem of three Bodies," Nos. 84 to 96 of Donkin's memoir "On a Class of Differential Equations &c." Part II. The author remarks that his differential equations No. 93 afford an example of the so-called elimination of the Nodes, quite different however (in that they are *merely* transformations of the original differential equations of the problem without any integrations) from that effected by Jacobi.

121. It may be right to refer again in this place to the concluding part of § 28 of Jacobi's memoir "Nova Theoria Multiplicatoris &c." (*antè*, No. 92), as bearing on the problem of three bodies.

122. Bour's "Mémoire sur le Problème des Trois Corps" (1856).—The author remarks that Bertrand's system of equations have lost the remarkable form and the properties which characterize the ordinary equations for the solution of a dynamical

problem. But by selecting eight new variables, functions of Bertrand's variables, the system may be brought back to the standard Hamiltonian form

$$\frac{dq_i}{dt}=\frac{dH}{dp_i},\ \frac{dp_i}{dt}=-\frac{dH}{dq_i},$$

or to the form adopted by M. Bour, of a partial differential equation

$$\Sigma\left(\frac{dH}{dq_i}\frac{d\zeta}{dp_i}-\frac{dH}{dp_i}\frac{d\zeta}{dq_i}\right)=0;$$

and guiding himself by a theorem in relation to canonical integrals obtained in his memoir of 1855 (see No. 66 of my former Report), he finds by a somewhat intricate analysis the expressions of the eight new variables $p_1, p_2, p_3, p_4, q_1, q_2, q_3, q_4$. The results ultimately obtained are of a very remarkable and interesting form, viz. $H =$ funct. $(p_1, p_2, p_3, p_4, q_1, q_2, q_3, q_4)$ is equal to the value it would have for motion in a plane, *plus* a term admitting of a simple geometrical interpretation; and he thus arrives at the following theorem as a *résumé* of the whole memoir, viz.,

"In order to integrate in the general case the problem of three bodies, it is sufficient to solve the case of motion in a plane, and then to take account of a disturbing function equal to the product of a constant depending on the areas by the sum of the moments of inertia of the bodies round a certain axis, divided by the square of the triangle formed by the three bodies."

123. It may be remarked that the only given integral of the system of eight equations is the integral of *Vis Viva*, $H =$ const., and that using this equation to eliminate one of the variables, and omitting the equation $(=dt)$, we have, as in the solutions of Jacobi and Bertrand, a system of six equations between seven variables. As the equations are in the standard dynamical form, no investigation is needed of the multiplier M, which is given by Jacobi's general theory, and consequently when any five integrals of the six equations are given, the remaining integral can be obtained by a quadrature.

In the case of three bodies moving in a plane, the solution takes a very simple form, which is given in the concluding paragraph of the memoir.

Transformation of Coordinates, Article Nos. 124 to 141.

124. It may be convenient to remark at once that two sets of rectangular coordinates may be related to each other properly or improperly, viz., the axes to which they belong (considered as drawn from the origin in the positive directions) may be either capable, or else incapable, of being brought into coincidence. The latter relation, although of equal generality with the former one, may for the most part be disregarded; for by merely reversing the directions of the one set of axes, the improper is converted into the proper relation.

125. In the memoir "Problema Algebraicum &c." (1770) Euler proposes to himself the question "Invenire novem numeros ita in quadratum disponendos

$$\begin{matrix} A, & B, & C \\ D, & E, & F \\ G, & H, & I \end{matrix}$$

ut satisfiat duodecem sequentibus conditionibus &c.", viz., substituting for A, B, C, &c. the ordinary letters

$$\begin{matrix} \alpha\,, & \beta\,, & \gamma\,, \\ \alpha', & \beta', & \gamma', \\ \alpha'', & \beta'', & \gamma'', \end{matrix}$$

the twelve conditions are

$$\begin{aligned} \alpha^2 + \alpha'^2 + \alpha''^2 &= 1, & \alpha\beta + \alpha'\beta' + \alpha''\beta'' &= 0, \\ \beta^2 + \beta'^2 + \beta''^2 &= 1, & \beta\gamma + \beta'\gamma' + \beta''\gamma'' &= 0, \\ \gamma^2 + \gamma'^2 + \gamma''^2 &= 1, & \gamma\alpha + \gamma'\alpha' + \gamma''\alpha'' &= 0, \end{aligned}$$

$$\begin{aligned} \alpha^2 + \beta^2 + \gamma^2 &= 1, & \alpha\alpha' + \beta\beta' + \gamma\gamma' &= 0, \\ \alpha'^2 + \beta'^2 + \gamma'^2 &= 1, & \alpha'\alpha'' + \beta'\beta'' + \gamma'\gamma'' &= 0, \\ \alpha''^2 + \beta''^2 + \gamma''^2 &= 1, & \alpha\alpha'' + \beta''\beta + \gamma''\gamma &= 0. \end{aligned}$$

And he remarks that this is in fact the problem of the transformation of coordinates, viz., if we have

$$\begin{aligned} X &= \alpha\, x + \beta\, y + \gamma\, z, \\ Y &= \alpha' x + \beta' y + \gamma' z, \\ Z &= \alpha'' x + \beta'' y + \gamma'' z, \end{aligned}$$

then the first equations are such as to give identically

$$X^2 + Y^2 + Z^2 = x^2 + y^2 + z^2.$$

126. Assuming the first six equations, he shows by a direct analytical process that $\alpha^2 = (\beta'\gamma'' - \beta''\gamma')^2$, or $\alpha = \pm(\beta'\gamma'' - \beta''\gamma')$; or taking the positive sign (for, as the numbers may be taken as well positively as negatively, there is nothing lost by doing so) $\alpha = \beta'\gamma'' - \beta''\gamma'$, which gives the system

$$\begin{aligned} \alpha\; &= \beta'\gamma'' - \beta''\gamma', & \beta\; &= \gamma'\alpha'' - \gamma''\alpha', & \gamma\; &= \alpha'\beta'' - \alpha''\beta', \\ \alpha' &= \beta''\gamma - \beta\gamma'', & \beta' &= \gamma''\alpha - \gamma\alpha'', & \gamma' &= \alpha''\beta - \alpha\beta'', \\ \alpha'' &= \beta\gamma' - \beta'\gamma, & \beta'' &= \gamma\alpha' - \gamma'\alpha, & \gamma'' &= \alpha\beta' - \alpha'\beta', \end{aligned}$$

and from these he deduces the second system of six equations. The inverse system of equations

$$\begin{aligned} X &= \alpha x + \alpha' y + \alpha'' z, \\ Y &= \beta x + \beta' y + \beta'' z, \\ Z &= \gamma x + \gamma' y + \gamma'' z \end{aligned}$$

is not explicitly referred to.

127. He then satisfies the equations by means of trigonometrical substitutions, viz., assuming $\alpha = \cos \zeta$, then $\alpha'^2 + \alpha''^2 = \sin^2 \zeta$, which is satisfied by $\alpha' = \sin \zeta \cos \eta$, $\alpha'' = \sin \zeta \sin \eta$, &c., and he thus obtains for the coefficients a set of values involving the angles ζ, η, θ, which are the same as those mentioned *post*, No. 130. And he shows how these formulæ may be obtained geometrically by three successive transformations of two coordinates only. The remainder of the memoir relates to the analogous problem of the transformation of four or more coordinates.

128. I have analysed so much of Euler's memoir in order to show that it contains nearly the whole of the ordinary theory of the transformation of coordinates; the only addition required is the equation

$$\begin{vmatrix} \alpha, & \beta, & \gamma \\ \alpha', & \beta', & \gamma' \\ \alpha'', & \beta'', & \gamma'' \end{vmatrix} = \pm 1,$$

where the sign + gives $\alpha = \beta'\gamma'' - \beta''\gamma'$, &c. (*ut supra*), but the sign − would give $\alpha = -(\beta'\gamma'' - \beta''\gamma')$, &c.

129. The distinction of the ambiguous sign is in fact the above-mentioned one of the proper and improper transformations; viz., for the sign + the two sets of axes can, for the sign − they cannot, be brought into coincidence: this very important remark was, I believe, first made by Jacobi in one of his early memoirs in *Crelle's Journal*, but I have lost the reference. As already mentioned, it is allowable to attend only to the proper transformation, and to consider the value of the determinant as being $= +1$; and this is in fact almost always done.

130. Euler's formulæ involving the three angles are those which are ordinarily made use of in the problem of rotation and the problems of physical astronomy generally.

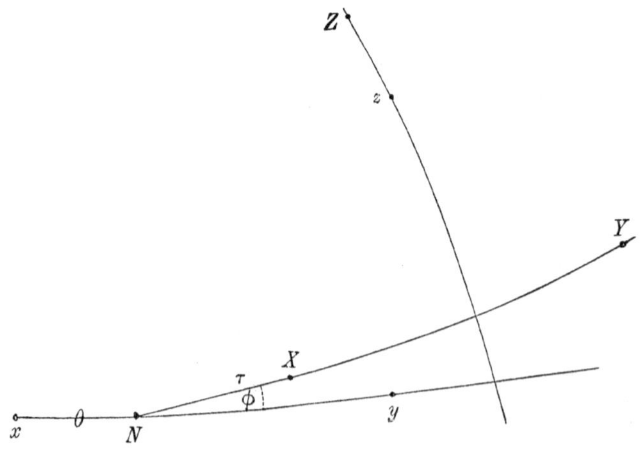

It is convenient to take them as in the figure, viz., θ, the longitude of node,

ϕ, the inclination, τ, the angular distance of X from node, and the formulæ of transformation then are

	X	Y	Z
x	$\cos\tau\cos\theta - \sin\tau\sin\theta\cos\phi$	$-\sin\tau\cos\theta - \cos\tau\sin\theta\cos\phi$	$\sin\theta\sin\phi$
y	$\cos\tau\sin\theta + \sin\tau\cos\theta\cos\phi$	$-\sin\tau\sin\theta + \cos\tau\cos\theta\cos\phi$	$-\cos\theta\sin\phi$
z	$\sin\tau\sin\phi$	$\cos\tau\sin\phi$	$\cos\phi$

The foregoing very convenient algorithm, viz., the employment of

	X	Y	Z
x	α	β	γ
y	α'	β'	γ'
z	α''	β''	γ''

to denote the system of equations

$$x = \alpha X + \beta Y + \gamma Z,$$
$$y = \alpha' X + \beta' Y + \gamma' Z,$$
$$z = \alpha'' X + \beta'' Y + \gamma'' Z,$$

is due to M. Lamé.

131. But previously to the foregoing investigations, viz., in the memoir "Du Mouvement de Rotation &c.," *Mém. de Berlin for* 1758 (pr. 1765), Euler had obtained incidentally a very elegant solution of the problem of the transformation of coordinates; this is in fact identical with the next mentioned one, the letters l, m, n; λ, μ, ν being used in the place of ζ, ζ', ζ''; η, η', η''.

132. In the memoir "Formulæ generales pro translatione &c." (1775), Euler gives the following formulæ for the transformation of coordinates, viz., if the position of the set of axes XYZ in reference to the set xyz is determined by

$$xX,\ yX,\ zX = 90° - \zeta,\ 90° - \zeta',\ 90° - \zeta'',\ \angle^{s}\, YXx,\ YXy,\ YXz = \eta,\ \eta',\ \eta'',$$

then the formulæ of transformation are

	X	Y	Z
x	$\sin\zeta$	$\cos\zeta\ \sin\eta$	$\cos\zeta\ \cos\eta$
y	$\sin\zeta'$	$\cos\zeta'\ \sin\eta'$	$\cos\zeta'\ \cos\eta'$
z	$\sin\zeta''$	$\cos\zeta''\ \sin\eta''$	$\cos\zeta''\ \cos\eta''$

with the following equations connecting the six angles, viz., if

$$-\Delta^2 = \cos(\eta' - \eta'')\cos(\eta'' - \eta)\cos(\eta - \eta'),$$

then

$$\tan\zeta = \frac{-\Delta}{\cos(\eta' - \eta'')}, \quad \tan\zeta' = \frac{-\Delta}{\cos(\eta'' - \eta)}, \quad \tan\zeta'' = \frac{-\Delta}{\cos(\eta - \eta')}.$$

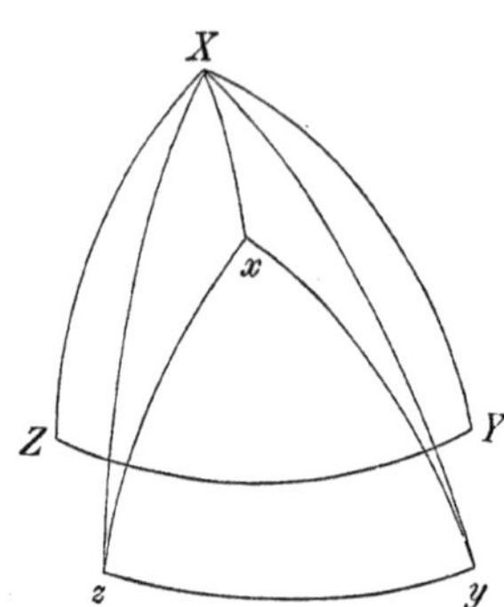

133. It is right to notice that these values of ζ, ζ', ζ'' give the twelve equations $\alpha^2 + \beta^2 + \gamma^2 = 1$, &c., but they do not give definitely $\alpha = \beta'\gamma'' - \beta''\gamma'$, &c., but only $\alpha = \pm(\beta'\gamma'' - \beta''\gamma')$; that is, in the formulæ in question the two sets of axes are not of necessity displacements the one of the other. In the same memoir Euler considers two sets of rectangular axes, and assuming that they are displacements the one of the other (this assumption is not made as explicitly as it should have been), he remarks that the one set may be made to coincide with the other set by means of a finite rotation about a certain axis (which may conveniently be termed the Resultant Axis). This consideration leads him to an equation which ought to be satisfied by the coefficients of transformation, but which he is not able to verify by means of the foregoing expressions in terms of ζ, ζ', ζ'', η, η', η''.

134. I remark that Euler's equation in fact is

$$\begin{vmatrix} \alpha - 1, & \beta, & \gamma \\ \alpha', & \beta' - 1, & \gamma' \\ \alpha'', & \beta'', & \gamma'' - 1 \end{vmatrix} = 0,$$

or, as it may be written,

$$\begin{vmatrix} \alpha, & \beta, & \gamma \\ \alpha', & \beta', & \gamma' \\ \alpha'', & \beta'', & \gamma'' \end{vmatrix} - (\beta'\gamma'' - \beta''\gamma') - (\gamma''\alpha - \gamma\alpha'') - (\alpha\beta' - \alpha'\beta) + \alpha + \beta' + \gamma'' - 1 = 0,$$

in which form it is an immediate consequence of the equations

$$\begin{vmatrix} \alpha, & \beta, & \gamma \\ \alpha', & \beta', & \gamma' \\ \alpha'', & \beta'', & \gamma'' \end{vmatrix} = 1, \quad \alpha = \beta'\gamma'' - \beta''\gamma', \text{ \&c.},$$

which are true for the proper, but not for the improper transformation.

135. In the undated addition to the memoir, Euler states the theorem of the resultant axis as follows:—"Theorema. Quomodocunque sphæra circa centrum suum convertatur, semper assignari potest diameter cujus directio in situ translato conveniat cum situ originali;" and he again endeavours to obtain a verification of the foregoing analytical theorem.

136. The theory of the Resultant Axis was further developed by Euler in the memoir "Nova Methodus Motum &c." (1775), and by Lexell in the memoir "Nonnulla theoremata generalia &c." (1775): the geometrical investigations are given more completely and in greater detail in Lexell's memoir. The result is contained in the following system of formulæ for the transformation of coordinates, viz., if α, β, γ are the inclinations of the resultant axis to the original set, and if ϕ is the rotation about the resultant axis, or say the resultant rotation, then we have

	X	Y	Z
x	$\cos^2\alpha + \sin^2\alpha\cos\phi$	$\cos\alpha\cos\beta(1-\cos\phi)+\cos\gamma\sin\phi$	$\cos\alpha\cos\gamma(1-\cos\phi)-\cos\beta\sin\phi$
y	$\cos\beta\cos\alpha(1-\cos\phi)-\cos\gamma\sin\phi$	$\cos^2\beta+\sin^2\beta\cos\phi$	$\cos\beta\cos\gamma(1-\cos\phi)+\cos\alpha\sin\phi$
z	$\cos\gamma\cos\alpha(1-\cos\phi)+\cos\beta\sin\phi$	$\cos\gamma\cos\beta(1-\cos\phi)-\cos\alpha\sin\phi$	$\cos^2\gamma+\sin^2\gamma\cos\phi$

Euler attempts, but not very successfully, to apply the formulæ to the dynamical problem of the rotation of a solid body: he does not introduce them into the differential equations, but only into the integral ones, and his results are complicated and inelegant. The further simplification effected by Rodrigues was in fact required.

137. Jacobi's paper, "Euleri formulæ &c." (1827), merely cites the last-mentioned result.

138. I find it stated in Lacroix's *Differential Calculus*, t. I. p. 533, that the following system for the transformation of coordinates was obtained by Monge (no reference is given in Lacroix), viz., the system being as above,

$$\begin{vmatrix} \alpha, & \beta, & \gamma, \\ \alpha', & \beta', & \gamma', \\ \alpha'', & \beta'', & \gamma'', \end{vmatrix}$$

and the quantities α, β', γ'' being arbitrary, then putting

$$1+\alpha+\beta'+\gamma''=M,$$
$$1+\alpha-\beta'-\gamma''=N,$$
$$1-\alpha+\beta'-\gamma''=P,$$
$$1-\alpha-\beta'+\gamma''=Q,$$

so that

$$M+N+P+Q=4,$$

we have

$$2\beta = \sqrt{NP} + \sqrt{MQ},\quad 2\gamma' = \sqrt{PQ} + \sqrt{MN},\quad 2\alpha'' = \sqrt{QN} + \sqrt{MP},$$
$$2\alpha' = \sqrt{NP} - \sqrt{MQ},\quad 2\beta'' = \sqrt{PQ} - \sqrt{MN},\quad 2\gamma = \sqrt{QN} - \sqrt{MP}.$$

These are formulæ very closely connected with those of Rodrigues.

139. The theory was perfected by Rodrigues in the valuable memoir "Des lois géométriques &c." (1840). Using for greater convenience λ, μ, ν in the place of his $\frac{1}{2}m$, $\frac{1}{2}n$, $\frac{1}{2}p$, he in effect writes

$$\tan\tfrac{1}{2}\phi\cos\alpha = \lambda,$$
$$\tan\tfrac{1}{2}\phi\cos\beta = \mu,$$
$$\tan\tfrac{1}{2}\phi\cos\gamma = \nu,$$

and this being so, the coefficients of transformation are

$$\begin{array}{lll} 1+\lambda^2-\mu^2-\nu^2, & 2(\lambda\mu+\nu), & 2(\lambda\nu-\mu), \\ 2(\mu\lambda-\nu), & 1-\lambda^2+\mu^2-\nu^2, & 2(\mu\nu+\lambda), \\ 2(\nu\lambda+\mu), & 2\nu\mu-\lambda, & 1-\lambda^2-\mu^2+\nu^2, \end{array}$$

all divided by the common denominator $1+\lambda^2+\mu^2+\nu^2$. Conversely, if the coefficients of transformation are as usual represented by

$$\begin{array}{lll} \alpha, & \beta, & \gamma, \\ \alpha', & \beta', & \gamma', \\ \alpha'', & \beta'', & \gamma'', \end{array}$$

then λ^2, μ^2, ν^2, λ, μ, ν are respectively equal to

$$\begin{array}{lll} 1+\alpha-\beta'-\gamma'', & 1-\alpha+\beta'-\gamma'', & 1-\alpha-\beta'+\gamma'', \\ \gamma'-\beta'', & \alpha''-\beta, & \beta-\alpha', \end{array}$$

each of them divided by $1+\alpha+\beta'+\gamma''$.

The memoir contains very elegant formulæ for the composition of finite rotations, and it will be again referred to in speaking of the kinematics of a solid body.

140. Sir W. R. Hamilton's first papers on the theory of quaternions were published in the years 1843 and 1844: the fundamental idea consists in the employment of the imaginaries i, j, k, which are such that

$$i^2 = j^2 = k^2 = -1,\quad jk = -kj = i,\quad ki = -ik = j,\quad ij = -ji = k,$$

whence also

$$\begin{aligned} (w+ix+jy+kz)(w'+ix'+jy'+kz') &= ww'-xx'-yy'-zz' \\ &\quad + i(wx'+w'x+yz'-y'z) \\ &\quad + j(wy'+w'y+zx'-z'x) \\ &\quad + k(wz'+w'z+xy'-x'y); \end{aligned}$$

so that representing the right-hand side by

$$W + iX + jY + kZ,$$

we have identically

$$W^2 + X^2 + Y^2 + Z^2 = (w^2 + x^2 + y^2 + z^2)(w'^2 + x'^2 + y'^2 + z'^2).$$

It is hardly necessary to remark that Sir W. R. Hamilton in his various publications on the subject, and in the *Lectures on Quaternions*, Dublin, 1853, has developed the theory in detail, and has made the most interesting applications of it to geometrical and dynamical questions; and although the first explicit application of it to the present question may have been made in my own paper next referred to, it seems clear that the whole theory was in its original conception intimately connected with the notion of rotation.

141. Cayley, "On certain Results relating to Quaternions" (1845).—It is shown that Rodrigues' transformation formula may be expressed in a very simple manner by means of quaternions; viz., we have

$$ix + jy + kz = (1 + i\lambda + j\mu + k\nu)^{-1}(iX + jY + kZ)(1 + i\lambda + j\mu + k\nu),$$

where developing the function on the right-hand side, and equating the coefficients of i, j, k, we obtain the formulæ in question. A subsequent paper, Cayley, "On the application of Quaternions to the Theory of Rotation" (1848), relates to the composition of rotations.

Principal Axes, and Moments of Inertia. Article Nos. 142—163.

142. The theorem of principal axes consists herein, that at any point of a solid body there exists a system of axes Ox, Oy, Oz, such that

$$\int yz\,dm = 0, \quad \int zx\,dm = 0, \quad \int xy\,dm = 0.$$

But this, the original form of the theorem, is a mere deduction from a general theory of the representation of the integrals

$$\int x^2dm, \quad \int y^2dm, \quad \int z^2dm, \quad \int yz\,dm, \quad \int zx\,dm, \quad \int xy\,dm$$

for any axes through the given origin by means of an ellipsoid depending on the values of these integrals corresponding to a given set of rectangular axes through the same origin.

143. If, for convenience, we write as follows, $M = \int dm$ the mass of the body, and

$$A' = \int x^2dm, \quad B' = \int y^2dm, \quad C' = \int z^2dm, \quad F' = \int yz\,dm, \; G' = \int zx\,dm, \quad H' = \int xy\,dm,$$

and moreover

$$A=\int(y^2+z^2)\,dm,\quad B=\int(z^2+x^2)\,dm,\quad C=\int(x^2+y^2)\,dm,$$

$$F=-\int yzdm,\quad G=-\int zxdm,\quad H=-\int xydm,\text{(}^1\text{)}$$

so that

$$A=B'+C',\quad B=C'+A',\quad C=A'+B',\quad F=-F',\quad G=-G',\quad H=-H',$$

then the ellipsoid which in the first instance presents itself for this purpose, and which Prof. Price has termed the Ellipsoid of Principal Axes, but which I would rather term the "Comomental Ellipsoid," is the ellipsoid

$$(A',\ B',\ C',\ F',\ G',\ H'\between x,\ y,\ z)^2=Mk^4,$$

where k is arbitrary, so that the absolute magnitude is not determined. But it is more usual, and in some respects better to consider in place thereof the "Momental Ellipsoid" (Cauchy, "Sur les Moments d'Inertie," *Exercices de Mathématique*, t. II. pp. 93—103, 1827),

$$(A,\ B,\ C,\ F,\ G,\ H\between x,\ y,\ z)^2=Mk^4,$$

or as it may also be written,

$$(A'+B'+C')(x^2+y^2+z^2)-(A',\ B',\ C',\ F',\ G',\ H'\between x,\ y,\ z)^2=Mk^4,$$

which shows that the two ellipsoids have their axes, and also their circular sections, coincident in direction.

144. And there is besides this a third ellipsoid, the "Ellipsoid of Gyration," which is the reciprocal of the momental ellipsoid in regard to the concentric sphere, radius k. The last-mentioned ellipsoid is given in magnitude, viz., if the body is referred to its principal axes, then putting $A=Ma^2$, $B=Mb^2$, $C=Mc^2$, the equation of the ellipsoid of gyration is

$$\frac{x^2}{a^2}+\frac{y^2}{b^2}+\frac{z^2}{c^2}=1.$$

The axes of any one of the foregoing ellipsoids coincide in direction with the principal axes of the body, and the magnitudes of the axes lead very simply to the values of the principal moments A, B, C.

145. The origin has so far been left arbitrary: in the dynamical applications, this origin is in the case of a solid body rotating about a fixed point, the fixed point; and in the case of a free body, the centre of gravity. But the values of the coefficients (A, B, C, F, G, H), or (A', B', C', F', G', H'), corresponding to any given origin whatever, are very easily expressed in terms of the coordinates of this origin, and the values of the corresponding coefficients for the centre of gravity as origin; or, what is the same thing, any one of the ellipsoids for the given origin may be geometrically con-

[1] I have ventured to make this change instead of writing as usual $F=\int yzdm$, &c.; as in most cases $F=G=H=0$, the formulæ affected by the alteration are not numerous.

structed by means of the ellipsoid for the centre of gravity. The geometrical theory, as regards the magnitudes of the axes, does not appear to have been anywhere explicitly enunciated; as regards their direction, it is comprised in the theorem that the directions at any point are the three rectangular directions at that point in regard to the ellipsoid of gyration for the centre of gravity[1], *post*, No. 159. The notion of the ellipsoids, and of the relation between the ellipsoids at a given point and those at the centre of gravity, once established, the theory of principal axes and moments of inertia becomes a purely geometrical one.

146. The existence of principal axes was first established by Segner in the work *Specimen Theoriæ Turbinum*, Halle (1755), where, however, it is remarked that Euler had said something on the subject in the {Berlin} *Memoirs* for 1749 and 1750 (*post*, No. 167), and had constructed a new mechanical principle, but without pursuing the question. Segner's course of investigation is in principle the same as that now made use of, viz. a principal axis is defined to be an axis such that when a body revolves round it the forces arising from the rotation have no tendency to alter the position of the axes. It is first shown that there are systems of axes x, y, z such that $\int xz dm = 0$, and then, in reference to such a set of axes, the position of a principal axis, say the axis of X, is determined by the conditions $\int XY dm = 0$, $\int XZ dm = 0$, viz. the unknown quantities being taken to be $t = \dfrac{\cos\alpha}{\cos\gamma}$, $\tau = \dfrac{\cos\beta}{\cos\gamma}$ (α, β, γ being the inclinations of the principal axis to those of x, y, z), and then putting $A = \int x^2 dm$, &c. ($F = 0$ by hypothesis), Segner's equations for the determination of t, τ, are

$$G't^2 + (C' - A')\,t - G' - H'\tau = 0,$$

$$(C' - B')\,\tau - G't\tau + H't = 0,$$

the second of which gives

$$\tau = \frac{H't}{B' - C' + G't},$$

and by means of it the first gives

$$G'^2t^3 - G'(A' - B')\,t^2 + \{(B' - C')(C' - A') - G'^2 - H'^2\}\,t + G'(B' - C') = 0,$$

which being a cubic equation shows that there are three principal axes; and it is afterwards proved that these are at right angles to each other.

147. To show the equivalence of Segner's solution to the modern one, I remark that if $u = \int X^2 dm$, we have

$$\begin{aligned}
(A' - u)\,t + H'\,\tau + G' &= 0,\\
B'\,t + (B' - u)\,\tau + F' &= 0,\\
G'\,t + F'\,\tau + C' - u &= 0,
\end{aligned}$$

[1] The rectangular directions at a point in regard to an ellipsoid are the directions of the axes of the circumscribed cone, or, what is the same thing, they are the directions of the normals to the three quadric surfaces, confocal with the given ellipsoid, which pass through the given point. The theory of confocal surfaces appears to have been first given by Chasles, Note XXXI. of the *Aperçu Historique* (1837).

whence

$$\begin{aligned} t^2 : \tau^2 : 1 : \tau : t : t\tau = \quad & B'C' - F'^2 - (B' + C')\,u + u^2 \\ : \; & C'A' - G'^2 - (C' + A')\,u + u^2 \\ : \; & A'B' - H'^2 - (A' + B')\,u + u^2 \\ : \; & G'H' - A'F' \quad + F'\,u \\ : \; & H'F' - B'G' \quad + G'\,u \\ : \; & F'G' - C'H' \quad + H'\,u, \end{aligned}$$

or putting therein $F'=0$,

$$\begin{aligned} t^2 : \tau^2 : 1 : \tau : t : t\tau = \quad & B'C' \qquad - (B' + C')\,u + u^2 \\ : \; & C'A' - G'^2 - (C' + A')\,u + u^2 \\ : \; & A'B' - H'^2 - (A' + B')\,u + u^2 \\ : \; & G'H' \\ : \; & - B'G' \qquad + G'\,u \\ : \; & - C'H' \qquad + H'\,u, \end{aligned}$$

by means of which Segner's equations may be verified. I have given this analysis, as the first solution of such a problem is a matter of interest.

148. There is little if anything added to Segner's results by the memoir, Euler, "Recherches sur la Connaissance Mécanique des Corps" (1758), which is introductory to the immediately following one on Rotation.

149. Relating to the theory of principal axes we have Binet's "Mémoire sur les axes conjugués &c.," (1813). The author proposes to make known the new systems of axes which he calls *conjugate axes*, which, when they are at right angles to each other, coincide with the principal axes; viz. considering the sum of the molecules each into its distance from a plane, such distance being measured in the direction of a line, then (the direction of the line being given) of all the planes which pass through a given point, there is one for which the sum in question is a minimum, and this plane is said to be *conjugate* to the given line, and from the notion of a line and conjugate plane he passes to that of a system of *conjugate* axes. The investigation (which is throughout an elegant one) is conducted analytically; the coordinates made use of are oblique ones, and the formulæ are thus rendered more complicated than they would have been; in referring to them it will be convenient to make the axes rectangular.

150. One of the results is the well-known equation

$$(A' - \Theta)(B' - \Theta)(C' - \Theta) - F'^2(A' - \Theta) - G'^2(B' - \Theta) - H'^2(C' - \Theta) + 2F'G'H' = 0;$$

which, if x_1, y_1, z_1 are the principal axes, has for its roots $\int x_1^2 dm$, $\int y_1^2 dm$, $\int z_1^2 dm$.

And the equations (1), p. 49, taking therein the original axes as rectangular, are

$$\left(\mathfrak{A}' - \frac{K'}{\Theta'}\right)\cos\alpha + \mathfrak{H}' \cos\beta + \mathfrak{G}' \cos\gamma = 0,$$

$$+\mathfrak{H}' \cos\alpha + \left(\mathfrak{B}' - \frac{K'}{\Theta'}\right)\cos\beta + \mathfrak{F}' \cos\gamma = 0,$$

$$+\mathfrak{G}' \cos\alpha + \mathfrak{F}' \cos\beta + \left(\mathfrak{C}' - \frac{K'}{\Theta'}\right)\cos\gamma = 0,$$

where $\mathfrak{A}'$, $\mathfrak{B}'$, $\mathfrak{C}'$, $\mathfrak{F}'$, $\mathfrak{G}'$, $\mathfrak{H}'$ denote the reciprocal coefficients, $\mathfrak{A}' = B'C' - F'^2$ &c., and K' is the discriminant $= A'B'C' - A'F'^2 - B'G'^2 - C'H'^2 + 2F'G'H'$: this is a symmetrical system of equations for finding $\cos\alpha : \cos\beta : \cos\gamma$, less simple however than the modern form (*post*, No. 154), the identity of which with Binet's may be shown without difficulty.

151. Another result (p. 57) is that if the original axes are principal axes, and if Ox, Oy, Oz are the principal axes through a point the coordinates whereof are f, g, h, and if $\Theta_1' = \text{(say)} \int x_1^2 dm$, then we have

$$\frac{f^2}{\Theta_1' - A'} + \frac{g^2}{\Theta_1' - B'} + \frac{h^2}{\Theta_1' - C'} = \frac{1}{M},$$

(in which I have restored the mass M, which is put equal to unity), so that if Θ_1' have a given constant value, the locus of the point is a quadric surface, the nature whereof will depend on the value of Θ_1. The surfaces in question are confocal with each other {and with the imaginary surface $\frac{x^2}{-A'} + \frac{y^2}{-B'} + \frac{z^2}{-C'} = \frac{1}{M}$, which is similar to the ellipsoid $\frac{x^2}{A'} + \frac{y^2}{B'} + \frac{z^2}{C'} = \frac{1}{M}$, which is the reciprocal of the comomental ellipsoid $A'x^2 + B'y^2 + C'z^2 = Mk^2$ in regard to a concentric sphere, radius k}. The author mentions the ellipsoid $\frac{x^2}{A'} + \frac{y^2}{B'} + \frac{z^2}{C'} = \frac{1}{M}$ (see p. 64), and he remarks that his conjugate axes are in fact conjugate axes in respect to this ellipsoid, and consequently that the principal axes are in direction the principal axes of this ellipsoid: it is noticeable that the ellipsoid thus incidentally considered is not the comomental ellipsoid itself, but, as just remarked, its reciprocal in regard to a concentric sphere.

152. Poisson, *Mécanique* (1st ed. 1811, and indeed 2nd ed. 1833), gives the theory of principal axes in a less complete form than in Binet's memoir; for the directions of the principal axes are obtained in anything but an elegant form.

153. Ampère's Memoir (1823).—The expression *permanent axis* is used in the place of principal axis, which is employed to designate a principal axis through the centre of gravity. The memoir contains a variety of very interesting geometrical theorems, which however, as no ellipsoid is made use of, can hardly be considered as exhibited in their proper connexion. The author arrives incidentally at certain conics, which are in fact the focal conics of the ellipsoid of gyration $\left(\frac{x^2}{A} + \frac{y^2}{B} + \frac{z^2}{C} = \frac{1}{M}\right)$ for the centre of gravity.

154. Cauchy, in the memoir "Sur les Momens d'Inertie" (1827), considers the momental ellipsoid $(A, B, C, F, G, H \between x, y, z)^2 = 1$, and employs it as well to prove the existence of the principal axes as to determine their direction, and also the magnitudes of the principal moments; the results are obtained in the simplest and best forms; viz. the direction cosines are given by

$$\begin{aligned}
(A-\Theta)\cos\alpha + H\cos\beta + G\cos\gamma &= 0,\\
H\cos\alpha + (B-\Theta)\cos\beta + F\cos\gamma &= 0,\\
G\cos\alpha + F\cos\beta + (C-\Theta)\cos\gamma &= 0,
\end{aligned}$$

where

$$(A-\Theta)(B-\Theta)(C-\Theta) - (A-\Theta)F^2 - (B-\Theta)G^2 - (C-\Theta)H^2 + 2FGH = 0,$$

Θ being one of the principal moments.

155. Poinsot, "Mémoire sur la Rotation" (1834), defines the "Central Ellipsoid" as an ellipsoid having for its axes the principal axes through the centre of gravity, the squares of the lengths being reciprocally proportional to the principal moments; and he remarks in passing that *the moment about any diameter of the ellipsoid is inversely proportional to the square of this diameter.* It is to be noticed that Poinsot speaks only of the ellipsoid having its centre at the centre of gravity, but that such ellipsoid may be constructed about any point whatever as centre, and that so generalized it is in fact the momental ellipsoid $Ax^2 + By^2 + Cz^2 = Mk^4$; and moreover that Poinsot defines his ellipsoid by reference to the principal axes.

156. Pirie, "On the Principal Axes &c." (1837), obtained analytically in a very elegant manner equations for determining the positions of the principal axes; viz. these are

$$\begin{aligned}
(A'-\Theta')\cos\alpha + H'\cos\beta + G'\cos\gamma &= 0,\\
H'\cos\alpha + (B'-\Theta')\cos\beta + F'\cos\gamma &= 0,\\
G'\cos\alpha + F'\cos\beta + (C'-\Theta')\cos\gamma &= 0,
\end{aligned}$$

where

$$(A'-\Theta')(B'-\Theta')(C'-\Theta') - (A'-\Theta')F'^2 - (B'-\Theta')G'^2 - (C'-\Theta')F'^2 + 2F'G'H' = 0\,;$$

viz. these are similar to those of Cauchy, only they belong to the comomental instead of the momental ellipsoid.

157. Maccullagh, in his Lectures of 1844 (see Haughton), considers the momental ellipsoid

$$(A, B, C, F, G, H \between x, y, z)^2 = Mk^4$$

(A, B, C, F, G, H *ut suprà*), which is such that the moment of inertia of the body with respect to any axis passing through the origin is proportional to the square of the radius vector of the ellipsoid; and from the geometrical theorem of the ellipsoid having principal axes he obtained the mechanical theorem of the existence of principal axes of the body; at least I infer that he did so, although the conclusion is not explicitly stated in Haughton's account; but in all this he had been anticipated by Cauchy. And afterwards, referring the ellipsoid to its principal axes, so that the equation is $Ax^2+By^2+Cz^2=Mk^4$, he writes $A=Ma^2$, $B=Mb^2$, $C=Mc^2$, which reduces the equation to $a^2x^2+b^2y^2+c^2z^2=k^4$, and he considers the reciprocal ellipsoid $\frac{x^2}{a^2}+\frac{y^2}{b^2}+\frac{z^2}{c^2}=1$, or, what is the same thing, $\frac{x^2}{A}+\frac{y^2}{B}+\frac{z^2}{C}=\frac{1}{M}$, which is the ellipsoid of gyration.

158. Thomson, "On the Principal Axes of a Solid Body" (1846), shows analytically that the principal axes coincide in direction with the axes of the momental ellipsoid

$$(A,\ B,\ C,\ F,\ G,\ H\between x,\ y,\ z)^2=Mk^4;$$

but the geometrical theorem might have been assumed: the investigation is really an investigation of the axes of this ellipsoid. And he remarks that the ellipsoid $(A',\ B',\ C',\ F',\ G',\ H'\between x,\ y,\ z)^2=Mk^4$ (the comomental ellipsoid) might equally well have been used for the purpose.

159. He obtains the before-mentioned theorem that the directions of the principal axes at any point are the rectangular directions in regard to the ellipsoid of gyration $\left(\frac{x^2}{A}+\frac{y^2}{B}+\frac{z^2}{C}=\frac{1}{M}\right)$ for the centre of gravity. And for determining the moments of inertia at the given point (say its coordinates are $\xi,\ \eta,\ \zeta$) he obtains the equation

$$\frac{\xi^2}{\xi^2+\eta^2+\zeta^2+\frac{A-P}{M}}+\frac{\eta^2}{\xi^2+\eta^2+\zeta^2+\frac{B-P}{M}}+\frac{\zeta^2}{\xi^2+\eta^2+\zeta^2+\frac{C-P}{M}}=1,$$

where the three roots of the cubic in P are the required moments. Analytically nothing can be more elegant, but, as already remarked, a geometrical construction for the magnitudes of these moments appears to be required.

160. Some very interesting geometrical results are obtained by considering the "equimomental surface", the locus of the points for which one of the moments of inertia is equal to a given quantity Π; the equation is of course

$$\frac{x^2}{x^2+y^2+z^2+\frac{A-\Pi}{M}}+\frac{y^2}{x^2+y^2+z^2+\frac{B-\Pi}{M}}+\frac{z^2}{x^2+y^2+z^2+\frac{C-\Pi}{M}}=1,$$

which includes Fresnel's wave-surface. In particular it is shown that the equimomental surface cuts any surface

$$\frac{x^2}{A+\Theta}+\frac{y^2}{B+\Theta}+\frac{z^2}{C+\Theta}=\frac{1}{M}$$

confocal with the ellipsoid of gyration in a spherical conic and a curve of curvature; a theorem which is also demonstrated, Cayley, "Note on a Geometrical Theorem &c." (1846).

161. Townsend, "On principal Axes &c." (1846).—This elaborate paper is contemporaneous, or nearly so, with Thomson's, and several of the conclusions are common to the two. From the character of the paper, I find it difficult to give an account of it; and I remark that, the theory of principal axes once brought into connexion with that of confocal surfaces, all ulterior developments belong more properly to the latter theory.

162. Haton de la Goupillière's two memoirs, "Sur la Théorie Nouvelle de la Géométrie des Masses" (1858), relate in a great measure to the theory of the integral $\int xydm$, and its variations according to the different positions of the two planes $x=0$ and $y=0$; the geometrical interpretations of the several results appear to be given with much care and completeness, but I have not examined them in detail. The author refers to the researches of Thomson and Townsend.

163. I had intended to show (but the paper has not been completed for publication) how the momental ellipsoid for any point of the body may be obtained from that for the centre of gravity by a construction depending on the "square potency" of a point in regard to the last-mentioned ellipsoid.

The Rotation of a solid body. Article Nos. 164—207.

164. It will be recollected that the problem is the same for a body rotating about a fixed point, and for the rotation of a free body about the centre of gravity; the case considered is that of a body not acted on by any forces. According to the ordinary analytical mode of treatment, the problem depends upon Euler's equations

$$Adp + (C-B)\,qrdt = 0,$$
$$Bdq + (A-C)\,rpdt = 0,$$
$$Cdr + (B-A)\,pqdt = 0,$$

for the determination of p, q, r, the angular velocities about the principal axes; considering p, q, r as known, we obtain by merely geometrical considerations a system of three differential equations of the first order for the determination of the position in space of the principal axes.

165. The solution of these, which constitutes the chief difficulty of the problem, is usually effected by referring the body to a set of axes fixed in space, the position whereof is not arbitrary, but depends on the initial circumstances of the motion; viz. the axis of z is taken to be perpendicular to the so-called *invariable plane.* The solution is obtained *without* this assumption both by Euler and Lagrange, although, as remarked by them, the formulæ are greatly simplified by making it; it is, on the other hand, made in the solution (which may be considered as the received one) by Poisson; and the results depending on it are the starting-point of the ulterior analytical developments of Rueb and Jacobi; the method of Poinsot is also based upon the consideration of the invariable plane.

166. D'Alembert's Principle, which affords a direct and general method for obtaining the equations of motion in any dynamical problem whatever, was given in his "Traité de Dynamique" (1743); and in his memoir of 1749 he applied it to the physical problem of the Precession of the Equinoxes, which is a special case of the problem of Rotation, the motion of rotation about the centre of gravity being in fact similar to that about a fixed point. But, as might be expected in the first attempt at the analytical treatment of so difficult a problem, the equations of motion are obtained in a cumbrous and unmanageable form.

167. They are obtained by Euler in the memoir "Découverte d'un Nouveau Principe de Mécanique," *Berlin Memoirs* for 1750 (1752) (written before the establishment of the theory of principal axes), in a perfectly elegant form, including the ordinary one already mentioned, and, in fact, reducible to it by merely putting the quantities F, G, H (which denote the integrals $\int yzdm$, &c.) equal to zero. But Euler does not in this memoir enter into the question of the integration of the equations.

168. The notion of principal axes having been suggested by Euler, and their existence demonstrated by Segner, we come to Euler's investigations contained in the memoirs "Du Mouvement de Rotation &c.," *Berlin Memoirs* for 1758 (printed 1765) and for 1760 (printed 1767), and the "Theoria Motus Corporum Solidorum &c." (1765). In the memoir of 1760, and in the "Theoria Motus," Euler employs $\aleph$, the angular velocity round the instantaneous axis, but not the resolved velocities $\aleph\cos\alpha$, $\aleph\cos\beta$, $\aleph\cos\gamma\,(=p,\ q,\ r)$: these quantities (there called x, y, z) are however employed in the memoir, *Berlin Memoirs* (1758), which must, I apprehend, have been written after the other, and in which at any rate the solution is developed with much greater completeness. It is in fact carried further than the ordinary solutions, and after the angular velocities p, q, r have been found, the remaining investigation for the position in space of the principal axes, conducted, (as above remarked), without the aid of the invariable plane, is one of great elegance.

169. In the last-mentioned memoir Euler starts from the equations given by d'Alembert's principle; viz. the impressed forces being put equal to zero, these are

$$\Sigma dm\left(y\frac{d^2z}{dt^2}-z\frac{d^2y}{dt^2}\right)=0,\ \ \&c.,$$

or, what is the same thing, using u, v, w to denote the velocities of an element in the directions of the axes fixed in space, these are

$$\Sigma dm\left(y\frac{dw}{dt}-z\frac{dv}{dt}\right)=0,$$

$$\Sigma dm\left(z\frac{du}{dt}-x\frac{dw}{dt}\right)=0,$$

$$\Sigma dm\left(x\frac{dv}{dt}-y\frac{dy}{dt}\right)=0.$$

It is assumed that at any moment the body revolves round an instantaneous axis, inclinations α, β, γ, with an angular velocity $\aleph$; this gives

$$u = \aleph(z\cos\beta - y\cos\gamma) = qz - ry,$$
$$v = \aleph(x\cos\gamma - z\cos\alpha) = rx - pz,$$
$$w = \aleph(y\cos\alpha - x\cos\beta) = px - qy,$$

if $\aleph\cos\alpha$, $\aleph\cos\beta$, $\aleph\cos\gamma$ are denoted by p, q, r. The values of du, dv, dw are obtained by differentiating these formulæ, treating p, q, r, x, y, z as variable, and replacing dx, dy, dz by udt, vdt, wdt respectively; in the resulting formulæ for $ydw - zdv$, &c., x, y, z are considered as denoting the coordinates of the element in regard to axes fixed in the body and moveable with it, but which at the moment under consideration coincide in position with the axes fixed in space. The expressions for $\Sigma(ydw - zdv)\,dm$ involve the integrals $A = \int (y^2 + z^2)\,dm$, &c., where the coordinates refer to axes fixed in the body; and if these are taken to be principal axes, the expression for $\Sigma(ydw - zdv)\,dm$ is $= Adp + (C - B)\,qrdt$, which gives the three equations

$$Adp + (C - B)\,qrdt = 0,$$
$$Bdq + (A - C)\,rpdt = 0,$$
$$Cdr + (B - A)\,pqdt = 0,$$

already referred to as Euler's equations.

170. Next, as regards the determination of the position in space of the principal axes: if about the fixed point we describe a sphere meeting the principal axes in x_1, y_1, z_1, and if P be an arbitrary point on the sphere and PQ an arbitrary direction through P, the quantities used to determine the positions of x_1, y_1, z_1 are the distances x_1P, y_1P, z_1P ($= l$, m, n) and the inclinations x_1PQ, y_1PQ, z_1PQ ($=\lambda$, μ, ν) of these arcs to the fixed direction PQ (it is to be observed that the sines and cosines of the differences of λ, μ, ν are given functions of the sines and cosines of l, m, n, and, moreover, that $\cos^2 l + \cos^2 m + \cos^2 n = 1$, so that the number of independent parameters is three). The above is Euler's definition; but if we consider a set of axes fixed in space meeting the sphere in the points X, Y, Z, then if the point X be taken for P, and the arc XY for PQ, it is at once seen that the angles used for determining the relative positions of the two sets of axes are the same as in Euler's memoir "Formulæ Generales &c.," 1775 (*ante*, No. 132), where the formulæ for this transformation of coordinates are considered apart from the dynamical theory.

Euler expresses the quantities p, q, r in terms of an auxiliary variable u, which is such that $du = pqrdt$; p, q, r are at once obtained in terms of u, and then t is given in terms of u by a quadrature. Euler employs also an auxiliary angle U, given in terms of u by a quadrature. And he obtains finite algebraical expressions in u, $\cos U$, $\sin U$ for the cosines or sines of l, m, n; s (the angular distance IP, if I denote the point in which the instantaneous axis meets the sphere), ϕ (the angle IPQ) and $\lambda - \phi$, $\mu - \phi$, $\nu - \phi$. The formulæ, although complicated, are extremely elegant, and they appear to have been altogether overlooked by subsequent writers.

171. Euler remarks, however, that the complexity of his solution is owing to the circumstance that the fixed point P is left arbitrary, and that they may be simplified by taking this point so that a certain relation $G - \mathfrak{D}^2 = 0$ may be satisfied between the constants of the solution; and he gives the far more simple formulæ corresponding to this assumption. This amounts to taking the point P *in the normal of the invariable plane,* and the resulting formulæ are in fact identical with the ordinary formulæ for the solution of the problem. The expression *invariable plane* is not used by Euler, and seems to have been first employed in Lagrange's memoir "Essai sur le Problème de Trois Corps," *Prix de l'Acad. de Berlin*, t. IX. (1772): the theory in reference to the solar system has been studied by Laplace, Poinsot, and others.

172. Lagrange's solution in the memoir of 1773 is substantially the same with that in the *Mécanique Analytique*. Only he starts from the integral equations of areas and of *Vis Viva*, but in the last-mentioned work from the equations of motion as expressed in the Lagrangian form by means of the *Vis Viva* function $T\left(=\tfrac{1}{2}\Sigma(x'^2+y'^2+z'^2)\,dm\right)$. The distinctive feature is that he does not refer the body to the principal axes but to any rectangular axes whatever fixed in the body: the expression for T consequently is $T = \tfrac{1}{2}(A, B, C, F, G, H \between p, q, r)^2$, instead of the more simple form

$$T = \tfrac{1}{2}(Ap^2 + Bq^2 + Cr^2),$$

which it assumes when the body is referred to its principal axes. And Lagrange effects the integration as well with this more general form of T, as without the simplification afforded by the invariable plane; the employment, however, of the more general form of T seems an unnecessary complication of the problem, and the formulæ are not worked out nearly so completely as in Euler's memoir. It may be observed that p, q, r are expressed as functions of the instantaneous velocity $\omega\,(=\sqrt{p^2+q^2+r^2})$, and thence t obtained by a quadrature as a function of ω.

173. Poisson's Memoir of 1809.—The problem is only treated incidentally for the sake of obtaining the expressions for the variations of the arbitrary constants; the results (depending, as already remarked, on the consideration of the invariable plane) are obtained and exhibited in a very compact form, and they have served as a basis for further developments; it will be proper to refer to them somewhat particularly. The Eulerian equations give, in the first place, the integrals

$$A\,p^2 + B\,q^2 + C\,r^2 = h,$$
$$A^2p^2 + B^2q^2 + C^2r^2 = k^2;$$

and then by means of these, p, q being expressed in terms of r, we have t in terms of r by a quadrature.

174. The position in space of the principal axes is determined by referring them, by means of the angles θ, ϕ, τ, to axes Ox, Oy, Oz fixed in space; if, to fix the ideas, we call the plane of xy the ecliptic (Ox being the origin of longitudes), and the plane of the two principal axes x_1y_1 the equator, then we have

θ, the longitude of node,
ϕ, the inclination,
τ, the hour-angle, or angular distance of Ox_1 from the node,

and α, β, γ the cosine inclinations of Ox_1, α', β', γ' those of Oy_1, and α'', β'', γ'' those of Oz_1 to Ox, Oy, Oz respectively are given functions of θ, ϕ, τ (the values of α'', β'', γ'' depending upon θ, ϕ only), we have

$$pdt = \sin\tau \sin\phi \, d\theta + \cos\tau \, d\phi,$$
$$qdt = \cos\tau \sin\phi \, d\theta - \sin\tau \, d\phi,$$
$$rdt = d\tau + \cos\phi \, d\theta.$$

175. A set of integrals is

$$Ap\alpha + Bq\beta + Cr\gamma = k\cos\lambda,$$
$$Ap\alpha' + Bq\beta' + Cr\gamma' = k\cos\mu,$$
$$Ap\alpha'' + Bq\beta'' + Cr\gamma'' = k\cos\nu,$$

equivalent to two independent equations, the values of λ, μ, ν being such that $\cos^2\lambda + \cos^2\mu + \cos^2\nu = 1$; but the position of the axis of z may be chosen so that the values on the right-hand sides become 0, 0, k; the axis of z is then perpendicular to the *invariable plane*, the condition in question serving as a definition. And the three equations then give

$$Ap = k\alpha'', \quad Bq = k\beta'', \quad Cr = k\gamma'',$$

where the values of α'', β'', γ'' in fact are

$$\alpha'' = \sin\tau \sin\phi, \quad \beta'' = \cos\tau \sin\phi, \quad \gamma'' = \cos\phi;$$

we have thus τ, ϕ in terms of r. And the equation $rdt = d\tau + \cos\phi d\theta$ then leads to the value of $d\theta$, or θ is determined as a function of r by a quadrature.

176. The constants of integration are h, k, l (the constant attached to t), g (the constant attached to θ); and two constants, say α the longitude of the node, and γ the inclination of the invariable plane in reference to an arbitrary plane of xy and origin x of longitudes therein. I remark in passing that Poisson obtains an elegant set of formulæ for the variations of the constants h, k, g, l, α, γ, not actually in the canonical form, but which may by a slight change be reduced to it.

177. Legendre considers the problem of Rotation in the *Exercices de Calcul Intégral*, t. II. (1817), and the *Théorie des Fonctions Elliptiques*, t. I. pp. 366—410 (1826). He does not employ the quantities p, q, r, but obtains *de novo* a set of differential equations of the second order involving the angles which determine the position of the principal axes with regard to the axes fixed in space: these angles are in fact (calling the plane of the fixed axes x, y the ecliptic) the longitude and latitude of *one* of the principal axes, and the azimuth from the meridian through such principal axis of an arbitrary axis fixed in the body and moveable with it. The solution is developed by means of the elliptic *integrals*. From the peculiar choice of variables there would, it would seem, be considerable labour in comparing the results with those of other writers, and there would be but little use in doing so.

178. Poinsot's "Théorie Nouvelle de la Rotation des Corps."—The 'Extrait' of the memoir was published in 1834, but the memoir itself was not published *in extenso* until the year 1851. The 'Extrait' contains, however, not only the fundamental theorem of the representation of the motion of a body about a fixed point by means of the momental ellipsoid rolling on a fixed tangent plane, but also the geometrical and mechanical reasonings by which this theorem is demonstrated; it establishes also the notions of the Poloid and Serpoloid curves; and it contains incidentally, and without any developments, a very important remark as to the representation of the motion by means of the rolling and sliding motion of an elliptic cone. The whole theory (including that of the last-mentioned representation of the motion) is in the memoir itself also analytically developed, but without the introduction of the elliptic and Jacobian functions: to form a complete theory, it would be necessary to incorporate the memoir with that of Jacobi.

179. The following is an outline of the 'Extrait':

The instantaneous motion of a body about a fixed point is a motion of rotation about an axis (the instantaneous axis); and hence the finite motion is as if there were a cone fixed in the body which rolls (without sliding) upon another cone fixed in space.

The instantaneous motion of a body in space is a motion of rotation about an axis combined with a translation in the direction of this axis: this remark is hardly required for Poinsot's purpose, and he does not further develope the theory of the motion of a body in space. The effect of a couple in a plane perpendicular to a principal axis is to turn the body about this axis with an angular velocity proportional to the moment of the couple divided by the moment of inertia about the axis.

And hence by resolving any couple into couples perpendicular to the principal axes, the effect of such couple may be calculated; but more simply by means of the central ellipsoid (momental ellipsoid $a^2x^2+b^2y^2+c^2z^2=k^4$, if A, B, $C=Ma^2$, Mb^2, Mc^2), viz., if the body is acted on by a couple in a tangent plane of the ellipsoid, the instantaneous axis passes through the point of contact; and reciprocally, given the instantaneous axis, the couple must act in the tangent plane.

180. Considering now a body rotating about a fixed point, and taking as the plane of the couple of impulsion a tangent plane of the ellipsoid, the instantaneous axis is initially the diameter through the point of contact; the centrifugal forces arising from the rotation produce however an accelerating couple, the effect whereof is continually to impress on the body a rotation which is compounded with that about the instantaneous axis, and thus to cause a variation in the position of this axis and in the angular velocity round it. The axis of the accelerating couple is always situate in the plane of the couple of impulsion.

181. Hence also

1°. Throughout the motion the angular velocity is proportional to the length of the instantaneous axis considered as a radius vector of the ellipsoid.

2°. The distance of the tangent plane from the centre is constant; that is, the tangent plane to the ellipsoid at the extremity of the instantaneous axis is a fixed plane in space.

Or, what is the same thing, the motion is such that the ellipsoid remains always in contact with a fixed plane, viz., the body revolves round the radius vector through the point of contact, the angular velocity being always proportional to the length of this radius vector.

It is right to remark that in Poinsot's theory the distance of this plane from the centre depends on the arbitrarily assumed magnitude of the central ellipsoid; the parallel plane through the centre is the invariable plane of the motion.

182. The motion is best understood by the consideration that it is implied in the theorem that the pole of the instantaneous axis describes on the ellipsoid a certain curve, "the Poloid," which is the locus of all the points for which the perpendicular on the tangent plane has a given constant value (the curve in question is easily seen to be the intersection of the ellipsoid by a concentric cone of the second order); and that the instantaneous axis describes on the fixed tangent plane a curve called "the Serpoloid," which is the locus of the points with which the several points of the poloid come successively in contact with the tangent plane, and is a species of undulating curve, viz., the radius vector as it moves through the angles θ to $\theta_1+2\Pi$, $\theta_1+2\Pi$ to $\theta_1+4\Pi$, &c. assumes continually the same series of values. This is in fact evident from the mode of generation; and it is moreover clear that the serpoloid is an algebraical or else a transcendental curve according as Π is or is not commensurable with π.

{Treating the poloid and serpoloid as cones instead of curves, the motion of the body is the rolling motion of the former upon the latter cone, which agrees with a previous remark.}

There is a very interesting special case where the perpendicular distance from the tangent plane is equal to the mean axis of the ellipse.

183. Poinsot remarks that the motion is such that {considering the plane of the couple of impulsion as drawn through the centre of the ellipsoid} the section of the ellipsoid is an ellipse variable in form but of constant magnitude, and that this leads to a new representation of the motion, viz., that it may be regarded as the *motion of an elliptic cone which rolls on the plane of the couple* {the invariable plane} *with a variable velocity, and which slides on this plane with a uniform velocity.*

184. The theory of the last-mentioned cone, say the "rolling and sliding cone," is developed in the memoir, *Liouville*, t. XVI. p. 303, in the chapter entitled "Nouvelle Image de la Rotation des Corps." If a, b, c signify as before (viz., A, B, $C=Ma^2$, Mb^2, Mc^2), and if h be the distance of the centre from Poinsot's fixed tangent plane ($h<a>c$), then the invariable axis describes in the body a cone the equation whereof is

$$(a^2-h^2)x^2+(b^2-h^2)y^2+(c^2-h^2)z^2=0;$$

the cone reciprocal to this, viz. the cone the equation whereof is

$$\frac{x^2}{a^2-h^2}+\frac{y^2}{b^2-h^2}+\frac{z^2}{c^2-h^2}=0,$$

is the "rolling and sliding cone." The generating line OT of this cone is perpendicular to the plane of the instantaneous axis OI, and of the invariable axis OG; and the analytical expressions for the rolling and sliding velocities follow from the geometrical consideration that the motion at any instant is a rotation round the instantaneous axis OI: that for the sliding velocity is the instantaneous angular velocity into the cosine of the angle IOG, which is in fact constant; that for the rolling velocity is given, but a further explanation of the geometrical signification is perhaps desirable.

185. I may in this place again refer to Cohen's memoir "On the Differential Coefficients and Determinants of Lines &c." (1862), the latter part of which contains an application of the method to finding Euler's equations for the motion of a rotating body.

186. Rueb in his memoir (1834) first applied the elliptic and Jacobian functions to the present problem. Starting from the equations

$$A\,p^2 + B\,q^2 + C\,r^2 = h,$$

$$A^2p^2 + B^2q^2 + C^2r^2 = l^2, \text{[1]}$$

and

$$dt = \frac{-Bdq}{(A-C)\,rp},$$

it is easy to perceive that by assuming q = a proper multiple of $\sin\xi$, the expression for dt takes the form $ndt = \dfrac{d\xi}{\sqrt{1-k^2\sin^2\xi}}$, so that writing $\xi = \text{am}\,u$, the integral equation is $nt-\epsilon = u$, or u is an angle varying directly as the time (and corresponding to the mean longitude, or, if we please, to the mean anomaly in the problem of elliptic motion). And then p, q, r are expressed as elliptic functions of u. The value of the modulus k, and that of n ($nt-\epsilon = u$ *ut suprà*) are

$$n = \sqrt{\frac{(B-C)(-l^2+Ah)}{ABC}},$$

$$k = \sqrt{\frac{(A-B)(l^2-Ch)}{ABC}},$$

and then

$$\left.\begin{aligned} p &= \pm\sqrt{\frac{l^2-Ch}{A\,.\,A-C}}\cos\text{am}\,u,\\ q &= -\sqrt{\frac{l^2-Ch}{B\,.\,B-C}}\sin\text{am}\,u,\\ r &= \sqrt{\frac{-l^2+Ah}{C\,.\,A-C}}\Delta\,\text{am}\,u. \end{aligned}\right\}$$

[1] l is Poisson's k, the constant of the principal area; it is the letter afterwards used by Jacobi; Rueb's letter is g. In quoting (*infrà*) the expressions for p, q, r, I have given them with Rueb's signs, but it would be too long to explain how the signs of the radicals are determined.

187. Substituting for p, q, r their values in terms of u, we have $d\theta$, and thence θ (the longitude of the node of the equator on the invariable plane) in the form

$$\theta = -\frac{l}{An}u + i\Pi(u, ia) \quad (i = \sqrt{-1}),$$

which by Jacobi's formulæ for the transformation of the elliptic integral of the third class becomes

$$\theta = \left(-\frac{l}{An} + iZ(ai)\right)u + \tfrac{1}{2}i\log\frac{\Theta(u-ai)}{\Theta(u+ai)},$$

which Rueb reduces to the real form

$$\theta = -n'u + \tan^{-1} W,$$

W being given in the form of a fraction, the numerator and denominator whereof are series in multiple sines and multiple cosines respectively of $\frac{\pi u}{2K}$.

188. Rueb investigates also the values in terms of u of the cosine inclinations of the instantaneous axis to the axes fixed in space; and he obtains a very elegant expression for the angle ζ, which is the angular distance from x of the projection on the plane of xy (the invariable plane) of the instantaneous axis; viz., this is

$$\zeta = \tan^{-1}\left(-\frac{ABn}{(A-B)\,l}\,\frac{\Delta\operatorname{am} u}{\sin\operatorname{am} u\cos\operatorname{am} u}\right) - \theta,$$

and there is throughout a careful discussion of the geometrical signification of the results.

189. The advance made was enormous; the result is that we have in terms of the time $\sin\tau\sin\phi$, $\cos\tau\sin\phi$, $\cos\phi$ (the cosine inclinations of the invariable axis to the principal axes), and also θ, the longitude of the node. The cosine inclinations of the axes of x and y to the principal axes could of course be obtained from these, but they would be of a very complicated and unmanageable form; *the reason of this is the presence in the expression for θ of the non-periodic term $-n'u$.* It will presently be seen how this difficulty was got over by Jacobi.

190. Briot's paper of 1842 contains an analytical demonstration of some of the theorems given in the 'Extrait' of Poinsot's memoir of 1834.

191. In Maccullagh's Lectures of 1844 (see Haughton, 1849; Maccullagh, 1847) the problem of the rotation of a solid body is treated in a mode somewhat similar to that of Poinsot; only the ellipsoid of gyration $\left(\frac{x^2}{a^2} + \frac{y^2}{b^2} + \frac{z^2}{c^2} = 1, \text{ if } A, B, C = Ma^2, Mb^2, Mc^2\right)$ is used instead of *the momental ellipsoid.* Thus, reciprocal to the poloid curve on the momental ellipsoid we have on the ellipsoid of gyration a curve all the points whereof are equidistant from the centre; such curve is of course the intersection of the ellipsoid by a concentric sphere, that is, it is a spherical conic; and the points of this spherical conic come successively to coincide with a fixed point on the invariable axis. This is a theorem stated in Art. VI. of Haughton's memoir: it may be added

that the several tangent planes of the ellipsoid of gyration at the points of the spherical conic as they come to coincide with the fixed point, form a cone reciprocal to Poinsot's serpoloid cone. It is clear that every theorem in the one theory has its reciprocal in the other theory; I have not particularly examined as to how far the reciprocal theorems have been stated in the two theories.

192. Cayley, "On the Motion of Rotation of a Solid Body" (1843).—The object was to apply to the solution of the problem Rodrigues' formulæ for the resultant rotation; viz., if the principal axes, considered as originally coinciding with the axes of x, y, z, can be brought into their actual position at the end of the time t by a rotation θ round an axis inclined at angles f, g, h to the axes of x, y, z, and if $\lambda = \tan\frac{1}{2}\theta\cos f$, $\mu = \tan\frac{1}{2}\theta\cos g$, $\nu = \tan\frac{1}{2}\theta\cos h$, then the principal axes are referred to the axes fixed in space by means of the quantities λ, μ, ν. And these are to be obtained from the equations

$$\begin{aligned}\kappa\, p dt &= 2\,(\quad d\lambda + \nu d\mu - \mu d\nu),\\ \kappa\, q dt &= 2\,(-\nu d\lambda + \quad d\mu + \lambda d\nu),\\ \kappa\, r dt &= 2\,(\quad \mu d\lambda - \lambda d\mu + \quad d\nu),\end{aligned}$$

where $\kappa = 1 + \lambda^2 + \mu^2 + \nu^2$, and p, q, r are to be considered as given functions of t, or of other the variable selected as the independent one. But for effecting the integration it was found necessary to take the axis of z as the invariable axis.

193. The solution by Jacobi, § 27 of the memoir "Theoria Novi Multiplicatoris" (1845), is given as an application of the general theory, the author remarking that, as well in this question as in the problem of the two fixed centres, he purposely employed a somewhat inartificial analysis, in order to show that the principle (of the Ultimate Multiplier) would lead to the result without any special artifices. The principal axes are referred to the axes fixed in space by the ordinary three angles (here called q_1, q_2, q_3), and the solution is carried so far as to give the integral equations, without any reduction of the integrals contained in them to elliptic integrals. The solution is, however, in so far remarkable that the integrations are effected without the aid of the invariable plane.

194. Cayley, "On the Rotation of a Solid Body &c." (1846).—It appeared desirable to obtain the solution by means of the quantities λ, μ, ν, *without the assistance of the invariable plane*, and Jacobi's discovery of the theorem of the Ultimate Multiplier induced me to resume the problem, and at least attempt to bring it so far as to obtain a differential equation of the first order between two variables only, the multiplier of which could be obtained theoretically by Jacobi's discovery. The choice of two new variables to which the equations of the problem led me, enabled me to effect this in a simple manner; and the differential equation which I finally obtained turned out to be integrable *per se*, so that the laborious process of finding the multiplier became unnecessary.

195. The new variables Ω, υ have the following geometrical significations: $\Omega = l\tan\frac{1}{2}\theta\cos I$, where l is the principal moment ($A^2p^2 + B^2q^2 + C^2r^2 = l^2$), θ (as before) the angle of resultant

rotation, and I is the inclination of the resultant axis to the invariable axis; and $v = l^2 \cos^2 \frac{1}{2} J$, where if we imagine a line AQ having the same position relatively to the axes fixed in space that the invariable axis has to the principal axes of the body, then J is the inclination of this line to the invariable axis. It is found that p, q, r are functions of v only, whereas λ, μ, ν contain besides the variable Ω. In obtaining these relations, there occurs a singular relation $\Omega^2 = \kappa v - l^2$, which may also be written $1 + \tan^2 \frac{1}{2}\theta \cos^2 I = \sec^2 \frac{1}{2}\theta \cos^2 \frac{1}{2} J$, where the geometrical significations of the quantities I, J have just been explained. The final results are that the time t, and the arc $\tan^{-1} \dfrac{\Omega}{l}$ are each of them expressible as the integrals of certain algebraical functions of v. There might be some interest in comparing the results with those of Euler's memoir of 1758, where the principal axes are also referred to an arbitrary system of axes fixed in space; but I was not then acquainted with Euler's memoir.

The concluding part of the paper relates to the determination of the variations of the constants in the disturbed problem.

196. Cayley, "Note on the Rotation of a Solid of Revolution" (1849), shows the simplification produced in the formulæ of the last-mentioned memoir in the case where two of the moments of inertia are equal, say $A = B$.

197. Jacobi's final solution of the problem of Rotation was given without demonstration in the letter to the Academy of Sciences at Paris; the demonstration is added in the memoir, *Crelle*, t. XXXIX. (1849). The fundamental idea consists in taking in the invariable plane, instead of the fixed axes xy, a set of axes xy revolving with uniform velocity, such that the angular distance of the axis of x from its initial position is precisely $= -n'u$; and consequently if θ' be the longitude of the node of the equator on the invariable plane, measured from the moveable axis of x as the origin of longitude, we have

$$\theta' = \theta + n'u = \frac{1}{2i} \log \frac{\Theta(u + ia)}{\Theta(u - ia)}, \quad (i = \sqrt{-1});$$

and in consequence of this form of the expression for $\theta' \left(= \dfrac{1}{2i} \text{ into a logarithmic function}\right)$ in passing to the trigonometrical functions $\sin\theta'$, $\cos\theta'$ the logarithm disappears altogether; and we have in a simple form the expressions for the actual functions $\sin\theta'$, $\cos\theta'$, through which θ' enters into the formulæ, and thus, Jacobi remarks, the barrier is cleared which stands in the way when the expression of an angle is reduced to an elliptic integral of the third class.

198. For the better expression of the results, Jacobi joins to the functions H, Θ, considered in the "Fundamenta Nova," the functions $\Theta_1 u = \Theta(K - u)$, $H_1(u) = H(K - u)$; so that

$$\sqrt{k} \sin \operatorname{am} u = \frac{Hu}{\Theta u}, \quad \sqrt{\frac{k}{k'}} \cos \operatorname{am} u = \frac{H_1 u}{\Theta u}, \quad \frac{1}{\sqrt{k'}} \Delta \operatorname{am} u = \frac{\Theta_1 u}{\Theta u},$$

and then considering the cosine inclinations of the principal axes to the invariable axis and the revolving axes in the invariable plane, these are all fractions which, neglecting constant factors, have the common denominator Θu; α'', β'', γ'' (as shown by Rueb's formulæ) have the numerators $H_1 u$, Hu, and $\Theta_1 u$ respectively; and α, α' have the numerators $H(u+ia) \pm H(u-ia)$, β, β' the numerators $H_1(u-ia) \pm H_1(u+ia)$, γ, γ' the numerators $\Theta(u+ia) \pm \Theta(u-ia)$ respectively: there are also expressions of a similar form for the angular velocities about the axis of x and y; that about the axis of z (the invariable axis) having, as was known, the constant value $\frac{h}{l}$. The memoir is also very valuable analytically, as completing the systems of formulæ given in the *Fundamenta Nova* in reference to elliptic integrals of the third class.

199. It is worth noticing how the results connect themselves with Poinsot's theorem of the rolling and sliding cone: the velocity of the rolling motion depends only upon the position, on the cone, of the line of contact, so that the same series of velocities recur after any number of complete revolutions of the cone; that is, the total angle described by the line of contact in consequence of the rolling motion, consists of a part varying directly with the time (or say varying as u) and a periodic part; the former part combines with the similar term arising from the sliding motion, and the two together give Jacobi's term $n'u$.

200. Somoff's memoir (1851), written after Jacobi's Note in the *Comptes Rendus*, but before the appearance of the memoir in *Crelle*, gives the demonstration of the greater part of Jacobi's results.

201. Booth's *Theory of Elliptic Integrals &c.* (1851) (contemporaneous with the publication of Poinsot's memoir of 1834) contains various interesting analytical developments, and, as an interpretation of them, the author obtains (p. 93) the theorem of the rolling and sliding cone. The investigations involve the elliptic integrals, but not the elliptic or Jacobian functions.

202. Richelot's two Notes (*Crelle*, tt. XLII. and XLIV.) relate to the solution of the problem of rotation given in his memoir "Eine neue Lösung &c." (1851). This is an application of Jacobi's theorem for the integration of a system of dynamical equations by means of the principal function S (see my "Report" of 1857, art. 34). Retaining Richelot's letters ϕ, ψ, θ, which signify

ψ, the longitude of the node,

θ, the inclination,

ϕ, the hour-angle,

the question is to find a complete solution of the partial differential equation

$$\begin{aligned} 0 = {} & \frac{1}{2A}\left\{\left(\frac{dV}{d\phi}\cos\theta + \frac{dV}{d\psi}\right)\frac{\sin\phi}{\sin\theta} - \frac{dV}{d\theta}\cos\phi\right\}^2 \\ & + \frac{1}{2B}\left\{\left(\frac{dV}{d\phi}\cos\theta + \frac{dV}{d\psi}\right)\frac{\cos\phi}{\sin\theta} + \frac{dV}{d\theta}\sin\phi\right\}^2 \\ & + \frac{1}{2C}\left(\frac{dV}{d\phi}\right)^2 \quad + \frac{dV}{dt}; \end{aligned}$$

that is, a solution involving (besides the constant attached to V by a mere addition) three arbitrary constants; these are t_1, ψ_1, ρ. Writing in the first place $V = W + tt_1 + \psi\psi_1$, the resulting equation for W may be satisfied by taking W, a function of ϕ and θ, without ψ or t; and it is sufficient to have a solution involving only a single arbitrary constant. This leads to a solution which is as follows:

$$V = tt_1 + \psi\psi_1 - \psi_1 \tan^{-1} \frac{\theta_1}{\sqrt{\rho^2 - \psi_1^2 - \theta_1^2}} + \rho \left\{ \tan^{-1} \frac{\psi_1\theta_1}{\rho\sqrt{\rho^2 - \psi_1^2 - \theta_1^2}} + \tan^{-1} \frac{\phi_1\theta_1}{\rho\sqrt{\rho^2 - \phi_1^2 - \theta_1^2}} \right\}$$

$$- \left(\frac{\rho^2}{C} + 2t_1\right) \int \frac{\phi_1^2 d\phi_1}{(\rho^2 - \phi_1^2) \sqrt{\left(\frac{1}{B} - \frac{1}{C}\right)\left(\phi_1^2 - \frac{\rho^2}{B} + 2t_1\right)\left(\left(\frac{1}{C} - \frac{1}{A}\right)\phi_1^2 + \frac{\rho^2}{A} + 2t_1\right)}},$$

where ϕ_1 and θ_1 are certain given functions of t_1, ψ_1, ρ, and of θ and ϕ. The solution of the dynamical problem is then obtained by putting the differential coefficients $\frac{dV}{dt_1}$, $\frac{dV}{d\psi_1}$, $\frac{dV}{d\rho}$ equal to arbitrary constants L, α, G respectively; the results are somewhat more simple than might be expected from the very complicated form of the function V. The foregoing results (although not by themselves very intelligible) will give an idea of the form in which the analytical solution in the first instance presents itself.

203. Richelot proceeds to remark that the solution in question, and the resulting integral equations of the problem, may be simplified in a peculiar manner by the method which he calls "the integration by the spherical triangle." For this purpose he introduces a spherical triangle, the sides and angles whereof are

$$\nu,\ \lambda,\ \mu;\ \mathrm{N},\ \Lambda,\ \mathrm{M},$$

and then assuming

$$\mathrm{N}\ \text{constant},\ \mathrm{M} = \pi - \theta$$

$$\left(\frac{1}{C} - \frac{1}{A}\right) \sin^2 (\phi + \nu) \sin^2 \Lambda + \left(\frac{1}{C} - \frac{1}{B}\right) \cos^2 (\phi + \nu) \sin^2 \Lambda = \frac{1}{C} + \frac{2t_1}{\rho^2},$$

where ρ and t_1 are constant, the solution is

$$V = t_1 t - \rho (\psi - \lambda) \cos \mathrm{N} - \rho \mathrm{M} + \rho \int \cos \Lambda \, d (\phi + \nu);$$

and that this expression leads to all the results almost without calculation.

204. I have quoted the foregoing results from the Note (*Crelle*, t. XLII.), having seen, but without having studied, the Memoir itself: the results appear very interesting and valuable ones; but they seem to require a more complete geometrical development than they have received in the Memoir; and I am not able to bring them into connexion with the other researches on the subject.

205. The solution, § 3 of Donkin's memoir "On a Class of Differential Equations &c." (part I. 1854), is given as an illustration of the general theory to which the memoir relates; it contains, however, some interesting geometrical developments in regard to the case ($A = B$) of two equal moments of inertia. I have not compared the results with those in my Note of 1849.

206. The solution of the rotation problem, § 66 of Jacobi's memoir "Nova Methodus &c." (1862), has for its object to show the complete analogy which exists between this problem and the problem of a body attracted to a fixed centre. The section is in fact headed "Solutio simultanea problematis de motu puncti versus centrum attracti atque problematis de rotatione &c."; and Jacobi, after noticing that Poisson, in his memoir of 1816 (*Mém. de l'Inst.* t. I.), had shown that the expressions for the variations of the elements in the two problems could be investigated by a common analysis, remarks, "Sed ipsa problemata duo imperturbata hic primum, quantum credo, amplexus sum." The solution is in fact as follows:—Suppose that in the one problem the position of the point in space, and in the other problem the position of the body in regard to the fixed axes, is determined in any manner by the quantities q_1, q_2, q_3. Let

$$\frac{dq_1}{dt}=q_1', \quad \frac{dq_2}{dt}=q_2', \quad \frac{dq_3}{dt}=q_3',$$

and expressing the *Vis Viva* function T in terms of $q_1, q_2, q_3, q_1', q_2', q_3'$, let

$$\frac{dT}{dq_1'}=p_1, \quad \frac{dT}{dq_2'}=p_2, \quad \frac{dT}{dq_3'}=p_3,$$

and let H be the value of T expressed in terms of $q_1, q_2, q_3, p_1, p_2, p_3$, so that $H=a$ is the integral of *Vis Viva* (this is merely the transformation to the Hamiltonian form). And let $H_1=a_1$, $\phi=a_1'$, $\psi=a_1''$ be the three integrals of areas (H, H_1, ϕ, ψ are functions of the variables only, not containing the arbitrary constants a, a_1, a_1', a_1''). Then, expressing

$$H, \ H_1, \ H_2 (=\sqrt{H_1^2+\phi^2+\psi^2})$$

in terms of $p_1, p_2, p_3, q_1, q_2, q_3$, and by means of the equations

$$H=a, \quad H_1=a_1, \quad H_2=a_2$$

(where $a_2=\sqrt{a_1^2+a_1'^2+a_1''^2}$) expressing p_1, p_2, p_3 in terms of q_1, q_2, q_3, we have $p_1dq_1+p_2dq_2+p_3dq_3$ a complete differential; and putting

$$\int(p_1dq_1+p_2dq_2+p_3dq_3)=V,$$

then (a, a_1, a_2, b, b_1, b_2 being arbitrary constants) we have

$$H=a, \quad H_1=a_1, \quad H_2=a_2,$$

$$\frac{dV}{da}=\int\left(\frac{dp_1}{da}dq_1+\frac{dp_2}{da}dq_2+\frac{dp_3}{da}dq_3\right)=t+b,$$

$$\frac{dV}{da_1}=\int\left(\frac{dp_1}{da_1}dq_1+\frac{dp_2}{da_1}dq_2+\frac{dp_3}{da_1}dq_3\right)=b_1,$$

$$\frac{dV}{da_2}=\int\left(\frac{dp_1}{da_2}dq_1+\frac{dp_2}{da_2}dq_2+\frac{dp_3}{da_2}dq_3\right)=b_2,$$

as the complete integrals of either problem, the last three of them being the final integrals.

And it is added that if in either problem we have $H+\Omega$ instead of H, the expressions for the variations of the elements assume the canonical forms $\frac{da}{dt}=-\frac{d\Omega}{db}$, $\frac{db}{dt}=\frac{d\Omega}{da}$, &c.

The solution is not further developed as regards the rotation problem, but it is so (§ 67) as regards the other problem.

207. It must, I think, be considered that a comprehensive memoir on the Problem of Rotation, embracing and incorporating all that has been done on the subject, is greatly needed.

Kinematics of a solid body. Article Nos. 208 to 215.

208. The general theorem in regard to the infinitesimal motions (rotations and translations) of a solid body is that these are compounded and resolved in the same way as if they were single forces and couples respectively. Thus any infinitesimal rotations and translations are resolvable into a rotation and a translation; the rotation is given as to its magnitude and as to the direction of its axis, but not as to the position of the axis (which may be any line in the given direction): the magnitude and direction of the translation depend on the assumed position of the axis of rotation; in particular this may be taken so that the translation shall be in the direction of the axis of rotation; and the magnitude of the rotation is then a minimum. I remark that the theorem as above stated presupposes the establishment of the theory of couples (of forces) which was first accomplished by Poinsot in his '*Élémens de Statique*,' 1st edit. 1804; it must have been, the whole or nearly the whole of it, familiar to Chasles at the date of his paper of 1830 next referred to (see also Note XXXIV. of the *Aperçu Historique,* 1837); and it is nearly the whole of it stated in the 'Extrait' of Poinsot's memoir on Rotation, 1834.

209. Chasles' paper in the *Bulletin Univ. des Sciences* for 1830.—The corresponding theorem is here given for the finite motions (rotations and translations) of a solid body as follows: viz. if any finite displacement be given to a free solid body in space, there exists always in the body a certain indefinite line which after the displacement remains in its original situation. The theorem is deduced from a more general one relating to two similar bodies. It may be otherwise stated thus: viz., any motions may be represented by a translation and a rotation (the order of the two being indifferent); the rotation is given as regards its magnitude and the direction of its axis, but not as to the position of the axis (which may be any line in the given direction); the magnitude and direction of the translation depend on the assumed position of the axis of rotation; in particular this may be taken so that the translation shall be in the direction of the axis of rotation; the magnitude of the translation is then a minimum.

It may be noticed that a translation may be represented as a couple of rotations; that is, two equal and opposite rotations about parallel axes.

210. It is part of the general theorem that any number of rotations about axes passing through one and the same point may be compounded into a single rotation about an axis through that point; this is, in fact, the theory of the "Resultant Axis" developed in Euler's and Lexell's memoirs of 1775.

211. The following properties are also given, viz., considering two similar solid bodies (or in particular any two positions of a solid body) and joining the corresponding points, the lines which pass through one and the same point form a cone of the second order; and the points of either body form on this cone a curve of the third order (skew cubic). And, reciprocally, the lines, intersections of corresponding planes, which lie in one and the same plane envelope a conic, and such planes of either body envelope a developable surface, which is such that any one of these planes meets it in a conic {or, what is the same thing, the planes envelope a developable surface of the fourth order}.

And also, given in space two equal bodies situate in any manner in respect to each other, then joining the points of the first body to the homologous points of the second body, the middle points of these lines form a body capable of an infinitesimal motion, each point of it along the line on which the same is situate.

212. The entire theory, as well of the infinitesimal as of the finite motions of a solid body, is carefully and successfully treated in Rodrigues' memoir "Des lois géométriques &c." (1840). It may be remarked that for the purpose of compounding together any rotations and translations, each rotation may be resolved into a rotation about a parallel axis and a couple of rotations, that is, a translation; the rotations are thus converted into rotations about axes through one and the same point, and these give rise to a single resultant rotation given as to its magnitude and the direction of the axis, but not as to the position of the axis (which is an arbitrary line in the given direction); the translations are then compounded together into a single translation, and finally the position of the axis of rotation is so determined that the translation shall be in the direction of this axis; the entire system is thus compounded (in accordance with Chasles' theorem) into a rotation and a translation in the direction of the axis of the rotation. The problem of the composition depends therefore on the composition of rotations about axes through one and the same point; that is, upon Euler's and Lexell's theory of the resultant axis. But, as already noticed, the analytical theory of the resultant axis was perfected by Rodrigues in the present memoir (see *ante*, 'Transformation of Coordinates,' Nos. 139—141, as to this memoir and the quaternion representation of the formulæ contained in it.

213. It was remarked in Poinsot's memoir of 1834 that every continuous motion of a solid body about a fixed point is the motion of a cone fixed in the body rolling upon another cone fixed in space. The corresponding theorem for the motion of a solid body in space is given

Cayley, "On the Geometrical Representation &c." (1846): viz. premising that a skew surface is said to be "deformed" if, considering the elements between consecutive generating lines as rigid, these elements be made in any manner to turn round and slide along the successive generating lines:—and that two skew surfaces can be made to roll and slide one upon the other, only if the one is a deformation of the other—and that then the rolling and sliding motions are perfectly determined—and that such a motion may be said to be a "gliding" motion: the theorem is that any motion whatever of a solid body in space may be represented as the gliding motion of one skew surface upon another skew surface of which it is the deformation.

214. The same paper contains the enunciation and analytical proof of the following theorem supplementary to that of Poinsot just referred to, viz. that when the motion of a solid body round a fixed point is represented as the rolling motion of one cone on another, then "the angular velocity round the line of contact (the instantaneous axis) is to the angular velocity of this line as the difference of the curvatures of the two cones at any point in this line is to the reciprocal of the distance of the point from the vertex."

215. There are a great number of theorems relating to the composition of forces and force-couples, which consequently relate also to infinitesimal rotations and translations. See, for instance, Chasles, "Théorèmes généraux &c." (1847), Möbius, "Lehrbuch der Statik" (1837), Steichen's Memoirs of 1853 and 1854, &c. Arising out of some theorems of Möbius in the "Statik," we have Sylvester's theory of the involution of six lines: viz. six lines (given in position) may be such that properly selected forces along them (or if we please, properly selected infinitesimal rotations round them) will counterbalance each other; or, what is the same thing, the six lines may be such that a system of forces, although satisfying for each of the six lines the condition moment $= 0$, will not of necessity be in equilibrium; such six lines are said to be in involution, and the geometrical theory is a very extensive and interesting one.

Miscellaneous Problems. Article Nos. 216 to 223.

216. As under the foregoing head, "Rotation round a fixed point," I have considered only the case of a body not acted upon by any forces, the case where the body is acted upon by any forces comes under the present head. The forces, whatever they are, may be considered as disturbing forces, and the problem be treated by the method of the variation of the elements; this is at any rate a *separate* part of the theory of rotation round a fixed point, and I have found it convenient to include it under the present head; the only case in which the forces have been treated as principal ones, seems to be that of a heavy body (a solid of revolution) rotating about a point not its centre of gravity. The case of a body suspended by a thread or resting on a plane comes under the present head, as also would (in some at least of the questions connected with it) the gyroscope. But none of these questions are here considered in any detail.

Rotation round a fixed point—Variation of the elements.

217. The forces acting on the body are treated as disturbing forces. Formulæ for the variations of the elements were first obtained by Poisson in the memoir "Sur la Variation des Constantes Arbitraires &c." (1809). The variations are expressed in terms

of the differential coefficients of the disturbing function in regard to the *elements*, and, as the author remarks, they are very similar in their form to, and can be rendered identical with, those for the variations of the elements in the theory of elliptic motion.

218. Cayley, "On the Rotation &c." (1846).—The latter part of the paper relates to the variations of the elements therein made use of, which are different from the ordinary ones.

219. Richelot, "Eine neue Lösung &c." (1851).—The form in which the integrals are obtained by means of a function V, satisfying a partial differential equation, leads at once to a canonical system for the variations of the elements; these formulæ are referred to in the introduction to the memoir, but they are not afterwards considered.

220. Cayley, "On the Rotation of a Solid Body" (1860).—The elements are those ordinarily made use of, with only a slight variation occasioned by the employment of the "departure" of the node. The course of the investigation consists in obtaining the variations in terms of the differential coefficients of the disturbing function in regard to *the coordinates* (formulæ which were thought interesting for their own sake), and in deducing therefrom those in terms of the differential coefficients in terms of the *elements*.

Other cases of the motion of a solid body.

221. In regard to a heavy solid of revolution rotating about a fixed point not its centre of gravity, we have

Poisson, "Mémoire sur un cas particulier &c." (1831), and the elaborate memoir

Lottner, "Reduction der Bewegung &c." (1855), where the solution is worked out by means of the Elliptic and Jacobian functions.

222. As regards a heavy solid suspended by a string,

Pagani, "Mémoire sur l'équilibre &c." (1839).

223. As regards the motion of a body resting on a fixed plane,

Cournot, "Mémoire sur le Mouvement &c." (1830 and 1832).

Puiseux, "Du Mouvement &c." (1848).

To these several others might doubtless be added; but the problems are so difficult, that the solutions cannot, it is probable, be obtained in any very complete form.

In conclusion, I can only regret that, notwithstanding the time which has elapsed since the present Report was undertaken, it is still—both as regards the omission of memoirs and works which should have been noticed, and the merely cursory examination of some of those which are mentioned—far from being as complete as I should have wished to make it. To have reproduced, to any much greater extent than has been done, the various mathematical investigations, would not have been proper, nor indeed practicable; at the same time, more especially as regards the subjects treated of in the second part of this Report, or say the kinematics and dynamics of a solid body, such a reproduction, incorporating and to some extent harmonizing the original researches, but without ignoring the points of view and methods of investigation of the several authors, would be a work which would well repay the labour of its accomplishment.

List of Memoirs and Works.

Ampère. Mémoire sur quelques propriétés nouvelles des axes permanens de rotation des corps, et des plans directeurs de ces axes. 4to. Paris, 1823.

———. Mémoire sur la Rotation. Mém. de l'Institut, t. v. 1834.

———. Mémoire sur les équations générales du mouvement. Liouv. t. I. pp. 211—228 (1836). (Written 1826.)

Anon. Note on the problem of falling bodies as affected by the earth's rotation. Camb. and Dubl. Math. Journ. t. III. pp. 206—208 (1848).

———. Remarks on the deviation of falling bodies to the east and south of the perpendicular, and corrections of a previously published paper on the same subject. Camb. and Dubl. Math. Journ. t. IV. pp. 96—105 (1849).

Baehr. Notice sur le mouvement du pendule ayant égard à la rotation de la terre. 4to. Middelbourg, 1853.

Bertrand. Mémoire sur l'intégration des équations différentielles de la Mécanique. Liouv. t. XVII. pp. 393—436 (1852).

———. Note sur le Gyroscope de M. Foucault. Liouv. t. I. 2 sér. (1856), pp. 379—382.

———. Mémoire sur quelques-unes des formes les plus simples que puissent présenter les équations différentielles du mouvement d'un point matériel. Liouv. t. II. 2 sér. (1857), pp. 113—140.

Bessel. Analytische Auflösung der Keplerschen Aufgabe. Berl. Abh. 1816—17, pp. 49—55. (Read July 1818.)

———. Ueber die Entwickelung der Functionen zweier Winkeln u und u' in Reihen welche nach der Cosinussen und Sinussen der Vielfachen von u und u' fortgehen. Berl. Abh. 1820—21, pp. 56—60. (Read June 1821.)

———. Untersuchung des Theils der planetarischen Störungen welcher aus der Bewegung der Sonne entsteht. Berl. Abh. 1824, pp. 1—52.

Binet. Mémoire sur la théorie des axes conjugués et des momens d'inertie des corps. Journ. Polyt. t. IX. (cah. 16), pp. 41—67 (1813). (Read May 1811.)

———. Note sur le mouvement du pendule simple en ayant égard à l'influence de la rotation diurne de la terre. Comptes Rendus, t. XXXII. (1851), pp. 157—160 and 197—205.

Bonnet. Note sur un théorème de Mécanique. Liouv. t. IX. p. 113 (1844), and Note IV. of t. II. of the last edition of the Méc. Anal., pp. 329—331 (1855).

Booth. Theory of Elliptic Integrals. 8vo. Lond. 1851.

Bour. Mémoire sur le problème des trois corps. Journ. École Polyt. t. XXI. (cah. 36), pp. 35—58 (1856).

Bravais. Mémoire sur l'influence qu'exerce la rotation de la terre sur le mouvement d'un pendule à oscillations coniques. Liouv. t. XIX. pp. 1—50 (1854).

———. Note sur une formule de Lagrange relative au mouvement pendulaire. Note VII. of t. II. of the last edition of the Méc. Anal., pp. 352—355 (1855).

Briot. Thèse sur le mouvement d'un corps solide autour d'un point fixe. Liouv. t. VII. pp. 70—84 (1842).

Cauchy. Sur les momens d'inertie. Exer. de Math. t. I. pp. 93—103 (1827).

———. Résumé d'un mémoire sur la Mécanique Céleste et sur un nouveau calcul appelé des limites. (Read at Turin Oct. 1831.) Exer. d'Anal. &c. t. II. pp. 41—109 (1841).

Cayley. On certain expansions in multiple sines and cosines. Camb. Math. Journ. t. III. pp. 162—167 (1842), [4].

———. On the motion of rotation of a solid body. Camb. Math. Journ. t. III. pp. 224—232 (1842), [5].

———. On certain results relating to quaternions. Phil. Mag. t. XXVI. (1845), p. 141, [20].

———. On the geometrical representation of the motion of a solid body. Camb. and Dubl. Math. Journ. t. I. pp. 164—167 (1846), [36].

———. On the rotation of a solid body round a fixed point. Camb. and Dubl. Math. Journ. t. I. pp. 167—173 and 264—274 (1846), [37].

———. Note on a geometrical theorem in Prof. Thomson's memoir on the principal axes of a solid body. Camb. and Dubl. Math. Journ. t. I. pp. 207, 208 (1846), [38].

———. On the application of quaternions to the theory of Rotation. Phil. Mag. t. XXXIII. (1848), p. 196, [68].

———. Note on the motion of rotation of a solid of revolution. Camb. and Dubl. Math. Journ. t. IV. pp. 268—271 (1849), [78].

———. Sur les déterminants gauches. Crelle, t. XXXVIII. (1849), pp. 93—96, [69].

———. Note on the theory of Elliptic Motion. Phil. Mag. t. XI. (1856), pp. 425—428, [199].

———. A demonstration of Sir W. R. Hamilton's theorem of the Isochronism of the Circular Hodograph. Phil. Mag. t. XIII. (1857), p. 427, [209].

———. Report on the recent progress of Theoretical Dynamics. Rep. Brit. Assoc. for 1857, pp. 1—42, [195].

Cayley. On Lagrange's solution of the problem of two fixed Centres. Quart. Math. Journ. t. II. pp. 76—82 (1858), [182].

———. Note on the expansion of the true anomaly. Quart. Math. Journ. t. II. pp. 229—232 (1858), [191].

———. Tables in the theory of Elliptic Motion. Mem. R. Astr. Soc. t. XXIX. (1860), pp. 191—306, [216].

———. A Memoir on the problem of the rotation of a solid body. Mem. R. Astr. Soc. t. XXIX. (1860), pp. 307—342, [217].

———. On Lambert's theorem for Elliptic Motion. Monthly Not. R. Astr. Soc. t. XXII. pp. 238—242, (1861), [222].

———. Note on a theorem of Jacobi's in relation to the problem of three bodies. Monthly Not. R. Astr. Soc. t. XI. pp. 76—79 (1861), [220].

Chasles. Note sur les propriétés générales du système de deux corps semblables entr'eux et placés d'une manière quelconque dans l'espace, et sur le déplacement fini ou infiniment petit d'un corps solide libre. (Read Feb. 1831.) Bull. Univ. des Sciences (Férussac), t. XIV. pp. 321—326.

———. Théorèmes généraux sur les systèmes de forces et leurs moments. Liouv. t. XII. pp. 213—224 (1847).

Clairaut. Théorie de la Lune déduite du seul principe de l'attraction réciproquement proportionnelle aux carrés des distances. 4to. St Pét. 1752, and Paris, 1765.

Cohen. On the Differential Coefficients and Determinants of Lines, and their Application to Theoretical Mechanics. Phil. Trans. t. 152 (1862), pp. 469—510.

Cotes. Harmonia mensurarum sive analysis et synthesis per rationum et angulorum mensuras promotæ; accedunt alia opuscula mathematica. 4to. Camb. 1722.

Cournot. Mémoire sur le mouvement d'un corps rigide soutenu par un plan fixe. Crelle, t. V. pp. 133—162 and 223—249 (1830); Suite, t. VIII. pp. 1—12 (1832).

Creedy. General and practical solution of Kepler's Problem. Quart. Math. Journ. t. I. pp. 259—271 (1855).

D'Alembert. Traité de Dynamique. Paris, 1743.

———. Recherches sur la précession des équinoxes et sur la nutation de l'axe de la terre. Mém. de Berl. (1749).

Desboves. Thèse sur le mouvement d'un point matériel attiré en raison inverse du carré des distances vers deux centres fixes. Liouv. t. XIII. pp. 369—396 (1848).

Donkin. On an application of the calculus of operations to the transformation of trigonometric series. Quart. Math. Journ. t. II. pp. 1—15 (1858).

———. On a class of Differential Equations, including those which occur in Dynamical Problems. Part I. Phil. Trans. t. CXLIV. (1854), pp. 71—113; Part II. t. CXLV. (1855), pp. 299—358.

Droop. On the Isochronism of the Circular Hodograph. Quart. Math. Journ. t. I. (1856), pp. 374—378.

Dumas. Ueber die Bewegung des Raumpendels mit Rucksicht auf die Rotation der Erde. Crelle, t. L. pp. 52—78 and 126—185 (1855).

Durège. Theorie der elliptischen Functionen. 8vo. Leipzig, 1861. (§ XX. reproduces some results on the spherical pendulum obtained in an unpublished memoir of 1849.)

Euler. Determinatio Orbitæ Cometæ anni 1742. Misc. Berl. t. VII. (1743), p. 1.

———. Theoria motuum planetarum et cometarum. 4to. Berl. 1744.

———. De motu corporis ad duo virium centra attracti. Nov. Comm. Petrop. t. X. for 1764, pub. 1766, pp. 207—242.

———. Problème : un corps étant attiré en raison réciproque carrée des distances vers deux points fixes donnés, trouver les cas où la courbe décrite par ce corps sera algébrique. Mém. de Berl. for 1760, pub. 1767, pp. 228—249.

———. De motu corporis ad duo centra virium fixa attracti. Nov. Comm. Petrop. t. XI. for 1765, pub. 1767, pp. 152—184.

———. Considerationes de motu corporum cœlestium. Nov. Comm. Petrop. t. X. for 1764, pub. 1766, pp. 544—558.

———. De motu rectilineo trium corporum se mutuo attrahentium. Nov. Comm. Petrop. t. XI. for 1765, pub. 1767, pp. 144—151.

———. De motu trium corporum se mutuo attrahentium super eâdem lineâ rectâ. Nov. Acta Petrop. t. III. (1776), pp. 126—141.

———. Problema algebraicum ob affectiones prorsus singulares memorabile. Nov. Comm. Petrop. t. XV. (1770), p. 75; Comm. Arith. Coll. t. I. pp. 427—443.

———. Formulæ generales pro translatione quâcunque corporum rigidorum. Nov. Comm. Petrop. t. XX. 1775, pp. 189—207.

———. Nova methodus motum corporum rigidorum determinandi. Nov. Comm. Petrop. t. XX. (1775), p. 208.

———. Recherches sur la précession des équinoxes et sur la nutation de l'axe de la terre. Mém. de Berl. t. V. for 1749, pub. 1751, pp. 326—338. (Euler mentions, t. VI., that this was written after he had seen D'Alembert's memoir.)

———. Découverte d'un nouveau principe de Mécanique. Mém. de Berl. t. VI. for 1750, pub. 1752, pp. 185—217.

———. Recherches sur la connaissance mécanique des corps. Mém. de Berl. for 1758, pub. 1767, pp. 132—153.

———. Du mouvement de rotation des corps solides autour d'une axe variable. Mém. de Berl. for 1758, pub. 1765, pp. 154—193.

Euler. Du mouvement d'un corps solide lorsqu'il tourne autour d'une axe mobile. Mém. de Berl. for 1760, pub. 1767, pp. 176—227.

———. Theoria motus corporum solidorum. 4to. Rostock, 1765.

Foucault. Démonstration physique du mouvement de rotation de la terre au moyen du pendule. Comptes Rendus, t. XXXII. (1851), pp. 135—138.

Gauss. Fundamental-Gleichungen für die Bewegung schwerer Körper auf der rotirenden Erde, 1804.

———. Theoria motus corporum cœlestium. 4to. Hamb. 1809. [*Werke*, t. VII. Gotha, 1871.]

Greatheed. Investigation of the general term of the expansion of the true anomaly in terms of the mean. Camb. Math. Journ. t. I. pp. 228—232 (1838).

Gudermann. De pendulis sphæricis et de curvis quæ ab ipsis describuntur sphæricis. Crelle, t. XXXVIII. pp. 185—215 (1849).

Hamilton, Sir W. R. A theorem of anthodographic isochronism. Proc. R. Irish Acad. 1847, t. III. pp. 465—466.

———. Lectures on Quaternions. 8vo. Dublin, &c. (1853).

Hansen. Fundamenta Nova investigationis orbitæ veræ quam Luna perlustrat. 4to. Gothæ, 1838.

———. Ermittelung der absoluten Störungen in Ellipsen von beliebigen Excentricität und Neigung. Gotha, 1843, pp. 1—167.

———. Entwickelung des Products einer Potenz des Radius-Vectors mit dem Sinus oder Cosinus eines Vielfachen der wahren Anomalie in Reihen die nach den Sinussen oder Cosinussen der Vielfachen der wahren excentrischen oder mittleren Anomalie fortschreiten. Abh. d. K. Sächs. Ges. zu Leipzig, t. II. pp. 183—281 (1853).

———. Entwickelung der negativen und ungeraden Potenzen der Quadratwurzel der Function $r^2 + r'^2 - 2rr' (\cos U \cos U' + \sin U \sin U' \cos J)$. Abh. d. K. Sächs. Ges. zu Leipzig, pp. 286—376 (1854).

———. Theorie der Pendelbewegung. 4to. Dantzig, 1856.

Haton de la Goupillière. Mémoire sur une théorie nouvelle de la géométrie des masses. Journ. École Polyt. t. XXI. (cah. 37), 1858, 1[r] Mémoire, pp. 35—72; 2[d] Mémoire, pp. 73—96.

Haughton. On the rotation of a solid body round a fixed point, being an account of the late Professor **Maccullagh's** lectures on that subject, Hilary Term, 1844, in Trinity College, Dublin; compiled by the Rev. S. Haughton. Trans. R. Irish Acad. t. XXII. (1849), pp. 1—18.

Jacobi. Euleri formulæ de transformatione coordinatarum. Crelle, t. II. pp. 188, 189 (1827).

Jacobi. Zur Theorie der Variations-Rechnung und der Differential-Gleichungen. Crelle, t. XVII. (1837), pp. 68—82.

———. Formulæ transformationis integralium definitorum. Crelle, t. XV. pp. 1—26 (1836).

———. De motu puncti singularis. Crelle, t. XXIV. pp. 5—27 (1842).

———. Elimination des nœuds dans le problème des trois corps. Crelle, t. XXVI. (1843), pp. 115—131.

———. Theoria novi multiplicatoris systemati æquationum differentialium vulgarium applicandi (§ 26, two centres). Crelle, t. XXIX. pp. 333—337 (1845).

———. Sur la rotation d'un corps. Extrait d'une lettre adressée à l'Académie des Sciences. Comptes Rendus, t. XXIX. p. 97; and Liouv. t. XIV. pp. 337—344 (1849).

———. ———. (With addition containing the demonstration of the formulæ.) Crelle, t. XXXIX. pp. 293—350 (1850).

———. Nova methodus æquationes differentiales partiales primi ordinis inter numerum variabilium quemcunque propositas integrandi (posthumous, edited by **A. Clebsch**). Crelle, t. LX. pp. 1—181 (1862).

Lagrange. Mécanique Analytique. 1st ed. 1788; 2nd ed. t. I. 1811; t. II. 1815; 3rd ed. 1855.

———. Sur une manière particulière d'exprimer le temps dans les sections coniques décrites par des forces tendantes au foyer et réciproquement proportionnelles aux carrés des distances. Mém. de Berlin for 1778; and Note V. of t. II. of the 3rd edition of the Méc. Anal. pp. 332—349.

———. Recherchès sur le mouvement d'un corps qui est attiré vers deux centres fixes. Premier Mémoire, où l'on suppose que l'attraction est en raison inverse des carrés des distances. Anc. Mém. de Turin, t. IV. (1766—1769), pp. 118—215.

———. ———. Second Mémoire, où l'on applique la méthode précédente à différentes hypothèses d'attraction. Anc. Mém. de Turin, t. IV. (1766—1769), pp. 215—243.

———. Nouvelle solution du problème du mouvement de rotation d'un corps. Mém. de Berl. for 1773.

Lambert. Insigniores Orbitæ Cometarum Proprietates. 8vo. Aug. 1765.

Laplace. Mécanique Céleste, t. I. 1799; t. II. 1799; t. III. 1802; t. IV. 1805; t. V. 1823.

———. Mémoire sur le développement de l'anomalie vraie et du rayon vecteur elliptique en séries ordonnées suivant les puissances de l'excentricité. Mém. de l'Inst. t. VI. (1823), pp. 63—80.

Lefort. Expressions numériques des intégrales définies qui se présentent quand on cherche les termes généraux des développements des coordonnés d'un planète dans son mouvement elliptique. Liouv. t. XI. pp. 142—152 (1846).

Legendre. Exercices de Calcul Intégrale, t. II. (containing the dynamical applications) 1817.

———. Traité des Fonctions Elliptiques, t. I. (1825) (but the dynamical applications are for the most part reproduced from the *Exercices*).

Leverrier. Annales de l'Observatoire de Paris, t. I. (1855).

Lexell. Theoremata nonnulla generalia de translatione corporum rigidorum. Nov. Comm. Petrop. t. XX. (1775), pp. 239—270.

Liouville. Sur l'intégrale $\int_0^\pi \cos i\,(u - x \sin u)\, du$. Liouv. t. VI. pp. 36, 37 (1841).

———. Extrait d'un mémoire sur un cas particulier du problème de trois corps. Liouv. t. VII. pp. 110—113 (1842).

———. Sur quelques cas particuliers où les équations du mouvement d'un point matériel peuvent s'intégrer. 3 Memoirs. Liouv. t. XI. pp. 345—379 (1846), t. XII. pp. 410—44 (1847), and t. XIV. pp. 257—300 (1849).

———. Mémoire sur un cas particulier du problème de trois corps. Conn. des Temps for 1845, and Liouv. t. I. (2 sér.), pp. 248—256 (1842).

———. Note ajoutée au mémoire de M. Serret (two revolving centres). Liouv. t. XIII. pp. 34—37 (1848).

Lottner. Reduction der Bewegung eines schweren um einen festen Punct rotirenden Revolutions-Körpers auf die elliptischen Transcendenten. Crelle, t. L. pp. 111—125 (1855).

———. Zur Theorie des Foucaultschen Pendelversuchs. Crelle, t. LII. pp. 52—58 (1856).

Maccullagh, *see* **Haughton.**

Möbius. Ueber die Zusammensetzung unendlich kleiner Drehungen. Crelle, t. XVIII. pp. 189—212 (1838).

———. Der barycentrische Calcul. 8vo. Leipzig, 1827.

———. Lehrbuch der Statik. 8vo. Leipzig, 1837.

———. Die Elemente der Mechanik des Himmels. 8vo. Leipzig, 1843.

Newton (Sir Isaac). Philosophiæ Naturalis Principia Mathematica. 4to. London, 1687.

Pagani. Démonstration d'un théorème de Lambert. Crelle, t. XV. pp. 350—352 (1834).

———. Mémoire sur l'équilibre d'un corps solide suspendu à un cordon flexible. Crelle, t. XXIX. pp. 185—204 (1839).

Painvin. Recherche du dernier multiplicateur pour deux formes spéciales et remarquables des équations différentielles du problème de trois corps. Liouv. t. XIX. pp. 88—111 (1854).

Poinsot. Mémoire sur la rotation. Extrait, 8vo. Paris, 1834.

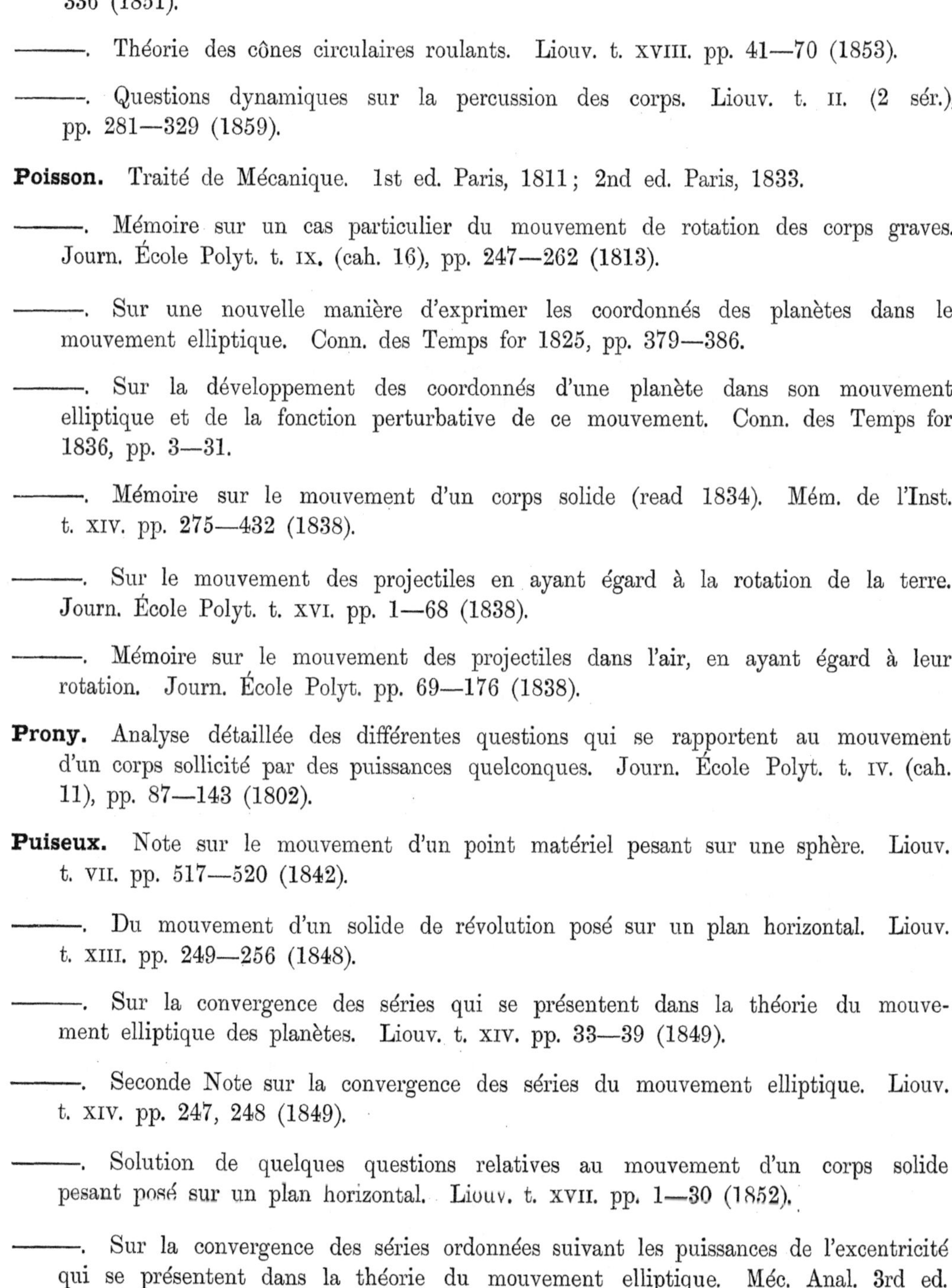

Poinsot. Théorie nouvelle de la rotation des corps. Liouv. t. XVI. pp. 9—130 and 289—336 (1851).

———. Théorie des cônes circulaires roulants. Liouv. t. XVIII. pp. 41—70 (1853).

———. Questions dynamiques sur la percussion des corps. Liouv. t. II. (2 sér.), pp. 281—329 (1859).

Poisson. Traité de Mécanique. 1st ed. Paris, 1811; 2nd ed. Paris, 1833.

———. Mémoire sur un cas particulier du mouvement de rotation des corps graves. Journ. École Polyt. t. IX. (cah. 16), pp. 247—262 (1813).

———. Sur une nouvelle manière d'exprimer les coordonnés des planètes dans le mouvement elliptique. Conn. des Temps for 1825, pp. 379—386.

———. Sur la développement des coordonnés d'une planète dans son mouvement elliptique et de la fonction perturbative de ce mouvement. Conn. des Temps for 1836, pp. 3—31.

———. Mémoire sur le mouvement d'un corps solide (read 1834). Mém. de l'Inst. t. XIV. pp. 275—432 (1838).

———. Sur le mouvement des projectiles en ayant égard à la rotation de la terre. Journ. École Polyt. t. XVI. pp. 1—68 (1838).

———. Mémoire sur le mouvement des projectiles dans l'air, en ayant égard à leur rotation. Journ. École Polyt. pp. 69—176 (1838).

Prony. Analyse détaillée des différentes questions qui se rapportent au mouvement d'un corps sollicité par des puissances quelconques. Journ. École Polyt. t. IV. (cah. 11), pp. 87—143 (1802).

Puiseux. Note sur le mouvement d'un point matériel pesant sur une sphère. Liouv. t. VII. pp. 517—520 (1842).

———. Du mouvement d'un solide de révolution posé sur un plan horizontal. Liouv. t. XIII. pp. 249—256 (1848).

———. Sur la convergence des séries qui se présentent dans la théorie du mouvement elliptique des planètes. Liouv. t. XIV. pp. 33—39 (1849).

———. Seconde Note sur la convergence des séries du mouvement elliptique. Liouv. t. XIV. pp. 247, 248 (1849).

———. Solution de quelques questions relatives au mouvement d'un corps solide pesant posé sur un plan horizontal. Liouv. t. XVII. pp. 1—30 (1852).

———. Sur la convergence des séries ordonnées suivant les puissances de l'excentricité qui se présentent dans la théorie du mouvement elliptique. Méc. Anal. 3rd ed. t. II. (Note I.), pp. 313—317 (1855).

Quet. Des mouvements relatifs en général, et spécialement des mouvements relatifs sur la terre. Liouv. t. XVIII. pp. 213—298 (1853).

Richelot. Bemerkung über einen Fall der Bewegung eines systems von materiellen Puncten. Crelle, t. XL. pp. 178—182 (1850).

———. Auszug eines Schreibens am Herrn Prof. Jacobi. Crelle, t. XLII. pp. 32—34 (1851).

———. Bemerkungen zur Theorie des Raumpendels. Crelle, t. XLII. pp. 233—238 (1853).

———. Eine neue Lösung des Problems der Rotation eines festen Körpers um einen Punct. (Auszug einer zu Berlin, 1851, erschienenen Schrift.) Crelle, t. XLIV. pp. 60—65.

Rodrigues. Des lois géométriques qui régissent les déplacements d'un système solide dans l'espace, et de la variation des coordonnés provenants de ces déplacements considérés indépendamment des causes qui peuvent les produire. Liouv. t. III. pp. 380—440 (1840).

Rueb. Specimen inaugurale de motu gyratorio corporis rigidi nullâ vi acceleratrice sollicitati. Utrecht, 1834.

Schellbach. Mathematische Miszellen. No. I.—IV. Ueber die Bewegung eines Puncts der von einem festen Puncte angezogen wird. Crelle, t. XLV. pp. 255—266 (1853).

Schubert. Astronomie Théorique. 4to. St Pét. (1822).

Segner. Specimen theoriæ turbinum. 4to. Halæ, 1755.

Serret. Thèse sur le mouvement d'un point attiré par deux centres fixes en raison inverse du carré des distances. Liouv. t. XIII. pp. 17—33 (1848).

———. Sur la solution particulière que peut admettre le problème du mouvement d'un corps attiré vers deux centres fixes par des forces réciproquement proportionnelles aux carrés des distances. Note III. to t. II. of the 3rd ed. of the Méc. Anal. pp. 327—331 (1855).

Sohncke. Motus corporum cœlestium in medio resistente. Crelle, t. X. pp. 23—40 (1833).

Somoff. Démonstration des formules de M. Jacobi relatives à la théorie de la rotation d'un corps solide. Crelle, t. XLII. pp. 95—116 (1851).

Stader. De orbitis et motibus puncti cujusdam corporei circa centrum attractivum aliis quam Neutonianâ attractionis legibus sollicitati. Crelle, t. XLVI. pp. 262—327 (1852).

Steichen. De la propriété fondamentale du mouvement cycloidal, et de sa liaison avec le principe de la composition des mouvements de rotation autour des axes parallèles et des axes qui se coupent. Crelle, t. XLVI. pp. 24—42 (1853).

———. Mémoire sur la question réciproque du centre de percussion. Crelle, t. XLVIII. pp. 1—68 (1854).

Steichen. Mémoire de Mécanique relatif au mouvement de rotation et au mouvement naissant des corps solides. Crelle, t. XLIII. pp. 161—244 (1852), and Addition t. XLIV. pp. 43—46 (1853).

Sylvester. Sur l'involution des lignes droites dans l'espace considérées comme des axes de rotation. Comptes Rendus, t. LII. (1861), p. 741.

———. Note sur l'involution de six lignes dans l'espace. Comptes Rendus, t. LII. (1861), p. 815.

Thomson. On the principal axes of a solid body. Camb. and Dubl. Math. Journ. t. I. pp. 127—133 and 195—206 (1846).

Tissot. Thèse de Mécanique. Liouv. t. XVII. pp. 88—116 (1852).

Townsend. On the principal axes of a solid body, their moments of inertia and distribution in space. Camb. and Dubl. Math. Journ. t. I. pp. 209—227, and t. II. pp. 19—42, 140—171 and 241—251 (1846—47).

Walton. Instantaneous lines and planes in a body revolving round a fixed point. Camb. and Dubl. Math. Journ. t. VII. pp. 111—113 (1852).

Whewell. A Treatise on Dynamics. 1st ed. Cambridge, 1823.

Worms. The Earth and its Mechanism—being an account of the various proofs of the rotation of the Earth, with a description of the instruments used in the experimental demonstrations: to which is added the theory of Foucault's Pendulum and Gyroscope. 8vo. London, 1862.

299.

MATHEMATICS, RECENT TERMINOLOGY IN.

[From the *English Cyclopædia*, vol. v. (1860), pp. 534—542.]

THE terms intended to be explained in the present article relate to subjects distinct indeed, but intimately connected together, as well logically as historically. *Determinants* were devised as a means to the solution of a system of simple equations, but the principle of their construction is contained in the rule of signs which belongs to the theory of *arrangements* (or permutations): this theory has been studied, as well for its own sake, as in reference to the theory of equations, and in it originated the notion of a *group*, the most outlying term of those which are here explained. Moreover, in a system of simple equations, if the coefficients arranged in the natural square order are considered apart by themselves, this leads to the theory of *matrices*, a theory which indeed might have preceded that of determinants; the matrix, is, so to speak, the matter of a determinant; the rule of signs giving the form. But when the rule of signs is applied to other matter, this leads to the function called *permutants;* these include *commutants* and *intermutants*, and also *Pfaffians*, which however were not originally so arrived at. The theory of elimination (according to one of the two ways in which it may be treated) is essentially dependent upon systems of linear equations, and is thus also connected with determinants. And all, or nearly all, the before-mentioned theories have an application to the theory of rational and integral homogeneous functions, or, as they have been termed, *forms* or *quantics;* they are thus connected with the "Calculus of Forms," and with "Quantics"; the last-mentioned expression (used as a singular), has been defined to denote the entire subject of rational and integral functions, and of the equations and loci to which those give rise. The theory of rational and integral functions was first studied in a general manner in the question of *linear transformations*, and it was this question which led to the discovery of the functions, called originally *hyperdeterminants*, but afterwards *invariants*, and of the more general functions called *covariants:* the theory of covariants is indeed the part which has been chiefly attended to of the Calculus of forms, or of Quantics.

The following list of terms may be convenient:

Rule of signs.
Group.
Determinant.
  Minor,
  Symmetric, skew, skew symmetric.
Commutant.
Pfaffian.
Permutant.
Intermutant.
Cumulant.
Matrix.
Resultant.
Discriminant.
Plexus.
Rational and integral functions (notation and nomenclature of).
  Quantic, quadric, cubic, &c., binary, ternary, &c., facient, tantipartite, lineo-linear.
Emanant.
Linear transformations.
  Modulus of transformation, unimodular.
Hyperdeterminant.
Invariant.
Covariant.
  Contravariant, peninvariant, seminvariant, quadrinvariant, quadricovariant, &c., catalecticant, canonisant.
Canonical form.
Bezoutic matrix, &c.
Tactinvariant, reciprocant.
Functional determinant, Jacobian, Hessian.
Concomitant, cogredient, contragredient.
Combinant.

Rule of Signs.—Any arrangement of a series of terms may be derived (and that in a variety of ways) from any other arrangement by successive interchanges of two terms; but in whatever way this is done, the number of interchanges will be constantly even or constantly odd; and the two arrangements are said to be of the same sign or of contrary signs accordingly. In particular, if any arrangement is selected as the primitive arrangement, and taken to be positive, then any other arrangement will be positive or negative according as it is derivable from the primitive arrangement by an even or an odd number of interchanges. The definition leads to the following theorem: any arrangement is positive or negative, according as the total number of times in which the several elements respectively precede (mediately or immediately) elements posterior to them in the primitive arrangement, is even or odd: it may be added, that the positive and negative arrangements are equal in number. Thus in the case of three terms, the primitive arrangement being 123; the positive arrangements are 123, 231, 312, the negative arrangements, 132, 213, 321: in the case of four terms, the primitive arrangement being 1234, the arrangements 1234, 2341, 3412, 4123 are respectively positive, negative, positive, negative; there are in all twelve positive and twelve negative arrangements.

GROUP.—The term was originally used as applied to substitutions only, but the more general use of the term is as follows: let θ be a symbol operating on any number of terms $x, y, z, ..$ and producing as the result of the operation the same number of new terms $X, Y, Z, ..$ (where $X, Y, Z ...$ may be each of them functions of all or any of the set, $x, y, z, ..$; if $X, Y, Z, ..$ are merely the terms, $x, y, x, ..$ in a different order, then θ is a substitution, which explains in what sense that term has just been used). Imagine a set of operative symbols $1, \theta, \phi, \chi ..$ (1, as an operative symbol denotes, of course, a symbol which leaves the operand unaltered) such that the result of the operation of *any* two symbols θ, ϕ (the same or different, and if different, then in either order) is identical with that of the operation of *some* symbol χ of the set; as thus, $\theta\phi(x, y, z..) = \theta(X, Y, Z, ..) = (X', Y', Z', ..) = \chi(x, y, z..)$, say, $\theta\phi = \chi$; then the symbols $1, \theta, \phi, \chi, ...$ form a *group*. It is to be remarked that 1 belongs to every group, and moreover, that if θ be any symbol of the group, then $\theta^2, \theta^3, \theta^4, ..$ belong to the group: the most simple form of a group (and when the number of terms is prime, the only form) is $1, \theta, \theta^2, ... \theta^{n-1} \{\theta^n = 1\}$. More generally, if there are n terms in the group, then every symbol θ of the group is an operation periodic of the order n (if not of an order a submultiple of n) and thus satisfies the symbolic equation $\theta^n = 1$. The symbols of the group are, so to speak, the symbolic n-th roots of unity, and as in the above-mentioned example, they may, whether n is prime or composite, form a system precisely analogous to that of the ordinary n-th roots of unity; but when n is composite, then upon two grounds this is not of necessity the case. 1°. The symbols of a group need not be convertible (thus $n = 6$, there is a group, $1, \beta, \beta^2, \alpha, \alpha\beta, \alpha\beta^2 [\alpha^2 = 1, \beta^3 = 1, \beta\alpha = \alpha\beta^2$ and $\therefore \beta^2\alpha = \alpha\beta$, this is in fact, the group of the substitutions of three things). 2°. There may be distinct n-th roots, thus $n = 4$, there is a group, $1, \alpha, \beta, \alpha\beta [\alpha^2 = 1, \beta^2 = 1, \alpha\beta = \beta\alpha]$, in which α, β are distinct square roots of (the symbolical) unity, and which is thus wholly different from the group, $1, \alpha, \alpha^2, \alpha^3 [\alpha^4 = 1]$.

The combination of a series of terms in the way of addition or subtraction, according to the rule of signs, gives rise to the class of functions called permutants, which include as a particular but the earliest discovered and most important case, the determinant:

DETERMINANT.—Imagine a square arrangement of terms, for example

$$\begin{vmatrix} a\ , & b\ , & c \\ a'\ , & b'\ , & c' \\ a'', & b'', & c'' \end{vmatrix}$$

and taking this as the primitive arrangement, permute in every possible way entire columns (or, what would give the same results, entire lines) and for each such arrangement form the product of the terms in the dexter diagonal (N.W. to S.E.) of the square, giving to such product the sign which belongs to the arrangement of the columns (or lines). The algebraical sum of these products is a determinant, and such determinant is or may be represented as above, by enclosing the terms within two vertical lines. Thus the developed value of the determinant in question is

$$ab'c'' - ab''c' + a'b''c + a''bc' - a''b'c - a'b''c.$$

The rule may be otherwise stated as follows: a determinant is the sum of a series of products each with its proper sign, such that in each product the factors are taken out of each line and out of each column, and if the factors are arranged according to the primitive arrangement of the columns in which they occur, then the sign is that corresponding to the resulting arrangement of the lines (or *vice versâ*): thus in the product $-ab''c'$, the factors a, b'', c' occur in the columns 1, 2, 3 (they are therefore arranged according to the primitive arrangement of the columns) and in the lines 1, 3, 2; such arrangement of the lines considered as derived from the primitive arrangement 1 2 3 is negative, and the product has therefore the sign $-$. A generalisation of this construction will be mentioned under the term commutant.

The word *resultant* was formerly used as synonymous with determinant, but it is now employed and is here explained in a more extended signification. The new synonym *eliminant* seems unnecessary.

A few of the numerous properties of determinants may be stated.

A determinant is a linear function (without constant term) of the terms in each of its columns, and also of the terms in each of its lines, or, more briefly expressed, it is a linear function of each column, and also of each line. Moreover, without altering the value of the determinant, the lines may be made columns, and the columns lines, and all the properties of the function exist equally with respect to the lines and to the columns. The absolute value of the determinant is not altered, but the sign is reversed, by an interchange of two columns, hence also if two columns become identical, the determinant vanishes. Moreover when the columns are permuted in any manner whatever, the absolute value is not altered, but the sign will be that corresponding to the arrangement of the columns. A determinant may be developed as a linear function of the terms in any line, thus

$$\begin{vmatrix} a, & b \\ a', & b' \end{vmatrix} = a \,|\, b' \,| - a' \,|\, b \,|,$$

$$\begin{vmatrix} a\,, & b\,, & c \\ a'\,, & b'\,, & c' \\ a'', & b'', & c'' \end{vmatrix} = a \begin{vmatrix} b'\,, & c' \\ b'', & c'' \end{vmatrix} + b \begin{vmatrix} c'\,, & a' \\ c'', & a'' \end{vmatrix} + c \begin{vmatrix} a'\,, & b' \\ a'', & b'' \end{vmatrix},$$

the signs being alternately positive and negative or else all positive, according as the number of columns is even or odd.

The square arrangement of terms out of which a determinant is formed, and generally any square or rectangular arrangement of terms, is called a matrix. Consider a determinant

$$\begin{vmatrix} a, & b, & c, & d \\ a', & b', & c', & d' \\ a'', & b'', & c'', & d'' \\ a''', & b''', & c''', & d''' \end{vmatrix}$$

and partitioning the lines in any manner, form with them the matrices

$$\begin{vmatrix} a, & b, & c, & d \\ a', & b', & c', & d' \end{vmatrix}, \quad \begin{vmatrix} a'', & b'', & c'', & d'' \\ a''', & b''', & c''', & d''' \end{vmatrix},$$

and out of these matrices, with complementary columns thereof, a sum of products

$$\Sigma \pm \begin{vmatrix} a, & b \\ a', & b' \end{vmatrix} \begin{vmatrix} c'', & d'' \\ c''', & d''' \end{vmatrix},$$

(the sign $\pm$ being that corresponding to the product $ab'c''d'''$ of the terms in the dexter diagonals of the factor determinants, considered as a term of the original determinant), the sum of all the products so obtained in the original determinant.

It has been mentioned that the determinant is a linear function of each column; hence if the terms of any column are $\rho\alpha$, $\rho'\alpha$,.. the determinant is equal to ρ times a determinant in which the corresponding column is a, a',... and similarly if the column is $a+b$, $a'+b'$,.. then the determinant is the sum of two other determinants in which the corresponding columns are a, a',... and b, b',... respectively. This property, in combination with some of those already mentioned, leads very simply to the rule for the multiplication of determinants; for example we have

$$\begin{vmatrix} \rho, & \sigma \\ \rho', & \sigma' \end{vmatrix} \begin{vmatrix} \alpha, & \beta \\ \alpha', & \beta' \end{vmatrix} = \begin{vmatrix} \rho\alpha+\sigma\beta, & \rho'\alpha+\sigma'\beta \\ \rho\alpha'+\sigma\beta', & \rho'\sigma'+\sigma'\beta' \end{vmatrix},$$

from which the law is obvious. The product might also be expressed, and although it appears less simple, there is an advantage in expressing it, in the form

$$\begin{vmatrix} \rho\alpha+\sigma\alpha', & \rho\beta+\sigma\beta' \\ \rho'\alpha+\sigma'\alpha', & \rho'\beta+\sigma'\beta' \end{vmatrix}.$$

If we omit simultaneously any line and any column of a determinant, and with the terms which are left form a determinant, the determinants so obtained are the *first minors* of the given determinant. A similar process, but omitting pairs, triads, &c. of lines and columns, gives the *second minors*, *third minors*, &c. of the given determinant. But the first minors are the most important, and are sometimes spoken of simply as the minors.

A determinant

$$\begin{vmatrix} a, & h, & g,.. \\ h, & b, & f \\ g, & f, & c \\ \vdots & & \end{vmatrix} \quad \text{or} \quad \begin{vmatrix} 11, & 12,.. \\ 21, & 22 \\ \vdots & \end{vmatrix},$$

where the corresponding terms on opposite sides of the dexter diagonal are equal to each other (say $rs = sr$) is said to be *symmetrical.*

But if the terms are equal in magnitude only, but have opposite signs (say $rs = -sr$, this relation not extending to the terms *in* the diagonal, for which $s = r$) the determinant is said to be *skew;* and if the relation extends to the case $s = r$, or what is the same thing, if the terms *in* the diagonal vanish, the determinant is said to be *skew symmetrical.* Skew determinants have an intimate connection with the functions called Pfaffians.

COMMUTANT.—The second rule for the construction of a determinant might have been thus stated, viz. for the determinant

$$\begin{vmatrix} 11, & 12,..\,1n \\ 21, & 22 \\ \vdots & \\ n1 & \end{vmatrix}$$

write down the expression

$$\begin{matrix} 1\,1 \\ 2\,2 \\ \vdots \\ n\,n \end{matrix}$$

and permute in every possible way the numbers in the first column, prefixing in each case the sign of the arrangement. Then reading off

$$\begin{matrix} \pm\, r\,1 \\ s\,2 \\ \vdots \\ z\,n. \end{matrix}$$

as meaning

$$\pm\, y1 \,.\, s2 \ldots zn$$

the sum of all the terms so obtained is in fact the determinant in question. The same result would be obtained by permuting the numbers in the second column instead of those in the first column. And moreover, if the numbers in both columns are permuted, the sign being the sign $\pm\,\pm$ compounded of the signs corresponding to the separate arrangements, the only difference is, that the determinant will be multiplied by the numerical factor $1 \,.\, 2 \,.\, 3 \ldots n$.

If instead of two we have three or more columns, the resulting function is a *commutant*. But a distinction is to be made according as the number of columns is even or odd. In the former case we may permute all but one of the columns, and it is indifferent which column is left unpermuted; and if all the columns are permuted, the effect is merely to introduce the numerical factor 1.2.3.... n. In the latter case, if all the columns are permuted, the result is zero, and it is therefore essential that one column should remain unpermuted; moreover, different results are obtained according to the column which is left unpermuted, and such column must therefore be distinguished; this is done by placing above it the mark †.

PFAFFIAN.—Suppose that the terms 12, 13, 21, &c., are such that $21 = -12$, and generally that $sr = -rs$, then the Pfaffians 1234, 123456, &c., are defined by means of the equations

$$1234 = 12\,.\,34 + 13\,.\,42 + 14\,.\,23,$$

$$123456 = 12\,.\,3456 + 13\,.\,4562 + 14\,.\,5623 + 15\,.\,6234 + 16\,.\,2345,$$

(where of course $3456 = 34\,.\,56 + 35\,.\,64 + 36\,.\,45$, and so for 4562, &c.)

and so on. The functions in question occur in the solution of an important problem (including that of partial differential equations of the first order and of any degree) known as Pfaff's problem, and were named accordingly.

It may be noticed that a skew symmetrical determinant of any odd order is equal to zero; but that a skew symmetrical determinant of any even order is the square of a Pfaffian, e.g. if $12 = -21$, &c., as above, then

$$\begin{vmatrix} 0, & 12, & 13, & 14 \\ 21, & 0, & 23, & 24 \\ 31, & 32, & 0, & 34 \\ 41, & 42, & 43, & 0 \end{vmatrix} = (12\,.\,34 + 13\,.\,42 + 14\,.\,23)^2.$$

PERMUTANT.—A very simple instance of a permutant is as follows, viz.: V_{123}, V_{213}, &c. being any quantities whatever, then the permutant $((V_{123}))$ denotes the sum

$$V_{123} + V_{231} + V_{312} - V_{132} - V_{213} - V_{321}$$

and in like manner for any number of permutable suffixes, or if instead of a single set of permutable suffixes we have two or more sets of such suffixes. It will be at once obvious how a permutant includes a determinant, commutant, or Pfaffian; thus, if V_{123} denotes $\alpha_1\beta_2\gamma_3$ and therefore $V_{213} = \alpha_2\beta_1\gamma_3$, &c., then we have a determinant, so if V_{1224} denotes $\alpha_{12}\,.\,\alpha_{34}$ where $\alpha_{21} = -\alpha_{12}$, we have a Pfaffian.

INTERMUTANT is a special form of permutant which need not be here further explained.

CUMULANT.—The name has been given to the function which is the numerator or denominator of a continued fraction. Such function may be exhibited (and indeed naturally presents itself) in the form of a determinant, thus the cumulant $(abcd)$ or numerator of the fractions $a+\frac{1}{b+}\frac{1}{c+}\frac{1}{d}$ is

$$\begin{vmatrix} a, & 1, & ., & . \\ -1, & b, & 1, & . \\ ., & -1, & c, & 1 \\ ., & ., & -1, & d \end{vmatrix},$$

and so for a greater number of terms. The developed expression is $abcd+ab+bc+bc+1$ which is formed from the product $abcd$ by successively omitting each product (cd, bc, ab), or set of products (cd, ab) of two consecutive letters; in like manner the cumulant $(abcde)$ is $abcde+abc+abe+a+c+e$.

MATRIX.—The term might be used to denote any arrangement of terms, but in a restricted sense it denotes a square or rectangular arrangement of terms, and it is thus employed in the theory of determinants.

To show further how the notion of a matrix is made use of, it may be remarked that a system of linear equations

$$\begin{aligned} \xi &= a\ x+b\ y+c\ z, \\ \eta &= a'\ x+b'\ y+c'z, \\ \zeta &= a''x+b''y+c''z \end{aligned}$$

is in the notation of matrices represented by

$$(\xi,\ \eta,\ \zeta)=\left(\begin{array}{ccc} a\ , & b\ , & c \\ a'\ , & b'\ , & c' \\ a'', & b'', & c'' \end{array}\right)(x,\ y,\ z).$$

The corresponding set of equations which give (x, y, z) in terms of (ξ, η, ζ) is represented by

$$(x,\ y,\ z)=\left(\left(\begin{array}{ccc} a\ , & b\ , & c \\ a'\ , & b'\ , & c' \\ a'', & c'', & c'' \end{array}\right)\right)^{-1}(\xi,\ \eta,\ \zeta),$$

and we have thus the definition of the *inverse* or *reciprocal* matrix: it follows from the theory of determinants that the terms of the reciprocal matrix are the first minor determinants formed out of the original matrix, each of them divided by the determinant formed out of the original matrix; but in writing down the expression some attention is required with respect to the arrangement and signs of the terms.

Similar considerations lead to the notion of *multiplying* or *compounding* together two or more matrices. As an instance of such composition, take

$$\left(\begin{array}{cc} \rho, & \sigma \\ \rho' & \sigma' \end{array}\right)\left(\begin{array}{cc} \alpha, & \beta \\ \alpha', & \beta' \end{array}\right)=\left(\begin{array}{cc} \rho\alpha+\sigma\alpha', & \rho\beta+\sigma'\beta' \\ \rho'\alpha+\sigma'\alpha', & \rho'\beta+\sigma'\beta \end{array}\right)$$

where it is to be observed that the lines of the first or *further* component matrix are compounded with the columns of the second or *nearer* component matrix to form the lines of the compound matrix. The words further, nearer, are used in reference to a set (x, y) which is, or may be considered to be, understood at the right of each side of the equation. A matrix may be compounded with itself once or oftener, giving rise to a positive power of such matrix; the notion of the negative powers is deducible from that of the inverse or reciprocal matrix, and the same process of generalisation as is employed for powers of a single quantity leads to the notion of the fractional powers of a matrix. As a definition of *addition*, matrices are added together by the addition of their corresponding terms, and as a particular case of the multiplication or composition of matrices we have the multiplication of a matrix by a single quantity, effected by multiplying by such quantity each term of the matrix; all these notions together lead to the notion of *functions* of a matrix.

As an instance of the employment of the notation of matrices for another purpose, take

$$\left(\begin{array}{ccc} a\ , & b\ , & c \\ a'\ , & b'\ , & c' \\ a'', & b'', & c'' \end{array}\right)(x,\ y,\ z)(\xi,\ \eta,\ \zeta)$$

used to denote the lineo-linear function

$$\begin{array}{r} (a\ x + b\ y + c\ z)\ \xi \\ (a'\ x + b'\ y + c'\ z)\ \eta \\ +\,(a''x + b''y + c''z)\ \zeta \end{array}$$

which includes

$$\left(\begin{array}{ccc} a, & h, & g \\ h, & b, & f \\ g, & f, & c \end{array}\right)(x,\ y,\ z)^2,$$

used to denote the quadric function

$$ax^2 + by^2 + cz^2 + 2fyz + 2gzx + 2hxy.$$

The last preceding notation is an instance of a *symmetrical* matrix: the terms *skew*, *skew symmetrical*, already explained with respect to determinants, apply also to matrices.

RESULTANT.—If there be a system of equations between the same number of unknown quantities (it is assumed that the several equations are of the form $U=0$, where U is a rational and integral homogeneous function), then the function of the coefficients which equated to zero expresses the result of the elimination of the unknown quantities from the several equations, or (what is the same thing) gives the condition for the existence of a set of values satisfying the equations simultaneously—is the *Resultant* of the equations, or of the functions which are thereby put equal to zero. In the case of two (non-homogeneous) equations involving a single unknown quantity, we may say more briefly that the resultant is the function which equated to zero gives the condition for the existence of a common root. In the particular case of a system of linear equations between as many unknown quantities, the resultant is the determinant formed with the coefficients of the equations.

DISCRIMINANT.—If in a system of equations the functions equated to zero are the derived functions of a single rational and integral homogeneous function with respect to each of the variables thereof, the resultant of the system is said to be the discriminant of the single function. The definition is easily made applicable to the case of a non-homogeneous function, the functions equated to zero are here the function itself and its derived functions with respect to each of the several variables. For a single function, it may be said that the discriminant is the function which equated to zero gives the condition for a pair of equal roots of the equation obtained by putting the function equal to zero.

To fix the precise value of the discriminant of a given function, it is assumed that the coefficient of some one selected term is $+1$. Thus, the discriminant of $ax^2 + 2\,bxy + cy^2$ is $ac - b^2$: that of

$$ax^3 + 3\,bx^2y + 3\,cxy^2 + dy^3 \text{ is } a^2d^2 - 6\,abcd + 4\,ac^3 + 4\,b^3d - 3\,b^2c^2.$$

In quadratic forms (in the theory of numbers) the expression $b^2 - ac$, which is the determinant $\begin{vmatrix} a, & b \\ b, & c \end{vmatrix}$ with the sign reversed, is called the determinant of the form $ax^2 + 2\,bxy + cy^2$. And in like manner for ternary quadratic forms, there is the same reversal of sign. It may be said as a convenient definition, that the determinant is the discriminant taken negatively.

PLEXUS.—It frequently happens, in problems of elimination and in other problems, that a given number of relations existing between a system of quantities can only be completely expressed by means of a greater number of equations. Thus, to take a very simple instance, if the unknown quantities x, y, are to be eliminated between the three equations $ax + by = 0$, $a'x + b'y = 0$, $a''x + b''y = 0$: this implies two relations between the coefficients a, b, a', b', a'', b''; but these relations cannot be completely expressed otherwise than by means of the three equations $ab' - a'b = 0$, $a'b'' - a''b' = 0$, $a''b - ab'' = 0$; for taking any two of these equations, e.g. the first and second, these would be satisfied by $a' = 0$, $b' = 0$, which however do not satisfy the third equation and are not a solution. Such a system of equations, or generally the system of equations required for the complete expression of the relations existing between a set of quantities (and which are *in general* more numerous than the relations themselves) is said to be a *Plexus*.

RATIONAL AND INTEGRAL FUNCTIONS (*Notation and Nomenclature of*).—A rational and integral homogeneous function, such as the function $ax^2 + 2bxy + cy^2$ is denoted by

$$(*)\,(x,\ y)^2$$

where the coefficients are only indicated by the asterisk, but are not expressed. A non-homogeneous rational and integral function is considered as derived from a homogeneous function by putting one of the variables thereof equal to unity, and is represented accordingly: thus $ax^2 + 2bx + c$ is denoted by

$$(*)(x,\ 1)^2.$$

But it is often proper to express the coefficients, and in regard to this the following distinction is made, namely

$$(a,\ b,\ c)\ (x,\ y)^2$$

denotes $ax^2 + 2bxy + cy^2$; and in like manner $(a, b, c, d)(x, y)^3$ denotes $ax^3 + 3bx^2y + 3cxy^2 + dy^3$, &c., the numerical coefficients being those of the successive powers of a binomial. But when such numerical coefficients are not to be inserted, this is denoted by an arrowhead, or other distinctive mark; thus $(a, b, c)\dagger (x, y)^2$ denotes $ax^2 + bxy + cy^2$. A rational and integral function of any order is termed a *quantic*, and a function of the orders two, three, four, five, &c., is termed a *quadric, cubic, quartic, quintic*, &c. respectively. The number of variables (the function being homogeneous) is denoted by the words *binary, ternary*, &c. As a correlative term to coefficients, the variables have been termed *facients.* A function which is linear in respect to several distinct sets of variables separately is said to be *tantipartite:* or, when there are two sets only, *lineo-linear.* Thus a determinant is a tantipartite function of the lines or of the columns; the function $axx' + bxy' + cx'y + dyy'$ is a lineo-linear function of (x, y) and (x', y'); a notation for it is

$$\left(\begin{array}{cc} a, & b \\ a', & b' \end{array}\right)(x,\ y)(x',\ y')$$

such as has been spoken of in regard to matrices.

EMANANT.—The development of an expression such as

$$(*)(\lambda x + \mu x',\ \lambda y + \mu y')^m$$

is naturally written under the form

$$\begin{array}{ll} (*)(x,\ y)^m & \lambda^m \\ +\frac{m}{1}(*)(x,\ y)^{m-2}(x',\ y') & \lambda^{m-1}\mu \\ \vdots & \\ +\quad (*)(x'\ y')^m & \mu^m, \end{array}$$

and the coefficients of the successive terms λ^m, $\lambda^{m-1}\mu$, &c. are said to be the emanants of the quantic $(*)(x, y)^m$. The coefficients of λ^m, or 0-th emanant, is the quantic itself, and the coefficient of μ^m, or ultimate emanant, is the quantic with (x', y') in the place of (x, y); but the intermediate emanants are functions of (x, y) and (x', y'), homogeneous in respect to the two sets separately. The coefficients may of course be expressed thus, the emanant (first emanant) of $(a, b, c)(x\ y)^2$ is $(a, b, c)(x, y)(x', y')$ which stands for $axx' + b(xy' + x'y) + cyy'$.

LINEAR TRANSFORMATIONS.—In this theory the variables of a function are supposed to be respectively linear functions of a new set of variables, so that the function is transformed into a similar function of these new variables, with of course altered values of the coefficients, and the question was to find the relations which existed between the original and new coefficients and the coefficients of the linear equations. The determinant composed of the coefficients of the linear equations is said to be the *modulus of transformation*, and when this determinant is unity the transformation is said to be *unimodular*. It was observed that a certain function of the coefficients, namely, the discriminant, possessed a remarkable property, found afterwards to belong to it as one of a class of functions called originally *hyperdeterminants*, but now *invariants*, and it was in this manner that the problem of linear transformation led to the general theory of covariants.

INVARIANTS.—An invariant is a function of the coefficients of a rational and integral homogeneous function or quantic, the characteristic property whereof is as follows: namely, if a linear transformation is effected on the quantic, then the new value of the invariant is to a factor *près* equal to the original value; the factor in question (or quotient of the two values) being a power of the modulus of transformation, and the two values being thus equal when the transformation is unimodular. The easiest example is afforded by the quadric function $(a, b, c)(x, y)^2$; effecting upon it a linear transformation, suppose that we have identically

$$(a, b, c)(\alpha x' + \beta y', \gamma x' + \delta y')^2 = (a', b', c')(x', y')^2$$

then it may be easily verified that $a'c' - b'^2 = (\alpha\delta - \beta\gamma)^2 (ac - b^2)$. The invariant $ac - b^2$ is however in this case nothing else than the discriminant; as another example take the quartic $(a, b, c, d, e)(x, y)^4$, for which $ae - 4bd + 3c^2$, $ace - ad^2 - b^2d + 2bcd - c^3$ are functions possessed of the like property of remaining to a factor *près* unaltered by the transformation, and are consequently invariants; it may be added that calling them I, J, respectively, the discriminant is here $= I^3 - 27J^2$, a rational and integral function of invariants of a lower degree.

COVARIANT.—Instead of a function of the coefficients only, we may have a function of the coefficients and variables, possessed of the like property of remaining unaltered to a factor *près* by the linear transformation: such function is termed a covariant. Thus, a covariant (the Hessian) of the quartic $(a, b, c, d, e)(x, y)^4$ is

$$(ac - b^2, 2ad - 2bc, ae + 2bd - 3c^2, 2be - 2cd, ce - d^2)(x, y)^4.$$

The quantic itself is one of its own covariants. The term covariant may be used in contradistinction to, or as including, invariant. The terms invariant, covariant, have been explained in reference to the simple case of a single quantic containing but one set of variables, but they apply equally to the case of a system of quantics, and to quantics which are homogeneous functions of two or more distinct sets of variables. There is one case which it is proper to mention; if in conjunction with a quantic $(*)(x, y, z, ..)^m$ we consider a linear function $\xi x + \eta y \ \zeta z + ..$, the invariants of the system are functions of the coefficients of the quantic, and of the coefficients $\xi, \eta, \zeta, ...$ of the linear function; and treating these as facients, the invariant is said to be a *contravariant* of the given quantic.

The foregoing definition gives the characteristic property of a covariant, but it does not directly show how the covariants of a given quantic are to be investigated. This is supplied as follows:—For any quantic with arbitrary coefficients, for example $(a, b, c, d)(x, y)^4$, there exist operators involving differentiations in respect to the coefficients, tantamount to the operators xd_y and yd_x in respect to the variables; thus the operator $ad_b + 2bd_c + 3cd_d$ is tantamount to yd_x, and $3bd_a + 2bd_c + cd_d$ is tantamount to xd_y. Or what is the same thing, denoting for shortness these operators by $\{yd_x\}$, $\{xd_y\}$ respectively, the quantic is reduced to zero by each of the operators $\{yd_x\} - yd_x$, $\{xd_y\} - xd_y$. Any function of the coefficients and the variables which, in like manner with the quantic itself, is reduced to zero by these two operators respectively, is said to be a *covariant;* or, if it contains the coefficients only, an *invariant* of the quantic. That the two definitions lead to the same result is of course a theorem to be proved.

The leading coefficient of a covariant, say the coefficient of x^m in any covariant of a binary quantic $(*)(x, y)^m$, possesses the property of being reduced to zero by the operator $\{yd_x\}$, and has been termed a *peninvariant* but a more appropriate term is *seminvariant.* An invariant is a function of a given degree in the coefficients, and a covariant is a function of a given degree in the coefficients and order in the variables, and they may be and are designated accordingly; thus, the above-mentioned invariants I, J of a binary quartic are called respectively the *quadrinvariant* and the *cubinvariant,* and the covariant of the same quartic is termed the *quadricovariant,* or if the distinction were required it would be termed the *quadricovariant quartic.* In these cases the designations are sufficient, but it is to be noticed that in general there is more than one invariant or covariant of the same degree or of the same degree and order, and that any such designation is only a generic, not a specific, name. An invariant or covariant may also be designated by a name referring to the mode of generation—for example, the discriminant. The name *catalecticant* denotes a certain invariant of a binary quantic of an even order: namely, it is a determinant, which, for the above-mentioned function, is

$$\begin{vmatrix} a, & b, & c \\ b, & c, & d \\ c, & d, & e \end{vmatrix}$$

(being in this particular case the cubinvariant), and the name *canonisant* denotes a certain covariant of a binary quantic of an odd order, namely, it is a determinant the terms whereof are linear functions of the coefficients, and which for the cubic $(a, b, c, d)\ (x, y)^3$ is

$$\begin{vmatrix} ax + by, & bx + cy \\ bx + cy, & cx + dy \end{vmatrix}$$

(being for the particular case the Hessian or quadricovariant).

Canonical Forms.—A binary quantic of an odd order $2m + 1$ admits of being expressed as a sum of $(m + 1)$ powers of linear functions, for example, the cubic $(a, b, c, d)(x, y)^3$ can be expressed in the form $(lx + my)^3 + (l'x + m'y)^3$—this is the *canonical form* of a binary function of an odd order. And there is in like manner a form (not admitting, however, of a simple definition) which is taken as the *canonical form* of a binary quantic of an even order. The catalecticant and the canonisant present themselves in the problem of the reduction of a binary quantic to the canonical form.

BEZOUTIC MATRIX.—If V, W are any two binary quantics of the same order m, and V', W' are what V, W become when the variables (x, y) of the two quantics are changed into (x', y'); then $(VW' - V'W) \div (xy' - x'y)$ is a rational and integral homogeneous function of the degree $m-1$ in each of the two sets $(x, y)(x', y')$, and the coefficients taken in their natural square arrangement constitute the Bezoutic matrix. The determinant formed out of this matrix is in fact the resultant of the two functions, or equated to zero it is the equation obtained by the elimination of the variables from the two equations $V=0$, $W=0$. If V, W are the derived functions of one and the same binary quantic of the order m, then the corresponding matrix, being of course of the order $m-2$, is the Bezoutoidal matrix, and the determinant is then the discriminant of the single quantic.

It would be too long to explain the allied terms Bezoutiant, Cobezoutiants, Bezoutoid, Cobezoutoids.

TACINVARIANT, RECIPROCANT.—A definition in the language of analytical geometry will be the most easily intelligible, and it can readily be converted into an analytical form and made applicable to any number of variables. The function of the coefficients which equated to zero expresses that the two curves $U=0$, $V=0$, touch each other, is an invariant, namely, it is the *tacinvariant* of the two functions U, V. And in particular, if, instead of the curve $V=0$, we have the line $\xi x + \eta y + \lambda z = 0$, then the function which equated to zero expresses that this line touches the curve $U=0$, is a contravariant, namely, it is the *reciprocant* of the function U.

FUNCTIONAL DETERMINANT, JACOBIAN, HESSIAN.—If V, W be quantics, then the determinant—

$$\begin{vmatrix} d_x V, & d_y V, \ldots \\ d_x W, & d_y W, \\ \vdots & \end{vmatrix}$$

is the functional determinant, or Jacobian, of the quantics V, $W\ldots$ And if V, $W, \ldots$ are the derived functions of $d_x U$, $d_y U, \ldots$ of one and the same quantic U, then the determinant in question is the *Hessian* of the single quantic: the Hessian is in fact to the Jacobian what the discriminant is to the resultant.

CONCOMITANT, COGREDIENT, CONTRAGREDIENT.—The theory of linear transformations has been considered from a different point of view; instead of the variables of a function being *put equal* to the linear functions of a new set of variables, they are considered as being *replaced* by a new set of variables, linear functions of the original variables. Two sets of variables may be so related that when the first set is thus replaced by a set of linear functions of themselves, the second set is also replaced by a set of linear functions of themselves, the coefficients of the two sets of linear functions being related together in a definite manner; this is *concomitance*, or rather it is (what is alone here spoken of) *simple concomitance*. The two most important kinds of concomitance are, 1st. *Congrediency*, that is, when the substitution on the second set of variables is identical with that upon the first set; 2nd. *Contragrediency*, that is, when the substitution on the second set of variables is the inverse or reciprocal one to that on the first set; it will make the notion of contragrediency clearer to remark that if the variables $x, y, ..$ and $\xi, \eta, ...$ are contragredients, then $x', y' ...$ (which are linear functions of $\xi, \eta ..$) are so related that $\xi' x' + \eta' y' + ..$ is $= \xi x + \eta y + ...$ It was from the consideration of contragredient variables that the notion of a contravariant was first derived, but as above remarked, the notion is really included in that of a covariant.

COMBINANT.—A combinant is a covariant (or invariant) of a set of quantics of the same order, which, besides being a covariant in the ordinary sense of the word, is, so to speak, a covariant *quoad* the system, that is, it remains to a factor *près* unaltered, when the quantics of the system are replaced by linear functions of themselves; the factor in question being a power of the determinant formed with the coefficients of the linear functions. For instance, if $U = (a, b, c)(x, y)^2$ and $U' = (a', b', c')(x, y)^2$, then $ac' - 2bb' + ca'$ is a function which, when for the original coefficients are substituted those of $\lambda U + \mu U'$, $\nu U + \rho U'$, is merely changed into $(\lambda\rho - \mu\nu)^2 (ac' - 2bb' + ca')$, and it is therefore a combinant. It would appear that the notion of a combinant might be extended to the case of a system of quantics not of the same order, and that the resultant of the system of quantics could be brought under the extended definition of a combinant, but this is a point which has not been considered.

The principal text-books on the foregoing subjects, are—on determinants:—Spottiswoode's *Elementary Theorems relating to Determinants*, 4to. London, 1851; Brioschi, *Teorica dei Determinanti*, 4to. Pavia, 1854, translated into French by Combescure and into German by Schellbach; Baltzer, *Ueber die Determinanten*, 8vo. Leipzig, 1857 (especially valuable for its references to the original sources). On elimination: Faà de Bruno, *Théorie générale de l'élimination*, 8vo. Paris, 1859. And extending to nearly all the subjects: Salmon, *Lessons introductory to the modern higher Algebra*, 8vo. Dublin, 1858. The memoirs on the different subjects are very numerous, and it was not thought expedient to give a list of them.

NOTES AND REFERENCES.

235. CONTAINS the demonstration and I think the first publication (1858) of Hermite's formula alluded to 135, for the reduction of an elliptic differential to the form

$$\frac{dr}{\sqrt{-J+zI-4z^3}},$$

which is in fact the Weierstrassian form, the theory of which has been of late so extensively developed.

241, 242. Figures of Poinsot's stellated Polyhedra are given, Fouché et Combe-rousse, *Traité de Geométrie Elementaire*, 8vo. Paris, 1866, and Dostor, "Propriétés générales des polyèdres reguliers etoilés," *Liouv.* t. V. (1879), pp. 209—226.

246. In connexion with this paper, On Contour and Slope Lines (1859), I refer to the earlier paper, Reech, "Démonstration d'une proprieté générale des surfaces fermées," *Jour. École Polyt.* Cah. 37 (1858), pp. 169—178: the contour lines are here considered with reference to a closed surface; the special object is the demonstration of the formula $I+S=M+2$, where I is the number of summits, S the number of imits (the letters I, S being thus interchanged) and M the number of cols. I refer also to the paper, Maxwell, "On Hills and Dales," *Phil. Mag.* t. XL. (1870), pp. 421—427, and *Works* (4to. Cambridge, 1890), t. II. No. XLIII.

259. It would have been proper to distinguish between *ab* and *ba*, and thus for instance to have presented the face-symbols of the polyhedron considered in the form *abcd*, *aefb*, *bfge*, *dcgh*, *adhe*, *ehgf* (read *ab*, *bc*, *cd*, *da*, &c.) so as to obtain therein each duad *once* in each of its two forms *ab* and *ba*, &c.: and the like as regards the vertex-symbols. And so as to the edge-symbols, instead of $ab=KL$, it would have been better to write $aKbL$, to be read, in like manner right-handedly, as a face- or vertex-symbol.

264. See the paper, Jenkins, "On Professor Cayley's Extension of Arbogast's Method of Derivations," *Amer. Math. Jour.* t. X. (1888), pp. 29—41.

268. In connexion herewith I refer to the memoirs Mc Clintock, "On the resolution of equations of the fifth degree," *Amer. Math. Jour.* t. VI. (1884), pp. 301—314, and "Analysis of Quintic Equations," *Amer. Math. Jour.* t. VIII. (1886), pp. 45—84: the author considers the *dexter resolvent* equation, which as he remarks is my equation

in ϕ, *ante*, p. 321; the *sinister resolvent* equation deduced from it by reversing the order of the coefficients (a, b, c, d, e, f), this is in fact the equation in χ obtained from my covariant equation, p. 323, for $\Phi = \frac{a}{U}(\phi x - \chi y)$ by writing therein $x = 0$ (viz. ψ, Mc Clintock $= \chi$, Cayley), but he afterwards modifies the form of this sinister resolvent; and a *central resolvent* equation for τ $\left(= \frac{\psi}{\phi},\ \text{Mc Clintock}\right) = \frac{\chi}{\phi}$, Cayley. We obtain this equation by writing in the equation for Φ, $x = \chi$, $y = \phi$, whence $\Phi = 0$, that is the equation becomes simply $AJ - 25D^2 = 0$, where in the covariants A, D, J the original variables x, y are to be replaced by χ, ϕ respectively: viz. the equation thus is

$$(a, b, c, d, e, f \between \chi, \phi)^5 . (J_0, J_1 \between \chi, \phi) - 25 \{(D_0, D_1, D_2, D_3 \between \chi, \phi)^3\}^2 = 0,$$

or what is the same thing

$$(a, b, c, d, e, f \between \tau, 1)^5 . (J_0, J_1 \between \tau, 1) - 25 \{(D_0, D_1, D_2, D_3 \between \tau, 1)^3\}^2 = 0,$$

which is the central resolvent equation for τ.

It is proper to remark that the foregoing expression $AJ - 25D^2$ for the last coefficient of the equation in Φ, which as appears p. 324 was not given in my original memoir, was in fact suggested to me by Mc Clintock's formula.

The equation in ϕ is

$$\phi^6 - 100 B\phi^4 + 2000 (3B^2 - 4H) \phi^2 - 800 A \sqrt{5Q'}\, \phi + AJ - 25D^2 = 0,$$

viz. getting rid of the radical, we have

$$\{\phi^6 - 100 B\phi^4 + 2000 (3B^2 - 4H) \phi^2 + AJ - 25D^2\}^2 - 320000 AQ'\phi^2 = 0,$$

a rational sextic equation in ϕ^2; and we infer that ϕ^2 is expressible as a rational function of τ. But the actual expression is obtained by Mc Clintock, and constitutes a very important and remarkable theorem; viz. we have

$$\frac{\phi^2}{500} = -\frac{(D_0, D_1, D_2, D_3 \between \tau, 1)^3}{(a, b, c, d, e, f \between \tau, 1)^5}, \quad = -\frac{(J_0, J_1 \between \tau, 1)}{25 (D_0, D_1, D_2, D_3 \between \tau, 1)^3},$$

the two expressions in τ being equal to each other in virtue of the foregoing equation in τ.

I verify this result in the case of the special quintic $(a, 0, 0, 0, e, f \between x, y)^5$. Writing with Mc Clintock $\phi^2 = w$, the equation in ϕ^2, or w, here is

$$(w^3 - 100\, aew^2 + 6000\, a^2e^2w + 40000\, a^3e^3)^2 - 320000\, (a^6f^4 + 256\, a^5e^5)\, w = 0;$$

and it is to be shown that, assuming as a definition of w the foregoing expressions in terms of τ, viz. for the form in question the expressions

$$\frac{w}{500} = \frac{ae^2\tau}{a\tau^5 + 5e\tau + f}, \quad = \frac{ae\tau - af}{25\tau},$$

(implying of course $(a\tau^5 + 5e\tau + f)(e\tau - f) - 25\, e^2\tau^2 = 0$ for the sextic equation in τ) the elimination of τ from these equations leads to the just mentioned sextic equation in w.

The two equations are

$$aw\tau^5 + 5e(w - 100\,ae)\tau + wf = 0, \quad \tau(w - 20\,ae) + 20\,af = 0,$$

or, if for greater convenience we write $20\,ae = \theta$, then

$$aw\tau^5 + 5e(w - \theta)\tau \qquad + wf = 0, \quad \tau(w - \theta) \qquad + 20\,af = 0,$$

we have

$$\tau = -\frac{20\,af}{w - \theta},$$

and thence

$$aw \cdot \frac{-3200000\,a^5 f^5}{(w - \theta)^5} - \frac{100\,aef(w - 5\theta)}{w - \theta} + wf = 0,$$

that is

$$w(w - \theta)^5 - 5\theta(w - 5\theta)(w - \theta)^4 - 3200000\,a^6 f^4 w = 0,$$

which should be identical with the before mentioned equation in w, that is with

$$(w^3 - 5\theta w^2 + 15\theta^2 w + 5\theta^3)^2 \qquad - (3200000 + 256\,\theta^5)\,w = 0,$$

and it is in fact at once seen that each of these equations is

$$\begin{aligned} 0 = \ & w^6 \\ & + w^5 \,.\, -10\,\theta \\ & + w^4 \,.\, \quad 55\,\theta^2 \\ & + w^3 \,.\, -140\,\theta^3 \\ & + w^2 \,.\, \quad 175\,\theta^4 \\ & + w \;\,.\, -106\,\theta^5 - 3200000\,a^6 f^4 \\ & + w^0 \,.\, \quad 25\,\theta^6\,; \end{aligned}$$

which completes the proof. The proof for the general form $(a, b, c, d, e, f \between x, y)^5$ is similar in principle, viz. treating for the moment ϕ^2 or w as a constant, we have in τ a quintic equation and a cubic equation, in each of which the coefficients contain w linearly; and the elimination of τ leads to the required sextic equation in w, but there would probably be considerable difficulty in effecting the calculations.

It thus appears that assuming the solution of the central resolvent equation for τ, $= \dfrac{\chi}{\phi}$, we also know ϕ: I recall that for the quintic equation whose roots are x_1, x_2, x_3, x_4, x_5, the significations of these quantities are

$$\tau = \frac{\chi}{\phi} = \frac{(12345) - (24135)}{12345 - 24135}, \quad \phi = 12345 - 24135,$$

where

$12345 = 12 + 23 + 34 + 45 + 51$, meaning thereby $x_1x_2 + x_2x_3 + x_3x_4 + x_4x_5 + x_5x_1$,
$(12345) = 123 + 234 + 345 + 451 + 512$ „ „ $x_1x_2x_3 + x_2x_3x_4 + x_3x_4x_5 + x_4x_5x_1 + x_5x_1x_2$,

viz. from the values just written down, we obtain by permutation of the roots the six values of each of the functions τ and ϕ.

By means of these data, or say of $t, = \tau + a^{-1}b$, and $v, = \dfrac{\phi}{10\sqrt{5}}$ (I write v instead of his $\sqrt{v}$) Mc Clintock completes in a very elegant manner the determination of the roots x_1, x_2, x_3, x_4, x_5 of the quintic equation: the solution contains also the coefficients $\gamma, \delta, \epsilon, \zeta$ which belong to the equation deprived of its second term, viz. for the definition of these, we have

$$(a, b, c, d, e, f \between x, 1)^5 = a\,(1, 0, \gamma, \delta, \epsilon, \zeta \between x + a^{-1}b, 1)^5 = 0.$$

I reproduce this solution: writing as usual

$$\begin{aligned}
5u_1 &= x_1 + \omega\, x_2 + \omega^2 x_3 + \omega^3 x_4 + \omega^4 x_5,\\
5u_2 &= x_1 + \omega^2 x_2 + \omega^4 x_3 + \omega\, x_4 + \omega^3 x_5,\\
5u_3 &= x_1 + \omega^3 x_2 + \omega\, x_3 + \omega^4 x_4 + \omega^2 x_5,\\
5u_4 &= x_1 + \omega^4 x_2 + \omega^3 x_3 + \omega^2 x_4 + \omega\, x_5,
\end{aligned}$$

(ω an imaginary fifth root of unity), we find the four "Eulerian" equations

$$\begin{aligned}
-2\gamma &= u_1u_4 + u_2u_3\\
-2\delta &= u_1^2u_3 + u_4^2u_2 + u_2^2u_1 + u_3^2u_4\\
-\epsilon + 4\gamma^2 - 3u_1u_2u_3u_4 &= u_1^3u_2 + u_4^2u_3 + u_2^2u_4 + u_3^2u_1\\
-\zeta - 20\gamma\delta &= u_1^5 + u_2^5 + u_3^5 + u_4^5 - 10\,(u_1^3u_3u_4 + u_4^3u_2u_1 + u_2^3u_1u_3 + u_3^3u_4u_2),
\end{aligned}$$

from which u_1, u_2, u_3, u_4 are to be obtained.

It is found that the first and second equations may be replaced by the two pairs of equations

$$u_1u_4 = -\gamma - v, \qquad u_2u_3 = -\gamma + v$$
$$u_1^2u_3 + u_4^2u_2 = -\delta - tv, \quad u_2^2u_1 + u_3^2u_4 = -\delta + tv.$$

As to the first of these pairs, we find

$$\begin{aligned}
25u_1u_4 = {} & x_1^2 + x_2^2 + x_3^2 + x_4^2 + x_5^2\\
& + (\omega + \omega^4)\,(x_1x_2 + x_2x_3 + x_3x_4 + x_4x_5 + x_5x_1)\\
& + (\omega^2 + \omega^3)\,(x_1x_3 + x_2x_4 + x_3x_5 + x_4x_1 + x_5x_2),
\end{aligned}$$

say this is

$$25u_1u_4 = \Sigma x^2 + (\omega + \omega^4)\,12345 + (\omega^2 + \omega^3)\,13524;$$

and then similarly

$$25u_2u_3 = \Sigma x^2 + (\omega^2 + \omega^3)\,12345 + (\omega + \omega^4)\,13524;$$

whence

$$25\,(u_1u_4 - u_2u_3) = (\omega + \omega^4 - \omega^2 - \omega^3)\,(12345 - 13524), = \sqrt{5}\,\phi, = 50\,v,$$

if as above $v = \dfrac{\phi}{10\sqrt{5}}$; that is

$$u_1u_4 - u_2u_3 = \quad 2v,$$

and combining herewith the equation

$$u_1u_4 + u_2u_3 = -2\gamma,$$

we have the first pair of equations. The second pair of equations is obtained by a similar process, but the work is longer. We have

$$\begin{aligned}
125\, u_1^2u_3 &= F + A\omega \;\; + B\omega^2 + C\omega^3 + D\omega^4 \\
125\, u_2^2u_1 &= F + A\omega^2 + B\omega^4 + C\omega \;\; + D\omega^3 \\
125\, u_3^2u_4 &= F + A\omega^3 + B\omega \;\; + C\omega^4 + D\omega^2 \\
125\, u_4^2u_2 &= F + A\omega^4 + B\omega^3 + C\omega^2 + D\omega
\end{aligned}$$

where

$$\begin{aligned}
F &= \Sigma x_1^3 + 2\Sigma x_1x_2x_3 \\
A &= \{24135\} + 2\,\{12345\} + 2\,(24135) \\
B &= \{54321\} + 2\,\{24135\} + 2\,(54321) \\
C &= \{12345\} + 2\,\{53142\} + 2\,(12345) \\
D &= \{53142\} + 2\,\{54321\} + 2\,(53142)
\end{aligned}$$

where

$$\{24135\} = x_2^2x_4 + x_4^2x_1 + x_1^2x_3 + x_3^2x_5 + x_5^2x_2, \text{ \&c.}:$$

and as before

$$(24135) = x_2x_4x_1 + x_4x_1x_3 + x_1x_3x_5 + x_3x_5x_2 + x_5x_2x_4, \text{ \&c.}$$

Hence

$$125\,(u_1^2u_3 + u_4^2u_2 - u_3^2u_4 - u_2^2u_1) = (\omega + \omega^4 - \omega^2 - \omega^3)(A + D - B - C) = \sqrt{5}\,(A + D - B - C).$$

Here

$$\begin{aligned}
A + D - B - C = &\quad \{12345\} - \{24135\} - \{53142\} + \{54321\} \\
&- 2\,[(12345) - (24135) - (53142) + (54123)],
\end{aligned}$$

where, substituting the values, the first line is

$x_1^2(x_2 - x_3 - x_4 + x_5) + x_2^2(x_3 - x_4 - x_5 + x_1) + x_3^2(x_4 - x_5 - x_1 + x_2) + x_4^2(x_5 - x_1 - x_2 + x_3) + x_5^2(x_1 - x_2 - x_3 + x_4)$, $=\Sigma'$ suppose: and the second line observing that (54123) and (53142) are equal to (12345) and (24135) respectively, is $=-4\,[(12345)-(24135)]$, $=-4\chi$: hence

$$A + D - B - C = \Sigma' - 4\chi.$$

But from the equations

$$a^{-1}\phi = 12345 - 24135, \; = x_1x_2 + x_2x_3 + x_3x_4 + x_4x_5 + x_5x_1 - x_2x_4 - x_4x_1 - x_1x_3 - x_3x_5 - x_5x_2,$$

and

$$-5b \qquad\qquad = x_1 + x_2 + x_3 + x_4 + x_5,$$

we easily obtain

$$-5a^{-1}b\phi = \Sigma' + \chi,$$

and the last result thus is

$$A + D - B - C = -5a^{-1}b\phi - 5\chi, \; = -5a^{-1}b\phi - 5\phi\tau, \; = -5\phi(a^{-1}b + \tau),$$

or substituting for ϕ and $a^{-1}b+\tau$ their values $10\sqrt{5}\,v$, and t, this is $= -50\sqrt{5}\,tv$. Hence the value of $125(u_1^2u_3 + u_4^2u_2 - u_3^2u_4 - u_2^2u_1)$ is $= -250\,tv$, that is we have

$$u_1^2u_3 + u_4^2u_2 - u_3^2u_4 - u_2^2u_1 = -2tv,$$

and combining herewith

$$u_1^2u_3 + u_4^2u_2 + u_3^2u_4 + u_2^2u_1 = -2\delta,$$

we have the second pair of equations.

We next write

$$\begin{aligned} 2u_1^2u_3 &= -\delta - tv + 2n_1, \\ 2u_4^2u_2 &= -\delta - tv - 2n_1, \\ 2u_2^2u_1 &= -\delta + tv + 2n_2, \\ 2u_3^2u_4 &= -\delta + tv - 2n_2, \end{aligned}$$

whence

$$\begin{aligned} 4n_1^2 &= (\delta + tv)^2 + 4(\gamma^2 - v^2)(\gamma + v), \\ 4n_2^2 &= (\delta - tv)^2 + 4(\gamma^2 - v^2)(\gamma - v), \end{aligned}$$

which equations determine n_1, n_2, so that $u_1^2u_3$, $u_4^2u_2$, $u_2^2u_1$ and $u_3^2u_4$ are now known; and we then have

$$\begin{aligned} u_1^5 &= (u_1^2u_3)^2(u_2^2u_1) \div (u_2u_3)^2, \\ u_2^5 &= (u_2^2u_1)^2(u_4^2u_2) \div (u_1u_4)^2, \\ u_4^5 &= (u_4^2u_2)^2(u_3^2u_4) \div (u_2u_3)^2, \\ u_3^2 &= (u_3^2u_4)^2(u_1^2u_3)^2 \div (u_1u_4)^4, \end{aligned}$$

which determine u_1, u_2, u_3, u_4. (Compare herewith the formulæ, $A = \dfrac{A'A''}{\beta\gamma}$, &c. p. 54 of my paper "On a solvible quintic Equation," *Amer. Math. Jour.* t. XIII. (1890), pp. 53—58.)

But, as Mc Clintock remarks, it is possible to obtain better formulæ: viz. these are

$$\begin{aligned} u_1^5 &= \tfrac{1}{4}r_1 + \tfrac{1}{4}r_2 + \sqrt{s_1 + s_2}, \\ u_2^5 &= \tfrac{1}{4}r_1 - \tfrac{1}{4}r_2 + \sqrt{s_1 - s_2}, \\ u_3^5 &= \tfrac{1}{4}r_1 - \tfrac{1}{4}r_2 - \sqrt{s_1 - s_2}, \\ u_4^5 &= \tfrac{1}{4}r_1 + \tfrac{1}{4}r_2 - \sqrt{s_1 + s_2}, \end{aligned}$$

where

$$\begin{aligned}
r_1 &= -\zeta + 20\,tv^2,\\
r_2 &= (\gamma^2 - v^2)^{-1}\,v^{-1}\left\{\begin{array}{l}(-\delta + 2\gamma t - t^3)\,v^4\\ +\{\delta\epsilon + 2\gamma^2\delta + (-2\gamma\epsilon + \delta^2 + 4\gamma^3)\,t + \gamma\delta t^2\}\,v^2\\ +\{\gamma^2\delta\epsilon - \gamma\delta^3\}\end{array}\right\},\\
s_1 &= \tfrac{1}{16}r_1^2 + \tfrac{1}{16}r_2^2 + \gamma^5 + 10\gamma^3v^2 + 5\gamma v^4,\\
s_2 &= \tfrac{1}{8}r_1r_2 \quad - 5\gamma^4v - 10\gamma^2v^3 - v^5.
\end{aligned}$$

To prove these results write for shortness

$2m_1 = -\delta - tv$, $2m_2 = -\delta + tv$, and (as above) $2n_1 = u_1^2u_3 - u_4^2u_2$, $2n_2 = u_2^2u_1 - u_3^2u_4$, then

$$u_1^2u_3 = m_1 + n_1,\quad u_4^2u_2 = m_1 - n_1,\quad u_2^2u_1 = m_2 + n_2,\quad u_3^2u_4 = m_2 - n_2\,;$$

we have $u_1u_2u_3u_4 = u_1u_4 \,.\, u_2u_3 = (-\gamma + v)(-\gamma - v),\ = \gamma^2 - v^2$, and the third Eulerian equation thus becomes

$$\epsilon = \gamma^2 + 3v^2 + (v^2 - \gamma^2)^{-1}\,\{u_1^4u_2^2\,u_3u_4 + u_4^4u_3^2\,u_1u_2 + u_2^4u_3^2\,u_1u_2 + u_3^4u_1^2\,u_2u_4\}.$$

The terms within the $\{\ \}$ are equal to

$$(m_1 + n_1)(m_2 + n_2)(-\gamma + v),\quad (m_1 - n_1)(m_2 - n_2)(-\gamma + v),\quad (m_1 - n_1)(m_2 + n_2)(-\gamma - v),$$
$$(m_1 + n_1)(m_2 - n_2)(-\gamma - v),$$

respectively, and their sum is $= -4\gamma m_1m_2 + 4vn_1n_2$.

Hence putting for shortness $p = 4vn_1n_2,\ = 4v\,(u_1^2u_3 - u_4^2u_2)(u_2^2u_1 - u_3^2u_4)$, the third equation becomes

$$\epsilon = \gamma^2 + 3v + (v^2 - \gamma^2)^{-1}\,(\gamma t^2v^2 - \gamma\delta^2 + p),$$

or, what is the same thing,

$$p = \gamma\delta^2 - \gamma t^2v^2 + (v^2 - \gamma^2)(\epsilon - \gamma^2 - 3v).$$

Again, writing the fourth Eulerian equation in the form

$$\Sigma u^5 = -\zeta - 20\gamma\delta + 10\,\{(m_1 + n_1)\,u_1u_4 + (m_1 - n_1)\,u_1u_4 + (m_2 + n_2)\,u_2u_3 + (m_2 - n_2)\,u_2u_3\},$$

the term in $\{\ \}$ is $2m_1u_1u_4 + 2m_2u_2u_3,\ = (-\delta - tv)(-\gamma + v) + (-\delta + tv)(-\gamma - v),\ = 2\gamma\delta - 2tv^2$, so that writing $\Sigma u^5 = r_1$, the equation becomes

$$r_1 = -\zeta - 20tv^2,$$

viz. this is the above-mentioned value of r_1.

We then have $r_2 = u_1^5 + u_4^5 - u_2^5 - u_3^5$, and for calculating this, we have

$$\begin{aligned}
(\gamma^2 - v^2)^2\,u_1^5 &= (m_1 + n_1)^2\,(m_2 + n_2)\,(\gamma - v)^2,\\
(\gamma^2 - v^2)^2\,u_4^5 &= (m_1 - n_1)^2\,(m_2 - n_2)\,(\gamma - v)^2,\\
(\gamma^2 - v^2)^2\,u_2^5 &= (m_2 + n_2)^2\,(m_1 - n_1)\,(\gamma + v)^2,\\
(\gamma^2 - v^2)\ \ u_3^5 &= (m_2 - n_2)^2\,(m_1 + n_1)\,(\gamma + v)^2,
\end{aligned}$$

and thence, instead of $4vn_1n_2$ which occurs on the right-hand side writing its value $=p$, we find

$$\begin{aligned}(\gamma^2-v^2)^2 r_2 = \;& (m_1+m_2)\,v^{-1}\,\{p\,(\gamma^2+v^2)-4m_1m_2\gamma\}\\ &+2\,(m_1-m_2)\quad\{m_1m_2\,(\gamma^2+v^2)-\;p\gamma\}\\ &+2\,(n_1{}^2m_2-n_2{}^2m_1)\quad(\gamma^2+v^2)\\ &-4\,(n_1{}^2m_1+n_2{}^2m_1)\quad\gamma v.\end{aligned}$$

We have

$$4n_1{}^2=(\delta+tv)^2+4\,(\gamma^2-v^2)\,(\gamma-v),\quad 4n_2{}^2=(\delta-tv)^2+4\,(\gamma^2-v^2)\,(\gamma+v),$$

p is given as above, and moreover $m_1+m_2=-\delta$, $m_1-m_2=-tv$, $m_1m_2=\frac{1}{4}(\delta^2-t^2v^2)$. Substituting these values and dividing out by $(\gamma^2-v^2)^2$, we find after some reductions the value given above for r_2.

Finally we have

$$4\,(\gamma-v)^5=-4\,(u_1u_4)^5=(u_1{}^5+u_4{}^5)^2-4\,(u_1u_4)^5-\tfrac{1}{2}\,(r_1+r_2)\,(u_1{}^5+u_4{}^5)=(u_1{}^5-u_4{}^5)^2-\tfrac{1}{2}\,(r_1+r_2)\,(u_1{}^5+u_4{}^5),$$

that is

$$4\,(\gamma-v)^5=4\,(s_1+s_2)-\tfrac{1}{4}\,(r_1+r_2)^2,$$

or say

$$s_1+s_2=(\tfrac{1}{4}r_1+\tfrac{1}{4}r_2)^2+(\gamma-v)^5,$$

and similarly

$$s_1-s_2=(\tfrac{1}{4}r_1-\tfrac{1}{4}r_2)^2+(\gamma+v)^5$$

which equations give the above-mentioned values of s_1 and s_2.

As to Jacobi's Memoir spoken of in the Addition, I refer to the paper Kronecker, "Ueber eine stelle in Jacobi's Aufsatz, Observatiunculæ ad theoriam æquationum pertinentes," *Crelle*, t. CVII. (1891), pp. 349—352, which incorporates some remarks of mine in regard thereto.

284, 294. The fundamental idea of these two papers is *not* that of "the six coordinates of a line," but (as indeed appears from the title) a somewhat different one, viz. I say that a curve in space may be represented by a homogeneous equation $V=0$ between six coordinates (p, q, r, s, t, u) such that $ps+qt+ru=0$; this equation being the equation of a cone of arbitrary vertex passing through the curve in question: taking x, y, z, w to be current point-coordinates and $\alpha, \beta, \gamma, \delta$ to be the point-coordinates of the arbitrary vertex, then p, q, r, s, t, u are the six determinants of the matrix

$$\left|\begin{matrix} x, & y, & z, & w \\ \alpha, & \beta, & \gamma, & \delta \end{matrix}\right|,$$

or, what is the same thing, we have

$$\begin{aligned} p&=\gamma y-\beta z, & s&=\delta x-\alpha w,\\ q&=\alpha z-\gamma x, & t&=\delta y-\beta w,\\ r&=\beta x-\alpha y, & u&=\delta z-\gamma w,\end{aligned}$$

values which satisfy $ps+qt+ru=0$. And I accordingly say that the equation of a line in space is

$$Ap+Bq+Cr+Fs+Gt+Hu=0,$$

viz. this is the equation of the cone of arbitrary vertex $(\alpha, \beta, \gamma, \delta)$ (that is, of the plane through the point $(\alpha, \beta, \gamma, \delta)$) which passes through the line in question. But I go on to say that if $(\alpha', \beta', \gamma', \delta'')$, $(\alpha'', \beta'', \gamma'', \delta'')$ are the coordinates of any two points on the given line, or if the line be given as the intersection of the two planes $ax+by+cz+dw=0$, $a'x+b'y+c'z+d'w=0$, then in the first case

$$\begin{aligned} A &= \alpha'\delta''-\alpha''\delta', & F &= \beta'\gamma''-\beta''\gamma', \\ B &= \beta'\delta''-\beta''\delta', & G &= \gamma'\alpha''-\gamma''\alpha', \\ C &= \gamma'\delta''-\gamma''\delta', & H &= \alpha'\beta''-\alpha''\beta', \end{aligned}$$

and in the second case

$$\begin{aligned} A &= bc'-b'c, & F &= ad'-a'd, \\ B &= ca'-c'a, & G &= bd'-b'd, \\ C &= ab'-a'b, & H &= cd'-c'd, \end{aligned}$$

so that in each case $AF+BG+CH=0$. I thus in effect, although not quite explicitly, define (A, B, C, F, G, H) as the "six coordinates of a line"; and after giving in terms of these quantities for any two lines the condition of the intersection of the two lines I say that any other problems in relation to the line, for instance...&c., may also be solved "by means of the new coordinates."

Plücker's Memoir "On a New Geometry of Space" is published *Phil. Trans.* t. CLV. (for 1865), pp. 725—791, the paper being received Dec. 22, 1864, and the Additional Note appended thereto, Dec. 11, 1865. My two papers are referred to in the foot-note p. 784, belonging to this additional note as follows: "In two remarkable papers 'On a New Analytical Representation of Curves in Space' published in the third and fifth volumes of the *Quarterly Journal of Mathematics,* Professor Cayley employed before me in order to represent cones the six coordinates of a right line depending upon any two of its points. Having lately only seen the papers I hasten to mention it now. But besides the coincidence referred to the leading views of Professor Cayley's paper and mine have nothing in common. On this occasion I may state that the principles upon which my paper is based were advanced by me nearly twenty years ago (Geometry of Space, No. 258), but this had entirely escaped my memory when I recurred to Geometry some time since."

In the work referred to, "System der Geometrie des Raumes u. s. w." (4to. Düsseldorf, 1846), No. 258, Plücker remarks that a straight line depends upon four constants, viz. its equations in point-coordinates being $x=\kappa z+\lambda$, $y=\mu z+\nu$, or in line-coordinates being $t=\kappa v+\lambda w$, $u=\mu v+\nu w$, then in either case the constants are $\kappa, \lambda, \mu, \nu$; and he defines these four quantities as the "four coordinates of a line."

The leading idea of Plücker's memoir appears in the first words thereof, "I. On linear Complexes of Right lines." He works at first with the *four* coordinates of a line; as long as these are arbitrary the line is any line whatever; but considering them as connected by a single equation, then there is he says "a Complex," considering them as connected by two equations "a Congruency", and considering them as connected by three equations "a Configuration" or ruled surface. The establishment of these notions of a Complex and a Congruency, and the general idea of regarding the line as an element in the Geometry of Space are absolutely Plücker's, there is no anticipation of them in my two papers. Later on five coordinates r, s, σ, ρ, $s\rho - r\sigma$ are introduced, but the *six* coordinates are first used in the Additional Note, viz. here instead of a non-homogeneous equation or equations between four coordinates we have a homogeneous equation or equations between six quantities connected by a homogeneous quadric equation. It is hardly necessary to refer to the posthumous work *Neue Geometrie des Raumes gegründet auf die Betrachtung der geraden Linie als Raumelement*, Leipzig, 1868 and 1869, or to later developments of the theory.

END OF VOL. IV.

CAMBRIDGE:
PRINTED BY C. J. CLAY, M.A. AND SONS,
AT THE UNIVERSITY PRESS.

SELECT SCIENTIFIC PUBLICATIONS OF

THE CAMBRIDGE UNIVERSITY PRESS.

The Collected Mathematical Papers of Arthur Cayley, Sc.D., F.R.S., Sadlerian Professor of Pure Mathematics in the University of Cambridge. Demy 4to. 10 vols. Vols. I. II. III. and IV. 25*s.* each. [Vol. V. *In the Press.*

The Electrical Researches of the Hon. H. Cavendish, F.R.S. Written between 1771 and 1781. Edited from the original MSS. in the possession of the Duke of Devonshire, K.G., by the late J. CLERK MAXWELL, F.R.S. Demy 8vo. 18*s.*

The Scientific Papers of the late Prof. J. Clerk Maxwell. Edited by W. D. NIVEN, M.A., formerly Fellow of Trinity College. In 2 vols. Royal 4to. £3. 3*s.* (net).

A History of the Theory of Elasticity and of the Strength of Materials, from Galilei to the present time. Vol. I. Galilei to Saint-Venant, 1639—1850. By the late I. TODHUNTER, Sc.D., F.R.S., edited and completed by Professor KARL PEARSON, M.A. Demy 8vo. 25*s.*

Vol. II. Saint-Venant to Sir WILLIAM THOMSON. By the same Editor. [*Nearly ready.*

A History of the Study of Mathematics at Cambridge. By W. W. ROUSE BALL, M.A., Fellow and Lecturer on Mathematics of Trinity College, Cambridge. Crown 8vo. 6*s.*

Mathematical and Physical Papers, by Sir W. THOMSON, LL.D., D.C.L., F.R.S., Professor of Natural Philosophy in the University of Glasgow. Collected from different Scientific Periodicals from May 1841, to the present time. Demy 8vo. Vol. I. 18*s.* Vol. II. 15*s.* Vol. III. 18*s.*

Mathematical and Physical Papers, by Sir G. G. STOKES, Sc.D., LL.D., F.R.S., Lucasian Professor of Mathematics in the University of Cambridge. Reprinted from the Original Journals and Transactions, with Additional Notes by the Author. Vol. I. Demy 8vo. 15*s.* Vol. II. 15*s.* [Vol. III. *In the Press.*

Catalogue of Scientific Papers compiled by the Royal Society of London: Vols. 1—6 for the years 1800—1863, Royal 4to. cloth (Vol. 1 in half morocco) £4 (net); half morocco £5. 5*s.* (net). Vols. 7—8 for the years 1864—1873, cloth £1. 11*s.* 6*d.* (net); half morocco £2. 5*s.* (net). Single volumes cloth 20*s.* or half-morocco 28*s.* (net). Vol. IX. New Series for the years 1874—1883. [*Nearly ready.*

Theory of Differential Equations. Part I. Exact Equations and Pfaff's Problem. By A. R. FORSYTH, Sc.D., F.R.S., Fellow of Trinity College, Cambridge. Demy 8vo. 12*s.*

A Treatise on Geometrical Optics. By R. S. HEATH, M.A., Professor of Mathematics in Mason Science College, Birmingham. Demy 8vo. 12*s.* 6*d.*

An Elementary Treatise on Geometrical Optics. By R. S. HEATH, M.A. Cr. 8vo. 5*s.*

A Treatise on Elementary Dynamics. By S. L. LONEY, M.A., late Fellow of Sidney Sussex College. New and Enlarged Edition. Crown 8vo. 7*s.* 6*d.*

Solutions to the Examples in a Treatise on Elementary Dynamics. By the same Author. [*In the Press.*

A Treatise on Analytical Statics. By E. J. ROUTH, Sc.D., F.R.S., Fellow of the University of London, Honorary Fellow of Peterhouse, Cambridge. Vol. I. Demy 8vo. 14*s.*

A Treatise on Plane Trigonometry. By E. W. HOBSON, M.A., Fellow and Lecturer of Christ's College, Cambridge. Demy 8vo. 12*s.*

A Catalogue of the Portsmouth Collection of Books and Papers written by or belonging to Sir ISAAC NEWTON. Demy 8vo. 5*s.*